5 STEPS TO A 5™

AP Calculus AB

2023

AP Calculus AB

2023

William Ma

New York Chicago San Francisco Athens London Madrid
Mexico City Milan New Delhi Singapore Sydney Toronto

1 2 3 4 5 6 7 8 9 LHS 27 26 25 24 23 22 (Cross-Platform Prep Course only)
1 2 3 4 5 6 7 8 9 LHS 27 26 25 24 23 22 (Elite Student Edition)

ISBN 978-1-264-39585-9 (Cross-Platform Prep Course only)
MHID 1-264-39585-X

e-ISBN 978-1-264-39808-9 (e-book Cross-Platform Prep Course only)
e-MHID 1-264-39808-5

ISBN 978-1-264-40579-4 (Elite Student Edition)
MHID 1-264-40579-0

e-ISBN 978-1-264-41211-2 (e-book Elite Student Edition)
e-MHID 1-264-41211-8

The series editor was Grace Freedson, and the project editor was Del Franz. Series design by Jane Tenenbaum.

Cross-Platform Prep Course

Single User License Terms and Conditions

CONTENTS

STEP 4 **Review the Knowledge You Need to Score High**

PREFACE

Congratulations! You are an AP Calculus student. Not too shabby! As you know, AP Calculus is one of the most challenging subjects in high school. You are studying mathematical ideas that helped change the world. Not that long ago, calculus was taught at the graduate level. Today, smart young people like yourself study calculus in high school. Most colleges will give you credit if you score a 3 or more on the AP Calculus exam.

So how do you do well on the AP Calculus exam? How do you get a 5? Well, you've already taken the first step. You're reading this book. The next thing you need to do is to make sure that you understand the materials and do the practice problems. In recent years, the AP Calculus exam has gone through many changes. For example, today the questions no longer stress long and tedious algebraic manipulations. Instead, you are expected to be able to solve a broad range of problems, including problems presented to you in the form of a graph, a chart, or a word problem. For many of the questions, you are also expected to use your calculator to find the solutions.

After having taught AP Calculus for many years and having spoken to students and other calculus teachers, we understand some of the difficulties that students might encounter with the AP Calculus exam. For example, some students have complained about not being able to visualize what the question was asking and other students said that even when the solution was given, they could not follow the steps. Under these circumstances, who wouldn't be frustrated? In this book, we have addressed these issues. Whenever possible, problems are accompanied by diagrams and solutions are presented in a step-by-step manner. The graphing calculator is used extensively whenever it is permitted. To make things even easier, this book begins with a chapter that reviews precalculus. So if you need to look up a formula, definition, or concept in precalculus, it is right here in the book. If you're familiar with these concepts, you might skip this chapter and begin with Chapter 6.

So how do you get a 5 on the AP Calculus exam?

Step 1: Set up your study program by selecting one of the three study plans in Chapter 2 of this book.

Step 2: Determine your test readiness by taking the diagnostic exam in Chapter 3.

Step 3: Develop strategies for success by learning the test-taking techniques offered in Chapter 4.

Step 4: Review the knowledge you need to score high by studying the subject material in Chapter 5 through Chapter 14.

Step 5: Build your test-taking confidence by taking the practice exams provided in this book.

As an old martial artist once said, "First you must understand. Then you must practice." Have fun and good luck!

ACKNOWLEDGMENTS

I could not have written this book without the help of the following people:

My high school calculus teacher, *Michael Cantor*, who taught me calculus.

Professor *Leslie Beebe*, who taught me how to write.

David Pickman, who fixed my computer and taught me Equation Editor.

Jennifer Tobin, who tirelessly edited many parts of the manuscript and with whom I look forward to coauthoring a math book in the future.

Robert Teseo and his calculus students who field-tested many of the problems.

Allison Litvack, *Rich Peck*, and *Liz Spiegel*, who proofread sections of the practice tests. And a special thanks to *Trisha Ho*, who edited some of the materials.

Mark Reynolds, who proofread part of the manuscript.

Maxine Lifsfitz, who offered many helpful comments and suggestions.

Grace Freedson, *Del Franz*, and *Sonam Arora* for all their assistance.

Sam Lee and *Derek Ma*, who were on 24-hour call for technical support.

My older daughter, *Janet*, for not killing me for missing one of her concerts.

My younger daughter, *Karen*, who helped me with many of the computer graphics.

My wife, *Mary*, who gave me many ideas for the book and who often has more confidence in me than I have in myself.

ABOUT THE AUTHOR

William Ma has taught calculus for many years. He received his BA and MA from Columbia University. He was the chairman of the Math Department at the Herricks School District on Long Island, New York, for many years before retiring. He also taught as adjunct instructor at Baruch College, Fordham University, and Columbia University. He is the author of several books, including test preparation books for the SAT, ACT, GMAT, and AP Calculus AB and BC. He is currently a math consultant.

INTRODUCTION: THE FIVE-STEP PROGRAM

How Is This Book Organized?

This book begins with an introduction to the five-step program followed by 14 chapters reflecting the five steps.

- Step 1 provides an overview of the AP Calculus AB exam and offers three study plans for preparing for this exam.
- Step 2 contains a diagnostic test with answers and explanations.
- Step 3 offers test-taking strategies for answering both multiple-choice and free-response questions, and for using a graphing calculator.
- Step 4 consists of 10 chapters providing a comprehensive review of all topics covered on the AP Calculus AB exam. At the end of each chapter (beginning with Chapter 5), you will find a set of practice problems with solutions, a set of cumulative review problems with solutions, and a rapid review section giving you the highlights of the chapter.
- Step 5 provides two full practice AP Calculus exams with answers, explanations, and worksheets to compute your score.

The book concludes with a summary of math formulas and theorems related to the AP Calculus exam. (*Please note that the exercises in this book are done with the TI-89 Graphing Calculator.*)

Introducing the Five-Step Preparation Program

This book is organized as a five-step program to prepare you to succeed in the AP Calculus AB exam. These steps are designed to provide you with vital skills, strategies, and the practice that can lead you to that perfect 5. Here are the five steps.

Step 1: Set Up Your Study Program

In this step, you will read an overview of the AP Calculus AB exam, including a summary of topics covered in the exam and a description of the format of the exam. You will also follow a process to help determine which of the following preparation programs is right for you:

- Full school year: September through May
- One semester: January through May
- Six weeks: basic training for the exam

Step 2: Determine Your Test Readiness

In this step, you will take a diagnostic multiple-choice exam in calculus. This pretest should give you an idea of how prepared you are to take the real exam before you begin to study for the actual AP Calculus AB exam.

Step 3: Develop Strategies for Success

In this step, you will learn strategies that will help you do your best on the exam. These strategies cover both the multiple-choice and free-response sections of the exam.

- Learn to read multiple-choice questions.
- Learn how to answer multiple-choice questions.
- Learn how to plan and write answers to the free-response questions.

Step 4: Review the Knowledge You Need to Score High

In this step, you will learn or review the material you need to know for the test. This review section takes up the bulk of this book. It contains:

- A comprehensive review of AP Calculus AB.
- A set of practice problems.
- A set of cumulative review problems beginning with Chapter 5.
- A rapid review summarizing the highlights of the chapter.

Step 5: Build Your Test-Taking Confidence

In this step, you will complete your preparation by testing yourself on practice exams. We have provided you with two complete AP Calculus AB exams, along with scoring guides for both of them. Although these practice exams are not reproduced questions from the actual AP calculus exams, they mirror both the material tested by AP and the way in which it is tested.

Finally, at the back of this book you will find additional resources to aid your preparation. These include:

- A brief bibliography.
- A list of websites related to the AP Calculus exam.
- A summary of formulas and theorems related to the AP Calculus exam.

Introduction to the Graphics Used in this Book

To emphasize particular skills and strategies, we use several icons throughout this book. An icon in the margin will alert you that you should pay particular attention to the accompanying text. We use these icons:

This icon points out a very important concept or fact that you should not pass over.

This icon calls your attention to a strategy that you may want to try.

This icon indicates a tip that you might find useful.

Set Up Your Study Plan

What You Need to Know About the AP Calculus AB Exam

IN THIS CHAPTER

Summary: Learn what topics are tested in the exam, what the format is, which calculators are allowed, and how the exam is graded.

Key Ideas

- The AP Calculus AB exam has 45 multiple-choice questions and 6 free-response questions. There are two types of questions, and each makes up 50% of the grade.
- Many graphing calculators are permitted on the exam, including the TI-98.
- You may bring up to two approved calculators for the exam.
- You may store programs in your calculator, and you are not required to clear the memories in your calculator for the exam.

1.1 What Is Covered on the AP Calculus AB Exam?

The AP Calculus AB exam covers the following topics:

- Functions, Limits and Graphs of Functions, Continuity
- Definition and Computation of Derivatives, Second Derivatives, Relationship between the Graphs of Functions and Their Derivatives, Applications of Derivatives
- Finding Antiderivatives, Definite Integrals, Applications of Integrals, Fundamental Theorem of Calculus, Numerical Approximations of Definite Integrals, and Separable Differential Equations

Students are expected to be able to solve problems that are expressed graphically, numerically, analytically, and verbally. For a more detailed description of the topics covered in the AP Calculus AB exam, visit the College Board AP website at: www.exploreap.org.

1.2 What Is the Format of the AP Calculus AB Exam?

The AP Calculus AB exam has two sections:

Section I contains 45 multiple-choice questions for which you are given 105 minutes to complete.

Section II contains 6 free-response questions for which you are given 90 minutes to complete.

The total time allotted for both sections is 3 hours and 15 minutes. Below is a summary of the different parts of each section.

Section I *Multiple-Choice*	Part A	30 Questions	No Calculator	60 Minutes
	Part B	15 Questions	Calculator	45 Minutes
Section II *Free-Response*	Part A	2 Questions	Calculator	30 Minutes
	Part B	4 Questions	No Calculator	60 Minutes

During the time allotted for Part B of Section II, students may continue to work on questions from Part A of Section II. However, they may not use a calculator at that time. Please note that you are not expected to be able to answer all the questions in order to receive a grade of 5. If you wish to see the specific instructions for each part of the test, visit the College Board website at: https://apstudent.collegeboard.org/apcourse/ap-calculus-ab/calculator-policy.

1.3 What Are the Advanced Placement Exam Grades?

Advanced Placement Exam grades are given on a 5-point scale with 5 being the highest grade. The grades are described below:

5 Extremely Well Qualified
4 Well Qualified
3 Qualified
2 Possibly Qualified
1 No Recommendation

How Is the AP Calculus AB Exam Grade Calculated?

- The exam has a total raw score of 108 points: 54 points for the multiple-choice questions in Section I and 54 points for the free-response questions for Section II.
- Each correct answer in Section I is worth 1.2 points; there is **no point deduction** for incorrect answers and no points are given for unanswered questions. For example, suppose your result in Section I is as follows:

Correct	Incorrect	Unanswered
40	5	0

Your raw score for Section I would be:

$40 \times 1.2 = 48$. Not a bad score!

- Each complete and correct solution for a question in Section II is worth 9 points.
- The total raw score for both Section I and II is converted to a 5-point scale. The cut-off points for each grade (1–5) vary from year to year. Visit the College Board website at: https://apstudent.collegeboard.org/exploreap/the-rewards/exam-scores for more information. Below is a rough estimate of the conversion scale:

Total Raw Score	Approximate AP Grade
80–108	5
65– 79	4
50– 64	3
36– 49	2
0– 35	1

Remember, these are approximate cut-off points.

1.4 Which Graphing Calculators Are Allowed for the Exam?

The following calculators are allowed:

CASIO	HEWLETT-PACKARD	TEXAS INSTRUMENTS
FX-6000 series	HP-9G	TI-73
FX-6200 series	HP-28 series	TI-80
FX-6300 series	HP-38G series	TI-81
FX-6500 series	HP-39 series	TI-82
FX-7000 series	HP-40G	TI-83
FX-7300 series	HP-48 series	TI-83 Plus
FX-7400 series	HP-49 series	TI-83 Plus Silver
FX-7500 series	HP-50 series	TI-84 Plus
FX-7700 series	HP Prime	TI-84 Plus SE
FX-7800 series		TI-84 Plus Silver
FX-8000 series	**RADIO SHACK**	TI-84 Plus C Silver
FX-8500 series	EC-4033	TI-84 Plus T
FX-8700 series	EC-4034	TI-84 Plus CE-T
FX-8800 series	EC-4037	TI-84 Plus CE Python
FX-9700 series		TI-84 Plus CE-T Python Ed.
FX-9750 series	**SHARP**	TI-85
FX-9860 series	EL-5200	TI-86
CFX-9800 series	EL-9200 series	TI-89
CFX-9850 series	EL-9300 series	TI-89 Titanium
CFX-9950 series	EL-9600 series (no stylus)	TI-Nspire
CFX-9970 series	EL-9900 series	TI-Nspire CX
FX 1.0 series		TI-Nspire CAS
Algebra FX 2.0 series	**OTHER**	TI-Nspire CX CAS
FX-CG-10 (PRIZM)	Datexx DS-883	TI-Nspire CM-C
FX-CG-20	Micronta	TI-Nspire CAS CX-C
FX-CG 500 (no stylus)	NumWorks	TI-Nspire CX-II CAS
FX-CG-50	Smart	TI-Nspire CX-T
Graph35 series		TI-Nspire CX-II
Graph75 series		TI-Nspire CX-T CAS
Graph85 series		TI-Nspire CX II-C CAS
Graph100 series		

For a more complete list, visit the College Board website at: https://apstudent.collegeboard.org/apcourse/ap-calculus-ab/calculator-policy. If you wish to use a graphing calculator that is not on the approved list, your teacher must obtain written permission from the ETS before April 1st of the testing year.

Calculators and Other Devices Not Allowed for the AP Calculus AB Exam

- TI-92 Plus, Voyage 200, and devices with QWERTY keyboards
- Non-graphing scientific calculators
- Laptop computers
- Pocket organizers, electronic writing pads, or pen-input devices
- Cellular phone calculators

Other Restrictions on Calculators

- You may bring up to two (but no more than two) approved graphing calculators to the exam.
- You may not share calculators with another student.
- You may store programs in your calculator.
- You are not required to clear the memories in your calculator for the exam.
- You may not use the memories of your calculator to store secured questions and take them out of the testing room.

How to Plan Your Time

IN THIS CHAPTER

Summary: The right preparation plan for you depends on your study habits and the amount of time you have before the test.

Key Idea

✪ Choose the study plan that is right for you.

2.1 Three Approaches to Preparing for the AP Calculus AB Exam

Overview of the Three Plans

No one knows your study habits, likes, and dislikes better than you. So you are the only one who can decide which approach you want and/or need to adopt to prepare for the Advanced Placement Calculus exam. Look at the brief profiles below. These may help you to place yourself in a particular prep mode.

You are a full-year prep student (Approach A) if:

1. You are the kind of person who likes to plan for everything far in advance . . . and I mean far . . . ;
2. You arrive at the airport 2 hours before your flight because, "you never know when these planes might leave early . . . ";
3. You like detailed planning and everything in its place;

4. You feel you must be thoroughly prepared;
5. You hate surprises.

You are a one-semester prep student (Approach B) if:

1. You get to the airport 1 hour before your flight is scheduled to leave;
2. You are willing to plan ahead to feel comfortable in stressful situations but are okay with skipping some details;
3. You feel more comfortable when you know what to expect, but a surprise or two is cool;
4. You're always on time for appointments.

You are a six-week prep student (Approach C) if:

1. You get to the airport just as your plane is announcing its final boarding;
2. You work best under pressure and tight deadlines;
3. You feel very confident with the skills and background you've learned in your AP Calculus class;
4. You decided late in the year to take the exam;
5. You like surprises;
6. You feel okay if you arrive 10–15 minutes late for an appointment.

2.2 Calendar for Each Plan

A Calendar for Approach A:
A Year-Long Preparation for the AP Calculus AB Exam

Although its primary purpose is to prepare you for the AP Calculus AB exam you will take in May, this book can enrich your study of calculus, your analytical skills, and your problem-solving techniques.

SEPTEMBER-OCTOBER (Check off the activities as you complete them.)

_______ Determine into which student mode you would place yourself.
_______ Carefully read Steps 1 and 2.
_______ Get on the Web and take a look at the AP website(s).
_______ Skim the Comprehensive Review section. (These areas will be part of your year-long preparation.)
_______ Buy a few highlighters.
_______ Flip through the entire book. Break the book in. Write in it. Toss it around a little bit. Highlight it.
_______ Get a clear picture of what your own school's AP Calculus curriculum is.
_______ Begin to use the book as a resource to supplement the classroom learning.
_______ Read and study Chapter 5—Review of Precalculus.
_______ Read and study Chapter 6—Limits and Continuity.
_______ Read and study Chapter 7—Differentiation.

NOVEMBER (The first ten weeks have elapsed.)

_______ Read and study Chapter 8—Graphs of Functions and Derivatives.
_______ Read and study Chapter 9— Applications of Derivatives.

DECEMBER

_______ Read and study Chapter 10—More Applications of Derivatives.
_______ Read and study Chapter 11—Integration.
_______ Review Chapters 6–8.

JANUARY (Twenty weeks have now elapsed.)

_______ Read and study Chapter 12—Definite Integrals.
_______ Review Chapters 9–11.

FEBRUARY

_______ Read and study Chapter 13—Areas and Volumes.
_______ Take the Diagnostic Test.
_______ Evaluate your strengths and weaknesses.
_______ Study appropriate chapters to correct weaknesses.

MARCH (Thirty weeks have now elapsed.)

_______ Read and study Chapter 14—More Applications of Definite Integrals.
_______ Review Chapters 12–14.

APRIL

_______ Take Practice Exam 1 in first week of April.
_______ Evaluate your strengths and weaknesses.
_______ Study appropriate chapters to correct weaknesses.
_______ Review Chapters 6–14.

MAY—First Two Weeks (THIS IS IT!)

_______ Take Practice Exam 2.
_______ Score yourself.
_______ Study appropriate chapters to correct weaknesses.
_______ Get a good night's sleep the night before the exam. Fall asleep knowing you are well prepared.

GOOD LUCK ON THE TEST!

A Calendar for Approach B:
A Semester-Long Preparation for the AP Calculus AB Exam

Working under the assumption that you've completed one semester of calculus studies, the following calendar will use those skills you've been practicing to prepare you for the May exam.

JANUARY

______ Carefully read Steps 1 and 2.
______ Read and study Chapter 6—Limits and Continuity.
______ Read and study Chapter 7—Differentiation.
______ Read and study Chapter 8—Graphs of Functions and Derivatives.
______ Read and study Chapter 9—Applications of Derivatives.

FEBRUARY

______ Read and study Chapter 5—Review of Precalculus.
______ Read and study Chapter 10—More Applications of Derivatives.
______ Read and study Chapter 11—Integration.
______ Take the Diagnostic Test.
______ Evaluate your strengths and weaknesses.
______ Study appropriate chapters to correct weaknesses.
______ Review Chapters 6–9.

MARCH (Ten weeks to go.)

______ Read and study Chapter 12—Definite Integrals.
______ Read and study Chapter 13—Areas and Volumes.
______ Read and study Chapter 14—More Applications of Definite Integrals.
______ Review Chapters 10–12.

APRIL

______ Take Practice Exam 1 in first week of April.
______ Evaluate your strengths and weaknesses.
______ Study appropriate chapters to correct weaknesses.
______ Review Chapters 6–14.

MAY—First Two Weeks (THIS IS IT!)

______ Take Practice Exam 2.
______ Score yourself.
______ Study appropriate chapters to correct weaknesses.
______ Get a good night's sleep the night before the exam. Fall asleep knowing you are well prepared.

GOOD LUCK ON THE TEST!

A Calendar for Approach C:
A Six-Week Preparation for the AP Calculus AB Exam

At this point, we are going to assume that you have been building your calculus knowledge base for more than six months. You will, therefore, use this book primarily as a specific guide to the AP Calculus AB exam.

Given the time constraints, now is not the time to try to expand your AP Calculus curriculum. Rather, it is time to limit and refine what you already do know.

APRIL 1st–15th

______ Skim Steps 1 and 2.
______ Skim Chapters 6–10.
______ Carefully go over the "Rapid Review" sections of Chapters 6–10.
______ Take the Diagnostic Test.
______ Evaluate your strengths and weaknesses.
______ Study appropriate chapters to correct weaknesses.

APRIL 16th–**MAY** 1st

______ Skim Chapters 11–14.
______ Carefully go over the "Rapid Review" sections of Chapters 11–14.
______ Complete Practice Exam 1.
______ Score yourself and analyze your errors.
______ Study appropriate chapters to correct weaknesses.

MAY—First Two Weeks (THIS IS IT!)

______ Complete Practice Exam 2.
______ Score yourself and analyze your errors.
______ Study appropriate chapters to correct weaknesses.
______ Get a good night's sleep the night before the exam. Fall asleep knowing you are well prepared.

GOOD LUCK ON THE TEST!

Summary of the Three Study Plans

MONTH	APPROACH A: SEPTEMBER PLAN	APPROACH B: JANUARY PLAN	APPROACH C: SIX-WEEK PLAN
September–October	Chapters 5–7		
November	Chapters 8 and 9		
December	Chapters 10 and 11 Review Chapters 6–8		
January	Chapter 12 Review Chapters 9–11	Chapters 6–9	
February	Chapter 13 Diagnostic Test	Chapters 5, 10–11 Diagnostic Test Review Chapters 6–9	
March	Chapter 14 Review Chapters 12–14	Chapters 12–14 Review Chapters 10–12	
April	Practice Exam 1 Review Chapters 5–14	Practice Exam 1 Review Chapters 5–14	Diagnostic Test Review Chapters 5–10 Practice Exam 1 Review Chapters 11–14
May	Practice Exam 2	Practice Exam 2	Practice Exam 2

Determine Your Test Readiness

Take a Diagnostic Exam

IN THIS CHAPTER

Summary: Get started in your review by working out the problems in the diagnostic exam. Use the answer sheet to record your answers. After you have finished working the problems, check your answers with the answer key. The problems in the diagnostic exam are presented in small groups matching the order of the review chapters. Your results should give you a good idea of how well prepared you are for the AP Calculus AB exam at this time. Note those chapters that you need to study the most, and spend more time on them. Good luck. You can do it.

Key Ideas

- Work out the problems in the diagnostic exam carefully.
- Check your work against the given answers.
- Determine your areas of strength and weakness.
- Identify and mark the pages that you must give special attention.

DIAGNOSTIC TEST ANSWER SHEET

1. ______
2. ______
3. ______
4. ______
5. ______
6. ______
7. ______
8. ______
9. ______
10. ______
11. ______
12. ______
13. ______
14. ______
15. ______
16. ______
17. ______
18. ______
19. ______
20. ______
21. ______
22. ______
23. ______
24. ______
25. ______
26. ______
27. ______
28. ______
29. ______
30. ______
31. ______
32. ______
33. ______
34. ______
35. ______
36. ______
37. ______
38. ______
39. ______
40. ______
41. ______
42. ______
43. ______
44. ______
45. ______
46. ______
47. ______
48. ______
49. ______
50. ______

3.1 Getting Started!

Taking the Diagnostic Test helps you assess your strengths and weaknesses as you begin preparing for the AP Calculus AB exam. The questions in the Diagnostic Test contain both multiple-choice and open-ended questions. They are arranged by topic and designed to review concepts tested on the AP Calculus AB exam. All questions in the Diagnostic Test should be done without the use of a graphing calculator, except in a few cases where you need to find the numerical value of a logarithmic or exponential function.

3.2 Diagnostic Test

Chapter 5

1. Write an equation of a line passing through the origin and perpendicular to the line $5x - 2y = 10$.

2. Solve the inequality $3|2x - 4| > 6$ graphically. Write the solution in interval notation.

3. Given $f(x) = x^2 + 1$, evaluate $\dfrac{f(3+h) - f(3)}{h}$.

4. Solve for x: $4 \ln x - 2 = 6$.

5. Given $f(x) = x^3 - 8$ and $g(x) = 2x - 2$, find $f(g(2))$.

Chapter 6

6. A function f is continuous on $[-2, 0]$, and some of the values of f are shown below.

x	-2	-1	0
f	4	b	4

If $f(x) = 2$ has no solution on $[-2, 0]$, then b could be

(A) 3

(B) 2

(C) 0

(D) -2

7. Evaluate $\lim_{x \to -\infty} \dfrac{\sqrt{x^2 - 4}}{2x}$.

8. If

$$h(x) = \begin{cases} \sqrt{x} & \text{if } x > 4 \\ x^2 - 12 & \text{if } x \le 4 \end{cases} \quad \text{find } \lim_{x \to 4} h(x).$$

9. If $f(x) = |2xe^x|$, what is the value of $\lim_{x \to 0^+} f'(x)$?

Chapter 7

10. Evaluate $\lim_{x \to \pi} \dfrac{e^x - e^\pi}{x^e - \pi^e}$.

11. Given the equation $y = (x+1)(x-3)^2$, what is the instantaneous rate of change of y at $x = -1$?

12. What is $\lim_{\Delta x \to 0} \dfrac{\tan\left(\frac{\pi}{4} + \Delta x\right) - \tan\left(\frac{\pi}{4}\right)}{\Delta x}$?

Chapter 8

13. The graph of f is shown in the figure below. Draw a possible graph of f' on (a, b).

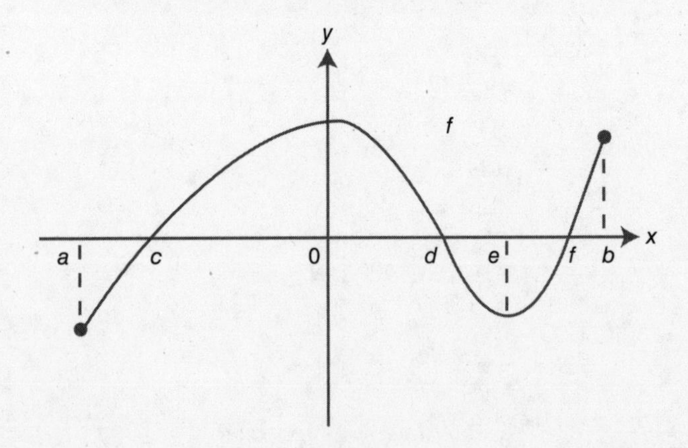

14. The graph of the function g is shown in the figure below. Which of the following is true for g on (a, b)?

I. g is monotonic on (a, b).
II. g' is continuous on (a, b).
III. $g'' > 0$ on (a, b).

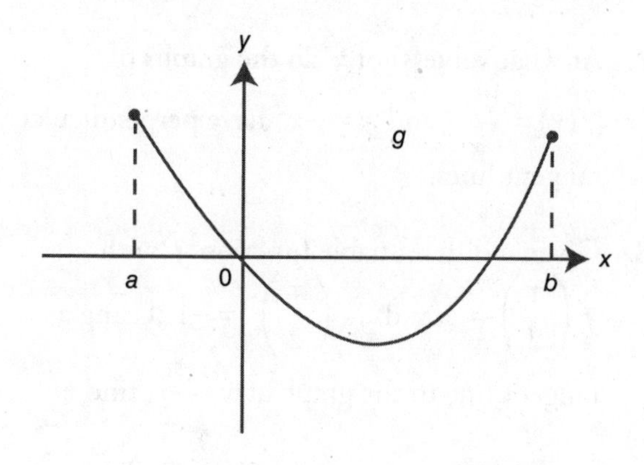

15. The graph of f is shown in the figure below. If f is twice differentiable, which of the following statements is true?

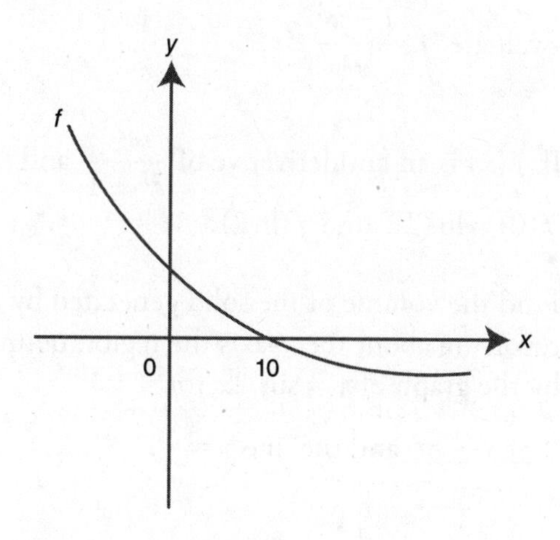

(A) $f(10) < f'(10) < f''(10)$
(B) $f''(10) < f'(10) < f(10)$
(C) $f'(10) < f(10) < f''(10)$
(D) $f'(10) < f''(10) < f(10)$

16. The graph of f', the derivative of f, is shown in the following figure. At what value(s) of x is the graph of f concave up?

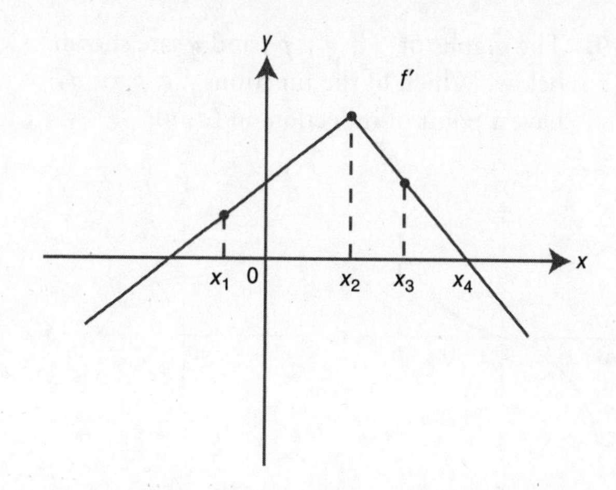

17. How many points of inflection does the graph of $y = \sin(x^2)$ have on the interval $[-\pi, \pi]$?

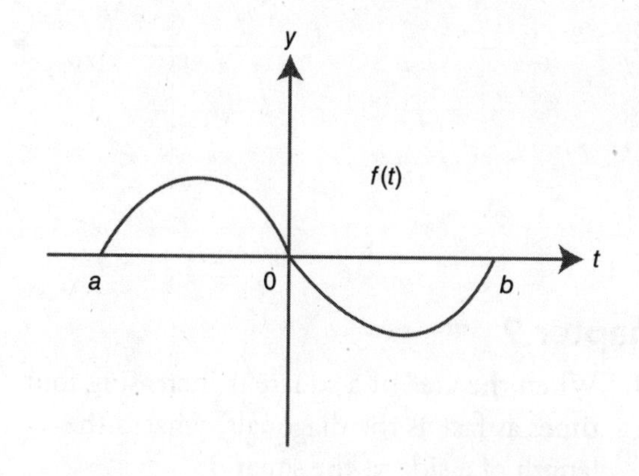

18. If $g(x) = \int_a^x f(t)dt$ and the graph of f is shown above, which of the graphs below is a possible graph of g?

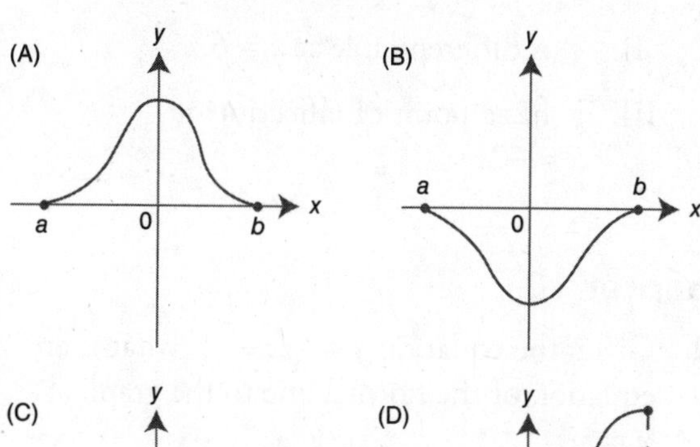

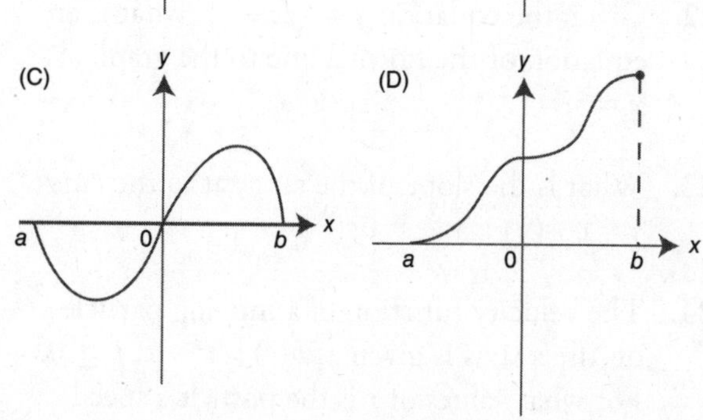

19. The graphs of f', g', p', and q' are shown below. Which of the functions f, g, p, or q have a point of inflection on (a, b)?

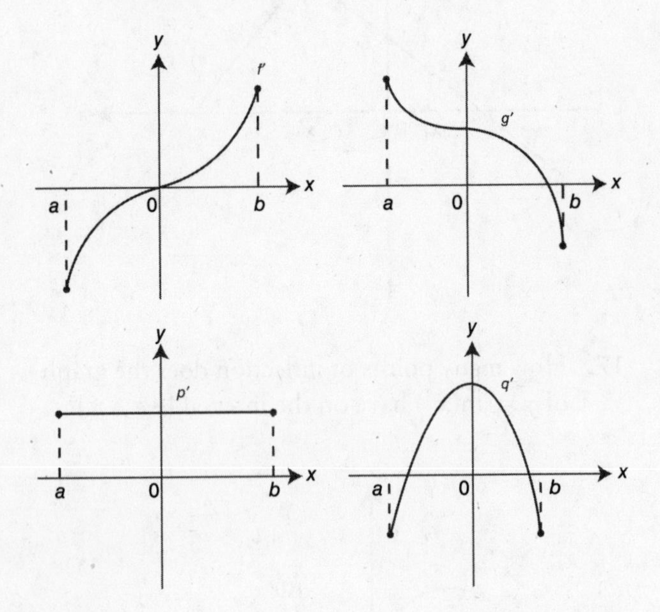

Chapter 9

20. When the area of a square is increasing four times as fast as the diagonals, what is the length of a side of the square?

21. If $g(x) = |x^2 - 4x - 12|$, which of the following statements about g is/are true?

I. g has a relative maximum at $x = 2$.

II. g is differentiable at $x = 6$.

III. g has a point of inflection at $x = -2$.

Chapter 10

22. Given the equation $y = \sqrt{x - 1}$, what is an equation of the normal line to the graph at $x = 5$?

23. What is the slope of the tangent to the curve $y = \cos(xy)$ at $x = 0$?

24. The velocity function of a moving particle on the x-axis is given as $v(t) = t^2 - t$, $t \geq 0$. For what values of t is the particle's speed decreasing?

25. The velocity function of a moving particle is $v(t) = \dfrac{t^3}{3} - 2t^2 + 5$ for $0 \leq t \leq 6$. What is the maximum acceleration of the particle on the interval $0 \leq t \leq 6$?

26. Write an equation of the normal line to the graph of $f(x) = x^3$ for $x \geq 0$ at the point where $f'(x) = 12$.

27. At what value(s) of x do the graphs of $f(x) = \dfrac{\ln x}{x}$ and $y = -x^2$ have perpendicular tangent lines?

28. Given a differentiable function f with $f\left(\dfrac{\pi}{2}\right) = 3$ and $f'\left(\dfrac{\pi}{2}\right) = -1$. Using a tangent line to the graph at $x = \dfrac{\pi}{2}$, find an approximate value of $f\left(\dfrac{\pi}{2} + \dfrac{\pi}{180}\right)$.

Chapter 11

29. Evaluate $\displaystyle\int \frac{1 - x^2}{x^2}\, dx$.

30. If $f(x)$ is an antiderivative of $\dfrac{e^x}{e^x + 1}$ and $f(0) = \ln(2)$, find $f(\ln 2)$.

31. Find the volume of the solid generated by revolving about the x-axis the region bounded by the graph of $y = \sin 2x$ for $0 \leq x \leq \pi$ and the line $y = \dfrac{1}{2}$.

Chapter 12

32. Evaluate $\displaystyle\int_1^4 \frac{1}{\sqrt{x}}\, dx$.

33. If $\displaystyle\int_{-1}^{k} (2x - 3)\, dx = 6$, find k.

34. If $h(x) = \displaystyle\int_{\pi/2}^{x} \sqrt{\sin t}\, dt$, find $h'(\pi)$.

35. If $f'(x) = g(x)$ and g is a continuous function for all real values of x, then $\int_0^2 g(3x)\,dx$ is

(A) $\frac{1}{3}f(6) - \frac{1}{3}f(0)$.
(B) $f(2) - f(0)$.
(C) $f(6) - f(0)$.
(D) $\frac{1}{3}f(0) - \frac{1}{3}f(6)$.

36. Evaluate $\int_\pi^x \sin(2t)\,dt$.

37. If a function f is continuous for all values of x, which of the following statements is/are always true?

I. $\int_a^c f(x)dx = \int_a^b f(x)dx + \int_b^c f(x)dx$

II. $\int_a^b f(x)dx = \int_a^c f(x)dx - \int_c^b f(x)dx$

III. $\int_b^c f(x)dx = \int_b^a f(x)dx - \int_c^a f(x)dx$

38. If $g(x) = \int_{\pi/2}^x 2\sin t\,dt$ on $\left[\frac{\pi}{2}, \frac{5\pi}{2}\right]$, find the value(s) of x where g has a local minimum.

Chapter 13

39. The graph of the velocity function of a moving particle is shown in the following figure. What is the total distance traveled by the particle during $0 \le t \le 6$?

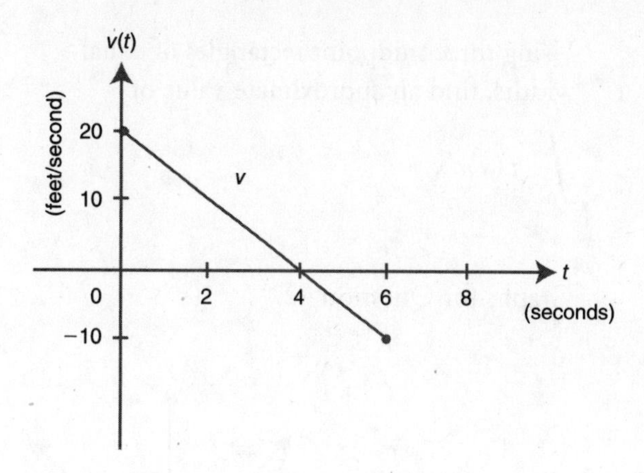

40. The graph of f consists of four line segments, for $-1 \le x \le 5$ as shown in the figure below. What is the value of $\int_{-1}^5 f(x)\,dx$?

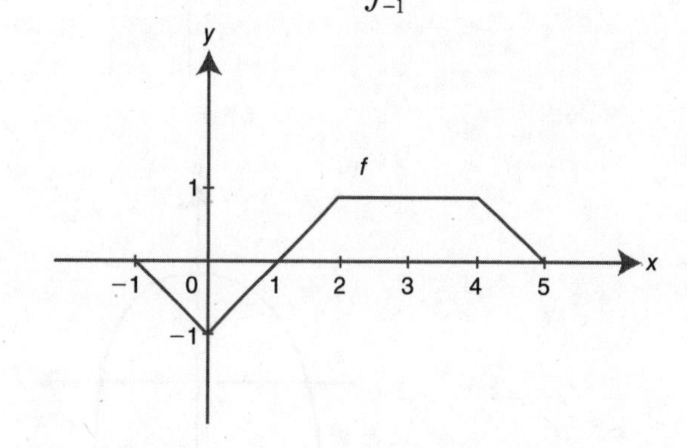

41. Find the area of the region enclosed by the graph of $y = x^2 - x$ and the x-axis.

42. If $\int_{-k}^k f(x)\,dx = 0$ for all real values of k, then which of the graphs shown on the next page could be the graph of f?

43. The area under the curve $y = \sqrt{x}$ from $x = 1$ to $x = k$ is 8. Find the value of k.

44. For $0 \le x \le 3\pi$, find the area of the region bounded by the graphs of $y = \sin x$ and $y = \cos x$.

45. Let f be a continuous function on [0, 6] that has selected values as shown below:

x	0	1	2	3	4	5	6
$f(x)$	1	2	5	10	17	26	37

Using three midpoint rectangles of equal widths, find an approximate value of

$$\int_0^6 f(x)dx.$$

Graphs for Question 42.

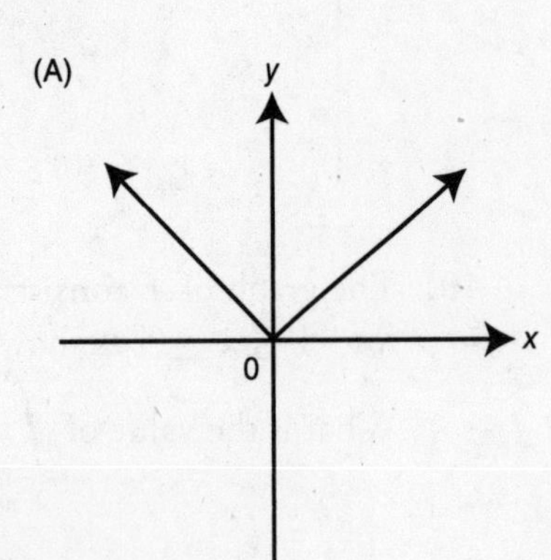

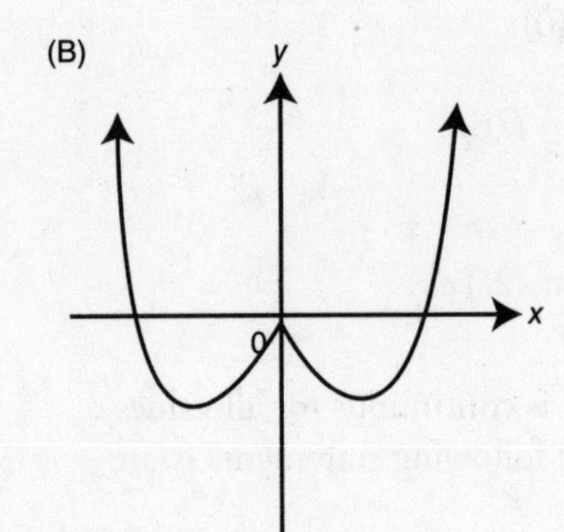

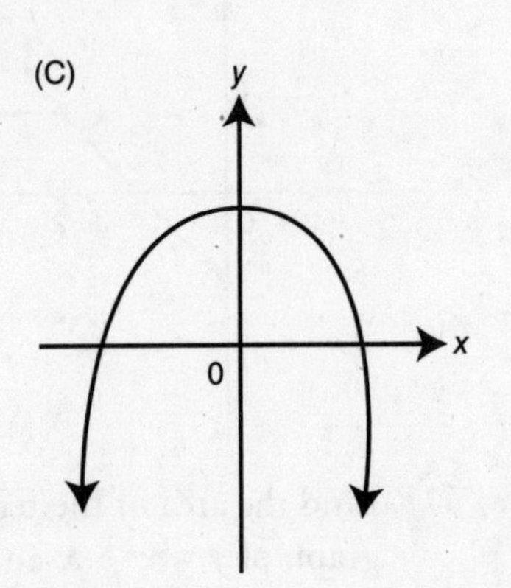

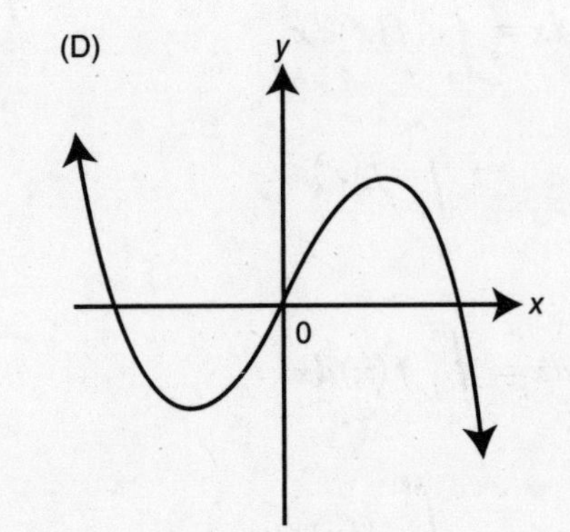

Chapter 14

46. What is the average value of the function $y = e^{-4x}$ on $[-\ln 2, \ln 2]$?

47. If $\frac{dy}{dx} = 2 \sin x$ and at $x = \pi$, $y = 2$, find a solution to the differential equation.

48. Water is leaking from a tank at the rate of $f(t) = 10 \ln(t+1)$ gallons per hour for $0 \leq t \leq 10$, where t is measured in hours. How many gallons of water have leaked from the tank after 5 hours?

49. Carbon-14 has a half-life of 5730 years. If y is the amount of Carbon-14 present and y decays according to the equation $\frac{dy}{dt} = ky$, where k is a constant and t is measured in years, find the value of k.

50. What is the volume of the solid whose base is the region enclosed by the graphs of $y = x^2$ and $y = x + 2$ and whose cross sections perpendicular to the x-axis are squares?

3.3 Answers to Diagnostic Test

1. $y = -\frac{2}{5}x$

2. $(-\infty, 1) \cup (3, \infty)$

3. $6 + h$

4. $x = e^2$

5. 0

6. A

7. $-1/2$

8. Does not exist.

9. 2

10. $\frac{e^{\pi-1}}{\pi^{e-1}}$

11. 16

12. 2

13. See the graph in Question 15 in the Solutions to Diagnostic Test.

14. II & III

15. C

16. $x < x_2$

17. 8

18. A

19. q

20. $2\sqrt{2}$

21. I

22. $y = -4x + 22$

23. 0

24. $\frac{1}{2} < t < 1$

25. 12

26. $y = \frac{-1}{12}x + \frac{49}{6}$

27. 1.370

28. 2.983

29. $\frac{-1}{x} - x + C$

30. $\ln 3$

31. 1.503

32. 2

33. $\{-2, 5\}$

34. 0

35. A

36. $\frac{-1}{2}\cos(2x) + \frac{1}{2}$

37. I & III

38. 2π

39. 50 feet

40. 2

41. $\frac{1}{6}$

42. D

43. $13^{2/3}$

44. 5.657

45. 76

46. $\frac{255}{128 \ln 2}$

47. $y = -2\cos x$

48. 57.506

49. $\frac{-\ln 2}{5730}$

50. $\frac{81}{10}$

3.4 Solutions to Diagnostic Test

Chapter 5

1. Write $5x - 2y = 10$ in slope-intercept form $y = mx + b$ and obtain $y = \frac{5}{2}x - 5$. The slope of the line $y = \frac{5}{2}x - 5$ is $\frac{5}{2}$. Therefore, the slope of the line perpendicular to $y = \frac{5}{2}x - 5$ is $-\frac{2}{5}$. Since the line perpendicular to $y = \frac{5}{2}x - 5$ also passes through the origin, its y-intercept is 0 and its equation is $y = -\frac{2}{5}x$.

2. Using your calculator, set $y_1 = 3|2x - 4|$ and $y_2 = 6$. Examine the figure below and note that $3|2x - 4| > 6$ when $x < 1$ or $x > 3$. Thus, the solution in interval notation is $(-\infty, 1) \cup (3, \infty)$.

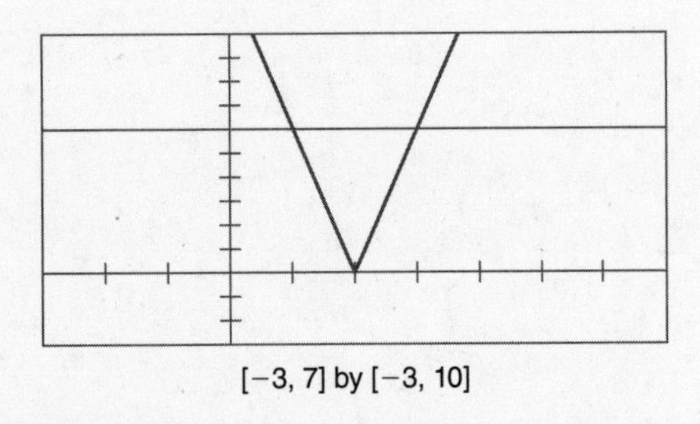

[−3, 7] by [−3, 10]

3. Since $f(x) = x^2 + 1$, $f(3) = 10$ and $f(3 + h) = (3 + h)^2 + 1 = 9 + 6h + h^2 + 1 = 10 + 6h + h^2$.
   The quotient $\frac{f(3+h) - f(3)}{h}$
   $= \frac{10 + 6h + h^2 - 10}{h} = \frac{6h + h^2}{h} = 6 + h$.

4. Simplify the equation $4 \ln x - 2 = 6$ and obtain $\ln x = 2$. (Note that e^x and $\ln x$ are inverse functions.) Thus, $e^{\ln x} = e^2$ or $x = e^2$.

5. Since $g(x) = 2x - 2$, $g(2) = 2(2) - 2 = 2$. Therefore, $f(g(2)) = f(2)$. Also $f(x) = x^3 - 8$. Thus, $f(2) = 2^3 - 8 = 0$.

Chapter 6

6. See the figure below.
   If $b = 2$, then $x = -1$ would be a solution for $f(x) = 2$.
   If $b = 0$, or -2, $f(x) = 2$ would have two solutions.
   Thus, $b = 3$, choice (A).

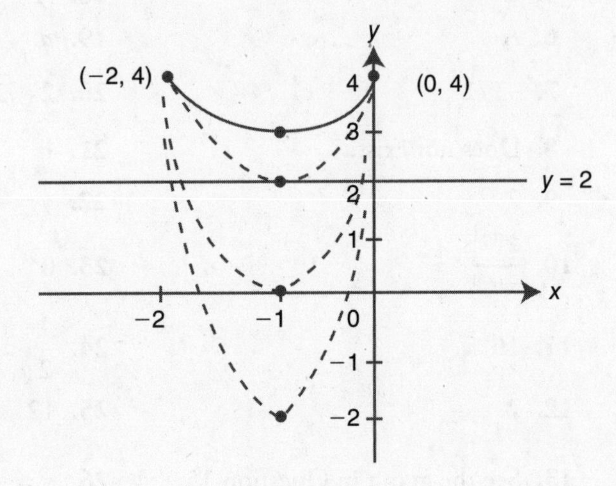

7. $\lim_{x \to -\infty} \frac{\sqrt{x^2 - 4}}{2x} = \lim_{x \to -\infty} \frac{\sqrt{x^2 - 4}/(-\sqrt{x^2})}{2x/(-\sqrt{x^2})}$

   (Note: as $x \to -\infty$, $x = -\sqrt{x^2}$.)

   $= \lim_{x \to -\infty} \frac{-\sqrt{(x^2 - 4)/x^2}}{2}$

   $= \lim_{x \to -\infty} \frac{-\sqrt{1 - (4/x^2)}}{2}$

   $= -\frac{\sqrt{1}}{2} = -\frac{1}{2}$

8. $h(x) = \begin{cases} \sqrt{x} & \text{if } x > 4 \\ x^2 - 12 & \text{if } x \leq 4 \end{cases}$

   $\lim_{x \to 4^+} h(x) = \lim_{x \to 4^+} \sqrt{x} = \sqrt{4} = 2$

   $\lim_{x \to 4^-} h(x) = \lim_{x \to 4^-} (x^2 - 12) = (4^2 - 12) = 4$

   Since $\lim_{x \to 4^+} h(x) \neq \lim_{x \to 4^-} h(x)$, thus $\lim_{x \to 4} h(x)$ does not exist.

9. $f(x) = \left|2xe^x\right| = \begin{cases} 2xe^x & \text{if } x \geq 0 \\ -2xe^x & \text{if } x < 0 \end{cases}$

If $x \geq 0$, $f'(x) = 2e^x + e^x(2x) = 2e^x + 2xe^x$

$\lim_{x \to 0^+} f'(x) = \lim_{x \to 0^+} (2e^x + 2xe^x) = 2e^0 + 0 = 2$

Chapter 7

10. By *L'Hôpital's* Rule, $\lim_{x \to \pi} \frac{e^x - e^\pi}{x^e - \pi^e} = \lim_{x \to \pi} \frac{e^x}{ex^{e-1}}$

$= \lim_{x \to \pi} \frac{e^{x-1}}{x^{e-1}} = \frac{e^{\pi-1}}{\pi^{e-1}}.$

11. $y = (x+1)(x-3)^2$;

$\frac{dy}{dx} = (1)(x-3)^2 + 2(x-3)(x+1)$

$= (x-3)^2 + 2(x-3)(x+1)$

$\left.\frac{dy}{dx}\right|_{x=-1} = (-1-3)^2 + 2(-1-3)(-1+1)$

$= (-4)^2 + 0 = 16$

12. $f'(x_1) = \lim_{\Delta x \to 0} \frac{f(x_1 + \Delta x) - f(x_1)}{\Delta x}$

Thus, $\lim_{\Delta x \to 0} \frac{\tan\left(\frac{\pi}{4} + \Delta x\right) - \tan\left(\frac{\pi}{4}\right)}{\Delta x}$

$= \frac{d}{dx}(\tan x)$ at $x = \frac{\pi}{4}$

$= \sec^2\left(\frac{\pi}{4}\right) = (\sqrt{2})^2 = 2.$

Chapter 8

13. See the graph shown at next column.

14. I. Since the graph of g is decreasing and then increasing, it is not monotonic.

II. Since the graph of g is a smooth curve, g' is continuous.

III. Since the graph of g is concave upward, $g'' > 0$.

Thus, only statements II and III are true.

15. The graph indicates that (1) $f(10) = 0$, (2) $f'(10) < 0$, since f is decreasing; and (3) $f''(10) > 0$, since f is concave upward. Thus, $f'(10) < f(10) < f''(10)$, choice (C).

Graph for Question 13.

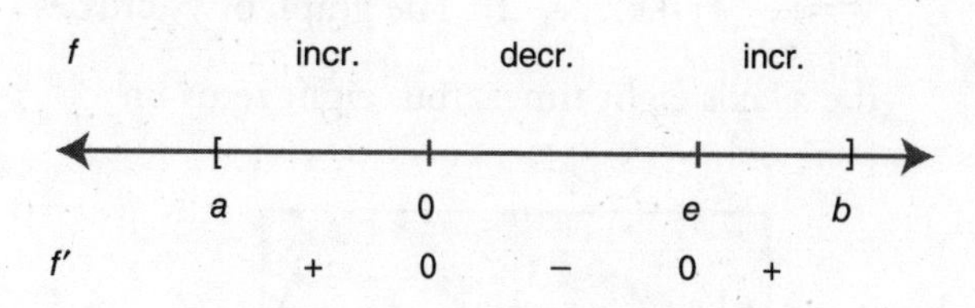

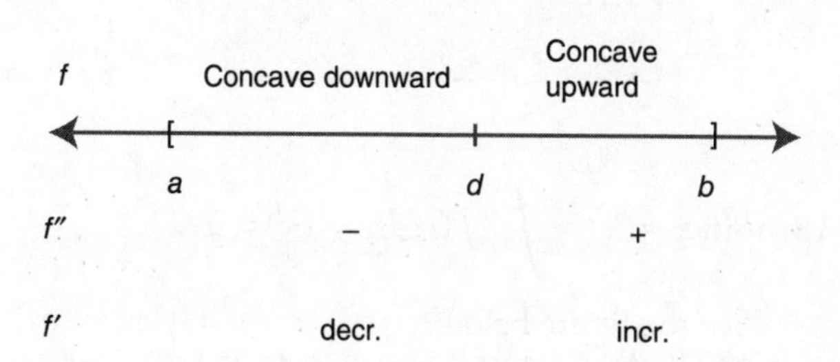

A possible graph of f'

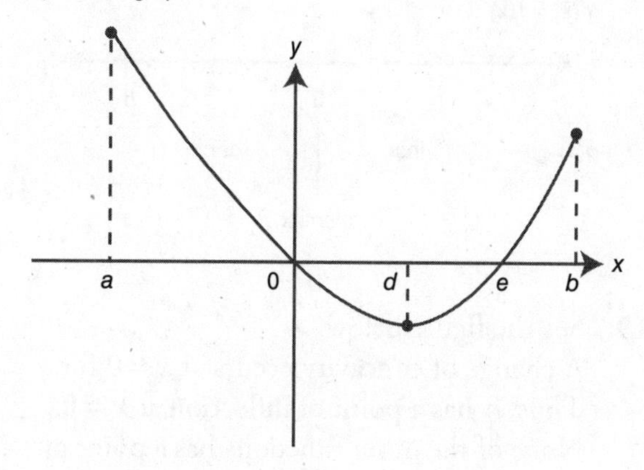

16. See the figure below.
The graph of f is concave upward for $x < x_2$.

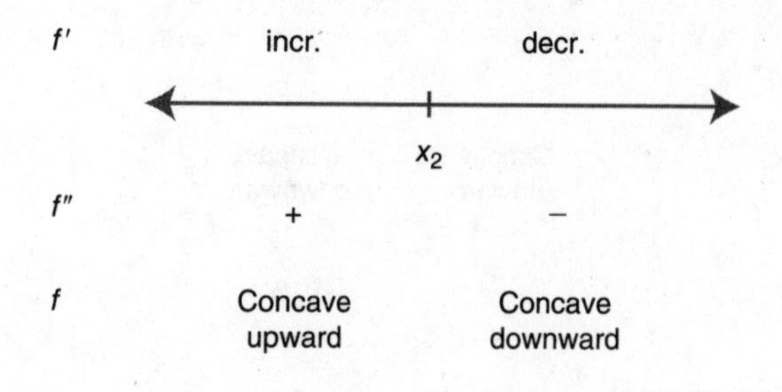

17. See the figure below.
Enter $y^1 = \sin(x^2)$. Using the [*Inflection*] function of your calculator, you obtain four points of inflection on $[0, \pi]$. The points of inflection occur at $x = 0.81$, 1.81, 2.52, and 3.07. Since $y_1 = \sin(x^2)$ is an even function,

there is a total of eight points of inflection on $[-\pi, \pi]$. An alternate solution is to enter $y^2 = \frac{d^2}{dx^2}(y_1(x), x, 2)$. The graph of y_2 crosses the x-axis eight times, thus eight zeros on $[-\pi, \pi]$.

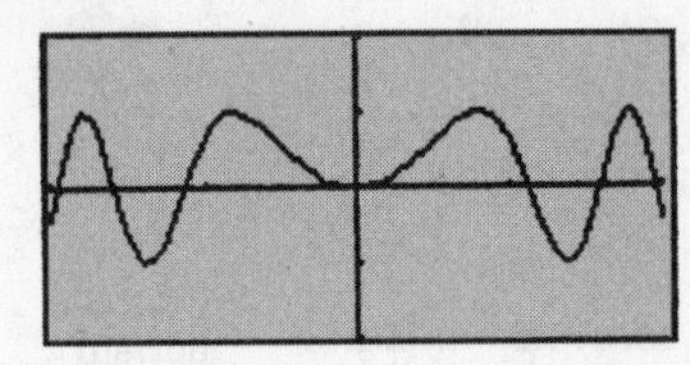

18. Since $g(x) = \int_a^x f(t)dt$, $g'(x) = f(x)$.

See the figure below.
The only graph that satisfies the behavior of g is choice (A).

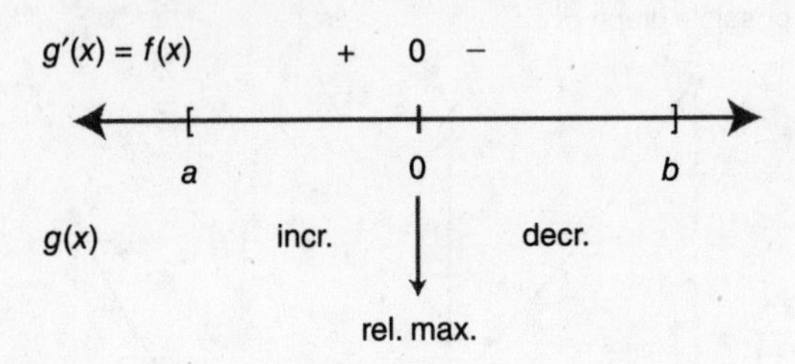

19. See the figure below.
A change of concavity occurs at $x = 0$ for q. Thus, q has a point of inflection at $x = 0$. None of the other functions has a point of inflection.

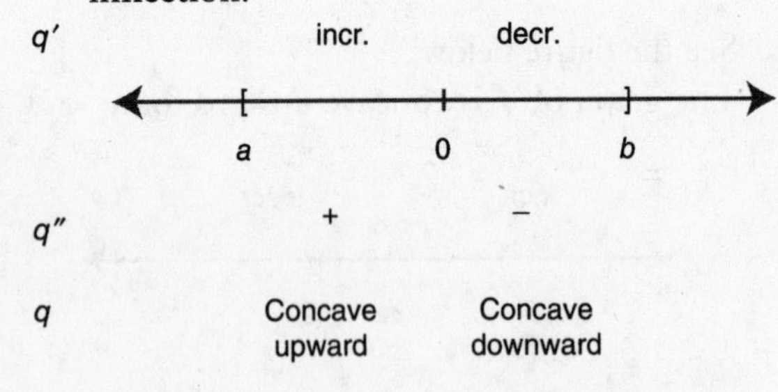

Chapter 9

20. Let z be the diagonal of a square. Area of a square $A = \frac{z^2}{2}$

$$\frac{dA}{dt} = \frac{2z}{2}\frac{dz}{dt} = z\frac{dz}{dt}.$$

Since $\frac{dA}{dt} = 4\frac{dz}{dt}$; $4\frac{dz}{dt} = z\frac{dz}{dt} \Rightarrow z = 4$.

Let s be a side of the square. Since the diagonal $z = 4$, $s^2 + s^2 = z^2$ or $2s^2 = 16$. Thus, $s^2 = 8$ or $s = 2\sqrt{2}$.

21. See the figure below.
The graph of g indicates that a relative maximum occurs at $x = 2$; g is not differentiable at $x = 6$, since there is a *cusp* at $x = 6$, and g does not have a point of inflection at $x = -2$, since there is no tangent line at $x = -2$. Thus, only statement I is true.

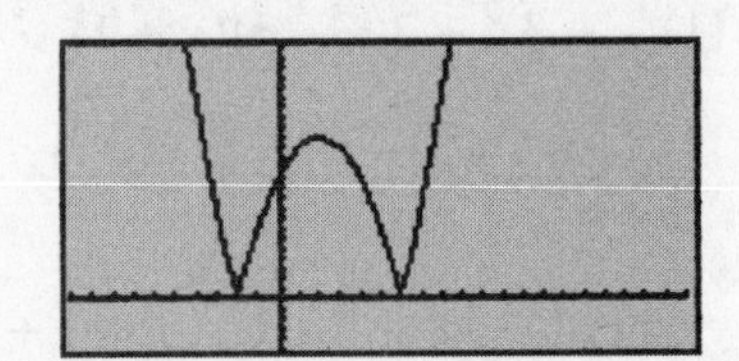

Chapter 10

22. $y = \sqrt{x - 1} = (x - 1)^{1/2}$;

$$\frac{dy}{dx} = \frac{1}{2}(x - 1)^{-1/2}$$

$$= \frac{1}{2(x - 1)^{1/2}}$$

$$\left.\frac{dy}{dx}\right|_{x=5} = \frac{1}{2(5 - 1)^{1/2}} = \frac{1}{2(4)^{1/2}} = \frac{1}{4}$$

At $x = 5$, $y = \sqrt{x - 1} = \sqrt{5 - 1}$
$= 2$; $(5, 2)$.

Slope of normal line = negative reciprocal of $\left(\frac{1}{4}\right) = -4$.

Equation of normal line:
$y - 2 = -4(x - 5) \Rightarrow y = -4(x - 5) + 2$ or
$y = -4x + 22$.

23. $y = \cos(xy)$; $\frac{dy}{dx}$

$$= [-\sin(xy)]\left(1y + x\frac{dy}{dx}\right)$$

$$\frac{dy}{dx} = -y\sin(xy) - x\sin(xy)\frac{dy}{dx}$$

$$\frac{dy}{dx} + x\sin(xy)\frac{dy}{dx} = -y\sin(xy)$$

$$\frac{dy}{dx}[1 + x\sin(xy)] = -y\sin(xy)$$

$$\frac{dy}{dx} = \frac{-y\sin(xy)}{1 + x\sin(xy)}$$

At $x = 0,\ y = \cos(xy) = \cos(0)$

$= 1;\ (0, 1)$

$$\left.\frac{dy}{dx}\right|_{x=0,y=1} = \frac{-(1)\sin(0)}{1 + 0\sin(0)} = \frac{0}{1} = 0.$$

Thus, the slope of the tangent at $x = 0$ is 0.

24. See the figure below.

$v(t) = t^2 - t$

Set $v(t) = 0 \Rightarrow t(t - 1) = 0$

$\Rightarrow t = 0$ or $t = 1$

$a(t) = v'(t) = 2t - 1.$

Set $a(t) = 0 \Rightarrow 2t - 1 = 0$ or $t = \frac{1}{2}$.

Since $v(t) < 0$ and $a(t) > 0$ on $\left(\frac{1}{2}, 1\right)$, the speed of the particle is decreasing on $\left(\frac{1}{2}, 1\right)$.

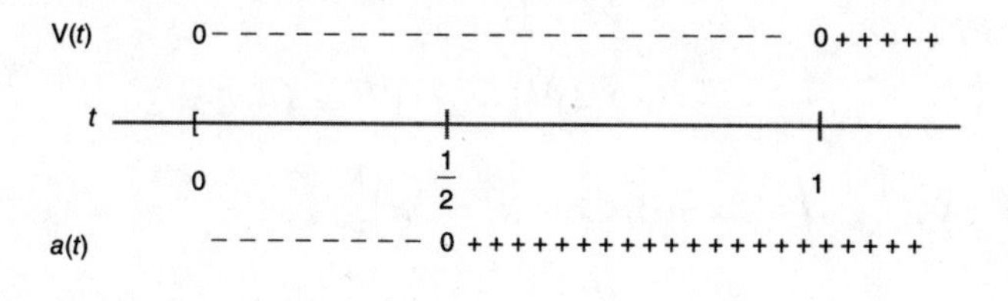

25. $v(t) = \frac{t^3}{3} - 2t^2 + 5$

$a(t) = v'(t) = t^2 - 4t$

See the figure below.
The graph indicates that for $0 \le t \le 6$, the maximum acceleration occurs at the endpoint $t = 6$. $a(t) = t^2 - 4t$ and $a(6) = 6^2 - 4(6) = 12$.

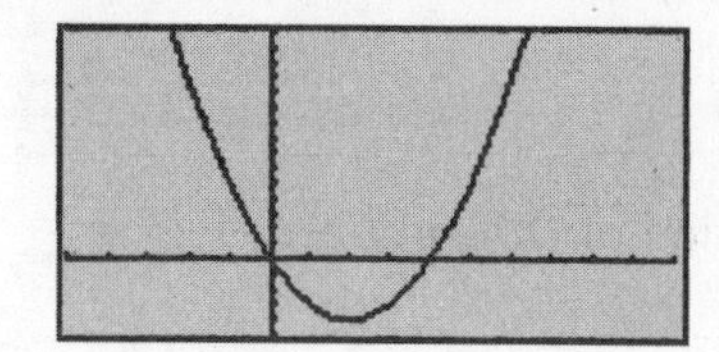

26. $y = x^3,\ x \ge 0;\ \frac{dy}{dx} = 3x^2$

$f'(x) = 12 \Rightarrow \frac{dy}{dx} = 3x^2 = 12$

$\Rightarrow x^2 = 4 \Rightarrow x = 2$

Slope of normal = negative reciprocal of slope of tangent $= -\frac{1}{12}$.

At $x = 2,\ y = x^3 = 2^3 = 8;\ (2, 8)$

$y - 8 = -\frac{1}{12}(x - 2).$

Equation of normal line: $\Rightarrow y = -\frac{1}{12}(x - 2) + 8$

or $y = -\frac{1}{12}x + \frac{49}{6}$.

27. $f(x) = \frac{\ln x}{x};\ f'x = \frac{(1/x)(x) - (1)\ln x}{x^2}$

$= \frac{1}{x^2} - \frac{\ln x}{x^2}$

$y = -x^2;\ \frac{dy}{dx} = -2x$

Perpendicular tangents

$\Rightarrow (f'(x))\left(\frac{dy}{dx}\right) = -1$

$\Rightarrow \left(\left(\frac{1}{x^2}\right) - \frac{\ln x}{x^2}\right)(-2x) = -1.$

Using the [*Solve*] function on your calculator, you obtain $x \approx 1.37015 \approx 1.370$.

28. $f\left(\frac{\pi}{2}\right)=3 \Rightarrow \left(\frac{\pi}{2}, 3\right)$ is on the graph.

$f'\left(\frac{\pi}{2}\right)=-1 \Rightarrow$ slope of the tangent at $x=\frac{\pi}{2}$

is -1.

Equation of tangent line: $y-3=$

$-1\left(x-\frac{\pi}{2}\right)$ or $y=-x+\frac{\pi}{2}+3$.

Thus, $f\left(\frac{\pi}{2}+\frac{\pi}{180}\right) \approx -\left(\frac{\pi}{2}+\frac{\pi}{180}\right)$

$+\frac{\pi}{2}+3$

$\approx 3-\frac{\pi}{180} \approx 2.98255$

$\approx 2.983.$

Chapter 11

29. $\int \frac{1-x^2}{x^2}dx=\int\left(\frac{1}{x^2}-\frac{x^2}{x^2}\right)dx$

$=\int\left(\frac{1}{x^2}-1\right)dx$

$=\int (x^{-2}-1)dx=\frac{x^{-1}}{-1}-x+C$

$=-\frac{1}{x}-x+C$

You can check the answer by differentiating your result.

30. Let $u=e^x+1$; $du=e^x dx$.

$f(x)=\int \frac{e^x}{e^x+1}dx=\int \frac{1}{u}du$

$=\ln|u|+C=\ln|e^x+1|+C$

$f(0)=\ln|e^0+1|+C=\ln(2)+C$

Since $f(0)=\ln 2 \Rightarrow \ln(2)+C=\ln 2$

$\Rightarrow C=0.$

Thus, $f(x)=\ln(e^x+1)$ and $f(\ln 2)$

$=\ln(e^{\ln 2}+1)=\ln(2+1)$

$=\ln 3.$

31. See the figure below.
To find the points of intersection, set

$\sin 2x=\frac{1}{2} \Rightarrow 2x=\sin^{-1}\left(\frac{1}{2}\right)$

$\Rightarrow 2x=\frac{\pi}{6}$ or $2x=\frac{5\pi}{6} \Rightarrow x=\frac{\pi}{12}$ or $x=\frac{5\pi}{12}$.

Volume of solid

$=\pi\int_{\pi/12}^{5\pi/12}\left[(\sin 2x)^2-\left(\frac{1}{2}\right)^2\right]dx.$

Using your calculator, you obtain:
Volume of solid $\approx (0.478306)\pi$
$\approx 1.50264 \approx 1.503.$

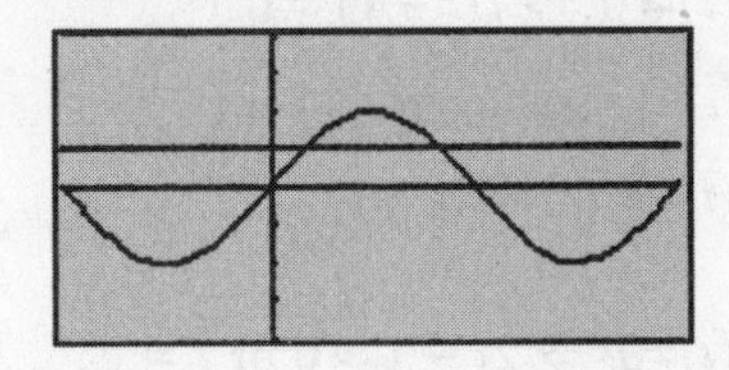

Chapter 12

32. $\int_1^4 \frac{1}{\sqrt{x}}dx=\int_1^4 x^{-1/2}dx=\left.\frac{x^{1/2}}{1/2}\right]_1^4$

$=\left.2x^{1/2}\right]_1^4$

$=2(4)^{1/2}-2(1)^{1/2}=4-2=2$

33. $\int_{-1}^{k}(2x-3)dx=\left.x^2-3x\right]_{-1}^{k}$

$=(k^2-3k)-((-1)^2$

$-3(-1))$

$=k^2-3k-(1+3)$

$=k^2-3k-4$

Set $k^2-3k-4=6 \Rightarrow k^2-3k-10=0$

$\Rightarrow (k-5)(k+2)=0 \Rightarrow k=5$ or $k=-2$.

You can check your answer by evaluating $\int_{-1}^{-2} (2x-3)dx$ and $\int_{-1}^{5} (2x-3)dx$.

34. $h(x) = \int_{\pi/2}^{\pi} \sqrt{\sin t}\, dt \Rightarrow h'(x) = \sqrt{\sin x}$

$h'(\pi) = \sqrt{\sin \pi} = \sqrt{0} = 0$

35. Let $u = 3x$; $du = 3dx$ or $\frac{du}{3} = dx$.

$$\int g(3x)dx = \int g(u)\frac{du}{3} = \frac{1}{3}\int g(u)du$$

$$= \frac{1}{3}f(u) + c = \frac{1}{3}f(3x) + c$$

$$\int_0^2 g(3x)dx = \frac{1}{3}[f(3x)]_0^2$$

$$= \frac{1}{3}f(6) - \frac{1}{3}f(0)$$

Thus, the correct choice is (A).

36. $\int_{\pi}^{x} \sin(2t)dt = \left[\frac{-\cos(2t)}{2}\right]_{\pi}^{x}$

$$= \frac{-\cos(2x)}{2} - \left(-\frac{\cos(2\pi)}{2}\right)$$

$$= -\frac{1}{2}\cos(2x) + \frac{1}{2}$$

37. I. $\int_a^c f(x)dx = \int_a^b f(x)dx + \int_b^c f(x)dx$

The statement is true, since the upper and lower limits of the integrals are in sequence, i.e., $a \to c = a \to b \to c$.

II. $\int_a^b f(x)dx = \int_a^c f(x)dx - \int_c^b f(x)dx$

$$= \int_a^c f(x)dx + \int_b^c f(x)dx$$

The statement is not always true.

III. $\int_b^c f(x)dx = \int_b^a f(x)dx - \int_c^a f(x)dx$

$$= \int_b^a f(x)dx + \int_a^c f(x)dx$$

The statement is true.
Thus, only statements I and III are true.

38. Since $g(x) = \int_{\pi/2}^{x} 2\sin t\, dt$, then $g'(x) = 2\sin x$.
Set $g'(x) = 0 \Rightarrow 2\sin x = 0 \Rightarrow x = \pi$ or 2π
$g''(x) = 2\cos x$ and $g''(\pi) = 2\cos\pi = -2$ and $g''(2\pi) = 1$.
Thus g has a local minimum at $x = 2\pi$. You can also approach the problem geometrically by looking at the area under the curve. See the figure below.

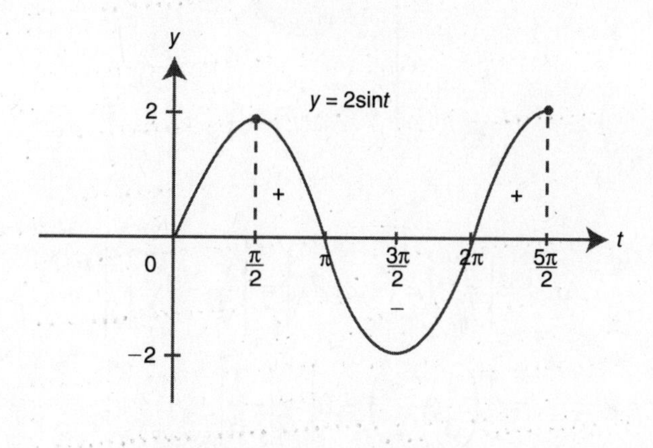

Chapter 13

39. Total distance $= \int_0^4 v(t)dt + \left|\int_4^6 v(t)dt\right|$

$$= \frac{1}{2}(4)(20) + \left|\frac{1}{2}(2)(-10)\right|$$

$$= 40 + 10 = 50 \text{ feet}$$

40. $\int_{-1}^{5} f(x)dx = \int_{-1}^{1} f(x)dx + \int_{1}^{5} f(x)dx$

$$= -\frac{1}{2}(2)(1) + \frac{1}{2}(2+4)(1)$$

$$= -1 + 3 = 2$$

41. To find points of intersection, set $y = x^2 - x = 0$
$\Rightarrow x(x-1) = 0 \Rightarrow x = 0$ or $x = 1$.
See the figure below.

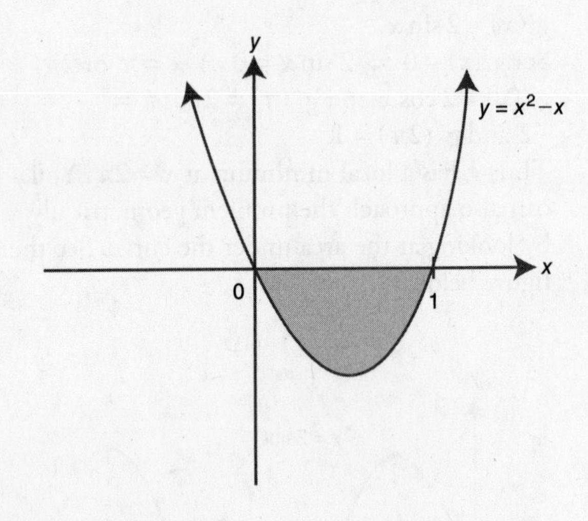

$$\text{Area} = \left|\int_0^1 (x^2 - x)\,dx\right| = \left|\left[\frac{x^3}{3} - \frac{x^2}{2}\right]_0^1\right|$$

$$= \left|\left(\frac{1}{3} - \frac{1}{2}\right) - 0\right| = \left|-\frac{1}{6}\right|$$

$$= \frac{1}{6}$$

42. $\int_{-k}^{k} f(x)\,dx = 0 \Rightarrow f(x)$ is an odd function, i.e., $f(x) = -f(-x)$. Thus, the graph in choice (D) is the only odd function.

43. Area $= \int_1^k \sqrt{x}\,dx = \int_1^k x^{1/2}\,dx$

$$= \left[\frac{x^{3/2}}{3/2}\right]_1^k$$

$$= \left[\frac{2}{3}x^{3/2}\right]_1^k = \frac{2}{3}k^{3/2} - \frac{2}{3}(1)^{3/2}$$

$$= \frac{2}{3}k^{3/2} - \frac{2}{3} = \frac{2}{3}(k^{3/2} - 1)$$

Since A = 8, set $\frac{2}{3}(k^{3/2}-1) = 8 \Rightarrow k^{3/2} - 1$
$= 12 \Rightarrow k^{3/2} = 13$ or $k = 13^{2/3}$.

44. See the figure below.

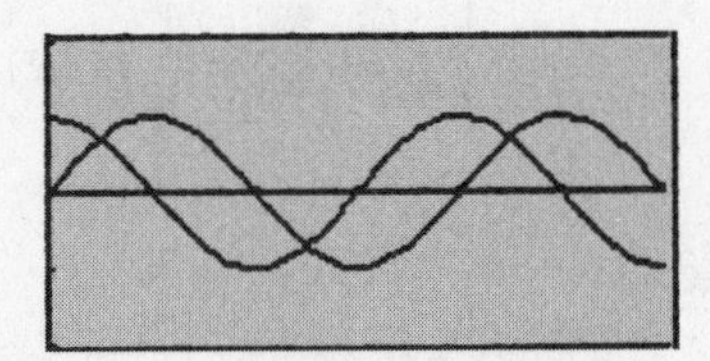

Using the [*Intersection*] function of the calculator, you obtain the intersection points at $x = 0.785398$, 3.92699, and 7.06858.

$$\text{Area} = \int_{0.785398}^{3.92699} (\sin x - \cos x)dx$$

$$+ \int_{3.92699}^{7.06858} (\cos x - \sin x)dx$$

$$= 2.82843 + 2.82843 \approx 5.65685$$

$$\approx 5.657$$

You can also find the area by:

$$\text{Area} = \int_{.785398}^{7.06858} |\sin x - \cos x|\,dx$$

$$\approx 5.65685 \approx 5.657$$

45. Width of a rectangle $= \dfrac{6-0}{3} = 2.$

Midpoints are $x = 1, 3,$ and 5 and $f(1) = 2,$ $f(3) = 10,$ and $f(5) = 26.$

$$\int_0^6 f(x)dx \approx 2(2+10+26) \approx 2(38) = 76$$

Chapter 14

46. Average value $= \dfrac{1}{\ln 2 - (-\ln 2)} \displaystyle\int_{-\ln 2}^{\ln 2} e^{-4x}\, dx.$

Let $u = -4x$; $du = -4dx$, or $\dfrac{-du}{4} = dx.$

$$\int e^{-4x} dx = \int e^{u}\left(\frac{-du}{4}\right) = \frac{-1}{4}e^{u} + C$$

$$= \frac{-1}{4}e^{-4x} + C$$

$$\text{Average value} = \frac{1}{2\ln 2}\left[\frac{e^{-4x}}{-4}\right]_{-\ln 2}^{\ln 2}$$

$$= \frac{1}{2\ln 2}\left[\left(\frac{e^{-4\ln 2}}{-4}\right) - \left(\frac{e^{-4(-\ln 2)}}{-4}\right)\right]$$

$$= \frac{1}{2\ln 2}\left[\frac{(e^{\ln 2})^{-4}}{-4} + \frac{(e^{\ln 2})^{4}}{4}\right]$$

$$= \frac{1}{2\ln 2}\left[\frac{2^{-4}}{-4} + \frac{2^{4}}{4}\right]$$

$$= \frac{1}{2\ln 2}\left(\frac{1}{-64} + 4\right)$$

$$= \frac{1}{2\ln 2}\left(\frac{255}{64}\right) = \frac{255}{128\ln 2}$$

47. $\dfrac{dy}{dx} = 2\sin x \Rightarrow dy = 2\sin x\, dx$

$$\int dy = \int 2\sin x\, dx \Rightarrow y = -2\cos x + C$$

At $x = \pi$, $y = 2 \Rightarrow 2 = -2\cos \pi + C$

$\Rightarrow 2 = (-2)(-1) + C$

$\Rightarrow 2 = 2 + C = 0.$

Thus, $y = -2\cos x.$

48. Amount of water leaked

$$= \int_0^5 10\ln(t+1)\, dt.$$

Using your calculator, you obtain $10(6\ln 6 - 5)$ which is approximately 57.506 gallons.

49. $\dfrac{dy}{dx} = ky \Rightarrow y = y_0e^{kt}$

Half-life $= 5730 \Rightarrow y = \dfrac{1}{2}y_0$

when $t = 5730.$

Thus, $\dfrac{1}{2}y_0 = y_0e^{k(5730)} \Rightarrow \dfrac{1}{2} = e^{5730k}.$

$$\ln\left(\frac{1}{2}\right) = \ln\left(e^{5730k}\right) \Rightarrow \ln\left(\frac{1}{2}\right) = 5730k$$

$$\ln 1 - \ln 2 = 5730k \Rightarrow -\ln 2 = 5730k$$

$$k = \frac{-\ln 2}{5730}$$

50. See the figure below.

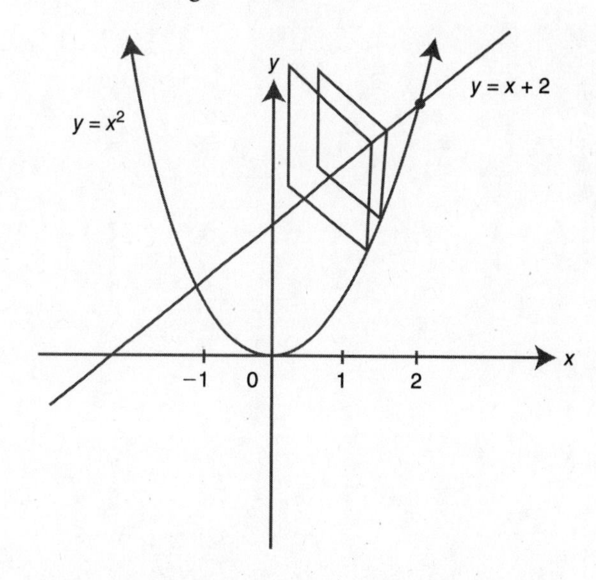

To find points of intersection, set $x^2 = x + 2$ $\Rightarrow x^2 - x - 2 = 0 \Rightarrow x = 2$ or $x = -1.$

Area of cross section $= ((x+2) - x^2)^2.$

Volume of solid, $V = \displaystyle\int_{-1}^{2} \left(x + 2 - x^2\right)^2 dx.$

Using your calculator, you obtain: $V = \dfrac{81}{10}.$

3.5 Calculate Your Score

Short-Answer Questions

Questions 1–50 for AP Calculus AB

Number of correct answers = ______________
raw score

AP Calculus AB Diagnostic Test

RAW SCORE	APPROXIMATE AP GRADE
43–50	5
36–42	4
29–35	3
18–28	2
0–17	1

Develop Strategies for Success

CHAPTER 4 How to Approach Each Question Type

How to Approach Each Question Type

IN THIS CHAPTER

Summary: Knowing and applying question-answering strategies helps you succeed on tests. This chapter provides you with many test-taking tips to help you earn a 5 on the AP Calculus exam.

Key Ideas

- Read each question carefully.
- Do not linger on a question. Time yourself accordingly.
- For multiple-choice questions, sometimes it is easier to work backward by trying each of the given choices. You will be able to eliminate some of the choices quickly.
- For free-response questions, always show sufficient work so that your line of reasoning is clear.
- Write legibly.
- Always use calculus notations instead of calculator syntax.
- If the question involves decimals, round your final answer to 3 decimal places unless the question indicates otherwise.
- Trust your instincts. Your first approach to solving a problem is usually the correct one.
- Get a good night's sleep the night before.

4.1 The Multiple-Choice Questions

- There are 45 multiple-choice questions for the AP Calculus AB exam. These questions are divided into Section I–Part A, which consists of 30 questions for which the use of a calculator is not permitted; and Section I–Part B with 15 questions, for which the use of a graphing calculator is allowed. The multiple-choice questions account for 50% of the grade for the whole test.
- Do the easy questions first because all multiple-choice questions are worth the same amount of credit. You have 60 minutes for the 30 questions in Section I–Part A and 45 minutes for the 15 questions in Section I–Part B. Do not linger on any one question. Time yourself accordingly.
- There is no partial credit for multiple-choice questions, and you do not need to show work to receive credit for the correct answer.

- Read the question carefully. If there is a graph or a chart, look at it carefully. For example, be sure to know if the given graph is that of $f(x)$ or $f'(x)$. Pay attention to the scale of the x and y axes and the unit of measurement.

- Never leave a question blank since there is **no penalty** for incorrect answers.
- If a question involves finding the derivative of a function, you must first find the derivative, and then see if you need to do additional work to get the final answer to the question. For example, if a question asks for an equation of the tangent line to a curve at a given point, you must first find the derivative, evaluate it at the given point (which gives you the slope of the line), and then proceed to find an equation of the tangent line. For some questions, finding the derivative of a given function (or sometimes, the antiderivative), is only the first step to solving the problem. It is not the final answer to the question. You might need to do more work to get the final answer.

- Sometimes, it is easier to work backward by trying each of the given choices as the final answer. Often, you will be able to eliminate some of the given choices quickly.
- If a question involves decimal numbers, do not round until the final answer, and at that point, the final answer is usually rounded to 3 decimal places. Look at the number of decimal places of the answers in the given choices.
- Trust your instincts. Usually your first approach to solving a problem is the correct one.

4.2 The Free-Response Questions

- There are 6 free-response questions in Section II: Part A consisting of 2 questions which allow the use of a calculator, and Part B with 4 questions which do not permit the use of a calculator. The 6 free-response questions account for 50% of the grade for the whole test.
- Read, read, read. Read the question carefully. Know what information is given, what quantity is being sought, and what additional information you need to find in order to answer the question.

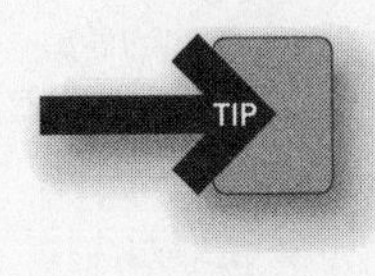

- Always show a sufficient amount of work so that your line of reasoning is clear. This is particularly important in determining partial credit. In general, use complete sentences to explain your reasoning. Include all graphs, charts, relevant procedures, and theorems. Clearly indicate all the important steps that you have taken in solving the problem. A correct answer with insufficient work will receive minimal credit.

- When appropriate, represent the given information in calculus notations. For example, if it is given that the volume of a cone is decreasing at 2 cm^3 per second, write $\frac{dV}{dt} = -2\ \text{cm}^3/\text{sec}$. Similarly, represent the quantity being sought in calculus notations. For example, if the question asks for the rate of change of the radius of the cone at 5 seconds, write "Find $\frac{dr}{dt}$ at $t = 5$ sec."

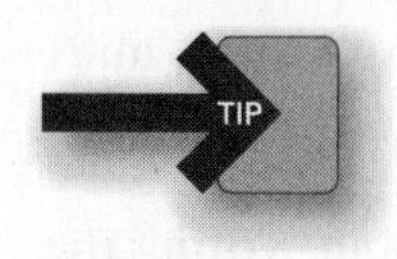

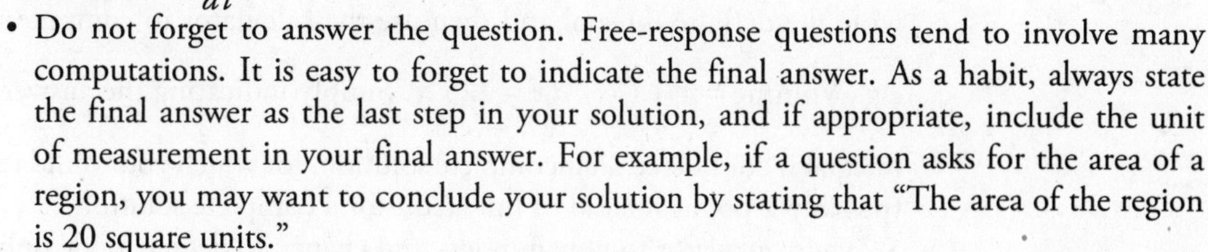

- Do not forget to answer the question. Free-response questions tend to involve many computations. It is easy to forget to indicate the final answer. As a habit, always state the final answer as the last step in your solution, and if appropriate, include the unit of measurement in your final answer. For example, if a question asks for the area of a region, you may want to conclude your solution by stating that "The area of the region is 20 square units."
- Do the easy questions first. Each of the 6 free-response questions is worth the same amount of credit. There is no penalty for an incorrect solution.
- Pay attention to the scales of the x and y axes, the unit of measurement, and the labeling of given charts and graphs. For example, be sure to know whether a given graph is that of $f(x)$ or $f'(x)$.
- When finding relative extrema or points of inflection, you must show the behavior of the function that leads to your conclusion. Simply showing a sign chart is not sufficient.

- Often a question has several parts. Sometimes, in order to answer a question in one part of the question, you might need the answer to an earlier part of the question. For example, to answer the question in part (b), you might need the answer in part (a). If you are not sure how to answer part (a), make an educated guess for the best possible answer and then use this answer to solve the problem in part (b). If your solution in part (b) uses the correct approach but your final answer is incorrect, you could still receive full or almost full credit for your work.
- As with solving multiple-choice questions, trust your instincts. Your first approach to solving a problem is usually the correct one.

4.3 Using a Graphing Calculator

- The use of a graphing calculator is permitted in Section I–Part B multiple-choice questions and in Section II–Part A free-response questions.
- You are permitted to use the following 4 built-in capabilities of your graphing calculator to obtain an answer:

  1. plotting the graph of a function
  2. finding the zeros of a function
  3. calculating numerically the derivative of a function
  4. calculating numerically the value of a definite integral

For example, if you have to find the area of a region, you need to show a definite integral. You may then proceed to use the calculator to produce the numerical value of the definite integral without showing any supporting work. All other capabilities of your calculator can only be used to *check* your answer. For example, you may not use the built-in [*Inflection*] function of your calculator to find points of inflection. You must use calculus showing derivatives and indicating a change of concavity.

- You may *not* use calculator syntax to substitute for calculus notations. For example, you may *not* write "Volume $= \int \left(\pi (5x)^\wedge 2, x, 0, 3\right) = 225\ \pi$"; instead, you need to write "Volume $= \pi \int_0^3 (5x)^2 dx = 225\ \pi$."
- When using a graphing calculator to solve a problem, you are required to write the setup that leads to the answer. For example, if you are finding the volume of a solid, you must write the definite integral and then use the calculator to compute the numerical value, e.g., Volume $= \pi \int_0^3 (5x)^2 dx = 225\ \pi$. Simply indicating the answer without writing the integral is considered an incomplete solution, for which you would receive minimal credit (possibly 1 point) instead of full credit for a complete solution.
- Set your calculator to radian mode, and change to degree mode only if necessary.
- If you are using a TI-89 graphing calculator, clear all previous entries for variables a through z before the AP Calculus exam.
- You are permitted to store computer programs in your calculator and use them in the AP Calculus exam. Your calculator memories will not be cleared.
- Using the [*Trace*] function to find points on a graph may not produce the required accuracy. Most graphing calculators have other built-in functions that can produce more accurate results. For example, to find the x-intercepts of a graph, use the [*Zero*] function, and to find the intersection point of two curves, use the [*Intersection*] function.
- When decimal numbers are involved, do not round until the final answer. Unless otherwise stated, your final answer should be accurate to three places after the decimal point.
- You may bring up to two calculators to the AP Calculus exam.
- Replace old batteries with new ones and make sure that the calculator is functioning properly before the exam.

4.4 Taking the Exam

What Do I Need to Bring to the Exam?

- Several Number 2 pencils.
- A good eraser and a pencil sharpener.
- Two black or blue pens.
- One or two approved graphing calculators with fresh batteries. (Be careful when you change batteries so that you don't lose your programs.)
- A watch.
- An admissions card or a photo I.D. card if your school or the test site requires it.
- Your Social Security number.
- Your school code number if the test site is not at your school.
- A simple snack *if the test site permits it.* (Don't eat anything you haven't eaten before. You might have an allergic reaction.)
- A light jacket if you know that the test site has strong air conditioning.
- Do *not* bring White Out or scrap paper.

Tips for Taking the Exam

General Tips

- Write legibly.
- Label all diagrams.
- Organize your solution so that the reader can follow your line of reasoning.
- Use complete sentences whenever possible. Always indicate what the final answer is.

More Tips

- Do easy questions first.
- Write out formulas and indicate all major steps.
- Never leave a question blank, especially a multiple-choice question, since there is no penalty for incorrect answers.
- Be careful to bubble in the right grid, especially if you skip a question.
- Move on. Don't linger on a problem too long. Make an educated guess.
- Go with your first instinct if you are unsure.

Still More Tips

- Indicate units of measure.
- Simplify numeric or algebraic expressions only if the question asks you to do so.
- Carry all decimal places and round only at the end.
- Round to 3 decimal places unless the question indicates otherwise.
- Watch out for different units of measure, e.g., the radius, r, is 2 feet, find $\frac{dr}{dt}$ in inches per second.
- Use calculus notations and not calculator syntax, e.g., write $\int x^2 dx$ and not $\int (x\wedge 2, x)$.
- Use only the four specified capabilities of your calculator to get your answer: plotting graphs, finding zeros, calculating numerical derivatives, and evaluating definite integrals. All other built-in capabilities can only be used to *check* your solution.
- Answer all parts of a question from Section II even if you think your answer to an earlier part of the question might not be correct.

Enough Already . . . Just 3 More Tips

- Be familiar with the instructions for the different parts of the exam. Review the practice exams in the back of this book.
- Get a good night's sleep the night before.
- Have a light breakfast before the exam.

Review the Knowledge You Need to Score High

Review of Precalculus

IN THIS CHAPTER

Summary: Many questions on the AP Calculus exam require the application of precalculus concepts. In this chapter, you will be guided through a summary of these precalculus concepts including the properties of lines, inequalities, and absolute values, and the behavior of functions and inverse functions. It is important that you review this chapter thoroughly. Your ability to solve the problems here is a prerequisite to doing well on the AP Calculus AB exam.

Key Ideas

- Writing an equation of a line
- Solving inequalities
- Working with functions and inverse functions
- Properties of trigonometric and inverse trigonometric functions
- Properties of exponential and logarithmic functions
- Properties of odd and even functions
- Behaviors of increasing and decreasing functions

5.1 Lines

Main Concepts: Slope of a Line, Equations of a Line, Parallel and Perpendicular Lines

Slope of a Line

Given two points $A(x_1, y_1)$ and $B(x_2, y_2)$, the *slope* of the line passing through the two given points is defined as

$$m = \frac{y_2 - y_1}{x_2 - x_1} \quad \text{where} \quad (x^2 - x_1) \neq 0$$

Note that if $(x_2 - x_1) = 0$, then $x_2 = x_1$, which implies that points A and B are on a vertical line parallel to the y-axis, and thus, the slope is *undefined.*

Example 1

Find the slope of the line passing through the points (3, 2) and (5, −4).

Using the definition $m = \frac{y_2 - y_1}{x_2 - x_1}$, the slope of the line is $m = \frac{-4 - 2}{5 - 3} = \frac{-6}{2} = -3$.

Example 2

Find the slope of the line passing through the points (−5, 3) and (2, 3).

The slope $m = \frac{3 - 3}{2 - (-5)} = \frac{0}{2 + 5} = \frac{0}{7} = 0$. This implies that the points (−5, 3) and (2, 3) are on a horizontal line parallel to the x-axis.

Example 3

Figure 5.1-1 is a summary of four different orientations of lines and their slopes:

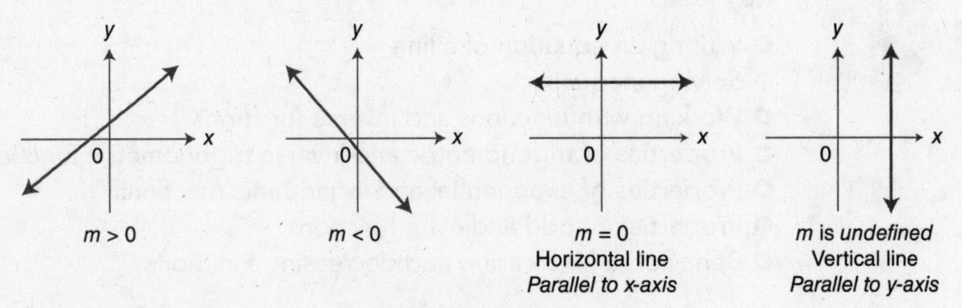

Figure 5.1-1

Equations of a Line

$y = mx + b$ — *Slope-intercept form* of a line where m is its slope and b is the y-intercept.

$y - y_1 = m(x - x_1)$ — *Point-slope form* of a line where m is the slope and (x_1, y_1) is a point on the line.

$Ax + By + C = 0$ — *General form* of a line where A, B, and C are constants and A and B are not *both* equal to 0.

Example 1

Write an equation of the line through the points (−2, 1) and (3, −9).

The slope of line passing through (−2, 1) and (3, −9) is $m = \frac{-9-1}{3-(-2)} = \frac{-10}{5} = -2$. Using the point-slope form and the point (−2, 1),

$$y - 1 = -2(x - (-2))$$

$$y - 1 = -2(x + 2) \text{ or } y = -2x - 3$$

An equation of the line is $y = -2x - 3$.

Example 2

An equation of a line l is $2x + 3y = 12$. Find the slope, the x-intercept, and the y-intercept of line l.

Begin by expressing the equation $2x + 3y = 12$ in *slope-intercept form.*

$$2x + 3y = 12$$

$$3y = -2x + 12$$

$$y = \frac{-2}{3}x + 4$$

Therefore, m, the slope of line l, is $\frac{-2}{3}$ and b, the y-intercept, is 4. To find the x-intercept, set $y = 0$ in the original equation $2x + 3y = 12$. Thus, $2x + 0 = 12$ and $x = 6$. The x-intercept of line l is 6.

Example 3

Equations of *vertical* and *horizontal* lines involve only a single variable. Figure 5.1-2 shows several examples:

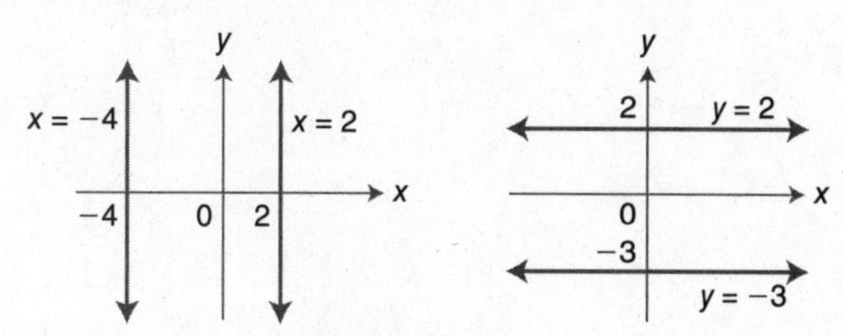

Figure 5.1-2

Parallel and Perpendicular Lines

Given two nonvertical lines l_1 and l_2 with slopes m_1 and m_2, as shown in Figure 5.1-3, respectively, they are parallel if and only if $m_1 = m_2$.

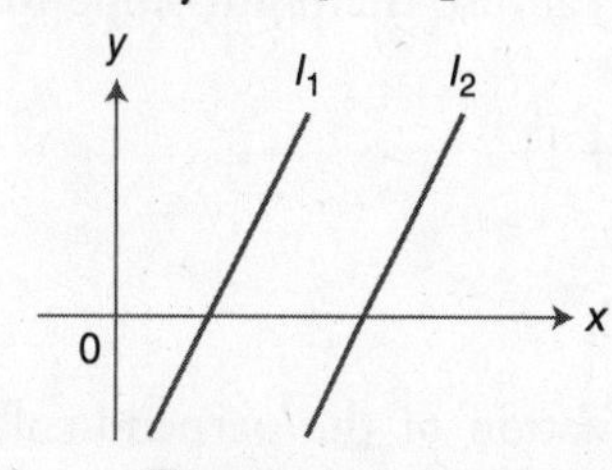

Figure 5.1-3

Lines l_1 and l_2 are perpendicular if and only if $m_1 m_2 = -1$. (See Figure 5.1-4.)

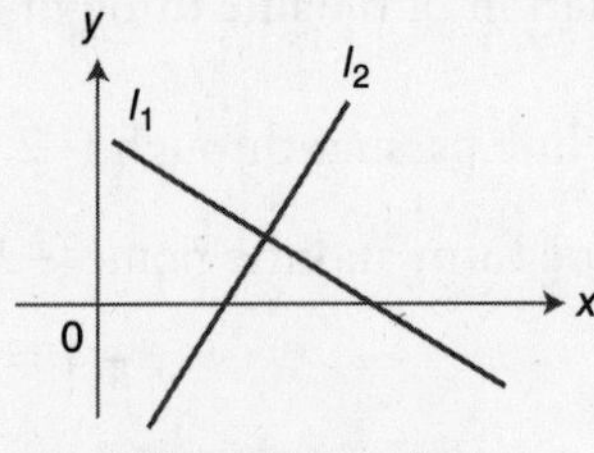

Figure 5.1-4

Example 1

Write an equation of the line through the point (–1, 3) and parallel to the line $3x - 2y = 6$. (See Figure 5.1-5.)

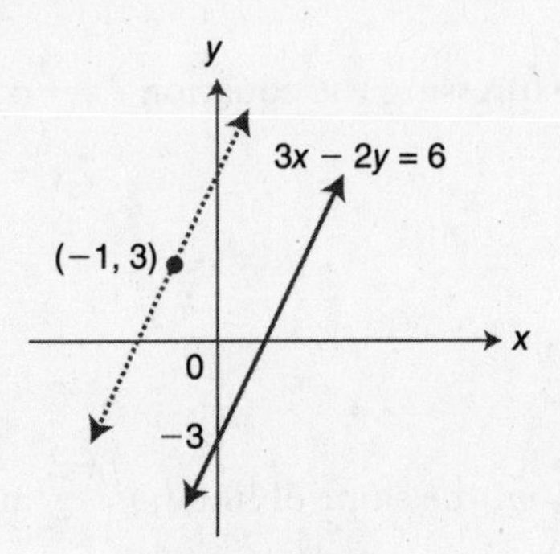

Figure 5.1-5

Begin by expressing $3x - 2y = 6$ in *slope-intercept form.*

$$3x - 2y = 6$$

$$-2y = -3x + 6$$

$$y = \frac{-3}{-2}x + \frac{6}{-2}$$

$$y = \frac{3}{2}x - 3$$

Therefore, the slope of the line $3x - 2y = 6$ is $m = \frac{3}{2}$, and the slope of the line parallel to the line $3x - 2y = 6$ is also $\frac{3}{2}$. Since the line parallel to $3x - 2y = 6$ passes through the point (–1, 3), you can use the point-slope form to obtain the equation $y - 3 = \frac{3}{2}(x - (-1))$ or $y - 3 = \frac{3}{2}(x + 1)$.

Example 2

Write an equation of the perpendicular bisector of the line segment joining the points $A(3, 0)$ and $B(-1, 4)$. (See Figure 5.1-6.)

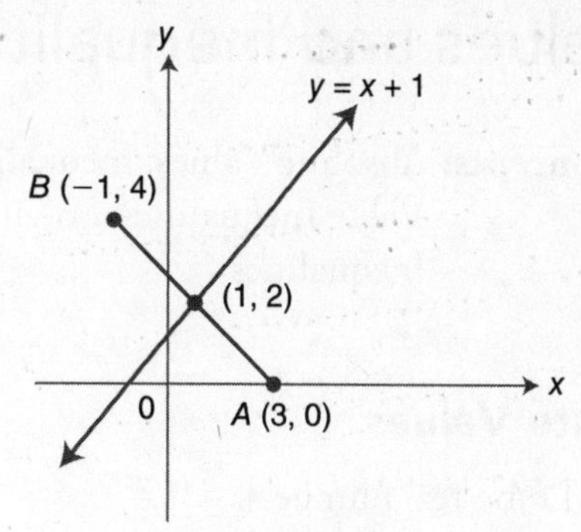

Figure 5.1-6

Begin by finding the midpoint of $\overline{AB}$. Midpoint $= \left(\frac{3+(-1)}{2}, \frac{0+4}{2}\right) = (1, 2)$. The slope of $\overline{AB}$ is $m = \frac{4-0}{-1-3} = -1$. Therefore, the perpendicular bisector of $\overline{AB}$ has a slope of 1. Since the perpendicular bisector of $\overline{AB}$ passes through the midpoint, you could use the point-slope form to obtain $y - 2 = 1(x - 1)$ or $y - 2 = x - 1$ or $y = x + 1$.

Example 3

Write an equation of the circle with center at $C(-2, 1)$ and tangent to line l having the equation $x + y = 5$. (See Figure 5.1-7.)

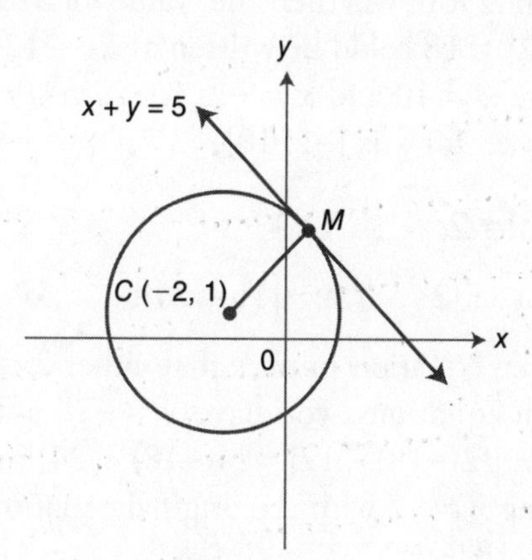

Figure 5.1-7

Let M be the point of tangency. Express the equation $x + y = 5$ in slope-intercept form to obtain $y = -x + 5$. Thus, the slope of line l is -1. Since $\overline{CM}$ is a radius drawn to the point of tangency, it is perpendicular to line l, and the slope of $\overline{CM}$ is 1. Using the point-slope formula, the equation of $\overline{CM}$ is $y - 1 = 1(x - (-2))$ or $y = x + 3$. To find the coordinates of point M, solve the two equations $y = -x + 5$ and $y = x + 3$ simultaneously. Thus, $-x + 5 = x + 3$ which is equivalent to $2 = 2x$ or $x = 1$.

Substituting $x = 1$ into $y = x + 3$, you have $y = 4$. Therefore, the coordinates of M are $(1, 4)$. Since $\overline{CM}$ is the radius of the circle, you should find the length of $\overline{CM}$ by using the distance formula $d = \sqrt{(x_2 - x_1)^2 + (y_2 - y_1)^2}$. Thus, $\overline{CM} = \sqrt{(1-(-2))^2 + (4-1)^2} = \sqrt{18}$. Now that you know both the radius of the circle ($r = \sqrt{18}$) and its center, $(-2, 1)$, use the formula $(x - h)^2 + (y - k)^2 = r^2$ to find an equation of the circle. Thus, an equation of the circle is $(x - (-2))^2 + (y - 1)^2 = 18$ or $(x + 2)^2 + (y - 1)^2 = 18$.

5.2 Absolute Values and Inequalities

Main Concepts: Absolute Values, Inequalities and the Real Number Line, Solving Absolute Value Inequalities, Solving Polynomial Inequalities, and Solving Rational Inequalities

Absolute Values

Let a and b be real numbers.

1. $|a| = \begin{cases} a, & \text{if } a \geq 0 \\ -a, & \text{if } a < 0 \end{cases}$
2. $|ab| = |a||b|$
3. $|a - b| = |b - a|$
4. $\sqrt{a^2} = |a| = \begin{cases} a, & \text{if } a \geq 0 \\ -a, & \text{if } a < 0 \end{cases}$

Example 1

Solve for x: $|3x - 12| = 18$.

Depending on whether the value of $(3x - 12)$ is positive or negative, the equation $|3x - 12| = 18$ could be written as $3x - 12 = 18$ or $3x - 12 = -18$. Solving both equations, you have $x = 10$ and $x = -2$. (*Be sure to check both answers in the original equation.*) The solution set for x is $\{-2, 10\}$.

Example 2

Solve for x: $|2x - 12| = |4x + 24|$.

The given equation implies that either $2x - 12 = 4x + 24$ or $2x - 12 = -(4x + 24)$. Solving both equations, you have $x = -18$ and $x = -2$. Checking $x = -18$ with the original equation: $|2(-18) - 12| = |4(-18) + 24|$ or $|-36 - 12| = |-72 + 24|$ or $|-48| = |-48|$. Checking $x = -2$ with the original equation, you have $|-16| = |16|$. Thus, the solution set for x is $\{18, -2\}$.

Example 3

Solve for x: $|11 - 3x| = 1 - x$.

Depending on the value of $(11 - 3x)$—whether it is greater than or less than 0—the given equation could be written as $11 - 3x = 1 - x$ or $11 - 3x = -(1 - x)$. Solving both equations, you have $x = 5$ and $x = 3$. Checking $x = 5$ with the original equation yields $|11 - 3(5)| = 1 - 5$ or $|-4| = -4$, which is *not* possible. Checking $x = 3$ with the original equation shows that $|11 - 3(3)| = 1 - 3$ or $|2| = -2$ which is also *not* possible. Thus, the solution for x is the empty set $\{\ \}$. You could also solve the equation by using a graphing calculator.

Enter $y_1 = |11 - 3x|$ and $y_2 = 1 - x$. The two graphs do not intersect; thus, there is no common solution. (See Figure 5.2-1.)

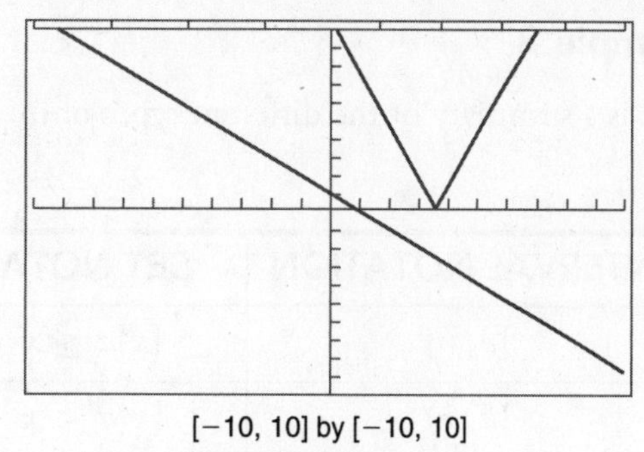

[−10, 10] by [−10, 10]

Figure 5.2-1

Inequalities and the Real Number Line

Properties of Inequalities

Let a, b, c, d, and k be real numbers:

1. **If $a < b$ and $b < c$, then $a < c$.** For example, $-7 < 2$ and $2 < 5 \Rightarrow -7 < 5$.
2. **If $a < b$ and $c < d$, then $a + c < b + d$.** For example, $5 < 7$ and $3 < 6 \Rightarrow 5 + 3 < 7 + 6$.
3. **If $a < b$ and $k > 0$, then $ak < bk$.** For example, $3 < 5$ and $2 > 0 \Rightarrow 3(2) < 5(2)$.
4. **If $a < b$ and $k < 0$, then $ak > bk$.** For example, $3 < 5$ and $-2 < 0 \Rightarrow 3(-2) > 5(-2)$.

Example 1

Solve the inequality $6 - 2x \leq 18$ and sketch the solution on the real number line.

Solving the inequality $6 - 2x \leq 18$ gives

$-2x \leq 12$

$x \geq -6$

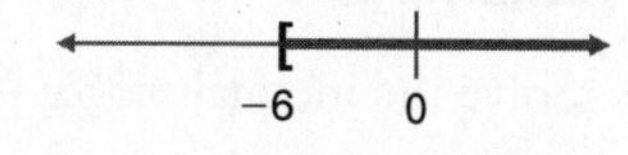

Therefore, the solution set is the interval $[-6, \infty)$ or, by expressing the solution set in set notation, $\{x | x \geq -6\}$.

Example 2

Solve the double inequality $-15 \leq 3x + 6 < 9$ and sketch the solution on the real number line.

Solving the double inequality $-15 \leq 3x + 6 < 9$ gives

$-21 \leq 3x < 3$

$-7 \leq x < 1$

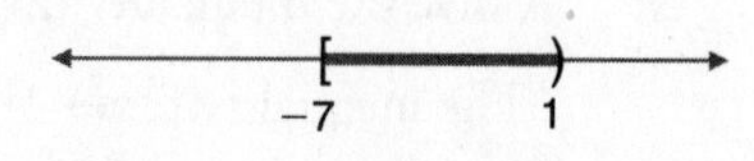

Therefore, the solution set is the interval $[-7, 1)$ or, by expressing the solution in set notation: $\{x | -7 \leq x < 1\}$.

Example 3

Here is a summary of the different types of intervals on a number line:

INTERVAL NOTATION	SET NOTATION	GRAPH
$[a, b]$	$\{x \mid a \leq x \leq b\}$	a b
(a, b)	$\{x \mid a < x < b\}$	a b
$[a, b)$	$\{x \mid a \leq x < b\}$	a b
$(a, b]$	$\{x \mid a < x \leq b\}$	a b
$[a, \infty)$	$\{x \mid x \geq a\}$	a
(a, ∞)	$\{x \mid x > a\}$	a
$(-\infty, b]$	$\{x \mid x \leq b\}$	b
$(-\infty, b)$	$\{x \mid x < b\}$	b
$(-\infty, \infty)$	$\{x \mid x \text{ is a real number}\}$	

Solving Absolute Value Inequalities

Let a be a real number such that $a \geq 0$.

$$|x| \geq a \Leftrightarrow (x \geq a \text{ or } x \leq -a) \text{ and } |x| > a \Leftrightarrow (x > a \text{ or } x < -a)$$
$$|x| \leq a \Leftrightarrow (-a \leq x \leq a) \quad \text{and } |x| < a \Leftrightarrow (-a < x < a)$$

Example 1

Solve the inequality $|3x - 6| \leq 15$ and sketch the solution on the real number line.

The given inequality is equivalent to

$$-15 \leq 3x - 6 \leq 15$$
$$-9 \leq 3x \leq 21$$
$$-3 \leq x \leq 7$$

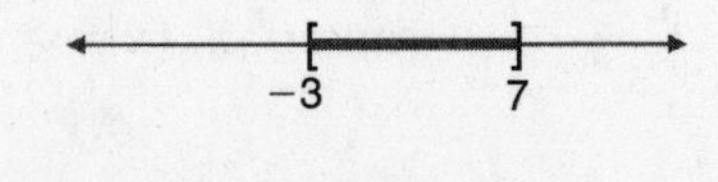

Therefore, the solution set is the interval $[-3, 7]$ or, in set notation, $\{x \mid -3 \leq x \leq 7\}$.

Example 2

Solve the inequality $|2x + 1| > 9$ and sketch the solution on the real number line.

The inequality $|2x + 1| > 9$ implies that

−5 4

$2x + 1 > 9$ or $2x + 1 < -9$

Solving the two inequalities in the above line, you have $x > 4$ or $x < -5$. Therefore, the solution set is the union of the two disjoint intervals $(x > 4) \cup (x < -5)$ or, by writing the solution in set notation, $\{x \mid (x > 4) \text{ or } (x < -5)\}$.

Example 3

Solve the inequality $|1 - 2x| \leq 7$ and sketch the solution on the real number line.

The inequality $|1 - 2x| \leq 7$ implies that

$$\begin{aligned} -7 \leq 1 - 2x &\leq 7 \\ -8 \leq -2x &\leq 6 \\ 4 \geq x &\geq -3 \\ -3 \leq x &\leq 4 \end{aligned}$$

Therefore, the solution set is the interval $[-3, 4]$ or, by writing the solution in set notation, $\{x | -3 \leq x \leq 4\}$. (See Figure 5.2-2.)

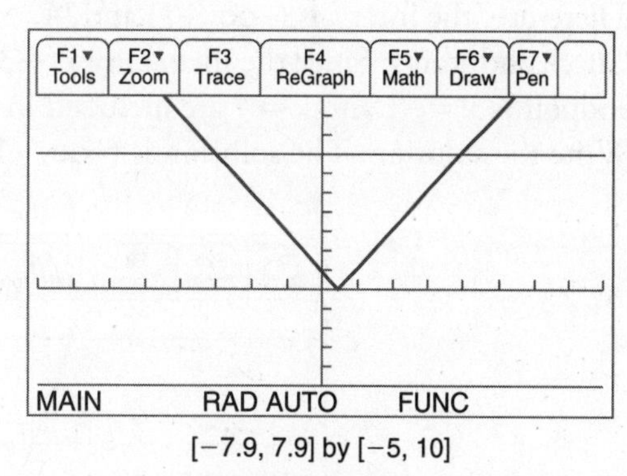

[−7.9, 7.9] by [−5, 10]

Figure 5.2-2

Note that you can solve an absolute value inequality by using a graphing calculator. For instance, in Example 3, enter $y_1 = |1 - 2x|$ and $y_2 = 7$. The graphs intersect at $x = -3$ and 4, and y_1 is below y_2 on the interval $(-3, 4)$. Since the inequality is $\leq$, the solution set is $[-3, 4]$.

Solving Polynomial Inequalities

1. Write the given inequality in standard form with the polynomial on the left and zero on the right.
2. Factor the polynomial, if possible.
3. Find all zeros of the polynomials.
4. Using the zeros on a number line, determine the test intervals.
5. Select an x-value from each interval and substitute it in the polynomial.
6. Check the *endpoints* of each interval with the inequality. Use a parenthesis if the endpoint is not included and a bracket if it is.
7. Write the solution to the inequality.

Example 1

Solve the inequality $x^2 - 3x \geq 4$.

1. Write in standard form: $x^2 - 3x - 4 \geq 0$
2. Factor the polynomial: $(x - 4)(x + 1)$
3. Find zeros: $(x - 4)(x + 1) = 0$ implies that $x = 4$ and $x = -1$.
4. Determine intervals:

$(-\infty, -1)$ and $(-1, 4)$ and $(4, \infty)$

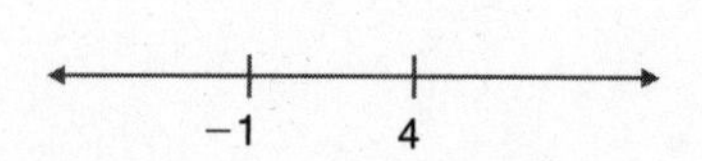

5. Select an x-value in each interval and evaluate the polynomial at that value:

INTERVAL	SELECTED x-VALUE	FACTOR $(x+1)$	FACTOR $(x-4)$	POLYNOMIAL $(x-4)(x+1)$
$(-\infty, -1)$	-2	$-$	$-$	$+$
$(-1, 4)$	0	$+$	$-$	$-$
$(4, \infty)$	6	$+$	$+$	$+$

Therefore, the intervals $(-\infty, -1)$ and $(4, \infty)$ make $(x-4)(x+1) > 0$.

6. Check endpoints: Since the inequality $x^2 - 3x - 4 \geq 0$ is greater than or equal to 0, both endpoints $x = -1$ and $x = 4$ are included in the solution.
7. Write the solution: The solution is $(-\infty, -1] \cup [4, \infty)$. (See Figure 5.2-3.)

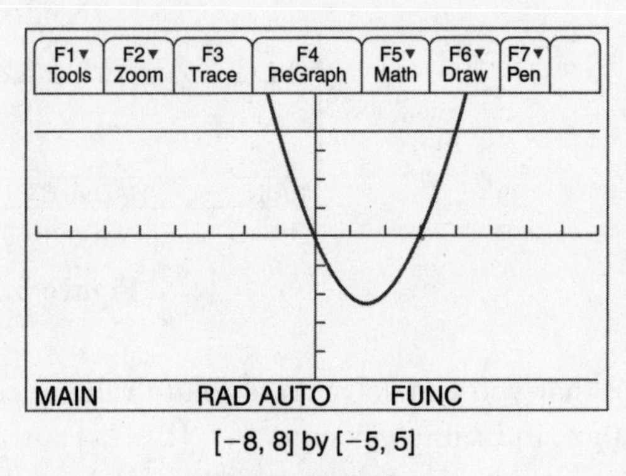

[−8, 8] by [−5, 5]

Figure 5.2-3

Note that the inequality $x^2 - 3x \geq 4$ could have been solved by using a graphing calculator. Enter $y_1 = x^2 - 3x$ and $y_2 = 4$. The graph of y_1 is above y_2 on $(-\infty, -1)$ and $(4, \infty)$. Since the inequality is $\geq$, the solution set is $(-\infty, -1]$ and $[4, \infty)$.

Example 2

Solve the inequality $x^3 - 9x < 0$, using a graphing calculator. (See Figure 5.2-4.)

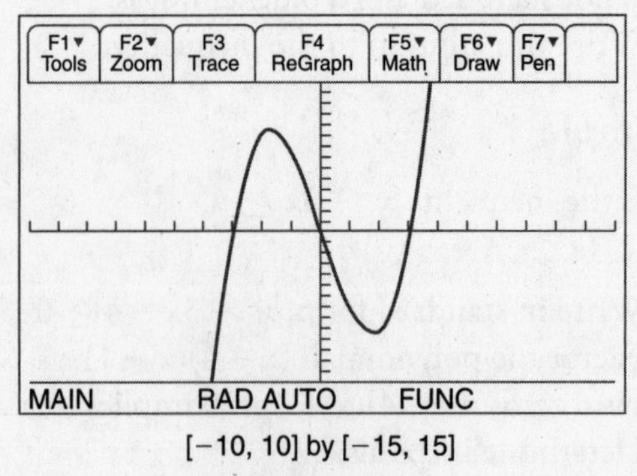

[−10, 10] by [−15, 15]

Figure 5.2-4

1. Enter $y = x^3 - 8x$ into your graphing calculator.
2. Find the zeros of y: $x = -3$, 0, and 3.
3. Determine the intervals on which $y < 0$: $(-\infty, -3)$ and $(0, 3)$.
4. Check whether the endpoints satisfy the inequality. Since the inequality is strictly less than 0, the endpoints are not included in the solution.
5. Write the solution to the inequality. The solution is $(-\infty, -3) \cup (0, 3)$.

Example 3

Solve the inequality $x^3 - 9x < 0$ algebraically.

1. Write in standard form: $x^3 - 9x < 0$ is already in standard form.
2. Factor the polynomial: $x(x - 3)(x + 3)$
3. Find zeros: $x(x - 3)(x + 3) = 0$ implies that $x = 0$, $x = 3$, and $x = -3$.
4. Determine the intervals:

   $(-\infty, -3)$, $(-3, 0)$, $(0, 3)$, and $(3, \infty)$

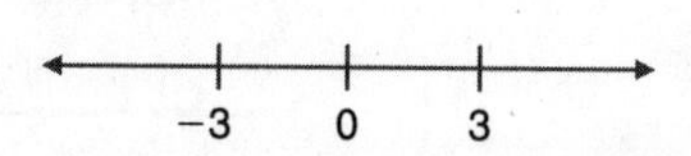

5. Select an x-value and evaluate the polynomial:

INTERVAL	SELECTED x-VALUE	FACTOR x	FACTOR (x + 3)	FACTOR (x − 3)	POLYNOMIAL x(x − 3)(x + 3)
$(-\infty, -3)$	−5	−	−	−	−
$(-3, 0)$	−1	−	+	−	+
$(0, 3)$	1	+	+	−	−
$(3, \infty)$	6	+	+	+	+

Therefore, the intervals $(-\infty, -3)$ and $(0, 3)$ make $x(x - 3)(x + 3) < 0$.

6. Check the endpoints: Since the inequality $x^3 - 9x < 0$ is strictly less than 0, none of the endpoints $x = -3$, 0, and 3 are included in the solution.
7. Write the solution: The solution is $(-\infty, -3) \cup (0, 3)$.

Solving Rational Inequalities

1. Rewrite the given inequality so that all the terms are on the left and only zero is on the right.
2. Find the least common denominator, and combine all the terms on the left into *a single fraction.*
3. Factor the numerator and the denominator, if possible.
4. Find all x-values for which the numerator or the denominator is zero.
5. Putting these x-values on a number line, determine the test intervals.
6. Select an x-value from each interval and substitute it in the fraction.
7. Check the *endpoints* of each interval with the inequality.
8. Write the solution to the inequality.

Example 1

Solve the inequality $\frac{2x-5}{x-3} \le 1$.

1. Rewrite: $\frac{2x-5}{x-3} - 1 \le 0$
2. Combine: $\frac{2x-5-x+3}{x-3} \le 0 \Leftrightarrow \frac{x-2}{x-3} \le 0$
3. Set the numerator and denominator equal to 0 and solve for x: $x = 2$ and 3.
4. Determine intervals:
   $(-\infty, 2)$, $(2, 3)$ and $(3, \infty)$

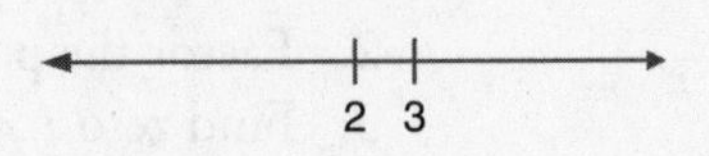

5. Select an x-value and evaluate the fraction:

INTERVAL	SELECTED x-VALUE	FACTOR $(x-2)$	FACTOR $(x-3)$	FRACTION $\frac{x-2}{x-3}$
$(-\infty, 2)$	0	−	−	+
$(2, 3)$	2.5	+	−	−
$(3, \infty)$	6	+	+	+

   Therefore, the interval $(2, 3)$ makes the fraction < 0.

6. Check the endpoints: At $x = 3$, the fraction is undefined. Thus, the only endpoint is $x = 2$. Since the inequality is less than or equal to 0, $x = 2$ is included in the solution.
7. Write the solution: The solution is the interval $[2, 3)$.

Example 2

Solve the inequality $\frac{2x-5}{x-3} \le 1$ by using a graphing calculator. (See Figure 5.2-5.)

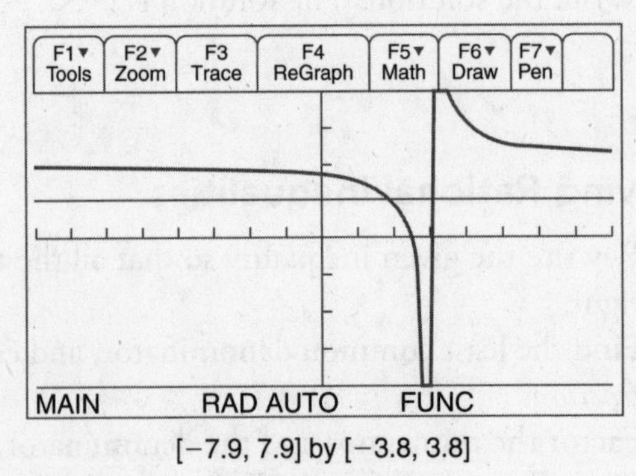

[−7.9, 7.9] by [−3.8, 3.8]

Figure 5.2-5

1. Enter $y_1 = \frac{2x-5}{x-3}$ and $y_2 = 1$.
2. Find the intersection points: $x = 2$. (Note that at $x = 3$, y_1 is undefined.)
3. Determine the intervals on which y_1 is below y_2: The interval is $(2, 3)$.

4. Check whether the endpoints satisfy the inequality. Since the inequality is less than or equal to 1, the endpoint at $x = 2$ is included in the solution.
5. Write the solution to the inequality. The solution is the interval [2, 3).

Example 3

Solve the inequality $\frac{1}{x} \geq x$ by using a graphing calculator. (See Figure 5.2-6.)

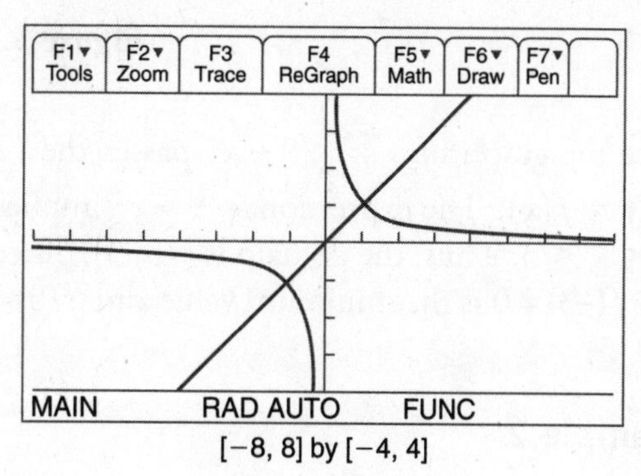

[−8, 8] by [−4, 4]

Figure 5.2-6

1. Enter $y_1 = \frac{1}{x}$ and $y_2 = x$.
2. Find the intersection points: $x = -1$ and $x = 1$. (Note that at $x = 0$, y_1 is undefined.)
3. Determine the intervals on which $y_1 \geq y_2$. The intervals are $(-\infty, -1)$ and $(0, 1)$.
4. Check whether the endpoints satisfy the inequality: Since the inequality is greater than or equal to x, the endpoints at $x = -1$ and $x = 1$ are included in the solution.
5. Write the solution to the inequality. The solution is the interval $(-\infty, -1] \cup (0, 1]$.

5.3 Functions

Main Concepts: Definition of a Function, Operations on Functions, Inverse Functions, Trigonometric and Inverse Trigonometric Functions, Exponential and Logarithmic Functions

Definition of a Function

A function f is a set of ordered pairs (x, y) in which for every x coordinate there is *one and only one* corresponding y coordinate. We write $f(x) = y$. The domain of f is the set of all possible values of x, and the range of f is the set of all values of y.

The Vertical Line Test

If all vertical lines pass through the graph of an equation at no more than one point, then the equation is a function.

Example 1

Given $y = \sqrt{9 - x^2}$, sketch the graph of the equation, determine if the equation is a function, and find the domain and range of the equation. (See Figure 5.3-1.)

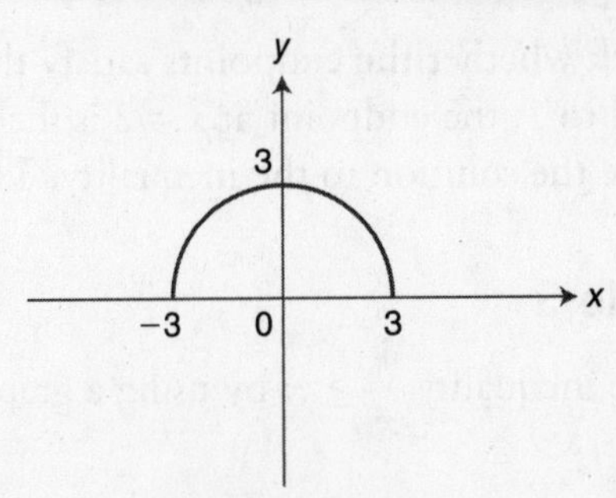

Figure 5.3-1

Since the graph of $y = \sqrt{9 - x^2}$ passes the vertical line test, the equation is a function. Let $y = f(x)$. The expression $\sqrt{9 - x^2}$ implies that $9 - x^2 \geq 0$. By inspection, note that $-3 \leq x \leq 3$. Thus, the domain is $[-3, 3]$. Since $f(x)$ is defined for all values of $x \in [-3, 3]$ and $f(-3) = 0$ is the minimum value and $f(0) = 3$ is the maximum value, the range of $f(x)$ is $[0, 3]$.

Example 2

Given $f(x) = x^2 - 4x$, find $f(-3)$, $f(-x)$, and $\dfrac{f(x+h) - f(x)}{h}$.

$$f(-3) = (-3)^2 - 4(-3) = 9 + 12 = 21$$

$$f(-x) = (-x)^2 - 4(-x) = x^2 + 4x$$

$$\frac{f(x+h) - f(x)}{h} = \frac{(x+h)^2 - 4(x+h) - (x^2 - 4x)}{h} = \frac{x^2 + 2hx + h^2 - 4x - 4h - x^2 + 4x}{h}$$

$$= \frac{2hx + h^2 - 4h}{h} = 2x + h - 4.$$

Operations on Functions

Let f and g be two given functions. Then for all x in the intersection of the domains of f and g, the *sum*, *difference*, *product*, and *quotient* of f and g, respectively, are defined as follows:

$$(f+g)(x) = f(x) + g(x)$$

$$(f-g)(x) = f(x) - g(x)$$

$$(fg)(x) = f(x) - g(x)$$

$$\left(\frac{f}{g}\right)(x) = \frac{f(x)}{g(x)}, \; g(x) \neq 0$$

The composition of f with g is $(f \circ g)(x) = f(g(x))$, where the domain of $f \circ g$ is the set containing all x in the domain of g for which $g(x)$ is in the domain of f.

Example 1

Given $f(x) = x^2 - 4$ and $g(x) = x - 5$, find

(a) $(f \circ g)(-1)$
(b) $(g \circ f)(-1)$
(c) $(f+g)(-3)$

(d) $(f - g)(1)$
(e) $(fg)(2)$
(f) $\left(\frac{f}{g}\right)(0)$
(g) $\left(\frac{f}{g}\right)(5)$
(h) $\left(\frac{g}{f}\right)(4)$

Solutions

(a) $(f \circ g)(x) = f(g(x)) = f(x-5) = (x-5)^2 - 4 = x^2 - 10x + 21$.
Thus $(f \circ g)(-1) = (-1)^2 - 10(-1) + 21 = 1 + 10 + 21 = 32$.
Or $(f \circ g)(-1) = f(g(-1)) = f(-6) = 32$.
(b) $(g \circ f)(x) = g(f(x)) = g(x^2 - 4) = (x^2 - 4) - 5 = x^2 - 9$.
Thus $(g \circ f)(-1) = (-1)^2 - 9 = 1 - 9 = -8$.
(c) $(f + g)(x) = (x^2 - 4) + (x - 5) = x^2 + x - 9$. Thus $(f + g)(-3) = -3$.
(d) $(f - g)(x) = (x^2 - 4) - (x - 5) = x^2 - x + 1$. Thus $(f - g)(1) = 1$.
(e) $(fg)(x) = (x^2 - 4)(x - 5) = x^3 - 5x^2 - 4x + 20$. Thus $(fg)(2) = 0$.
(f) $\left(\frac{f}{g}\right)(x) = \frac{x^2 - 4}{x - 5}$, $x \neq 5$. Thus $\left(\frac{f}{g}\right)(0) = \frac{4}{5}$.
(g) Since $g(5) = 0$, $x = 5$ is *not* in the domain of $\left(\frac{f}{g}\right)$ and $\left(\frac{f}{g}\right)(5)$ is *undefined.*
(h) $\left(\frac{g}{f}\right)(x) = \frac{x - 5}{x^2 - 4}$, $x \neq 2$ or -2. Thus $\left(\frac{g}{f}\right)(4) = -\frac{1}{12}$.

Example 2

Given $h(x) = \sqrt{x}$ and $k(x) = \sqrt{9 - x^2}$:

(a) find $\left(\frac{h}{k}\right)(x)$ and indicate its domain and
(b) find $\left(\frac{k}{h}\right)(x)$ and indicate its domain.

Solutions

(a) $\left(\frac{h}{k}\right)(x) = \frac{\sqrt{x}}{\sqrt{9 - x^2}}$

The domain of $h(x)$ is $[0, \infty)$, and the domain of $k(x)$ is $[-3, 3]$.
The intersection of the two domains is $[0, 3]$. However, $k(3) = 0$.

Therefore, the domain of $\left(\frac{h}{k}\right)$ is $[0, 3)$.

Note that $\frac{\sqrt{x}}{\sqrt{9 - x^2}}$ is not equivalent to $\sqrt{\frac{x}{9 - x^2}}$ outside of the domain $[0, 3)$.

(b) $\left(\frac{k}{h}\right)(x) = \frac{\sqrt{9 - x^2}}{\sqrt{x}}$

The intersection of the two domains is $[0, 3]$. However, $h(0) = 0$.

Therefore, the domain of $\left(\frac{k}{h}\right)$ is $(0, 3]$.

Example 3

Given the graphs of functions $f(x)$ and $g(x)$ in Figures 5.3-2 and 5.3-3,

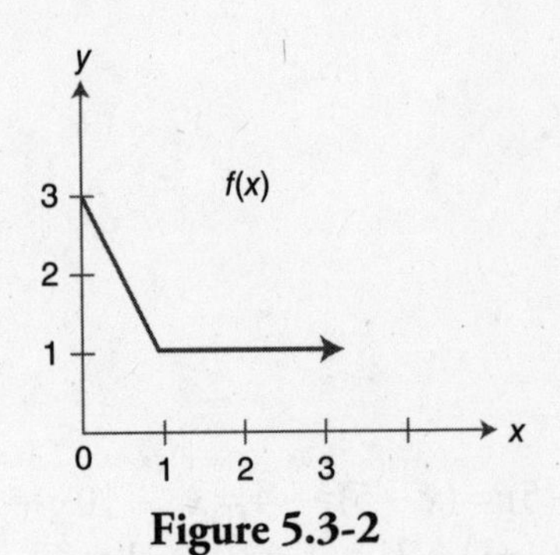

Figure 5.3-2

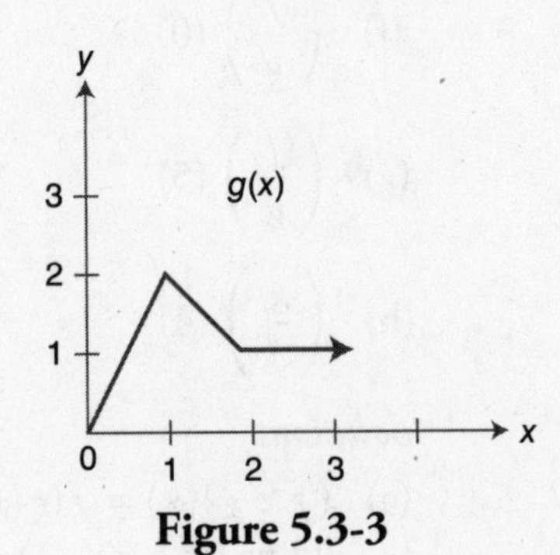

Figure 5.3-3

find:

(a) $(f+g)(1)$
(b) $(fg)(0)$
(c) $\left(\frac{f}{g}\right)(0)$
(d) $f(g(3))$

Solutions

(a) $(f+g)(1) = f(1) + g(1) = 3$
(b) $(fg)(0) = f(0)g(0) = 3(0) = 0$
(c) $\left(\frac{f}{g}\right)(0) = \frac{f(0)}{g(0)} = \frac{3}{0}$ undefined
(d) $f(g(3)) = f(1) = 1$

Inverse Functions

Given a function f, the inverse of f(if it exists) is a function g such that $f(g(x)) = x$ for every x in the domain of g and $g(f(x)) = x$ for every x in the domain of f. The function g is written as f^{-1}. Thus, $f(f^{-1}(x)) = x$ and $f^{-1}(f(x)) = x$. The graphs of f and f^{-1} in Figure 5.3-4 are *reflections* of each other in the line $y = x$. The point (a, b) is on the graph of f if and only if the point (b, a) is on the graph of f^{-1}.

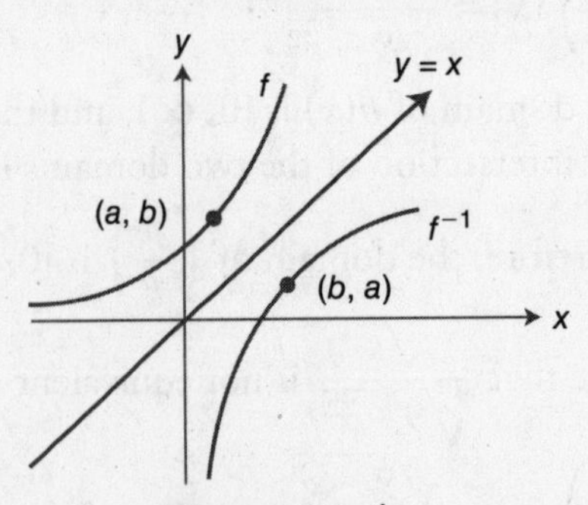

Figure 5.3-4

A function f is *one-to-one* if for any two points x_1 and x_2 in the domain such that $x_1 \neq x_2$, then $f(x_1) \neq f(x_2)$.

Equivalent Statements

Given a function f:

1. The function f has an inverse.
2. The function f is one-to-one.
3. Every horizontal line passes through the graph of f no more than once.
4. The function f is monotonic—strictly increasing or decreasing.

To find the inverse of a function f:

1. Check if f has an inverse, that is, if f is one-to-one or passes the horizontal line test.
2. Replace $f(x)$ by y.
3. Interchange the variables x and y.
4. Solve for y.
5. Replace y by $f^{-1}(x)$.
6. Indicate the domain of $f^{-1}(x)$ as the range of $f(x)$.
7. Verify $f^{-1}(x)$ by checking if $f(f^{-1}(x)) = f^{-1}(f(x)) = x$.

Example 1

Given the graph of $f(x)$ in Figure 5.3-5, find:

(a) $f^{-1}(0)$
(b) $f^{-1}(1)$
(c) $f^{-1}(3)$

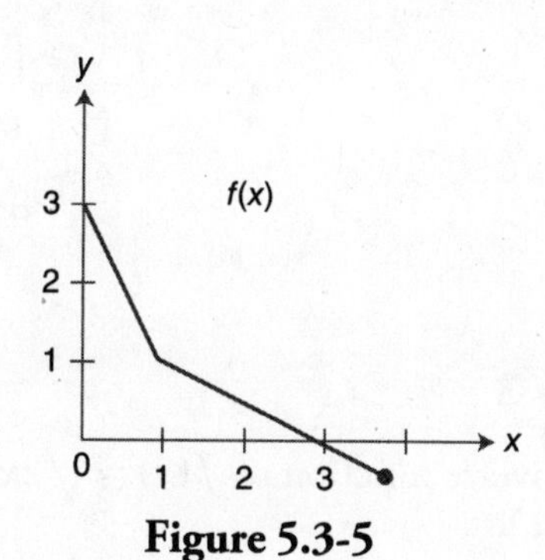

Figure 5.3-5

Solutions

(a) By inspection, $f(3) = 0$. Thus, $f^{-1}(0) = 3$.
(b) Since $f(1) = 1$, thus, $f^{-1}(1) = 1$.
(c) Since $f(0) = 3$, therefore, $f^{-1}(3) = 0$.

Example 2

Determine if the given functions have an inverse:

(a) $f(x) = x^3 + x - 2$
(b) $f(x) = x^3 - 2x + 1$

Solutions

(a) By inspection, the graph of $f(x)=x^3+x-2$ in Figure 5.3-6 is strictly increasing, which implies that $f(x)$ is one-to-one. (You could also use the horizontal line test.) Therefore, $f(x)$ has an inverse function.

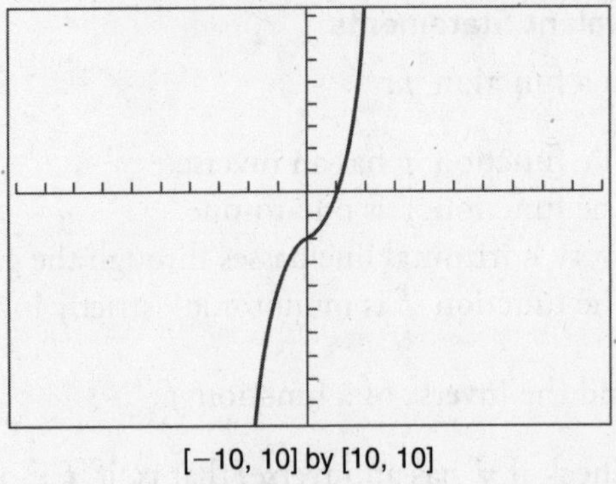

[−10, 10] by [10, 10]

Figure 5.3-6

(b) By inspection, the graph of $f(x) = x^3 - 2x + 1$ in Figure 5.3-7 fails the horizontal line test. Thus, $f(x)$ has no inverse function.

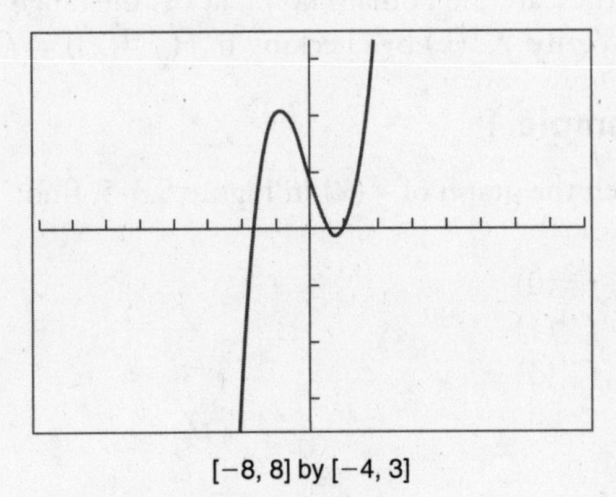

[−8, 8] by [−4, 3]

Figure 5.3-7

Example 3

Find the inverse function of $f(x) = \sqrt{2x - 1}$. (See Figure 5.3-8.)

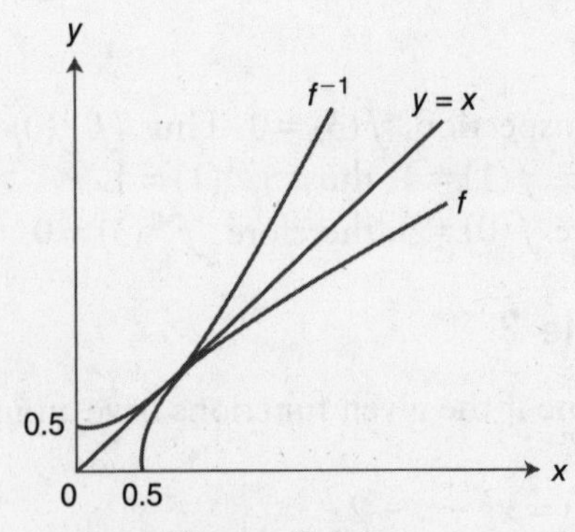

Figure 5.3-8

1. Since $f(x)$ is a strictly increasing function, the inverse function exists.
2. Let $y = f(x)$. Thus, $y = \sqrt{2x - 1}$.
3. Interchange x and y. You have $x = \sqrt{2y - 1}$.

4. Solve for y. Thus, $y = \dfrac{x^2+1}{2}$.
5. Replace y by $f^{-1}(x)$. You have $f^{-1}(x) = \dfrac{x^2+1}{2}$.
6. Since the range of $f(x)$ is $[0, \infty)$, the domain of $f^{-1}(x)$ is $[0, \infty)$.
7. Verify $f^{-1}(x)$ by checking:
   Since $x > 0$, $\sqrt{x^2} = x$,

$$f(f^{-1}(x)) = f\left(\frac{x^2+1}{2}\right) = \sqrt{2\left(\frac{x^2+1}{2}\right) - 1} = x$$

$$f^{-1}(f(x)) = f^{-1}(\sqrt{2x-1}) = \frac{(\sqrt{2x-1})^2+1}{2} = x$$

Trigonometric and Inverse Trigonometric Functions

There are six basic trigonometric functions and six inverse trigonometric functions. Their graphs are illustrated in Figures 5.3-9 to 5.3-20.

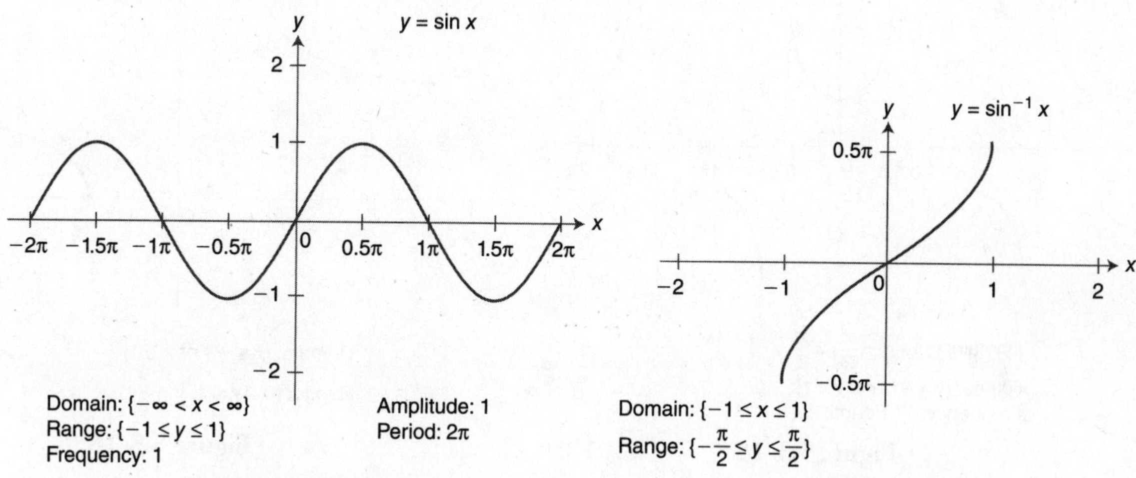

Figure 5.3-9

Figure 5.3-10

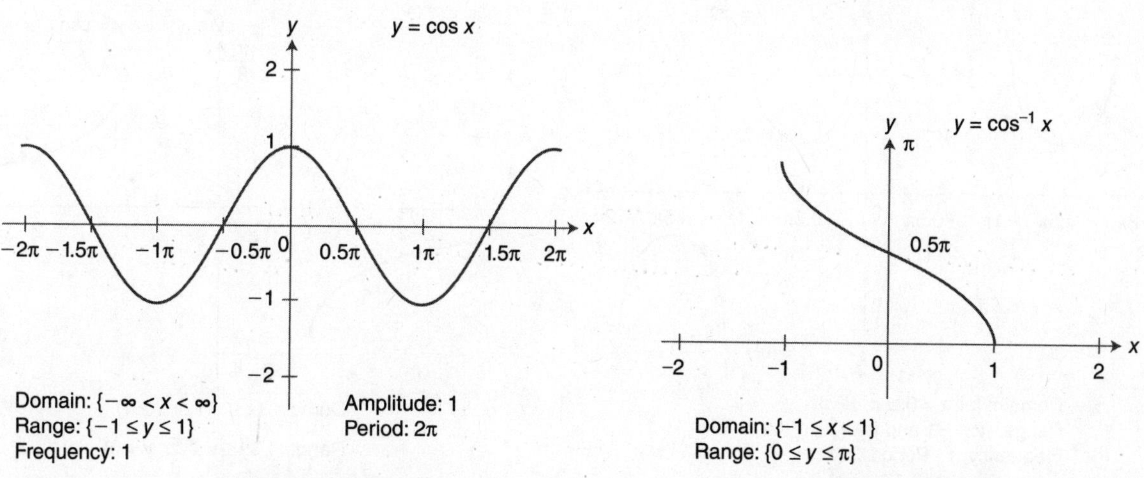

Figure 5.3-11

Figure 5.3-12

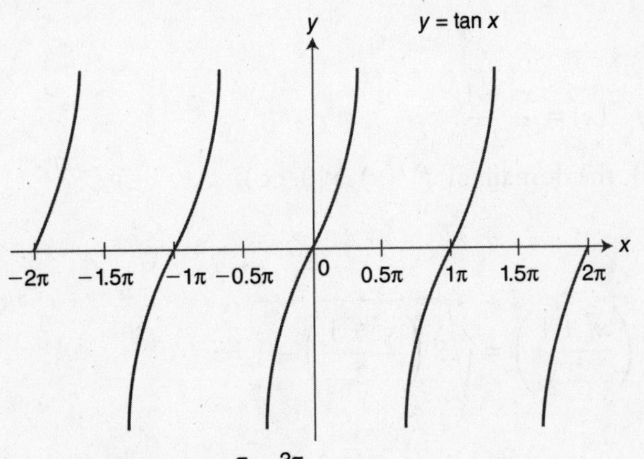

Domain: {all $x \neq \pm\frac{\pi}{2}, \pm\frac{3\pi}{2}, \ldots$}

Range: $\{-\infty < y < \infty\}$

Frequency: 1 Period: π

Figure 5.3-13

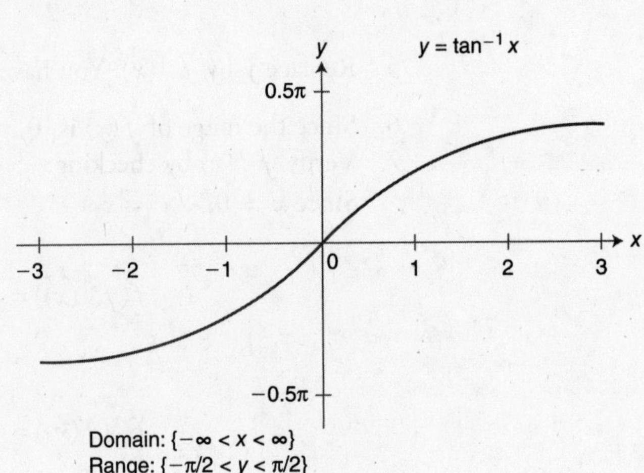

Domain: $\{-\infty < x < \infty\}$

Range: $\{-\pi/2 < y < \pi/2\}$

Figure 5.3-14

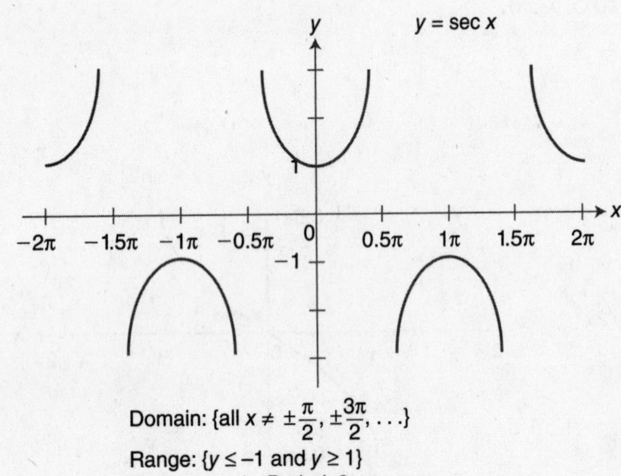

Domain: {all $x \neq \pm\frac{\pi}{2}, \pm\frac{3\pi}{2}, \ldots$}

Range: $\{y \leq -1 \text{ and } y \geq 1\}$

Frequency: 1 Period: 2π

Figure 5.3-15

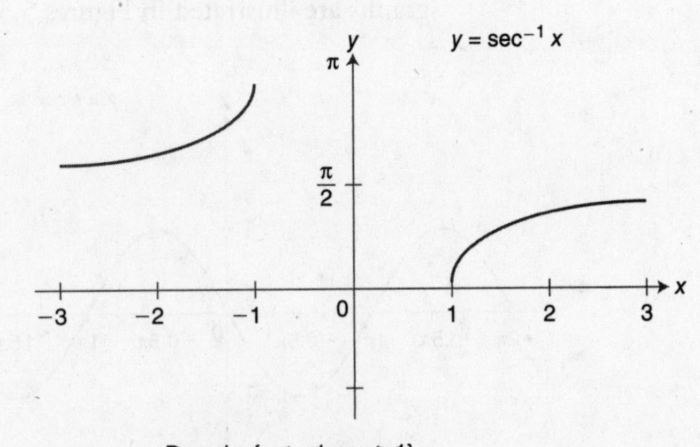

Domain: $\{x \leq -1 \text{ or } x \geq 1\}$

Range: $\{0 \leq y \leq \pi, y \neq \frac{\pi}{2}\}$

Figure 5.3-16

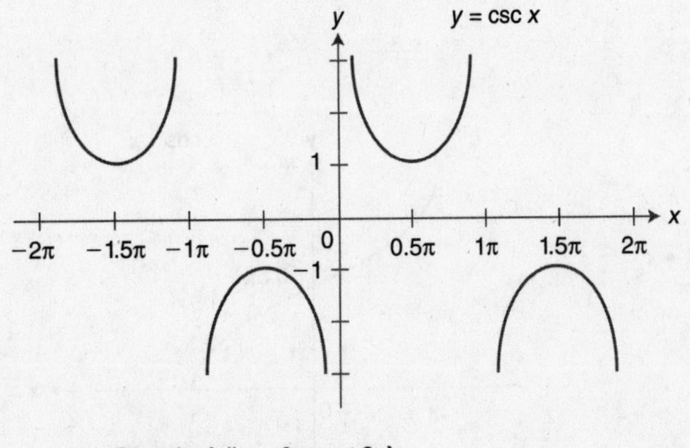

Domain: {all $x \neq 0, \pm\pi, \pm 2\pi$}

Range: $\{y \leq -1 \text{ and } y \geq 1\}$

Frequency: 1 Period: 2π

Figure 5.3-17

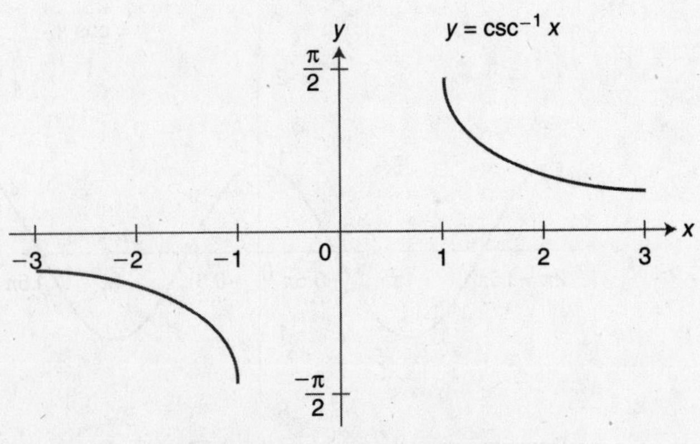

Domain: $\{x \leq -1 \text{ or } x \geq 1\}$

Range: $\{\frac{-\pi}{2} \leq y \leq \frac{\pi}{2}, y \neq 0\}$

Figure 5.3-18

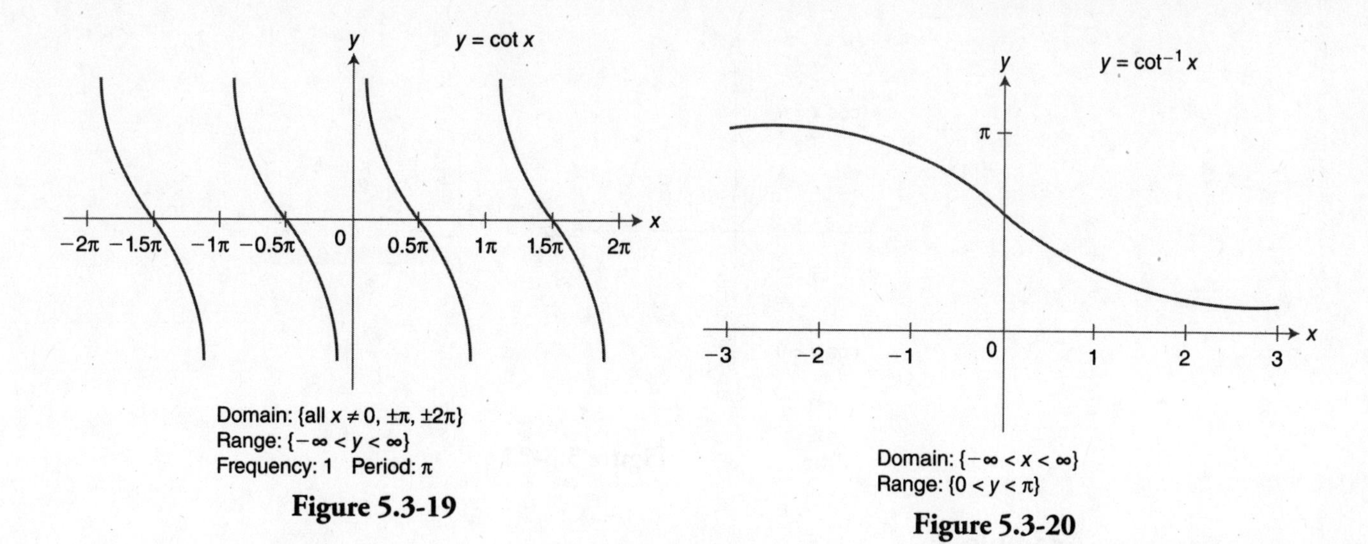

Figure 5.3-19

Figure 5.3-20

Formulas for using a calculator to get $\sec^{-1} x$, $\csc^{-1} x$, and $\cot^{-1} x$:

$\sec^{-1} x = \cos^{-1}(1/x)$

$\csc^{-1} x = \sin^{-1}(1/x)$

$\cot^{-1} x = \pi/2 - \tan^{-1} x$

Example 1

Sketch the graph of the function $y = 3 \sin 2x$. Indicate its domain, range, amplitude, period, and frequency.

The domain is all real numbers. The range is $[-3, 3]$. The amplitude is 3, which is the coefficient of $\sin 2x$. The frequency is 2, the coefficient of x, and the period is $(2\pi) \div$ (the frequency), thus $2\pi \div 2 = \pi$. (See Figure 5.3-21.)

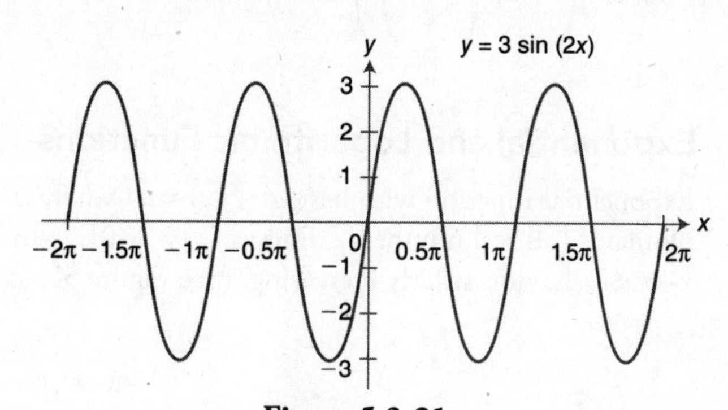

Figure 5.3-21

Example 2

Solve the equation $\cos x = -0.5$ if $0 \leq x \leq 2\pi$.

Note that $\cos(\pi/3) = 0.5$ and that the cosine is negative in the second and third quadrants. Since $\cos x = -0.5$, x must be in the second or third quadrant with a reference angle of $\pi/3$. In the second quadrant, $x = \pi - (\pi/3) = 2\pi/3$ and in the third quadrant, $x = \pi + (\pi/3) = 4\pi/3$. Thus $x = 2\pi/3$ or $4\pi/3$. (See Figure 5.3-22.)

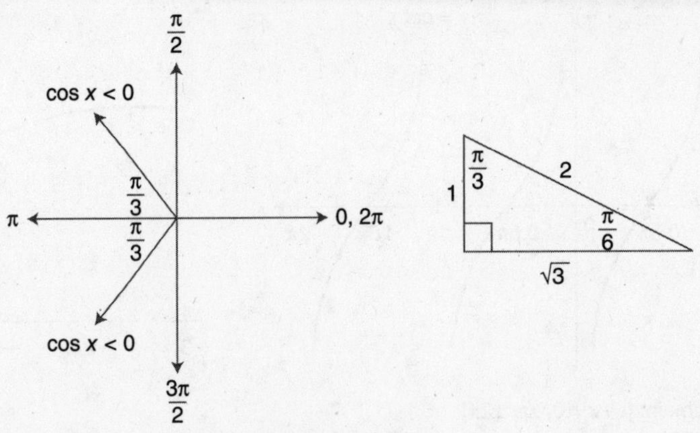

Figure 5.3-22

Example 3

Evaluate $\tan^{-1}(3)$.

Using your graphing calculator, enter $\tan^{-1}(3)$. The result is 1.2490457724. Note that the range of $\tan^{-1} x$ is $(-\pi/2, \pi/2)$ and $-\pi/2 \le 1.2490457724 \le \pi/2$. Thus $\tan^{-1}(3) \approx 1.2490457724$.

Example 4

Evaluate $\sin\left(\cos^{-1}\left(\frac{\sqrt{3}}{2}\right)\right)$.

Note that $\cos^{-1}\left(\frac{\sqrt{3}}{2}\right) = \frac{\pi}{6}$, and thus, $\sin\left(\frac{\pi}{6}\right) = 0.5$. Or you could use a calculator and enter $\sin\left(\cos^{-1}\left(\frac{\sqrt{3}}{2}\right)\right)$ and get 0.5.

Exponential and Logarithmic Functions

Exponential function with base a: $f(x) = a^x$ where $a > 0$ and $a \neq 1$.
Domain: {all real numbers}. Range: $\{y \mid y > 0\}$. y-Intercept: (0, 1). Horizontal asymptote: x-axis. Behavior: strictly increasing. (See Figure 5.3-23.)

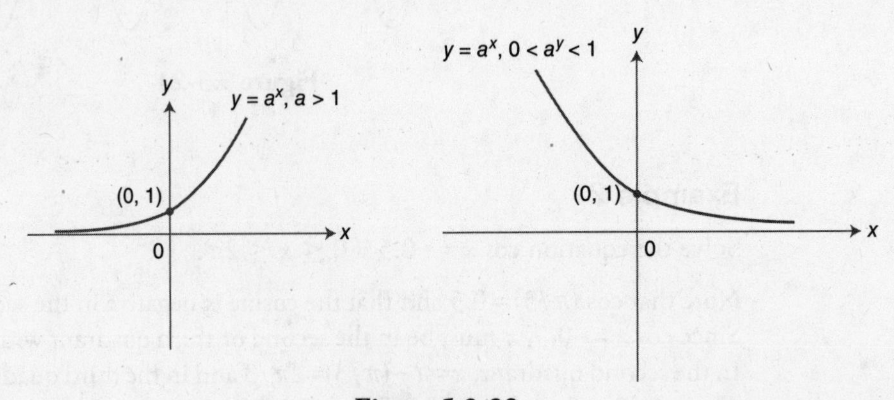

Figure 5.3-23

Properties of Exponents

Given $a > 0$, $b > 0$, and x and y are real numbers, then

$$a^x \cdot a^y = a^{x+y}$$

$$a^x \div a^y = a^{x-y}$$

$$(a^x)^y = a^{xy}$$

$$(ab)^x = a^x \cdot b^x$$

$$\left(\frac{a}{b}\right)^x = \frac{a^x}{b^x}$$

Logarithmic function with base a: $y = \log_a x$ if and only if $a^y = x$ where $x > 0$, $a > 0$, and $a \neq 1$. (See Figure 5.3-24.)

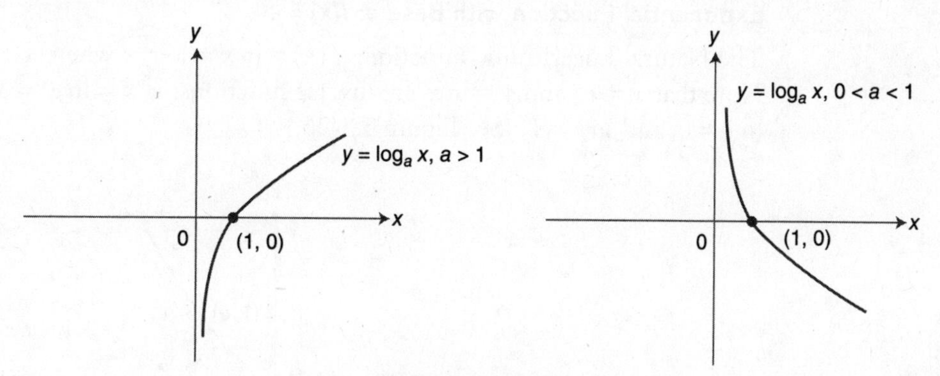

Figure 5.3-24

Domain: $\{x | x > 0\}$. Range: $\{$all real numbers$\}$. x-Intercept: (1, 0). Vertical asymptote: y-axis. Behavior: strictly increasing.

Note that $y = \log_a x$ and $y = a^x$ are inverse functions (that is, $\log_a(a^x) = a^{\log_a x} = x$). (See Figure 5.3-25.)

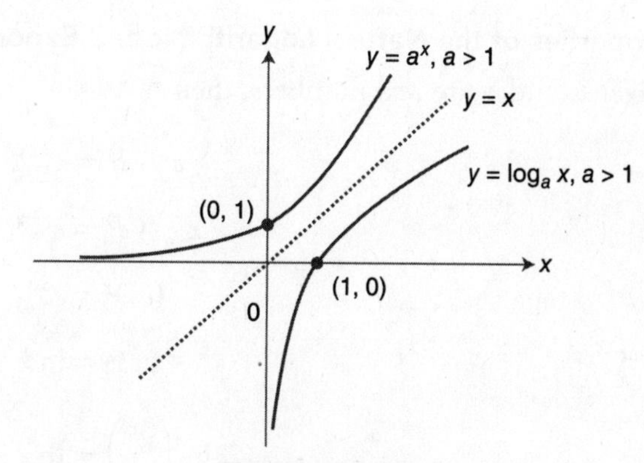

Figure 5.3-25

Properties of Logarithms

Given that x, y, and a are positive numbers with $a \neq 1$ and n is a real number, then

$$\log_a (xy) = \log_a x + \log_a y$$

$$\log_a \left(\frac{x}{y}\right) = \log_a x - \log_a y$$

$$\log_a x^n = n \log_a x$$

Note that $\log_a 1 = 0$, $\log_a a = 1$, and $\log_a a^x = x$.

The Natural Base *e*

$e \approx 2.71828182846\ldots$

The expression $\left(1 + \frac{1}{x}\right)^x$ approaches the number e as x gets larger and larger. An equivalent expression is $(1+h)^{1/h}$. The expression $(1+h)^{1/h}$ also approaches e as h approaches 0.

Exponential Function with Base *e*: $f(x) = e^x$

The Natural Logarithmic Function: $f(x) = \ln x = \log_e x$ where $x > 0$.
Note that $y = e^x$ and $y = \ln x$ are inverse functions: $e^{\ln x} = \ln e^x = x$. Also note that $e^0 = 1$, $\ln 1 = 0$, and $\ln e = 1$. (See Figure 5.3-26.)

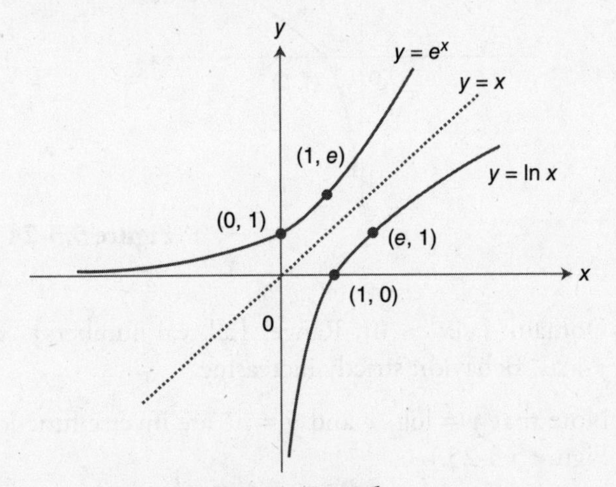

Figure 5.3-26

Properties of the Natural Logarithmic and Exponential Functions

Given x and y are real numbers, then

$$e^x \cdot e^y = e^{x+y}$$

$$e^x \div e^y = e^{x-y}$$

$$(e^x)^y = e^{xy}$$

$$\ln xy = \ln x + \ln y$$

$$\ln \left(\frac{x}{y}\right) = \ln x - \ln y$$

$$\ln x^n = n \ln x$$

Change of Base Formula

$\log_a x = \dfrac{\ln x}{\ln a}$ where $a > 0$ and $a \neq 1$

Example 1

Sketch the graph of $f(x) = \ln(x - 2)$. Note that the domain of $f(x)$ is $\{x | x > 2\}$ and that $f(3) = \ln(1) = 0$; thus, the x-intercept is 3. (See Figure 5.3-27.)

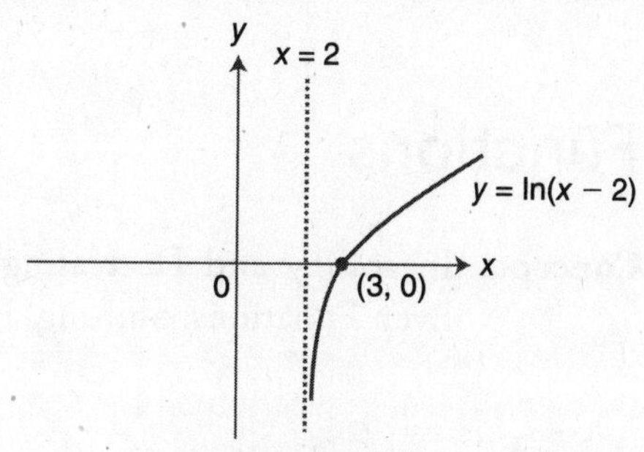

Figure 5.3-27

Example 2

Evaluate:

(a) $\log_2 8$

(b) $\log_5 \dfrac{1}{25}$

(c) $\ln e^5$

Solutions

(a) Let $n = \log_2 8$ and thus $2^n = 8 = 2^3$. Therefore, $n = 3$.

(b) Let $n = \log_5 \dfrac{1}{25}$, and thus $5^n = \dfrac{1}{25} = 5^{-2}$. Therefore, $n = -2$.

(c) You know that $y = e^x$ and $y = \ln x$ are inverse functions. Thus, $\ln e^5 = 5$.

Example 3

Express $\ln(x(2x + 5)^3)$ as the sum and multiplication of logarithms.

$\ln(x(2x + 5)^3) = \ln x + \ln(2x + 5)^3 = \ln x + 3\ln(2x + 5)$

Example 4

Solve $2e^{x+1} = 18$ to the nearest thousandth.

$$\begin{aligned}
2e^{x+1} &= 18 \\
e^{x+1} &= 9 \\
\ln(e^{x+1}) &= \ln 9 \\
x + 1 &= \ln 9 \\
x &= 1.197
\end{aligned}$$

Example 5

Solve $3 \ln 2x = 12$ to the nearest thousandth.

$$3 \ln 2x = 12$$
$$\ln 2x = 4$$
$$e^{\ln 2x} = e^4$$
$$2x = e^4$$
$$x = \frac{e^4}{2} = 27.299$$

5.4 Graphs of Functions

Main Concepts: Increasing and Decreasing Functions; Intercepts and Zeros; Odd and Even Functions; Shifting, Reflecting, and Stretching Graphs

Increasing and Decreasing Functions

Given a function f defined on an interval:

- f is increasing on an interval if $f(x_1) < f(x_2)$ whenever $x_1 < x_2$ for any x_1 and x_2 in the interval.
- f is decreasing on an interval if $f(x_1) > f(x_2)$ whenever $x_1 < x_2$ for any x_1 and x_2 in the interval.
- f is constant on an interval if $f(x_1) = f(x_2)$ for any x_1 and x_2 in the interval.

A function value $f(c)$ is called a *relative minimum* of f if there exists an interval (a, b) in the domain of f containing c such that $f(c) \leq f(x)$ for all $x \in (a, b)$.

A function value $f(c)$ is called a *relative maximum* of f if there exists an interval (a, b) in the domain of f containing c such that $f(c) \geq f(x)$ for all $x \in (a, b)$.

(See Figure 5.4-1.)

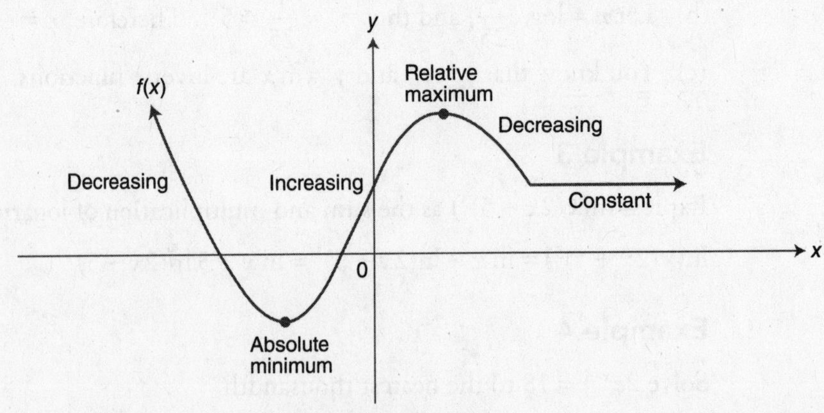

Figure 5.4-1

See the following examples. Using your graphing calculator, determine the intervals over which the given function is increasing, decreasing, or constant. Indicate any relative minimum and maximum values of the function.

Example 1

$f(x) = x^3 - 3x + 2$

The function $f(x) = x^3 - 3x + 2$ is increasing on $(-\infty, -1)$ and $(1, \infty)$ and decreasing on $(-1, 1)$. A relative minimum value of the function is 0, occurring at the point $(1, 0)$, and a relative maximum value of 4 is located at the point $(-1, 4)$. (See Figure 5.4-2.)

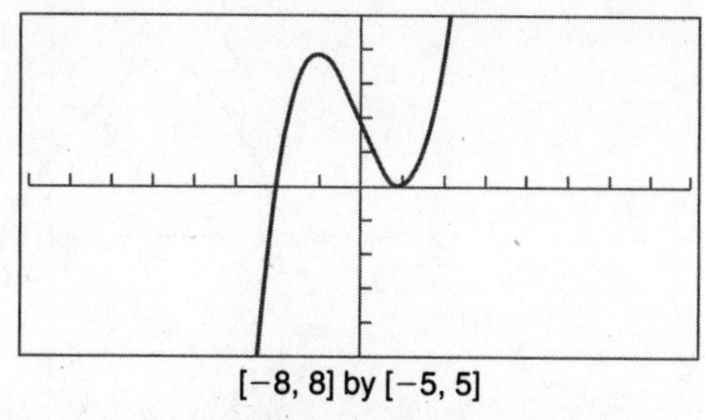

[−8, 8] by [−5, 5]

Figure 5.4-2

Example 2

$g(x) = (x - 1)^3$

Note that $g(x) = (x - 1)^3$ is increasing for the entire domain $(-\infty, \infty)$, and it has no relative minimum or relative maximum values. (See Figure 5.4-3.)

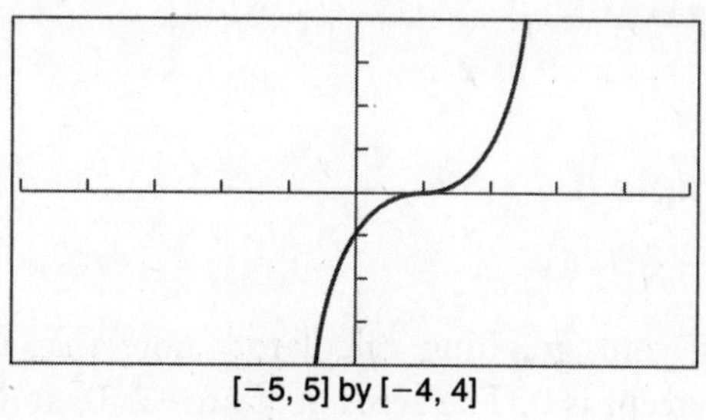

[−5, 5] by [−4, 4]

Figure 5.4-3

Example 3

$$f(x) = \frac{x}{x - 2}$$

The function f is decreasing on the intervals $(-\infty, 2)$ and $(2, \infty)$, and it has no relative minimum or relative maximum values. (See Figure 5.4-4.)

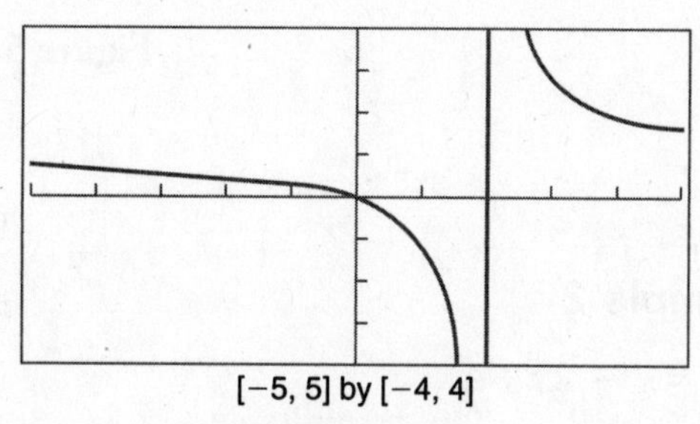

[−5, 5] by [−4, 4]

Figure 5.4-4

Intercepts and Zeros

Given a function f, if $f(a) = 0$, then the point $(a, 0)$ is an x-intercept of the graph of the function, and the number a is called a *zero* of the function.

If $f(0) = b$, then b is the y-intercept of the graph of the function. (See Figure 5.4-5.)

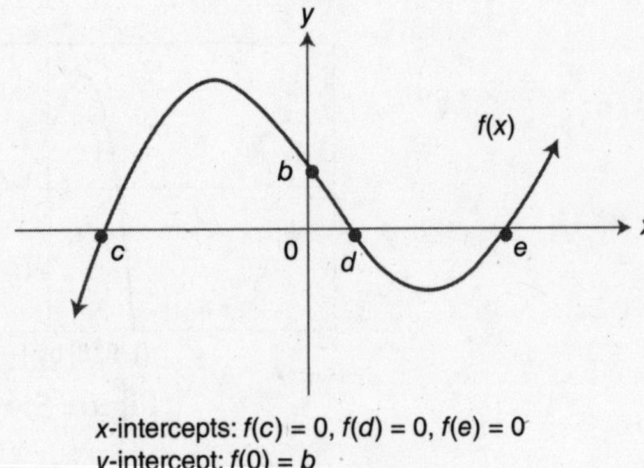

Figure 5.4-5

Note that to find the x-intercepts or zeros of a function, you should set $f(x) = 0$; and to find the y-intercept, let x be 0 (i.e., find $f(0)$).

In the examples below, find the x-intercepts, y-intercept, and zeros of the given function if they exist.

Example 1

$f(x) = x^3 - 4x$

Using your graphing calculator, note that the x-intercepts are −2, 0, and 2, and the y-intercept is 0. The zeros of f are −2, 0, and 2. (See Figure 5.4-6.)

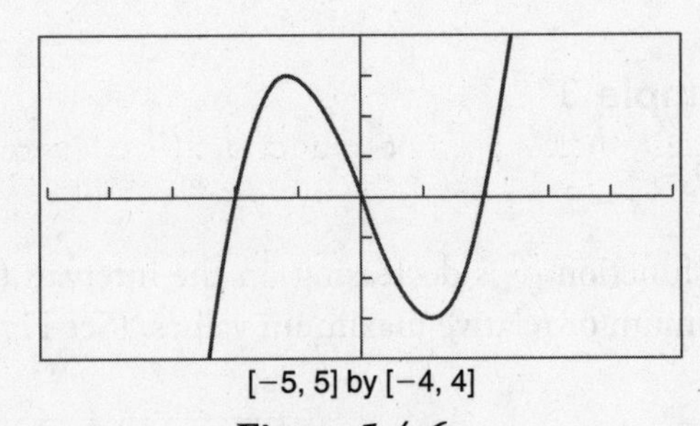

Figure 5.4-6

Example 2

$f(x) = x^2 - 2x + 4$

Using your calculator, you see that the y-intercept is (0, 4) and the function f has no x-intercept or zeros. (See Figure 5.4-7.)

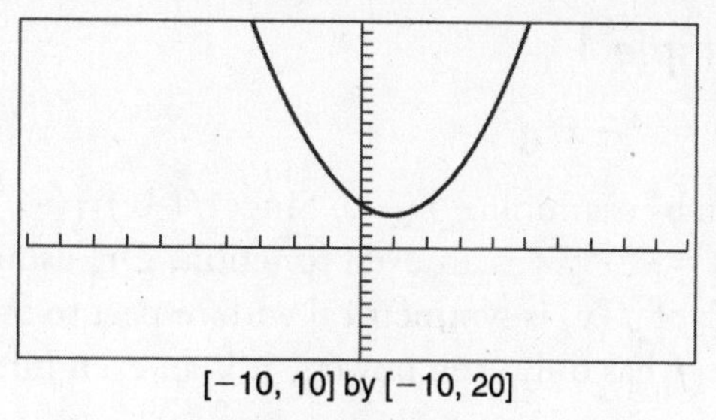

Figure 5.4-7

Odd and Even Functions

A function f is an even function if $f(-x) = f(x)$ for all x in the domain. The graph of an even function is symmetrical with respect to the y-axis. If a point (a, b) is on the graph, so is the point $(-a, b)$. If a function is a polynomial with only even powers, then it is an even function. (See Figure 5.4-8.)

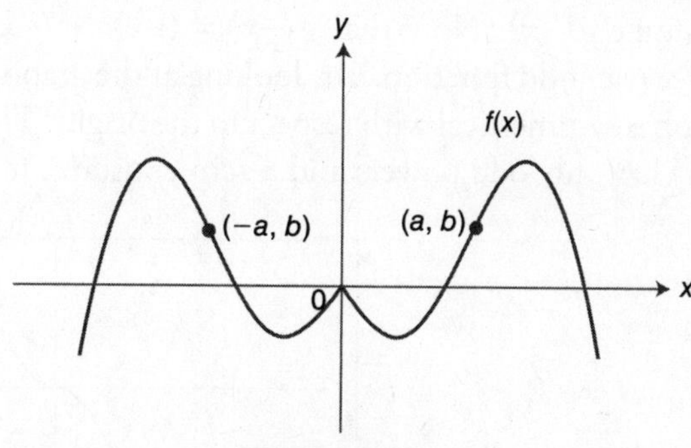

Figure 5.4-8

A function f is an odd function if $f(-x) = -f(x)$ for all x in the domain. The graph of an odd function is symmetrical with respect to the origin. If a point (a, b) is on the graph, so is the point $(-a, -b)$. If a function is a polynomial with only odd powers and a zero constant, then it is an odd function. (See Figure 5.4-9.)

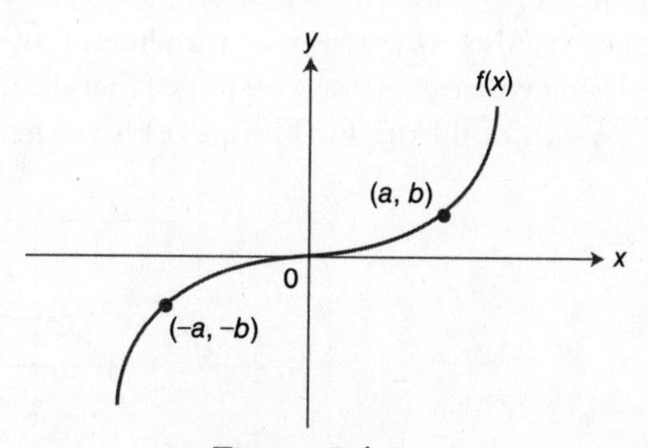

Figure 5.4-9

For the following examples, determine if the given functions are even, odd, or neither.

Example 1

$f(x) = x^4 - x^2$

Begin by examining $f(-x)$. Since $f(-x) = (-x)^4 - (-x)^2 = x^4 - x^2$, $f(-x) = f(x)$. Therefore, $f(x) = x^4 - x^2$ is an even function. Or, using your graphing calculator, you see that the graph of $f(x)$ is symmetrical with respect to the y-axis. Thus, $f(x)$ is an even function. Or, since f has only even powers, it is an even function. (See Figure 5.4-10.)

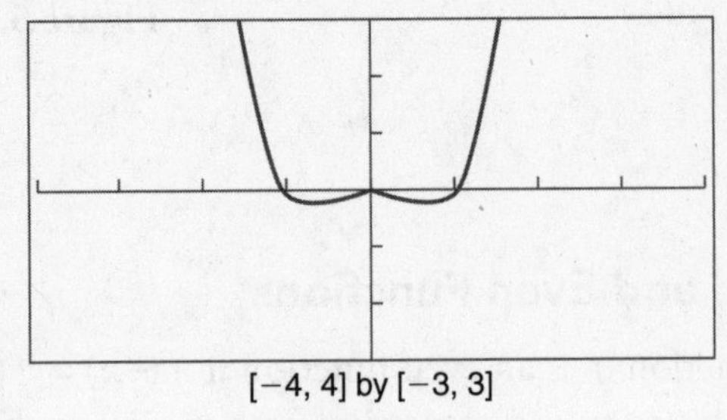

[−4, 4] by [−3, 3]

Figure 5.4-10

Example 2

$g(x) = x^3 + x$

Examine $g(-x)$. Note that $g(-x) = (-x)^3 + (-x) = -x^3 - x = -g(x)$. Therefore, $g(x) = x^3 + x$ is an odd function. Or, looking at the graph of $g(x)$ in your calculator, you see that the graph is symmetrical with respect to the origin. Therefore, $g(x)$ is an odd function. Or, since $g(x)$ has only odd powers and a zero constant, it is an odd function. (See Figure 5.4-11.)

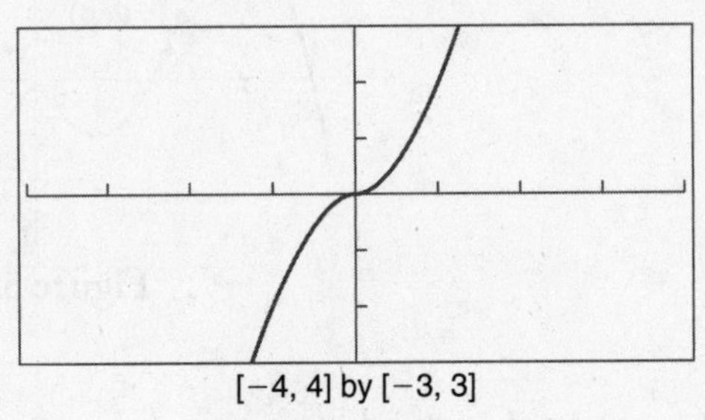

[−4, 4] by [−3, 3]

Figure 5.4-11

Example 3

$h(x) = x^3 + 1$

Examine $h(-x)$. Since $h(-x) = (-x)^3 + 1 = -x^3 + 1$, $h(-x) \neq h(x)$, which indicates that $h(x)$ is not even. Also, $-h(x) = -x^3 - 1$; therefore, $h(-x) \neq -h(x)$, which implies that $h(x)$ is not odd. Using your calculator, you notice that the graph of $h(x)$ is not symmetrical with respect to the y-axis or the origin. Thus, $h(x)$ is neither even nor odd. (See Figure 5.4-12.)

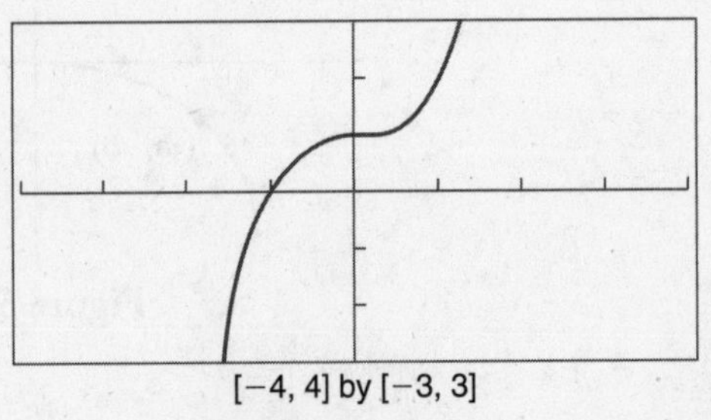

[−4, 4] by [−3, 3]

Figure 5.4-12

Shifting, Reflecting, and Stretching Graphs

Vertical and Horizontal Shifts

Given $y = f(x)$ and $a > 0$, the graph of $y = f(x) + a$ is a vertical shift of the graph of $y = f(x)$ by a units upward. And $y = f(x) - a$ is a vertical shift of the graph of $y = f(x)$ by a units downward. (See Figure 5.4-13.)

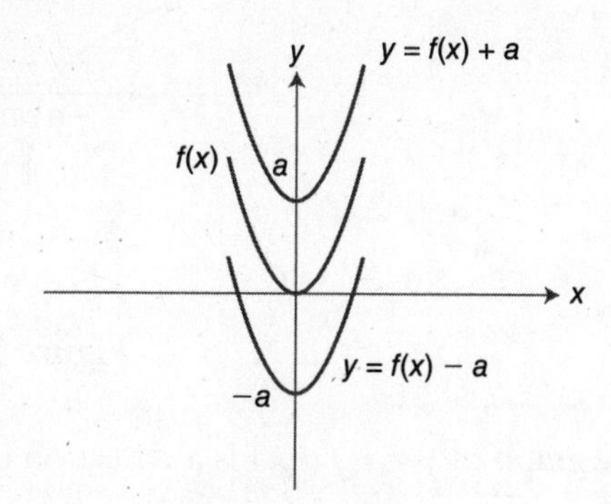

Figure 5.4-13

$y = f(x - a)$ is a horizontal shift of the graph of $y = f(x)$ by a units to the right. $y = f(x + a)$ is a horizontal shift of the graph of $y = f(x)$ by a units to the left. (See Figure 5.4-14.)

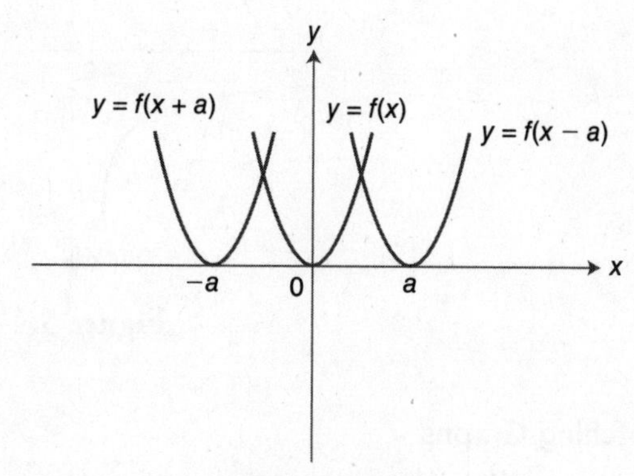

Figure 5.4-14

Reflections About the x-Axis, y-Axis, and the Origin

Given $y = f(x)$, then the graph of $y = -f(x)$ is a reflection of the graph of $y = f(x)$ about the x-axis. (See Figure 5.4-15.)

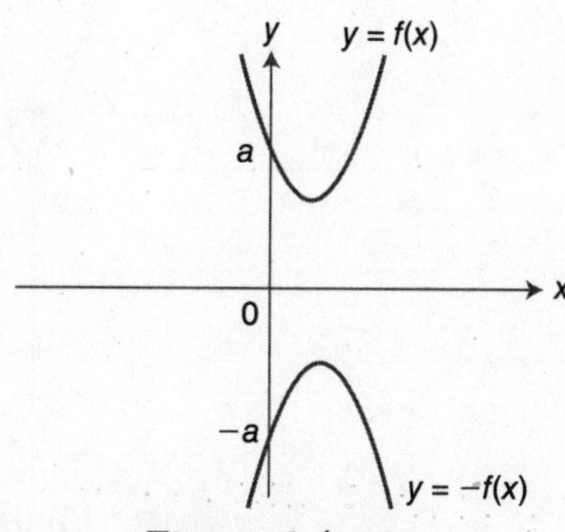

Figure 5.4-15

The graph of $y = f(-x)$ is a reflection of the graph of $y = f(x)$ about the y-axis. (See Figure 5.4-16.)

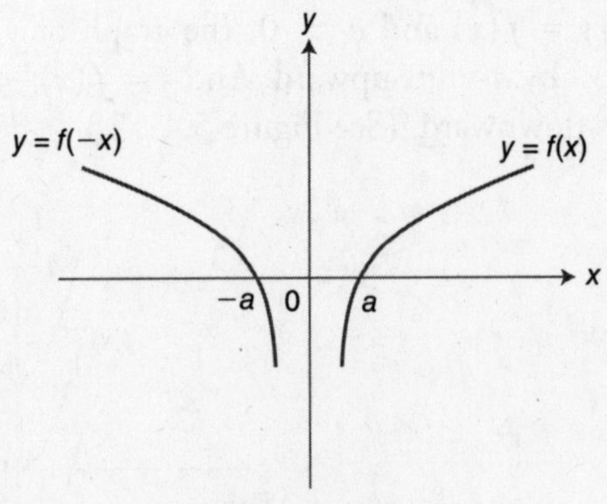

Figure 5.4-16

The graph of $y = -f(-x)$ is a reflection of the graph of $y = f(x)$ about the origin. (See Figure 5.4-17.)

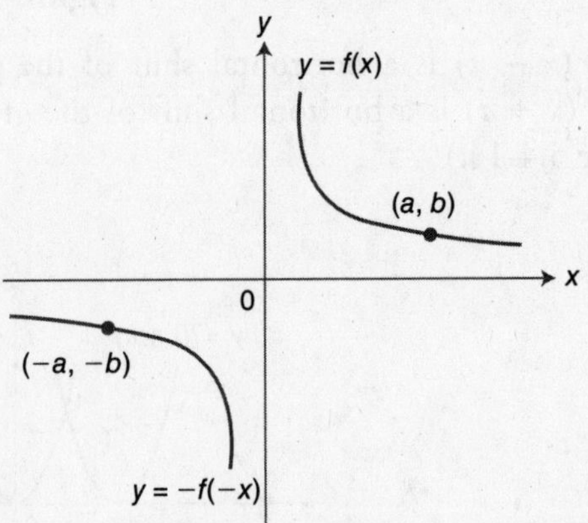

Figure 5.4-17

Stretching Graphs

Given $y = f(x)$, the graph of $y = af(x)$, where $a > 1$ is a vertical stretch of the graph of $y = f(x)$, and $y = af(x)$, where $0 < a < 1$ is a vertical shrink of the graph of $y = f(x)$. (See Figure 5.4-18.)

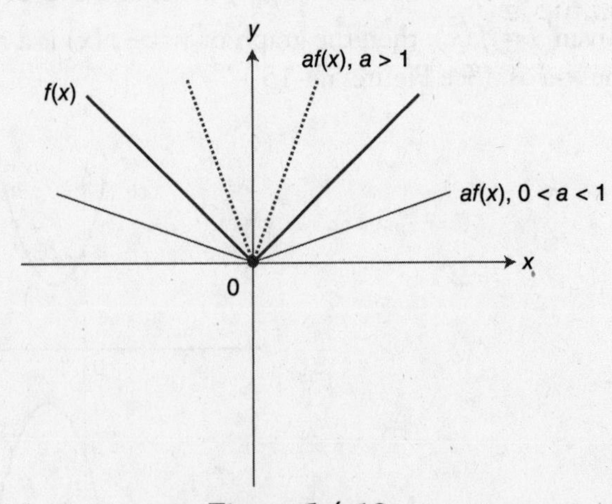

Figure 5.4-18

Example 1

Sketch the graphs of the given functions and verify your results with your graphing calculator: $f(x) = x^2$, $g(x) = 2x^2$, and $p(x) = (x - 3)^2 + 2$.

Note that $g(x)$ is a vertical stretch of $f(x)$ and that $p(x)$ is a horizontal shift of $f(x)$ by 3 units to the right followed by a vertical shift of 2 units upward. (See Figure 5.4-19.)

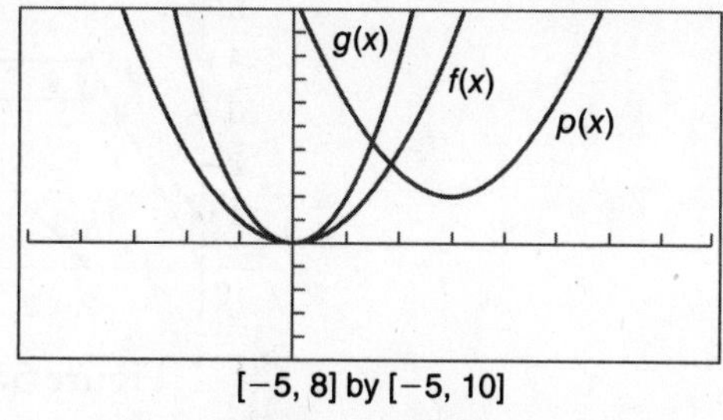

Figure 5.4-19

Example 2

Figure 5.4-20 contains the graphs of $f(x) = x^3$, $h(x)$, and $g(x)$. Find an equation for $h(x)$ and an equation for $g(x)$. (See Figure 5.4-20.)

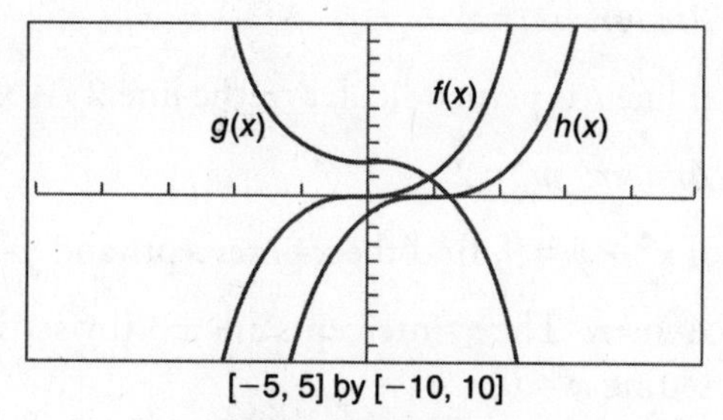

Figure 5.4-20

The graph of $h(x)$ is a horizontal shift of the graph of $f(x)$ by 1 unit to the right. Therefore, $h(x) = (x - 1)^3$. The graph of $g(x)$ is a reflection of the graph of $f(x)$ about the x-axis followed by a vertical shift of 2 units upward. Thus, $g(x) = -x^3 + 2$.

Example 3

Given $f(x)$ as shown in Figure 5.4-21, sketch the graphs of $f(x-2)$, $f(x)+1$, and $2f(x)$.

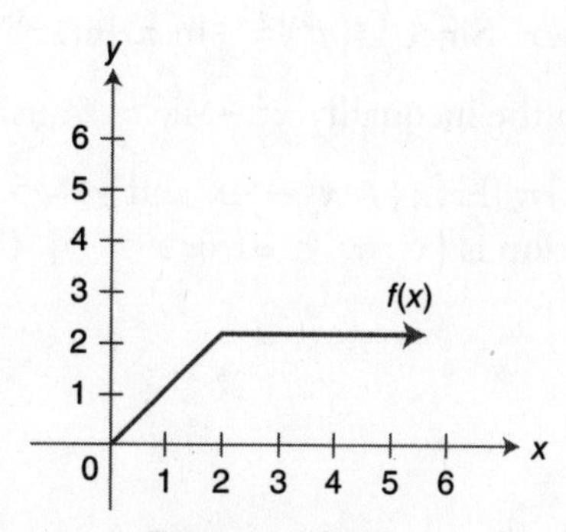

Figure 5.4-21

Note that (a) $f(x-2)$ is a horizontal shift of $f(x)$ by 2 units to the right, (b) $f(x)+1$ is a vertical shift of $f(x)$ by 1 unit upward, and (c) $2f(x)$ is a vertical stretch of $f(x)$ by a factor of 2. (See Figure 5.4-22.)

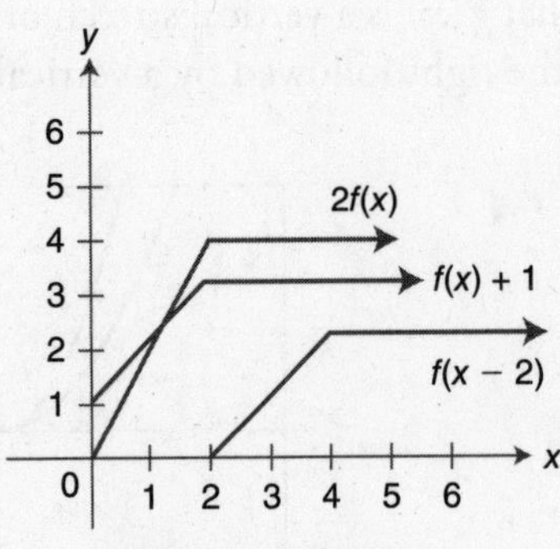

Figure 5.4-22

5.5 Rapid Review

1. If line l is parallel to the line $y - 3x = 2$, find m_l, the slope of line l.

   Answer: $m_l = 3$

2. If line l is perpendicular to the line $2y + x = 6$, find m_l.

   Answer: $m_l = 2$

3. If $x^2 + y = 9$, find the x-intercepts and y-intercept.

   Answer: The x-intercepts are ± 3 (by setting $y = 0$), and the y-intercept is 9 (by setting $x = 0$).

4. Simplify (a) $\ln(e^{3x})$ and (b) $e^{\ln(2x)}$.

   Answer: Since $y = \ln x$ and $y = e^x$ are inverse functions, $\ln(e^{3x}) = 3x$ and $e^{\ln(2x)} = 2x$.

5. Simplify $\ln\left(\frac{1}{x}\right)$.

   Answer: Since $\ln\left(\frac{a}{b}\right) = \ln(a) - \ln(b)$, $\ln\left(\frac{1}{x}\right) = \ln(1) - \ln x = -\ln x$.

6. Simplify $\ln(x^3)$.

   Answer: Since $\ln(a^b) = b \ln a$, $\ln(x^3) = 3 \ln x$.

7. Solve the inequality $x^2 - 4x > 5$, using your calculator.

   Answer: Let $y_1 = x^2 - 4x$ and $y_2 = 5$. Look at the graph and see where y_1 is above y_2. Solution is $\{x : x < -1 \text{ or } x > 5\}$. (See Figure 5.5-1.)

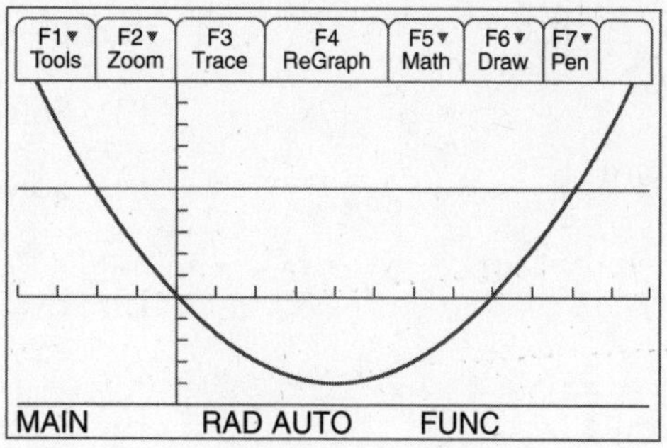

[−2, 6] by [−5, 10]

Figure 5.5-1

8. Evaluate $\sin\left(\frac{\pi}{6}\right)$, $\tan\left(\frac{\pi}{4}\right)$, and $\cos\left(\frac{\pi}{6}\right)$.

   Answer: $\sin\left(\frac{\pi}{6}\right) = \frac{1}{2}$, $\tan\left(\frac{\pi}{4}\right) = 1$, and $\cos\left(\frac{\pi}{6}\right) = \frac{\sqrt{3}}{2}$.

9. $f(x) = \dfrac{\sqrt{x^2-1}}{x-2}$.

   Answer: The domain of f is $\{x : |x| \geq 1 \text{ and } x \neq 2\}$.

10. Is the function $f(x) = x^4 - x^2$ even, odd, or neither?

    Answer: $f(x)$ is an even function since the exponents of x are all even.

5.6 Practice Problems

Part A—The use of a calculator is not allowed.

1. Write an equation of a line passing through the point (−2, 5) and parallel to the line $3x - 4y + 12 = 0$.

2. The vertices of a triangle are $A(-2, 0)$, $B(0, 6)$, and $C(4, 0)$. Find an equation of a line containing the median from vertex A to $\overline{BC}$.

3. Write an equation of a circle whose center is at (2, −3) and tangent to the line $y = -1$.

4. Solve for x: $|x - 2| = 2x + 5$.

5. Solve the inequality $|6 - 3x| < 18$ and sketch the solution on the real number line.

6. Given $f(x) = x^2 + 3x$, find $\dfrac{f(x+h) - f(x)}{h}$ in simplest form.

7. Determine which of the following equations represent y as a function of x:

   (1) $xy = -8$ (2) $4x^2 + 9y^2 = 36$
   (3) $3x^2 - y = 1$ (4) $y^2 - x^2 = 4$

8. If $f(x) = x^2$ and $g(x) = \sqrt{25 - x^2}$, find $(f \circ g)(x)$ and indicate its domain.

9. Given the graphs of f and g in Figures 5.6-1 and 5.6-2, evaluate:

   (1) $(f - g)(2)$ (2) $(f \circ g)(1)$ (3) $(g \circ f)(0)$

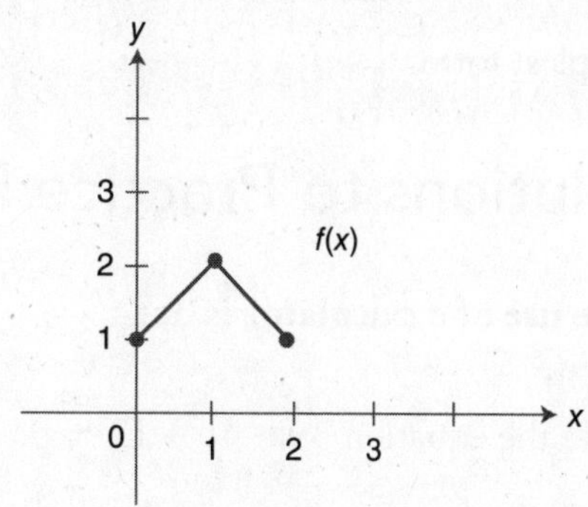

Figure 5.6-1

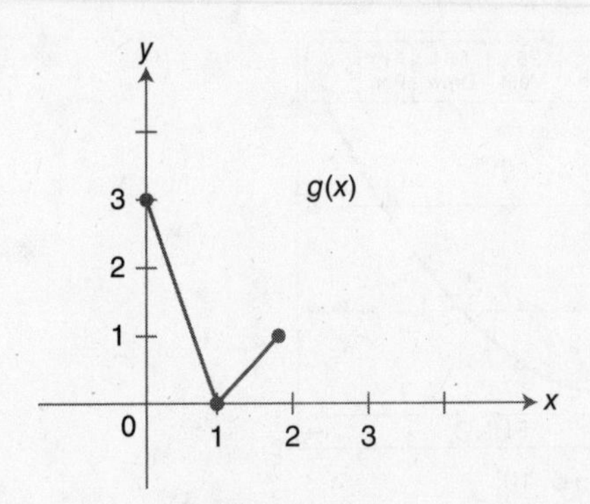

Figure 5.6-2

10. Find the inverse of the function $f(x) = x^3 + 1$.

11. Sketch the graph of the equation $y = 3\cos\left(\frac{1}{2}x\right)$ in the interval $-2\pi \le x \le 2\pi$ and indicate the amplitude, frequency, and period.

12. On the same set of axes, sketch the graphs of:
(1) $y = \ln x$ (2) $y = \ln(-x)$
(3) $y = -\ln(x + 3)$

Part B—Calculators are permitted.

13. Solve the inequality $|2x + 4| \le 10$.

14. Solve the inequality $x^3 - 2x > 1$.

15. Evaluate $\tan\left(\arccos\frac{\sqrt{2}}{2}\right)$.

16. Solve for x to the nearest thousandth: $e^{2x} - 6e^x + 5 = 0$.

17. Solve for x to the nearest thousandth: $3\ln 2x - 3 = 12$.

18. Solve the inequality $\frac{2x - 1}{x + 1} \le 1$.

19. Determine if the function $f(x) = -2x^4 + x^2 + 5$ is even, odd, or neither.

20. Given the function $f(x) = x^4 - 4x^3$, determine the intervals over which the function is increasing, decreasing, or constant. Find all zeros of $f(x)$, and indicate any relative minimum and maximum values of the function.

5.7 Cumulative Review Problems

21. Given a linear function $y = f(x)$, with $f(2) = 4$ and $f(-4) = 10$, find $f(x)$.

22. Solve the inequality $x^3 - x \ge 0$ graphically.

23. If $f(x) = \frac{1}{x}$, $x \ne 0$, evaluate $\frac{f(x + h) - f(x)}{h}$ and express the answer in simplest form.

24. Given $g(x) = 3x - 12$, find $g^{-1}(3)$.

25. Write an equation of the tangent line to the graph of $x^2 + y^2 = 25$ at the point $(4, -3)$.

5.8 Solutions to Practice Problems

Part A—The use of a calculator is not allowed.

1. Rewrite the equation $3x - 4y + 12 = 0$ in $y = mx + b$ form: $y = \frac{3}{4}x + 3$. Thus, the slope of the line is $\frac{3}{4}$. Since line l is parallel to this line, the slope of line l must also be $\frac{3}{4}$. Line l also passes through the point $(-2, 5)$. Therefore, an equation of line l is $y - 5 = \frac{3}{4}(x + 2)$.

2. Let M be the midpoint of $\overline{BC}$. Using the midpoint formula, you will find the

coordinates of M to be (2, 3). The slope of median $\overline{AM}$ is $\frac{3}{4}$. Thus, an equation of $\overline{AM}$ is $y - 3 = \left(\frac{3}{4}\right)(x - 2)$.

3. Since the circle is tangent to the line $y = -1$, the radius of the circle is 2 units. Therefore, the equation of the circle is $(x - 2)^2 + (y + 3)^2 = 4$.

4. The two derived equations are $x - 2 = 2x + 5$ and $x - 2 = -2x - 5$. From $x - 2 = 2x + 5$, $x = -7$ and from $x - 2 = -2x - 5$, $x = -1$. However, substituting $x = -7$ into the original equation $|x - 2| = 2x + 5$ results in $9 = -9$, which is not possible. Thus, the only solution is -1.

5. The inequality $|6 - 3x| < 18$ is equivalent to $-18 < 6 - 3x < 18$. Thus, $-24 < -3x < 12$. Dividing through by -3 and reversing the inequality sign, you have $8 > x > -4$ or $-4 < x < 8$.

6. Since $f(x + h) = (x + h)^2 + 3(x + h)$, the expression $\frac{f(x + h) - f(x)}{h}$ is equivalent to $\frac{[(x + h)^2 + 3(x + h)] - [x^2 + 3x]}{h}$

$$= \frac{(x^2 + 2xh + h^2 + 3x + 3h) - x^2 - 3x}{h}$$

$$= \frac{2xh + h^2 + 3h}{h}$$

$$= 2x + h + 3$$

7. The graph of equation (2) $4x^2 + 9y^2 = 36$ is an ellipse, and the graph of equation (4) $y^2 - x^2 = 4$ is a hyperbola intersecting the y-axis at two distinct points. Both of these graphs fail the vertical line test. Only the graphs of equations (1) $xy = -8$ and (3) $3x^2 - y = 1$ (which are a hyperbola in the second and fourth quadrants and a parabola, respectively) pass the vertical line test. Thus, only (1) $xy = -8$ and (3) $3x^2 - y = 1$ are functions.

8. The domain of $g(x)$ is $-5 \le x \le 5$, and the domain of $f(x)$ is the set of all real numbers. Therefore, the domain of $(f \circ g)(x) = \left(\sqrt{25 - x^2}\right)^2 = 25 - x^2$ is the interval $-5 \le x \le 5$.

9. From the graph, $(f - g)(2) = f(2) - g(2) = 1 - 1 = 0$, $(f \circ g)(1) = f(g(1)) = f(0) = 1$, and $(g \circ f)(0) = g(f(0)) = g(1) = 0$.

10. Let $y = f(x)$ and, thus, $y = x^3 + 1$. Switch x and y to obtain $x = y^3 + 1$. Solve for y, and you will have $y = (x - 1)^{1/3}$. Thus, $f^{-1}(x) = (x - 1)^{1/3}$.

11. The amplitude is 3, frequency is $1/2$, and period is 4π. (See Figure 5.8-1.)

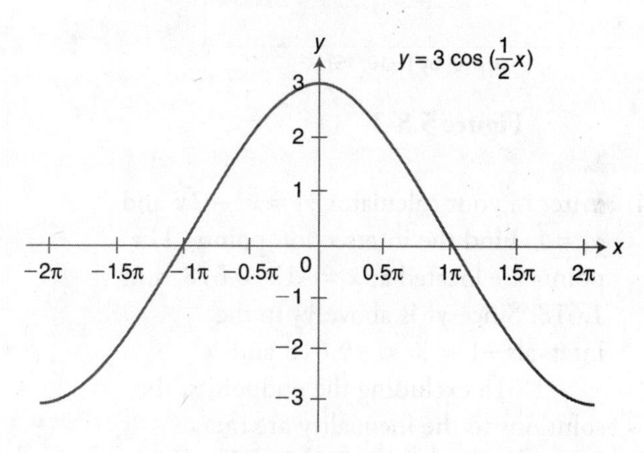

Figure 5.8-1

12. Note that (1) $y = \ln x$ is the graph of the natural logarithmic function. (2) $y = \ln(-x)$ is the reflection about the y-axis. (3) $y = -\ln(x + 3)$ is a horizontal shift 2 units to the left followed by a reflection about the x-axis. (See Figure 5.8-2.)

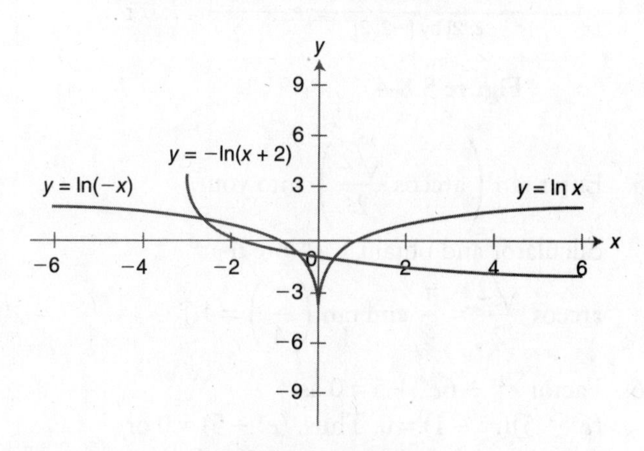

Figure 5.8-2

Part B—Calculators are permitted.

13. Enter into your calculator $y_1 = |2x + 4|$ and $y_2 = 10$. Locate the intersection points. They occur at $x = -7$ and 3. Note that y_1 is below y_2 from $x = -7$ to 3. Since the inequality is less than or equal to, the solution is $-7 \le x \le 3$. (See Figure 5.8-3.)

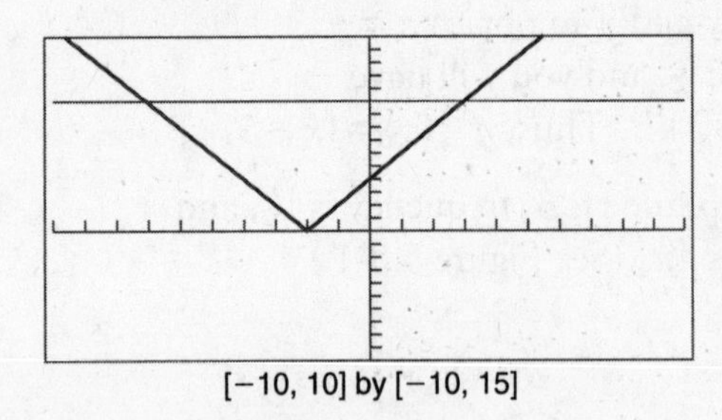

[−10, 10] by [−10, 15]

Figure 5.8-3

14. Enter in your calculator $y_1 = x^3 - 2x$ and $y_2 = 1$. Find the intersection points. The points are located at $x = -1, -0.618$, and 1.618. Since y_1 is above y_2 in the intervals $-1 < x < -0.618$ and $x > 1.618$ excluding the endpoints, the solutions to the inequality are the intervals $-1 < x < -0.618$ and $x > 1.618$. (See Figure 5.8-4.)

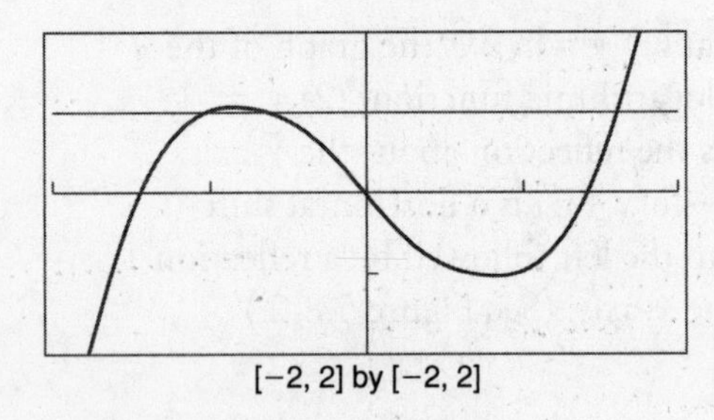

[−2, 2] by [−2, 2]

Figure 5.8-4

15. Enter $\tan\left(\arccos\dfrac{\sqrt{2}}{2}\right)$ into your calculator and obtain 1. (Note that $\arccos\dfrac{\sqrt{2}}{2} = \dfrac{\pi}{4}$ and $\tan\left(\dfrac{\pi}{4}\right) = 1$.)

16. Factor $e^{2x} - 6e^x + 5 = 0$ as $(e^x - 5)(e^x - 1) = 0$. Thus, $(e^x - 5) = 0$ or $(e^x - 1) = 0$, resulting in $e^x = 5$ and $e^x = 1$. Taking the natural log of both sides yields $\ln(e^x) = \ln 5 \approx 1.609$ and $\ln(e^x) = \ln 1 = 0$. Therefore to the nearest thousandth, $x = 1.609$ or 0. (Note that $\ln(e^x) = x$.)

17. The equation $3 \ln 2x - 3 = 12$ is equivalent to $\ln 2x = 5$. Therefore, $e^{\ln 2x} = e^5$, $2x = e^5 \approx 148.413159$, and $x \approx 74.207$.

18. Enter $y_1 = \dfrac{2x - 1}{x + 1}$ and $y_2 = 1$ into your calculator. Note that y_1 is below $y_2 = 1$ on the interval $(-1, 2)$. Since the inequality is $\le$, which includes the endpoint at $x = 2$, the solution is $(-1, 2]$. (See Figure 5.8-5.)

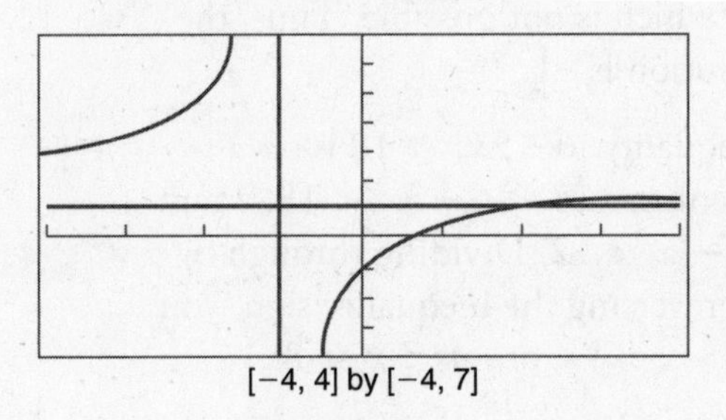

[−4, 4] by [−4, 7]

Figure 5.8-5

19. Examine $f(-x)$ and $f(-x) = -2(-x)^4 + (-x)^2 + 5 = -2x^4 + x^2 + 5 = f(x)$. Therefore, $f(x)$ is an even function. (Note that the graph of $f(x)$ is symmetrical with respect to the y-axis; thus, $f(x)$ is an even function.) (See Figure 5.8-6.)

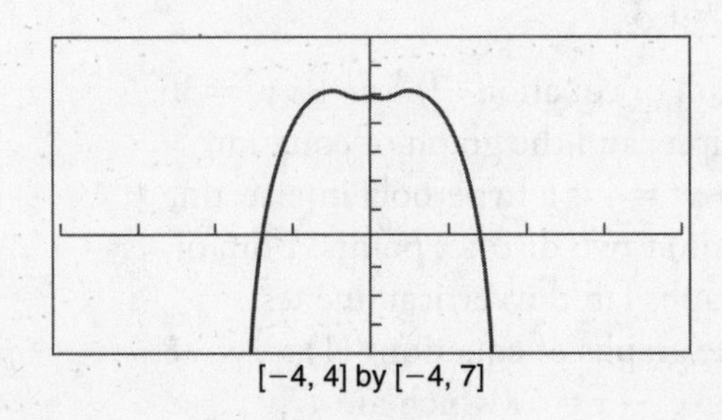

[−4, 4] by [−4, 7]

Figure 5.8-6

20. Enter $y_1 = x^4 - 4x^3$ into your calculator and examine the graph. Note that the graph is decreasing on the interval $(-\infty, 3)$ and increasing on $(3, \infty)$. The function crosses the x-axis at 0 and 4. Thus, the zeros of the function are 0 and

4. There is one relative minimum point at (3, −27). Thus, the relative minimum value for the function is −27. There is no relative maximum. (See Figure 5.8-7.)

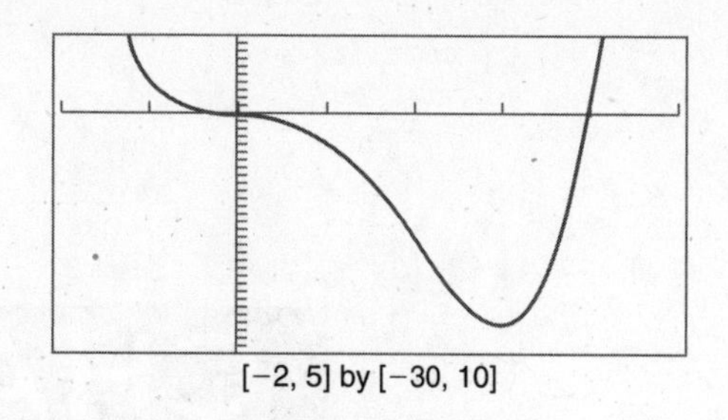
[−2, 5] by [−30, 10]

Figure 5.8-7

5.9 Solutions to Cumulative Review Problems

21. The notation $f(2) = 4$ means that when $x = 2$, $y = 4$, and thus, the point (2, 4) is on the graph of $f(x)$. Similarly, $f(-4) = 10$ implies that the point (−4, 10) is also on the graph. Since $f(x)$ is a linear function, its graph is a line. The slope of a line, m, is defined as $m = \dfrac{y_2 - y_1}{x_2 - x_1}$. Thus, $m = \dfrac{10-4}{-4-2} = \dfrac{6}{-6} = -1$. Using the point slope of a line $y - y_1 = m(x - x_1)$, you have $y - 4 = -1(x - 2)$ or $y = -x + 6$.

22. Enter into your calculator $y_1 = x^3 - x$ and examine the graph. (See Figure 5.9-1.)

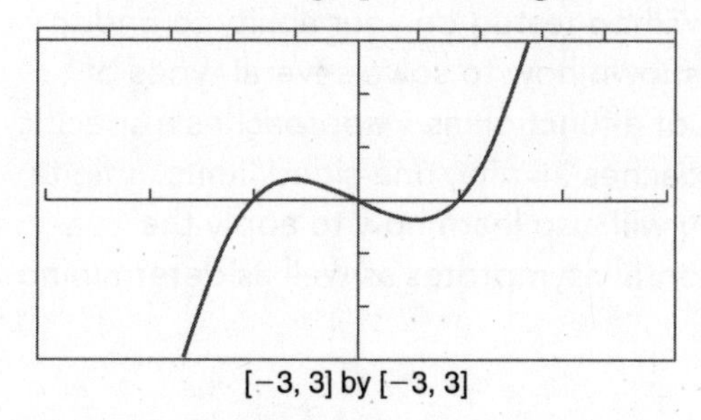
[−3, 3] by [−3, 3]

Figure 5.9-1

Note that $f(x) \geq 0$ on the intervals $[-1, 0]$ and $[1, \infty)$. Therefore, the solution to $x^3 - x \geq 0$ is $-1 \leq x \leq 0$ or $x \geq 1$.

23. Since $f(x) = \dfrac{1}{x}$, $\dfrac{f(x+h) - f(x)}{h} = \dfrac{\frac{1}{x+h} - \frac{1}{x}}{h}$ and the lowest common denominator (LCD) of $\dfrac{1}{x+h}$ and $\dfrac{1}{x}$ is $x(x+h)$. Multiplying the numerator and denominator of the complex fraction by the LCD, you have $\dfrac{\frac{1}{x+h} - \frac{1}{x}}{h} \cdot \dfrac{x(x+h)}{x(x+h)}$, which is equivalent to $\dfrac{x - (x+h)}{xh(x+h)}$ or $\dfrac{-h}{xh(x+h)}$ or $\dfrac{-1}{x(x+h)}$.

24. Begin by finding $g^{-1}(x)$. Rewrite $g(x) = 3x - 12$ as $y = 3x - 12$. Switch x and y, and you have $x = 3y - 12$. Solving for y, you have $y = \dfrac{x+12}{3}$. Substitute $g^{-1}(x)$ for y. Thus, $g^{-1}(x) = \dfrac{x+12}{3}$ and

$g^{-1}(3) = \dfrac{3+12}{3} = 5$.

25. The slope of the line segment joining the origin (0, 0) and the point of tangency (4, −3) is $m = \dfrac{-3-0}{4-0} = \dfrac{-3}{4}$. Since this line segment is perpendicular to the tangent line, the slope of the tangent line is $\dfrac{4}{3}$. Using the point-slope form of a line, you have $y - y_1 = m(x - x_1)$, or $y - (-3) = \dfrac{4}{3}(x - 4)$ or $y = \dfrac{4}{3}x - \dfrac{25}{3}$. (See Figure 5.9-2.)

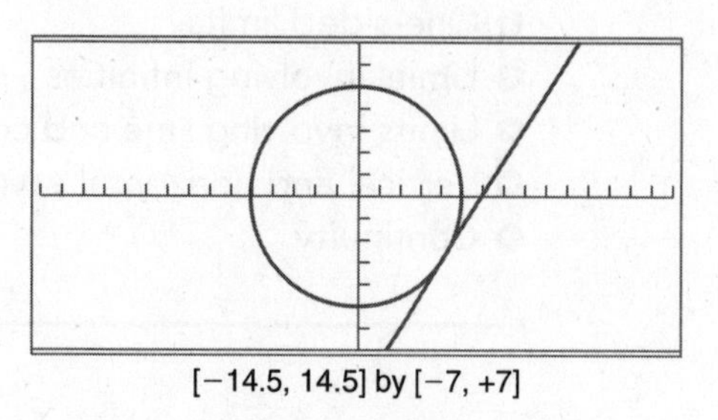
[−14.5, 14.5] by [−7, +7]

Figure 5.9-2

Big Idea 1: Limits
Limits and Continuity

IN THIS CHAPTER

Summary: On the AP Calculus AB exam, you will be tested on your ability to find the limit of a function. In this chapter, you will be shown how to solve several types of limit problems, which include finding the limit of a function as x approaches a specific value, finding the limit of a function as x approaches infinity, one-sided limits, infinite limits, and limits involving sine and cosine. You will also learn how to apply the concepts of limits to finding vertical and horizontal asymptotes as well as determining the continuity of a function.

Key Ideas

- Definition of the limit of a function
- Properties of limits
- Evaluating limits as x approaches a specific value
- Evaluating limits as x approaches $\pm$ infinity
- One-sided limits
- Limits involving infinities
- Limits involving sine and cosine
- Vertical and horizontal asymptotes
- Continuity

6.1 The Limit of a Function

Main Concepts: Definition and Properties of Limits, Evaluating Limits, One-Sided Limits, Squeeze Theorem

Definition and Properties of Limits

Definition of Limit

Let f be a function defined on an open interval containing a, except possibly at a itself. Then $\lim_{x\to a} f(x) = L$ (read as the limit of $f(x)$ as x approaches a is L) if for any $\varepsilon > 0$, there exists a $\delta > 0$ such that $|f(x) - L| < \varepsilon$ whenever $|x - a| < \delta$.

Properties of Limits

Given $\lim_{x\to a} f(x) = L$ and $\lim_{x\to a} g(x) = M$ and L, M, a, c, and n are real numbers, then:

1. $\lim_{x\to a} c = c$
2. $\lim_{x\to a} [cf(x)] = c \lim_{x\to a} f(x) = cL$
3. $\lim_{x\to a} [f(x) \pm g(x)] = \lim_{x\to a} f(x) \pm \lim_{x\to a} g(x) = L + M$
4. $\lim_{x\to a} [f(x) \cdot g(x)] = \lim_{x\to a} f(x) \cdot \lim_{x\to a} g(x) = L \cdot M$
5. $\lim_{x\to a} \dfrac{f(x)}{g(x)} = \dfrac{\lim_{x\to a} f(x)}{\lim_{x\to a} g(x)} = \dfrac{L}{M}, \; M \neq 0$
6. $\lim_{x\to a} [f(x)]^n = \left(\lim_{x\to a} f(x)\right)^n = L^n$

Evaluating Limits

If f is a continuous function on an open interval containing the number a, then $\lim_{x\to a} f(x) = f(a)$.

Common techniques in evaluating limits are:

1. Substituting directly
2. Factoring and simplifying
3. Multiplying the numerator and denominator of a rational function by the conjugate of either the numerator or denominator
4. Using a graph or a table of values of the given function

Example 1

Find the limit: $\lim_{x\to 5} \sqrt{3x + 1}$.

Substituting directly: $\lim_{x\to 5} \sqrt{3x + 1} = \sqrt{3(5) + 1} = 4$.

Example 2

Find the limit: $\lim_{x \to \pi} 3x \sin x$.

Using the product rule, $\lim_{x \to \pi} 3x \sin x = \left(\lim_{x \to \pi} 3x\right)\left(\lim_{x \to \pi} \sin x\right) = (3\pi)(\sin \pi) = (3\pi)(0) = 0.$

Example 3

Find the limit: $\lim_{t \to 2} \frac{t^2 - 3t + 2}{t - 2}$.

Factoring and simplifying: $\lim_{t \to 2} \frac{t^2 - 3t + 2}{t - 2} = \lim_{t \to 2} \frac{(t-1)(t-2)}{(t-2)}$

$$= \lim_{t \to 2} (t - 1) = (2 - 1) = 1.$$

(Note that had you substituted $t = 2$ directly in the original expression, you would have obtained a zero in both the numerator and denominator.)

Example 4

Find the limit: $\lim_{x \to b} \frac{x^5 - b^5}{x^{10} - b^{10}}$.

Factoring and simplifying: $\lim_{x \to b} \frac{x^5 - b^5}{x^{10} - b^{10}} = \lim_{x \to b} \frac{x^5 - b^5}{(x^5 - b^5)(x^5 + b^5)}$

$$= \lim_{x \to b} \frac{1}{x^5 + b^5} = \frac{1}{b^5 + b^5} = \frac{1}{2b^5}.$$

Example 5

Find the limit: $\lim_{t \to 0} \frac{\sqrt{t+2} - \sqrt{2}}{t}$.

Multiplying both the numerator and the denominator by the conjugate of the numerator, $\left(\sqrt{t+2} + \sqrt{2}\right)$, yields $\lim_{t \to 0} \frac{\sqrt{t+2} - \sqrt{2}}{t} \left(\frac{\sqrt{t+2} + \sqrt{2}}{\sqrt{t+2} + \sqrt{2}}\right)$

$$= \lim_{t \to 0} \frac{t + 2 - 2}{t\left(\sqrt{t+2} + \sqrt{2}\right)}$$

$$= \lim_{t \to 0} \frac{t}{t\left(\sqrt{t+2} + \sqrt{2}\right)} = \lim_{t \to 0} \frac{1}{\left(\sqrt{t+2} + \sqrt{2}\right)} = \frac{1}{\sqrt{0+2} + \sqrt{2}} = \frac{1}{2\sqrt{2}}$$

$$= \frac{1}{2\sqrt{2}}\left(\frac{\sqrt{2}}{\sqrt{2}}\right) = \frac{\sqrt{2}}{4}.$$

(Note that substituting 0 directly into the original expression would have produced a 0 in both the numerator and denominator.)

Example 6

Find the limit: $\lim_{x\to 0} \frac{3\sin 2x}{2x}$.

Enter $y1 = \frac{3\sin 2x}{2x}$ in the calculator. You see that the graph of $f(x)$ approaches 3 as x approaches 0. Thus, the $\lim_{x\to 0} \frac{3\sin 2x}{2x} = 3$. (Note that had you substituted $x = 0$ directly in the original expression, you would have obtained a zero in both the numerator and denominator.) (See Figure 6.1-1.)

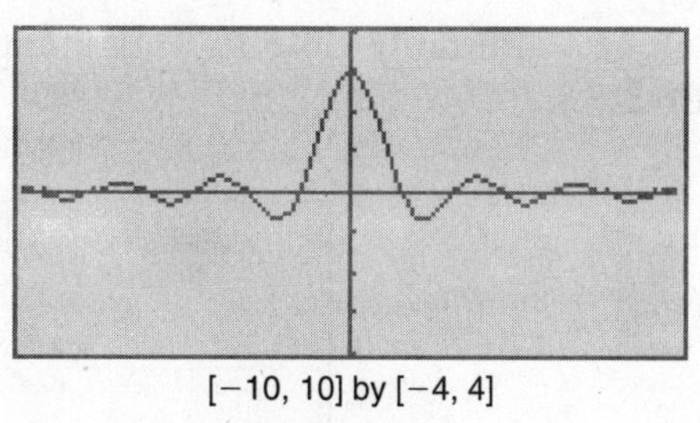

[−10, 10] by [−4, 4]

Figure 6.1-1

Example 7

Find the limit: $\lim_{x\to 3} \frac{1}{x-3}$.

Enter $y1 = \frac{1}{x-3}$ into your calculator. You notice that as x approaches 3 from the right, the graph of $f(x)$ goes higher and higher and that as x approaches 3 from the left, the graph of $f(x)$ goes lower and lower. Therefore, $\lim_{x\to 3} \frac{1}{x-3}$ is undefined. (See Figure 6.1-2.)

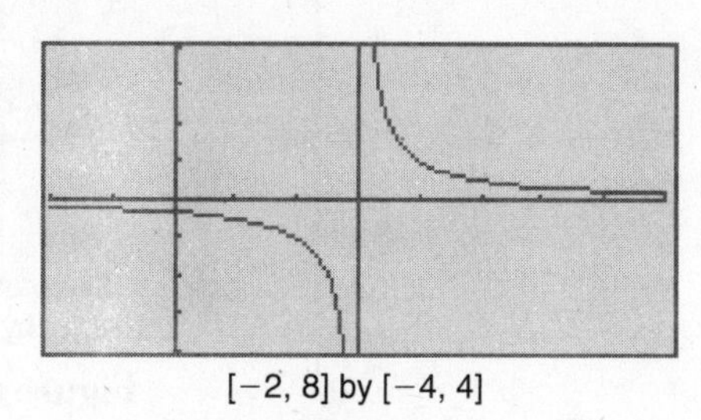

[−2, 8] by [−4, 4]

Figure 6.1-2

- Always indicate what the final answer is, e.g., "The maximum value of f is 5." Use complete sentences whenever possible.

One-Sided Limits

Let f be a function and let a be a real number. Then the right-hand limit: $\lim_{x\to a^+} f(x)$ represents the limit of f as x approaches a from the right, and the left-hand limit: $\lim_{x\to a^-} f(x)$ represents the limit of f as x approaches a from the left.

Existence of a Limit

Let f be a function and let a and L be real numbers. Then the two-sided limit: $\lim_{x\to a} f(x) = L$ if and only if the one-sided limits exist and $\lim_{x\to a^+} f(x) = \lim_{x\to a^-} f(x) = L$.

Example 1

Given $f(x) = \dfrac{x^2 - 2x - 3}{x - 3}$, find the limits: (a) $\lim\limits_{x \to 3^+} f(x)$, (b) $\lim\limits_{x \to 3^-} f(x)$, and (c) $\lim\limits_{x \to 3} f(x)$.

Substituting $x = 3$ into $f(x)$ leads to a 0 in both the numerator and denominator. Factor $f(x)$ as $\dfrac{(x-3)(x+1)}{(x-3)}$, which is equivalent to $(x + 1)$ where $x \neq 3$. Thus, (a) $\lim\limits_{x \to 3^+} f(x) = \lim\limits_{x \to 3^+} (x + 1) = 4$, (b) $\lim\limits_{x \to 3^-} f(x) = \lim\limits_{x \to 3^-} (x + 1) = 4$, and (c) since the one-sided limits exist and are equal, $\lim\limits_{x \to 3^+} f(x) = \lim\limits_{x \to 3^-} f(x) = 4$; therefore, the two-sided limit $\lim\limits_{x \to 3} f(x)$ exists and $\lim\limits_{x \to 3} f(x) = 4$. (Note that $f(x)$ is undefined at $x = 3$, but the function gets arbitrarily close to 4 as x approaches 3. Therefore, the limit exists.) (See Figure 6.1-3.)

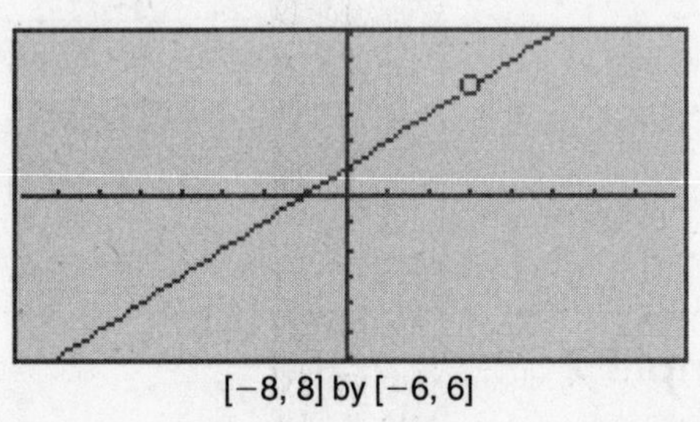

[−8, 8] by [−6, 6]

Figure 6.1-3

Example 2

Given $f(x)$ as illustrated in the accompanying diagram (See Figure 6.1-4), find the limits: (a) $\lim\limits_{x \to 0^-} f(x)$, (b) $\lim\limits_{x \to 0^+} f(x)$, and (c) $\lim\limits_{x \to 0} f(x)$.

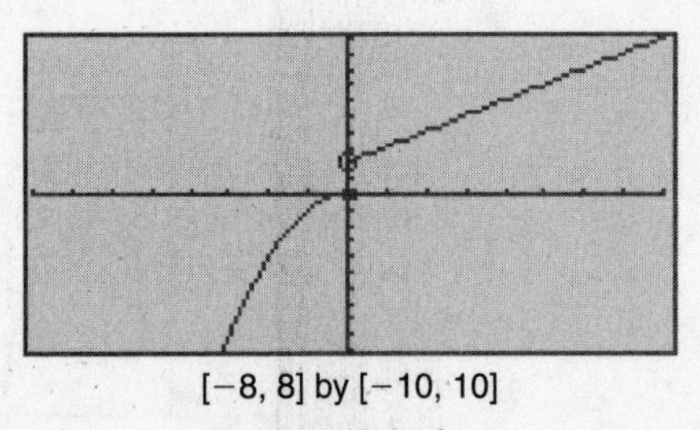

[−8, 8] by [−10, 10]

Figure 6.1-4

(a) As x approaches 0 from the left, $f(x)$ gets arbitrarily close to 0. Thus, $\lim\limits_{x \to 0^-} f(x) = 0$.

(b) As x approaches 0 from the right, $f(x)$ gets arbitrarily close to 2. Therefore, $\lim\limits_{x \to 0^+} f(x) = 2$. Note that $f(0) \neq 2$.

(c) Since $\lim\limits_{x \to 0^+} f(x) \neq \lim\limits_{x \to 0^-} f(x)$, $\lim\limits_{x \to 0} f(x)$ does not exist.

Example 3

Given the greatest integer function $f(x) = [x]$, find the limits: (a) $\lim\limits_{x \to 1^+} f(x)$, (b) $\lim\limits_{x \to 1^-} f(x)$, and (c) $\lim\limits_{x \to 1} f(x)$.

(a) Enter $y1 = \text{int}(x)$ in your calculator. You see that as x approaches 1 from the right, the function stays at 1. Thus, $\lim\limits_{x \to 1^+} [x] = 1$. Note that $f(1)$ is also equal to 1.

(b) As x approaches 1 from the left, the function stays at 0. Therefore, $\lim_{x\to 1^-}[x]=0$. Notice that $\lim_{x\to 1^-}[x]\neq f(1)$.

(c) Since $\lim_{x\to 1^-}[x]\neq \lim_{x\to 1^+}[x]$, therefore, $\lim_{x\to 1}[x]$ does not exist. (See Figure 6.1-5.)

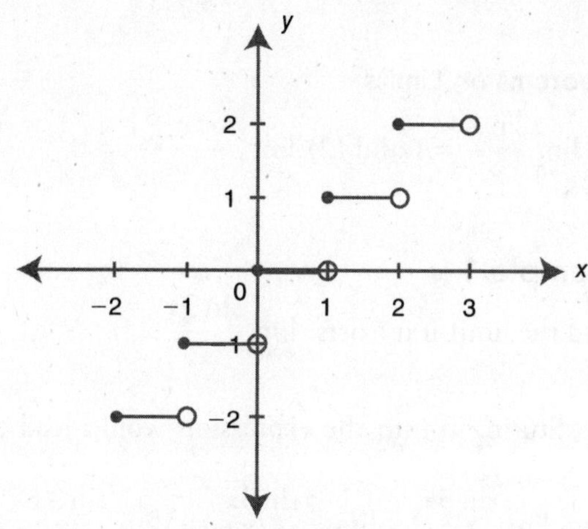

Figure 6.1-5

Example 4

Given $f(x)=\dfrac{|x|}{x}$, $x\neq 0$, find the limits: (a) $\lim_{x\to 0^+} f(x)$, (b) $\lim_{x\to 0^-} f(x)$, and (c) $\lim_{x\to 0} f(x)$.

(a) From inspecting the graph, $\lim_{x\to 0^+} = \dfrac{|x|}{x} = 1$, (b) $\lim_{x\to 0^-} = \dfrac{|x|}{x} = -1$,

and (c) since $\lim_{x\to 0^+}\dfrac{|x|}{x}\neq \lim_{x\to 0^-}\dfrac{|x|}{x}$, therefore, $\lim_{x\to 0} = \dfrac{|x|}{x}$ does not exist. (See Figure 6.1-6.)

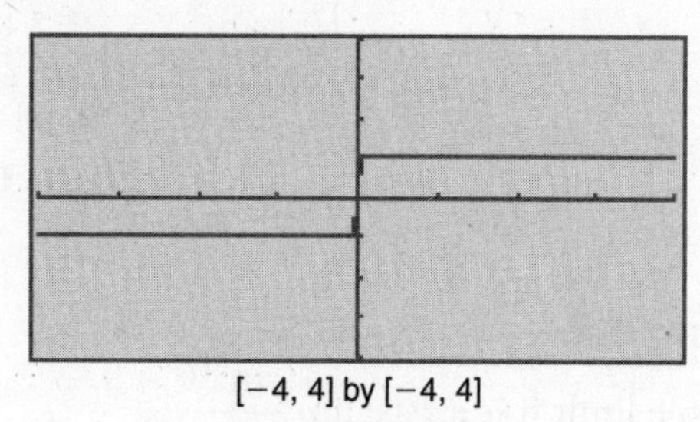

[−4, 4] by [−4, 4]

Figure 6.1-6

Example 5

If $f(x)=\begin{cases} e^{2x} & \text{for } -4\leq x<0 \\ xe^{x} & \text{for } \ \ 0\leq x\leq 4 \end{cases}$, find $\lim_{x\to 0} f(x)$.

$\lim_{x\to 0^+} f(x)=\lim_{x\to 0^+} xe^{x}=0$ and $\lim_{x\to 0^-} f(x)=\lim_{x\to 0^-} e^{2x}=1$.

Thus, $\lim_{x\to 0} f(x)$ does not exist.

- Remember $\ln(e)=1$ and $e^{\ln 3}=3$ since $y=\ln x$ and $y=e^{x}$ are inverse functions.

Squeeze Theorem

If f, g, and h are functions defined on some open interval containing a such that $g(x) \leq f(x) \leq h(x)$ for all x in the interval except possibly at a itself, and $\lim_{x \to a} g(x) = \lim_{x \to a} h(x) = L$, then $\lim_{x \to a} f(x) = L$.

Theorems on Limits

(1) $\lim_{x \to 0} \frac{\sin x}{x} = 1$ and (2) $\lim_{x \to 0} \frac{\cos x - 1}{x} = 0$

Example 1

Find the limit if it exists: $\lim_{x \to 0} \frac{\sin 3x}{x}$.

Substituting 0 into the expression would lead to 0/0. Rewrite $\frac{\sin 3x}{x}$ as $\frac{3}{3} \cdot \frac{\sin 3x}{x}$ and thus, $\lim_{x \to 0} \frac{\sin 3x}{x} = \lim_{x \to 0} \frac{3 \sin 3x}{3x} = 3 \lim_{x \to 0} \frac{\sin 3x}{3x}$. As x approaches 0, so does $3x$. Therefore, $3 \lim_{x \to 0} \frac{\sin 3x}{3x} = 3 \lim_{3x \to 0} \frac{\sin 3x}{3x} = 3(1) = 3$. (Note that $\lim_{3x \to 0} \frac{\sin 3x}{3x}$ is equivalent to $\lim_{x \to 0} \frac{\sin x}{x}$ by replacing $3x$ by x.) Verify your result with a calculator. (See Figure 6.1-7.)

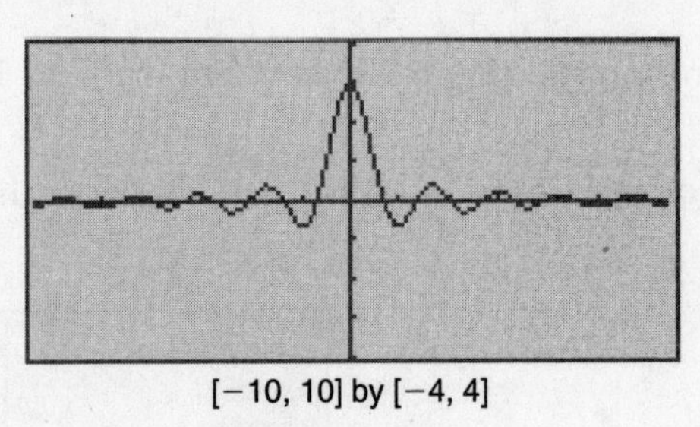

[−10, 10] by [−4, 4]

Figure 6.1-7

Example 2

Find the limit if it exists: $\lim_{h \to 0} \frac{\sin 3h}{\sin 2h}$.

Rewrite $\frac{\sin 3h}{\sin 2h}$ as $\frac{3\left(\frac{\sin 3h}{3h}\right)}{2\left(\frac{\sin 2h}{2h}\right)}$. As h approaches 0, so do $3h$ and $2h$. Therefore,

$$\lim_{h \to 0} \frac{\sin 3h}{\sin 2h} = \frac{3 \lim_{3h \to 0} \frac{\sin 3h}{3h}}{2 \lim_{2h \to 0} \frac{\sin 2h}{2h}} = \frac{3(1)}{2(1)} = \frac{3}{2}.$$

(Note that substituting $h = 0$ into the original expression would have produced 0/0.) Verify your result with a calculator. (See Figure 6.1-8.)

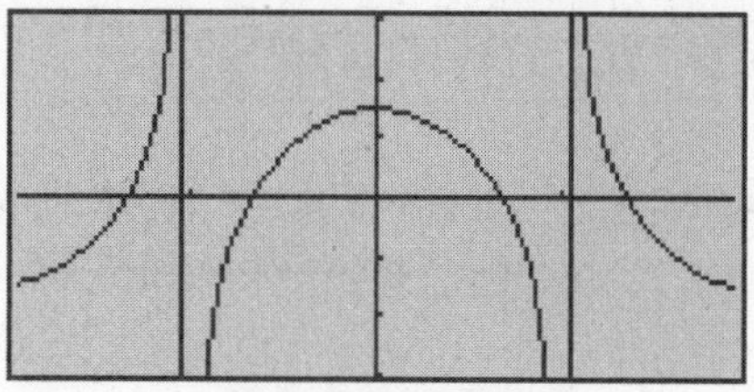

[−3, 3] by [−3, 3]

Figure 6.1-8

Example 3

Find the limit if it exists: $\lim_{y\to 0} \dfrac{y^2}{1-\cos y}$.

Substituting 0 in the expression would lead to 0/0. Multiplying both the numerator and denominator by the conjugate (1 + cos y) produces $\lim_{y\to 0} \dfrac{y^2}{1-\cos y} \cdot \dfrac{(1+\cos y)}{(1+\cos y)}$

$$= \lim_{y\to 0} \frac{y^2(1+\cos y)}{1-\cos^2 y} = \lim_{y\to 0} \frac{y^2(1+\cos y)}{\sin^2 y} = \lim_{y\to 0} \frac{y^2}{\sin^2 y} \cdot \lim_{y\to 0} (1+\cos^2 y)$$

$$= \lim_{y\to 0} \left(\frac{y}{\sin y}\right)^2 \cdot \lim_{y\to 0} (1+\cos^2 y) = \left(\lim_{y\to 0} \frac{y}{\sin y}\right)^2 \cdot \lim_{y\to 0} (1+\cos^2 y) = (1)^2(1+1) = 2.$$

(Note that $\lim_{y\to 0} \dfrac{y}{\sin y} = \lim_{y\to 0} \dfrac{1}{\frac{\sin y}{y}} = \dfrac{\lim_{y\to 0}(1)}{\lim_{y\to 0} \frac{\sin y}{y}} = \dfrac{1}{1} = 1$.) Verify your result with a calculator. (See Figure 6.1-9.)

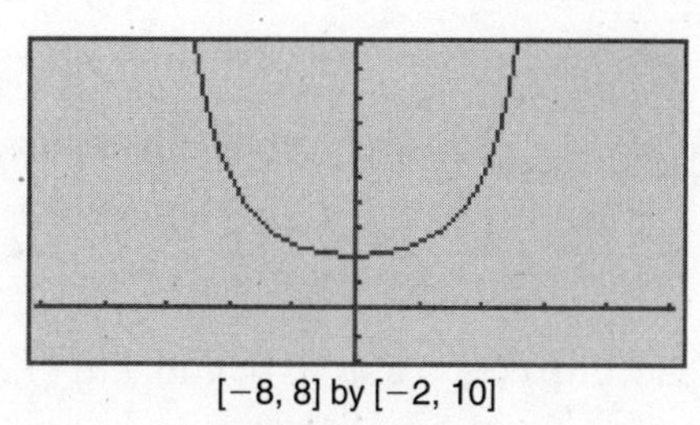

[−8, 8] by [−2, 10]

Figure 6.1-9

Example 4

Find the limit if it exists: $\lim_{x\to 0} \dfrac{3x}{\cos x}$.

Using the quotient rule for limits, you have $\lim_{x\to 0} \dfrac{3x}{\cos x} = \dfrac{\lim_{x\to 0}(3x)}{\lim_{x\to 0}(\cos x)} = \dfrac{0}{1} = 0$.

Verify your result with a calculator. (See Figure 6.1-10.)

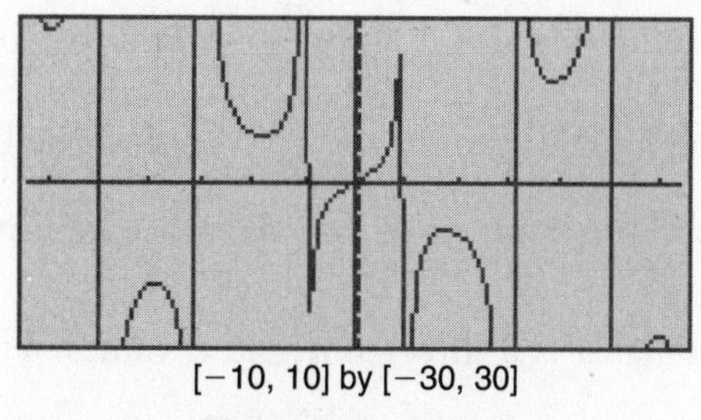

[−10, 10] by [−30, 30]

Figure 6.1-10

6.2 Limits Involving Infinities

Main Concepts: Infinite Limits (as $x \to a$), Limits at Infinity (as $x \to \infty$), Horizontal and Vertical Asymptotes

Infinite Limits (as x ⟶ a)

If f is a function defined at every number in some open interval containing a, except possibly at a itself, then

(1) $\lim_{x \to a} f(x) = \infty$ means that $f(x)$ increases without bound as x approaches a.
(2) $\lim_{x \to a} f(x) = -\infty$ means that $f(x)$ decreases without bound as x approaches a.

Limit Theorems

(1) If n is a positive integer, then

(a) $\lim_{x \to 0^+} \frac{1}{x^n} = \infty$

(b) $\lim_{x \to 0^-} \frac{1}{x^n} = \begin{cases} \infty & \text{if } n \text{ is even} \\ -\infty & \text{if } n \text{ is odd} \end{cases}$

(2) If the $\lim_{x \to a} f(x) = c$, $c > 0$, and $\lim_{x \to a} g(x) = 0$, then

$$\lim_{x \to a} \frac{f(x)}{g(x)} = \begin{cases} \infty & \text{if } g(x) \text{ approaches 0 through positive values} \\ -\infty & \text{if } g(x) \text{ approaches 0 through negative values} \end{cases}$$

(3) If the $\lim_{x \to a} f(x) = c$, $c < 0$, and $\lim_{x \to a} g(x) = 0$, then

$$\lim_{x \to a} \frac{f(x)}{g(x)} = \begin{cases} -\infty & \text{if } g(x) \text{ approaches 0 through positive values} \\ \infty & \text{if } g(x) \text{ approaches 0 through negative values} \end{cases}$$

(Note that limit theorems 2 and 3 hold true for $x \to a^+$ and $x \to a^-$.)

Example 1

Evaluate the limit: (a) $\lim_{x \to 2^+} \frac{3x - 1}{x - 2}$ and (b) $\lim_{x \to 2^-} \frac{3x - 1}{x - 2}$.

(a) The limit of the numerator is 5 and the limit of the denominator is 0 through positive values. Thus, $\lim_{x \to 2^+} \frac{3x - 1}{x - 2} = \infty$. (b) The limit of the numerator is 5 and the limit of the denominator is 0 through negative values. Therefore, $\lim_{x \to 2^-} \frac{3x - 1}{x - 2} = -\infty$. Verify your result with a calculator. (See Figure 6.2-1.)

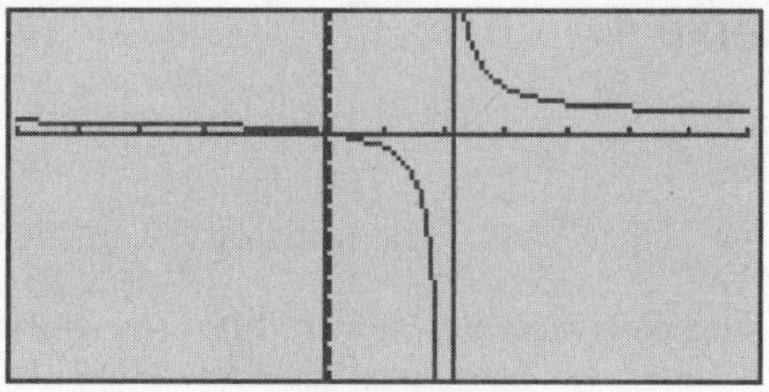

[−5, 7] by [−40, 20]

Figure 6.2-1

Example 2

Find: $\lim_{x \to 3^-} \dfrac{x^2}{x^2 - 9}$.

Factor the denominator obtaining $\lim_{x \to 3^-} \dfrac{x^2}{x^2 - 9} = \lim_{x \to 3^-} \dfrac{x^2}{(x-3)(x+3)}$. The limit of the numerator is 9 and the limit of the denominator is $(0)(6) = 0$ through negative values. Therefore, $\lim_{x \to 3^-} \dfrac{x^2}{x^2 - 9} = -\infty$. Verify your result with a calculator. (See Figure 6.2-2.)

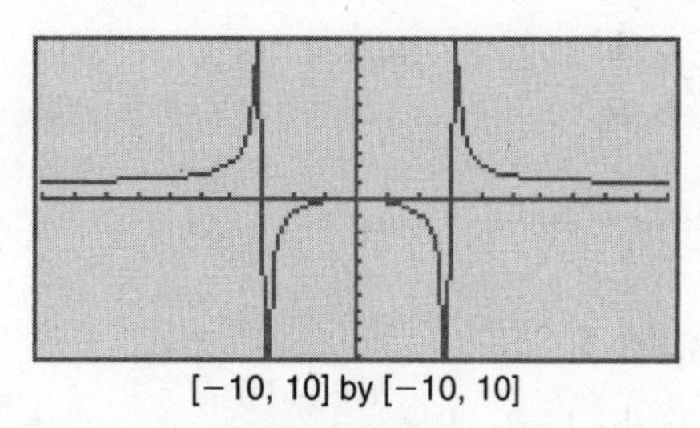

[−10, 10] by [−10, 10]

Figure 6.2-2

Example 3

Find: $\lim_{x \to 5^-} \dfrac{\sqrt{25 - x^2}}{x - 5}$.

Substituting 5 into the expression leads to 0/0. Factor the numerator $\sqrt{25 - x^2}$ into $\sqrt{(5-x)(5+x)}$. As $x \to 5^-$, $(x - 5) < 0$. Rewrite $(x - 5)$ as $-(5 - x)$. As $x \to 5^-$, $(5 - x) > 0$ and thus, you may express $(5 - x)$ as $\sqrt{(5-x)^2} = \sqrt{(5-x)(5-x)}$. Therefore, $(x - 5) = -(5 - x) = -\sqrt{(5-x)(5-x)}$. Substituting these equivalent expressions into the original problem, you have $\lim_{x \to 5^-} \dfrac{\sqrt{25 - x^2}}{x - 5} = \lim_{x \to 5^-} \dfrac{\sqrt{(5-x)(5+x)}}{\sqrt{(5-x)(5-x)}} = -\lim_{x \to 5^-} \sqrt{\dfrac{(5-x)(5+x)}{(5-x)(5-x)}} = -\lim_{x \to 5^-} \sqrt{\dfrac{(5+x)}{(5-x)}}$. The limit of the numerator is 10 and the limit of the denominator is 0 through positive values. Thus, the $\lim_{x \to 5^-} \dfrac{\sqrt{25 - x^2}}{x - 5} = -\infty$.

Example 4

Find: $\lim_{x \to 2^-} \frac{[x]-x}{2-x}$, where $[x]$ is the greatest integer value of x.

As $x \to 2^-$, $[x] = 1$. The limit of the numerator is $(1 - 2) = -1$. As $x \to 2^-$, $(2 - x) = 0$ through positive values. Thus, $\lim_{x \to 2^-} \frac{[x]-x}{2-x} = -\infty$.

- Do easy questions first. The easy ones are worth the same number of points as the hard ones.

Limits at Infinity (as x → ±∞)

If f is a function defined at every number in some interval (a, ∞), then $\lim_{x \to \infty} f(x) = L$ means that L is the limit of $f(x)$ as x increases without bound.
If f is a function defined at every number in some interval $(-\infty, a)$, then $\lim_{x \to -\infty} f(x) = L$ means that L is the limit of $f(x)$ as x decreases without bound.

Limit Theorem

If n is a positive integer, then

(a) $\lim_{x \to \infty} \frac{1}{x^n} = 0$

(b) $\lim_{x \to -\infty} \frac{1}{x^n} = 0$

Example 1

Evaluate the limit: $\lim_{x \to \infty} \frac{6x-13}{2x+5}$.

Divide every term in the numerator and denominator by the highest power of x (in this case, it is x), and obtain:

$$\lim_{x \to \infty} \frac{6x-13}{2x+5} = \lim_{x \to \infty} \frac{6-\frac{13}{x}}{2+\frac{5}{x}} = \frac{\lim_{x \to \infty}(6) - \lim_{x \to \infty}\frac{13}{x}}{\lim_{x \to \infty}(2) + \lim_{x \to \infty}\left(\frac{5}{x}\right)} = \frac{\lim_{x \to \infty}(6) - 13\lim_{x \to \infty}\left(\frac{1}{x}\right)}{\lim_{x \to \infty}(2) + 5\lim_{x \to \infty}\left(\frac{1}{x}\right)}$$

$$= \frac{6-13(0)}{2+5(0)} = 3.$$

Verify your result with a calculator. (See Figure 6.2-3.)

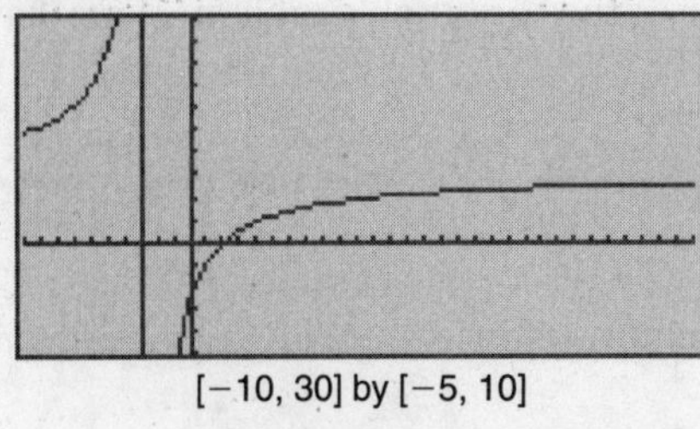

[−10, 30] by [−5, 10]

Figure 6.2-3

Example 2

Evaluate the limit: $\lim_{x\to-\infty} \frac{3x-10}{4x^3+5}$.

Divide every term in the numerator and denominator by the highest power of x. In this case, it is x^3. Thus, $\lim_{x\to-\infty} \frac{3x-10}{4x^3+5} = \lim_{x\to-\infty} \frac{\frac{3}{x^2}-\frac{10}{x^3}}{4+\frac{5}{x^3}} = \frac{0-0}{4+0} = 0.$

Verify your result with a calculator. (See Figure 6.2-4.)

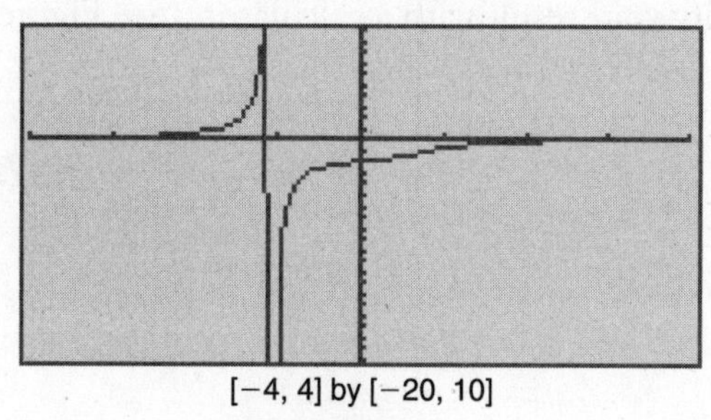

[−4, 4] by [−20, 10]

Figure 6.2-4

Example 3

Evaluate the limit: $\lim_{x\to\infty} \frac{1-x^2}{10x+7}$.

Divide every term in the numerator and denominator by the highest power of x. In this case, it is x^2. Therefore, $\lim_{x\to\infty} \frac{1-x^2}{10x+7} = \lim_{x\to\infty} \frac{\frac{1}{x^2}-1}{\frac{10}{x}+\frac{7}{x^2}} = \frac{\lim_{x\to\infty}\left(\frac{1}{x^2}\right)-\lim_{x\to\infty}(1)}{\lim_{x\to\infty}\left(\frac{10}{x}\right)+\lim_{x\to\infty}\frac{7}{x^2}}$. The limit of the numerator is -1 and the limit of the denominator is 0. Thus, $\lim_{x\to\infty} \frac{1-x^2}{10x+7} = -\infty$.

Verify your result with a calculator. (See Figure 6.2-5.)

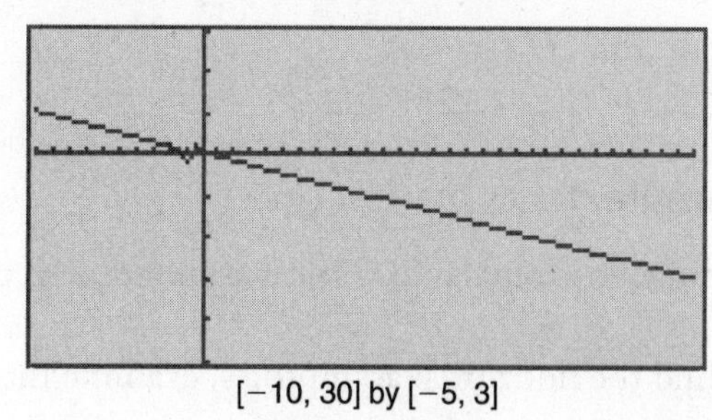

[−10, 30] by [−5, 3]

Figure 6.2-5

Example 4

Evaluate the limit: $\lim_{x\to-\infty} \frac{2x+1}{\sqrt{x^2+3}}$.

As $x \to -\infty$, $x < 0$ and thus, $x = -\sqrt{x^2}$. Divide the numerator and denominator by x (not x^2 since the denominator has a square root). Thus, you have $\lim_{x\to-\infty} \frac{2x+1}{\sqrt{x^2+3}} = \lim_{x\to-\infty} \frac{\frac{2x+1}{x}}{\frac{\sqrt{x^2+3}}{x}}$.

Replacing the x below $\sqrt{x^2+3}$ with $(-\sqrt{x^2})$, you have $\lim_{x\to-\infty} \frac{2x+1}{\sqrt{x^2+3}} = \lim_{x\to-\infty} \frac{\frac{2x+1}{x}}{\frac{\sqrt{x^2+3}}{-\sqrt{x^2}}}$.

Thus, $\lim_{x\to-\infty} \frac{2+\frac{1}{x}}{-\sqrt{1+\frac{3}{x^2}}} = \frac{\lim_{x\to-\infty}(2) - \lim_{x\to-\infty}\frac{1}{x}}{-\sqrt{\lim_{x\to-\infty}(1) + \lim_{x\to-\infty}\left(\frac{3}{x^2}\right)}} = \frac{2}{-1} = -2.$

Verify your result with a calculator. (See Figure 6.2-6.)

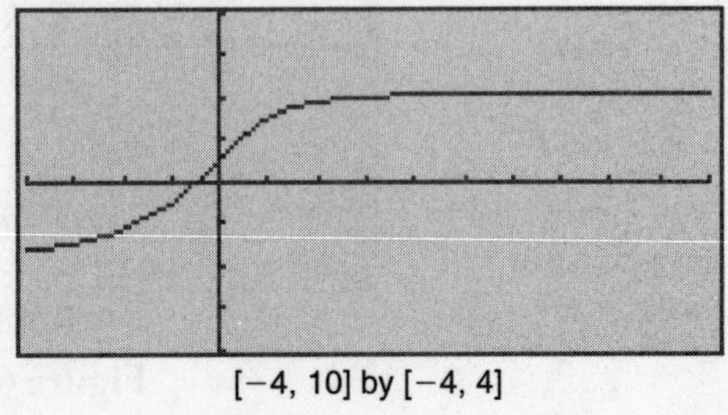
[−4, 10] by [−4, 4]

Figure 6.2-6

- Remember that $\ln\left(\frac{1}{x}\right) = \ln(1) - \ln x = -\ln x$ and $y = e^{-x} = \frac{1}{e^x}$.

Horizontal and Vertical Asymptotes

A line $y = b$ is called a *horizontal asymptote* for the graph of a function f if either $\lim_{x\to\infty} f(x) = b$ or $\lim_{x\to-\infty} f(x) = b$.

A line $x = a$ is called a *vertical asymptote* for the graph of a function f if either $\lim_{x\to a^+} f(x) = +\infty$ or $\lim_{x\to a^-} f(x) = +\infty$.

Example 1

Find the horizontal and vertical asymptotes of the function $f(x) = \frac{3x+5}{x-2}$.

To find the horizontal asymptotes, examine the $\lim_{x\to\infty} f(x)$ and the $\lim_{x\to-\infty} f(x)$.

The $\lim_{x\to\infty} f(x) = \lim_{x\to\infty} \frac{3x+5}{x-2} = \lim_{x\to\infty} \frac{3+\frac{5}{x}}{1-\frac{2}{x}} = \frac{3}{1} = 3$, and the $\lim_{x\to-\infty} f(x) = \lim_{x\to-\infty} \frac{3x+5}{x-2} =$

$\lim_{x\to-\infty} \frac{3+\frac{5}{x}}{1-\frac{2}{x}} = \frac{3}{1} = 3.$

Thus, $y = 3$ is a horizontal asymptote.

To find the vertical asymptotes, look for x values such that the denominator $(x - 2)$ would be 0, in this case, $x = 2$. Then examine:

(a) $\lim_{x\to 2^+} f(x) = \lim_{x\to 2^+} \frac{3x+5}{x-2} = \frac{\lim_{x\to 2^+}(3x+5)}{\lim_{x\to 2^+}(x-2)}$, the limit of the numerator is 11 and the limit of the denominator is 0 through positive values, and thus, $\lim_{x\to 2^+} \frac{3x+5}{x-2} = \infty$.

(b) $\lim_{x\to 2^-} f(x) = \lim_{x\to 2^-} \frac{3x+5}{x-2} = \frac{\lim_{x\to 2^-}(3x+5)}{\lim_{x\to 2^-}(x-2)}$, the limit of the numerator is 11 and the limit of the denominator is 0 through negative values, and thus, $\lim_{x\to 2^-} \frac{3x+5}{x-2} = -\infty$.

Therefore, $x = 2$ is a vertical asymptote.

Example 2

Using your calculator, find the horizontal and vertical asymptotes of the function $f(x) = \frac{x}{x^2-4}$.

Enter $y1 = \frac{x}{x^2-4}$. The graph shows that as $x \to \pm\infty$, the function approaches 0, thus $\lim_{x\to\infty} f(x) = \lim_{x\to-\infty} f(x) = 0$. Therefore, a horizontal asymptote is $y = 0$ (or the x-axis).

For vertical asymptotes, you notice that $\lim_{x\to 2^+} f(x) = \infty$, $\lim_{x\to 2^-} f(x) = -\infty$, and $\lim_{x\to -2^+} f(x) = \infty$, $\lim_{x\to -2^-} f(x) = -\infty$. Thus, the vertical asymptotes are $x = -2$ and $x = 2$. (See Figure 6.2-7.)

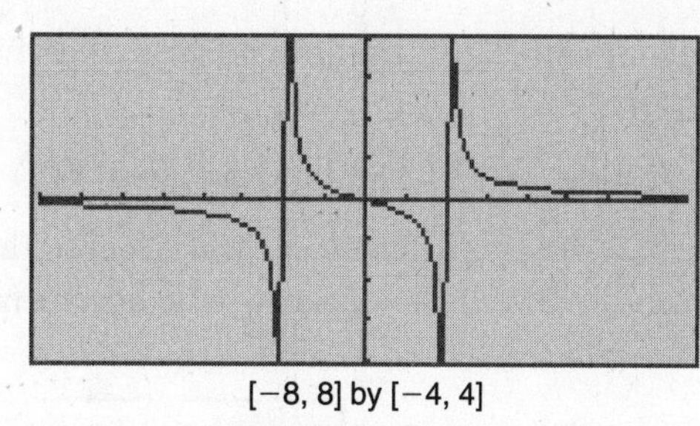

[−8, 8] by [−4, 4]

Figure 6.2-7

Example 3

Using your calculator, find the horizontal and vertical asymptotes of the function $f(x) = \frac{x^3+5}{x}$.

Enter $y1 = \frac{x^3+5}{x}$. The graph of $f(x)$ shows that as x increases in the first quadrant, $f(x)$ goes higher and higher without bound. As x moves to the left in the second quadrant, $f(x)$ again goes higher and higher without bound. Thus, you may conclude that $\lim_{x\to\infty} f(x) = \infty$ and $\lim_{x\to-\infty} f(x) = \infty$ and thus, $f(x)$ has no horizontal asymptote. For vertical asymptotes, you notice that $\lim_{x \to 0^+} f(x) = \infty$, and $\lim_{x \to 0^-} f(x) = -\infty$. Therefore, the line $x = 0$ (or the y-axis) is a vertical asymptote. (See Figure 6.2-8.)

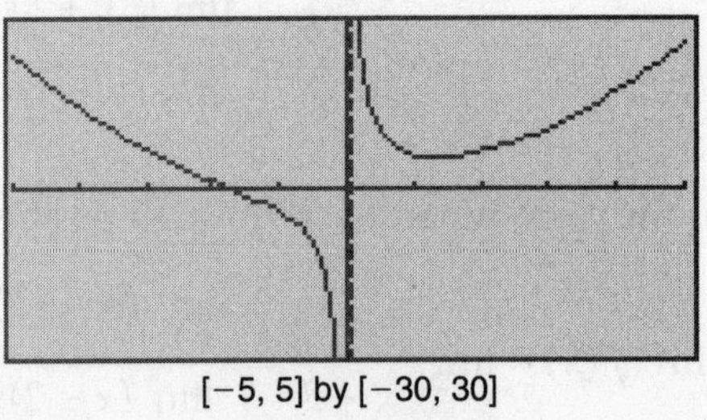

[−5, 5] by [−30, 30]

Figure 6.2-8

Relationship between the limits of rational functions as $x \to \infty$ and horizontal asymptotes:

Given $f(x) = \frac{p(x)}{q(x)}$, then:

(1) If the degree of $p(x)$ is the same as the degree of $q(x)$, then $\lim_{x \to \infty} f(x) = \lim_{x \to -\infty} f(x) = \frac{a}{b}$, where a is the coefficient of the highest power of x in $p(x)$ and b is the coefficient of the highest power of x in $q(x)$. The line $y = \frac{a}{b}$ is a horizontal asymptote. See Example 1 on page 96.

(2) If the degree of $p(x)$ is smaller than the degree of $q(x)$, then $\lim_{x \to \infty} f(x) = \lim_{x \to -\infty} f(x) = 0$. The line $y = 0$ (or the x-axis) is a horizontal asymptote. See Example 2 on page 97.

(3) If the degree of $p(x)$ is greater than the degree of $q(x)$, then $\lim_{x \to \infty} f(x) = \pm\infty$ and $\lim_{x \to -\infty} f(x) = \pm\infty$. Thus, $f(x)$ has no horizontal asymptote. (See Example 3 on page 97.)

Example 4

Using your calculator, find the horizontal asymptotes of the function $f(x) = \frac{2 \sin x}{x}$.

Enter $y1 = \frac{2 \sin x}{x}$. The graph shows that $f(x)$ oscillates back and forth about the x-axis. As $x \to \pm\infty$, the graph gets closer and closer to the x-axis, which implies that $f(x)$ approaches 0. Thus, the line $y = 0$ (or the x-axis) is a horizontal asymptote. (See Figure 6.2-9.)

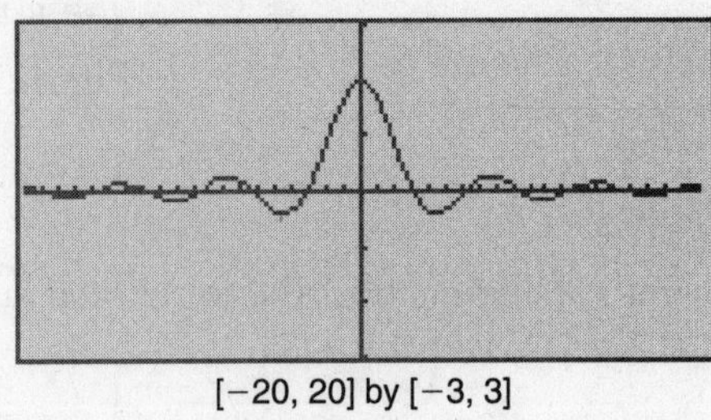

[−20, 20] by [−3, 3]

Figure 6.2-9

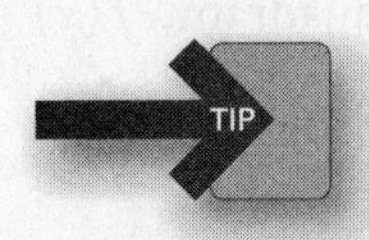

- When entering a rational function into a calculator, use parentheses for both the numerator and denominator, e.g., $(x - 2) + (x + 3)$.

6.3 Continuity of a Function

Main Concepts: Continuity of a Function at a Number, Continuity of a Function over an Interval, Theorems on Continuity

Continuity of a Function at a Number

A function f is said to be continuous at a number a if the following three conditions are satisfied:

1. $f(a)$ exists
2. $\lim_{x \to a} f(x)$ exists
3. $\lim_{x \to a} f(x) = f(a)$

The function f is said to be discontinuous at a if one or more of these three conditions are not satisfied and a is called the point of discontinuity.

Continuity of a Function over an Interval

A function is continuous over an interval if it is continuous at every point in the interval.

Theorems on Continuity

1. If the functions f and g are continuous at a, then the functions $f + g$, $f - g$, $f \cdot g$, f/g, and $g(a) \neq 0$, are also continuous at a.
2. A polynomial function is continuous everywhere.
3. A rational function is continuous everywhere, except at points where the denominator is zero.
4. Intermediate Value Theorem: If a function f is continuous on a closed interval $[a, b]$ and k is a number with $f(a) \leq k \leq f(b)$, then there exists a number c in $[a, b]$ such that $f(c) = k$.

Example 1

Find the points of discontinuity of the function $f(x) = \dfrac{x+5}{x^2 - x - 2}$.

Since $f(x)$ is a rational function, it is continuous everywhere, except at points where the denominator is 0. Factor the denominator and set it equal to 0: $(x - 2)(x + 1) = 0$. Thus $x = 2$ or $x = -1$. The function $f(x)$ is undefined at $x = -1$ and at $x = 2$. Therefore, $f(x)$ is discontinuous at these points. Verify your result with a calculator. (See Figure 6.3-1.)

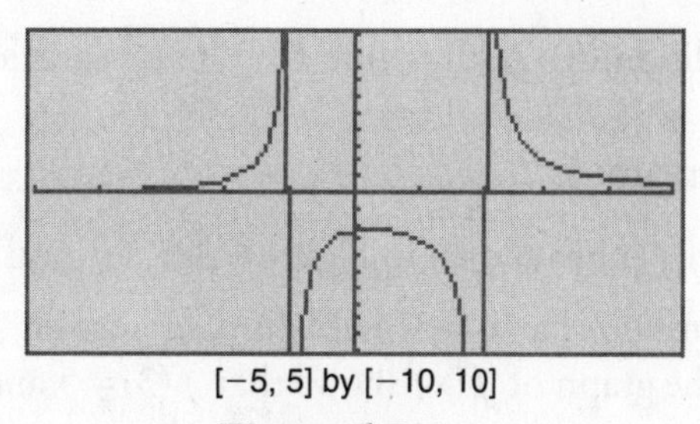

[−5, 5] by [−10, 10]

Figure 6.3-1

Example 2

Determine the intervals on which the given function is continuous:

$$f(x) = \begin{cases} \dfrac{x^2+3x-10}{x-2}, & x \neq 2 \\ 10, & x = 2 \end{cases}$$

Check the three conditions of continuity at $x = 2$:

Condition 1: $f(2) = 10$.

Condition 2: $\lim\limits_{x\to 2} \dfrac{x^2+3x-10}{x-2} = \lim\limits_{x\to 2} \dfrac{(x+5)(x-2)}{x-2} = \lim\limits_{x\to 2} (x+5) = 7$.

Condition 3: $f(2) \neq \lim\limits_{x\to 2} f(x)$. Thus, $f(x)$ is discontinuous at $x = 2$.

The function is continuous on $(-\infty, 2)$ and $(2, \infty)$. Verify your result with a calculator. (See Figure 6.3-2.)

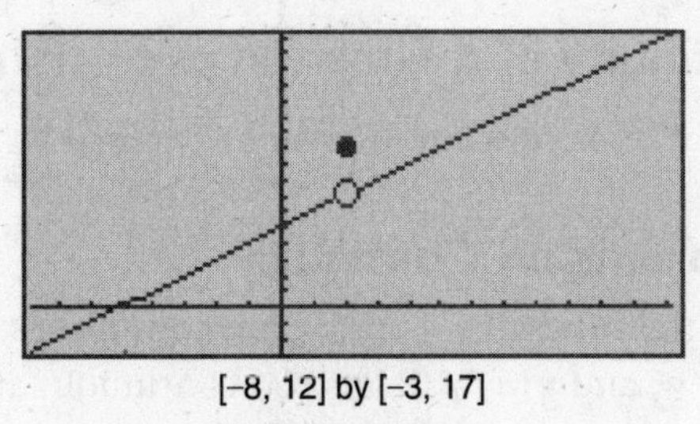

[–8, 12] by [–3, 17]

Figure 6.3-2

- Remember that $\dfrac{d}{dx}\left(\dfrac{1}{x}\right) = -\dfrac{1}{x^2}$ and $\displaystyle\int \frac{1}{x}\,dx = \ln|x| + C$.

Example 3

For what value of k is the function $f(x) = \begin{cases} x^2 - 2x, & x \leq 6 \\ 2x + k, & x > 6 \end{cases}$ continuous at $x = 6$?

For $f(x)$ to be continuous at $x = 6$, it must satisfy the three conditions of continuity:

Condition 1: $f(6) = 6^2 - 2(6) = 24$.

Condition 2: $\lim\limits_{x\to 6^-} (x^2 - 2x) = 24$; thus $\lim\limits_{x\to 6^-} (2x + k)$ must also be 24 in order for the $\lim\limits_{x\to 6} f(x)$ to equal 24. Thus, $\lim\limits_{x\to 6^-} (2x + k) = 24$ which implies $2(6) + k = 24$ and $k = 12$. Therefore, if $k = 12$,

Condition (3): $f(6) = \lim\limits_{x\to 6} f(x)$ is also satisfied.

Example 4

Given $f(x)$ as shown in Figure 6.3-3, (a) find $f(3)$ and $\lim\limits_{x\to 3} f(x)$, and (b) determine if $f(x)$ is continuous at $x = 3$? Explain your answer.

(a) The graph of $f(x)$ shows that $f(3) = 5$ and the $\lim\limits_{x\to 3} f(x) = 1$. (b) Since $f(3) \neq \lim\limits_{x\to 3} f(x)$, $f(x)$ is discontinuous at $x = 3$.

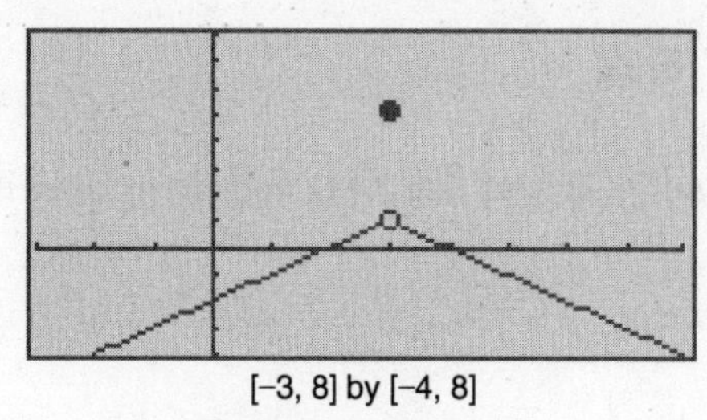

Figure 6.3-3

Example 5

If $g(x) = x^2 - 2x - 15$, using the Intermediate Value Theorem show that $g(x)$ has a root in the interval [1, 7].

Begin by finding $g(1)$ and $g(7)$, and $g(1) = -16$ and $g(7) = 20$. If $g(x)$ has a root, then $g(x)$ crosses the x-axis, i.e., $g(x) = 0$. Since $-16 \leq 0 \leq 20$, by the Intermediate Value Theorem, there exists at least one number c in [1, 7] such that $g(c) = 0$. The number c is a root of $g(x)$.

Example 6

A function f is continuous on [0, 5], and some of the values of f are shown below.

x	0	3	5
f	−4	b	−4

If $f(x) = -2$ has no solution on [0, 5], then b could be

(A) 1 (B) 0 (C) −2 (D) −5

If $b = -2$, then $x = 3$ would be a solution for $f(x) = -2$.

If $b = 0$, 1, or 3, $f(x) = -2$ would have two solutions for $f(x) = -2$.

Thus, $b = -5$, choice (D). (See Figure 6.3-4.)

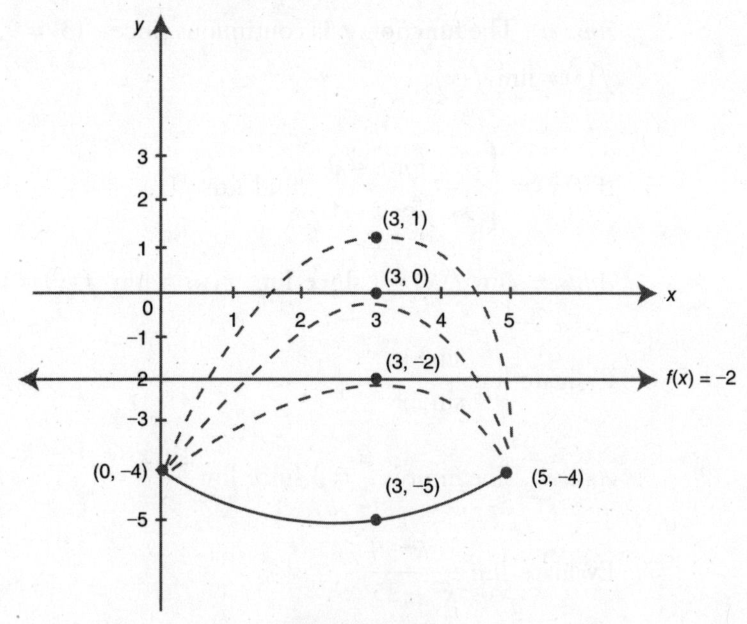

Figure 6.3-4

6.4 Rapid Review

1. Find $f(2)$ and $\lim_{x\to 2} f(x)$ and determine if f is continuous at $x = 2$. (See Figure 6.4-1.)

   Answer: $f(2) = 2$, $\lim_{x\to 2} f(x) = 4$, and f is discontinuous at $x = 2$.

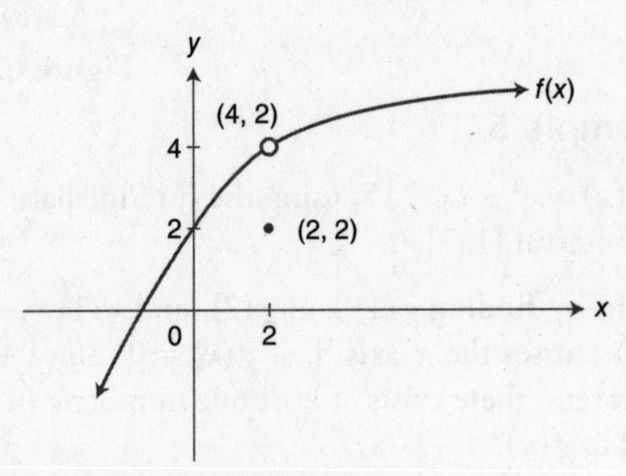

Figure 6.4-1

2. Evaluate $\lim_{x\to a} \dfrac{x^2 - a^2}{x - a}$.

   Answer: $\lim_{x\to a} \dfrac{(x+a)(x-a)}{x-a} = 2a$.

3. Evaluate $\lim_{x\to\infty} \dfrac{1 - 3x^2}{x^2 + 100x + 99}$.

   Answer: The limit is -3 since the polynomials in the numerator and denominator have the same degree.

4. Determine if $f(x) = \begin{cases} x+6 & \text{for } x < 3 \\ x^2 & \text{for } x \geq 3 \end{cases}$ is continuous at $x = 3$.

   Answer: The function f is continuous since $f(3) = 9$, $\lim_{x\to 3^+} f(x) = \lim_{x\to 3^-} f(x) = 9$, and $f(3) = \lim_{x\to 3} f(x)$.

5. If $f(x) = \begin{cases} e^x & \text{for } x \neq 0 \\ 5 & \text{for } x = 0 \end{cases}$, find $\lim_{x\to 0} f(x)$.

   Answer: $\lim_{x\to 0} f(x) = 1$ since $\lim_{x\to 0^+} f(x) = \lim_{x\to 0^-} f(x) = 1$.

6. Evaluate $\lim_{x\to 0} \dfrac{\sin 6x}{\sin 2x}$.

   Answer: The limit is $\dfrac{6}{2} = 3$ since $\lim_{x\to 0} \dfrac{\sin x}{x} = 1$.

7. Evaluate $\lim_{x\to 5^-} \dfrac{x^2}{x^2 - 25}$.

   Answer: The limit is $-\infty$ since $(x^2 - 25)$ approaches 0 through negative values.

8. Find the vertical and horizontal asymptotes of $f(x) = \dfrac{1}{x^2 - 25}$.

Answer: The vertical asymptotes are $x = \pm 5$, and the horizontal asymptote is $y = 0$, since $\lim\limits_{x \to \pm\infty} f(x) = 0$.

6.5 Practice Problems

Part A—The use of a calculator is not allowed.

Find the limits of the following:

1. $\lim\limits_{x \to 0} (x - 5) \cos x$
2. If $b \neq 0$, evaluate $\lim\limits_{x \to b} \dfrac{x^3 - b^3}{x^6 - b^6}$.
3. $\lim\limits_{x \to 0} \dfrac{2 - \sqrt{4 - x}}{x}$
4. $\lim\limits_{x \to \infty} \dfrac{5 - 6x}{2x + 11}$
5. $\lim\limits_{x \to -\infty} \dfrac{x^2 + 2x - 3}{x^3 + 2x^2}$
6. $\lim\limits_{x \to \infty} \dfrac{3x^2}{5x + 8}$
7. $\lim\limits_{x \to -\infty} \dfrac{3x}{\sqrt{x^2 - 4}}$
8. If $f(x) = \begin{cases} e^x & \text{for } 0 \leq x < 1 \\ x^2 e^x & \text{for } 1 \leq x \leq 5 \end{cases}$, find $\lim\limits_{x \to 1} f(x)$.
9. $\lim\limits_{x \to \infty} \dfrac{e^x}{1 - x^3}$
10. $\lim\limits_{x \to 0} \dfrac{\sin 3x}{\sin 4x}$
11. $\lim\limits_{x \to 3^+} \dfrac{\sqrt{t^2 - 9}}{t - 3}$
12. The graph of a function f is shown in Figure 6.5-1. Which of the following statements is/are true?

    I. $\lim\limits_{x \to 4^-} f(x) = 5$.

    II. $\lim\limits_{x \to 4} f(x) = 2$.

    III. $x = 4$ is not in the domain of f.

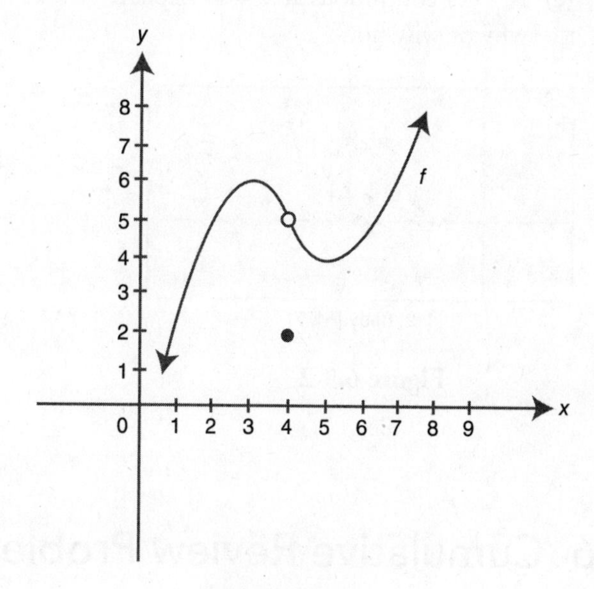

Figure 6.5-1

Part B—Calculators are allowed.

13. Find the horizontal and vertical asymptotes of the graph of the function $f(x) = \dfrac{1}{x^2 + x - 2}$.
14. Find the limit: $\lim\limits_{x \to 5^+} \dfrac{5 + [x]}{5 - x}$ when $[x]$ is the greatest integer of x.
15. Find all x-values where the function $f(x) = \dfrac{x + 1}{x^2 + 4x - 12}$ is discontinuous.
16. For what value of k is the function $g(x) = \begin{cases} x^2 + 5, & x \leq 3 \\ 2x - k, & x > 3 \end{cases}$ continuous at $x = 3$?
17. Determine if $f(x) = \begin{cases} \dfrac{x^2 + 5x - 14}{x - 2}, & \text{if } x \neq 2 \\ 12, & \text{if } x = 2 \end{cases}$ is continuous at $x = 2$. Explain why or why not.

18. Given $f(x)$ as shown in Figure 6.5-2, find

   (a) $f(3)$.
   (b) $\lim_{x \to 3^+} f(x)$.
   (c) $\lim_{x \to 3^-} f(x)$.
   (d) $\lim_{x \to 3} f(x)$.
   (e) Is $f(x)$ continuous at $x = 3$? Explain why or why not.

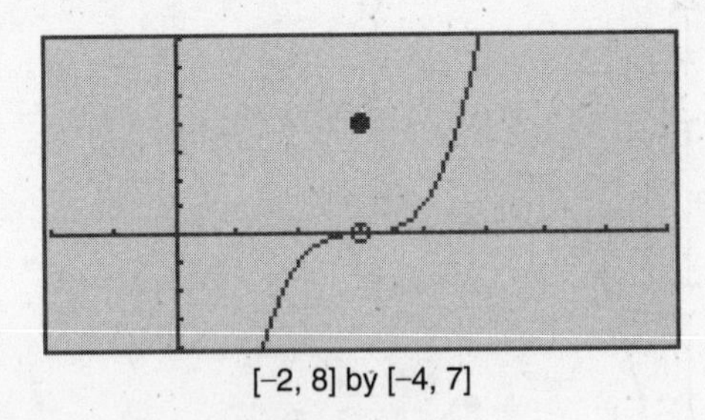

[-2, 8] by [-4, 7]

Figure 6.5-2

19. A function f is continuous on [−2, 2], and some of the values of f are shown below:

x	−2	0	2
$f(x)$	3	b	4

If f has only one root, r, on the closed interval [−2, 2], and $r \neq 0$, then a possible value of b is

(A) −2 (B) −1 (C) 0 (D) 1

20. Evaluate $\lim_{x \to 0} \dfrac{1 - \cos x}{\sin^2 x}$.

6.6 Cumulative Review Problems

21. Write an equation of the line passing through the point (2, −4) and perpendicular to the line $3x - 2y = 6$.

22. The graph of a function f is shown in Figure 6.6-1. Which of the following statements is/are true?

   I. $\lim_{x \to 4^-} f(x) = 3$.
   II. $x = 4$ is not in the domain of f.
   III. $\lim_{x \to 4} f(x)$ does not exist.

23. Evaluate $\lim_{x \to 0} \dfrac{|3x - 4|}{x - 2}$.

24. Find $\lim_{x \to 0} \dfrac{\tan x}{x}$.

25. Find the horizontal and vertical asymptotes of $f(x) = \dfrac{x}{\sqrt{x^2 + 4}}$.

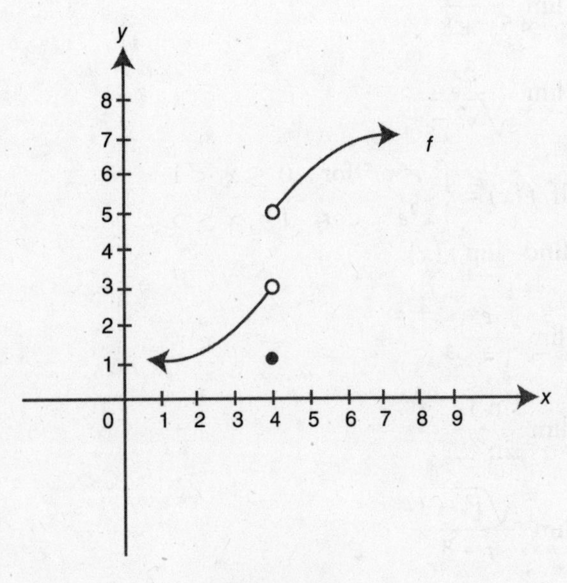

Figure 6.6-1

6.7 Solutions to Practice Problems

Part A—The use of a calculator is not allowed.

1. Using the product rule,
$\lim_{x\to 0}(x-5)(\cos x)=$
$\left[\lim_{x\to 0}(x-5)\right]\left[\lim_{x\to 0}(\cos x)\right]$
$=(0-5)(\cos 0)=(-5)(1)=-5.$
(Note that $\cos 0 = 1$.)

2. Rewrite $\lim_{x\to b}\frac{x^3-b^3}{x^6-b^6}$ as
$\lim_{x\to b}\frac{x^3-b^3}{(x^3-b^3)(x^3+b^3)}=\lim_{x\to b}\frac{1}{x^3+b^3}.$
Substitute $x = b$ and obtain $\frac{1}{b^3+b^3}=\frac{1}{2b^3}.$

3. Substituting $x = 0$ into the expression $\frac{2-\sqrt{4-x}}{x}$ leads to $0/0$, which is an indeterminate form. Thus, multiply both the numerator and denominator by the conjugate $(2+\sqrt{4-x})$ and obtain

$$\lim_{x\to 0}\frac{2-\sqrt{4-x}}{x}\left(\frac{2+\sqrt{4-x}}{2+\sqrt{4-x}}\right)$$
$$=\lim_{x\to 0}\frac{4-(4-x)}{x\left(2+\sqrt{4-x}\right)}$$
$$=\lim_{x\to 0}\frac{x}{x\left(2+\sqrt{4-x}\right)}$$
$$=\lim_{x\to 0}\frac{1}{\left(2+\sqrt{4-x}\right)}$$
$$=\frac{1}{\left(2+\sqrt{4-(0)}\right)}=\frac{1}{4}.$$

4. Since the degree of the polynomial in the numerator is the same as the degree of the polynomial in the denominator,
$\lim_{x\to\infty}\frac{5-6x}{2x+11}=-\frac{6}{2}=-3.$

5. Since the degree of the polynomial in the numerator is 2 and the degree of the polynomial in the denominator is 3,
$\lim_{x\to-\infty}\frac{x^2+2x-3}{x^3+2x^2}=0.$

6. The degree of the monomial in the numerator is 2, and the degree of the binomial in the denominator is 1. Thus,
$\lim_{x\to\infty}\frac{3x^2}{5x+8}=\infty.$

7. Divide every term in both the numerator and denominator by the highest power of x. In this case, it is x. Thus, you have
$\lim_{x\to-\infty}\frac{\frac{3x}{x}}{\frac{\sqrt{x^2-4}}{x}}$. As $x\to-\infty$, $x=-\sqrt{x^2}$.
Since the denominator involves a radical, rewrite the expression as
$$\lim_{x\to-\infty}\frac{\frac{3x}{x}}{\frac{\sqrt{x^2-4}}{-\sqrt{x^2}}}=\lim_{x\to-\infty}\frac{3}{-\sqrt{1-\frac{4}{x^2}}}$$
$$=\frac{3}{-\sqrt{1-0}}=-3.$$

8. $\lim_{x\to 1^+}f(x)=\lim_{x\to 1^+}(x^2e^x)=e$ and $\lim_{x\to 1^-}f(x)=\lim_{x\to 1^-}(e^x)=e$. Thus, $\lim_{x\to 1}f(x)=e$.

9. $\lim_{x\to\infty}e^x=\infty$ and $\lim_{x\to\infty}(1-x^3)=-\infty$.
However, as $x\to\infty$, the rate of increase of e^x is much greater than the rate of decrease of $(1-x^3)$. Thus,
$\lim_{x\to\infty}\frac{e^x}{1-x^3}=-\infty.$

10. Divide both numerator and denominator by x and obtain $\lim_{x\to 0}\frac{\frac{\sin 3x}{x}}{\frac{\sin 4x}{x}}$. Now rewrite the limit as $\lim_{x\to 0}\frac{3\frac{\sin 3x}{3x}}{4\frac{\sin 4x}{4x}}=\frac{3}{4}\lim_{x\to 0}\frac{\frac{\sin 3x}{3x}}{\frac{\sin 4x}{4x}}.$
As x approaches 0, so do $3x$ and $4x$.

Thus, you have

$$\frac{3}{4}\frac{\displaystyle\lim_{3x\to 0}\frac{\sin 3x}{3x}}{\displaystyle\lim_{4x\to 0}\frac{\sin 4x}{4x}}=\frac{3(1)}{4(1)}=\frac{3}{4}.$$

11. As $t \to 3^+$, $(t-3) > 0$, and thus $(t-3)=\sqrt{(t-3)^2}$. Rewrite the limit as

$$\lim_{t\to 3^+}\frac{\sqrt{(t-3)(t+3)}}{\sqrt{(t-3)^2}}=\lim_{t\to 3^+}\frac{\sqrt{(t+3)}}{\sqrt{(t-3)}}.$$

The limit of the numerator is $\sqrt{6}$, and the denominator is approaching 0 through positive values. Thus, $\displaystyle\lim_{t\to 3^+}\frac{\sqrt{t^2-9}}{t-3}=\infty$.

12. The graph of f indicates that:
    I. $\lim_{x\to 4^-} f(x)=5$ is true.
    II. $\lim_{x\to 4} f(x)=2$ is false.
    (The $\lim_{x\to 4} f(x)=5$.)
    III. "$x=4$ is not in the domain of f" is false since $f(4)=2$.

Part B—Calculators are allowed.

13. Examining the graph in your calculator, you notice that the function approaches the x-axis as $x\to\infty$ or as $x\to-\infty$. Thus, the line $y=0$ (the x-axis) is a horizontal asymptote. As x approaches 1 from either side, the function increases or decreases without bound. Similarly, as x approaches -2 from either side, the function increases or decreases without bound. Therefore, $x=1$ and $x=-2$ are vertical asymptotes. (See Figure 6.7-1.)

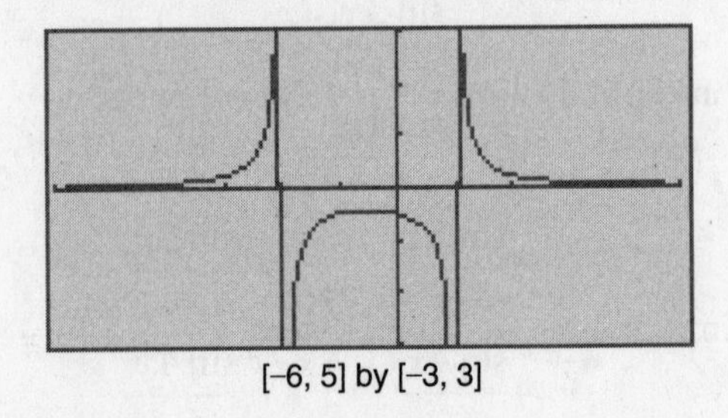

[-6, 5] by [-3, 3]

Figure 6.7-1

14. As $x\to 5^+$, the limit of the numerator $(5+[5])$ is 10 and as $x\to 5^+$, the denominator approaches 0 through negative values. Thus, the $\displaystyle\lim_{x\to 5^+}\frac{5+[x]}{5-x}=-\infty$.

15. Since $f(x)$ is a rational function, it is continuous everywhere except at values where the denominator is 0. Factoring and setting the denominator equal to 0, you have $(x+6)\ (x-2)=0$. Thus, the function is discontinuous at $x=-6$ and $x=2$. Verify your result with a calculator. (See Figure 6.7-2.)

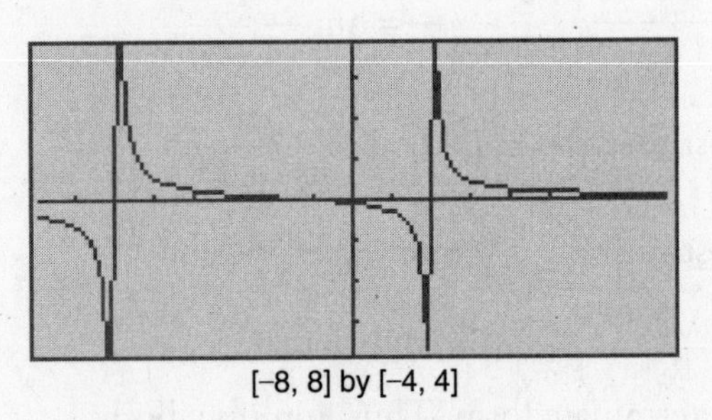

[-8, 8] by [-4, 4]

Figure 6.7-2

16. In order for $g(x)$ to be continuous at $x=3$, it must satisfy the three conditions of continuity:
    (1) $g(3)=3^2+5=14$,
    (2) $\lim_{x\to 3^+}(x^2+5)=14$, and
    (3) $\lim_{x\to 3^-}(2x-k)=6-k$, and the two one-sided limits must be equal in order for $\lim_{x\to 3} g(x)$ to exist. Therefore, $6-k=14$ and $k=-8$.
    Now, $g(3)=\lim_{x\to 3} g(x)$ and condition 3 is satisfied.

17. Checking with the three conditions of continuity:
    (1) $f(2)=12$,
    (2) $\displaystyle\lim_{x\to 2}\frac{x^2+5x-14}{x-2}=\lim_{x\to 2}\frac{(x+7)(x-2)}{x-2}=\lim_{x\to 2}(x+7)=9$, and
    (3) $f(2)\neq\lim_{x\to 2}(x+7)$. Therefore, $f(x)$ is discontinuous at $x=2$.

18. The graph indicates that (a) $f(3) = 4$, (b) $\lim_{x\to 3^+} f(x) = 0$, (c) $\lim_{x\to 3^-} f(x) = 0$, (d) $\lim_{x\to 3} f(x) = 0$, and (e) therefore, $f(x)$ is not continuous at $x = 3$ since $f(3) \neq \lim_{x\to 3} f(x)$.

19. (See Figure 6.7-3.) If $b = 0$, then $r = 0$, but r cannot be 0. If $b = -3, -2$, or -1, f would have more than one root. Thus $b = 1$. Choice (D).

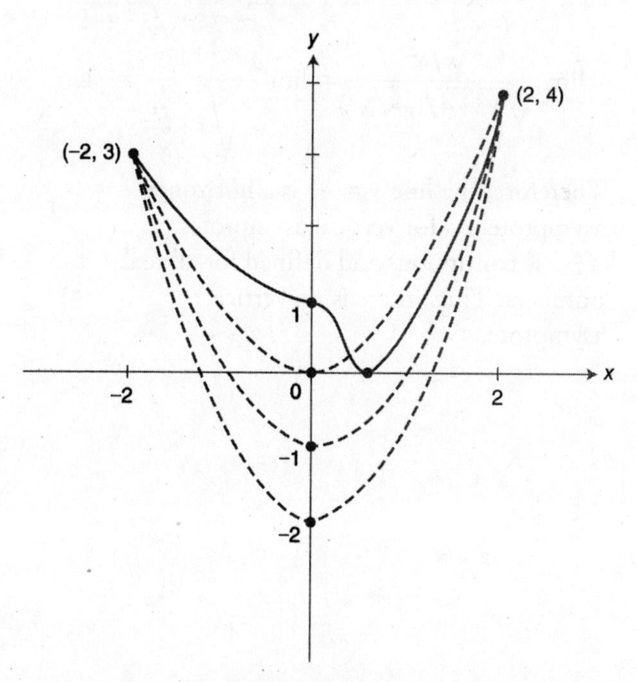

Figure 6.7-3

20. Substituting $x = 0$ would lead to 0/0. Substitute $(1 - \cos^2 x)$ in place of $\sin^2 x$ and obtain

$$\lim_{x\to 0} \frac{1 - \cos x}{\sin^2 x} = \lim_{x\to 0} \frac{1 - \cos x}{(1 - \cos^2 x)}$$

$$= \lim_{x\to 0} \frac{1 - \cos x}{(1 - \cos x)(1 + \cos x)}$$

$$= \lim_{x\to 0} \frac{1}{(1 + \cos x)}$$

$$= \frac{1}{1+1} = \frac{1}{2}.$$

Verify your result with a calculator. (See Figure 6.7-4)

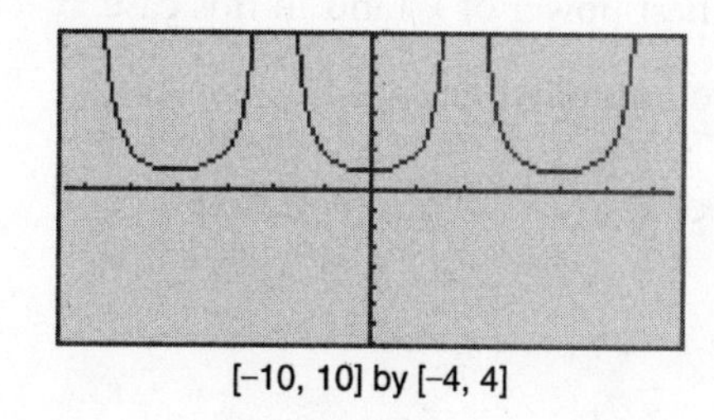

[-10, 10] by [-4, 4]

Figure 6.7-4

6.8 Solutions to Cumulative Review Problems

21. Rewrite $3x - 2y = 6$ in $y = mx + b$ form, which is $y = \frac{3}{2}x - 3$. The slope of this line whose equation is $y = \frac{3}{2}x - 3$ is $m = \frac{3}{2}$. Thus, the slope of a line perpendicular to this line is $m = -\frac{2}{3}$. Since the perpendicular line passes through the point $(2, -4)$, therefore, an equation of the perpendicular line is $y - (-4) = -\frac{2}{3}(x - 2)$, which is equivalent to $y + 4 = -\frac{2}{3}(x - 2)$.

22. The graph indicates that $\lim_{x\to 4^-} f(x) = 3$, $f(4) = 1$, and $\lim_{x\to 4} f(x)$ does not exist. Therefore, only statements I and III are true.

23. Substituting $x = 0$ into $\frac{|3x-4|}{x-2}$, you obtain $\frac{4}{-2} = -2$.

24. Rewrite $\lim_{x \to 0} \frac{\tan x}{x}$ as $\lim_{x \to 0} \frac{\sin x/\cos x}{x}$, which is equivalent to $\lim_{x \to 0} \frac{\sin x}{x \cos x}$, which is equal to
$\lim_{x \to 0} \frac{\sin x}{x} \cdot \lim_{x \to 0} \frac{1}{\cos x} = (1)(1) = 1$.

25. To find horizontal asymptotes, examine the $\lim_{x \to \infty} f(x)$ and the $\lim_{x \to -\infty} f(x)$. The $\lim_{x \to \infty} f(x) = \lim_{x \to \infty} \frac{x}{\sqrt{x^2+4}}$. Dividing by the highest power of x (and in this case, it is x), you obtain $\lim_{x \to \infty} \frac{x/x}{\sqrt{x^2+4}/x}$. As $x \to \infty$, $x = \sqrt{x^2}$. Thus, you have

$$\lim_{x \to \infty} \frac{x/x}{\sqrt{x^2+4}/\sqrt{x^2}} = \lim_{x \to \infty} \frac{1}{\sqrt{\frac{x^2+4}{x^2}}}$$

$= \lim_{x \to \infty} \frac{1}{\sqrt{1+\frac{4}{x^2}}} = 1$. Thus, the line $y = 1$ is a horizontal asymptote.

The $\lim_{x \to -\infty} f(x) = \lim_{x \to -\infty} \frac{x}{\sqrt{x^2+4}}$.

As $x \to -\infty$, $x = -\sqrt{x^2}$. Thus, $\lim_{x \to -\infty} \frac{x}{\sqrt{x^2+4}}$

$$= \lim_{x \to -\infty} \frac{x/x}{\sqrt{x^2+4}/(-\sqrt{x^2})} = \lim_{x \to -\infty} \frac{1}{-\sqrt{1+\frac{4}{x^2}}} = -1.$$

Therefore, the line $y = -1$ is a horizontal asymptote. As for vertical asymptotes, $f(x)$ is continuous and defined for all real numbers. Thus, there is no vertical asymptote.

Big Idea 2: Derivatives
Differentiation

IN THIS CHAPTER

Summary: The derivative of a function is often used to find rates of change. It is also related to the slope of a tangent line. On the AP Calculus AB exam, many questions involve finding the derivative of a function. In this chapter, you will learn different techniques for finding a derivative which include using the Power Rule, Product and Quotient Rules, Chain Rule, and Implicit Differentiation. You will also learn to find the derivatives of trigonometric, exponential, logarithmic, and inverse functions.

Key Ideas

- Definition of the derivative of a function
- Power Rule, Product and Quotient Rules, and Chain Rule
- Derivatives of trigonometric, exponential, and logarithmic functions
- Derivatives of inverse functions
- Implicit Differentiation
- Higher order derivatives
- *L'Hôpital's* Rule for Indeterminate Forms

7.1 Derivatives of Algebraic Functions

Main Concepts: Definition of the Derivative of a Function; Power Rule; The Sum, Difference, Product, and Quotient Rules; The Chain Rule

Definition of the Derivative of a Function

The derivative of a function f, written as f', is defined as

$$f'(x) = \lim_{h \to 0} \frac{f(x+h) - f(x)}{h},$$

if this limit exists. (Note that $f'(x)$ is read as f prime of x.)
Other symbols of the derivative of a function are:

$$D_x f,\ \frac{d}{dx} f(x),\ \text{and if } y = f(x),\ y',\ \frac{dy}{dx},\ \text{and } D_x y.$$

Let m_{tangent} be the slope of the tangent to a curve $y = f(x)$ at a point on the curve. Then,

$$m_{\text{tangent}} = f'(x) = \lim_{h \to 0} \frac{f(x+h) - f(x)}{h}$$

$$m_{\text{tangent}}(\text{at } x = a) = f'(a) = \lim_{h \to 0} \frac{f(a+h) - f(a)}{h} \text{ or } \lim_{x \to a} \frac{f(x) - f(a)}{x - a}.$$

(See Figure 7.1-1.)

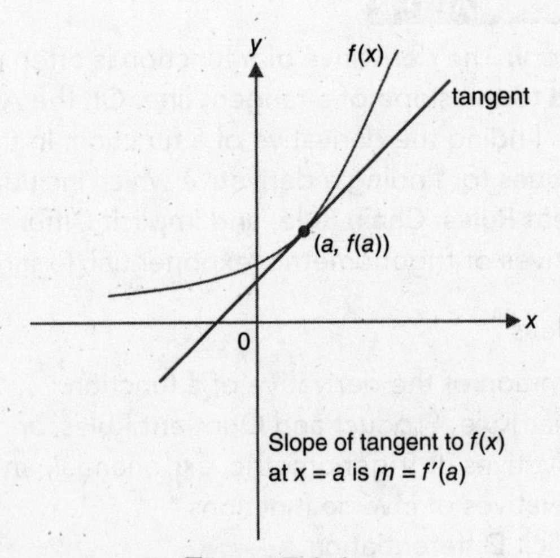

Figure 7.1-1

Given a function f, if $f'(x)$ exists at $x = a$, then the function f is said to be differentiable at $x = a$. If a function f is differentiable at $x = a$, then f is continuous at $x = a$. (Note that the converse of the statement is not necessarily true, i.e., if a function f is continuous at $x = a$, then f may or may not be differentiable at $x = a$.) Here are several examples of functions that are not differentiable at a given number $x = a$. (See Figures 7.1-2–7.1-5.)

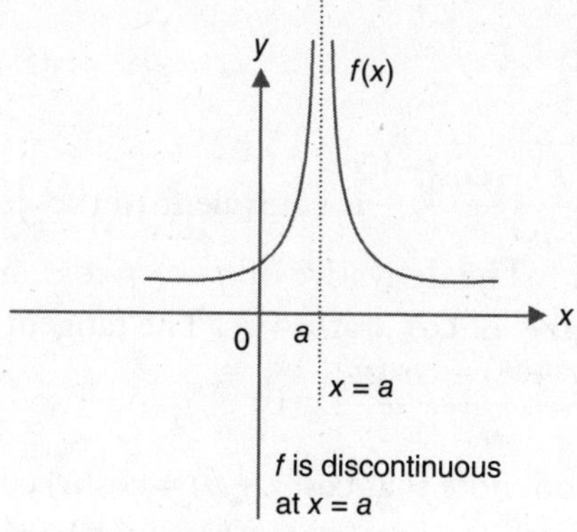

Figure 7.1-2

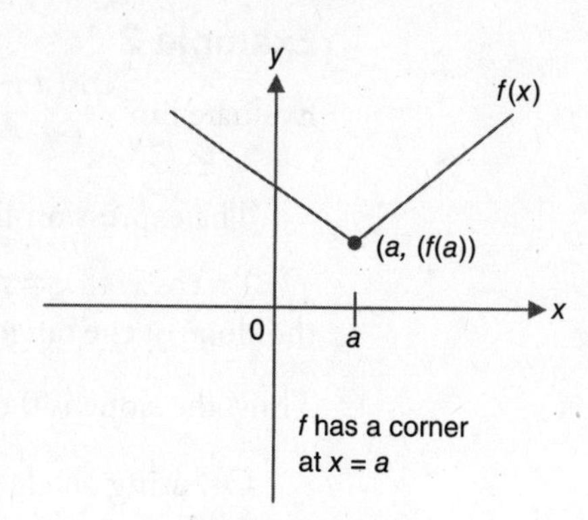

Figure 7.1-3

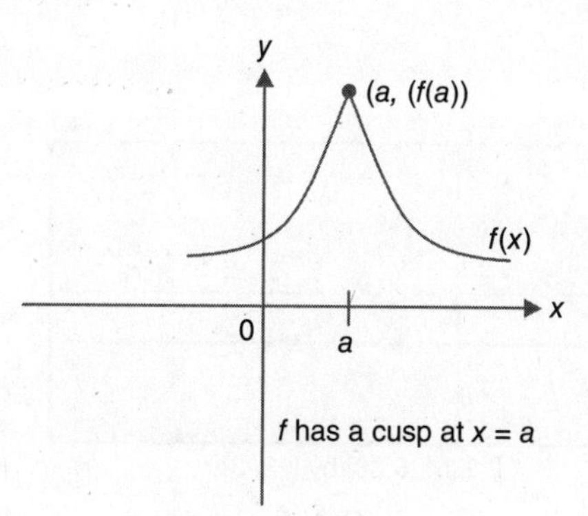

Figure 7.1-4

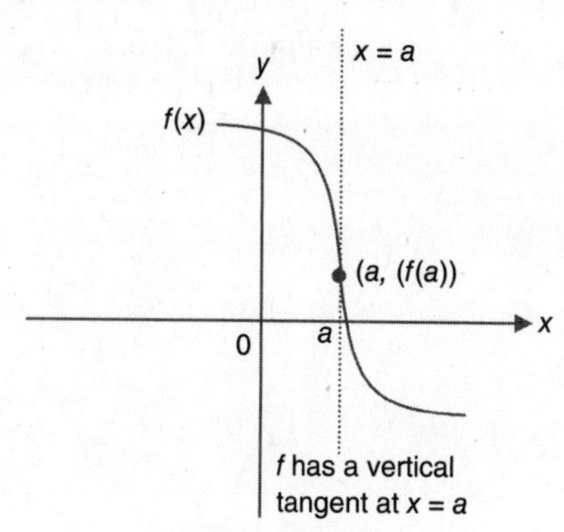

Figure 7.1-5

Example 1

If $f(x) = x^2 - 2x - 3$, find (a) $f'(x)$ using the definition of derivative, (b) $f'(0)$, (c) $f'(1)$, and (d) $f'(3)$.

(a) Using the definition of derivative, $f'(x) = \lim_{h \to 0} \frac{f(x+h) - f(x)}{h}$

$$= \lim_{h \to 0} \frac{[(x+h)^2 - 2(x+h) - 3] - [x^2 - 2x - 3]}{h}$$

$$= \lim_{h \to 0} \frac{[x^2 + 2xh + h^2 - 2x - 2h - 3] - [x^2 - 2x - 3]}{h}$$

$$= \lim_{h \to 0} \frac{2xh + h^2 - 2h}{h}$$

$$= \lim_{h \to 0} \frac{h(2x + h - 2)}{h}$$

$$= \lim_{h \to 0} (2x + h - 2) = 2x - 2.$$

(b) $f'(0) = 2(0) - 2 = -2$, (c) $f'(1) = 2(1) - 2 = 0$, and (d) $f'(3) = 2(3) - 2 = 4$.

Example 2

Evaluate $\lim_{h \to 0} \frac{\cos(\pi + h) - \cos(\pi)}{h}$.

The expression $\lim_{h \to 0} \frac{\cos(\pi + h) - \cos(\pi)}{h}$ is equivalent to the derivative of the function $f(x) = \cos x$ at $x = \pi$, i.e., $f'(\pi)$. The derivative of $f(x) = \cos x$ at $x = \pi$ is equivalent to the slope of the tangent to the curve of cos x at $x = \pi$. The tangent is parallel to the x-axis. Thus, the slope is 0 or $\lim_{h \to 0} \frac{\cos(\pi + h) - \cos(\pi)}{h} = 0$.

Or, using an algebraic method, note that $\cos(a + b) = \cos(a)\cos(b) - \sin(a)\sin(b)$. Then rewrite $\lim_{h \to 0} \frac{\cos(\pi + h) - \cos(\pi)}{h} = \lim_{h \to 0} \frac{\cos(\pi)\cos(h) - \sin(\pi)\sin(h) - \cos(\pi)}{h} =$ $\lim_{h \to 0} \frac{-\cos(h) - (-1)}{h} = \lim_{h \to 0} \frac{-\cos(h) + 1}{h} = \lim_{h \to 0} \frac{-[\cos(h) - 1]}{h} = -\lim_{h \to 0} \frac{[\cos(h) - 1]}{h} = 0$.

(See Figure 7.1-6.)

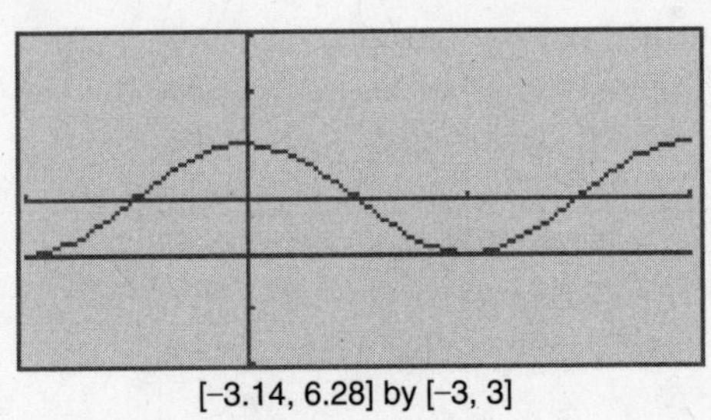

[-3.14, 6.28] by [-3, 3]

Figure 7.1-6

Example 3

If the function $f(x) = x^{2/3} + 1$, find all points where f is not differentiable.

The function $f(x)$ is continuous for all real numbers, and the graph of $f(x)$ forms a "cusp" at the point (0, 1). Thus, $f(x)$ is not differentiable at $x = 0$. (See Figure 7.1-7.)

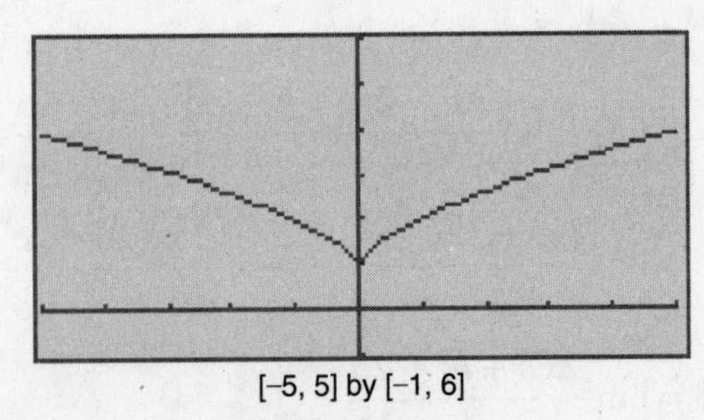

[-5, 5] by [-1, 6]

Figure 7.1-7

Example 4

Using a calculator, find the derivative of $f(x) = x^2 + 4x$ at $x = 3$.

There are several ways to find $f'(3)$, using a calculator. One way is to use the [*nDeriv*] function of the calculator. From the main Home screen, select *F3-Calc* and then select [*nDeriv*]. Enter [*nDeriv*] $(x^2 + 4x,\ x)|x = 3$. The result is 10.

- Always write out all formulas in your solutions.

Power Rule

If $f(x) = c$ where c is a constant, then $f'(x) = 0$.
If $f(x) = x^n$ where n is a real number, then $f'(x) = nx^{n-1}$.
If $f(x) = cx^n$ where c is a constant and n is a real number, then $f'(x) = cnx^{n-1}$.

Summary of Derivatives of Algebraic Functions

$$\frac{d}{dx}(c) = 0, \quad \frac{d}{dx}(x^n) = nx^{n-1}, \quad \text{and} \quad \frac{d}{dx}(cx^n) = cnx^{n-1}$$

Example 1

If $f(x) = 2x^3$, find (a) $f'(x)$, (b) $f'(1)$, and (c) $f'(0)$.
Note that (a) $f'(x) = 6x^2$, (b) $f'(1) = 6(1)^2 = 6$, and (c) $f'(0) = 0$.

Example 2

If $y = \frac{1}{x^2}$, find (a) $\frac{dy}{dx}$ and (b) $\frac{dy}{dx}\big|_{x=0}$ (which represents $\frac{dy}{dx}$ at $x = 0$).

Note that (a) $y = \frac{1}{x^2} = x^{-2}$ and thus, $\frac{dy}{dx} = -2x^{-3} = \frac{-2}{x^3}$ and (b) $\frac{dy}{dx}\big|_{x=0}$ does not exist because the expression $\frac{-2}{0}$ is undefined.

Example 3

Here are several examples of algebraic functions and their derivatives:

FUNCTION	WRITTEN IN cx^n FORM	DERIVATIVE	DERIVATIVE WITH POSITIVE EXPONENTS
$3x$	$3x^1$	$3x^0 = 3$	3
$-5x^7$	$-5x^7$	$-35x^6$	$-35x^6$
$8\sqrt{x}$	$8x^{\frac{1}{2}}$	$4x^{-\frac{1}{2}}$	$\frac{4}{x^{\frac{1}{2}}}$ or $\frac{4}{\sqrt{x}}$
$\frac{1}{x^2}$	x^{-2}	$-2x^{-3}$	$\frac{-2}{x^3}$
$\frac{-2}{\sqrt{x}}$	$\frac{-2}{x^{\frac{1}{2}}} = -2x^{-\frac{1}{2}}$	$x^{-\frac{3}{2}}$	$\frac{1}{x^{\frac{3}{2}}}$ or $\frac{1}{\sqrt{x^3}}$
4	$4x^0$	0	0
π^2	$(\pi^2)x^0$	0	0

Example 4

Using a calculator, find $f'(x)$ and $f'(3)$ if $f(x)=\frac{1}{\sqrt{x}}$.

There are several ways of finding $f'(x)$ and $f'(9)$ using a calculator. One way is to use the *d* [*Differentiate*] function. Go to the Home screen. Select *F3-Calc* and then select *d* [*Differentiate*]. Enter $d(1/\sqrt{(x)},\ x)$. The result is $f'(x)=\frac{-1}{2x^{\frac{3}{2}}}$. To find $f'(3)$, enter $d(1/\sqrt{(x)},\ x)|x=3$. The result is $f'(3)=\frac{-1}{54}$.

The Sum, Difference, Product, and Quotient Rules

If u and v are two differentiable functions, then

$$\frac{d}{dx}(u\pm v)=\frac{du}{dx}\pm\frac{dv}{dx} \qquad \text{Sum and Difference Rules}$$

$$\frac{d}{dx}(uv)=v\frac{du}{dx}+u\frac{dv}{dx} \qquad \text{Product Rule}$$

$$\frac{d}{dx}\left(\frac{u}{v}\right)=\frac{v\frac{du}{dx}-u\frac{dv}{dx}}{v^2},\ v\neq 0 \qquad \text{Quotient Rule}$$

Summary of Sum, Difference, Product, and Quotient Rules

$$(u\pm v)'=u'\pm v' \qquad (uv)'=u'v+v'u \qquad \left(\frac{u}{v}\right)'=\frac{u'v-v'u}{v^2}$$

Example 1

Find $f'(x)$ if $f(x)=x^3-10x+5$.

Using the sum and difference rules, you can differentiate each term and obtain $f'(x)=3x^2-10$. Or using your calculator, select the *d* [*Differentiate*] function and enter $d(x^3-10x+5,\ x)$ and obtain $3x^2-10$.

Example 2

If $y=(3x-5)(x^4+8x-1)$, find $\frac{dy}{dx}$.

Using the product rule $\frac{d}{dx}(uv)=v\frac{du}{dx}+u\frac{dv}{dx}$, let $u=(3x-5)$ and $v=(x^4+8x-1)$.

Then $\frac{dy}{dx}=(3)(x^4+8x-1)+(4x^3+8)(3x-5)=(3x^4+24x-3)+(12x^4-20x^3+24x-40)=15x^4-20x^3+48x-43$. Or you can use your calculator and enter $d((3x-5)(x^4+8x-1),\ x)$ and obtain the same result.

Example 3

If $f(x)=\frac{2x-1}{x+5}$, find $f'(x)$.

Using the quotient rule $\left(\frac{u}{v}\right)'=\frac{u'v-v'u}{v^2}$, let $u=2x-1$ and $v=x+5$. Then $f'(x)=\frac{(2)(x+5)-(1)(2x-1)}{(x+5)^2}=\frac{2x+10-2x+1}{(x+5)^2}=\frac{11}{(x+5)^2},\ x\neq-5$. Or you can use your calculator and enter $d((2x-1)/(x+5),\ x)$ and obtain the same result.

Example 4

Using your calculator, find an equation of the tangent to the curve $f(x) = x^2 - 3x + 2$ at $x = 5$.

Find the slope of the tangent to the curve at $x = 5$ by entering $d(x^2 - 3x + 2,\ x)|x = 5$. The result is 7. Compute $f(5) = 12$. Thus, the point (5, 12) is on the curve of $f(x)$. An equation of the line whose slope $m = 7$ and passing through the point (5, 12) is $y - 12 = 7(x - 5)$.

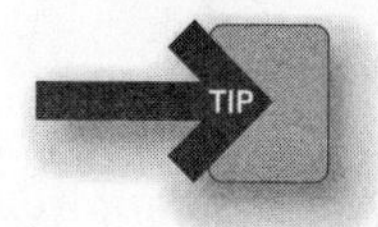

- Remember that $\frac{d}{dx} \ln x = \frac{1}{x}$ and $\int \ln x\, dx = x \ln x - x + c$. The integral formula is not usually tested in the AB exam.

The Chain Rule

If $y = f(u)$ and $u = g(x)$ are differentiable functions of u and x respectively, then $\frac{d}{dx}[f(g(x))] = f'(g(x)) \cdot g'(x)$ or $\frac{dy}{dx} = \frac{dy}{du} \cdot \frac{du}{dx}$.

Example 1

If $y = (3x - 5)^{10}$, find $\frac{dy}{dx}$.

Using the chain rule, let $u = 3x - 5$ and thus, $y = u^{10}$. Then, $\frac{dy}{du} = 10u^9$ and $\frac{du}{dx} = 3$. Since $\frac{dy}{dx} = \frac{dy}{du} \cdot \frac{du}{dx}$, $\frac{dy}{dx} = \left(10u^9\right)(3) = 10(3x - 5)^9(3) = 30(3x - 5)^9$. Or you can use your calculator and enter $d((3x - 5)^{10},\ x)$ and obtain the same result.

Example 2

If $f(x) = 5x\sqrt{25 - x^2}$, find $f'(x)$.

Rewrite $f(x) = 5x\sqrt{25 - x^2}$ as $f(x) = 5x(25 - x^2)^{1/2}$. Using the product rule, $f'(x) = (25 - x^2)^{1/2}\frac{d}{dx}(5x) + (5x)\frac{d}{dx}(25 - x^2)^{1/2} = 5(25 - x^2)^{1/2} + (5x)\frac{d}{dx}(25 - x^2)^{1/2}$.

To find $\frac{d}{dx}(25 - x^2)^{1/2}$, use the chain rule and let $u = 25 - x^2$.

Thus, $\frac{d}{dx}(25 - x^2)^{1/2} = \frac{1}{2}(25 - x^2)^{-1/2}(-2x) = \frac{-x}{(25 - x^2)^{1/2}}$. Substituting this quantity back into $f'(x)$, you have $f'(x) = 5(25 - x^2)^{1/2} + (5x)\left(\frac{-x}{(25 - x^2)^{1/2}}\right) = \frac{5(25 - x^2) - 5x^2}{(25 - x^2)^{1/2}} = \frac{125 - 10x^2}{(25 - x^2)^{1/2}}$. Or you can use your calculator and enter $d(5x\sqrt{25 - x^2},\ x)$ and obtain the same result.

Example 3

If $y = \left(\frac{2x-1}{x^2}\right)^3$, find $\frac{dy}{dx}$.

Using the chain rule, let $u = \left(\frac{2x-1}{x^2}\right)$. Then $\frac{dy}{dx} = 3\left(\frac{2x-1}{x^2}\right)^2 \frac{d}{dx}\left(\frac{2x-1}{x^2}\right)$.

To find $\frac{d}{dx}\left(\frac{2x-1}{x^2}\right)$, use the quotient rule.

Thus, $\frac{d}{dx}\left(\frac{2x-1}{x^2}\right) = \frac{(2)(x^2)-(2x)(2x-1)}{(x^2)^2} = \frac{-2x^2+2x}{x^4}$. Substituting this quantity back into $\frac{dy}{dx} = 3\left(\frac{2x-1}{x^2}\right)^2 \frac{d}{dx}\left(\frac{2x-1}{x^2}\right) = 3\left(\frac{2x-1}{x^2}\right)^2 \frac{-2x^2+2x}{x^4} = \frac{-6(x-1)(2x-1)^2}{x^7}$.

An alternate solution is to use the product rule and rewrite $y = \left(\frac{2x-1}{x^2}\right)^3$ as $y = \frac{(2x-1)^3}{(x^2)^3} = \frac{(2x-1)^3}{x^6}$ and use the quotient rule.

Another approach is to express $y = (2x-1)^3(x^{-6})$ and use the product rule. Of course, you can always use your calculator if you are permitted to do so.

7.2 Derivatives of Trigonometric, Inverse Trigonometric, Exponential, and Logarithmic Functions

Main Concepts: Derivatives of Trigonometric Functions, Derivatives of Inverse Trigonometric Functions, Derivatives of Exponential and Logarithmic Functions

Derivatives of Trigonometric Functions

Summary of Derivatives of Trigonometric Functions

$$\frac{d}{dx}(\sin x) = \cos x \qquad \frac{d}{dx}(\cos x) = -\sin x$$

$$\frac{d}{dx}(\tan x) = \sec^2 x \qquad \frac{d}{dx}(\cot x) = -\csc^2 x$$

$$\frac{d}{dx}(\sec x) = \sec x \tan x \qquad \frac{d}{dx}(\csc x) = -\csc x \cot x$$

Note that the derivatives of *cosine*, *cotangent*, and *cosecant* all have a negative sign.

Example 1

If $y = 6x^2 + 3\sec x$, find $\frac{dy}{dx}$.

$\frac{dy}{dx} = 12x + 3\sec x \tan x$.

Example 2

Find $f'(x)$ if $f(x) = \cot(4x - 6)$.
Using the chain rule, let $u = 4x - 6$. Then $f'(x) = [-\csc^2(4x-6)][4] = -4\csc^2(4x-6)$.

Or using your calculator, enter $d(1/\tan(4x-6),\ x)$ and obtain $\dfrac{-4}{\sin^2(4x-6)}$, which is an equivalent form.

Example 3

Find $f'(x)$ if $f(x) = 8\sin(x^2)$.
Using the chain rule, let $u = x^2$. Then $f'(x) = [8\cos(x^2)][2x] = 16x\cos(x^2)$.

Example 4

If $y = \sin x \cos(2x)$, find $\dfrac{dy}{dx}$.

Using the product rule, let $u = \sin x$ and $v = \cos(2x)$.

Then $\dfrac{dy}{dx} = \cos x\cos(2x) + [-\sin(2x)](2)(\sin x) = \cos x\cos(2x) - 2\sin x\sin(2x)$.

Example 5

If $y = \sin[\cos(2x)]$, find $\dfrac{dy}{dx}$.

Using the chain rule, let $u = \cos(2x)$. Then

$$\frac{dy}{dx} = \frac{dy}{du}\cdot\frac{du}{dx} = \cos[\cos(2x)]\frac{d}{dx}[\cos(2x)].$$

To evaluate $\dfrac{d}{dx}[\cos(2x)]$, use the chain rule again by making another u-substitution, this time for $2x$. Thus, $\dfrac{d}{dx}[\cos(2x)] = [-\sin(2x)]2 = -2\sin(2x)$. Therefore, $\dfrac{dy}{dx}\cos[\cos(2x)](-2\sin(2x)) = -2\sin(2x)\cos[\cos(2x)]$.

Example 6

Find $f'(x)$ if $f(x) = 5x\csc x$.
Using the product rule, let $u = 5x$ and $v = \csc x$. Then $f'(x) = 5\csc x + (-\csc x\cot x)(5x) = 5\csc x - 5x(\csc x)(\cot x)$.

Example 7

If $y = \sqrt{\sin x}$, find $\dfrac{dy}{dx}$.

Rewrite $y = \sqrt{\sin x}$ as $y = (\sin x)^{1/2}$. Using the chain rule, let $u = \sin x$. Thus, $\dfrac{dy}{dx} = \dfrac{1}{2}(\sin x)^{-1/2}(\cos x) = \dfrac{\cos x}{2(\sin x)^{1/2}} = \dfrac{\cos x}{2\sqrt{\sin x}}$.

Example 8

If $y = \frac{\tan x}{1 + \tan x}$, find $\frac{dy}{dx}$.

Using the quotient rule, let $u = \tan x$ and $v = (1 + \tan x)$. Then,

$$\frac{dy}{dx} = \frac{(\sec^2 x)(1 + \tan x) - (\sec^2 x)(\tan x)}{(1 + \tan x)^2}$$

$$= \frac{\sec^2 x + (\sec^2 x)(\tan x) - (\sec^2 x)(\tan x)}{(1 + \tan x)^2}$$

$$= \frac{\sec^2 x}{(1 + \tan x)^2}, \text{ which is equivalent to } \frac{\frac{1}{(\cos x)^2}}{1 + \left(\frac{\sin x}{\cos x}\right)^2}$$

$$= \frac{\frac{1}{(\cos x)^2}}{\left(\frac{\cos x + \sin x}{\cos x}\right)^2} = \frac{1}{(\cos x + \sin x)^2}.$$

Note that for all of the above exercises, you can find the derivatives by using a calculator, provided that you are permitted to do so.

Derivatives of Inverse Trigonometric Functions

Summary of Derivatives of Inverse Trigonometric Functions

Let u be a differentiable function of x, then

$$\frac{d}{dx}\sin^{-1} u = \frac{1}{\sqrt{1-u^2}}\frac{du}{dx}, \ |u| < 1 \qquad \frac{d}{dx}\cos^{-1} u = \frac{-1}{\sqrt{1-u^2}}\frac{du}{dx}, \ |u| < 1$$

$$\frac{d}{dx}\tan^{-1} u = \frac{1}{1+u^2}\frac{du}{dx} \qquad \frac{d}{dx}\cot^{-1} u = \frac{-1}{1+u^2}\frac{du}{dx}$$

$$\frac{d}{dx}\sec^{-1} u = \frac{1}{|u|\sqrt{u^2-1}}\frac{du}{dx}, \ |u| > 1 \qquad \frac{d}{dx}\csc^{-1} u = \frac{-1}{|u|\sqrt{u^2-1}}\frac{du}{dx}, \ |u| > 1.$$

Note that the derivatives of $\cos^{-1} x$, $\cot^{-1} x$, and $\csc^{-1} x$ all have a "−1" in their numerators.

Example 1

If $y = 5\sin^{-1}(3x)$, find $\frac{dy}{dx}$.

Let $u = 3x$. Then $\frac{dy}{dx} = (5)\frac{1}{\sqrt{1-(3x)^2}}\frac{du}{dx} = \frac{5}{\sqrt{1-(3x)^2}}(3) = \frac{15}{\sqrt{1-9x^2}}$.

Or using a calculator, enter $d[5\sin^{-1}(3x),\ x]$ and obtain the same result.

Example 2

Find $f'(x)$ if $f(x) = \tan^{-1}\sqrt{x}$.

Let $u = \sqrt{x}$. Then $f'(x) = \dfrac{1}{1+(\sqrt{x})^2}\dfrac{du}{dx} = \dfrac{1}{1+x}\left(\dfrac{1}{2}x^{-\frac{1}{2}}\right) = \dfrac{1}{1+x}\left(\dfrac{1}{2\sqrt{x}}\right)$

$$= \frac{1}{2\sqrt{x}(1+x)}.$$

Example 3

If $y = \sec^{-1}(3x^2)$, find $\dfrac{dy}{dx}$.

Let $u = 3x^2$. Then $\dfrac{dy}{dx} = \dfrac{1}{|3x^2|\sqrt{(3x^2)^2-1}}\dfrac{du}{dx} = \dfrac{1}{3x^2\sqrt{9x^4-1}}(6x) = \dfrac{2}{x\sqrt{9x^4-1}}$.

Example 4

If $y = \cos^{-1}\left(\dfrac{1}{x}\right)$, find $\dfrac{dy}{dx}$.

Let $u = \left(\dfrac{1}{x}\right)$. Then $\dfrac{dy}{dx} = \dfrac{-1}{\sqrt{1-\left(\dfrac{1}{x}\right)^2}}\dfrac{du}{dx}$.

Rewrite $u = \left(\dfrac{1}{x}\right)$ as $u = x^{-1}$. Then $\dfrac{du}{dx} = -1x^{-2} = \dfrac{-1}{x^2}$.

Therefore, $\dfrac{dy}{dx} = \dfrac{-1}{\sqrt{1-\left(\dfrac{1}{x}\right)^2}}\dfrac{du}{dx} = \dfrac{-1}{\sqrt{1-\left(\dfrac{1}{x}\right)^2}}\dfrac{-1}{x^2} = \dfrac{1}{\sqrt{\dfrac{x^2-1}{x^2}}(x^2)}$

$$= \frac{1}{\dfrac{\sqrt{x^2-1}}{|x|}(x^2)} = \frac{1}{|x|\sqrt{x^2-1}}.$$

Note that for all of the above exercises, you can find the derivatives by using a calculator, provided that you are permitted to do so.

Derivatives of Exponential and Logarithmic Functions

Summary of Derivatives of Exponential and Logarithmic Functions

Let u be a differentiable function of x, then

$$\frac{d}{dx}(e^u) = e^u\frac{du}{dx} \qquad \frac{d}{dx}(a^u) = a^u \ln a\frac{du}{dx}, \quad a > 0 \;\&\; a \neq 1$$

$$\frac{d}{dx}(\ln u) = \frac{1}{u}\frac{du}{dx}, \quad u > 0 \qquad \frac{d}{dx}(\log_a u) = \frac{1}{u\ln a}\frac{du}{dx}, \quad a > 0 \;\&\; a \neq 1.$$

For the following examples, find $\dfrac{dy}{dx}$ and verify your result with a calculator.

Example 1

$y = e^{3x} + 5xe^3 + e^3$

$\frac{dy}{dx} = (e^{3x})(3) + 5e^3 + 0 = 3e^{3x} + 5e^3$ (Note that e^3 is a constant.)

Example 2

$y = xe^x - x^2e^x$

Using the product rule for both terms, you have

$$\frac{dy}{dx} = (1)e^x + (e^x)x - \left[(2x)e^x + (e^x)x^2\right] = e^x + xe^x - 2xe^x - x^2e^x = e^x - xe^x - x^2e^x$$
$$= -x^2e^x - xe^x + e^x = e^x(-x^2 - x + 1).$$

Example 3

$y = 3^{\sin x}$

Let $u = \sin x$. Then, $\frac{dy}{dx} = (3^{\sin x})(\ln 3)\frac{du}{dx} = (3^{\sin x})(\ln 3)\cos x = (\ln 3)(3^{\sin x})\cos x$.

Example 4

$y = e^{(x^3)}$

Let $u = x^3$. Then, $\frac{dy}{dx} = \left[e^{(x^3)}\right]\frac{du}{dx} = \left[e^{(x^3)}\right]3x^2 = 3x^2e^{(x^3)}$.

Example 5

$y = (\ln x)^5$

Let $u = \ln x$. Then, $\frac{dy}{dx} = 5(\ln x)^4\frac{du}{dx} = 5(\ln x)^4\left(\frac{1}{x}\right) = \frac{5(\ln x)^4}{x}$.

Example 6

$y = \ln(x^2 + 2x - 3) + \ln 5$

Let $u = x^2 + 2x - 3$. Then, $\frac{dy}{dx} = \frac{1}{x^2 + 2x - 3}\frac{du}{dx} + 0 = \frac{1}{x^2 + 2x - 3}(2x + 2) = \frac{2x + 2}{x^2 + 2x - 3}$.

(Note that $\ln 5$ is a constant. Thus, the derivative of $\ln 5$ is 0.)

Example 7

$y = 2x\ln x + x$

Using the product rule for the first term,

you have $\frac{dy}{dx} = (2)\ln x + \left(\frac{1}{x}\right)(2x) + 1 = 2\ln x + 2 + 1 = 2\ln x + 3$.

Example 8

$y = \ln(\ln x)$

Let $u = \ln x$. Then $\dfrac{dy}{dx} = \dfrac{1}{\ln x}\dfrac{du}{dx} = \dfrac{1}{\ln x}\left(\dfrac{1}{x}\right) = \dfrac{1}{x \ln x}$.

Example 9

$y = \log_5(2x + 1)$

Let $u = 2x + 1$. Then $\dfrac{dy}{dx} = \dfrac{1}{(2x+1)\ln 5}\dfrac{du}{dx} = \dfrac{1}{(2x+1)\ln 5} \cdot (2) = \dfrac{2}{(2x+1)\ln 5}$.

Example 10

Write an equation of the line tangent to the curve of $y = e^x$ at $x = 1$.

The slope of the tangent to the curve $y = e^x$ at $x = 1$ is equivalent to the value of the derivative of $y = e^x$ evaluated at $x = 1$. Using your calculator, enter $d(e \wedge (x),\ x)|x = 1$ and obtain e. Thus, $m = e$, the slope of the tangent to the curve at $x = 1$. At $x = 1$, $y = e^1 = e$, and thus, the point on the curve is $(1, e)$. Therefore, the equation of the tangent is $y - e = e(x - 1)$ or $y = ex$. (See Figure 7.2-1.)

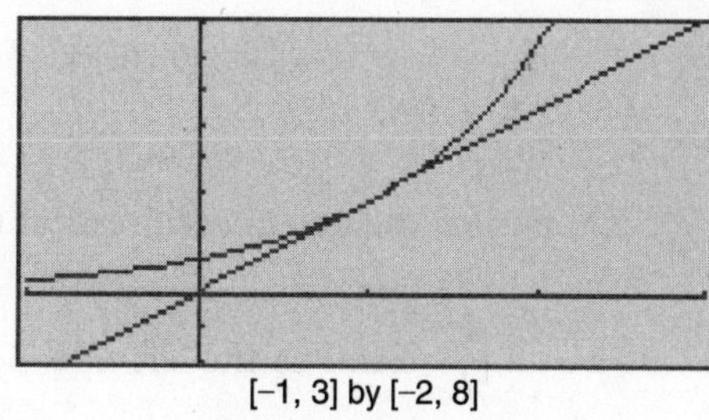

[-1, 3] by [-2, 8]

Figure 7.2-1

- Never leave a multiple-choice question blank. There is no penalty for incorrect answers.

7.3 Implicit Differentiation

Main Concept: Procedure for Implicit Differentiation

Procedure for Implicit Differentiation

Given an equation containing the variables x and y for which you cannot easily solve for y in terms of x, you can find $\dfrac{dy}{dx}$ by doing the following:

Steps

1. Differentiate each term of the equation with respect to x.
2. Move all terms containing $\dfrac{dy}{dx}$ to the left side of the equation and all other terms to the right side.
3. Factor out $\dfrac{dy}{dx}$ on the left side of the equation.
4. Solve for $\dfrac{dy}{dx}$.

Example 1

Find $\frac{dy}{dx}$ if $y^2 - 7y + x^2 - 4x = 10$.

Step 1: Differentiate each term of the equation with respect to x. (Note that y is treated as a function of x.) $2y\frac{dy}{dx} - 7\frac{dy}{dx} + 2x - 4 = 0$

Step 2: Move all terms containing $\frac{dy}{dx}$ to the left side of the equation and all other terms to the right side: $2y\frac{dy}{dx} - 7\frac{dy}{dx} = -2x + 4$.

Step 3: Factor out $\frac{dy}{dx}$: $\frac{dy}{dx}(2y - 7) = -2x + 4$.

Step 4: Solve for $\frac{dy}{dx}$: $\frac{dy}{dx} = \frac{-2x + 4}{(2y - 7)}$.

Example 2

Find $\frac{dy}{dx}$ if $x^3 + y^3 = 6xy$.

Step 1: Differentiate each term with respect to x: $3x^2 + 3y^2\frac{dy}{dx} = (6)y + \left(\frac{dy}{dx}\right)(6x)$.

Step 2: Move all $\frac{dy}{dx}$ terms to the left side: $3y^2\frac{dy}{dx} - 6x\frac{dy}{dx} = 6y - 3x^2$.

Step 3: Factor out $\frac{dy}{dx}$: $\frac{dy}{dx}(3y^2 - 6x) = 6y - 3x^2$.

Step 4: Solve for $\frac{dy}{dx}$: $\frac{dy}{dx} = \frac{6y - 3x^2}{3y^2 - 6x} = \frac{2y - x^2}{y^2 - 2x}$.

Example 3

Find $\frac{dy}{dx}$ if $(x + y)^2 - (x - y)^2 = x^5 + y^5$.

Step 1: Differentiate each term with respect to x:

$$2(x + y)\left(1 + \frac{dy}{dx}\right) - 2(x - y)\left(1 - \frac{dy}{dx}\right) = 5x^4 + 5y^4\frac{dy}{dx}.$$

Distributing $2(x + y)$ and $-2(x - y)$, you have

$$2(x + y) + 2(x + y)\frac{dy}{dx} - 2(x - y) + 2(x - y)\frac{dy}{dx} = 5x^4 + 5y^4\frac{dy}{dx}.$$

Step 2: Move all $\frac{dy}{dx}$ terms to the left side:

$$2(x + y)\frac{dy}{dx} + 2(x - y)\frac{dy}{dx} - 5y^4\frac{dy}{dx} = 5x^4 - 2(x + y) + 2(x - y).$$

Step 3: Factor out $\frac{dy}{dx}$:

$$\frac{dy}{dx}[2(x+y)+2(x-y)-5y^4]=5x^4-2x-2y+2x-2y$$

$$\frac{dy}{dx}[2x+2y+2x-2y-5y^4]=5x^4-4y$$

$$\frac{dy}{dx}[4x-5y^4]=5x^4-4y.$$

Step 4: Solve for $\frac{dy}{dx}$: $\frac{dy}{dx}=\frac{5x^4-4y}{4x-5y^4}$.

Example 4

Write an equation of the tangent to the curve $x^2+y^2+19=2x+12y$ at (4, 3).
The slope of the tangent to the curve at (4, 3) is equivalent to the derivative $\frac{dy}{dx}$ at (4, 3).

Using implicit differentiation, you have:

$$2x+2y\frac{dy}{dx}=2+12\frac{dy}{dx}$$

$$2y\frac{dy}{dx}-12\frac{dy}{dx}=2-2x$$

$$\frac{dy}{dx}(2y-12)=2-2x$$

$$\frac{dy}{dx}=\frac{2-2x}{2y-12}=\frac{1-x}{y-6} \text{ and } \left.\frac{dy}{dx}\right|_{(4,3)}=\frac{1-4}{3-6}=1.$$

Thus, the equation of the tangent is $y-3=(1)(x-4)$ or $y-3=x-4$.

Example 5

Find $\frac{dy}{dx}$, if $\sin(x+y)=2x$.

$$\cos(x+y)\left(1+\frac{dy}{dx}\right)=2$$

$$1+\frac{dy}{dx}=\frac{2}{\cos(x+y)}$$

$$\frac{dy}{dx}=\frac{2}{\cos(x+y)}-1$$

7.4 Approximating a Derivative

Given a continuous and differentiable function, you can find the approximate value of a derivative at a given point numerically. Here are two examples.

Example 1

The graph of a function f on [0, 5] is shown in Figure 7.4-1. Find the approximate value of $f'(3)$.

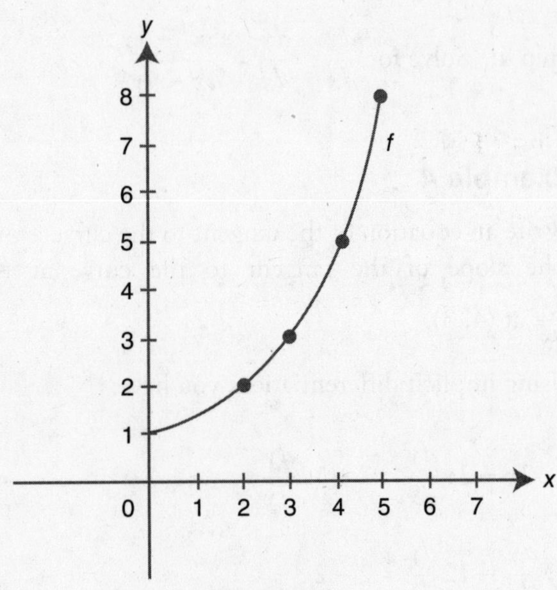

Figure 7.4-1

Since $f'(3)$ is equivalent to the slope of the tangent to $f(x)$ at $x = 3$, there are several ways you can find its approximate value.

Method 1: Use the slope of the line segment joining the points at $x = 3$ and $x = 4$.

$$f(3) = 3 \text{ and } f(4) = 5$$

$$m = \frac{f(4) - f(3)}{4 - 3} = \frac{5 - 3}{4 - 3} = 2$$

Method 2: Use the slope of the line segment joining the points at $x = 2$ and $x = 3$.

$$f(2) = 2 \text{ and } f(3) = 3$$

$$m = \frac{f(3) - f(2)}{3 - 2} = \frac{3 - 2}{3 - 2} = 1$$

Method 3: Use the slope of the line segment joining the points at $x = 2$ and $x = 4$.

$f(2) = 2$ and $f(4) = 5$

$$m = \frac{f(4) - f(2)}{4 - 2} = \frac{5 - 2}{4 - 2} = \frac{3}{2}$$

Note that $\frac{3}{2}$ is the average of the results from methods 1 and 2.

Thus, $f'(3) \approx 1, 2,$ or $\frac{3}{2}$ depending on which line segment you use.

Example 2

Let f be a continuous and differentiable function. Selected values of f are shown below. Find the approximate value of f' at $x = 1$.

x	−2	−1	0	1	2	3
f	1	0	1	1.59	2.08	2.52

You can use the difference quotient $\frac{f(a+h) - f(a)}{h}$ to approximate $f'(a)$.

$$\text{Let } h = 1; \quad f'(1) \approx \frac{f(2) - f(1)}{2 - 1} \approx \frac{2.08 - 1.59}{1} \approx 0.49.$$

$$\text{Let } h = 2; \quad f'(1) \approx \frac{f(3) - f(1)}{3 - 1} \approx \frac{2.52 - 1.59}{2} \approx 0.465.$$

Or, you can use the symmetric difference quotient $\frac{f(a+h) - f(a-h)}{2h}$ to approximate $f'(a)$.

$$\text{Let } h = 1; \quad f'(1) \approx \frac{f(2) - f(0)}{2 - 0} \approx \frac{2.08 - 1}{2} \approx 0.54.$$

$$\text{Let } h = 2; \quad f'(1) \approx \frac{f(3) - f(-1)}{3 - (-1)} \approx \frac{2.52 - 0}{4} \approx 0.63.$$

Thus, $f'(3) \approx 0.49, 0.465, 0.54,$ or 0.63 depending on your method.

Note that f is decreasing on $(-2, -1)$ and increasing on $(-1, 3)$. Using the symmetric difference quotient with $h = 3$ would not be accurate. (See Figure 7.4-2.)

[−2, 4] by [−2, 4]

Figure 7.4-2

- Remember that the $\lim_{x \to 0} \frac{\sin 6x}{\sin 2x} = \frac{6}{2} = 3$ because the $\lim_{x \to 0} \frac{\sin x}{x} = 1$.

7.5 Derivatives of Inverse Functions

Let f be a one-to-one differentiable function with inverse function f^{-1}. If $f'(f^{-1}(a)) \neq 0$, then the inverse function f^{-1} is differentiable at a and $(f^{-1})'(a) = \frac{1}{f'(f^{-1}(a))}$. (See Figure 7.5-1.)

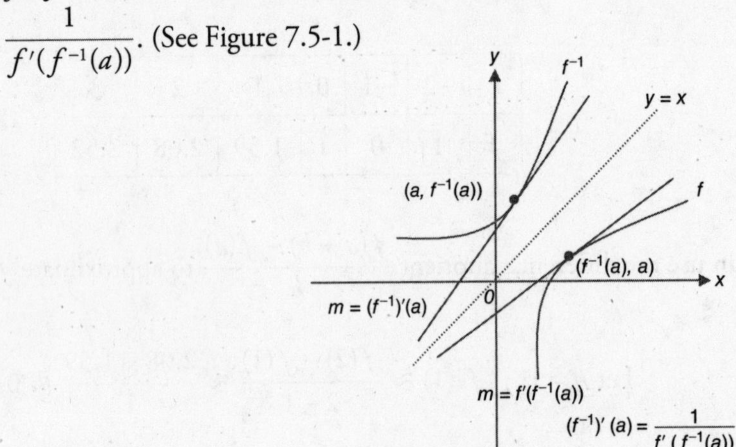

Figure 7.5-1

If $y = f^{-1}(x)$ so that $x = f(y)$, then $\frac{dy}{dx} = \frac{1}{dx/dy}$ with $\frac{dx}{dy} \neq 0$.

Example 1

If $f(x) = x^3 + 2x - 10$, find $(f^{-1})'(x)$.

Step 1: Check if $(f^{-1})'(x)$ exists. $f'(x) = 3x^2 + 2$ and $f'(x) > 0$ for all real values of x. Thus, $f(x)$ is strictly increasing which implies that $f(x)$ is $1 - 1$. Therefore, $(f^{-1})'(x)$ exists.

Step 2: Let $y = f(x)$ and thus $y = x^3 + 2x - 10$.

Step 3: Interchange x and y to obtain the inverse function $x = y^3 + 2y - 10$.

Step 4: Differentiate with respect to y: $\frac{dx}{dy} = 3y^2 + 2$.

Step 5: Apply formula $\frac{dy}{dx} = \frac{1}{dx/dy}$.

$$\frac{dy}{dx} = \frac{1}{dx/dy} = \frac{1}{3y^2 + 2}. \text{ Thus, } (f^{-1})'(x) = \frac{1}{3y^2 + 2}.$$

Example 2

Example 1 could have been done by using implicit differentiation.

Step 1: Let $y = f(x)$, and thus $y = x^3 + 2x - 10$.

Step 2: Interchange x and y to obtain the inverse function $x = y^3 + 2y - 10$.

Step 3: Differentiate each term implicitly with respect to x.

$$\frac{d}{dx}(x) = \frac{d}{dx}(y^3) + \frac{d}{dx}(2y) - \frac{d}{dx}(-10)$$

$$1 = 3y^2\frac{dy}{dx} + 2\frac{dy}{dx} - 0$$

Step 4: Solve for $\frac{dy}{dx}$.

$$1 = \frac{dy}{dx}(3y^2 + 2)$$

$$\frac{dy}{dx} = \frac{1}{3y^2 + 2}. \text{ Thus, } (f^{-1})'(x) = \frac{1}{3y^2 + 2}.$$

Example 3

If $f(x) = 2x^5 + x^3 + 1$, find (a) $f(1)$ and $f'(1)$ and (b) $(f^{-1})(4)$ and $(f^{-1})'(4)$.
Enter $y1 = 2x^5 + x^3 + 1$. Since $y1$ is strictly increasing, $f(x)$ has an inverse.

(a) $f(1) = 2(1)^5 + (1)^3 + 1 = 4$
$f'(x) = 10x^4 + 3x^2$
$f'(1) = 10(1)^4 + 3(1)^2 = 13$

(b) Since $f(1) = 4$ implies the point (1, 4) is on the curve $f(x) = 2x^5 + x^3 + 1$, therefore, the point (4, 1) (which is the reflection of (1, 4) on $y = x$) is on the curve $(f^{-1})(x)$. Thus, $(f^{-1})(4) = 1$.

$$(f^{-1})'(4) = \frac{1}{f'(1)} = \frac{1}{13}$$

Example 4

If $f(x) = 5x^3 + x + 8$, find $(f^{-1})'(8)$.
Enter $y1 = 5x^3 + x + 8$. Since $y1$ is strictly increasing near $x = 8$, $f(x)$ has an inverse near $x = 8$.
Note that $f(0) = 5(0)^3 + 0 + 8 = 8$, which implies the point (0, 8) is on the curve of $f(x)$. Thus, the point (8, 0) is on the curve of $(f^{-1})(x)$.

$$f'(x) = 15x^2 + 1$$

$$f'(0) = 1$$

Therefore, $(f^{-1})'(8) = \frac{1}{f'(0)} = \frac{1}{1} = 1$.

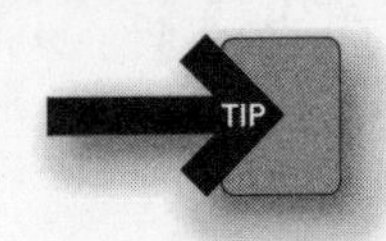

- You do not have to answer every question correctly to get a 5 on the AP Calculus AB exam. But always select an answer to a multiple-choice question. There is no penalty for incorrect answers.

7.6 Higher Order Derivatives

If the derivative f' of a function f is differentiable, then the derivative of f' is the second derivative of f represented by f'' (reads as f double prime). You can continue to differentiate f as long as there is differentiability.

Some of the Symbols of Higher Order Derivatives

$$f'(x), f''(x), f'''(x), f^{(4)}(x)$$

$$\frac{dy}{dx}, \frac{d^2y}{dx^2}, \frac{d^3y}{dx^3}, \frac{d^4y}{dx^4}$$

$$y', y'', y''', y^{(4)}$$

$$D_x(y), D_x^2(y), D_x^3(y), D_x^4(y)$$

Note that $\frac{d^2y}{dx^2} = \frac{d}{dx}\left(\frac{dy}{dx}\right)$ or $\frac{dy'}{dx}$.

Example 1

If $y = 5x^3 + 7x - 10$, find the first four derivatives.

$$\frac{dy}{dx} = 15x^2 + 7; \frac{d^2y}{dx^2} = 30x; \frac{d^3y}{dx^3} = 30; \frac{d^4y}{dx^4} = 0$$

Example 2

If $f(x) = \sqrt{x}$, find $f''(4)$.

Rewrite: $f(x) = \sqrt{x} = x^{1/2}$ and differentiate: $f'(x) = \frac{1}{2}x^{-1/2}$.

Differentiate again:

$$f''(x) = -\frac{1}{4}x^{-3/2} = \frac{-1}{4x^{3/2}} = \frac{-1}{4\sqrt{x^3}} \text{ and } f''(4) = \frac{-1}{4\sqrt{4^3}} = -\frac{1}{32}.$$

Example 3

If $y = x \cos x$, find y''.

Using the product rule, $y' = (1)(\cos x) + (x)(-\sin x) = \cos x - x \sin x$

$$y'' = -\sin x - [(1)(\sin x) + (x)(\cos x)]$$

$$= -\sin x - \sin x - x \cos x$$

$$= -2 \sin x - x \cos x.$$

Or, you can use a calculator and enter $d[x^* \cos x, x, 2]$ and obtain the same result.

7.7 *L'Hôpital's* Rule for Indeterminate Forms

Let lim represent one of the limits: $\lim_{x\to c}$, $\lim_{x\to c^+}$, $\lim_{x\to c^-}$, $\lim_{x\to\infty}$, or $\lim_{x\to-\infty}$. Suppose $f(x)$ and $g(x)$ are differentiable and $g'(x)\neq 0$ near c, except possibly at c, and suppose $\lim f(x)=0$ and $\lim g(x)=0$. Then the $\lim \frac{f(x)}{g(x)}$ is an indeterminate form of the type $\frac{0}{0}$. Also, if $\lim f(x)=\pm\infty$ and $\lim g(x)=\pm\infty$, then the $\lim \frac{f(x)}{g(x)}$ is an indeterminate form of the type $\frac{\infty}{\infty}$. In both cases, $\frac{0}{0}$ and $\frac{\infty}{\infty}$, *L'Hôpital's* Rule states that $\lim \frac{f(x)}{g(x)} = \lim \frac{f'(x)}{g'(x)}$.

Example 1

Find $\lim_{x\to 0} \frac{1-\cos x}{x^2}$, if it exists.

Since $\lim_{x\to 0}(1-\cos x)=0$ and $\lim_{x\to 0}(x^2)=0$, this limit is an indeterminate form. Take the derivatives, $\frac{d}{dx}(1-\cos x)=\sin x$ and $\frac{d}{dx}(x^2)=2x$. By *L'Hôpital's* Rule, $\lim_{x\to 0}\frac{1-\cos x}{x^2} = \lim_{x\to 0}\frac{\sin x}{2x} = \frac{1}{2}\lim_{x\to 0}\frac{\sin x}{x} = \frac{1}{2}$.

Example 2

Find $\lim_{x\to\infty} x^3 e^{-x^2}$, if it exists.

Rewriting $\lim_{x\to\infty} x^3 e^{-x^2}$ as $\lim_{x\to\infty}\left(\frac{x^3}{e^{x^2}}\right)$ shows that the limit is an indeterminate form, since $\lim_{x\to\infty}(x^3)=\infty$ and $\lim_{x\to\infty}\left(e^{x^2}\right)=\infty$. Differentiating and applying *L'Hôpital's* Rule means that $\lim_{x\to\infty}\left(\frac{x^3}{e^{x^2}}\right)=\lim_{x\to\infty}\left(\frac{3x^2}{2xe^{x^2}}\right)=\frac{3}{2}\lim_{x\to\infty}\left(\frac{x}{e^{x^2}}\right)$. Unfortunately, this new limit is also indeterminate. However, it is possible to apply *L'Hôpital's* Rule again, so $\frac{3}{2}\lim_{x\to\infty}\left(\frac{x}{e^{x^2}}\right)$ equals to $\frac{3}{2}\lim_{x\to\infty}\left(\frac{1}{2xe^{x^2}}\right)$. This expression approaches zero as x becomes large, so $\lim_{x\to\infty} x^3 e^{-x^2}=0$.

7.8 Rapid Review

1. If $y=e^{x^3}$, find $\frac{dy}{dx}$.

   Answer: Using the chain rule, $\frac{dy}{dx}=\left(e^{x^3}\right)(3x^2)$.

2. Evaluate $\lim_{h\to 0}\frac{\cos\left(\frac{\pi}{6}+h\right)-\cos\left(\frac{\pi}{6}\right)}{h}$.

   Answer: The limit is equivalent to $\frac{d}{dx}\cos x\bigg|_{x=\frac{\pi}{6}} = -\sin\left(\frac{\pi}{6}\right) = -\frac{1}{2}$.

3. Find $f'(x)$ if $f(x) = \ln(3x)$.

   Answer: $f'(x) = \dfrac{1}{3x}(3) = \dfrac{1}{x}$.

4. Find the approximate value of $f'(3)$. (See Figure 7.8-1.)

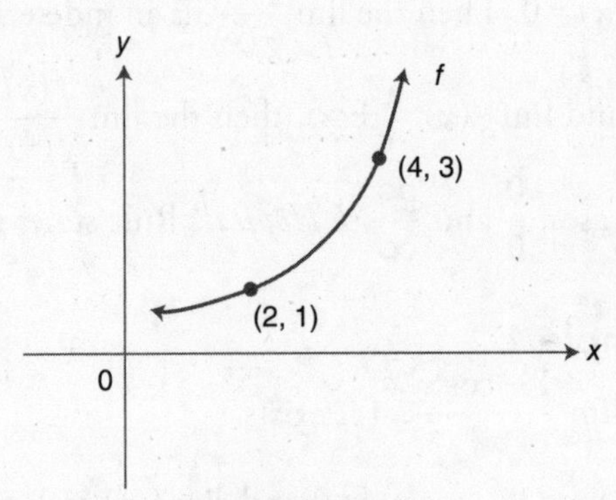

Figure 7.8-1

   Answer: Using the slope of the line segment joining (2, 1) and (4, 3), $f'(3) = \dfrac{3-1}{4-2} = 1$.

5. Find $\dfrac{dy}{dx}$ if $xy = 5x^2$.

   Answer: Using implicit differentiation, $1y + x\dfrac{dy}{dx} = 10x$. Thus, $\dfrac{dy}{dx} = \dfrac{10x - y}{x}$.

   Or simply solve for y leading to $y = 5x$ and thus, $\dfrac{dy}{dx} = 5$.

6. If $y = \dfrac{5}{x^2}$, find $\dfrac{d^2y}{dx^2}$.

   Answer: Rewrite $y = 5x^{-2}$. Then, $\dfrac{dy}{dx} = -10x^{-3}$ and $\dfrac{d^2y}{dx^2} = 30x^{-4} = \dfrac{30}{x^4}$.

7. Using a calculator, write an equation of the line tangent to the graph $f(x) = -2x^4$ at the point where $f'(x) = -1$.

   Answer: $f'(x) = -8x^3$. Using a calculator, enter [*Solve*] $[-8x\char`^3 = -1,\ x]$ and obtain $x = \dfrac{1}{2} \Rightarrow f'\left(\dfrac{1}{2}\right) = -1$. Using the calculator $f\left(\dfrac{1}{2}\right) = -\dfrac{1}{8}$. Thus, tangent is $y + \dfrac{1}{8} = -1\left(x - \dfrac{1}{2}\right)$.

8. $\lim\limits_{x\to 2} \dfrac{x^2 + x - 6}{x^2 - 4}$

   Answer: Since $\dfrac{x^2 + x - 6}{x^2 - 4} \to \dfrac{0}{0}$, consider $\lim\limits_{x\to 2} \dfrac{2x+1}{2x} = \dfrac{5}{4}$.

9. $\lim\limits_{x\to\infty} \dfrac{\ln x}{x}$

   Answer: Since $\dfrac{\ln x}{x} \to \dfrac{\infty}{\infty}$, consider $\lim\limits_{x\to\infty} \dfrac{1/x}{1} = \lim\limits_{x\to\infty} \dfrac{1}{x} = 0$.

7.9 Practice Problems

Part A—The use of a calculator is not allowed.

Find the derivative of each of the following functions.

1. $y = 6x^5 - x + 10$

2. $f(x) = \dfrac{1}{x} + \dfrac{1}{\sqrt[3]{x^2}}$

3. $y = \dfrac{5x^6 - 1}{x^2}$

4. $y = \dfrac{x^2}{5x^6 - 1}$

5. $f(x) = (3x - 2)^5(x^2 - 1)$

6. $y = \sqrt{\dfrac{2x+1}{2x-1}}$

7. $y = 10\cot(2x - 1)$

8. $y = 3x \sec(3x)$

9. $y = 10 \cos[\sin(x^2 - 4)]$

10. $y = 8\cos^{-1}(2x)$

11. $y = 3e^5 + 4xe^x$

12. $y = \ln(x^2 + 3)$

Part B—Calculators are allowed.

13. Find $\dfrac{dy}{dx}$, if $x^2 + y^3 = 10 - 5xy$.

14. The graph of a function f on [1, 5] is shown in Figure 7.9-1. Find the approximate value of $f'(4)$.

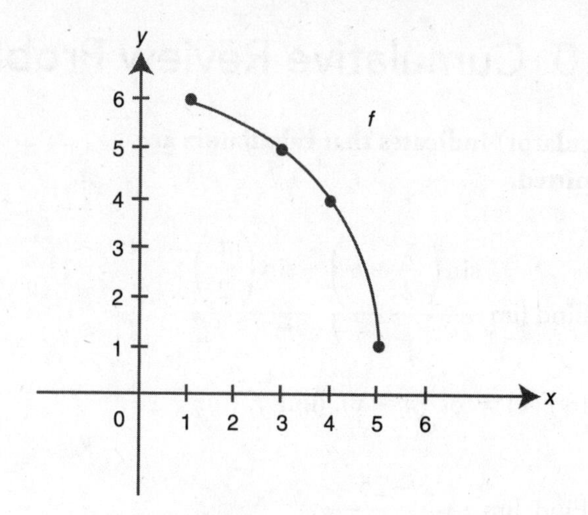

Figure 7.9-1

15. Let f be a continuous and differentiable function. Selected values of f are shown below. Find the approximate value of f' at $x = 2$.

x	−1	0	1	2	3
f	6	5	6	9	14

16. If $f(x) = x^5 + 3x - 8$, find $(f^{-1})'(-8)$.

17. Write an equation of the tangent to the curve $y = \ln x$ at $x = e$.

18. If $y = 2x \sin x$, find $\dfrac{d^2y}{dx^2}$ at $x = \dfrac{\pi}{2}$.

19. If the function $f(x) = (x - 1)^{2/3} + 2$, find all points where f is not differentiable.

20. Write an equation of the normal line to the curve $x \cos y = 1$ at $\left(2, \dfrac{\pi}{3}\right)$.

21. $\lim\limits_{x\to 3} \dfrac{x^2 - 3x}{x^2 - 9}$

22. $\lim\limits_{x\to 0^+} \dfrac{\ln(x+1)}{\sqrt{x}}$

23. $\lim\limits_{x\to 0} \dfrac{e^x - 1}{\tan 2x}$

24. $\lim\limits_{x\to 0} \dfrac{\cos(x) - 1}{\cos(2x) - 1}$

25. $\lim\limits_{x\to \infty} \dfrac{5x + 2\ln x}{x + 3\ln x}$

7.10 Cumulative Review Problems

(Calculator) indicates that calculators are permitted.

26. Find $\lim_{h \to 0} \frac{\sin\left(\frac{\pi}{2}+h\right)-\sin\left(\frac{\pi}{2}\right)}{h}$.

27. If $f(x)=\cos^2(\pi - x)$, find $f'(0)$.

28. Find $\lim_{x \to \infty} \frac{x-25}{10+x-2x^2}$.

29. (Calculator) Let f be a continuous and differentiable function. Selected values of f are shown below. Find the approximate value of f' at $x = 2$.

x	0	1	2	3	4	5
f	3.9	4	4.8	6.5	8.9	11.8

30. (Calculator) If $f(x) = \begin{cases} \frac{x^2-9}{x-3}, & x \neq 3, \\ 3, & x = 3 \end{cases}$

determine if $f(x)$ is continuous at $(x = 3)$. Explain why or why not.

7.11 Solutions to Practice Problems

Part A—The use of a calculator is not allowed.

1. Applying the power rule, $\frac{dy}{dx} = 30x^4 - 1$.

2. Rewrite $f(x) = \frac{1}{x} + \frac{1}{\sqrt[3]{x^2}}$ as $f(x) = x^{-1} + x^{-2/3}$. Differentiate: $f'(x) = -x^{-2} - \frac{2}{3}x^{-5/3} = -\frac{1}{x^2} - \frac{2}{3\sqrt[3]{x^5}}$.

3. Rewrite

$y = \frac{5x^6-1}{x^2}$ as $y = \frac{5x^6}{x^2} - \frac{1}{x^2} = 5x^4 - x^{-2}$.

Differentiate:

$\frac{dy}{dx} = 20x^3 - (-2)x^{-3} = 20x^3 + \frac{2}{x^3}$.

An alternate method is to differentiate

$y = \frac{5x^6-1}{x^2}$ directly, using the quotient rule.

4. Applying the quotient rule,

$$\frac{dy}{dx} = \frac{(2x)(5x^6-1)-(30x^5)(x^2)}{(5x^6-1)^2}$$

$$= \frac{10x^7 - 2x - 30x^7}{(5x^6-1)^2}$$

$$= \frac{-20x^7 - 2x}{(5x^6-1)^2} = \frac{-2x(10x^6+1)}{(5x^6-1)^2}.$$

5. Applying the product rule, $u = (3x-2)^5$ and $v = (x^2-1)$, and then the chain rule,

$$f'(x) = [5(3x-2)^4(3)][x^2-1] + [2x] \times [(3x-2)^5]$$

$$= 15(x^2-1)(3x-2)^4 + 2x(3x-2)^5$$

$$= (3x-2)^4[15(x^2-1) + 2x(3x-2)]$$

$$= (3x-2)^4[15x^2 - 15 + 6x^2 - 4x]$$

$$= (3x-2)^4(21x^2 - 4x - 15).$$

6. Rewrite $y = \sqrt{\frac{2x+1}{2x-1}}$ as

$y = \left(\frac{2x+1}{2x-1}\right)^{1/2}$. Applying first the chain rule and then the quotient rule,

$$\frac{dy}{dx} = \frac{1}{2}\left(\frac{2x+1}{2x-1}\right)^{-1/2} \times \left[\frac{(2)(2x-1)-(2)(2x+1)}{(2x-1)^2}\right]$$

$$= \frac{1}{2}\frac{1}{\left(\frac{2x+1}{2x-1}\right)^{1/2}}\left[\frac{-4}{(2x-1)^2}\right]$$

$$= \frac{1}{2}\frac{1}{\frac{(2x+1)^{1/2}}{(2x-1)^{1/2}}}\left[\frac{-4}{(2x-1)^2}\right]$$

$$= \frac{-2}{(2x+1)^{1/2}(2x-1)^{3/2}}.$$

Note that $\left(\frac{2x+1}{2x-1}\right)^{1/2} = \frac{(2x+1)^{1/2}}{(2x-1)^{1/2}}$, if $\frac{2x+1}{2x-1} > 0$, which implies $x < -\frac{1}{2}$ or $x > \frac{1}{2}$.

An alternate method of solution is to write $y = \frac{\sqrt{2x+1}}{\sqrt{2x-1}}$ and use the quotient rule.

Another method is to write $y = (2x+1)^{1/2}(2x-1)^{1/2}$ and use the product rule.

7. Let $u = 2x - 1$,

$$\frac{dy}{dx} = 10[-\csc^2(2x-1)](2)$$
$$= -20\csc^2(2x-1).$$

8. Using the product rule,

$$\frac{dy}{dx} = (3[\sec(3x)]) + [\sec(3x)\ \tan(3x)](3)[3x]$$

$$= 3\sec(3x) + 9x\sec(3x)\tan(3x)$$

$$= 3\sec(3x)[1 + 3x\tan(3x)].$$

9. Using the chain rule, let $u = \sin(x^2 - 4)$.

$$\frac{dy}{dx} = 10(-\sin[\sin(x^2-4)])[\cos(x^2-4)](2x)$$
$$= -20x\cos(x^2-4)\sin[\sin(x^2-4)]$$

10. Using the chain rule, let $u = 2x$.

$$\frac{dy}{dx} = 8\left(\frac{-1}{\sqrt{1-(2x)^2}}\right)(2) = \frac{-16}{\sqrt{1-4x^2}}$$

11. Since $3e^5$ is a constant, its derivative is 0.

$$\frac{dy}{dx} = 0 + (4)(e^x) + (e^x)(4x)$$
$$= 4e^x + 4xe^x = 4e^x(1+x)$$

12. Let $u = (x^2+3)$, $\frac{dy}{dx} = \left(\frac{1}{x^2+3}\right)(2x)$

$$= \frac{2x}{x^2+3}.$$

Part B—Calculators are allowed.

13. Using implicit differentiation, differentiate each term with respect to x.

$$2x + 3y^2\frac{dy}{dx} = 0 - \left[(5)(y) + \frac{dy}{dx}(5x)\right]$$

$$2x + 3y^2\frac{dy}{dx} = -5y - 5x\frac{dy}{dx}$$

$$3y^2\frac{dy}{dx} + 5x\frac{dy}{dx} = -5y - 2x$$

$$\frac{dy}{dx} = (3y^2 + 5x) = -5y - 2x$$

$$\frac{dy}{dx} = \frac{-5y-2x}{3y^2+5x} \text{ or } \frac{dy}{dx} = \frac{-(2x+5y)}{5x+3y^2}$$

14. Since $f'(4)$ is equivalent to the slope of the tangent to $f(x)$ at $x = 4$, there are several ways you can find its approximate value.

Method 1: Use the slope of the line segment joining the points at $x = 4$ and $x = 5$.

$f(5) = 1$ and $f(4) = 4$

$$m = \frac{f(5) - f(4)}{5 - 4}$$

$$= \frac{1 - 4}{1} = -3$$

Method 2: Use the slope of the line segment joining the points at $x = 3$ and $x = 4$.

$f(3) = 5$ and $f(4) = 4$

$$m = \frac{f(4) - f(3)}{4 - 3}$$

$$= \frac{4 - 5}{4 - 3} = -1$$

Method 3: Use the slope of the line segment joining the points at $x = 3$ and $x = 5$.

$f(3) = 5$ and $f(5) = 1$

$$m = \frac{f(5) - f(3)}{5 - 3}$$

$$= \frac{1 - 5}{5 - 3} = -2$$

Note that -2 is the average of the results from methods 1 and 2. Thus, $f'(4) \approx -3, -1,$ or -2 depending on which line segment you use.

15. You can use the difference quotient $\frac{f(a+h) - f(a)}{h}$ to approximate $f'(a)$.

Let $h = 1$; $f'(2) \approx \frac{f(3) - f(2)}{3 - 2} \approx \frac{14 - 9}{3 - 2} \approx 5.$

Or, you can use the symmetric difference quotient $\frac{f(a+h) - f(a-h)}{2h}$ to approximate $f'(a)$.

Let $h = 1$; $f'(2) \approx \frac{f(3) - f(1)}{2 - 0} \approx \frac{14 - 6}{2} \approx 4.$

Thus, $f'(2) \approx 4$ or 5 depending on your method.

16. Enter $y1 = x^5 + 3x - 8$. The graph of $y1$ is strictly increasing. Thus, $f(x)$ has an inverse. Note that $f(0) = -8$. Thus, the point $(0, -8)$ is on the graph of $f(x)$, which implies that the point $(-8, 0)$ is on the graph of $f^{-1}(x)$.

$f'(x) = 5x^4 + 3$ and $f'(0) = 3$.

Since $(f^{-1})'(-8) = \frac{1}{f'(0)}$, thus

$(f^{-1})'(-8) = \frac{1}{3}$.

17. $\frac{dy}{dx} = \frac{1}{x}$ and $\left.\frac{dy}{dx}\right|_{x=e} = \frac{1}{e}$

Thus, the slope of the tangent to $y = \ln x$ at $x = e$ is $\frac{1}{e}$. At $x = e$, $y = \ln x = \ln e = 1$, which means the point $(e, 1)$ is on the curve of $y = \ln x$. Therefore, an equation of the tangent is $y - 1 = \frac{1}{e}(x - e)$ or $y = \frac{x}{e}$.

(See Figure 7.11-1.)

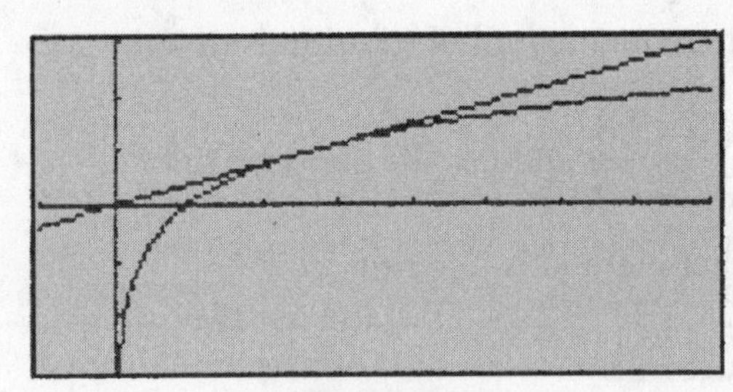

[-1.8] by [-3, 3]

Figure 7.11-1

18. $\frac{dy}{dx} = (2)(\sin x) + (\cos x)(2x) =$

$2 \sin x + 2x \cos x$

$$\frac{d^2y}{dx^2} = 2 \cos x + [(2)(\cos x) + (-\sin x)(2x)]$$

$$= 2 \cos x + 2 \cos x - 2x \sin x$$

$$= 4 \cos x - 2x \sin x$$

$$\left.\frac{d^2y}{dx^2}\right|_{x=\pi/2} = 4\cos\left(\frac{\pi}{2}\right) - 2\left(\frac{\pi}{2}\right)\left(\sin\left(\frac{\pi}{2}\right)\right)$$

$$= 0 - 2\left(\frac{\pi}{2}\right)(1) = -\pi$$

Or, using a calculator, enter $d(2x - \sin(x), x, 2)\ x = \frac{\pi}{2}$ and obtain $-\pi$.

19. Enter $y1 = (x-1)^{2/3} + 2$ in your calculator. The graph of $y1$ forms a cusp at $x = 1$. Therefore, f is not differentiable at $x = 1$.

20. Differentiate with respect to x:

$$(1)\cos y + \left[(-\sin y)\frac{dy}{dx}\right](x) = 0$$

$$\cos y - x\sin y\frac{dy}{dx} = 0$$

$$\frac{dy}{dx} = \frac{\cos y}{x\sin y}$$

$$\left.\frac{dy}{dx}\right|_{x=2, y=\pi/3} = \frac{\cos(\pi/3)}{(2)\sin(\pi/3)}$$

$$= \frac{1/2}{2(\sqrt{3}/2)} = \frac{1}{2\sqrt{3}}.$$

Thus, the slope of the tangent to the curve at $(2, \pi/3)$ is $m = \frac{1}{2\sqrt{3}}$. The slope of the normal line to the curve at $(2, \pi/3)$ is $m = -\frac{2\sqrt{3}}{1} = -2\sqrt{3}$. Therefore, an equation of the normal line is $y - \pi/3 = -2\sqrt{3}(x-2)$.

21. $$\lim_{x\to 3}\frac{x^2-3x}{x^2-9} = \lim_{x\to 3}\frac{2x-3}{2x} = \frac{1}{2}$$

22. $$\lim_{x\to 0^+}\frac{\ln(x+1)}{\sqrt{x}} = \lim_{x\to 0^+}\frac{1/(x+1)}{1/(2\sqrt{x})}$$

$$= \lim_{x\to 0^+}\frac{2\sqrt{x}}{x+1} = 0$$

23. $$\lim_{x\to 0}\frac{e^x-1}{\tan 2x} = \lim_{x\to 0}\frac{e^x}{2\sec^2 2x} = \frac{1}{2}$$

24. $$\lim_{x\to 0}\frac{\cos(x)-1}{\cos(2x)-1} = \lim_{x\to 0}\frac{-\sin x}{-2\sin(2x)}$$

$$= \lim_{x\to 0}\frac{-\cos x}{-4\cos(2x)} = \frac{1}{4}$$

25. $$\lim_{x\to\infty}\frac{5x+2\ln x}{x+3\ln x} = \lim_{x\to\infty}\frac{5+(2/x)}{1+(3/x)} = 5$$

7.12 Solutions to Cumulative Review Problems

26. The expression

$$\lim_{h\to 0}\frac{\sin\left(\frac{\pi}{2}+h\right) - \sin\left(\frac{\pi}{2}\right)}{h}$$ is

the derivative of $\sin x$ at $x = \pi/2$, which is the slope of the tangent to $\sin x$ at $x = \pi/2$. The tangent to $\sin x$ at $x = \pi/2$ is parallel to the x-axis.

Therefore, the slope is 0, i.e.,

$$\lim_{h\to 0}\frac{\sin\left(\frac{\pi}{2}+h\right) - \sin\left(\frac{\pi}{2}\right)}{h} = 0.$$

An alternate method is to expand $\sin\left(\frac{\pi}{2}+h\right)$ as

$$\sin\left(\frac{\pi}{2}\right)\cos h + \cos\left(\frac{\pi}{2}\right)\sin h.$$

Thus, $\lim_{h\to 0}\frac{\sin\left(\frac{\pi}{2}+h\right) - \sin\left(\frac{\pi}{2}\right)}{h} =$

$$\lim_{h\to 0}\frac{\sin\left(\frac{\pi}{2}\right)\cos h + \cos\left(\frac{\pi}{2}\right)\sin h - \sin\left(\frac{\pi}{2}\right)}{h}$$

$$= \lim_{h\to 0} \frac{\sin\left(\frac{\pi}{2}\right)[\cos h - 1] + \cos\left(\frac{\pi}{2}\right)\sin h}{h}$$

$$= \lim_{h\to 0} \sin\left(\frac{\pi}{2}\right)\left(\frac{\cos h - 1}{h}\right)$$

$$- \lim_{h\to 0} \cos\left(\frac{\pi}{2}\right)\left(\frac{\sin h}{h}\right)$$

$$= \sin\left(\frac{\pi}{2}\right)\lim_{h\to 0}\left(\frac{\cos h - 1}{h}\right)$$

$$- \cos\left(\frac{\pi}{2}\right)\lim_{h\to 0}\left(\frac{\sin h}{h}\right)$$

$$= \left[\sin\left(\frac{\pi}{2}\right)\right] 0 + \cos\left(\frac{\pi}{2}\right)(1)$$

$$= \cos\left(\frac{\pi}{2}\right) = 0.$$

27. Using the chain rule, let $u = (\pi - x)$.
Then, $f'(x) = 2\cos(\pi - x)[-\sin(\pi - x)](-1)$

$$= 2\cos(\pi - x)\sin(\pi - x)$$

$$f'(0) = 2\cos\pi \sin\pi = 0.$$

28. Since the degree of the polynomial in the denominator is greater than the degree of the polynomial in the numerator, the limit is 0.

29. You can use the difference quotient $\frac{f(a+h) - f(a)}{h}$ to approximate $f'(a)$.

Let $h = 1$; $f'(2) \approx \frac{f(3) - f(2)}{3 - 2}$

$$\approx \frac{6.5 - 4.8}{1} \approx 1.7.$$

Let $h = 2$; $f'(2) \approx \frac{f(4) - f(2)}{4 - 2}$

$$\approx \frac{8.9 - 4.8}{2} \approx 2.05.$$

Or, you can use the symmetric difference quotient $\frac{f(a+h) - f(a-h)}{2h}$ to approximate $f'(a)$.

Let $h = 1$; $f'(2) \approx \frac{f(3) - f(1)}{3 - 1}$

$$\approx \frac{6.5 - 4}{2} \approx 1.25.$$

Let $h = 2$; $f'(2) \approx \frac{f(4) - f(0)}{4 - 0}$

$$\approx \frac{8.9 - 3.9}{4} \approx 1.25.$$

Thus, $f'(2) = 1.7$, 2.05, or 1.25 depending on your method.

30. (See Figure 7.12-1.) Checking the three conditions of continuity:

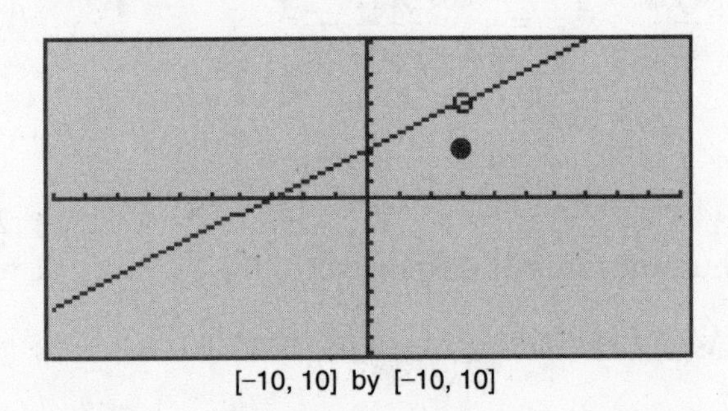

[−10, 10] by [−10, 10]

Figure 7.12-1

(1) $f(3) = 3$

(2) $\lim_{x\to 3} \frac{x^2 - 9}{x - 3} = \lim_{x\to 3}\left(\frac{(x+3)(x-3)}{(x-3)}\right)$

$$= \lim_{x\to 3}(x+3) = (3) + 3 = 6$$

(3) Since $f(3) \neq \lim_{x\to 3} f(x)$, $f(x)$ is discontinuous at $x = 3$.

Big Idea 2: Derivatives
Graphs of Functions and Derivatives

IN THIS CHAPTER

Summary: Many questions on the AP Calculus AB exam involve working with graphs of a function and its derivatives. In this chapter, you will learn how to use derivatives both algebraically and graphically to determine the behavior of a function. Applications of Rolle's Theorem, the Mean Value Theorem, and the Extreme Value Theorem are also shown.

Key Ideas

- Rolle's Theorem, Mean Value Theorem, and Extreme Value Theorem
- Test for Increasing and Decreasing Functions
- First and Second Derivative Tests for Relative Extrema
- Test for Concavity and Point of Inflection
- Curve Sketching
- Graphs of Derivatives

8.1 Rolle's Theorem, Mean Value Theorem, and Extreme Value Theorem

Main Concepts: Rolle's Theorem, Mean Value Theorem, Extreme Value Theorem

- Set your calculator to Radians and change it to Degrees if/when you need to. Do not forget to change it back to Radians after you have finished using it in Degrees.

Rolle's Theorem

If f is a function that satisfies the following three conditions:

1. f is continuous on a closed interval $[a, b]$
2. f is differentiable on the open interval (a, b)
3. $f(a) = f(b) = 0$

then there exists a number c in (a, b) such that $f'(c) = 0$. (See Figure 8.1-1.)

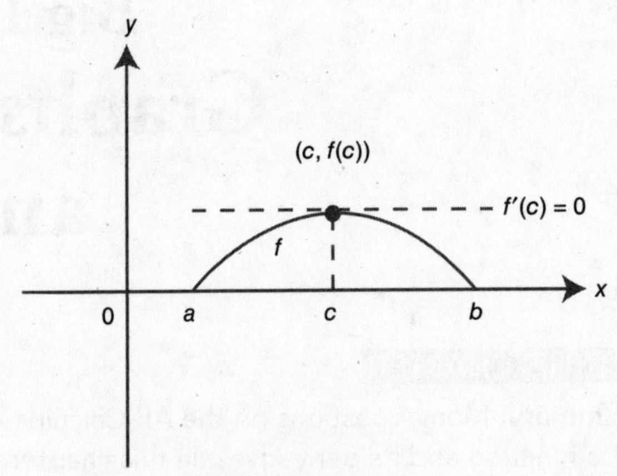

Figure 8.1-1

Note that if you change condition 3 from $f(a) = f(b) = 0$ to $f(a) = f(b)$, the conclusion of Rolle's Theorem is still valid.

Mean Value Theorem

If f is a function that satisfies the following conditions:

1. f is continuous on a closed interval $[a, b]$
2. f is differentiable on the open interval (a, b)

then there exists a number c in (a, b) such that $f'(c) = \dfrac{f(b) - f(a)}{b - a}$. (See Figure 8.1-2.)

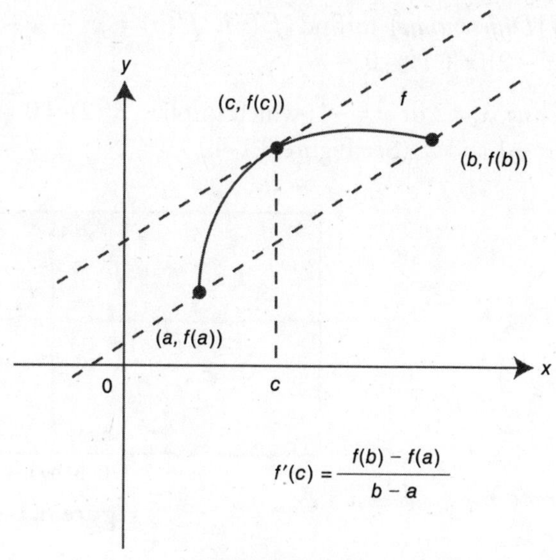

Figure 8.1-2

Example 1

If $f(x) = x^2 + 4x - 5$, show that the hypotheses of Rolle's Theorem are satisfied on the interval $[-4, 0]$ and find all values of c that satisfy the conclusion of the theorem. Check the three conditions in the hypotheses of Rolle's Theorem:

(1) $f(x) = x^2 + 4x - 5$ is continuous everywhere since it is polynomial.
(2) The derivative $f'(x) = 2x + 4$ is defined for all numbers and thus is differentiable on $(-4, 0)$.
(3) $f(0) = f(-4) = -5$. Therefore, there exists a c in $(-4, 0)$ such that $f'(c) = 0$. To find c, set $f'(x) = 0$. Thus, $2x + 4 = 0 \Rightarrow x = -2$, i.e., $f'(-2) = 0$. (See Figure 8.1-3.)

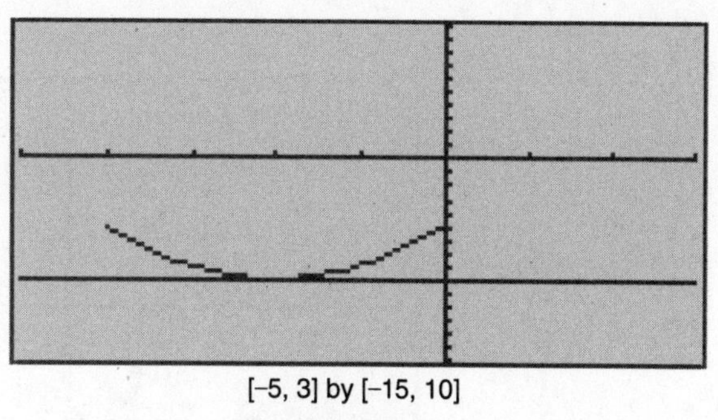

[–5, 3] by [–15, 10]

Figure 8.1-3

Example 2

Let $f(x) = \dfrac{x^3}{3} - \dfrac{x^2}{2} - 2x + 2$. Using Rolle's Theorem, show that there exists a number c in the domain of f such that $f'(c) = 0$. Find all values of c.

Note $f(x)$ is a polynomial and thus $f(x)$ is continuous and differentiable everywhere. Enter $y1 = \dfrac{x^3}{3} - \dfrac{x^2}{2} - 2x + 2$. The zeros of $y1$ are approximately -2.3, 0.9, and 2.9 i.e., $f(-2.3) = f(0.9) = f(2.9) = 0$. Therefore, there exists at least one c in the interval $(-2.3, 0.9)$ and at least one c in the interval $(0.9, 2.9)$ such that $f'(c) = 0$. Use *d* [*Differentiate*] to find $f'(x)$: $f'(x) = x^2 - x - 2$. Set $f'(x) = 0 \Rightarrow x^2 - x - 2 = 0$ or $(x - 2)(x + 1) = 0$.

Thus, $x = 2$ or $x = -1$, which implies $f'(2) = 0$ and $f'(-1) = 0$. Therefore, the values of c are -1 and 2. (See Figure 8.1-4.)

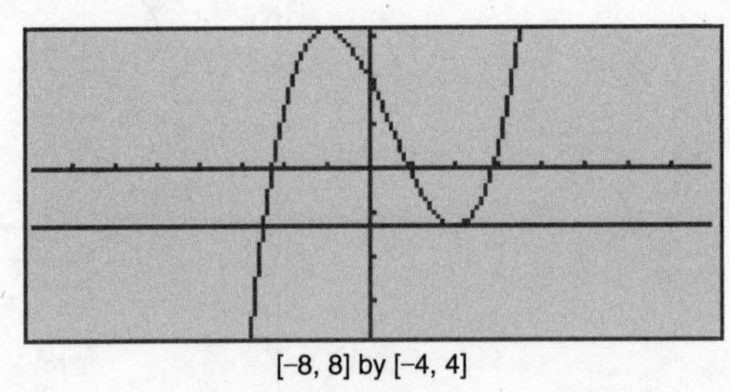

[−8, 8] by [−4, 4]

Figure 8.1-4

Example 3

The points $P(1, 1)$ and $Q(3, 27)$ are on the curve $f(x) = x^3$. Using the Mean Value Theorem, find c in the interval (1, 3) such that $f'(c)$ is equal to the slope of the secant $\overline{PQ}$.

The slope of secant $\overline{PQ}$ is $m = \frac{27 - 1}{3 - 1} = 13$. Since $f(x)$ is defined for all real numbers, $f(x)$ is continuous on [1, 3]. Also $f'(x) = 3x^2$ is defined for all real numbers. Thus, $f(x)$ is differentiable on (1, 3). Therefore, there exists a number c in (1, 3) such that $f'(c) = 13$.

Set $f'(c) = 13 \Rightarrow 3(c)^2 = 13$ or $c^2 = \frac{13}{3}$, $c = \pm\sqrt{\frac{13}{3}}$. Since only $\sqrt{\frac{13}{3}}$ is in the interval (1, 3), $c = \sqrt{\frac{13}{3}}$. (See Figure 8.1-5.)

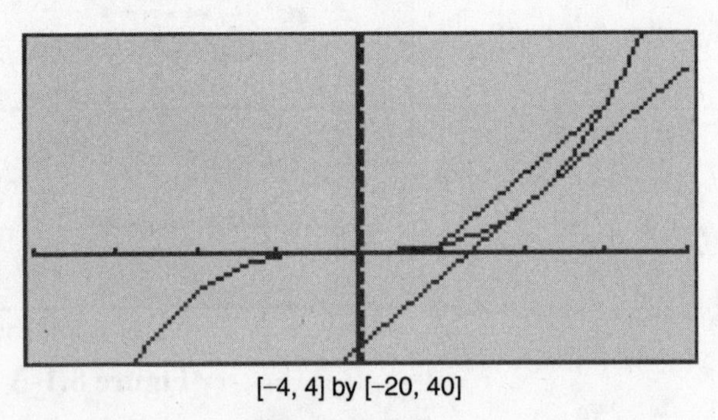

[−4, 4] by [−20, 40]

Figure 8.1-5

Example 4

Let f be the function $f(x) = (x - 1)^{2/3}$. Determine if the hypotheses of the Mean Value Theorem are satisfied on the interval [0, 2], and if so, find all values of c that satisfy the conclusion of the theorem.

Enter $y1 = (x - 1)^{2/3}$. The graph $y1$ shows that there is a cusp at $x = 1$. Thus, $f(x)$ is not differentiable on (0, 2), which implies there may or may not exist a c in (0, 2) such that $f'(c) = \frac{f(2) - f(0)}{2 - 0}$. The derivative $f'(x) = \frac{2}{3}(x - 1)^{-1/3}$ and $\frac{f(2) - f(0)}{2 - 0} = \frac{1 - 1}{2} = 0$.

Set $\frac{2}{3}(x-1)^{1/3}=0 \Rightarrow x=1$. Note that f is not differentiable $(a+x=1)$. Therefore, c does not exist. (See Figure 8.1-6.)

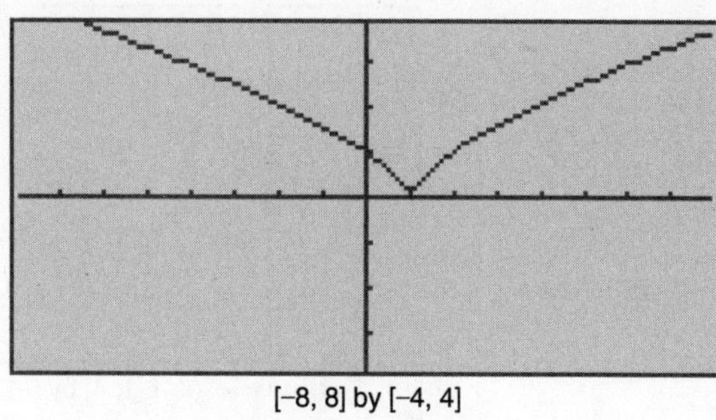

[–8, 8] by [–4, 4]

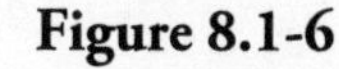

Figure 8.1-6

- The formula for finding the area of an equilateral triangle is $area = \frac{s^2\sqrt{3}}{4}$ where s is the length of a side. You might need this to find the volume of a solid whose cross sections are equilateral triangles.

Extreme Value Theorem

If f is a continuous function on a closed interval $[a, b]$, then f has both a maximum and a minimum value on the interval.

Example 1

If $f(x)=x^3+3x^2-1$, find the maximum and minimum values of f on $[-2, 2]$. Since $f(x)$ is a polynomial, it is a continuous function everywhere. Enter $y1=x^3+3x^2-1$. The graph of $y1$ indicates that f has a minimum of -1 at $x=0$ and a maximum value of 19 at $x=2$. (See Figure 8.1-7.)

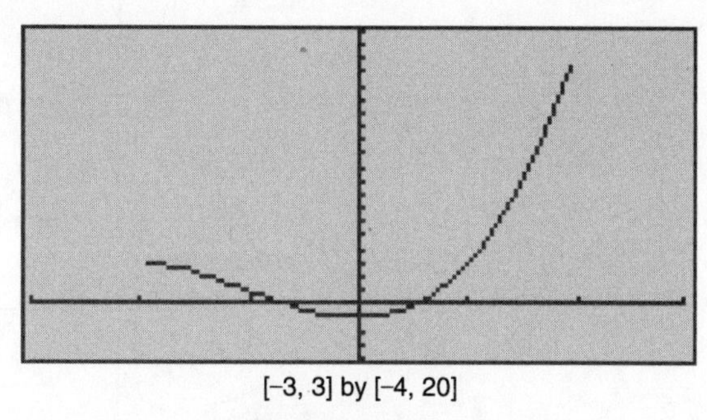

[–3, 3] by [–4, 20]

Figure 8.1-7

Example 2

If $f(x)=\frac{1}{x^2}$, find any maximum and minimum values of f on $[0, 3]$. Since $f(x)$ is a rational function, it is continuous everywhere except at values where the denominator is 0. In this case, at $x=0$, $f(x)$ is undefined. Since $f(x)$ is not continuous on $[0, 3]$, the Extreme Value Theorem may not be applicable. Enter $y1=\frac{1}{x^2}$. The graph of $y1$ shows that as $x \to 0^+$, $f(x)$ increases without bound (i.e., $f(x)$ goes to infinity). Thus, f has no maximum value. The minimum value occurs at the endpoint $x = 3$ and the minimum value is $\frac{1}{9}$. (See Figure 8.1-8.)

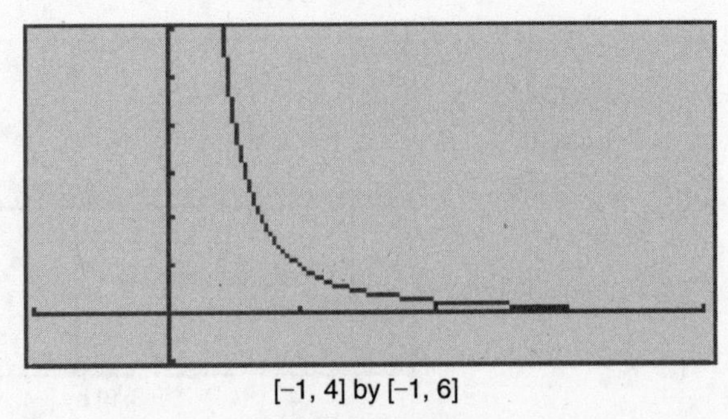

[−1, 4] by [−1, 6]

Figure 8.1-8

8.2 Determining the Behavior of Functions

Main Concepts: Test for Increasing and Decreasing Functions, First Derivative Test and Second Derivative Test for Relative Extrema, Test for Concavity and Points of Inflection

Test for Increasing and Decreasing Functions

Let f be a continuous function on the closed interval $[a, b]$ and differentiable on the open interval (a, b).

1. If $f'(x) > 0$ on (a, b), then f is increasing on $[a, b]$.
2. If $f'(x) < 0$ on (a, b), then f is decreasing on $[a, b]$.
3. If $f'(x) = 0$ on (a, b), then f is constant on $[a, b]$.

Definition: Let f be a function defined at a number c. Then c is a critical number of f if either $f'(c) = 0$ or $f'(c)$ does not exist. (See Figure 8.2-1.)

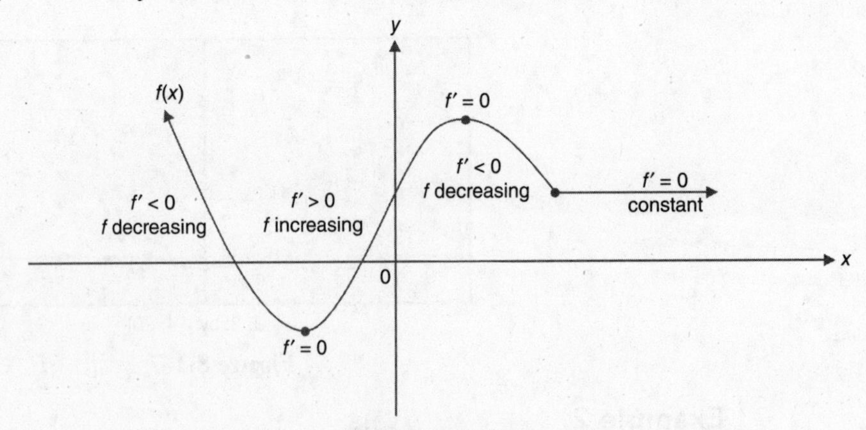

Figure 8.2-1

Example 1

Find the critical numbers of $f(x) = 4x^3 + 2x^2$.

To find the critical numbers of $f(x)$, you have to determine where $f'(x) = 0$ and where $f'(x)$ does not exist. Note $f'(x) = 12x^2 + 4x$, and $f'(x)$ is defined for all real numbers. Let $f'(x) = 0$ and thus $12x^2 + 4x = 0$, which implies $4x(3x + 1) = 0 \Rightarrow x = -1/3$ or $x = 0$. Therefore, the critical numbers of f are 0 and $-1/3$. (See Figure 8.2-2.)

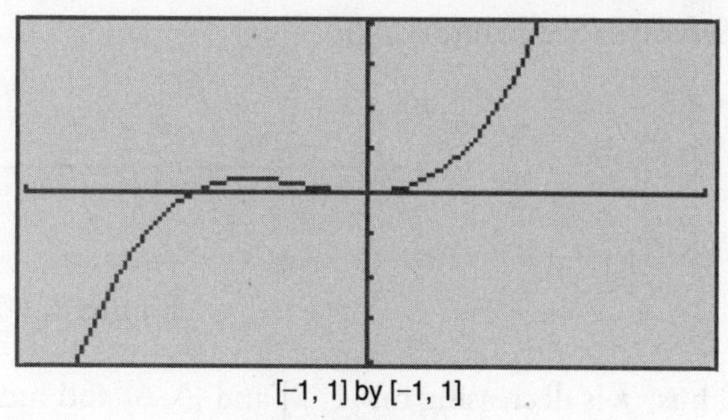

[−1, 1] by [−1, 1]

Figure 8.2-2

Example 2

Find the critical numbers of $f(x)=(x-3)^{2/5}$.

$f'(x) = \frac{2}{5}(x-3)^{-3/5} = \frac{2}{5(x-3)^{3/5}}$. Note that $f'(x)$ is undefined at $x = 3$ and that $f'(x) \neq 0$. Therefore, 3 is the only critical number of f. (See Figure 8.2-3.)

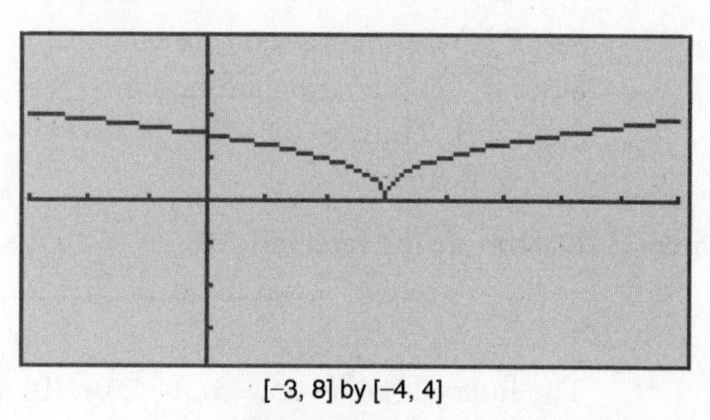

[−3, 8] by [−4, 4]

Figure 8.2-3

Example 3

The graph of f' on (1, 6) is shown in Figure 8.2-4. Find the intervals on which f is increasing or decreasing.

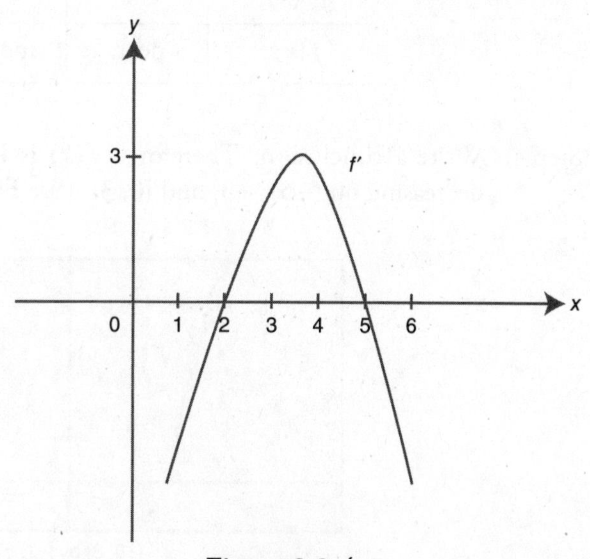

Figure 8.2-4

Solution: (See Figure 8.2-5.)

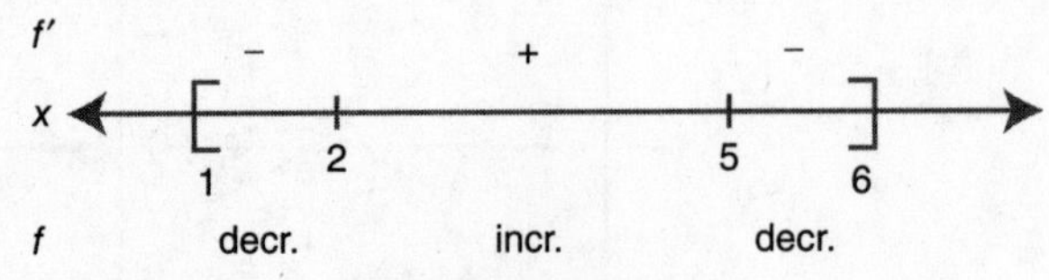

Figure 8.2-5

Thus, f is decreasing on [1, 2] and [5, 6] and increasing on [2, 5].

Example 4

Find the open intervals on which $f(x) = (x^2 - 9)^{2/3}$ is increasing or decreasing.

Step 1: Find the critical numbers of f.

$$f'(x) = \frac{2}{3}(x^2 - 9)^{-1/3}(2x) = \frac{4x}{3(x^2 - 9)^{1/3}}$$

Set $f'(x) = 0 \Rightarrow 4x = 0$ or $x = 0$.
Since $f'(x)$ is a rational function, $f'(x)$ is undefined at values where the denominator is 0. Thus, set $x^2 - 9 = 0 \Rightarrow x = 3$ or $x = -3$. Therefore, the critical numbers are -3, 0, and 3.

Step 2: Determine the intervals.

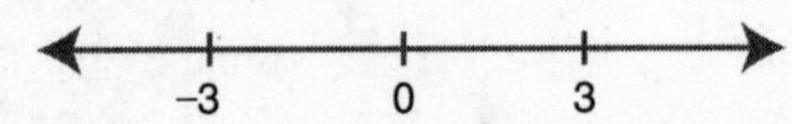

The intervals are $(-\infty, -3)$, $(-3, 0)$, $(0, 3)$, and $(3, \infty)$.

Step 3: Set up a table.

INTERVALS	$(-\infty, -3)$	$(-3, 0)$	$(0, 3)$	$(3, \infty)$
Test Point	-5	-1	1	5
$f'(x)$	$-$	$+$	$-$	$+$
$f(x)$	decr.	incr.	decr.	incr.

Step 4: Write a conclusion. Therefore, $f(x)$ is increasing on $[-3, 0]$ and $[3, \infty)$ and decreasing on $(-\infty, -3]$ and $[0, 3]$. (See Figure 8.2-6.)

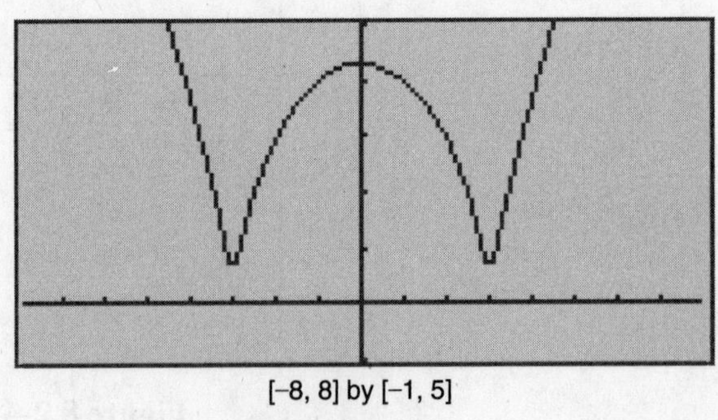

[–8, 8] by [–1, 5]

Figure 8.2-6

Example 5

The derivative of a function f is given as $f'(x) = \cos(x^2)$. Using a calculator, find the values of x on $\left[-\frac{\pi}{2}, \frac{\pi}{2}\right]$ such that f is increasing. (See Figure 8.2-7.)

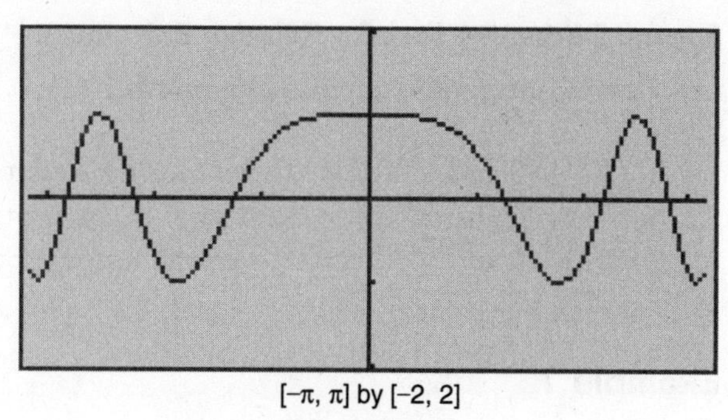

Figure 8.2-7

Using the [*Zero*] function of the calculator, you obtain $x = 1.25331$ is a zero of f' on $\left[0, \frac{\pi}{2}\right]$. Since $f'(x) = \cos(x^2)$ is an even function, $x = -1.25331$ is also a zero on $\left[-\frac{\pi}{2}, 0\right]$. (See Figure 8.2-8.)

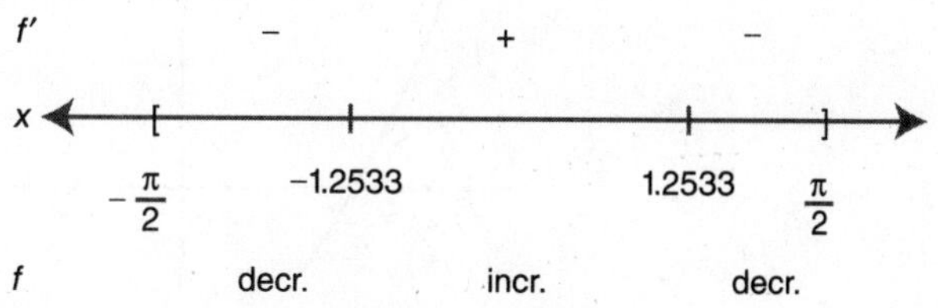

Figure 8.2-8

Thus, f is increasing on $[-1.2533, 1.2533]$.

- Be sure to bubble in the right grid. You have to be careful in filling in the bubbles, especially when you skip a question.

First Derivative Test and Second Derivative Test for Relative Extrema

First Derivative Test for Relative Extrema

Let f be a continuous function and c be a critical number of f. (See Figure 8.2-9.)

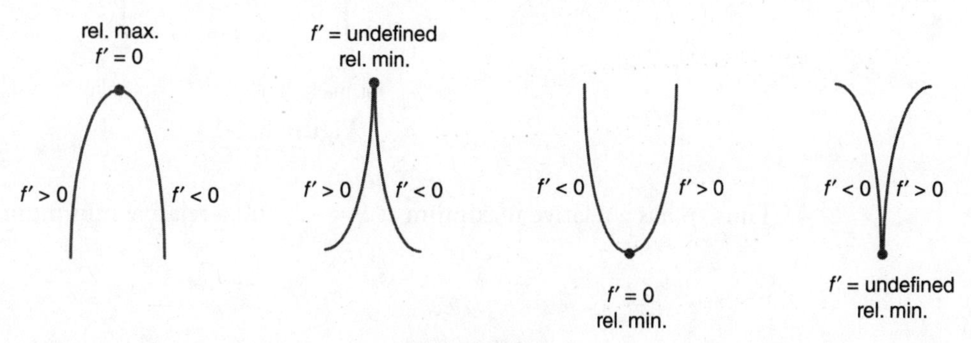

Figure 8.2-9

1. If $f'(x)$ changes from positive to negative at $x = c$ ($f' > 0$ for $x < c$ and $f' < 0$ for $x > c$), then f has a relative maximum at c.
2. If $f'(x)$ changes from negative to positive at $x = c$ ($f' < 0$ for $x < c$ and $f' > 0$ for $x > c$), then f has a relative minimum at c.

Second Derivative Test for Relative Extrema

Let f be a continuous function at a number c.

1. If $f'(c) = 0$ and $f''(c) < 0$, then $f(c)$ is a relative maximum.
2. If $f'(c) = 0$ and $f''(c) > 0$, then $f(c)$ is a relative minimum.
3. If $f'(c) = 0$ and $f''(c) = 0$, then the test is inconclusive. Use the First Derivative Test.

Example 1

The graph of f', the derivative of a function f, is shown in Figure 8.2-10. Find the relative extrema of f.

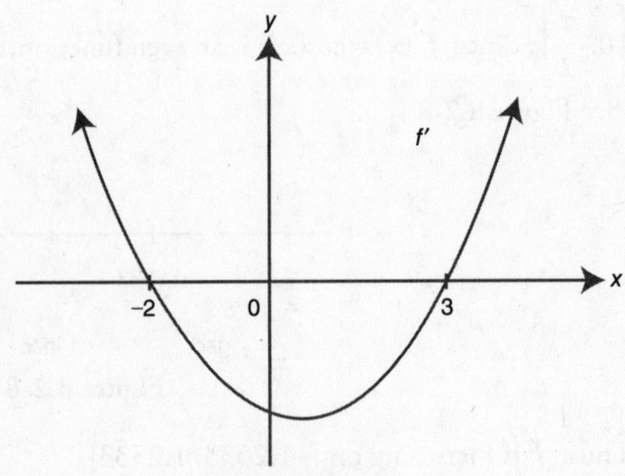

Figure 8.2-10

Solution: (See Figure 8.2-11.)

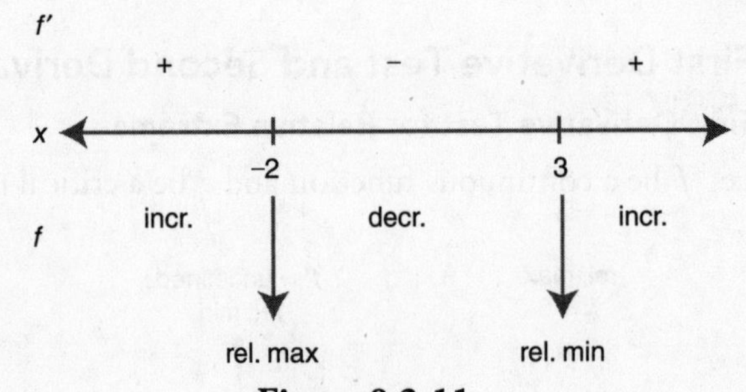

Figure 8.2-11

Thus, f has a relative maximum at $x = -2$, and a relative minimum at $x = 3$.

Example 2

Find the relative extrema for the function $f(x) = \frac{x^3}{3} - x^2 - 3x$.

Step 1: Find $f'(x)$.

$$f'(x) = x^2 - 2x - 3$$

Step 2: Find all critical numbers of $f(x)$.
Note that $f'(x)$ is defined for all real numbers.
Set $f'(x) = 0$: $x^2 - 2x - 3 = 0 \Rightarrow (x-3)(x+1) = 0 \Rightarrow x = 3$ or $x = -1$.

Step 3: Find $f''(x)$: $f''(x) = 2x - 2$.

Step 4: Apply the Second Derivative Test.
$f''(3) = 2(3) - 2 = 4 \Rightarrow f(3)$ is a relative minimum.
$f''(-1) = 2(-1) - 2 = -4 \Rightarrow f(-1)$ is a relative maximum.
$f(3) = \frac{3^3}{3} - (3)^2 - 3(3) = -9$ and $f(-1) = \frac{5}{3}$.
Therefore, -9 is a relative minimum value of f and $\frac{5}{3}$ is a relative maximum value.
(See Figure 8.2-12.)

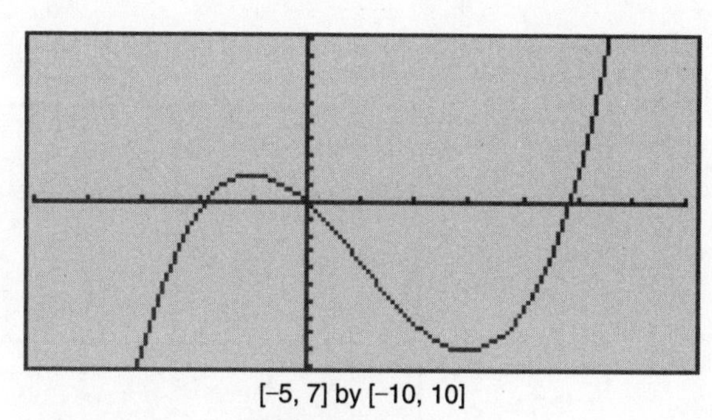

[-5, 7] by [-10, 10]

Figure 8.2-12

Example 3

Find the relative extrema for the function $f(x) = (x^2 - 1)^{2/3}$.

Using the First Derivative Test

Step 1: Find $f'(x)$.

$$f'(x) = \frac{2}{3}(x^2 - 1)^{-1/3}(2x) = \frac{4x}{3(x^2 - 1)^{1/3}}.$$

Step 2: Find all critical numbers of f.
Set $f'(x) = 0$. Thus, $4x = 0$ or $x = 0$.
Set $x^2 - 1 = 0$. Thus, $f'(x)$ is undefined at $x = 1$ and $x = -1$. Therefore, the critical numbers are -1, 0, and 1.

Step 3: Determine the intervals.

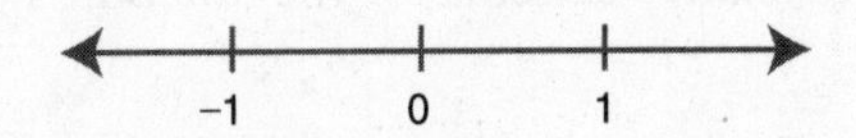

The intervals are $(-\infty, -1)$, $(-1, 0)$, $(0, 1)$, and $(1, \infty)$.

Step 4: Set up a table.

INTERVALS	$(-\infty, -1)$	$X = -1$	$(-1, 0)$	$X = 0$	$(0, 1)$	$X = 1$	$(1, \infty)$
Test Point	-2		$-1/2$		$1/2$		2
$f'(x)$ $f(x)$	$-$ decr.	undefined rel. min.	$+$ incr.	0 rel. max.	$-$ decr.	undefined rel. min.	$+$ incr.

Step 5: Write a conclusion.
Using the First Derivative Test, note that $f(x)$ has a relative maximum at $x = 0$ and relative minimums at $x = -1$ and $x = 1$.

Note that $f(-1) = 0$, $f(0) = 1$, and $f(1) = 0$. Therefore, 1 is a relative maximum value and 0 is a relative minimum value. (See Figure 8.2-13.)

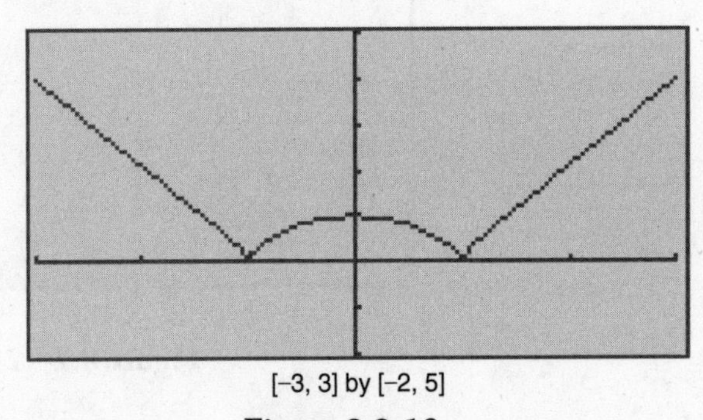

[−3, 3] by [−2, 5]

Figure 8.2-13

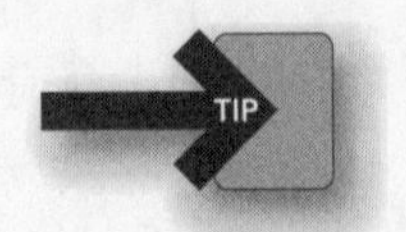

- Do not forget the constant, C, when you write the antiderivative after evaluating an indefinite integral, e.g., $\int \cos x dx = \sin x + C$.

Test for Concavity and Points of Inflection

Test for Concavity

Let f be a differentiable function.

1. If $f'' > 0$ on an interval I, then f is concave upward on I.
2. If $f'' < 0$ on an interval I, then f is concave downward on I.

(See Figures 8.2-14 and 8.2-15.)

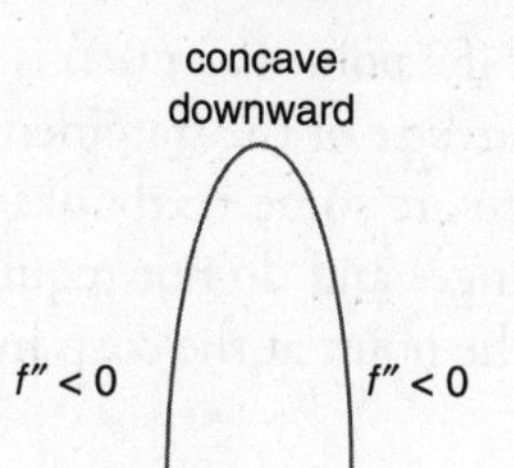

Figure 8.2-14

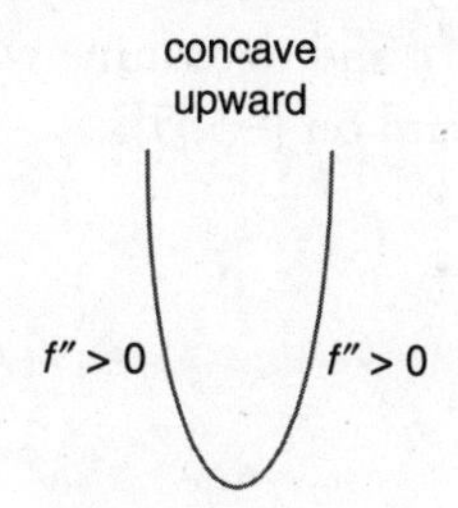

Figure 8.2-15

Points of Inflection

A point P on a curve is a point of inflection if:

1. the curve has a tangent line at P, and
2. the curve changes concavity at P (from concave upward to downward or from concave downward to upward).

(See Figures 8.2-16–8.2-18.)

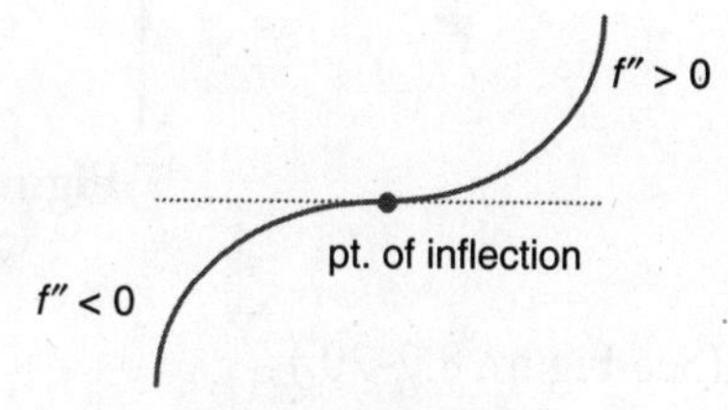

Figure 8.2-16

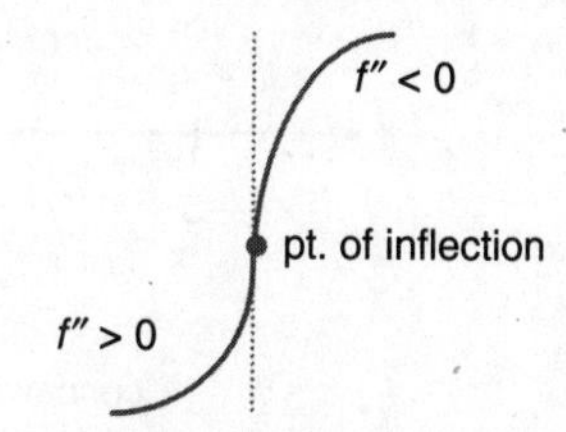

Figure 8.2-17

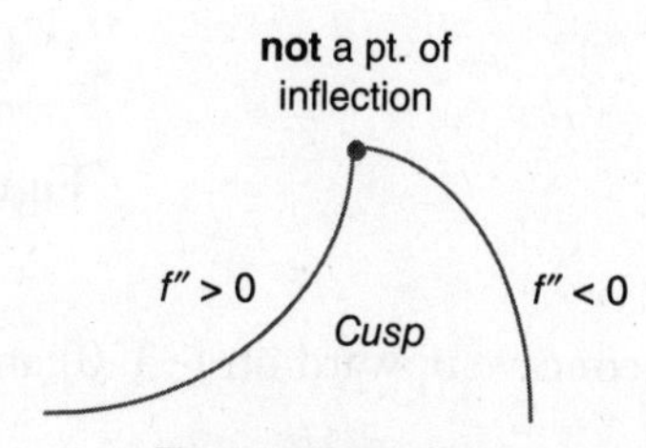

Figure 8.2-18

Note that if a point $(a, f(a))$ is a point of inflection, then $f''(c) = 0$ or $f''(c)$ does not exist. (The converse of the statement is not necessarily true.)

Note there are some textbooks that define a point of inflection as a point where the concavity changes and do not require the existence of a tangent at the point of inflection. In that case, the point at the cusp in Figure 8.2-18 would be a point of inflection.

Example 1

The graph of f', the derivative of a function f, is shown in Figure 8.2-19. Find the points of inflection of f and determine where the function f is concave upward and where it is concave downward on $[-3, 5]$.

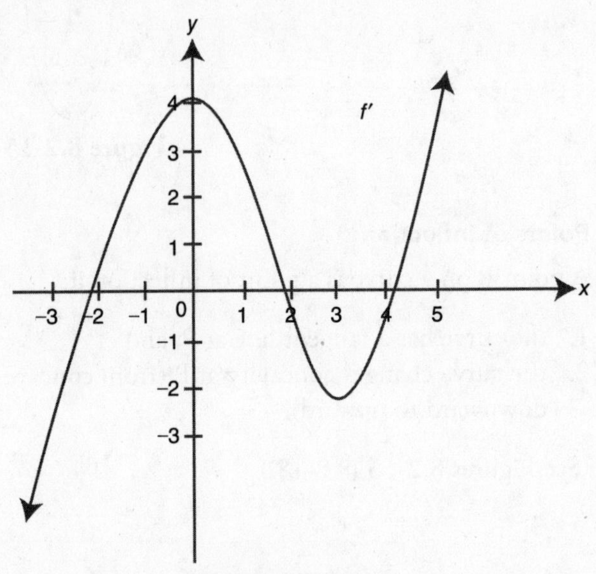

Figure 8.2-19

Solution: (See Figure 8.2-20.)

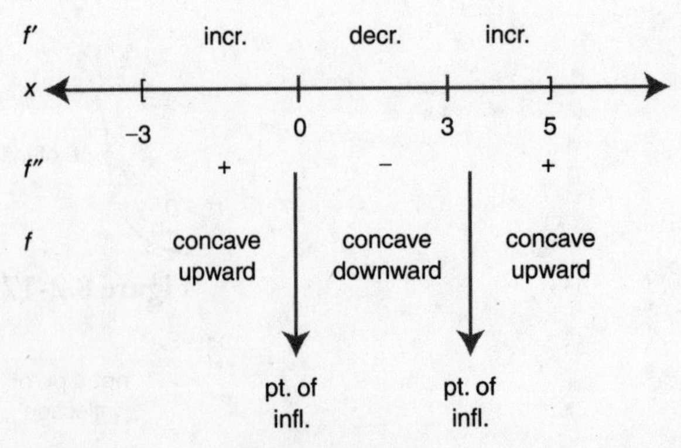

Figure 8.2-20

Thus, f is concave upward on $[-3, 0)$ and $(3, 5]$, and is concave downward on $(0, 3)$.

There are two points of inflection: one at $x = 0$ and the other at $x = 3$.

Example 2

Using a calculator, find the values of x at which the graph of $y = x^2e^x$ changes concavity.

Enter $y1 = x\hat{}2 * e\hat{}x$ and $y2 = d(y1(x), x, 2)$. The graph of $y2$, the second derivative of y, is shown in Figure 8.2-21. Using the [*Zero*] function, you obtain $x = -3.41421$ and $x = -0.585786$. (See Figures 8.2-21 and 8.2-22.)

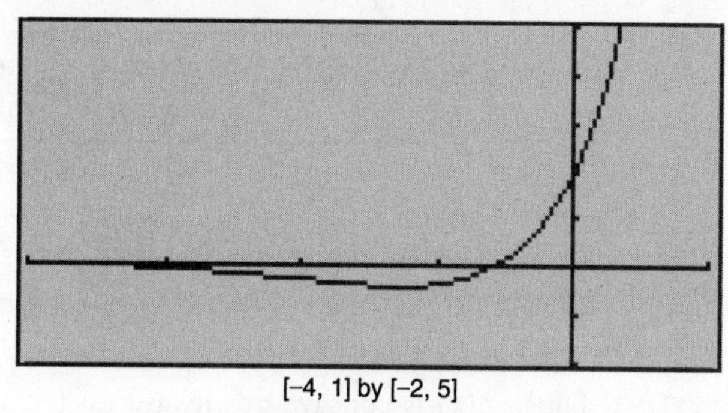

[–4, 1] by [–2, 5]

Figure 8.2-21

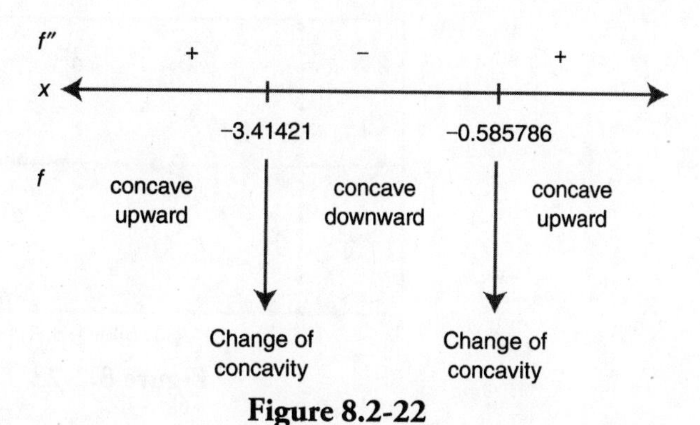

Figure 8.2-22

Thus, f changes concavity at $x = -3.41421$ and $x = -0.585786$.

Example 3

Find the points of inflection of $f(x) = x^3 - 6x^2 + 12x - 8$ and determine the intervals where the function f is concave upward and where it is concave downward.

Step 1: Find $f'(x)$ and $f''(x)$.

$$f'(x) = 3x^2 - 12x + 12$$
$$f''(x) = 6x - 12$$

Step 2: Set $f''(x) = 0$.

$$6x - 12 = 0$$
$$x = 2$$

Note that $f''(x)$ is defined for all real numbers.

Step 3: Determine the intervals.

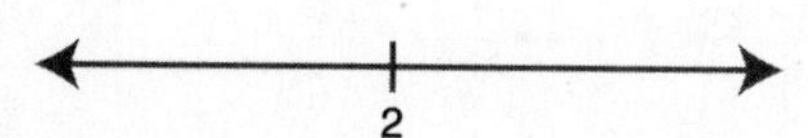

The intervals are $(-\infty, 2)$ and $(2, \infty)$.

Step 4: Set up a table.

INTERVALS	$(-\infty, 2)$	$X = 2$	$(2, \infty)$
Test Point	0		5
$f''(x)$	–	0	+
$f(x)$	concave downward	point of inflection	concave upward

Since $f(x)$ has change of concavity at $x = 2$, the point $(2, f(2))$ is a point of inflection. $f(2) = (2)^3 - 6(2)^2 + 12(2) - 8 = 0$.

Step 5: Write a conclusion.
Thus, $f(x)$ is concave downward on $(-\infty, 2)$, concave upward on $(2, \infty)$, and $f(x)$ has a point of inflection at $(2, 0)$. (See Figure 8.2-23.)

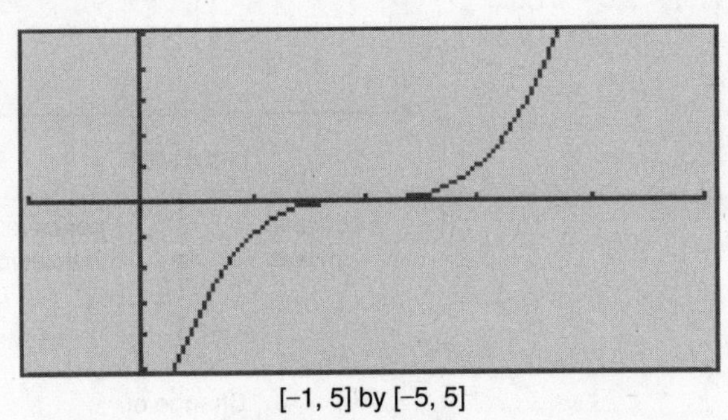

[–1, 5] by [–5, 5]

Figure 8.2-23

Example 4

Find the points of inflection of $f(x) = (x - 1)^{2/3}$ and determine the intervals where the function f is concave upward and where it is concave downward.

Step 1: Find $f'(x)$ and $f''(x)$.

$$f'(x) = \frac{2}{3}(x-1)^{-1/3} = \frac{2}{3(x-1)^{1/3}}$$

$$f''(x) = -\frac{2}{9}(x-1)^{-4/3} = \frac{-2}{9(x-1)^{4/3}}$$

Step 2: Find all values of x where $f''(x) = 0$ or $f''(x)$ is undefined.
Note that $f''(x) \neq 0$ and that $f''(1)$ is undefined.

Step 3: Determine the intervals.

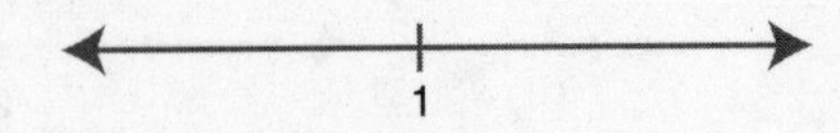

The intervals are $(-\infty, 1)$ and $(1, \infty)$.

Step 4: Set up a table.

INTERVALS	$(-\infty, 1)$	$X = 1$	$(1, \infty)$
Test Point	0		2
$f''(x)$	–	undefined	–
$f(x)$	concave downward	no change of concavity	concave downward

Note that since $f(x)$ has no change of concavity at $x = 1$, f does not have a point of inflection.

Step 5: Write a conclusion.
Therefore, $f(x)$ is concave downward on $(-\infty, \infty)$ and has no point of inflection. (See Figure 8.2-24.)

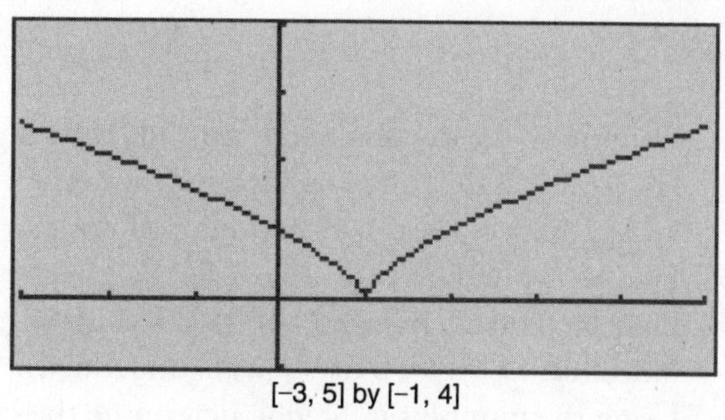

[-3, 5] by [-1, 4]
Figure 8.2-24

Example 5

The graph of f is shown in Figure 8.2-25, and f is twice differentiable. Which of the following statements is true?

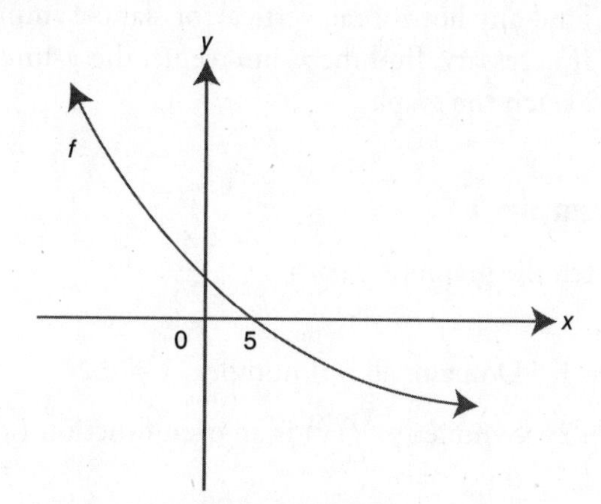

Figure 8.2-25

(A) $f(5) < f'(5) < f''(5)$
(B) $f''(5) < f'(5) < f(5)$
(C) $f'(5) < f(5) < f''(5)$
(D) $f'(5) < f''(5) < f(5)$

The graph indicates that (1) $f(5) = 0$; (2) $f'(5) < 0$, since f is decreasing; and (3) $f''(5) > 0$, since f is concave upward. Thus, $f'(5) < f(5) < f''(5)$, choice (C).

- Move on. Do not linger on a problem too long. Make an educated guess. You can earn many more points from other problems.

8.3 Sketching the Graphs of Functions

Main Concepts: Graphing without Calculators, Graphing with Calculators

Graphing without Calculators

General Procedure for Sketching the Graph of a Function

Steps:

1. Determine the domain and if possible the range of the function $f(x)$.
2. Determine if the function has any symmetry, i.e., if the function is even ($f(x) = f(-x)$), odd ($f(x) = -f(-x)$), or periodic ($f(x+p) = f(x)$).
3. Find $f'(x)$ and $f''(x)$.
4. Find all critical numbers ($f'(x) = 0$ or $f'(x)$ is undefined) and possible points of inflection ($f''(x) = 0$ or $f''(x)$ is undefined).
5. Using the numbers in Step 4, determine the intervals on which to analyze $f(x)$.
6. Set up a table using the intervals, to
   (a) determine where $f(x)$ is increasing or decreasing.
   (b) find relative and absolute extrema.
   (c) find points of inflection.
   (d) determine the concavity of $f(x)$ on each interval.
7. Find any horizontal, vertical, or slant asymptotes.
8. If necessary, find the x-intercepts, the y-intercepts, and a few selected points.
9. Sketch the graph.

Example 1

Sketch the graph of $f(x) = \dfrac{x^2 - 4}{x^2 - 25}$.

Step 1: Domain: all real numbers $x \neq \pm 5$.

Step 2: Symmetry: $f(x)$ is an even function ($f(x) = f(-x)$); symmetrical with respect to the y-axis.

Step 3: $$f'(x) = \frac{(2x)(x^2-25)-(2x)(x^2-4)}{(x^2-25)^2} = \frac{-42x}{(x^2-25)^2}$$

$$f''(x) = \frac{-42(x^2-25)^2 - 2(x^2-25)(2x)(-42x)}{(x^2-25)^4} = \frac{42(3x^2+25)}{(x^2-25)^3}$$

Step 4: Critical numbers:
$f'(x) = 0 \Rightarrow -42x = 0$ or $x = 0$
$f'(x)$ is undefined at $x = \pm 5$ which are not in the domain.
Possible points of inflection:
$f''(x) \neq 0$ and $f''(x)$ is undefined at $x = \pm 5$ which are not in the domain.

Step 5: Determine the intervals:

−5 0 5

Intervals are $(-\infty, -5)$, $(-5, 0)$, $(0, 5)$, and $(5, \infty)$.

Step 6: Set up a table:

INTERVALS	$(-\infty, -5)$	$X = -5$	$(-5, 0)$	$X = 0$	$(0, 5)$	$X = 5$	$(5, \infty)$
$f(x)$		undefined		4/25		undefined	
$f'(x)$	+	undefined	+	0	–	undefined	–
$f''(x)$	+	undefined	–	–	–	undefined	+
conclusion	incr. concave upward		incr. concave downward	rel. max.	decr. concave downward		decr. concave upward

Step 7: Vertical asymptote: $x = 5$ and $x = -5$
Horizontal asymptote: $y = 1$

Step 8: y-intercept: $\left(0, \dfrac{4}{25}\right)$
x-intercept: $(-2, 0)$ and $(2, 0)$
(See Figure 8.3-1.)

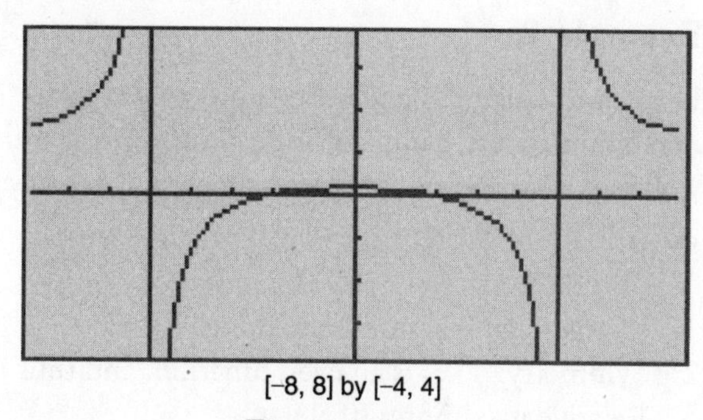

[-8, 8] by [-4, 4]

Figure 8.3-1

Graphing with Calculators

Example 1

Using a calculator, sketch the graph of $f(x) = -x^{5/3} + 3x^{2/3}$ indicating all relative extrema, points of inflection, horizontal and vertical asymptotes, intervals where $f(x)$ is increasing or decreasing, and intervals where $f(x)$ is concave upward or downward.

Steps:

1. Domain: all real numbers; Range: all real numbers
2. No symmetry
3. Relative maximum: (1.2, 2.03)
   Relative minimum: (0, 0)
   Points of inflection: (–0.6, 2.56)
4. No asymptote
5. $f(x)$ is decreasing on $(-\infty, 0]$, $[1.2, \infty)$ and increasing on $(0, 1.2)$.
6. Evaluating $f''(x)$ on either side of the point of inflection (–0.6, 2.56)

$$d\left(-x \wedge \left(\frac{5}{3}\right) + 3 * x \wedge \left(\frac{2}{3}\right), x, 2\right) \quad x = -2 \rightarrow 0.19$$

$$d\left(-x \wedge \left(\frac{5}{3}\right) + 3 * x \wedge \left(\frac{2}{3}\right), x, 2\right) \quad x = -1 \rightarrow -4.66$$

$\Rightarrow$ $f(x)$ is concave upward on $(-\infty, -0.6)$ and concave downward on $(-0.6, \infty)$. (See Figure 8.3-2.)

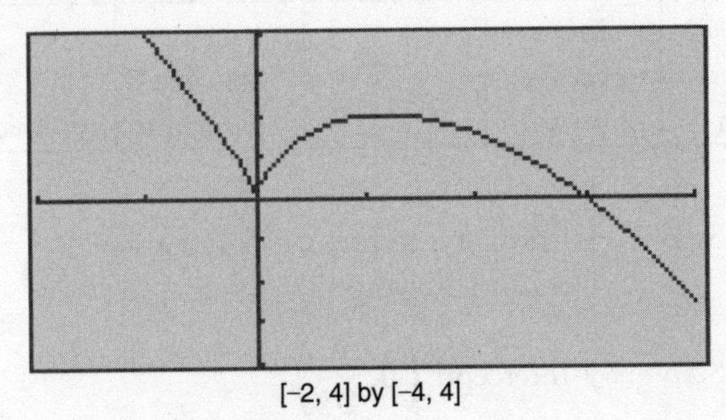

[–2, 4] by [–4, 4]

Figure 8.3-2

Example 2

Using a calculator, sketch the graph of $f(x) = e^{-x^2/2}$, indicating all relative minimum and maximum points; points of inflection; vertical and horizontal asymptotes; and intervals on which $f(x)$ is increasing, decreasing, concave upward, or concave downward.

Steps:

1. Domain: all real numbers; Range (0, 1]
2. Symmetry: $f(x)$ is an even function, and thus is symmetrical with respect to the y-axis.
3. Relative maximum: (0, 1)
   No relative minimum
   Points of inflection: (–1, 0.6) and (1, 0.6)
4. $y = 0$ is a horizontal asymptote; no vertical asymptote.
5. $f(x)$ is increasing on $(-\infty, 0]$ and decreasing on $[0, \infty)$.
6. $f(x)$ is concave upward on $(-\infty, -1)$ and $(1, \infty)$; and concave downward on $(-1, 1)$.

(See Figure 8.3-3.)

[–4, 4] by [–1, 2]

Figure 8.3-3

- When evaluating a definite integral, you do not have to write a constant C, e.g., $\int_1^3 2x\,dx = x^2\Big|_1^3 = 8$. Notice, no C.

8.4 Graphs of Derivatives

The functions f, f', and f'' are interrelated, and so are their graphs. Therefore, you can usually infer from the graph of one of the three functions (f, f', or f'') and obtain information about the other two. Here are some examples.

Example 1

The graph of a function f is shown in Figure 8.4-1. Which of the following is true for f on (a, b)?

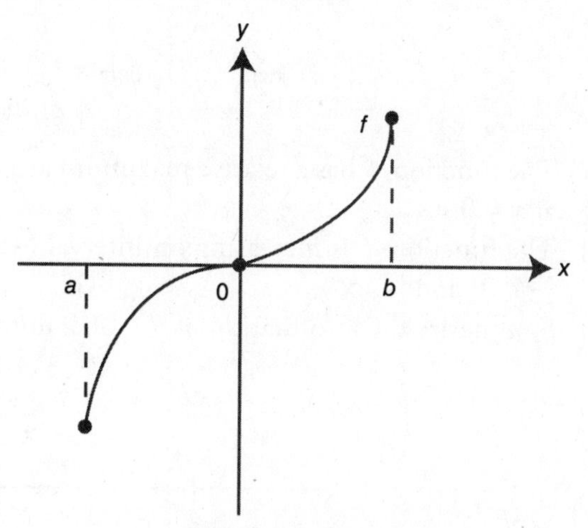

Figure 8.4-1

I. $f' \geq 0$ on (a, b)
II. $f'' > 0$ on (a, b)

Solution:

I. Since f is strictly increasing, $f' \geq 0$ on (a, b) is true.
II. The graph is concave downward on $(a, 0)$ and upward on $(0, b)$. Thus, $f'' > 0$ on $(0, b)$ only. Therefore, only statement I is true.

Example 2

Given the graph of f' in Figure 8.4-2, find where the function f: (a) has its relative maximum(s) or relative minimums, (b) is increasing or decreasing, (c) has its point(s) of inflection, (d) is concave upward or downward, and (e) if $f(-2) = f(2) = 1$ and $f(0) = -3$, draw a sketch of f.

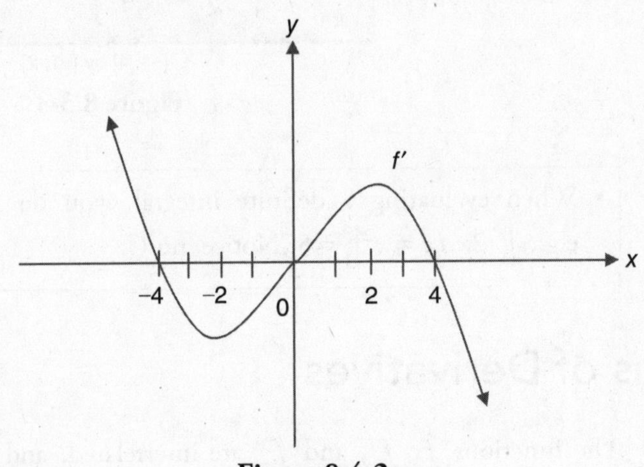

Figure 8.4-2

(a) Summarize the information of f' on a number line:

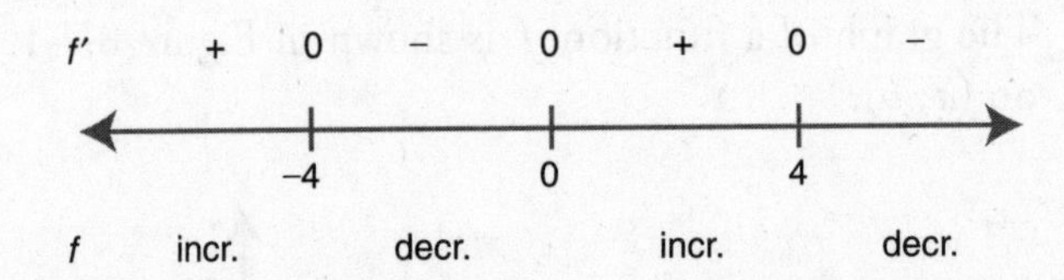

The function f has a relative maximum at $x = -4$ and at $x = 4$, and a relative minimum at $x = 0$.

(b) The function f is increasing on interval $(-\infty, -4]$ and $[0, 4]$, and f is decreasing on $[-4, 0]$ and $[4, \infty)$.

(c) Summarize the information of f'' on a number line:

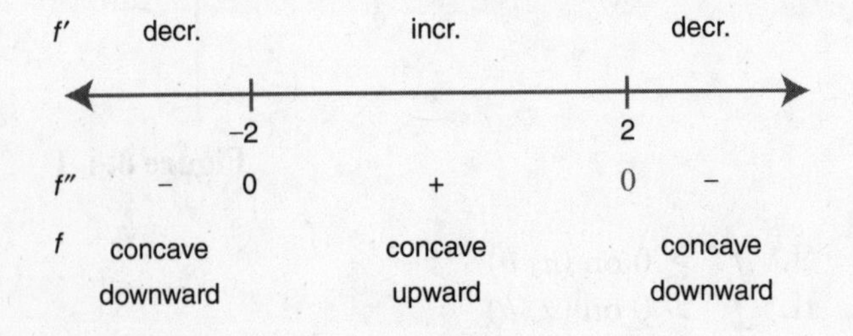

A change of concavity occurs at $x = -2$ and at $x = 2$ and f' exists at $x = -2$ and at $x = 2$, which implies that there is a tangent line to the graph of f at $x = -2$ and at $x = 2$. Therefore, f has a point of inflection at $x = -2$ and at $x = 2$.

(d) The graph of f is concave upward on the interval $(-2, 2)$ and concave downward on $(-\infty, -2)$ and $(2, \infty)$.

(e) A sketch of the graph of f is shown in Figure 8.4-3.

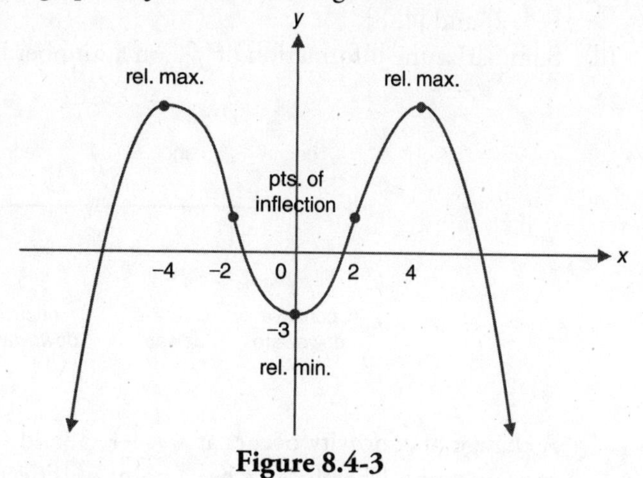

Figure 8.4-3

Example 3

Given the graph of f' in Figure 8.4-4, find where the function f (a) has a horizontal tangent, (b) has its relative extrema, (c) is increasing or decreasing, (d) has a point of inflection, and (e) is concave upward or downward.

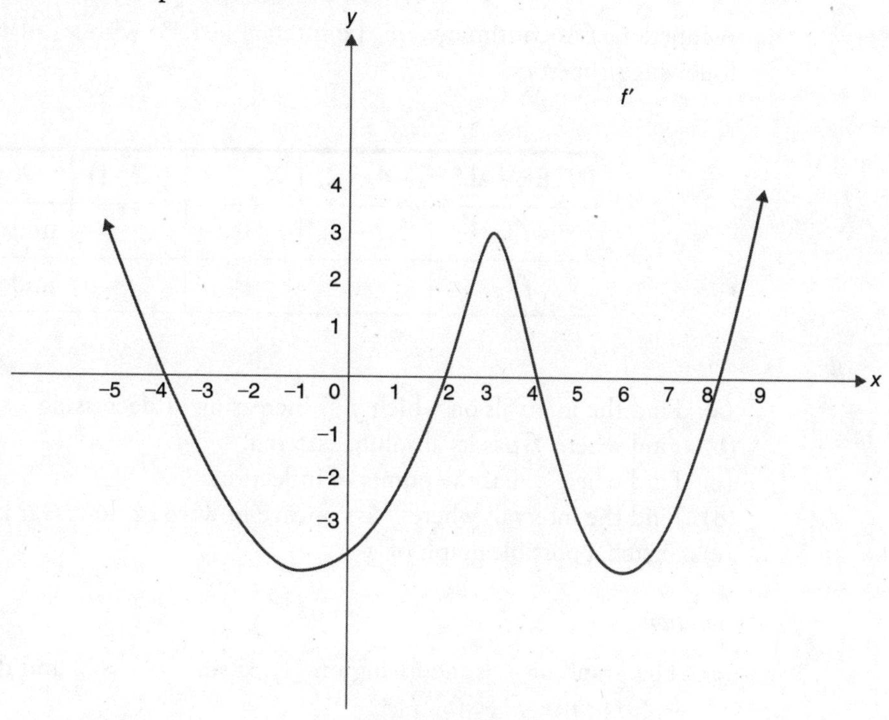

Figure 8.4-4

(a) $f'(x) = 0$ at $x = -4$, 2, 4, 8. Thus, f has a horizontal tangent at these values.

(b) Summarize the information of f' on a number line:

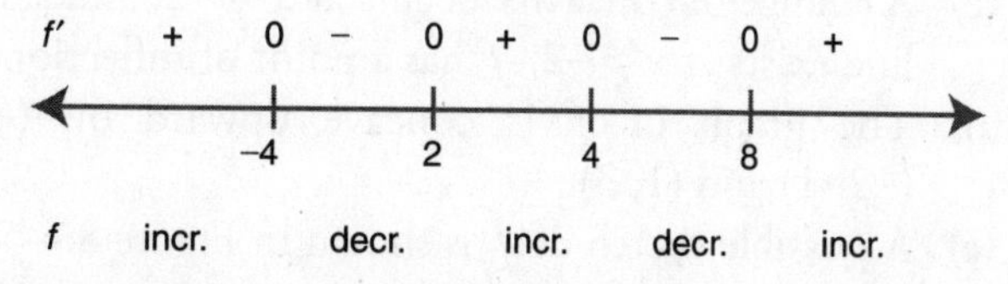

The First Derivative Test indicates that f has relative maximums at $x = -4$ and 4; and f has relative minimums at $x = 2$ and 8.

(c) The function f is increasing on $(-\infty, -4]$, $[2, 4]$, and $[8, \infty)$ and is decreasing on $[-4, 2]$ and $[4, 8]$.
(d) Summarize the information of f'' on a number line:

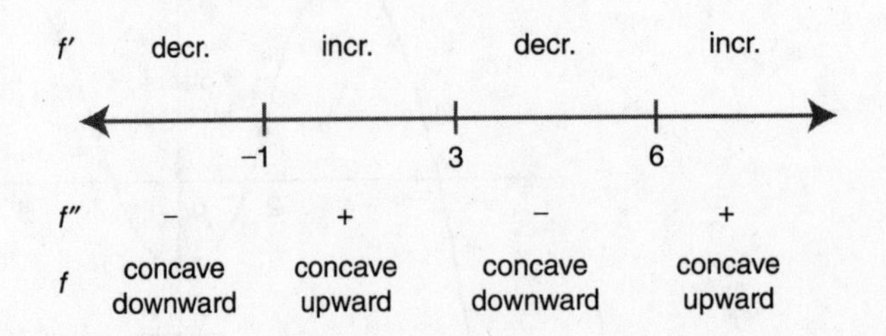

A change of concavity occurs at $x = -1$, 3, and 6. Since $f'(x)$ exists, f has a tangent at every point. Therefore, f has a point of inflection at $x = -1$, 3, and 6.
(e) The function f is concave upward on the interval $(-1, 3)$ and $(6, \infty)$ and concave downward on $(-\infty, -1)$ and $(3, 6)$.

Example 4

A function f is continuous on the interval $[-4, 3]$ with $f(-4) = 6$ and $f(3) = 2$ and the following properties:

INTERVALS	(–4, –2)	X = –2	(–2, 1)	X = 1	(1, 3)
f'	–	0	–	undefined	+
f''	+	0	–	undefined	–

(a) Find the intervals on which f is increasing or decreasing.
(b) Find where f has its absolute extrema.
(c) Find where f has the points of inflection.
(d) Find the intervals where f is concave upward or downward.
(e) Sketch a possible graph of f.

Solution:

(a) The graph of f is increasing on $[1, 3]$ since $f' > 0$ and decreasing on $[-4, -2]$ and $[-2, 1]$ since $f' < 0$.
(b) At $x = -4$, $f(x) = 6$. The function decreases until $x = 1$ and increases back to 2 at $x = 3$. Thus, f has its absolute maximum at $x = -4$ and its absolute minimum at $x = 1$.
(c) A change of concavity occurs at $x = -2$, and since $f'(-2) = 0$, which implies a tangent line exists at $x = -2$, f has a point of inflection at $x = -2$.
(d) The graph of f is concave upward on $(-4, -2)$ and concave downward on $(-2, 1)$ and $(1, 3)$.
(e) A possible sketch of f is shown in Figure 8.4-5.

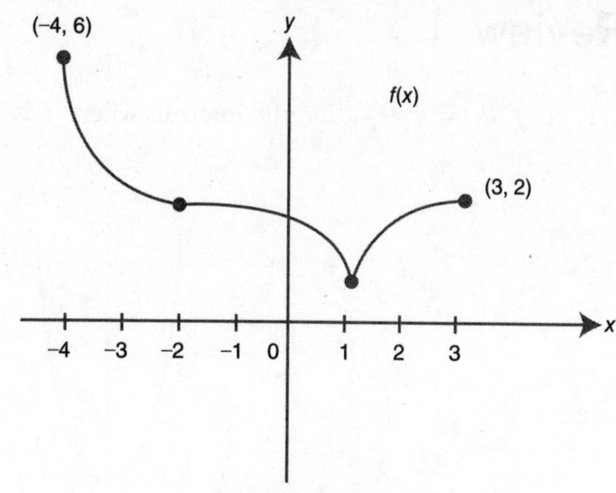

Figure 8.4-5

Example 5

If $f(x) = |\ln(x+1)|$, find $\lim_{x \to 0^-} f'(x)$. (See Figure 8.4-6.)

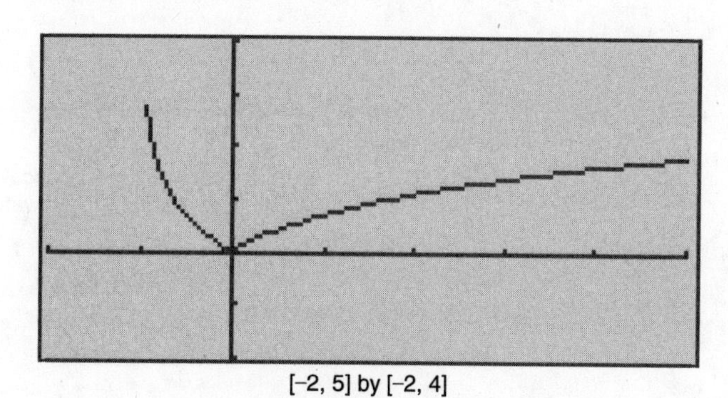

[−2, 5] by [−2, 4]

Figure 8.4-6

The domain of f is $(-1, \infty)$.

$$f(0) = |\ln(0+1)| = |\ln(1)| = 0$$

$$f(x) = |\ln(x+1)| = \begin{cases} \ln(x+1) & \text{if } x \geq 0 \\ -\ln(x+1) & \text{if } x < 0 \end{cases}$$

$$\text{Thus, } f'(x) = \begin{cases} \dfrac{1}{x+1} & \text{if } x \geq 0 \\ -\dfrac{1}{x+1} & \text{if } x < 0 \end{cases}$$

Therefore, $\lim_{x \to 0^-} f'(x) = \lim_{x \to 0^-} \left(-\dfrac{1}{x+1}\right) = -1.$

8.5 Rapid Review

1. If $f'(x) = x^2 - 4$, find the intervals where f is decreasing. (See Figure 8.5-1.)

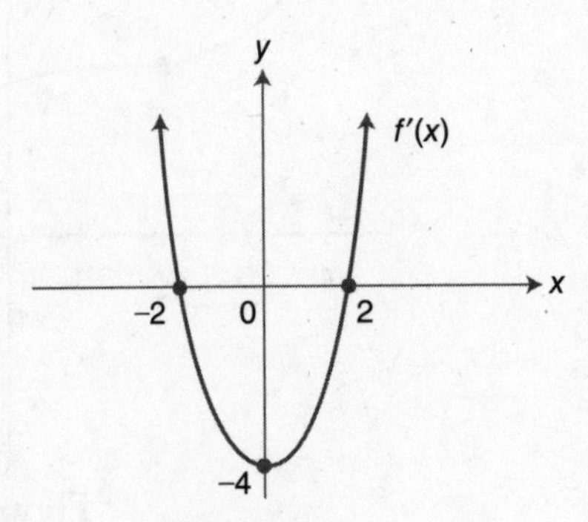

Figure 8.5-1

Answer: Since $f'(x) < 0$ if $-2 < x < 2$, f is decreasing on $(-2, 2)$.

2. If $f''(x) = 2x - 6$ and f' is continuous, find the values of x where f has a point of inflection. (See Figure 8.5-2.)

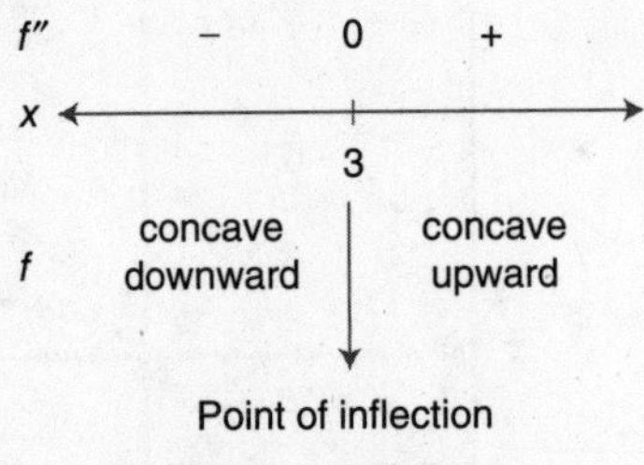

Figure 8.5-2

Answer: Thus, f has a point of inflection at $x = 3$.

3. Find the values of x where f has change of concavity. (See Figure 8.5-3.)

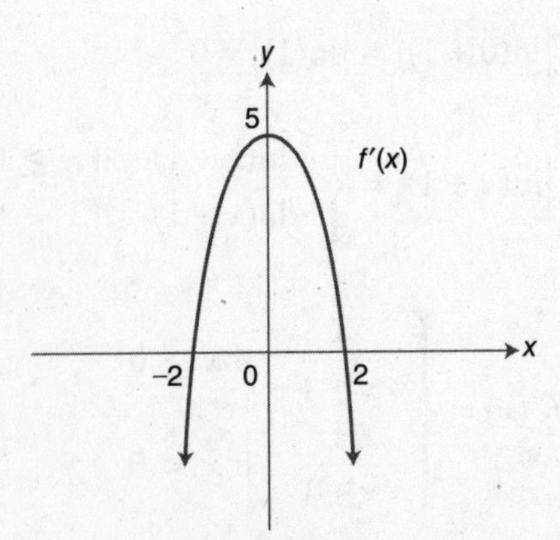

Figure 8.5-3

Answer: f has a change of concavity at $x = 0$. (See Figure 8.5-4.)

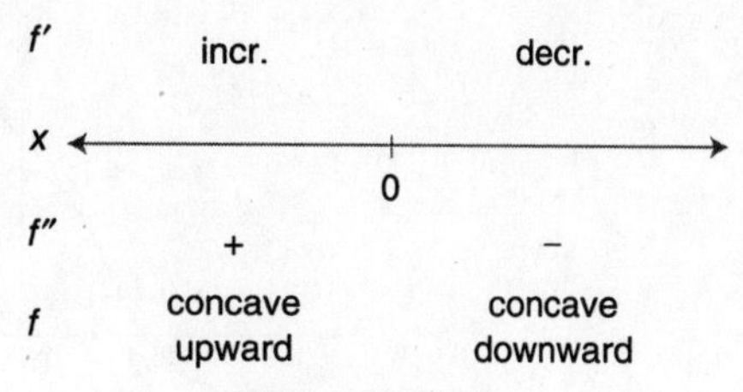

Figure 8.5-4

4. Find the values of x where f has a relative minimum. (See Figure 8.5-5.)

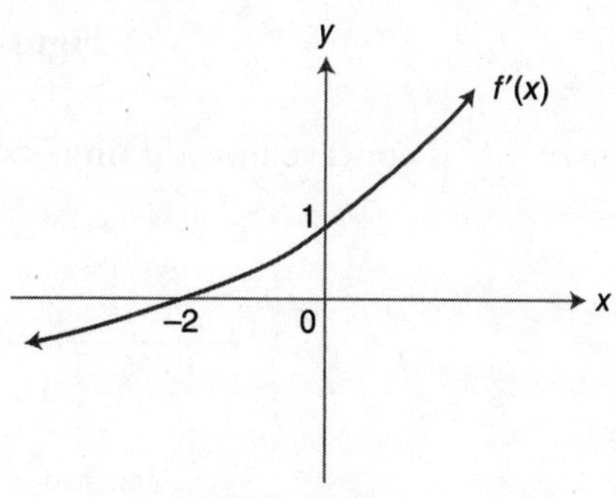

Figure 8.5-5

Answer: f has a relative minimum at $x = -2$. (See Figure 8.5-6.)

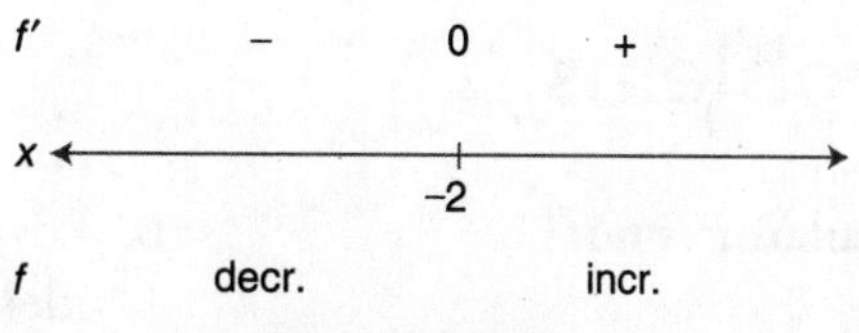

Figure 8.5-6

5. Given f is twice differentiable, arrange $f(10)$, $f'(10)$, $f''(10)$ from smallest to largest. (See Figure 8.5-7.)

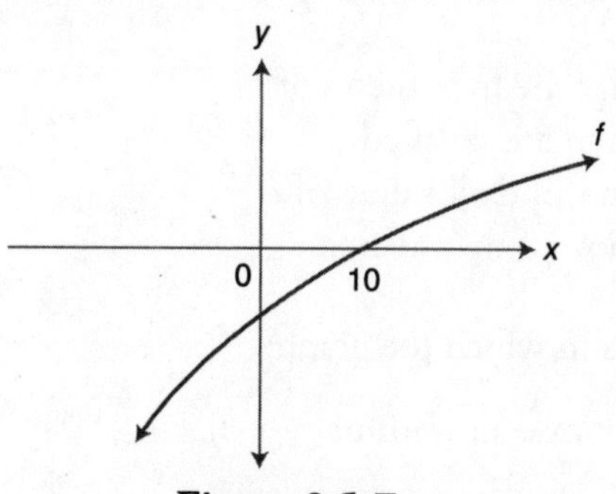

Figure 8.5-7

Answer: $f(10) = 0$, $f'(10) > 0$ since f is increasing, and $f''(10) < 0$ since f is concave downward. Thus, the order is $f''(10)$, $f(10)$, $f'(10)$.

6. Find the values of x where f' is concave upward. (See Figure 8.5-8.)

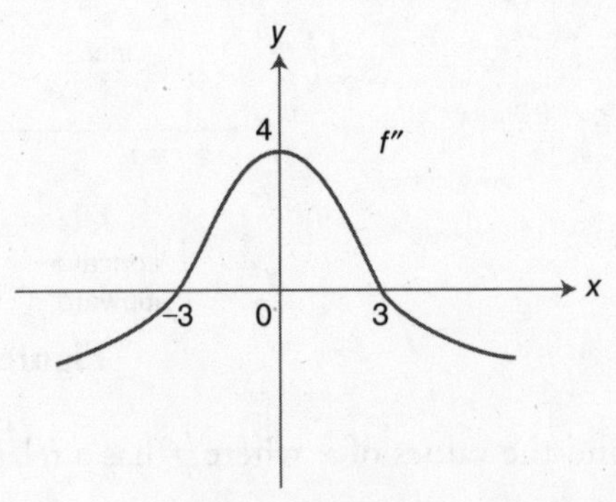

Figure 8.5-8

Answer: f' is concave upward on $(-\infty, 0)$. (See Figure 8.5-9.)

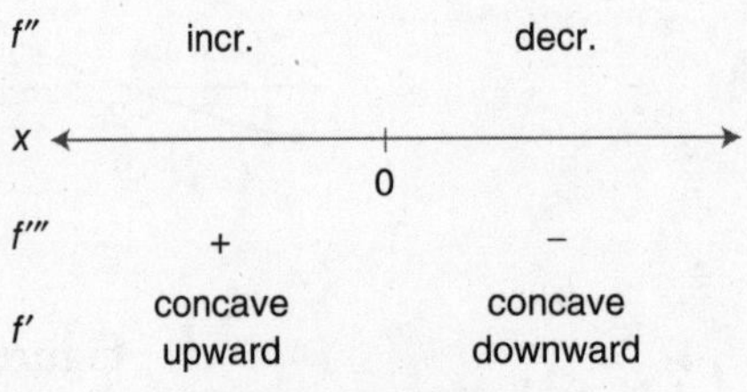

Figure 8.5-9

8.6 Practice Problems

Part A—The use of a calculator is not allowed.

1. If $f(x) = x^3 - x^2 - 2x$, show that the hypotheses of Rolle's Theorem are satisfied on the interval $[-1, 2]$ and find all values of c that satisfy the conclusion of the theorem.

2. Let $f(x) = e^x$. Show that the hypotheses of the Mean Value Theorem are satisfied on the interval $[0, 1]$ and find all values of c that satisfy the conclusion of the theorem.

3. Determine the intervals in which the graph of $f(x) = \dfrac{x^2 + 9}{x^2 - 25}$ is concave upward or downward.

4. Given $f(x) = x + \sin x \; 0 \le x \le 2\pi$, find all points of inflection of f.

5. Show that the absolute minimum of $f(x) = \sqrt{25 - x^2}$ on $[-5, 5]$ is 0 and the absolute maximum is 5.

6. Given the function f in Figure 8.6-1, identify the points where:
(a) $f' < 0$ and $f'' > 0$.
(b) $f' < 0$ and $f'' < 0$.
(c) $f' = 0$.
(d) f'' does not exist.

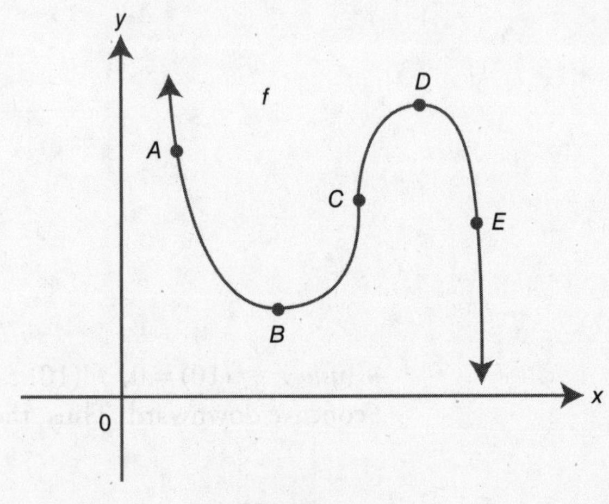

Figure 8.6-1

7. Given the graph of f'' in Figure 8.6-2, determine the values of x at which the function f has a point of inflection.

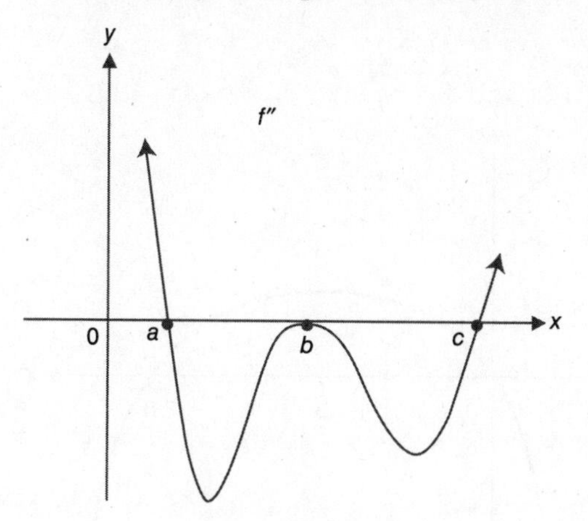

Figure 8.6-2

8. If $f''(x) = x^2(x+3)(x-5)$, find the values of x at which the graph of f has a change of concavity.

9. The graph of f' on $[-3, 3]$ is shown in Figure 8.6-3. Find the values of x on $[-3, 3]$ such that (a) f is increasing and (b) f is concave downward.

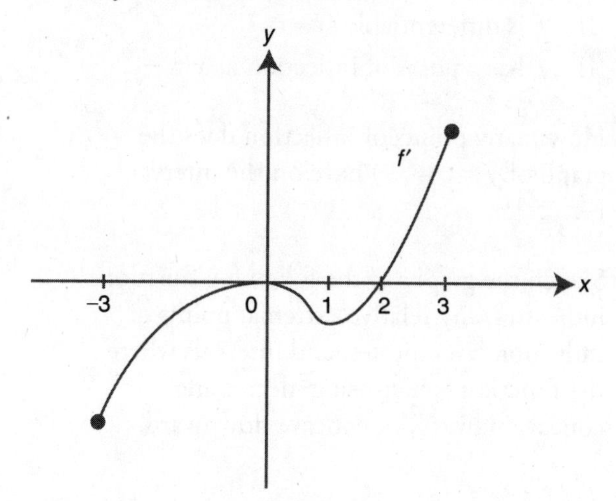

Figure 8.6-3

10. The graph of f is shown in Figure 8.6-4 and f is twice differentiable. Which of the following has the largest value?

 (a) $f(-1)$
 (b) $f'(-1)$
 (c) $f''(-1)$
 (d) $f(-1)$ and $f'(-1)$

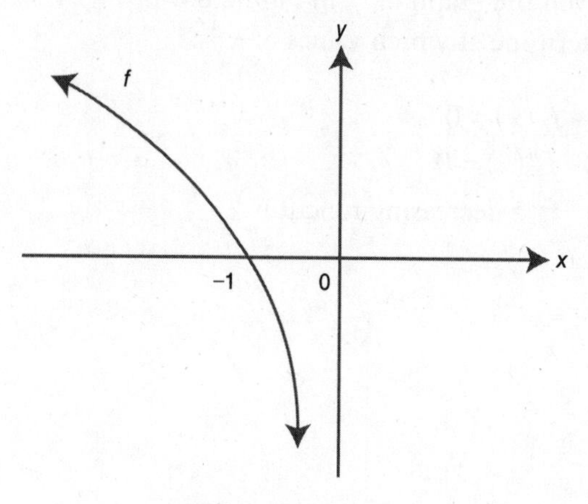

Figure 8.6-4

Sketch the graphs of the following functions indicating any relative and absolute extrema, points of inflection, intervals on which the function is increasing, decreasing, concave upward, or concave downward.

11. $f(x) = x^4 - x^2$

12. $f(x) = \dfrac{x+4}{x-4}$

Part B—Calculators are allowed.

13. Given the graph of f' in Figure 8.6-5, determine at which of the four values of x (x_1, x_2, x_3, x_4) f has:

 (a) the largest value.
 (b) the smallest value.
 (c) a point of inflection.
 (d) and at which of the four values of x does f'' have the largest value.

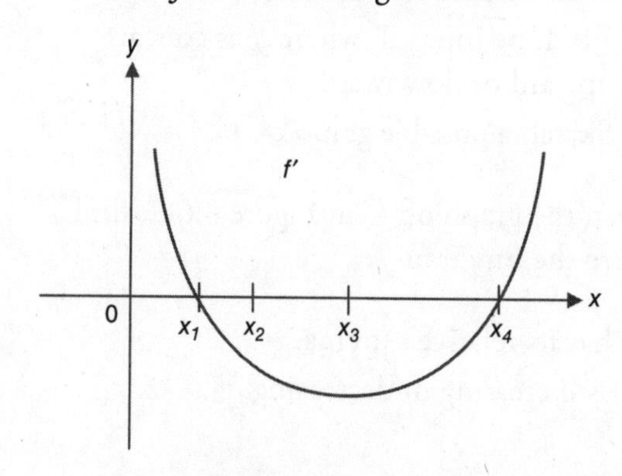

Figure 8.6-5

14. Given the graph of f in Figure 8.6-6, determine at which values of x is:

(a) $f'(x) = 0$
(b) $f''(x) = 0$
(c) f' a decreasing function

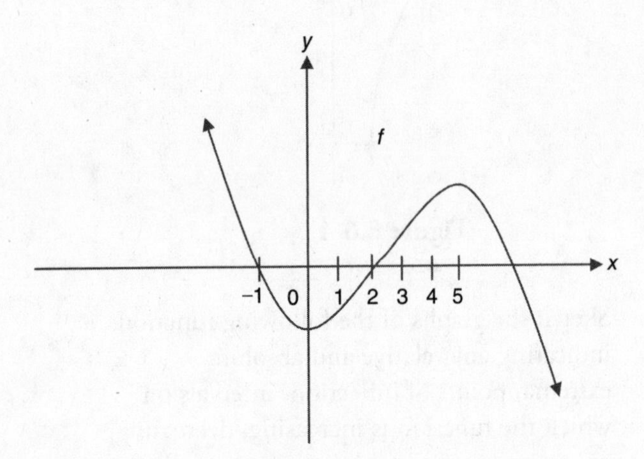

Figure 8.6-6

15. A function f is continuous on the interval $[-2, 5]$ with $f(-2) = 10$ and $f(5) = 6$ and the following properties:

INTERVALS	(−2, 1)	X = 1	(1, 3)	X = 3	(3, 5)
f'	+	0	−	undefined	+
f''	−	0	−	undefined	+

(a) Find the intervals on which f is increasing or decreasing.
(b) Find where f has its absolute extrema.
(c) Find where f has points of inflection.
(d) Find the intervals where f is concave upward or downward.
(e) Sketch a possible graph of f.

16. Given the graph of f' in Figure 8.6-7, find where the function f:

(a) has its relative extrema.
(b) is increasing or decreasing.
(c) has its point(s) of inflection.
(d) is concave upward or downward.
(e) if $f(0) = 1$ and $f(6) = 5$, draw a sketch of f.

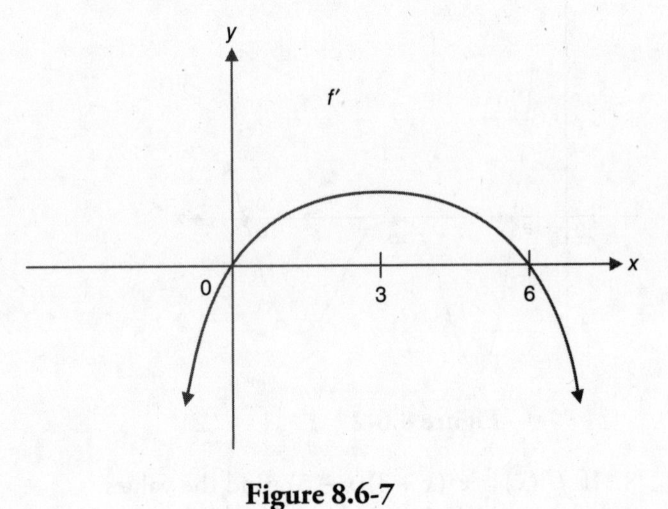

Figure 8.6-7

17. If $f(x) = |x^2 - 6x - 7|$, which of the following statements about f are true?

I. f has a relative maximum at $x = 3$.
II. f is differentiable at $x = 7$.
III. f has a point of inflection at $x = -1$.

18. How many points of inflection does the graph of $y = \cos(x^2)$ have on the interval $[-\pi, \pi]$?

Sketch the graphs of the following functions indicating any relative extrema, points of inflection, asymptotes, and intervals where the function is increasing, decreasing, concave upward, or concave downward.

19. $f(x) = 3e^{-x^2/2}$

20. $f(x) = \cos x \sin^2 x \ [0, 2\pi]$

8.7 Cumulative Review Problems

(Calculator) indicates that calculators are permitted.

21. Find $\frac{dy}{dx}$ if $(x^2 + y^2)^2 = 10xy$.

22. Evaluate $\lim_{x \to 0} \frac{\sqrt{x+9} - 3}{x}$.

23. Find $\frac{d^2y}{dx^2}$ if $y = \cos(2x) + 3x^2 - 1$.

24. (Calculator) Determine the value of k such that the function

$$f(x) = \begin{cases} x^2 - 1, & x \leq 1 \\ 2x + k, & x > 1 \end{cases}$$ is continuous

for all real numbers.

25. A function f is continuous on the interval $[-1, 4]$ with $f(-1) = 0$ and $f(4) = 2$ and the following properties:

INTERVALS	(–1, 0)	X = 0	(0, 2)	X = 2	(2, 4)
f'	+	undefined	+	0	–
f''	+	undefined	–	0	–

(a) Find the intervals on which f is increasing or decreasing.
(b) Find where f has its absolute extrema.
(c) Find where f has points of inflection.
(d) Find intervals on which f is concave upward or downward.
(e) Sketch a possible graph of f.

8.8 Solutions to Practice Problems

Part A—The use of a calculator is not allowed.

1. Condition 1: Since $f(x)$ is a polynomial, it is continuous on $[-1, 2]$.
Condition 2: Also, $f(x)$ is differentiable on $(-1, 2)$ because $f'(x) = 3x^2 - 2x - 2$ is defined for all numbers in $[-1, 2]$.
Condition 3: $f(-1) = f(2) = 0$. Thus, $f(x)$ satisfies the hypotheses of Rolle's Theorem, which means there exists a c in $[-1, 2]$ such that $f'(c) = 0$. Set $f'(x) = 3x^2 - 2x - 2 = 0$. Solve $3x^2 - 2x - 2 = 0$, using the quadratic formula and obtain $x = \frac{1 \pm \sqrt{7}}{3}$. Thus, $x \approx 1.215$ or -0.549 and both values are in the interval $(-1, 2)$. Therefore, $c = \frac{1 \pm \sqrt{7}}{3}$.

2. Condition 1: $f(x) = e^x$ is continuous on $[0, 1]$.
Condition 2: $f(x)$ is differentiable on $(0, 1)$ since $f'(x) = e^x$ is defined for all numbers in $[0, 1]$.
Thus, there exists a number c in $[0, 1]$ such that $f'(c) = \frac{e^1 - e^0}{1 - 0} = (e - 1)$.
Set $f'(x) = e^x = (e - 1)$. Thus, $e^x = (e - 1)$. Take ln of both sides.
$\ln(e^x) = \ln(e - 1) \Rightarrow x = \ln(e - 1)$.
Thus, $x \approx 0.541$, which is in the interval $(0, 1)$. Therefore, $c = \ln(e - 1)$.

3. $f(x) = \frac{x^2 + 9}{x^2 - 25}$,

$$f'(x) = \frac{2x(x^2 - 25) - (2x)(x^2 + 9)}{(x^2 - 25)^2}$$

$$= \frac{-68x}{(x^2 - 25)^2}, \text{ and}$$

$f''(x)$

$$= \frac{-68(x^2-25)^2 - 2(x^2-25)(2x)(-68x)}{(x^2-25)^4}$$

$$= \frac{68(3x^2+25)}{(x^2-25)^3}.$$

Set $f'' > 0$. Since $(3x^2+25) > 0$, $\Rightarrow (x^2-25)^3 > 0 \Rightarrow x^2 - 25 > 0$, $x < -5$ or $x > 5$. Thus, $f(x)$ is concave upward on $(-\infty, -5)$ and $(5, \infty)$ and concave downward on $(-5, 5)$.

4. Step 1: $f(x) = x + \sin x$,
   $f'(x) = 1 + \cos x$,
   $f'' = -\sin x$.

   Step 2: Set $f''(x) = 0 \Rightarrow -\sin x = 0$ or $x = 0, \pi, 2\pi$.

   Step 3: Check the intervals.

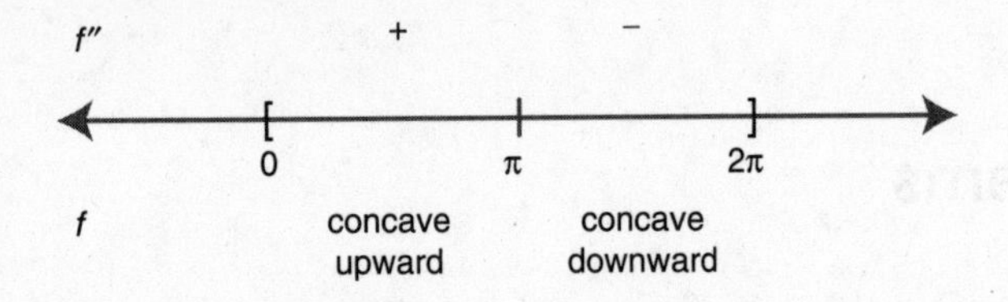

   Step 4: Check for tangent line: At $x = \pi$, $f'(x) = 1 + (-1) \Rightarrow 0$ there is a tangent line at $x = \pi$.

   Step 5: Thus, (π, π) is a point of inflection.

5. Step 1: Rewrite $f(x)$ as $f(x) = (25 - x^2)^{1/2}$.

   Step 2: $f'(x) = \frac{1}{2}(25-x^2)^{-1/2}(-2x)$

   $$= \frac{-x}{(25-x^2)^{1/2}}$$

   Step 3: Find critical numbers. $f'(x) = 0$; at $x = 0$; and $f'(x)$ is undefined at $x = \pm 5$.

   Step 4:

   $f''(x)$

   $$= \frac{(-1)\sqrt{(25-x^2)} - \dfrac{(-2x)(-x)}{2\sqrt{(25-x^2)}}}{(25-x^2)}$$

   $$= \frac{-1}{(25-x^2)^{1/2}} - \frac{x^2}{(25-x^2)^{3/2}}$$

   $f'(0) = 0$ and $f''(0) = \frac{1}{5}$ (and $f(0) = 5$) $\Rightarrow (0, 5)$ is a relative maximum. Since $f(x)$ is continuous on $[-5, 5]$, $f(x)$ has both a maximum and a minimum value on $[-5, 5]$ by the Extreme Value Theorem. And since the point (0,5) is the only relative extremum, it is an absolute extremum. Thus, (0,5) is an absolute maximum point, and 5 is the maximum value. Now we check the endpoints, $f(-5) = 0$ and $f(5) = 0$. Therefore, $(-5, 0)$ and $(5, 0)$ are the lowest points for f on $[-5, 5]$. Thus, 0 is the absolute minimum value.

6. (a) Point A $f' < 0 \Rightarrow$ decreasing and $f'' > 0 \Rightarrow$ concave upward.
   (b) Point E $f' < 0 \Rightarrow$ decreasing and $f'' < 0 \Rightarrow$ concave downward.
   (c) Points B and D $f' = 0 \Rightarrow$ horizontal tangent.
   (d) Point C f'' does not exist $\Rightarrow$ vertical tangent.

7. A change in concavity $\Rightarrow$ a point of inflection. At $x = a$, there is a change of concavity; f'' goes from positive to negative $\Rightarrow$ concavity changes from upward to downward. At $x = c$, there is a change of concavity; f'' goes from negative to positive $\Rightarrow$ concavity changes from downward to upward. Therefore, f has two points of inflection, one at $x = a$ and the other at $x = c$.

8. Set $f''(x) = 0$. Thus, $x^2(x+3)(x-5) = 0 \Rightarrow x = 0$, $x = -3$ or $x = 5$. (See Figure 8.8-1.)

   Thus, f has a change of concavity at $x = -3$ and at $x = 5$.

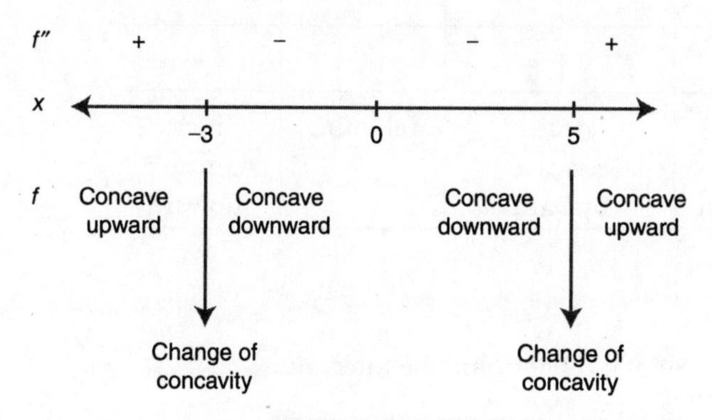

Figure 8.8-1

9. See Figure 8.8-2.
   Thus, f is increasing on [2, 3] and concave downward on (0, 1).

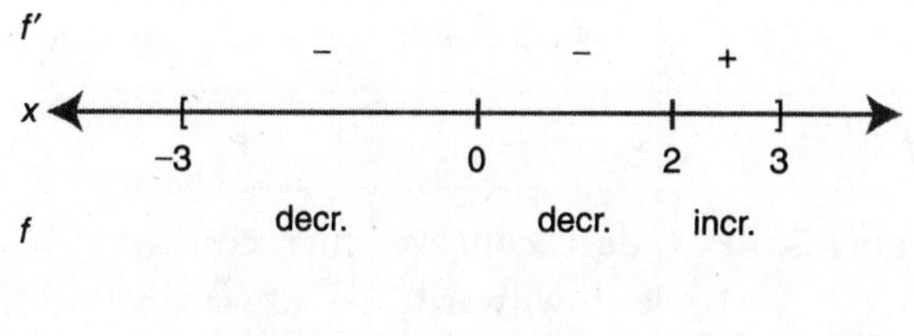

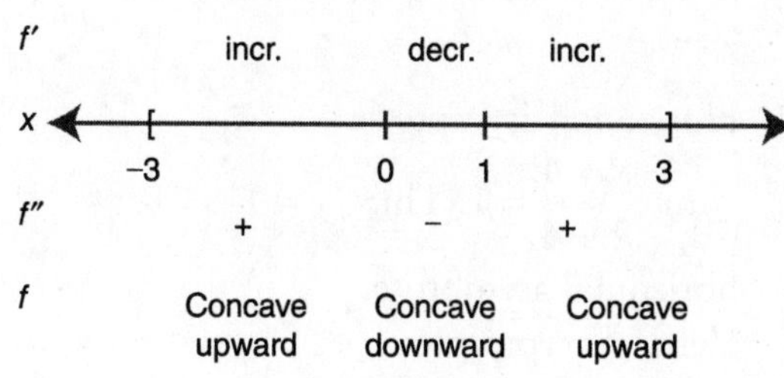

Figure 8.8-2

10. The correct answer is (A).
    $f(-1) = 0$; $f'(0) < 0$ since f is decreasing and $f''(-1) < 0$ since f is concave downward. Thus, $f(-1)$ has the largest value.

11. Step 1: Domain: all real numbers.

    Step 2: Symmetry: Even function ($f(x) = f(-x)$); symmetrical with respect to the y-axis.

    Step 3: $f'(x) = 4x^3 - 2x$ and $f''(x) = 12x^2 - 2$.

    Step 4: Find the critical numbers: $f'(x)$ is defined for all real numbers. Set $f'(x) = 4x^3 - 2x = 0 \Rightarrow 2x(2x^2 - 1) = 0 \Rightarrow x = 0$ or $x = \pm\sqrt{1/2}$.
    Possible points of inflection: $f''(x)$ is defined for all real numbers. Set $f''(x) = 12x^2 - 2 = 0 \Rightarrow 2(6x^2 - 1) = 0 \Rightarrow x = \pm\sqrt{1/6}$.

    Step 5: Determine the intervals:

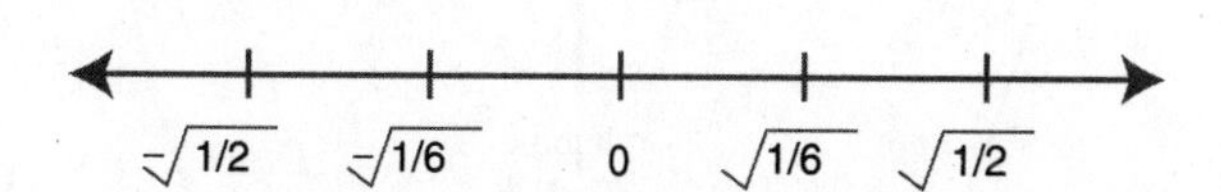

    The intervals are: $(-\infty, -\sqrt{1/2})$, $(-\sqrt{1/2}, -\sqrt{1/6})$, $(-\sqrt{1/6}, 0)$, $(0, \sqrt{1/6})$, $(\sqrt{1/6}, \sqrt{1/2})$, and $(\sqrt{1/2}, \infty)$.
    Since $f'(x)$ is symmetrical with respect to the y-axis, you only need to examine half of the intervals.

    Step 6: Set up a table (See Table 8.8-1). The function has an absolute minimum value of (−1/4) and no absolute maximum value.

Table 8.8-1

INTERVALS	$X=0$	$(0, \sqrt{1/6})$	$X=\sqrt{1/6}$	$(\sqrt{1/6}, \sqrt{1/2})$	$X=\sqrt{1/2}$	$(\sqrt{1/2}, \infty)$
$f(x)$	0		−5/36		−1/4	
$f'(x)$	0	−	−	−	0	+
$f''(x)$	−	−	0	+	+	+
conclusion	rel. max.	decr. concave downward	decr. pt. of inflection	decr. concave upward	rel. min.	incr. concave upward

Step 7: Sketch the graph. (See Figure 8.8-3.)

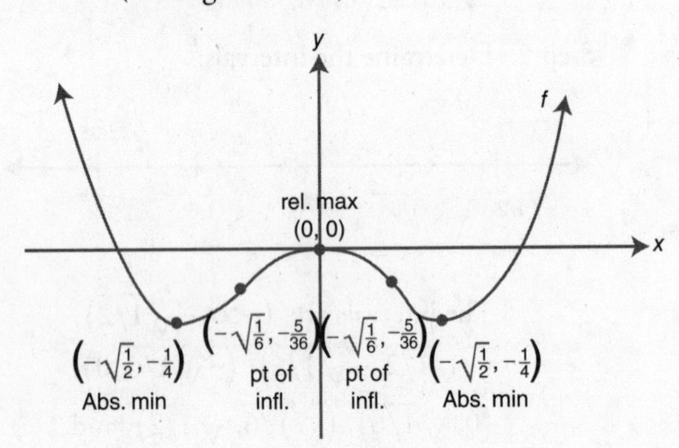

Figure 8.8-3

12. Step 1: Domain: all real numbers $x \neq 4$.

Step 2: Symmetry: none.

Step 3: Find f' and f''.

$$f'(x) = \frac{(1)(x-4)-(1)(x+4)}{(x-4)^2}$$

$$= \frac{-8}{(x-4)^2}, \quad f''(x) = \frac{16}{(x-4)^3}$$

Step 4: Find the critical numbers:
$f'(x) \neq 0$ and
$f'(x)$ is undefined at $x = 4$.

Step 5: Determine the intervals.

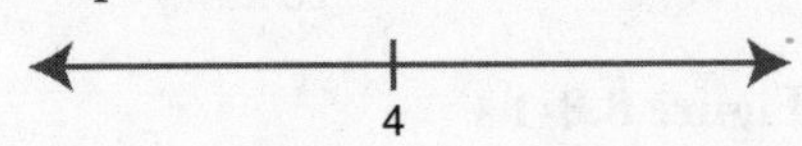

Intervals are $(-\infty, 4)$ and $(4, \infty)$.

Step 6: Set up a table as below:

INTERVALS	$(-\infty, 4)$	$(4, \infty)$
f'	−	−
f''	−	+
conclusion	decr. concave downward	incr. concave upward

Step 7: Horizontal asymptote:
$\lim_{x\to\pm\infty} \frac{x+4}{x-4} = 1$. Thus, $y = 1$ is a horizontal asymptote.
Vertical asymptote:
$\lim_{x\to 4^+} \frac{x+4}{x-4} = \infty$ and
$\lim_{x\to 4^-} \frac{x+4}{x-4} = -\infty$; Thus, $x = 4$ is a vertical asymptote.

Step 8: Determine the intercepts:
x-intercept: Set $f'(x) = 0$
$\Rightarrow x + 4 = 0$; $x = -4$.
y-intercept: Set $x = 0$
$\Rightarrow f(x) = -1$.

Step 9: Sketch the graph.
(See Figure 8.8-4.)

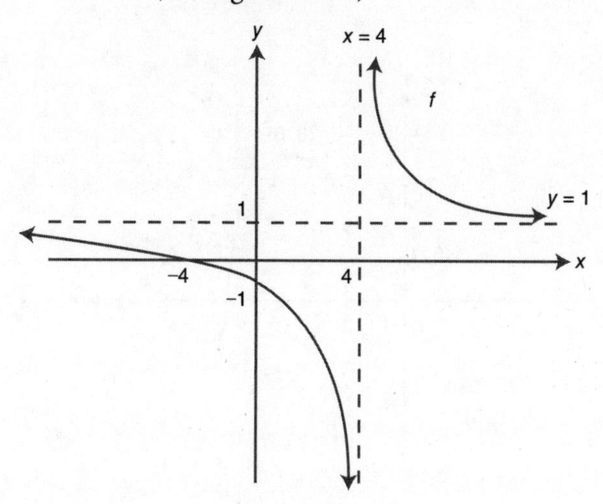

Figure 8.8-4

13. (a)

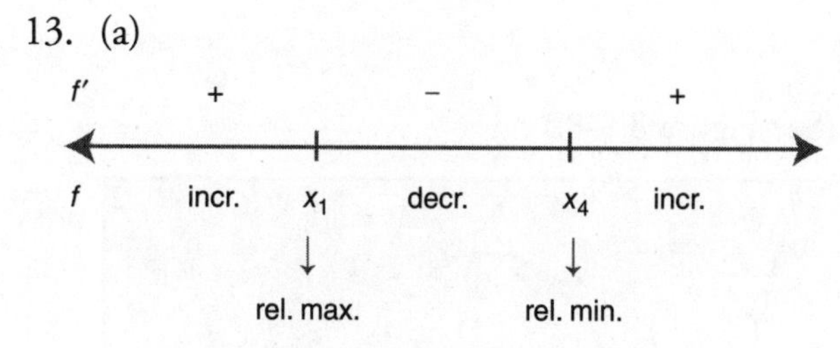

The function f has the largest value (of the four choices) at $x = x_1$. (See Figure 8.8-5.)

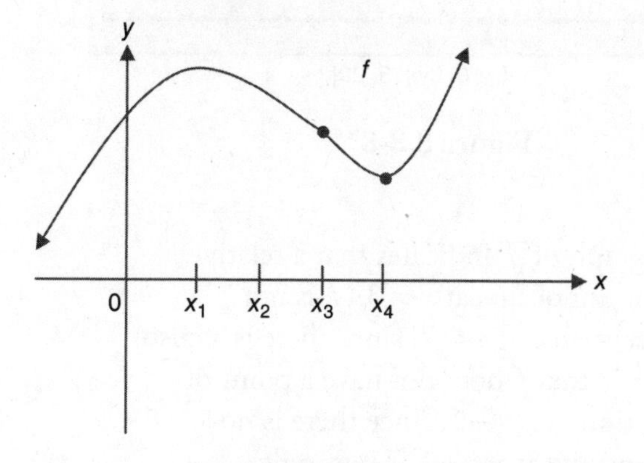

Figure 8.8-5

(b) And f has the smallest value at $x = x_4$.
(c)

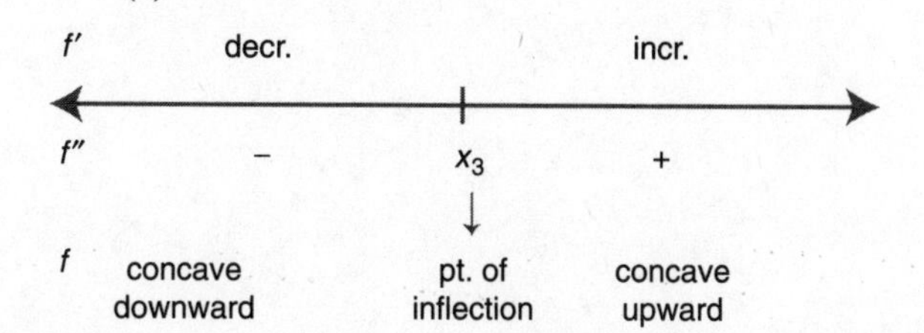

A change of concavity occurs at $x = x_3$, and $f'(x_3)$ exists, which implies there is a tangent to f at $x = x_3$. Thus, at $x = x_3$, f has a point of inflection.

(d) The function f'' represents the slope of the tangent to f'. The slope of the tangent to f' is the largest at $x = x_4$.

14. (a) Since $f'(x)$ represents the slope of the tangent, $f'(x) = 0$ at $x = 0$, and $x = 5$.

(b) At $x = 2$, f has a point of inflection, which implies that if $f''(x)$ exists then $f''(x) = 0$. Since $f'(x)$ is differentiable for all numbers in the domain, $f''(x)$ exists, and $f''(x) = 0$ at $x = 2$.

(c) Since the function f is concave downward on $(2, \infty)$, $f'' < 0$ on $(2, \infty)$, which implies f' is decreasing on $(2, \infty)$.

15. (a) The function f is increasing on the intervals $(-2, 1)$ and $(3, 5)$ and decreasing on $(1, 3)$.

(b) The absolute maximum occurs at $x = 1$, since it is a relative maximum, $f(1) > f(-2)$ and $f(5) < f(-2)$. Similarly, the absolute minimum occurs at $x = 3$, since it is a relative minimum, and $f(3) < f(5) < f(-2)$.

(c) No point of inflection. (Note that at $x = 3$, f has a cusp.)

Note that some textbooks define a *point of inflection* as a point where the concavity changes and do not require the existence of a tangent. In that case, at $x = 3$, f has a point of inflection.

(d) Concave upward on $(3, 5)$ and concave downward on $(-2, 3)$.

(e) A possible graph is shown in Figure 8.8-6.

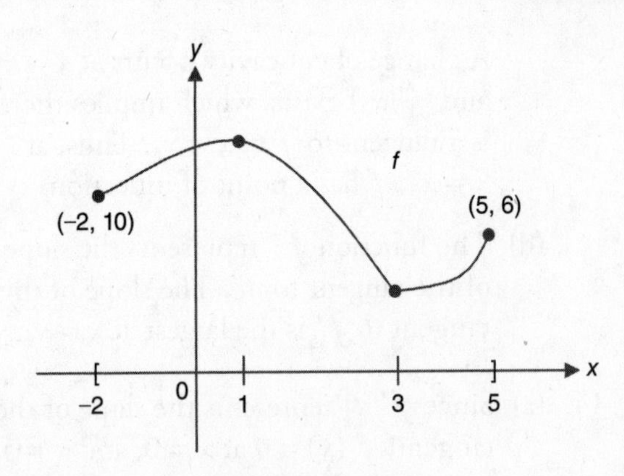

Figure 8.8-6

16. (a)

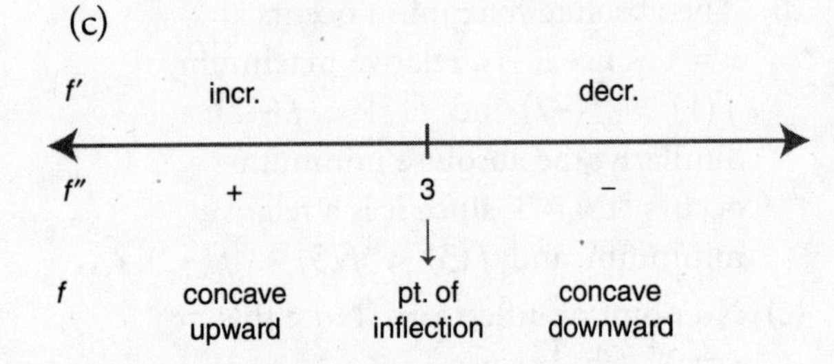

The function f has its relative minimum at $x = 0$ and its relative maximum at $x = 6$.

(b) The function f is increasing on $[0, 6]$ and decreasing on $(-\infty, 0]$ and $[6, \infty)$.

(c)

f′ incr. decr.

f″ + 3 –

f concave upward pt. of inflection concave downward

Since $f'(3)$ exists and a change of concavity occurs at $x = 3$, f has a point of inflection at $x = 3$.

(d) Concave upward on $(-\infty, 3)$ and downward on $(3, \infty)$.

(e) Sketch a graph. (See Figure 8.8-7.)

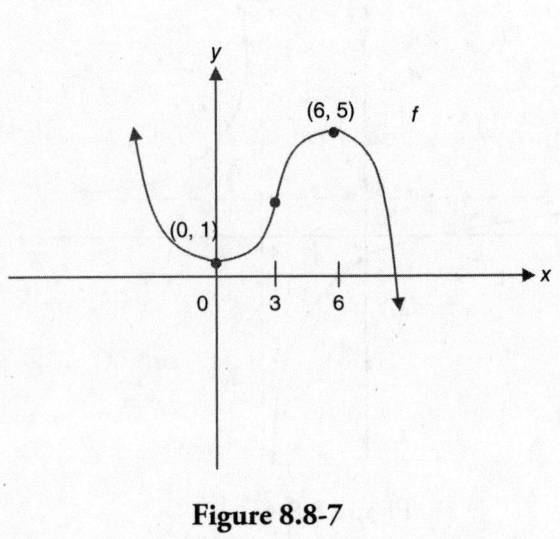

Figure 8.8-7

17. (See Figure 8.8-8.)

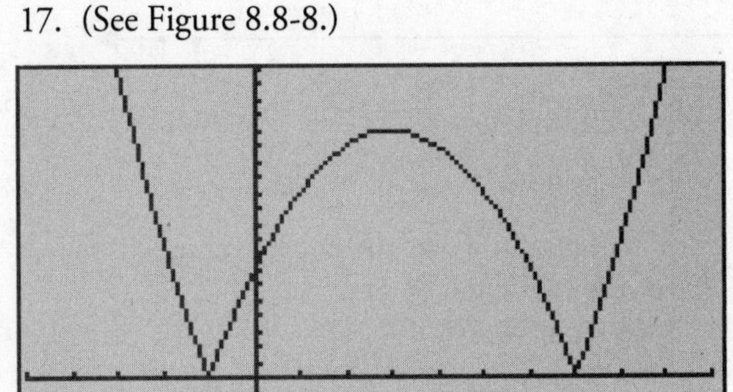

[–5, 10] by [–5, 20]

Figure 8.8-8

The graph of f indicates that a relative maximum occurs at $x = 3$, f is not differentiable at $x = 7$, since there is a cusp at $x = 7$ and f does not have a point of inflection at $x = -1$, since there is no tangent line at $x = -1$. Thus, only statement I is true.

18. (See Figure 8.8-9.)

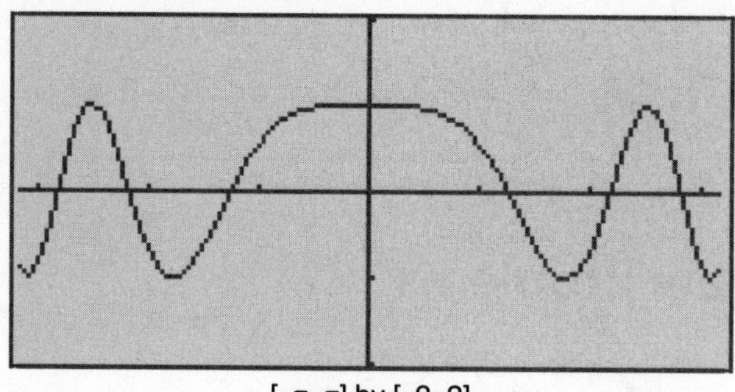

$[-\pi, \pi]$ by $[-2, 2]$

Figure 8.8-9

Enter $y1 = \cos(x^2)$
Using the [*Inflection*] function of your calculator, you obtain three points of inflection on $[0, \pi]$. The points of inflection occur at $x = 1.35521, 2.1945,$ and 2.81373. Since $y_1 = \cos(x^2)$ is an even function, there is a total of 6 points of inflection on $[-\pi, \pi]$. An alternate solution is to enter
$y2 = \frac{d^2}{dx^2}\left(y_1(x), x, 2\right)$. The graph of y_2 indicates that there are 6 zeros on $[-\pi, \pi]$.

19. Enter $y1 = 3 * e \wedge (-x \wedge 2/2)$. Note that the graph has a symmetry about the y-axis. Using the functions of the calculator, you will find:

(a) a relative maximum point at (0, 3), which is also the absolute maximum point;
(b) points of inflection at (−1, 1.819) and (1, 1.819);
(c) $y = 0$ (the x-axis) a horizontal asymptote;
(d) y_1 increasing on $(-\infty, 0]$ and decreasing on $[0, \infty)$; and
(e) y_1 concave upward on $(-\infty, -1)$ and $(1, \infty)$ and concave downward on (−1, 1). (See Figure 8.8-10.)

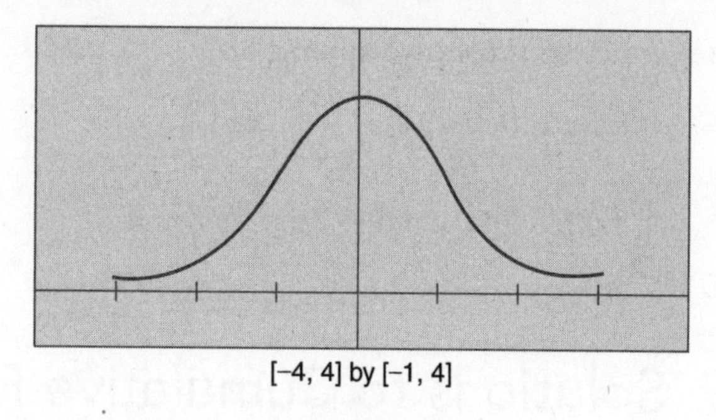

[−4, 4] by [−1, 4]

Figure 8.8-10

20. (See Figure 8.8-11.) Enter $y1 = \cos(x) * (\sin(x)) \wedge 2$. A fundamental domain of y_1 is $[0, 2\pi]$. Using the functions of the calculator, you will find:

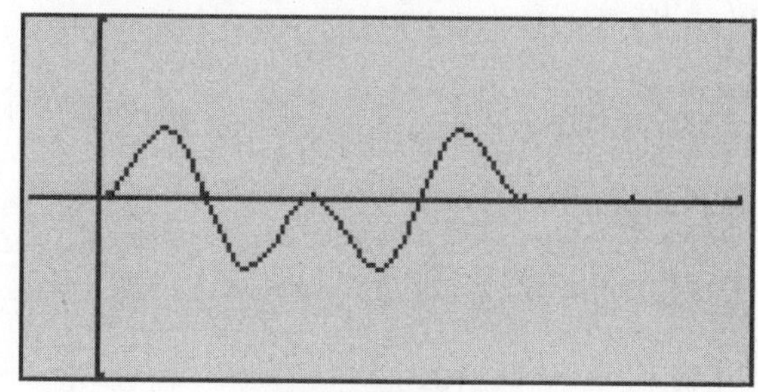

[−1, 9.4] by [−1, 1]

Figure 8.8-11

(a) relative maximum points at (0.955, 0.385), $(\pi, 0)$, and (5.328, 0.385), and relative minimum points at (2.186, −0.385) and (4.097, −0.385);
(b) points of inflection at (0.491, 0.196), $\left(\frac{\pi}{2}, 0\right)$, (2.651, −0.196), (3.632, −0.196), $\left(\frac{3\pi}{2}, 0\right)$, and (5.792, 0.196);
(c) no asymptote;
(d) function is increasing on intervals (0, 0.955), $(2.186, \pi)$, and (4.097, 5.328), and decreasing on intervals (0.955, 2.186), $(\pi, 4.097)$, and $(5.328, 2\pi)$;

(e) function is concave upward on intervals $(0, 0.491)$, $\left(\frac{\pi}{2}, 2.651\right)$, $\left(3.632, \frac{3\pi}{2}\right)$, and $(5.792, 2\pi)$, and concave downward on the intervals $\left(0.491, \frac{\pi}{2}\right)$, $(2.651, 3.632)$, and $\left(\frac{3\pi}{2}, 5.792\right)$.

8.9 Solutions to Cumulative Review Problems

21. $(x^2+y^2)^2 = 10xy$

$$2\left(x^2+y^2\right)\left(2x+2y\frac{dy}{dx}\right) = 10y + (10x)\frac{dy}{dx}$$

$$4x\left(x^2+y^2\right)+4y\left(x^2+y^2\right)\frac{dy}{dx} = 10y + (10x)\frac{dy}{dx}$$

$$4y\left(x^2+y^2\right)\frac{dy}{dx} - (10x)\frac{dy}{dx} = 10y - 4x\left(x^2+y^2\right)$$

$$\frac{dy}{dx}\left(4y\left(x^2+y^2\right)-10x\right) = 10y - 4x\left(x^2+y^2\right)$$

$$\frac{dy}{dx} = \frac{10y-4x\left(x^2+y^2\right)}{4y\left(x^2+y^2\right)-10x} = \frac{5y-2x\left(x^2+y^2\right)}{2y\left(x^2+y^2\right)-5x}$$

22. Substituting $x = 0$ in the expression $\frac{\sqrt{x+9}-3}{x}$ leads to $\frac{0}{0}$, an indeterminant form. Apply *L'Hôpital's* Rule and you have

$$\lim_{x\to 0}\frac{\frac{1}{2}(x+9)^{-\frac{1}{2}}(1)}{1}, \text{ or } \frac{1}{2}(0+9)^{-\frac{1}{2}} = \frac{1}{6}.$$

Alternatively,

$$\lim_{x\to 0}\frac{\sqrt{x+9}-3}{x} = \lim_{x\to 0}\frac{\left(\sqrt{x+9}-3\right)}{x}\cdot\frac{\left(\sqrt{x+9}+3\right)}{\left(\sqrt{x+9}+3\right)}$$

$$= \lim_{x\to 0}\frac{(x+9)-9}{x\left(\sqrt{x+9}+3\right)}$$

$$= \lim_{x\to 0}\frac{x}{x\left(\sqrt{x+9}+3\right)}$$

$$= \lim_{x\to 0}\frac{1}{\sqrt{x+9}+3} = \frac{1}{\sqrt{0+9}+3}$$

$$= \frac{1}{3+3} = \frac{1}{6}$$

23. $y = \cos(2x) + 3x^2 - 1$

$$\frac{dy}{dx} = [-\sin(2x)](2) + 6x = -2\sin(2x) + 6x$$

$$\frac{d^2y}{dx^2} = -2(\cos(2x))(2) + 6 = -4\cos(2x) + 6$$

24. (Calculator) The function f is continuous everywhere for all values of k except possibly at $x = 1$. Checking with the three conditions of continuity at $x = 1$:

(1) $f(1) = (1)^2 - 1 = 0$

(2) $\lim_{x\to 1^+}(2x+k) = 2+k$, $\lim_{x\to 1^-}\left(x^2-1\right) = 0$; thus, $2 + k = 0 \Rightarrow k = -2$. Since $\lim_{x\to 1^+} f(x) = \lim_{x\to 1^-} f(x) = 0$, therefore, $\lim_{x\to 1} f(x) = 0$.

(3) $f(1) = \lim_{x\to 1} f(x) = 0$. Thus, $k = -2$.

25. (a) Since $f' > 0$ on $(-1, 0)$ and $(0, 2)$, the function f is increasing on the intervals $[-1, 0]$ and $[0, 2]$. Since $f' < 0$ on $(2, 4)$, f is decreasing on $[2, 4]$.

(b) The absolute maximum occurs at $x = 2$, since it is a relative maximum and it is the only relative extremum on $(-1, 4)$. The absolute minimum occurs at $x = -1$, since $f(-1) < f(4)$ and the function has no relative minimum on $[-1, 4]$.

(c) A change of concavity occurs at $x = 0$. However, $f'(0)$ is undefined, which implies f may or may not have a tangent at $x = 0$. Thus, f may or may not have a point of inflection at $x = 0$.

(d) Concave upward on $(-1, 0)$ and concave downward on $(0, 4)$.

(e) A possible graph is shown in Figure 8.9-1.

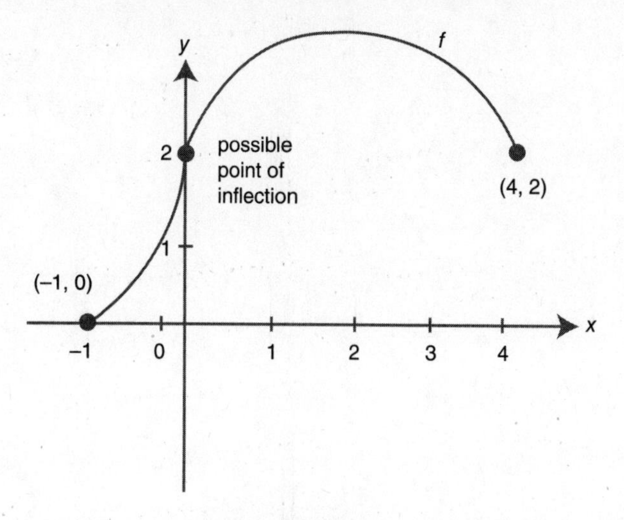

Figure 8.9-1

Big Idea 2: Derivatives
Applications of Derivatives

IN THIS CHAPTER

Summary: Two of the most common applications of derivatives involve solving related rate problems and applied maximum and minimum problems. In this chapter, you will learn the general procedures for solving these two types of problems and to apply these procedures to examples. Both related rate and applied maximum and minimum problems appear often on the AP Calculus AB exam.

Key Ideas

- General Procedure for Solving Related Rate Problems
- Common Related Rate Problems
- Inverted Cone, Shadow, and Angle of Elevation Problems
- General Procedure for Solving Applied Maximum and Minimum Problems
- Distance, Area, Volume, and Business Problems

9.1 Related Rate

Main Concepts: General Procedure for Solving Related Rate Problems, Common Related Rate Problems, Inverted Cone (Water Tank) Problem, Shadow Problem, Angle of Elevation Problem

General Procedure for Solving Related Rate Problems

1. Read the problem and, if appropriate, draw a diagram.
2. Represent the given information and the unknowns by mathematical symbols.
3. Write an equation involving the rate of change to be determined. (If the equation contains more than one variable, it may be necessary to reduce the equation to one variable.)
4. Differentiate each term of the equation with respect to time.
5. Substitute all known values and known rates of change into the resulting equation.
6. Solve the resulting equation for the desired rate of change.
7. Write the answer and indicate the units of measure.

Common Related Rate Problems

Example 1

When the area of a square is increasing twice as fast as its diagonals, what is the length of a side of the square?

Let z represent the diagonal of the square. The area of a square is $A = \frac{z^2}{2}$.

$$\frac{dA}{dt} = 2z\frac{dz}{dt}\left(\frac{1}{2}\right) = z\frac{dz}{dt}$$

Since $\frac{dA}{dt} = 2\frac{dz}{dt}$, $2\frac{dz}{dt} = z\frac{dz}{dt} \Rightarrow z = 2.$

Let s be a side of the square. Since the diagonal $z = 2$, then $s^2 + s^2 = z^2$
$\Rightarrow 2s^2 = 4 \Rightarrow s^2 = 4 \Rightarrow s^2 = 2$ or $s = \sqrt{2}$.

Example 2

Find the surface area of a sphere at the instant when the rate of increase of the volume of the sphere is nine times the rate of increase of the radius.

Volume of a sphere: $V = \frac{4}{3}\pi r^3$; Surface area of a sphere: $S = 4\pi r^2$.

$$V = \frac{4}{3}\pi r^3; \quad \frac{dV}{dt} = 4r^2\frac{dr}{dt}$$

Since $\frac{dV}{dt} = 9\frac{dr}{dt}$, you have $9\frac{dr}{dt} = 4\pi r^2\frac{dr}{dt}$ or $9 = 4\pi r^2$.

Since $S = 4\pi r^2$, the surface area is $S = 9$ square units.

Note: At $9 = 4\pi r^2$, you could solve for r and obtain $r^2 = \frac{9}{4\pi}$ or $r = \frac{3}{2}\frac{1}{\sqrt{\pi}}$. You could then substitute $r = \frac{3}{2}\frac{1}{\sqrt{\pi}}$ into the formula for surface area $S = 4\pi r^2$ and obtain 9. These steps are of course correct but not necessary.

Example 3

The height of a right circular cone is always three times the radius. Find the volume of the cone at the instant when the rate of increase of the volume is twelve times the rate of increase of the radius.

Let r, h be the radius and height of the cone, respectively.

Since $h = 3r$, the volume of the cone $V = \frac{1}{3}\pi r^2 h = \frac{1}{3}\pi r^2 (3r) = \pi r^3$.

$$V = \pi r^3; \quad \frac{dV}{dt} = 3\pi r^2 \frac{dr}{dt}.$$

When $\frac{dV}{dt} = 12\frac{dr}{dt}$, $12\frac{dr}{dt} = 3\pi r^2 \frac{dr}{dt} \Rightarrow 4 = \pi r^2 \Rightarrow r = \frac{2}{\sqrt{\pi}}$.

Thus, $V = \pi r^3 = \pi \left(\frac{2}{\sqrt{\pi}}\right)^3 = \pi \left(\frac{8}{\pi\sqrt{\pi}}\right) = \frac{8}{\sqrt{\pi}}$.

- Go with your first instinct if you are unsure. Usually that is the correct one.

Inverted Cone (Water Tank) Problem

A water tank is in the shape of an inverted cone. The height of the cone is 10 meters, and the diameter of the base is 8 meters as shown in Figure 9.1-1. Water is being pumped into the tank at the rate of 2 m^3/min. How fast is the water level rising when the water is 5 meters deep? (See Figure 9.1-1.)

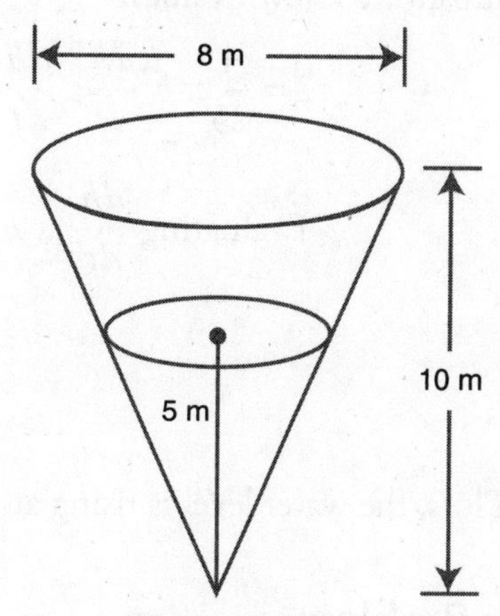

Figure 9.1-1

Solution:

Step 1: Define the variables. Let V be the volume of water in the tank; h be the height of the water level at t minutes; r be the radius of the surface of the water at t minutes; and t be the time in minutes.

Step 2: Given: $\frac{dV}{dt} = 2$ m^3/min. Height = 10 m, diameter = 8 m.

Find: $\frac{dh}{dt}$ at $h = 5$.

Step 3: Set up an equation: $V = \frac{1}{3}\pi r^2 h$.

Using similar triangles, you have $\frac{4}{10} = \frac{r}{h} \Rightarrow 4h = 10r$; or $r = \frac{2h}{5}$.

(See Figure 9.1-2.)

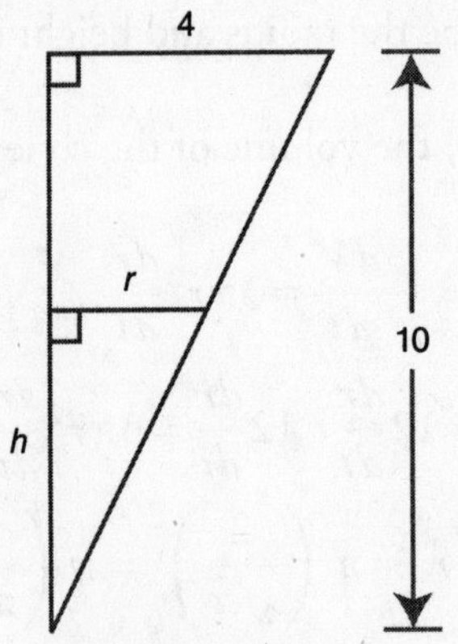

Figure 9.1-2

Thus, you can reduce the equation to one variable:

$$V = \frac{1}{3}\pi\left(\frac{2h}{5}\right)^2 h = \frac{4}{75}\pi h^3.$$

Step 4: Differentiate both sides of the equation with respect to t.

$$\frac{dV}{dt} = \frac{4}{75}\pi(3)h^2\frac{dh}{dt} = \frac{4}{25}\pi h^2\frac{dh}{dt}$$

Step 5: Substitute known values.

$$2 = \frac{4}{25}\pi h^2\frac{dh}{dt};\quad \frac{dh}{dt} = \left(\frac{25}{2}\right)\frac{1}{\pi h^2}\ \text{m/min}$$

$$\text{Evaluating } \frac{dh}{dt} \text{ at } h = 5;\quad \left.\frac{dh}{dt}\right|_{h=5} = \left(\frac{25}{2}\right)\frac{1}{\pi(5)^2}\ \text{m/min}$$

$$= \frac{1}{2\pi}\ \text{m/min.}$$

Step 6: Thus, the water level is rising at $\frac{1}{2\pi}$ m/min when the water is 5 m high.

Shadow Problem

A light on the ground 100 feet from a building is shining at a 6-foot-tall man walking away from the light and toward the building at the rate of 4 ft/sec. How fast is his shadow on the building becoming shorter when he is 40 feet from the building? (See Figure 9.1-3.)

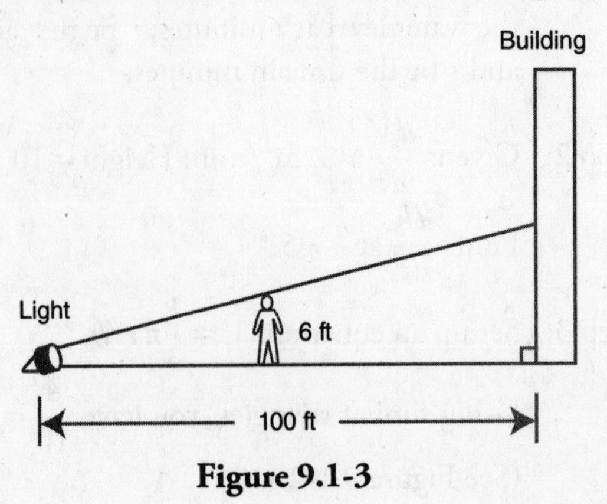

Figure 9.1-3

Solution:

Step 1: Let s be the height of the man's shadow; x be the distance between the man and the light; and t be the time in seconds.

Step 2: Given: $\frac{dx}{dt} = 4$ ft/sec; the man is 6 ft tall; distance between light and building is 100 ft. Find $\frac{ds}{dt}$ at $x = 60$.

Step 3: (See Figure 9.1-4.) Writing an equation using similar triangles, you have:

$$\frac{6}{s} = \frac{x}{100};\; s = \frac{600}{x} = 600x^{-1}$$

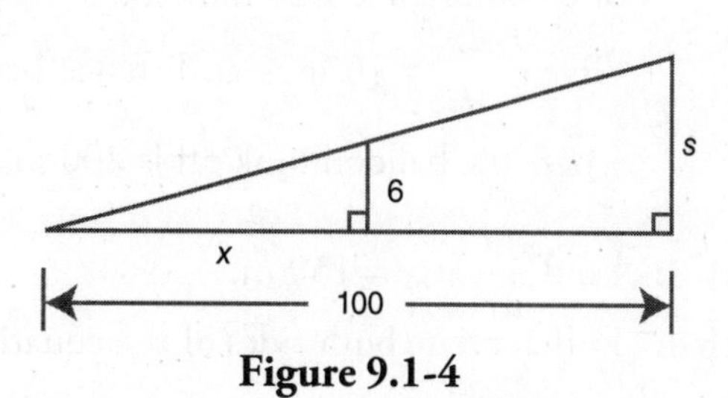

Figure 9.1-4

Step 4: Differentiate both sides of the equation with respect to t.

$$\frac{ds}{dt} = (-1)(600)x^{-2}\frac{dx}{dt} = \frac{-600}{x^2}\frac{dx}{dt} = \frac{-600}{x^2}(4) = \frac{-2400}{x^2} \text{ ft/sec}$$

Step 5: Evaluate $\frac{ds}{dt}$ at $x = 60$.

Note that when the man is 40 ft from the building, x (distance from the light) is 60 ft.

$$\left.\frac{ds}{dt}\right|_{x=60} = \frac{-2400}{(60)^2} \text{ ft/sec} = -\frac{2}{3} \text{ ft/sec}$$

Step 6: The height of the man's shadow on the building is changing at $-\frac{2}{3}$ ft/sec.

- Indicate units of measure, e.g., the velocity is 5 m/sec *or* the volume is 25 in^3.

Angle of Elevation Problem

A camera on the ground 200 meters away from a hot air balloon (also on the ground) records the balloon rising into the sky at a constant rate of 10 m/sec. How fast is the camera's angle of elevation changing when the balloon is 150 m in the air? (See Figure 9.1-5.)

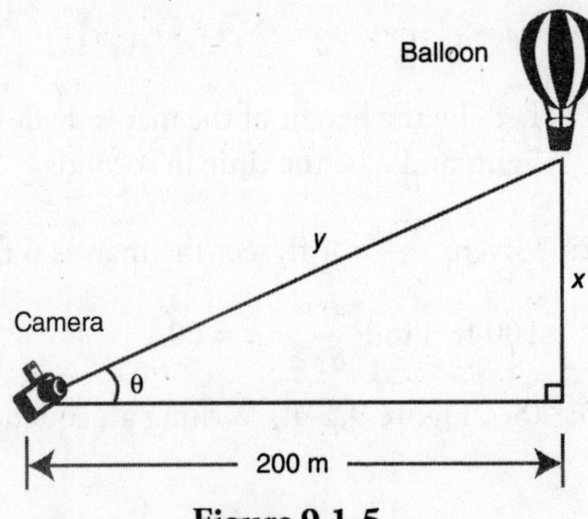

Figure 9.1-5

Step 1: Let x be the distance between the balloon and the ground; θ be the camera's angle of elevation; and t be the time in seconds.

Step 2: Given: $\frac{dx}{dt} = 10$ m/sec; distance between camera and the point on the ground where the balloon took off is 200 m, $\tan\theta = \frac{x}{200}$.

Step 3: Find $\frac{d\theta}{dt}$ at $x = 150$ m.

Step 4: Differentiate both sides of the equation with respect to t.

$$\sec^2\theta \frac{d\theta}{dt} = \frac{1}{200}\frac{dx}{dt}; \quad \frac{d\theta}{dt} = \frac{1}{200}\left(\frac{1}{\sec^2\theta}\right)(10) = \frac{1}{20\sec^2\theta}.$$

Step 5: $\sec\theta = \frac{y}{200}$ and at $x = 150$.

Using the Pythagorean Theorem: $y^2 = x^2 + (200)^2$

$$y^2 = (150)^2 + (200)^2$$

$$y = \pm 250.$$

Since $y > 0$, then $y = 250$. Thus, $\sec\theta = \frac{250}{200} = \frac{5}{4}$.

$$\text{Evaluating } \left.\frac{d\theta}{dt}\right|_{x=150} = \frac{1}{20\sec^2\theta} = \frac{1}{20\left(\frac{5}{4}\right)^2} \text{ radian/sec}$$

$$= \frac{1}{20\left(\frac{5}{4}\right)^2} = \frac{1}{20\left(\frac{25}{16}\right)} = \frac{1}{\frac{125}{4}} = \frac{4}{125} \text{ radian/sec}$$

or .032 radian/sec

= 1.833 deg/sec.

Step 6: The camera's angle of elevation changes at approximately 1.833 deg/sec when the balloon is 150 m in the air.

9.2 Applied Maximum and Minimum Problems

Main Concepts: General Procedure for Solving Applied Maximum and Minimum Problems, Distance Problem, Area and Volume Problems, Business Problems

General Procedure for Solving Applied Maximum and Minimum Problems

Steps:

1. Read the problem carefully and if appropriate, draw a diagram.
2. Determine what is given and what is to be found, and represent these quantities by mathematical symbols.
3. Write an equation that is a function of the variable representing the quantity to be maximized or minimized.
4. If the equation involves other variables, reduce the equation to a single variable that represents the quantity to be maximized or minimized.
5. Determine the appropriate interval for the equation (i.e., the appropriate domain for the function) based on the information given in the problem.
6. Differentiate to obtain the first derivative and to find critical numbers.
7. Apply the First Derivative Test or the Second Derivative Test by finding the second derivative.
8. Check the function values at the endpoints of the interval.
9. Write the answer(s) to the problem and, if given, indicate the units of measure.

Distance Problem

Find the shortest distance between the point A (19, 0) and the parabola $y = x^2 - 2x + 1$.

Solution:

Step 1: Draw a diagram. (See Figure 9.2-1.)

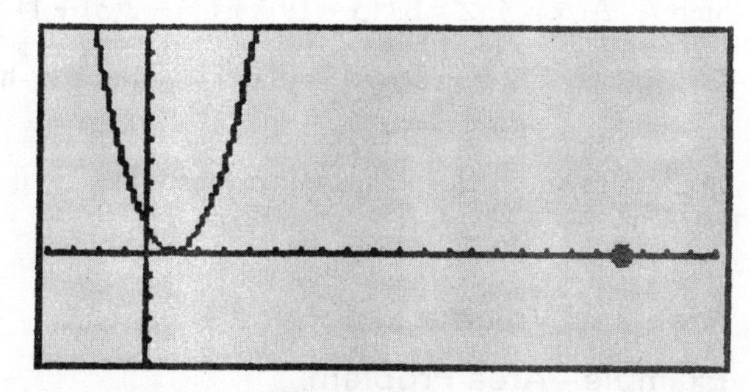

Figure 9.2-1

Step 2: Let $P(x, y)$ be the point on the parabola and let Z represent the distance between points $P(x, y)$ and $A(19, 0)$.

Step 3: Using the distance formula,

$$Z=\sqrt{(x-19)^2+(y-0)^2}=\sqrt{(x-19)^2+(x^2-2x+1-0)^2}$$
$$=\sqrt{(x-19)^2+\left((x-1)^2\right)^2}=\sqrt{(x-19)^2+(x-1)^4}.$$

(Special case: In distance problems, the distance and the square of the distance have the same maximum and minimum points.) Thus, to simplify computations, let $L=Z^2=(x-19)^2+(x-1)^4$. The domain of L is $(-\infty,\infty)$.

Step 4: Differentiate: $\frac{dL}{dx}=2(x-19)(1)+4(x-1)^3(1)$
$=2x-38+4x^3-12x^2+12x-4=4x^3-12x^2+14x-42$
$=2(2x^3-6x^2+7x-21).$

$\frac{dL}{dx}$ is defined for all real numbers.

Set $\frac{dL}{dx}=0$; $2x^3-6x^2+7x-21=0$. The factors of 21 are $\pm1, \pm3, \pm7$, and ± 21.

Using Synthetic Division, $2x^3-6x^2+7x-21=(x-3)(2x^2+7)=0 \Rightarrow x=3$.

Thus, the only critical number is $x=3$.

(Note: Step 4 could have been done using a graphing calculator.)

Step 5: Apply the First Derivative Test.

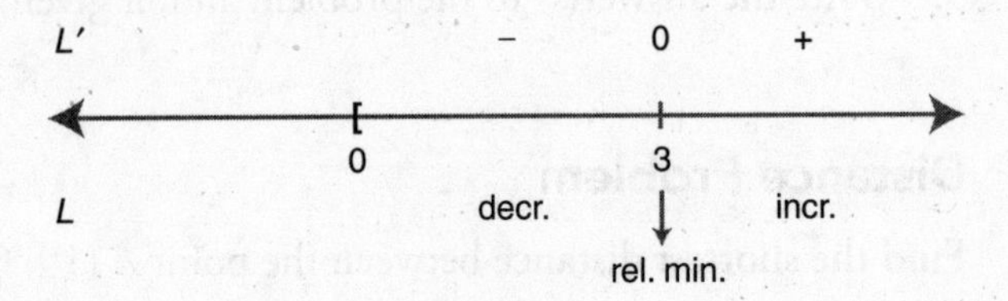

Step 6: Since $x=3$ is the only relative minimum point in the interval, it is the absolute minimum.

Step 7: At $x=3$, $Z=\sqrt{(3-19)^2+(3^2-2(3)+1)^2}=\sqrt{(-16)^2+(4)^2}$
$=\sqrt{272}=\sqrt{16}\sqrt{17}=4\sqrt{17}$. Thus, the shortest distance is $4\sqrt{17}$.

• Simplify numeric or algebraic expressions only if the question asks you to do so.

Area and Volume Problems

Example—Area Problem

The graph of $y=-\frac{1}{2}x+2$ encloses a region with the x-axis and y-axis in the first quadrant. A rectangle in the enclosed region has a vertex at the origin and the opposite vertex on the graph of $y=-\frac{1}{2}x+2$. Find the dimensions of the rectangle so that its area is a maximum.

Solution:

Step 1: Draw a diagram. (See Figure 9.2-2.)

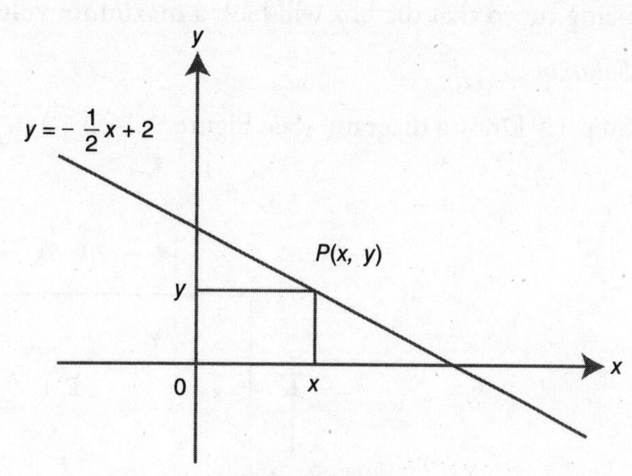

Figure 9.2-2

Step 2: Let $P(x, y)$ be the vertex of the rectangle on the graph of $y = -\frac{1}{2}x + 2$.

Step 3: Thus, the area of the rectangle is:

$$A = xy \text{ or } A = x\left(-\frac{1}{2}x + 2\right) = -\frac{1}{2}x^2 + 2x.$$

The domain of A is $[0, 4]$.

Step 4: Differentiate:

$$\frac{dA}{dx} = -x + 2.$$

Step 5: $\frac{dA}{dx}$ is defined for all real numbers.

Set $\frac{dA}{dx} = 0 \Rightarrow -x + 2 = 0$; $x = 2$.

$A(x)$ has one critical number $x = 2$.

Step 6: Apply the Second Derivative Test:

$\frac{d^2A}{dx^2} = -1 \Rightarrow A(x)$ has a relative maximum point at $x = 2$; $A(2) = 2$.

Since $x = 2$ is the only relative maximum, it is the absolute maximum. (Note that at the endpoints: $A(0) = 0$ and $A(4) = 0$.)

Step 7: At $x = 2$, $y = -\frac{1}{2}(2) + 2 = 1$.

Therefore, the length of the rectangle is 2, and its width is 1.

Example—Volume Problem (with calculator)

If an open box is to be made using a square sheet of tin, 20 inches by 20 inches, by cutting a square from each corner and folding the sides up, find the length of a side of the square being cut so that the box will have a maximum volume.

Solution:

Step 1: Draw a diagram. (See Figure 9.2-3.)

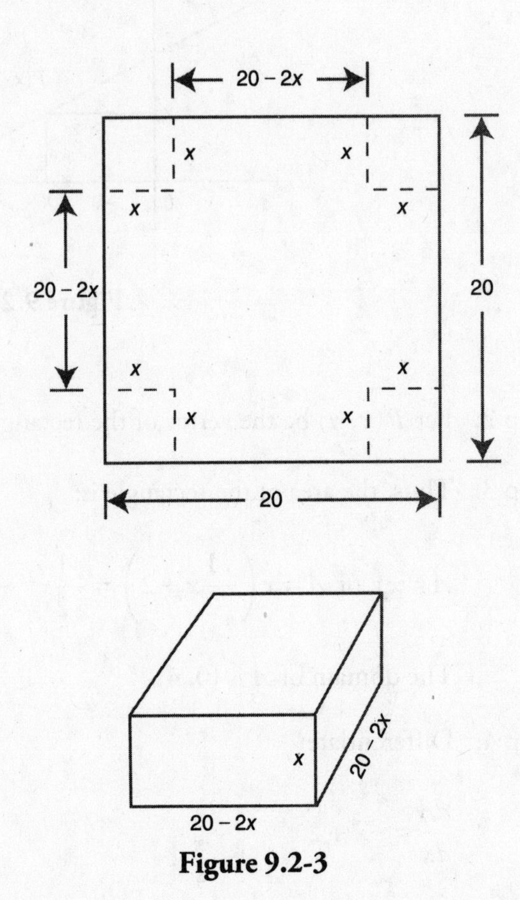

Figure 9.2-3

Step 2: Let x be the length of a side of the square to be cut from each corner.

Step 3: The volume of the box is $V(x) = x(20 - 2x)(20 - 2x)$.
The domain of V is $[0, 10]$.

Step 4: Differentiate $V(x)$.
Enter $d(x * (20 - 2x) * (20 - 2x),\ x)$ and we have $4(x - 10)(3x - 10)$.

Step 5: $V'(x)$ is defined for all real numbers.
Set $V'(x) = 0$ by entering: [*Solve*] $(4(x - 10)(3x - 10) = 0,\ x)$, and obtain $x = 10$ or $x = \frac{10}{3}$. The critical numbers of $V(x)$ are $x = 10$ and $x = \frac{10}{3}$. $V(10) = 0$ and $V\left(\frac{10}{3}\right) = 592.59$. Since $V(10) = 0$, you need to test only $x = \frac{10}{3}$.

Step 6: Apply the Second Derivative Test. Enter $d(x*(20-2x)*(20-2x), x, 2)|x=\frac{10}{3}$ and obtain –80. Thus, $V\left(\frac{10}{3}\right)$ is a relative maximum. Since it is the only relative maximum on the interval, it is the absolute maximum. (Note at the other endpoint $x=0$, $V(0)=0$.)

Step 7: Therefore, the length of a side of the square to be cut is $x=\frac{10}{3}$.

- The formula for the average value of a function f from $x=a$ to $x=b$ is $\frac{1}{b-a}\int_a^b f(x)dx$.

Business Problems

Summary of Formulas

1. $P=R-C$: Profit = Revenue – Cost
2. $R=xp$: Revenue = (Units Sold)(Price Per Unit)
3. $\overline{C}=\frac{C}{x}$: Average Cost $=\frac{\text{Total Cost}}{\text{Units produced/Sold}}$
4. $\frac{dR}{dx}$: Marginal Revenue ≈ Revenue from selling one more unit
5. $\frac{dP}{dx}$: Marginal Profit ≈ Profit from selling one more unit
6. $\frac{dC}{dx}$: Marginal Cost ≈ Cost of producing one more unit

Example 1

Given the cost function $C(x)=100+8x+0.1x^2$, (a) find the marginal cost when $x=50$; and (b) find the marginal profit at $x=50$, if the price per unit is \$20.

Solution:

(a) Marginal cost is $C'(x)$. Enter $d(100+8x+0.1x^2, x)|x=50$ and obtain \$18.

(b) Marginal profit is $P'(x)$

$P=R-C$

$P=20x-(100+8x+0.1x^2)$. Enter $d(20x-(100+8x+0.1x\wedge 2, x)|x=50$ and obtain 2.

- Carry all decimal places and round only at the final answer. Round to 3 decimal places unless the question indicates otherwise.

Example 2

Given the cost function $C(x) = 500 + 3x + 0.01x^2$ and the demand function (the price function) $p(x) = 10$, find the number of units produced in order to have maximum profit.

Solution:

Step 1: Write an equation.
Profit = Revenue − Cost
$P = R - C$
Revenue = (Units Sold)(Price Per Unit)
$R = xp(x) = x(10) = 10x$
$P = 10x - (500 + 3x + 0.01x^2)$

Step 2: Differentiate.
Enter $d(10x - (500 + 3x + 0.01x\text{^}2, x))$ and obtain $7 - 0.02x$.

Step 3: Find critical numbers.
Set $7 - 0.02x = 0 \Rightarrow x = 350$.
Critical number is $x = 350$.

Step 4: Apply the Second Derivative Test.
Enter $d(10x - (500 + 3x + 0.01x\text{^}2), x, 2)|x = 350$ and obtain -0.02.
Since $x = 350$ is the only relative maximum, it is the absolute maximum.

Step 5: Write a solution.
Thus, producing 350 units will lead to maximum profit.

9.3 Rapid Review

1. Find the instantaneous rate of change at $x = 5$ of the function $f(x) = \sqrt{2x - 1}$.

Answer: $f(x) = \sqrt{2x - 1} = (2x - 1)^{1/2}$

$f'(x) = \frac{1}{2}(2x - 1)^{-1/2}(2) = (2x - 1)^{-1/2}$

$f'(5) = \frac{1}{3}$

2. If h is the diameter of a circle and h is increasing at a constant rate of 0.1 cm/sec, find the rate of change of the area of the circle when the diameter is 4 cm.

Answer: $A = \pi r^2 = \pi\left(\frac{h}{2}\right)^2 = \frac{1}{4}\pi h^2$

$\frac{dA}{dt} = \frac{1}{2}\pi h\frac{dh}{dt} = \frac{1}{2}\pi(4)(0.1) = 0.2\pi \text{ cm}^2/\text{sec}$

3. The radius of a sphere is increasing at a constant rate of 2 inches per minute. In terms of the surface area, what is the rate of change of the volume of the sphere?

Answer: $V = \frac{4}{3}\pi r^3$; $\frac{dV}{dt} = 4\pi r^2\frac{dr}{dt}$ since $S = \pi r^2$, $\frac{dV}{dt} = 28 \text{ in.}^3/\text{min}$

4. **Using your calculator, find the shortest distance between the point (4, 0) and the line $y = x$. (See Figure 9.3-1.)**

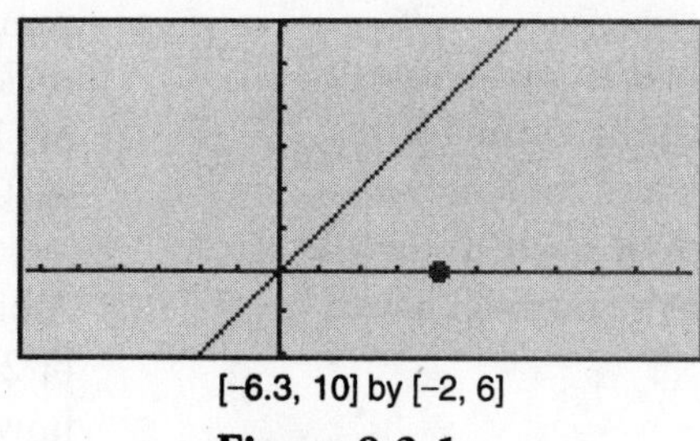

[-6.3, 10] by [-2, 6]

Figure 9.3-1

Answer:

$S = \sqrt{(x-4)^2 + (y-0)^2} = \sqrt{(x-4)^2 + x^2}$

Enter $y_1 = \sqrt{(x-4)^2 + x^2}$ and $y_2 = d(y_1(x), x)$.

Use the [*Zero*] function for y_2 and obtain $x = 2$. Note that when $x < 2$, $y_2 < 0$, which means y_1 is decreasing and when $x > 2$, $y_2 > 0$, which means y_1 is increasing, and thus at $x = 2$, y_1 is a minimum. Use the [*Value*] function for y_1 at $x = 2$ and obtain $y_1 = 2.82843$. Thus, the shortest distance is approximately 2.828.

9.4 Practice Problems

Part A—The use of a calculator is not allowed.

1. A spherical balloon is being inflated. Find the volume of the balloon at the instant when the rate of increase of the surface area is eight times the rate of increase of the radius of the sphere.

2. A 13-foot ladder is leaning against a wall. If the top of the ladder is sliding down the wall at 2 ft/sec, how fast is the bottom of the ladder moving away from the wall when the top of the ladder is 5 feet from the ground? (See Figure 9.4-1.)

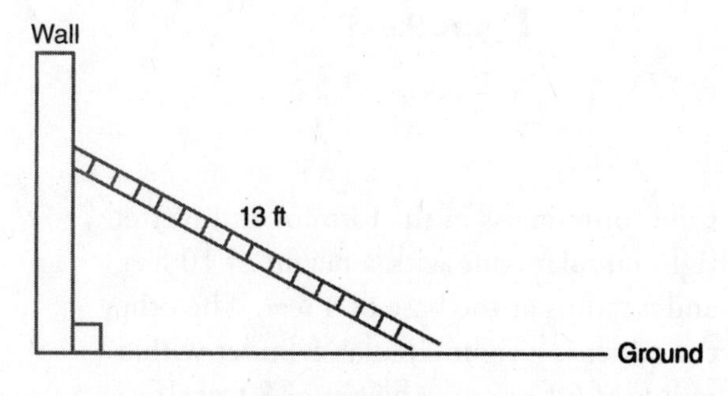

Figure 9.4-1

3. Air is being pumped into a spherical balloon at the rate of 100 cm³/sec. How fast is the diameter increasing when the radius is 5 cm?

4. A woman 5 feet tall is walking away from a streetlight hung 20 feet from the ground at the rate of 6 ft/sec. How fast is her shadow lengthening?

5. A water tank in the shape of an inverted cone has a height of 18 feet and a base radius of 12 feet. If the tank is full and the water is drained at the rate of 4 ft³/min, how fast is the water level dropping when the water level is 6 feet high?

6. Two cars leave an intersection at the same time. The first car is going due east at the rate of 40 mph and the second is going due south at the rate of 30 mph. How fast is the distance between the two cars increasing when the first car is 120 miles from the intersection?

7. If the perimeter of an isosceles triangle is 18 cm, find the maximum area of the triangle.

8. Find a number in the interval (0, 2) such that the sum of the number and its reciprocal is the absolute minimum.

9. An open box is to be made using a piece of cardboard 8 cm by 15 cm by cutting a square from each corner and folding the sides up. Find the length of a side of the square being cut so that the box will have a maximum volume.

10. What is the shortest distance between the point $\left(2, -\frac{1}{2}\right)$ and the parabola $y = -x^2$?

11. If the cost function is $C(x) = 3x^2 + 5x + 12$, find the value of x such that the average cost is a minimum.

12. A man with 200 meters of fence plans to enclose a rectangular piece of land using a river on one side and a fence on the other three sides. Find the maximum area that the man can obtain.

Part B—Calculators are allowed.

13. A trough is 10 meters long and 4 meters wide. (See Figure 9.4-2.) The two sides of the trough are equilateral triangles. Water is pumped into the trough at 1 m³/min. How fast is the water level rising when the water is 2 meters high?

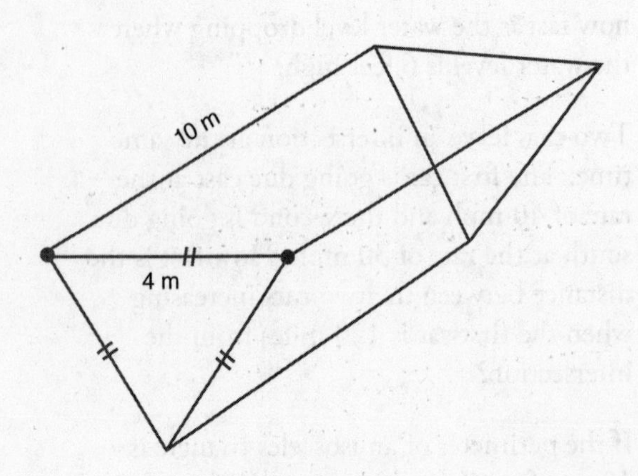

Figure 9.4-2

14. A rocket is sent vertically up in the air with the position function $s = 100t^2$ where s is measured in meters and t in seconds. A camera 3000 m away is recording the rocket. Find the rate of change of the angle of elevation of the camera 5 sec after the rocket went up.

15. A plane lifts off from a runway at an angle of 20°. If the speed of the plane is 300 mph, how fast is the plane gaining altitude?

16. Two water containers are being used. (See Figure 9.4-3.)

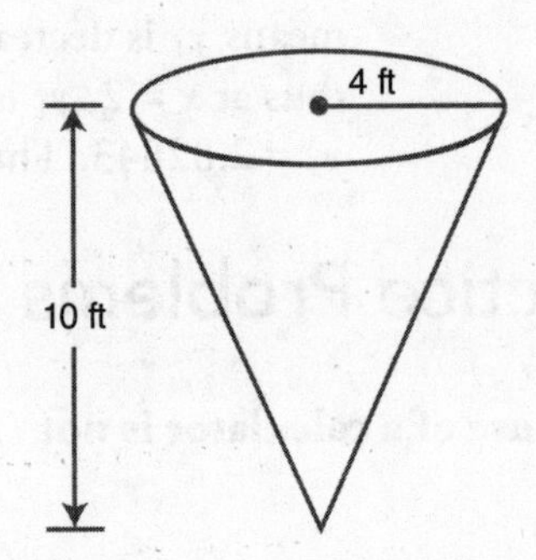

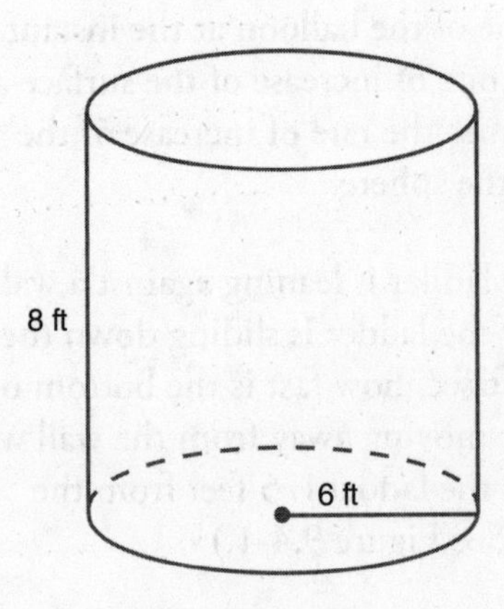

Figure 9.4-3

One container is in the form of an inverted right circular cone with a height of 10 feet and a radius at the base of 4 feet. The other container is a right circular cylinder with a radius of 6 feet and a height of 8 feet. If water is being drained from the conical container into the cylindrical container at the rate of 15 ft^3/min, how fast is the water level falling in the conical tank when the water level in the conical tank is 5 feet high? How fast is the water level rising in the cylindrical container?

17. The wall of a building has a parallel fence that is 6 feet high and 8 feet from the wall. What is the length of the shortest ladder that passes over the fence and leans on the wall? (See Figure 9.4-4.)

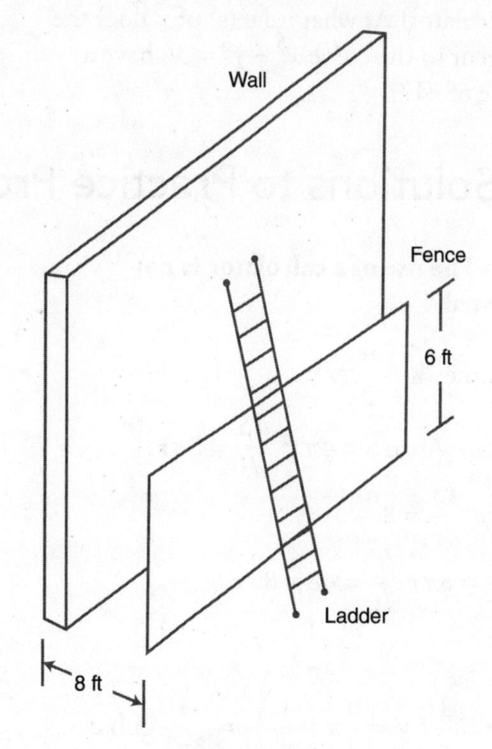

Figure 9.4-4

18. Given the cost function $C(x) = 2500 + 0.02x + 0.004x^2$, find the product level such that the average cost per unit is a minimum.

19. Find the maximum area of a rectangle inscribed in an ellipse whose equation is $4x^2 + 25y^2 = 100$.

20. A right triangle is in the first quadrant with a vertex at the origin and the other two vertices on the x- and y-axes. If the hypotenuse passes through the point (0.5, 4), find the vertices of the triangle so that the length of the hypotenuse is minimum.

9.5 Cumulative Review Problems

(Calculator) indicates that calculators are permitted.

21. If $y = \sin^2(\cos(6x - 1))$, find $\frac{dy}{dx}$.

22. Evaluate $\lim_{x \to \infty} \frac{100/x}{-4 + x + x^2}$.

23. The graph of f' is shown in Figure 9.5-1. Find where the function f: (a) has its relative extrema or absolute extrema; (b) is increasing or decreasing; (c) has its point(s) of inflection; (d) is concave upward or downward; and (e) if $f(3) = -2$. Draw a possible sketch of f.

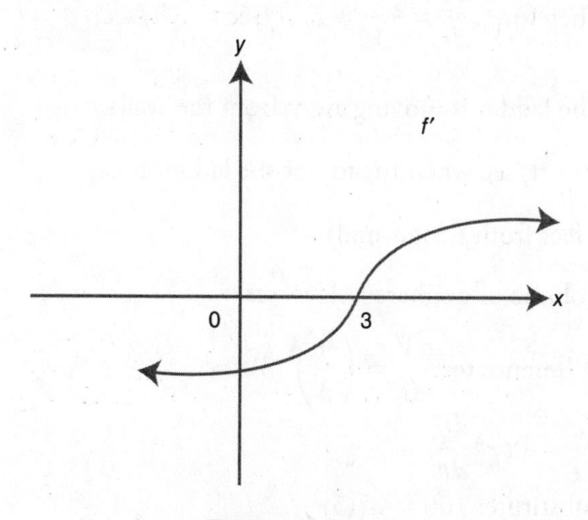

Figure 9.5-1

24. (Calculator) At what value(s) of x does the tangent to the curve $x^2 + y^2 = 36$ have a slope of -1?

25. (Calculator) Find the shortest distance between the point (1, 0) and the curve $y = x^3$.

9.6 Solutions to Practice Problems

Part A—The use of a calculator is not allowed.

1. Volume: $V = \frac{4}{3}\pi r^3$;

   Surface Area: $S = 4\pi r^2 \; \frac{dS}{dt} = 8\pi r \frac{dr}{dt}$.

   Since $\frac{dS}{dt} = 8\frac{dr}{dt}$,

   $8\frac{dr}{dt} = 8\pi r \frac{dr}{dt} \Rightarrow 8 = 8\pi r$

   or $r = \frac{1}{\pi}$.

   At $r = \frac{1}{\pi}$, $V = \frac{4}{3}\pi\left(\frac{1}{\pi}\right)^3 = \frac{4}{3\pi^2}$ cubic units.

2. Pythagorean Theorem yields $x^2 + y^2 = (13)^2$.

   Differentiate: $2x\frac{dx}{dt} + 2y\frac{dy}{dt} = 0 \Rightarrow \frac{dy}{dt} = \frac{-x}{y}\frac{dx}{dt}$.

   At $x = 5$, $(5)^2 + y^2 = 13^2 \Rightarrow y = \pm 12$, since $y > 0$, $y = 12$.

   Therefore, $\frac{dy}{dt} = -\frac{5}{12}(-2)$ ft/sec$= \frac{5}{6}$ ft/sec.

   The ladder is moving away from the wall at $\frac{5}{6}$ ft/sec when the top of the ladder is 5 feet from the ground.

3. Volume of a sphere is $V = \frac{4}{3}\pi r^3$.

   Differentiate: $\frac{dV}{dt} = \left(\frac{4}{3}\right)(3)\pi r^2 \frac{dr}{dt} = 4\pi r^2 \frac{dr}{dt}$.

   Substitute: $100 = 4\pi(5)^2 \frac{dr}{dt} \Rightarrow \frac{dr}{dt} = \frac{1}{\pi}$ cm/sec.

   Let x be the diameter. Since $x = 2r$, $\frac{dx}{dt} = 2\frac{dr}{dt}$.

   Thus, $\left.\frac{dx}{dt}\right|_{r=5} = 2\left(\frac{1}{\pi}\right)$ cm/sec $= \frac{2}{\pi}$ cm/sec. The diameter is increasing at $\frac{2}{\pi}$ cm/sec when the radius is 5 cm.

4. (See Figure 9.6-1.) Using similar triangles, with y the length of the shadow, you have:

   $\frac{5}{20} = \frac{y}{y + x} \Rightarrow 20y = 5y + 5x \Rightarrow$

   $15y = 5x$ or $y = \frac{x}{3}$.

   Differentiate:

   $\frac{dy}{dt} = \frac{1}{3}\frac{dx}{dt} \Rightarrow \frac{dy}{dt} = \frac{1}{3}(6)$

   $= 2$ ft/sec.

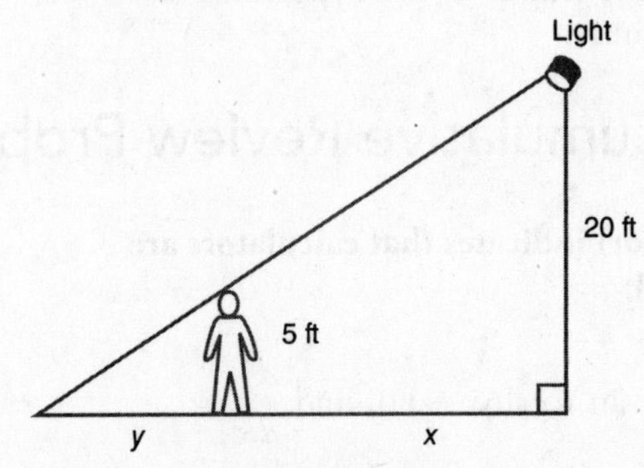

Figure 9.6-1

5. (See Figure 9.6-2.) Volume of a cone $V = \frac{1}{3}\pi r^2 h$.

   Using similar triangles, you have

   $\frac{12}{18} = \frac{r}{h} \Rightarrow 2h = 3r$ or $r = \frac{2}{3}h$, thus

   reducing the equation to

   $V = \frac{1}{3}\pi\left(\frac{2}{3}h\right)^2 (h) = \frac{4\pi}{27}h^3$.

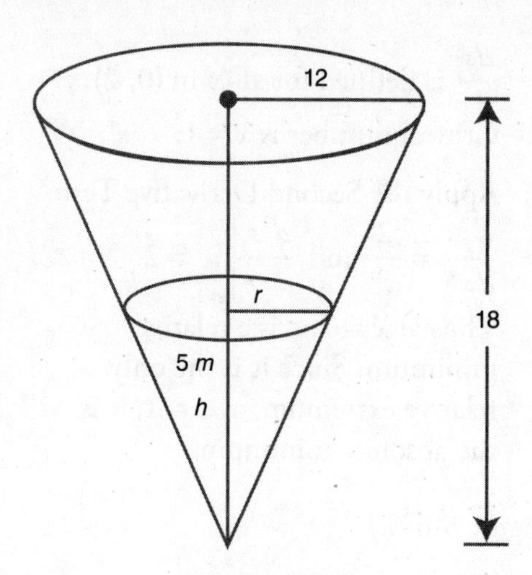

Figure 9.6-2

Differentiate: $\frac{dV}{dt} = \frac{4}{9}\pi h^2 \frac{dh}{dt}$.

Substitute known values:

$-4 = \frac{4\pi}{9}(6)^2 \frac{dh}{dt} \Rightarrow -4 = 16\pi \frac{dh}{dt}$ or

$\frac{dh}{dt} = -\frac{1}{4\pi}$ ft/min. The water level is

dropping at $\frac{1}{4\pi}$ ft/min when $h = 6$ ft.

6. (See Figure 9.6-3.)

Step 1: Using the Pythagorean Theorem, you have $x^2 + y^2 = z^2$. You also have $\frac{dx}{dt} = 40$ and $\frac{dy}{dt} = 30$.

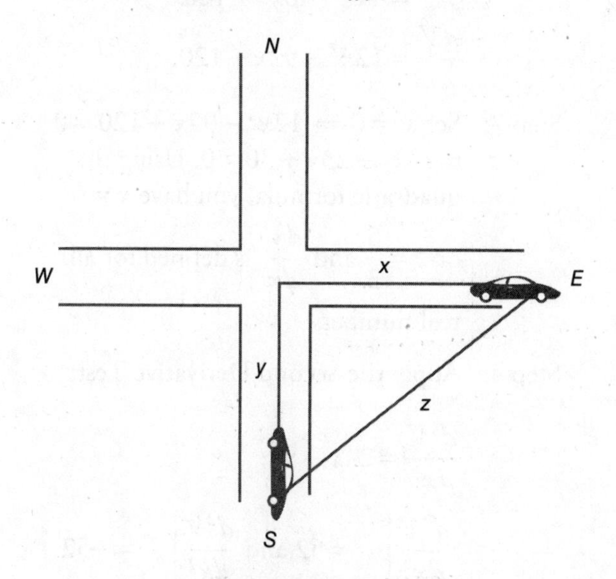

Figure 9.6-3

Step 2: Differentiate:

$2x\frac{dx}{dt} + 2y\frac{dy}{dt} = 2z\frac{dz}{dt}$.

At $x = 120$, both cars have traveled 3 hours and thus, $y = 3(30) = 90$. By the Pythagorean Theorem, $(120)^2 + (90)^2 = z^2 \Rightarrow z = 150$.

Step 3: Substitute all known values into the equation:

$2(120)(40) + 2(90)(30) = 2(150)\frac{dz}{dt}$.

Thus $\frac{dz}{dt} = 50$ mph.

Step 4: The distance between the two cars is increasing at 50 mph at $x = 120$.

7. (See Figure 9.6-4.)

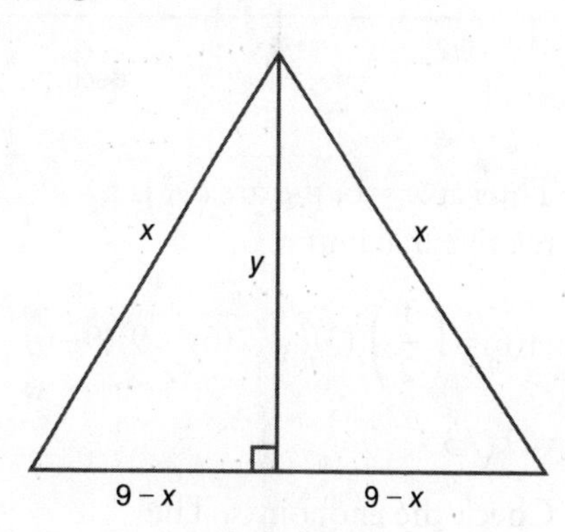

Figure 9.6-4

Step 1: Applying the Pythagorean Theorem, you have

$x^2 = y^2 + (9 - x)^2 \Rightarrow$

$y^2 = x^2 - (9 - x)^2 =$

$x^2 - \left(81 - 18x + x^2\right) =$

$18x - 81 = 9(2x - 9)$, or

$y = \pm\sqrt{9(2x - 9)} =$

$\pm 3\sqrt{(2x - 9)}$ since $y > 0$,

$y = 3\sqrt{(2x - 9)}$.

The area of the triangle

$A = \frac{1}{2}\left(3\sqrt{2x - 9}\right)(18 - 2x) =$

$\left(3\sqrt{2x - 9}\right)(9 - x) =$

$3(2x - 9)^{1/2}(9 - x)$.

Step 2: $\frac{dA}{dx} = \frac{3}{2}(2x-9)^{-1/2}(2)(9-x)$

$+(-1)(3)(2x-9)^{1/2}$.

$= \frac{3(9-x)-3(2x-9)}{\sqrt{2x-9}}$

$= \frac{54-9x}{\sqrt{2x-9}}$

Step 3: Set $\frac{dA}{dx} = 0 \Rightarrow 54 - 9x = 0;\ x = 6$.

$\frac{dA}{dx}$ is undefined at $x = \frac{9}{2}$. The critical numbers are $\frac{9}{2}$ and 6.

Step 4: First Derivative Test:

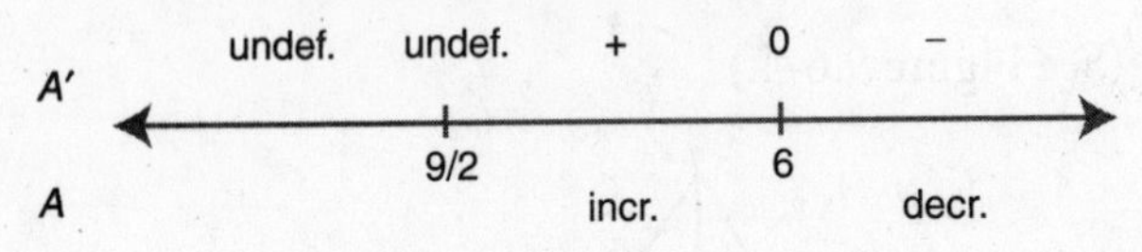

Thus at $x = 6$, the area A is a relative maximum.

$A(6) = \left(\frac{1}{2}\right)(3)(\sqrt{2(6)-9})(9-6)$

$= 9\sqrt{3}$

Step 5: Check the endpoints. The domain of A is $[9/2, 9]$. $A(9/2) = 0$; and $A(9) = 0$. Therefore, the maximum area of an isosceles triangle with the perimeter of 18 cm is $9\sqrt{3}$ cm^2. (Note that at $x = 6$, the triangle is an equilateral triangle.)

8. Step 1: Let x be the number and $\frac{1}{x}$ be its reciprocal.

Step 2: $s = x + \frac{1}{x}$ with $0 < x < 2$.

Step 3: $\frac{ds}{dx} = 1 + (-1)x^{-2} = 1 - \frac{1}{x^2}$

Step 4: Set $\frac{ds}{dx} = 0 \Rightarrow 1 - \frac{1}{x^2} = 0$
$\Rightarrow x = \pm 1$, since the domain is $(0, 2)$, thus $x = 1$.

$\frac{ds}{dx}$ is defined for all x in $(0, 2)$. Critical number is $x = 1$.

Step 5: Apply the Second Derivative Test:
$\frac{d^2s}{dx^2} = \frac{2}{x^3}$ and $\left.\frac{d^2s}{dx^2}\right|_{x=1} = 2$.
Thus at $x = 1$, s is a relative minimum. Since it is the only relative extremum, at $x = 1$, it is the absolute minimum.

9. (See Figure 9.6-5.)

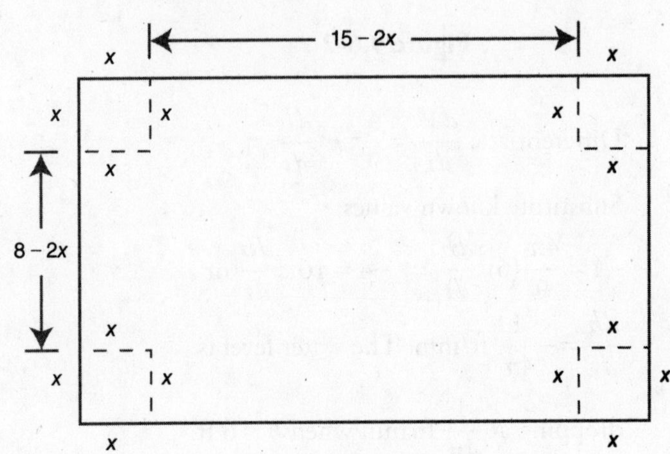

Figure 9.6-5

Step 1: Volume: $V = x(8-2x)(15-2x)$ with $0 \le x \le 4$.

Step 2: Differentiate: Rewrite as
$V = 4x^3 - 46x^2 + 120x$
$\frac{dV}{dx} = 12x^2 - 92x + 120$.

Step 3: Set $V = 0 \Rightarrow 12x^2 - 92x + 120 = 0$
$\Rightarrow 3x^2 - 23x + 30 = 0$. Using the quadratic formula, you have $x = 6$ or $x = \frac{5}{3}$ and $\frac{dV}{dx}$ is defined for all real numbers.

Step 4: Apply the Second Derivative Test:

$\frac{d^2V}{dx^2} = 24x - 92$;

$\left.\frac{d^2V}{dx^2}\right|_{x=6} = 52$ and $\left.\frac{d^2V}{dx^2}\right|_{x=\frac{5}{3}} = -52$.

Thus at $x = \frac{5}{3}$ is a relative maximum.

Step 5: Check the endpoints. At $x = 0$, $V = 0$ and at $x = 4$, $V = 0$. Therefore, at $x = \frac{5}{3}$, V is the absolute maximum.

10. (See Figure 9.6-6.)

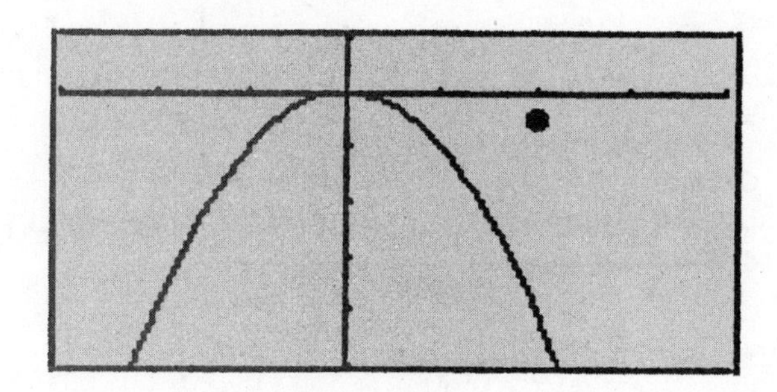

Figure 9.6-6

Step 1: Distance Formula:

$$Z = \sqrt{(x-2)^2 + \left(y - \left(-\frac{1}{2}\right)\right)^2}$$

$$= \sqrt{(x-2)^2 + \left(-x^2 + \frac{1}{2}\right)^2}$$

$$= \sqrt{x^2 - 4x + 4 + x^4 - x^2 + \frac{1}{4}}$$

$$= \sqrt{x^4 - 4x + \frac{17}{4}}$$

Step 2: Let $S = Z^2$, since S and Z have the same maximums and minimums.

$S = x^4 - 4x + \frac{17}{4}$; $\frac{dS}{dx} = 4x^3 - 4$

Step 3: Set $\frac{dS}{dx} = 0$; $x = 1$ and $\frac{dS}{dx}$ is defined for all real numbers.

Step 4: Apply the Second Derivative Test:

$\frac{d^2S}{dx^2} = 12x^2$ and $\left.\frac{d^2S}{dx^2}\right|_{x=1} = 12.$

Thus at $x = 1$, Z has a minimum, and since it is the only relative extremum, it is the absolute minimum.

Step 5: At $x = 1$,

$$Z = \sqrt{(1)^4 - 4(1) + \frac{17}{4}}$$

$$= \sqrt{\frac{5}{4}}.$$

Therefore, the shortest distance is $\sqrt{\frac{5}{4}}$.

11. Step 1: Average Cost:

$$\overline{C} = \frac{C(x)}{x} = \frac{3x^2 + 5x + 12}{x}$$

$$= 3x + 5 + \frac{12}{x}.$$

Step 2: $\frac{d\overline{C}}{dx} = 3 - 12x^{-2} = 3 - \frac{12}{x^2}$

Step 3: Set $\frac{d\overline{C}}{dx} = 0 \Rightarrow 3 - \frac{12}{x^2} = 0 \Rightarrow$

$3 = \frac{12}{x^2} \Rightarrow x = \pm 2$. Since $x > 0$, $x = 2$ and $\overline{C}(2) = 17$. $\frac{d\overline{C}}{dx}$ is undefined at $x = 0$, which is not in the domain.

Step 4: Apply the Second Derivative Test:

$\frac{d^2\overline{C}}{dx^2} = \frac{24}{x^3}$ and $\left.\frac{d^2\overline{C}}{dx^2}\right|_{x=2} = 3$

Thus at $x = 2$, the average cost is a minimum.

12. (See Figure 9.6-7.)

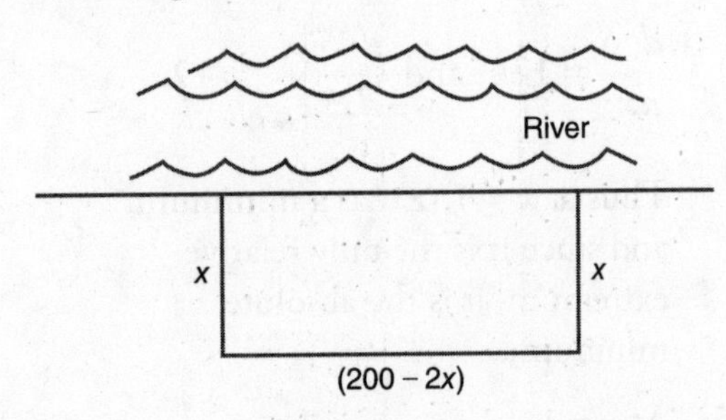

Figure 9.6-7

Step 1: Area:
$A = x(200 - 2x) = 200x - 2x^2$
with $0 \le x \le 100$.

Step 2: $A'(x) = 200 - 4x$

Step 3: Set $A'(x) = 0 \Rightarrow 200 - 4x = 0$;
$x = 50$.

Step 4: Apply the Second Derivative Test:
$A''(x) = -4$; thus at $x = 50$, the area is a relative maximum.
$A(50) = 5000\text{ m}^2$.

Step 5: Check the endpoints.
$A(0) = 0$ and $A(100) = 0$;
therefore at $x = 50$, the area is the absolute maximum and 5000 m^2 is the maximum area.

Part B—Calculators are allowed.

13. Step 1: Let h be the height of the trough and 4 be a side of one of the two equilateral triangles. Thus, in a 30–60 right triangle, $h = 2\sqrt{3}$.

Step 2: Volume:
$V = (\text{area of the triangle}) \cdot 10$

$$= \left[\frac{1}{2}(h)\left(\frac{2}{\sqrt{3}}h\right)\right] 10 = \frac{10}{\sqrt{3}}h^2.$$

Step 3: Differentiate with respect to t.

$$\frac{dV}{dt} = \left(\frac{10}{\sqrt{3}}\right)(2)h\frac{dh}{dt}$$

Step 4: Substitute known values:

$$1 = \frac{20}{\sqrt{3}}(2)\frac{dh}{dt};$$

$$\frac{dh}{dt} = \frac{\sqrt{3}}{40}\text{ m/min}.$$

The water level is rising $\frac{\sqrt{3}}{40}$ m/min when the water level is 2 m high.

14. (See Figure 9.6-8.)

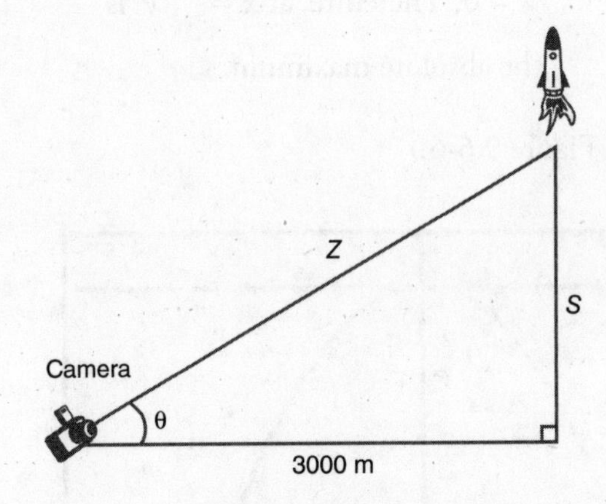

Figure 9.6-8

Step 1: $\tan\theta = S/3000$

Step 2: Differentiate with respect to t.

$$\sec^2\theta\frac{d\theta}{dt} = \frac{1}{3000}\frac{dS}{dt};$$

$$\frac{d\theta}{dt} = \frac{1}{3000}\left(\frac{1}{\sec^2\theta}\right)\frac{dS}{dt}$$

$$= \frac{1}{3000}\left(\frac{1}{\sec^2\theta}\right)(200t)$$

Step 3: At $t = 5$; $S = 100(5)^2 = 2500$;
Thus, $Z^2 = (3000)^2 + (2500)^2 = 15{,}250{,}000$. Therefore,
$Z = \pm 500\sqrt{61}$, since $Z > 0$,
$Z = 500\sqrt{61}$. Substitute known values into the equation:

$$\frac{d\theta}{dt} = \frac{1}{3000}\left(\frac{1}{\frac{500\sqrt{61}}{3000}}\right)^2(1000),$$

since $\sec\theta = \dfrac{Z}{3000}$.

$\frac{d\theta}{dt} = 0.197$ radian/sec. The angle of elevation is changing at 0.197 radian/sec, 5 seconds after liftoff.

15. (See Figure 9.6-9.)

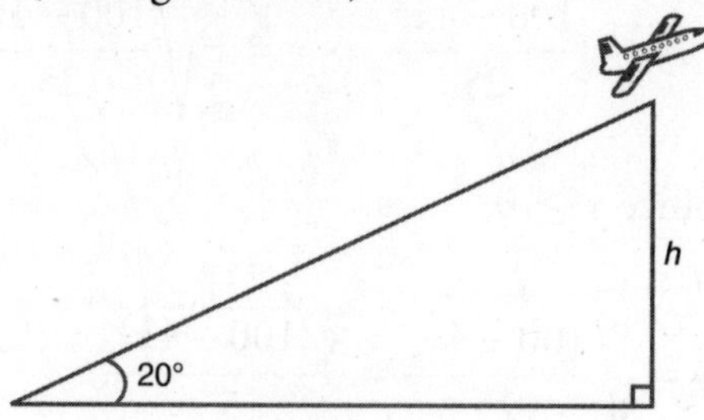

Figure 9.6-9

$\text{Sin } 20° = \frac{h}{300t}$

$h = (\sin 20°)300t;$

$\frac{dh}{dt} = (\sin 20°)(300) \approx 102.606$ mph.

The plane is gaining altitude at 102.606 mph.

16. $V_{\text{cone}} = \frac{1}{3}\pi r^2 h$

Similar triangles: $\frac{4}{10} = \frac{r}{h} \Rightarrow 5r = 2h$ or $r = \frac{2h}{5}$.

$V_{\text{cone}} = \frac{1}{3}\pi\left(\frac{2h}{5}\right)^2 h = \frac{4\pi}{75}h^3;$

$\frac{dV}{dt} = \frac{4\pi}{75}(3)h^2\frac{dh}{dt}.$

Substitute known values:

$-15 = \frac{4\pi}{25}(5)^2\frac{dh}{dt};$

$-15 = 4\pi\frac{dh}{dt}; \frac{dh}{dt} = \frac{-15}{4\pi} \approx -1.19$ ft/min.

The water level in the cone is falling at $\frac{-15}{4\pi}$ ft/min ≈ -1.19 ft/ min when the water level is 5 feet high.

$V_{\text{cylinder}} = \pi R^2 H = \pi(6)2H = 36\pi H.$

$\frac{dV}{dt} = 36\pi\frac{dH}{dt}; \frac{dH}{dt} = \frac{1}{36\pi}\frac{dV}{dt};$

$\frac{dH}{dt} = \frac{1}{36\pi}(15) = \frac{5}{12\pi}$ ft/min

≈ 0.1326 ft/min or 1.592 in/min.

The water level in the cylinder is rising at $\frac{5}{12\pi}$ ft/min=0.133 ft/min.

17. Step 1: Let x be the distance of the foot of the ladder from the higher wall. Let y be the height of the point where the ladder touches the higher wall. The slope of the ladder is $m = \frac{y-6}{0-8}$ or $m = \frac{6-0}{8-x}$. Thus,

$\frac{y-6}{-8} = \frac{6}{8-x} \Rightarrow (y-6)(8-x)$

$= -48$

$\Rightarrow 8y - xy - 48 + 6x = -48$

$\Rightarrow y(8-x) = -6x \Rightarrow y = \frac{-6x}{8-x}.$

Step 2: Pythagorean Theorem:

$l^2 = x^2 + y^2 = x^2 + \left(\frac{-6x}{8-x}\right)^2$

Since $l > 0$, $l = \sqrt{x^2 + \left(\frac{-6x}{8-x}\right)^2}$, $x > 8$.

Step 3: Enter $y1 = \sqrt{\{x\verb|^|2 + [(-6 * x)/(8-x)]\verb|^|2\}}$.

The graph of y_1 is continuous on the interval $x > 8$. Use the [*Minimum*] function of the calculator and obtain $x = 14.604$; $y = 19.731$. Thus, the minimum value of l is 19.731 or the shortest ladder is approximately 19.731 feet.

18. Step 1: Average Cost $\overline{C} = \frac{C}{x}$; thus, $\overline{C}(x)$

$= \frac{2500 + 0.02x + 0.004x^2}{x}$

$= \frac{2500}{x} + 0.02 + 0.004x.$

Step 2: Enter: $y1 = \frac{2500}{x} + .02 + .004 * x$

Step 3: Use the [*Minimum*] function in the calculator and obtain $x = 790.6$.

Step 4: Verify the result with the First Derivative Test. Enter $y2 = d(2500/x + .02 + .004x, x)$; Use the [*Zero*] function and obtain $x = 790.6$. Thus $\frac{dC}{dx} = 0$; at $x = 790.6$.
Apply the First Derivative Test:

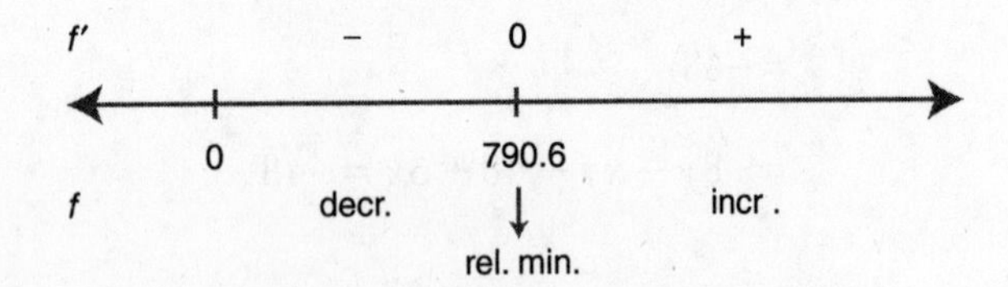

Thus the minimum average cost per unit occurs at $x = 790.6$. (The graph of the average cost function is shown in Figure 9.6-10.)

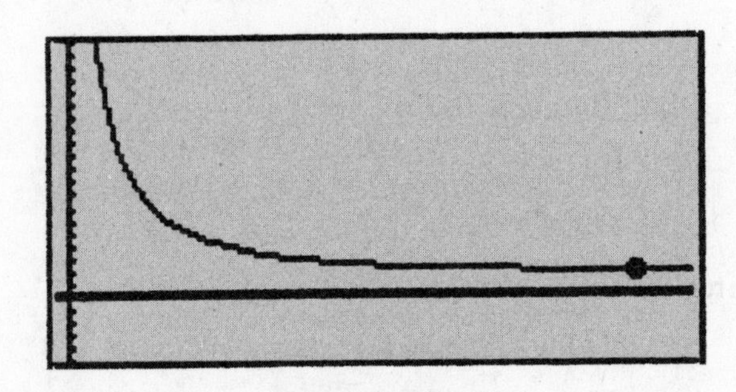

Figure 9.6-10

19. (See Figure 9.6-11.)

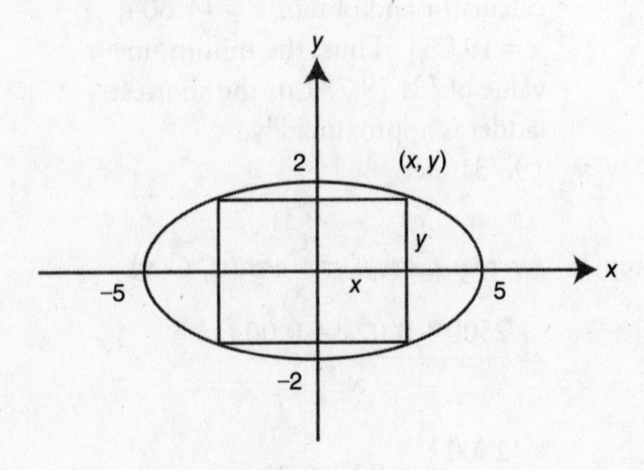

Figure 9.6-11

Step 1: Area $A = (2x)(2y)$; $0 \le x \le 5$ and $0 \le y \le 2$.

Step 2: $4x^2 + 25y^2 = 100$;
$25y^2 = 100 - 4x^2$.

$$y^2 = \frac{100 - 4x^2}{25} \Rightarrow y = \pm\sqrt{\frac{100 - 4x^2}{25}}$$

Since $y \ge 0$

$$y = \sqrt{\frac{100 - 4x^2}{25}} = \frac{\sqrt{100 - 4x^2}}{5}.$$

Step 3: $A = (2x)\left(\frac{2}{5}\right)\left(\sqrt{100 - 4x^2}\right)$

$$= \frac{4x}{5}\sqrt{100 - 4x^2}$$

Step 4: Enter $y1 = \frac{4x}{5}\sqrt{100 - 4x^2}$
Use the [*Maximum*] function and obtain $x = 3.536$ and $y_1 = 20$.

Step 5: Verify the result with the First Derivative Test.
Enter

$$y2 = d\left(\frac{4x}{5}\sqrt{100 - 4x^2}, x\right).$$

Use the [*Zero*] function and obtain $x = 3.536$.
Note that:

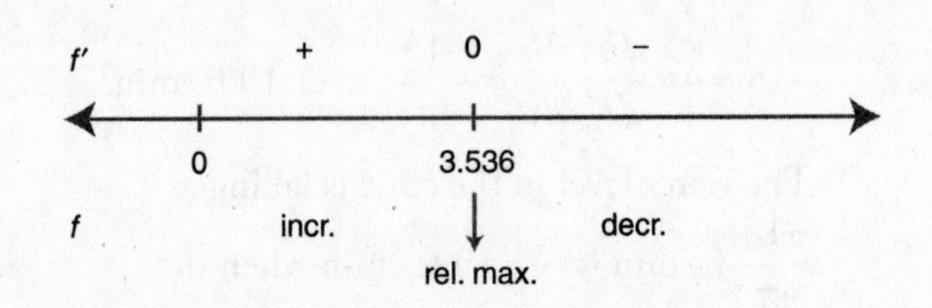

The function f has only one relative extremum. Thus, it is the absolute extremum. Therefore, at $x = 3.536$, the area is 20 and the area is the absolute maxima.

20. (See Figure 9.6-12.)

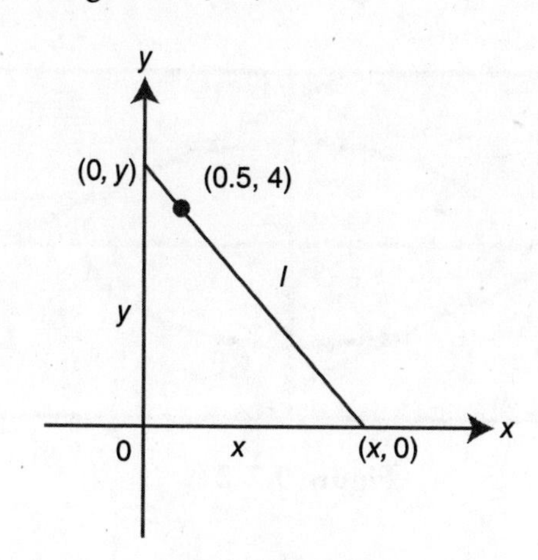

Figure 9.6-12

Step 1: Distance formula:
$l^2 = x^2 + y^2; x > 0.5$ and $y > 4$.

Step 2: The slope of the hypotenuse:
$$m = \frac{y-4}{0-0.5} = \frac{-4}{x-0.5}$$
$$\Rightarrow (y-4)(x-0.5) = 2$$
$$\Rightarrow xy - 0.5y - 4x + 2 = 2$$
$$y(x-0.5) = 4x$$
$$y = \frac{4x}{x-0.5}.$$

Step 3: $l^2 = x^2 + \left(\frac{4x}{x-0.5}\right)^2;$
$$l = \pm\sqrt{x^2 + \left(\frac{4x}{x-0.5}\right)^2}$$

Since $l > 0$, $l = \sqrt{x^2 + \left(\frac{4x}{x-0.5}\right)^2}$.

Step 4: Enter $y1 = \sqrt{x^2 + \left(\frac{4x}{x-0.5}\right)^2}$ and use the [*Minimum*] function of the calculator and obtain $x = 2.5$.

Step 5: Apply the First Derivative Test. Enter $y2 = d(y1(x), x)$ and use the [*Zero*] function and obtain $x = 2.5$.
Note that:

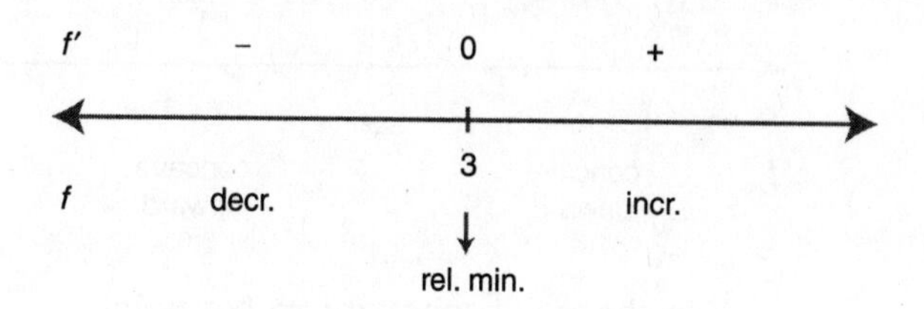

Since f has only one relative extremum, it is the absolute extremum.

Step 6: Thus, at $x = 2.5$, the length of the hypotenuse is the shortest. At $x = 2.5$, $y = \frac{4(2.5)}{2.5-0.5} = 5$. The vertices of the triangle are (0, 0), (2.5, 0), and (0, 5).

9.7 Solutions to Cumulative Review Problems

21. Rewrite: $y = [\sin(\cos(6x-1))]^2$

Thus, $\frac{dy}{dx} = 2\,[\sin(\cos(6x-1))]$
$\times [\cos(\cos(6x-1))]$
$\times [-\sin(6x-1)]\,(6)$
$= -12\sin(6x-1)$
$\times [\sin(\cos(6x-1))]$
$\times [\cos(\cos(6x-1))]$.

22. As $x \to \infty$, the numerator $\frac{100}{x}$ approaches 0 and the denominator increases without bound (i.e., ∞).

Thus, the $\lim\limits_{x\to\infty} \frac{100/x}{-4+x+x^2} = 0$.

23. (a) Summarize the information of f' on a number line.

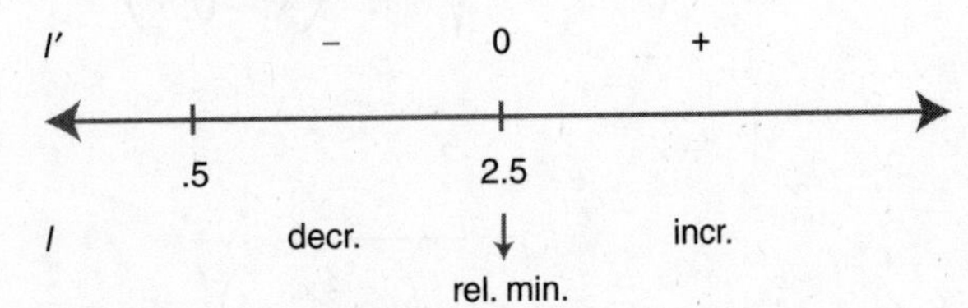

Since f has only one relative extremum, it is the absolute extremum. Thus, at $x = 3$, it is an absolute minimum.

(b) The function f is decreasing on the interval $(-\infty, 3)$ and increasing on $(3, \infty)$.

(c)

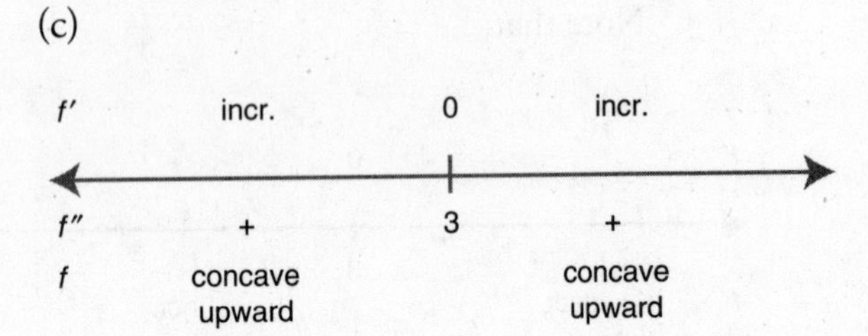

No change of concavity ⇒ No point of inflection.

(d) The function f is concave upward for the entire domain $(-\infty, \infty)$.

(e) Possible sketch of the graph for $f(x)$. (See Figure 9.7-1.)

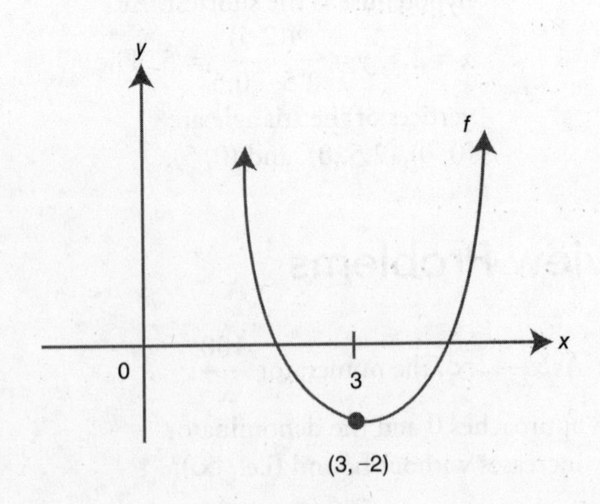

Figure 9.7-1

24. (Calculator) (See Figure 9.7-2.)

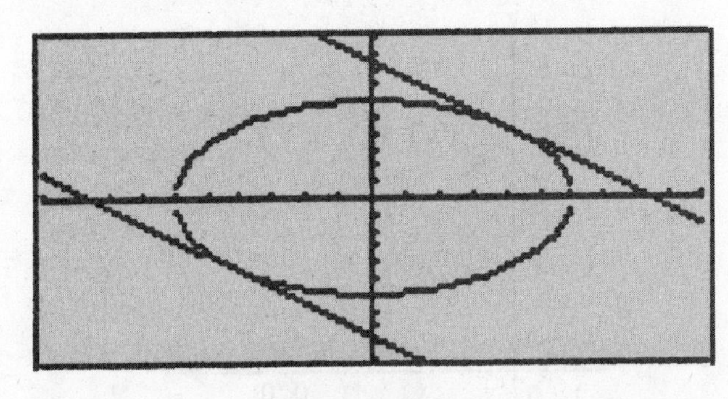

Figure 9.7-2

Step 1: Differentiate:
$2x + 2y\dfrac{dy}{dx} = 0 \Rightarrow \dfrac{dy}{dx} = -\dfrac{x}{y}$.

Step 2: Set $\dfrac{dy}{dx} = -1 \Rightarrow \dfrac{-x}{y} = -1 \Rightarrow$
$y = x$.

Step 3: Solve for y: $x^2 + y^2 = 36 \Rightarrow$
$y^2 = 36 - x^2$;
$y = \pm\sqrt{36 - x^2}$.

Step 4: Thus, $y = x \Rightarrow \pm\sqrt{36 - x^2} = x \Rightarrow 36 - x^2 = x^2 \Rightarrow$
$36 = 2x^2$ or $x = \pm 3\sqrt{2}$.

25. (Calculator) (See Figure 9.7-3.)

Step 1: Distance formula:
$z = \sqrt{(x-1)^2 + (x^3)^2} = \sqrt{(x-1)^2 + x^6}$.

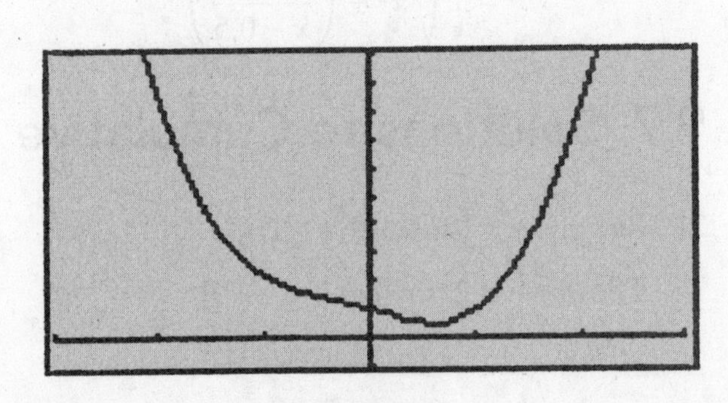

Figure 9.7-3

Step 2: Enter: $y1 = \sqrt{((x-1)\text{^}2 + x\text{^}6)}$.
Use the [*Minimum*] function of the calculator and obtain $x = .65052$ and $y1 = .44488$. Verify the result with the First Deriative Test. Enter $y2 = d(y1(x), x)$ and use the [*Zero*] function and obtain $x = .65052$.

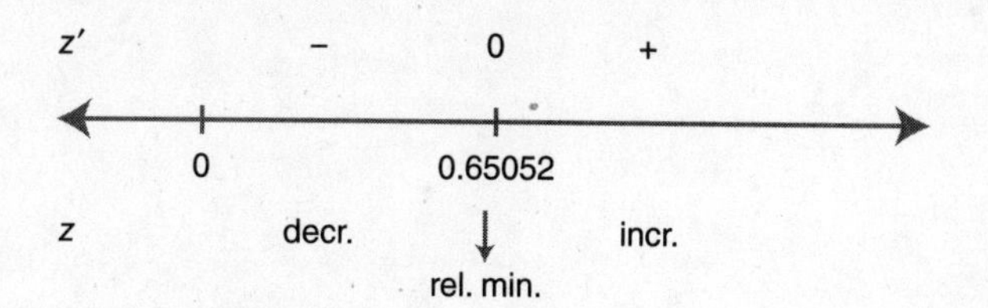

Thus, the shortest distance is approximately 0.445.

Big Idea 2: Derivatives
More Applications of Derivatives

IN THIS CHAPTER

Summary: Finding an equation of a tangent is one of the most common questions on the AP Calculus AB exam. In this chapter, you will learn how to use derivatives to find an equation of a tangent and to use the tangent line to approximate the value of a function at a specific point. You will also learn to apply derivatives to solve rectilinear motion problems.

Key Ideas

- Tangent and Normal Lines
- Linear Approximations
- Motion Along a Line

10.1 Tangent and Normal Lines

Main Concepts: Tangent Lines, Normal Lines

Tangent Lines

If the function y is differentiable at $x = a$, then the slope of the tangent line to the graph of y at $x = a$ is given as $m_{(\text{tangent at } x = a)} = \left.\frac{dy}{dx}\right|_{x=a}$.

Types of Tangent Lines

Horizontal Tangents: $\left(\dfrac{dy}{dx}=0\right)$. (See Figure 10.1-1.)

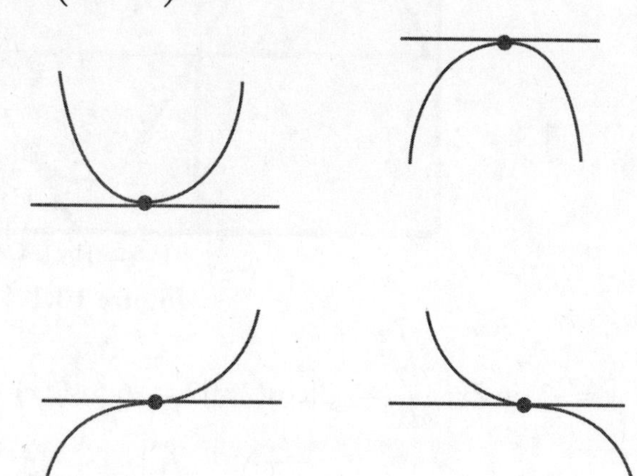

Figure 10.1-1

Vertical Tangents: $\left(\dfrac{dy}{dx} \text{ does not exist but } \dfrac{dx}{dy}=0\right)$. (See Figure 10.1-2.)

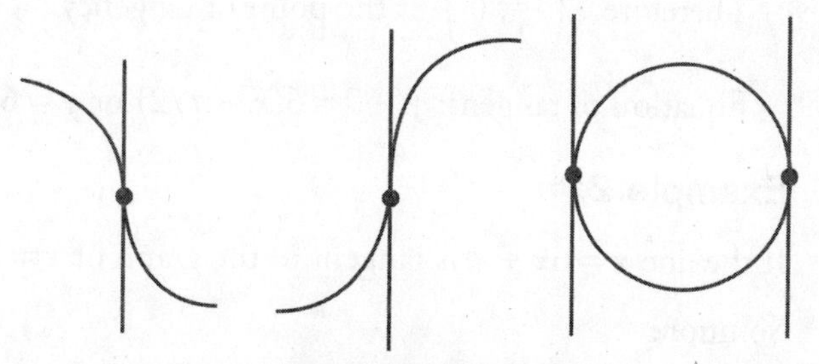

Figure 10.1-2

Parallel Tangents: $\left(\left.\dfrac{dy}{dx}\right|_{x=a}=\left.\dfrac{dy}{dx}\right|_{x=c}\right)$. (See Figure 10.1-3.)

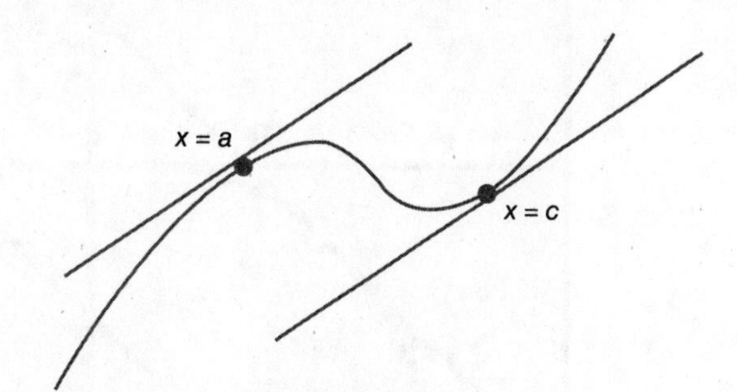

Figure 10.1-3

Example 1

Write an equation of the line tangent to the graph of $y=-3\sin 2x$ at $x=\dfrac{\pi}{2}$. (See Figure 10.1-4.)

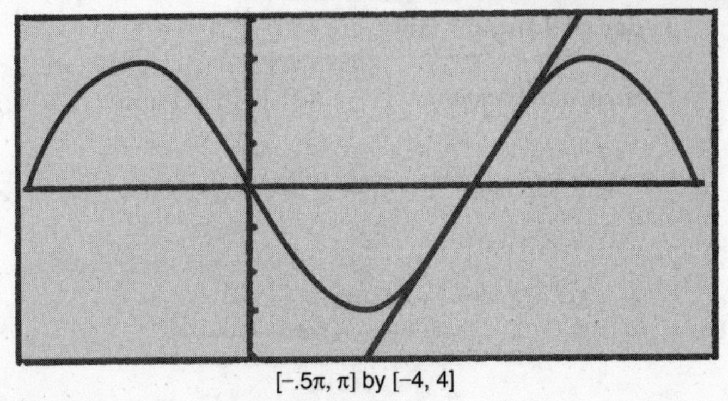

[−.5π, π] by [−4, 4]

Figure 10.1-4

$y = -3\sin 2x;\ \dfrac{dy}{dx} = -3[\cos(2x)]2 = -6\cos(2x)$

Slope of tangent $\left(\text{at } x = \dfrac{\pi}{2}\right)$: $\left.\dfrac{dy}{dx}\right|_{x=\pi/2} = -6\cos\,[2(\pi/2)] = -6\cos\pi = 6.$

Point of tangency: At $x = \dfrac{\pi}{2},\ y = -3\sin(2x)$

$= -3\sin[2(\pi/2)] = -3\sin(\pi) = 0.$

Therefore, $\left(\dfrac{\pi}{2}, 0\right)$ is the point of tangency.

Equation of tangent: $y - 0 = 6(x - \pi/2)$ or $y = 6x - 3\pi$.

Example 2

If the line $y = 6x + a$ is tangent to the graph of $y = 2x^3$, find the value(s) of a.

Solution:

$y = 2x^3;\ \dfrac{dy}{dx} = 6x^2$. (See Figure 10.1-5.)

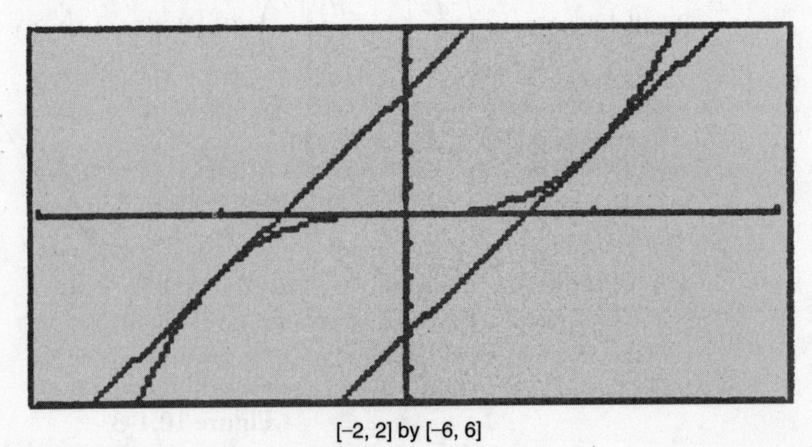

[−2, 2] by [−6, 6]

Figure 10.1-5

The slope of the line $y = 6x + a$ is 6.

Since $y = 6x + a$ is tangent to the graph of $y = 2x^3$, thus $\dfrac{dy}{dx} = 6$ for some values of x.

Set $6x^2 = 6 \Rightarrow x^2 = 1$ or $x = \pm 1$.

At $x = -1$, $y = 2x^3 = 2(-1)^3 = -2$; $(-1, -2)$ is a tangent point.
Thus, $y = 6x + a \Rightarrow -2 = 6(-1) + a$ or $a = 4$.

At $x = 1$, $y = 2x^3 = 2(1)^3 = 2$; $(1, 2)$ is a tangent point.
Thus, $y = 6x + a \Rightarrow 2 = 6(1) + a$ or $a = -4$.

Therefore, $a = \pm 4$.

Example 3

Find the coordinates of each point on the graph of $y^2 - x^2 - 6x + 7 = 0$ at which the tangent line is vertical. Write an equation of each vertical tangent. (See Figure 10.1-6.)

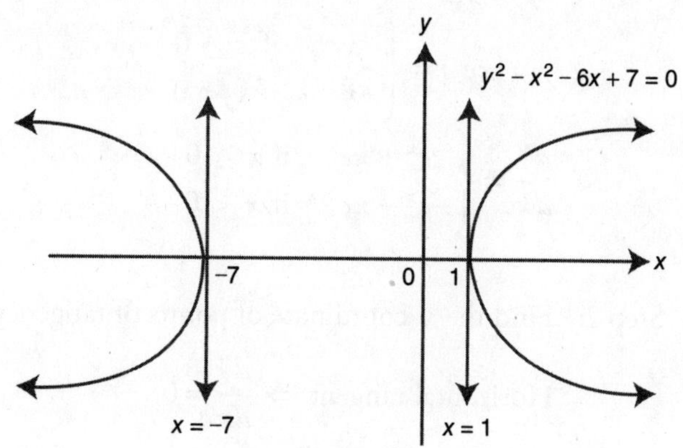

Figure 10.1-6

Step 1: Find $\frac{dy}{dx}$.

$$y^2 - x^2 - 6x + 7 = 0$$

$$2y\frac{dy}{dx} - 2x - 6 = 0$$

$$\frac{dy}{dx} = \frac{2x + 6}{2y} = \frac{x + 3}{y}$$

Step 2: Find $\frac{dx}{dy}$.

Vertical tangent $\Rightarrow \frac{dx}{dy} = 0$.

$$\frac{dx}{dy} = \frac{1}{dy/dx} = \frac{1}{(x + 3)/y} = \frac{y}{x + 3}$$

Set $\frac{dx}{dy} = 0 \Rightarrow y = 0$.

Step 3: Find points of tangency.
At $y = 0$, $y^2 - x^2 - 6x + 7 = 0$ becomes $-x^2 - 6x + 7 = 0 \Rightarrow x^2 + 6x - 7 = 0$
$\Rightarrow (x+7)(x-1) = 0 \Rightarrow x = -7$ or $x = 1$.
Thus, the points of tangency are $(-7, 0)$ and $(1, 0)$.

Step 4: Write equations for vertical tangents.
$x = -7$ and $x = 1$.

Example 4

Find all points on the graph of $y = |xe^x|$ at which the graph has a horizontal tangent.

Step 1: Find $\frac{dy}{dx}$.

$$y = |xe^x| = \begin{cases} xe^x & \text{if } x \geq 0 \\ -xe^x & \text{if } x < 0 \end{cases}$$

$$\frac{dy}{dx} = \begin{cases} e^x + xe^x & \text{if } x \geq 0 \\ -e^x - xe^x & \text{if } x < 0 \end{cases}$$

Step 2: Find the x-coordinate of points of tangency.

Horizontal tangent $\Rightarrow \frac{dy}{dx} = 0$.

If $x \geq 0$, set $e^x + xe^x = 0 \Rightarrow e^x(1+x) = 0 \Rightarrow x = -1$ but $x \geq 0$, therefore, no solution.

If $x < 0$, set $-e^x - xe^x = 0 \Rightarrow -e^x(1+x) = 0 \Rightarrow x = -1$.

Step 3: Find points of tangency.
At $x = -1$, $y = -xe^x = -(-1)e^{-1} = \frac{1}{e}$.
Thus at the point $(-1, 1/e)$, the graph has a horizontal tangent. (See Figure 10.1-7.)

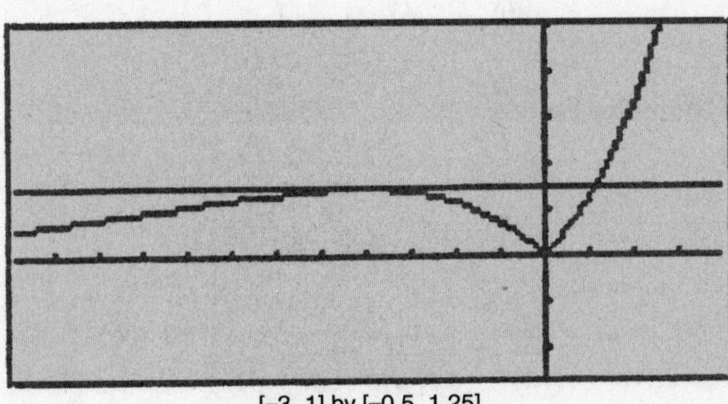

[−3, 1] by [−0.5, 1.25]

Figure 10.1-7

Example 5

Using your calculator, find the value(s) of x to the nearest hundredth at which the slope of the line tangent to the graph of $y = 2 \ln (x^2 + 3)$ is equal to $-\frac{1}{2}$.

(See Figures 10.1-8 and 10.1-9.)

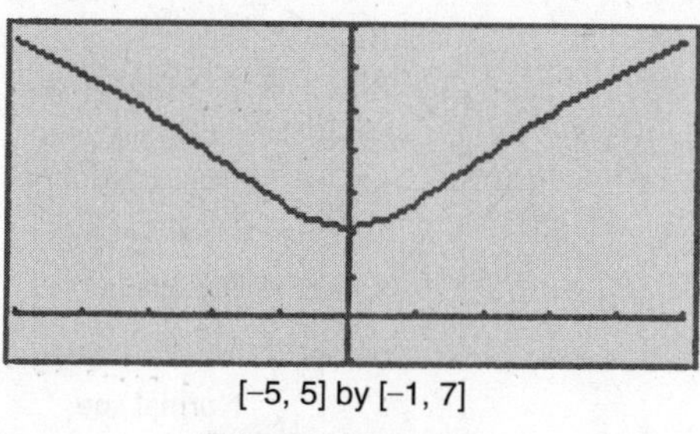

[−5, 5] by [−1, 7]

Figure 10.1-8

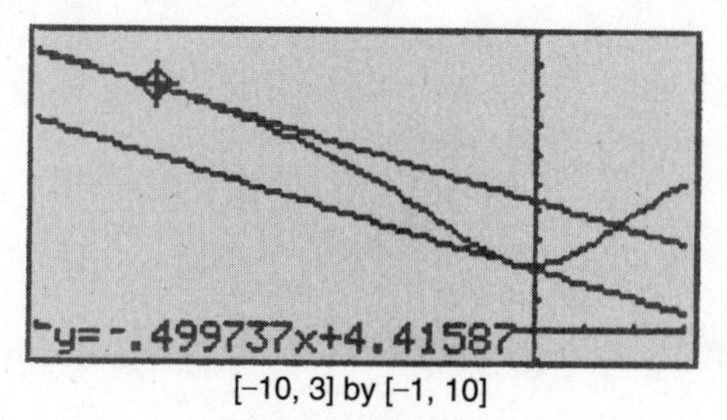

[−10, 3] by [−1, 10]

Figure 10.1-9

Step 1: Enter $y1 = 2 * \ln (x \wedge 2 + 3)$.

Step 2: Enter $y2 = d(y_1(x), x)$ and enter $y_3 = -\frac{1}{2}$.

Step 3: Using the [*Intersection*] function of the calculator for y_2 and y_3, you obtain $x = -7.61$ or $x = -0.39$.

Example 6

Using your calculator, find the value(s) of x at which the graphs of $y = 2x^2$ and $y = e^x$ have parallel tangents.

Step 1: Find $\frac{dy}{dx}$ for both $y = 2x^2$ and $y = e^x$.

$$y = 2x^2; \frac{dy}{dx} = 4x$$

$$y = e^x; \frac{dy}{dx} = e^x$$

Step 2: Find the x-coordinate of the points of tangency. Parallel tangents $\Rightarrow$ slopes are equal.

Set $4x = e^x \Rightarrow 4x - e^x = 0$.

Using the [*Solve*] function of the calculator, enter [*Solve*] $(4x - e^\wedge(x) = 0,\ x)$ and obtain $x = 2.15$ and $x = 0.36$.

- Watch out for different units of measure, e.g., the radius, r, is 2 feet, find $\frac{dr}{dt}$ in inches per second.

Normal Lines

The normal line to the graph of f at the point (x_1, y_1) is the line perpendicular to the tangent line at (x_1, y_1). (See Figure 10.1-10.)

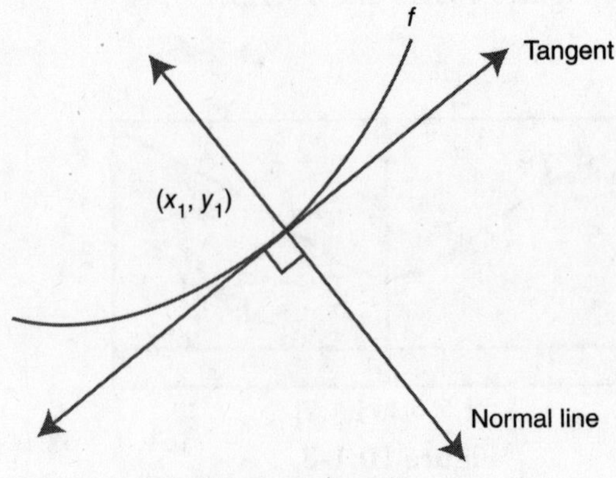

Figure 10.1-10

Note that the slope of the normal line and the slope of the tangent line at any point on the curve are negative reciprocals, provided that both slopes exist.

$(m_{\text{normal line}})(m_{\text{tangent line}}) = -1.$

Special Cases:
(See Figure 10.1-11.)
At these points, $m_{\text{tangent}} = 0$; but m_{normal} does not exist.

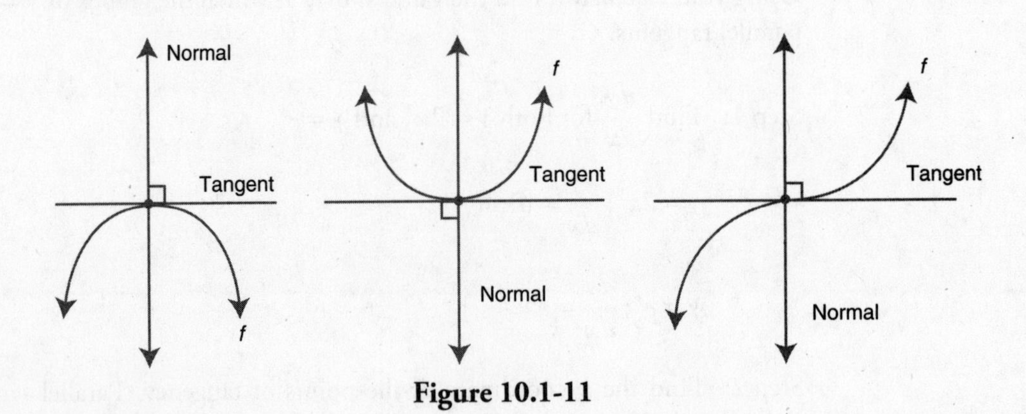

Figure 10.1-11

(See Figure 10.1-12.)
At these points, m_{tangent} does not exist; however, $m_{\text{normal}} = 0$.

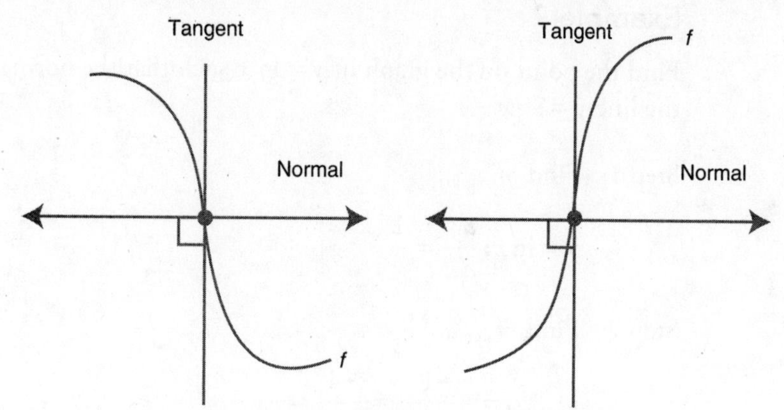

Figure 10.1-12

Example 1

Write an equation for each normal line to the graph of $y = 2 \sin x$ for $0 \leq x \leq 2\pi$ that has a slope of $\frac{1}{2}$.

Step 1: Find m_{tangent}.

$$y = 2 \sin x; \quad \frac{dy}{dx} = 2 \cos x$$

Step 2: Find m_{normal}.

$$m_{\text{normal}} = -\frac{1}{m_{\text{tangent}}} = -\frac{1}{2 \cos x}$$

Set $m_{\text{normal}} = \frac{1}{2} \Rightarrow -\frac{1}{2 \cos x} = \frac{1}{2} \Rightarrow \cos x = -1$

$\Rightarrow x = \cos^{-1}(-1)$ or $x = \pi$. (See Figure 10.1-13.)

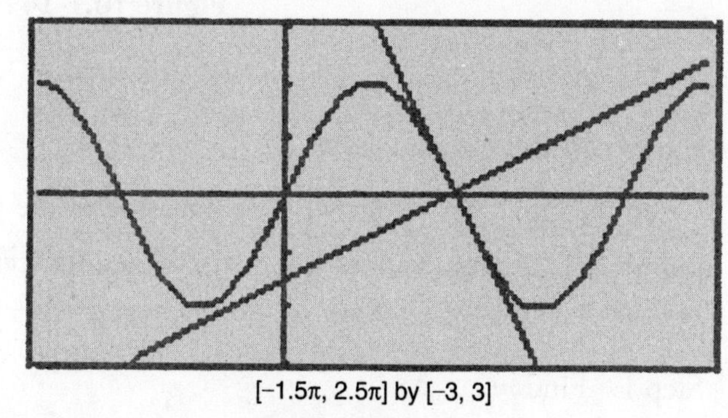

[−1.5π, 2.5π] by [−3, 3]

Figure 10.1-13

Step 3: Write equation of normal line.

At $x = \pi$, $y = 2 \sin x = 2(0) = 0$; $(\pi, 0)$.

Since $m = \frac{1}{2}$, equation of normal is:

$$y - 0 = \frac{1}{2}(x - \pi) \text{ or } y = \frac{1}{2}x - \frac{\pi}{2}.$$

Example 2

Find the point on the graph of $y = \ln x$ such that the normal line at this point is parallel to the line $y = -ex - 1$.

Step 1: Find m_{tangent}.

$$y = \ln x; \frac{dy}{dx} = \frac{1}{x}$$

Step 2: Find m_{normal}.

$$m_{\text{normal}} = \frac{-1}{m_{\text{tangent}}} = \frac{-1}{1/x} = -x$$

Slope of $y = -ex - 1$ is $-e$.

Since normal line is parallel to the line $y = -ex - 1$, set $m_{\text{normal}} = -e \Rightarrow -x = -e$ or $x = e$.

Step 3: Find the point on the graph. At $x = e$, $y = \ln x = \ln e = l$. Thus, the point of the graph of $y = \ln x$ at which the normal is parallel to $y = -ex - 1$ is $(e, 1)$. (See Figure 10.1-14.)

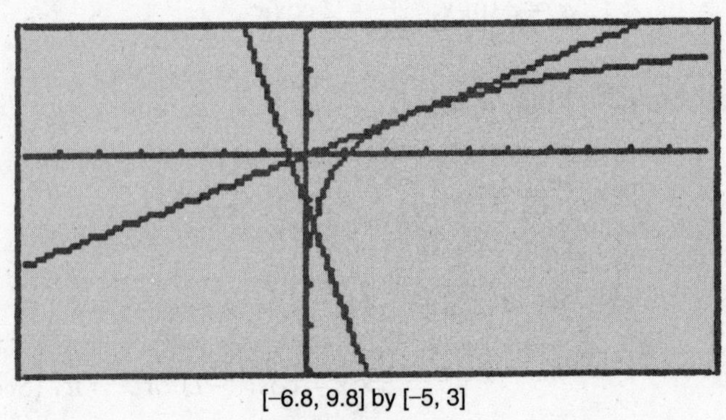

[-6.8, 9.8] by [-5, 3]

Figure 10.1-14

Example 3

Given the curve $y = \frac{1}{x}$: (a) write an equation of the normal line to the curve $y = \frac{1}{x}$ at the point (2, 1/2), and (b) does this normal line intersect the curve at any other point? If yes, find the point.

Step 1: Find m_{tangent}.

$$y = \frac{1}{x}; \frac{dy}{dx} = (-1)(x^{-2}) = -\frac{1}{x^2}$$

Step 2: Find m_{normal}.

$$m_{\text{normal}} = \frac{-1}{m_{\text{tangent}}} = \frac{-1}{-1/x^2} = x^2$$

At (2, 1/2), $m_{\text{normal}} = 2^2 = 4$.

Step 3: Write equation of normal line.

$m_{\text{normal}} = 4;\ (2, 1/2)$

Equation of normal line: $y - \frac{1}{2} = 4(x - 2)$, or $y = 4x - \frac{15}{2}$.

Step 4: Find other points of intersection.

$y = \frac{1}{x};\ y = 4x - \frac{15}{2}$

Using the [*Intersection*] function of your calculator, enter $y1 = \frac{1}{x}$ and $y2 = 4x - \frac{15}{2}$ and obtain $x = -0.125$ and $y = -8$. Thus, the normal line intersects the graph of $y = \frac{1}{x}$ at the point $(-0.125, -8)$ as well.

- Remember that $\int 1dx = x + C$ and $\frac{d}{dx}(1) = 0$.

10.2 Linear Approximations

Main Concepts: Tangent Line Approximation, Estimating the *n*th Root of a Number, Estimating the Value of a Trigonometric Function of an Angle

Tangent Line Approximation (or Linear Approximation)

An equation of the tangent line to a curve at the point $(a, f(a))$ is:
$y = f(a) + f'(a)(x - a)$, providing that f is differentiable at a. (See Figure 10.2-1.)
Since the curve of $f(x)$ and the tangent line are close to each other for points near $x = a$, $f(x) \approx f(a) + f'(a)(x - a)$.

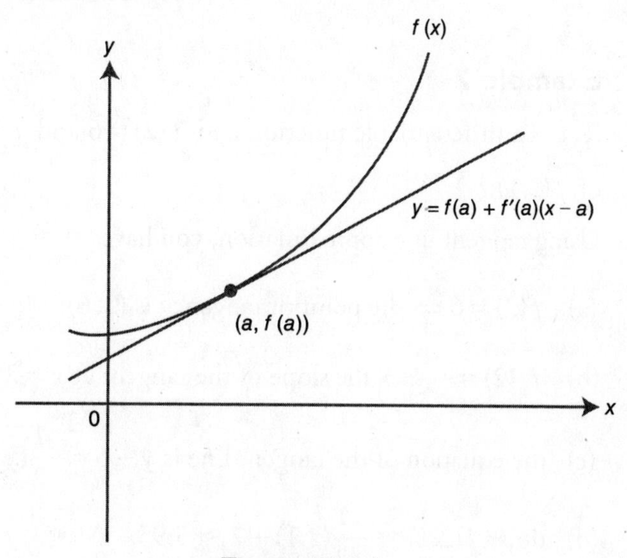

Figure 10.2-1

Example 1

Write an equation of the tangent line to $f(x)=x^3$ at (2, 8). Use the tangent line to find the approximate values of $f(1.90)$ and $f(2.01)$.

Differentiate $f(x)$: $f'(x)=3x^2$; $f'(2)=3(2)^2=12$. Since f is differentiable at $x=2$, an equation of the tangent at $x=2$ is:

$y=f(2)+f'(2)(x-2)$

$y=(2)^3+12(x-2)=8+12x-24=12x-16$

$f(1.90)\approx 12(1.90)-16=6.80$

$f(2.01)\approx 12(2.01)-16=8.12$. (See Figure 10.2-2.)

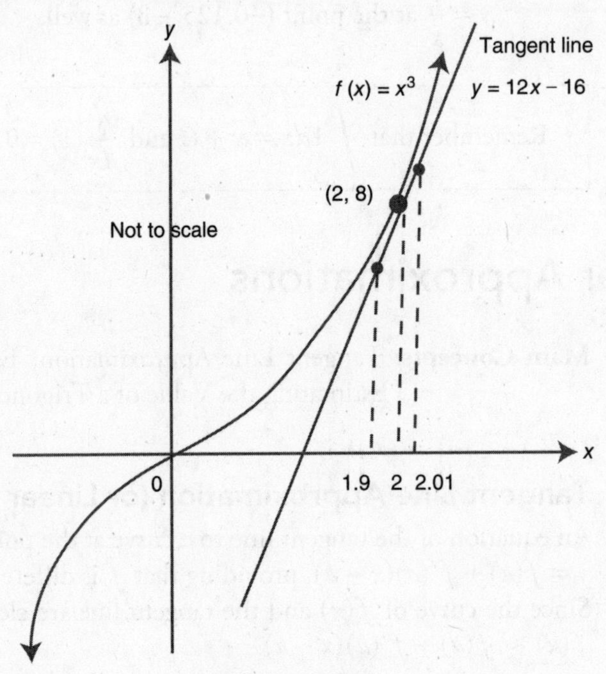

Figure 10.2-2

Example 2

If f is a differentiable function and $f(2)=6$ and $f'(2)=-\frac{1}{2}$, find the approximate value of $f(2.1)$.

Using tangent line approximation, you have

(a) $f(2)=6 \Rightarrow$ the point of tangency is $(2, 6)$;

(b) $f'(2)=-\frac{1}{2} \Rightarrow$ the slope of the tangent at $x=2$ is $m=-\frac{1}{2}$;

(c) the equation of the tangent line is $y-6=-\frac{1}{2}(x-2)$ or $y=-\frac{1}{2}x+7$;

(d) thus, $f(2.1)\approx -\frac{1}{2}(2.1)+7\approx 5.95$.

Example 3

The slope of a function at any point (x, y) is $-\frac{x+1}{y}$. The point (3, 2) is on the graph of f.

(a) Write an equation of the line tangent to the graph of f at $x = 3$.

(b) Use the tangent line in part (a) to approximate $f(3.1)$.

(a) Let $y = f(x)$, then $\frac{dy}{dx} = -\frac{x+1}{y}$

$$\left.\frac{dy}{dx}\right|_{x=3,\ y=2} = -\frac{3+1}{2} = -2.$$

Equation of tangent line: $y - 2 = -2(x - 3)$ or $y = -2x + 8$.

(b) $f(3.1) \approx -2(3.1) + 8 \approx 1.8$

Estimating the *n*th Root of a Number

Another way of expressing the tangent line approximation is:
$f(a + \Delta x) \approx f(a) + f'(a)\Delta x$, where Δx is a relatively small value.

Example 1

Find the approximate value of $\sqrt{50}$ using linear approximation.

Using $f(a + \Delta x) \approx f(a) + f'(a)\Delta x$, let $f(x) = \sqrt{x}$; $a = 49$ and $\Delta x = 1$.

Thus, $f(49 + 1) \approx f(49) + f'(49)(1) \approx \sqrt{49} + \frac{1}{2}(49)^{-1/2}(1) \approx 7 + \frac{1}{14} \approx 7.0714$.

Example 2

Find the approximate value of $\sqrt[3]{62}$ using linear approximation.

Let $f(x) = x^{1/3}$, $a = 64$, $\Delta x = -2$. Since $f'(x) = \frac{1}{3}x^{-2/3} = \frac{1}{3x^{2/3}}$ and

$f'(64) = \frac{1}{3(64)^{2/3}} = \frac{1}{48}$, you can use $f(a + \Delta x) \approx f(a) + f'(a)\Delta x$. Thus, $f(62) =$ $f(64 - 2) \approx f(64) + f'(64)(-2) \approx 4 + \frac{1}{48}(-2) \approx 3.958$.

- Use calculus notations and not calculator syntax, e.g., write $\int x^2 dx$ and not $\int (x\text{^}2, x)$.

Estimating the Value of a Trigonometric Function of an Angle

Example

Approximate the value of $\sin 31°$.

Note that you must express the angle measurement in radians before applying linear approximations. $30° = \frac{\pi}{6}$ radians and $1° = \frac{\pi}{180}$ radians.

Let $f(x) = \sin x$, $a = \frac{\pi}{6}$ and $\Delta x = \frac{\pi}{180}$.

Since $f'(x) = \cos x$ and $f'\left(\frac{\pi}{6}\right) = \cos\left(\frac{\pi}{6}\right) = \frac{\sqrt{3}}{2}$, you can use linear approximations:

$$f\left(\frac{\pi}{6} + \frac{\pi}{180}\right) \approx f\left(\frac{\pi}{6}\right) + f'\left(\frac{\pi}{6}\right)\left(\frac{\pi}{180}\right)$$

$$\approx \sin\frac{\pi}{6} + \left[\cos\left(\frac{\pi}{6}\right)\right]\left(\frac{\pi}{180}\right)$$

$$\approx \frac{1}{2} + \frac{\sqrt{3}}{2}\left(\frac{\pi}{180}\right) = 0.515.$$

10.3 Motion Along a Line

Main Concepts: Instantaneous Velocity and Acceleration, Vertical Motion, Horizontal Motion

Instantaneous Velocity and Acceleration

Position Function: $s(t)$

Instantaneous Velocity: $v(t) = s'(t) = \frac{ds}{dt}$

If particle is moving to the right →, then $v(t) > 0$.
If particle is moving to the left ←, then $v(t) < 0$.

Acceleration: $a(t) = v'(t) = \frac{dv}{dt}$ or $a(t) = s''(t) = \frac{d^2s}{dt^2}$

Instantaneous Speed: $|v(t)|$

Example 1

The position function of a particle moving on a straight line is $s(t) = 2t^3 - 10t^2 + 5$. Find (a) the position, (b) instantaneous velocity, (c) acceleration, and (d) speed of the particle at $t = 1$.

Solution:

(a) $s(1) = 2(1)^3 - 10(1)^2 + 5 = -3$

(b) $v(t) = s'(t) = 6t^2 - 20t$

$v(1) = 6(1)^2 - 20(1) = -14$

(c) $a(t) = v'(t) = 12t - 20$

$a(1) = 12(1) - 20 = -8$

(d) Speed $= |v(t)| = |v(1)| = 14$

Example 2

The velocity function of a moving particle is $v(t) = \frac{t^3}{3} - 4t^2 + 16t - 64$ for $0 \le t \le 7$.

What is the minimum and maximum acceleration of the particle on $0 \le t \le 7$?

$$v(t) = \frac{t^3}{3} - 4t^2 + 16t - 64$$

$$a(t) = v'(t) = t^2 - 8t + 16$$

(See Figure 10.3-1.) The graph of $a(t)$ indicates that:

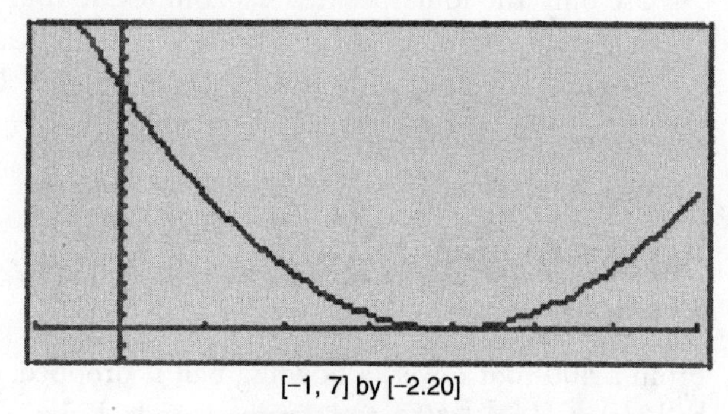

[−1, 7] by [−2.20]

Figure 10.3-1

(1) The minimum acceleration occurs at $t = 4$ and $a(4) = 0$.
(2) The maximum acceleration occurs at $t = 0$ and $a(0) = 16$.

Example 3

The graph of the velocity function is shown in Figure 10.3-2.

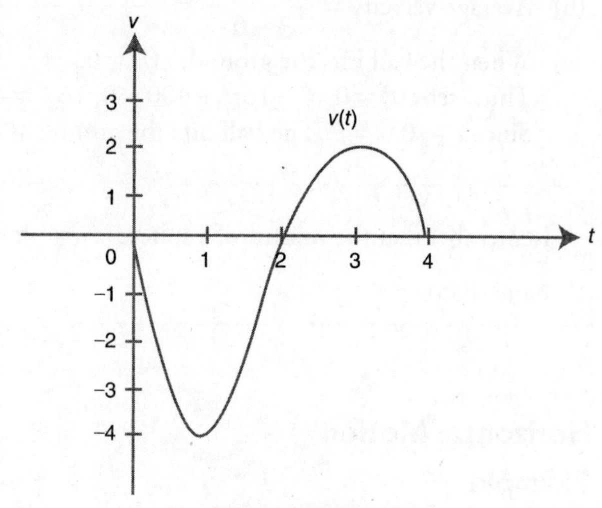

Figure 10.3-2

(a) When is the acceleration 0?
(b) When is the particle moving to the right?
(c) When is the speed the greatest?

Solution:

(a) $a(t) = v'(t)$ and $v'(t)$ is the slope of the tangent line to the graph of v. At $t = 1$ and $t = 3$, the slope of the tangent is 0.
(b) For $2 < t < 4$, $v(t) > 0$. Thus, the particle is moving to the right during $2 < t < 4$.
(c) Speed $= |v(t)|$ at $t = 1$, $v(t) = -4$.
Thus, speed at $t = 1$ is $|-4| = 4$, which is the greatest speed for $0 \leq t \leq 4$.

- Use only the four specified capabilities of your calculator to get your answer: plotting graphs, finding zeros, calculating numerical derivatives, and evaluating definite integrals. All other built-in capabilities can only be used to *check* your solution.

Vertical Motion

Example

From a 400-foot tower, a bowling ball is dropped. The position function of the bowling ball $s(t) = -16t^2 + 400$, $t \geq 0$ is in seconds. Find:

(a) the instantaneous velocity of the ball at $t = 2$ seconds.
(b) the average velocity for the first 3 seconds.
(c) when the ball will hit the ground.

Solution:

(a) $v(t) = s'(t) = -32t$
$v(2) = 32(2) = -64$ ft/sec
(b) Average velocity $= \dfrac{s(3) - s(0)}{3 - 0} = \dfrac{(-16(3)^2 + 400) - (0 + 400)}{3} = -48$ ft/sec.
(c) When the ball hits the ground, $s(t) = 0$.
Thus, set $s(t) = 0 \Rightarrow -16t^2 + 400 = 0$; $16t^2 = 400$; $t = \pm 5$.
Since $t \geq 0$, $t = 5$. The ball hits the ground at $t = 5$ sec.

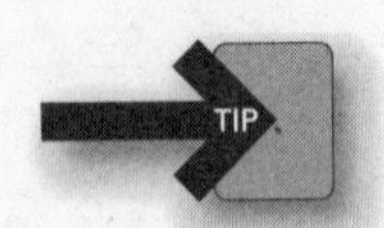

- Remember that the volume of a sphere is $v = \frac{4}{3}\pi r^3$ and that the surface area is $s = 4\pi r^2$.
Note that $v' = s$.

Horizontal Motion

Example

The position function of a particle moving in a straight line is $s(t) = t^3 - 6t^2 + 9t - 1$, $t \geq 0$. Describe the motion of the particle.

Step 1: Find $v(t)$ and $a(t)$.
$v(t) = 3t^2 - 12t + 9$

$a(t) = 6t - 12$

Step 2: Set $v(t)$ and $a(t) = 0$.

Set $v(t) = 0 \Rightarrow 3t^2 - 12t + 9 = 0 \Rightarrow 3(t^2 - 4t + 3) = 0$
$\Rightarrow 3(t-1)(t-3) = 0$ or $t = 1$ or $t = 3$.

Set $a(t) = 0 \Rightarrow 6t - 12 = 0 \Rightarrow 6(t-2) = 0$ or $t = 2$.

Step 3: Determine the directions of motion. (See Figure 10.3-3.)

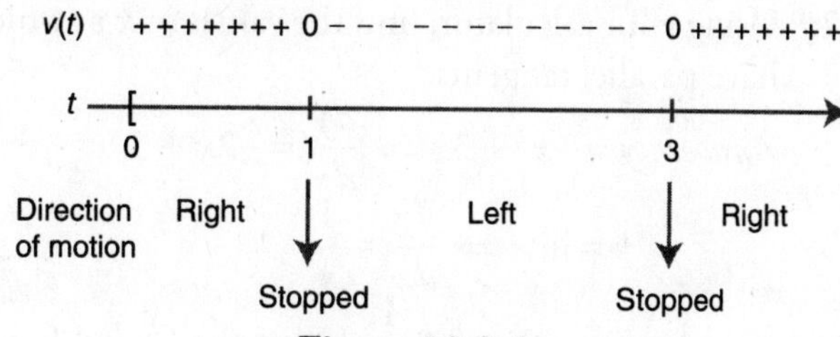

Figure 10.3-3

Step 4: Determine acceleration. (See Figure 10.3-4.)

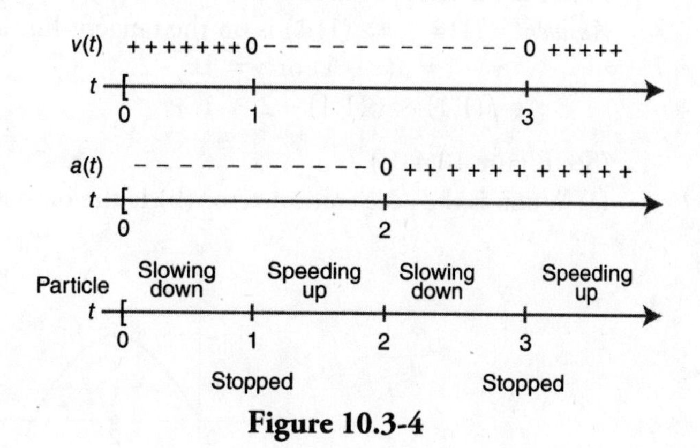

Figure 10.3-4

Step 5: Draw the motion of the particle. (See Figure 10.3-5.)
$s(0) = -1$, $s(1) = 3$, $s(2) = 1$ and $s(3) = -1$

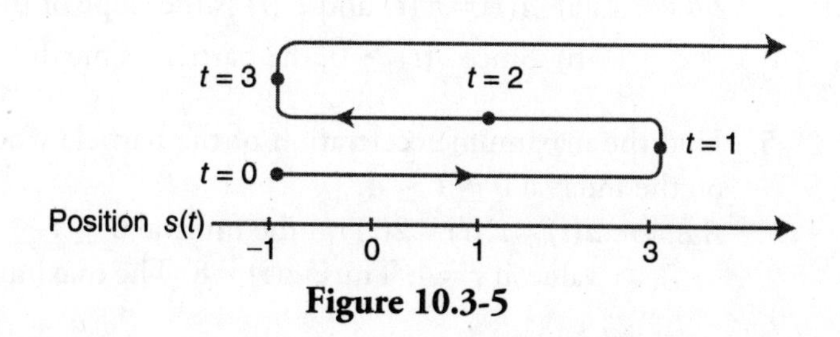

Figure 10.3-5

At $t = 0$, the particle is at -1 and moving to the right. It slows down and stops at $t = 1$ and at $t = 3$. It reverses direction (moving to the left) and speeds up until it reaches 1 at $t = 2$. It continues moving left but slows down and stops at -1 at $t = 3$. Then it reverses direction (moving to the right) again and speeds up indefinitely. (Note that "speeding up" is defined as when $|v(t)|$ increases and "slowing down" is defined as when $|v(t)|$ decreases.)

10.4 Rapid Review

1. Write an equation of the normal line to the graph $y = e^x$ at $x = 0$.

   Answer: $\left.\frac{dy}{dx}\right|_{x=0} e^x = e^x|_{x=0} = e^0 = 1 \Rightarrow m_{\text{normal}} = -1$

   At $x = 0$, $y = e^0 = 1 \Rightarrow$ you have the point (0, 1).
   Equation of normal line: $y - 1 = -1(x - 0)$ or $y = -x + 1$.

2. Using your calculator, find the values of x at which the function $y = -x^2 + 3x$ and $y = \ln x$ have parallel tangents.

   Answer: $y = -x^2 + 3x \Rightarrow \frac{dy}{dx} = -2x + 3$

   $y = \ln x \Rightarrow \frac{dy}{dx} = \frac{1}{x}$

   Set $-2x + 3 = \frac{1}{x}$. Using the [*Solve*] function on your calculator, enter

   [*Solve*] $\left(-2x + 3 = \frac{1}{x}, x\right)$ and obtain $x = 1$ or $x = \frac{1}{2}$.

3. Find the linear approximation of $f(x) = x^3$ at $x = 1$ and use the equation to find $f(1.1)$.

   Answer: $f(1) = 1 \Rightarrow (1, 1)$ is on the tangent line and $f'(x) = 3x^2 \Rightarrow f'(1) = 3$.
   $y - 1 = 3(x - 1)$ or $y = 3x - 2$.
   $f(1.1) \approx 3(1.1) - 2 \approx 1.3$

4. (See Figure 10.4-1.)
   (a) When is the acceleration zero? (b) Is the particle moving to the right or left?

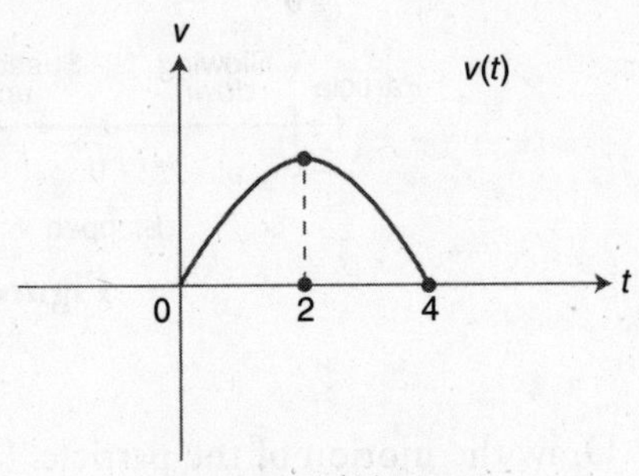

Figure 10.4-1

   Answer: (a) $a(t) = v'(t)$ and $v'(t)$ is the slope of the tangent. Thus, $a(t) = 0$ at $t = 2$.
   (b) Since $v(t) \geq 0$, the particle is moving to the right.

5. Find the maximum acceleration of the particle whose velocity function is $v(t) = t^2 + 3$ on the interval $0 \leq t \leq 4$.

   Answer: $a(t) = v'(t) = 2(t)$ on the interval $0 \leq t \leq 4$, $a(t)$ has its maximum value at $t = 4$. Thus, $a(t) = 8$. The maximum acceleration is 8.

10.5 Practice Problems

Part A—The use of a calculator is not allowed.

1. Find the linear approximation of $f(x) = (1+x)^{1/4}$ at $x = 0$ and use the equation to approximate $f(0.1)$.

2. Find the approximate value of $\sqrt[3]{28}$ using linear approximation.

3. Find the approximate value of $\cos 46°$ using linear approximation.

4. Find the point on the graph of $y = |x^3|$ such that the tangent at the point is parallel to the line $y - 12x = 3$.

5. Write an equation of the normal line to the graph of $y = e^x$ at $x = \ln 2$.

6. If the line $y - 2x = b$ is tangent to the graph $y = -x^2 + 4$, find the value of b.

7. If the position function of a particle is $s(t) = \frac{t^3}{3} - 3t^2 + 4$, find the velocity and position of the particle when its acceleration is 0.

8. The graph in Figure 10.5-1 represents the distance in feet covered by a moving particle in t seconds. Draw a sketch of the corresponding velocity function.

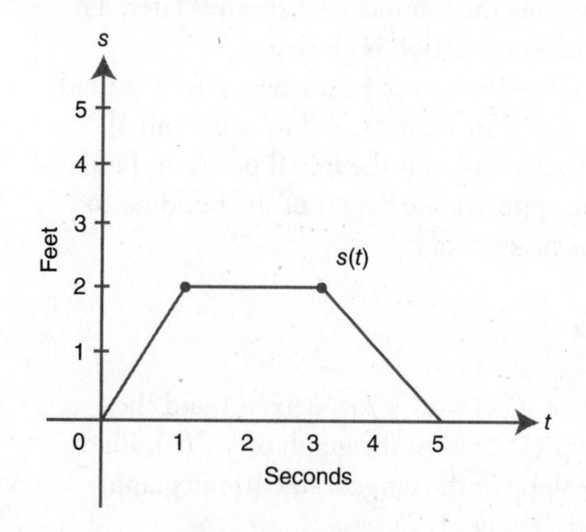

Figure 10.5-1

9. The position function of a moving particle is shown in Figure 10.5-2.

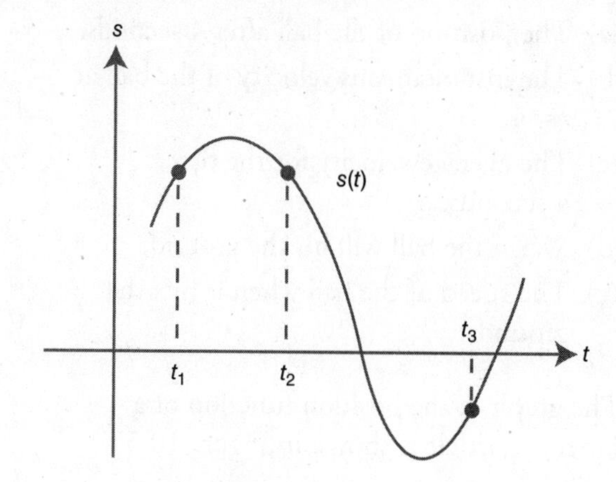

Figure 10.5-2

For which value(s) of $t(t_1, t_2, t_3)$ is:

(a) the particle moving to the left?
(b) the acceleration negative?
(c) the particle moving to the right and slowing down?

10. The velocity function of a particle is shown in Figure 10.5-3.

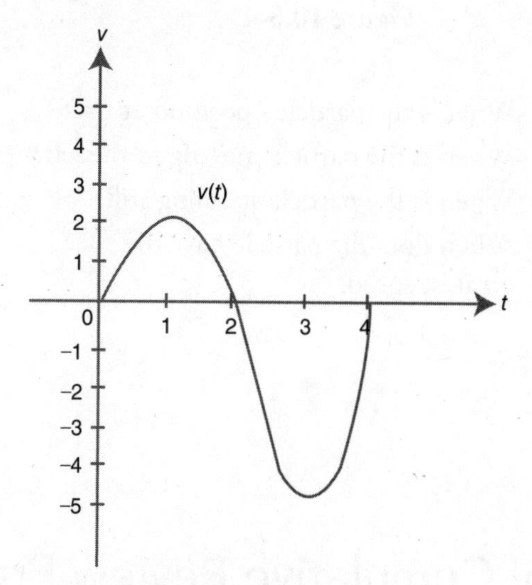

Figure 10.5-3

(a) When does the particle reverse direction?
(b) When is the acceleration 0?
(c) When is the speed the greatest?

11. A ball is dropped from the top of a 640-foot building. The position function of the ball is $s(t) = -16t^2 + 640$, where t is measured in seconds and $s(t)$ is in feet. Find:

    (a) The position of the ball after 4 seconds.
    (b) The instantaneous velocity of the ball at $t = 4$.
    (c) The average velocity for the first 4 seconds.
    (d) When the ball will hit the ground.
    (e) The speed of the ball when it hits the ground.

12. The graph of the position function of a moving particle is shown in Figure 10.5-4.

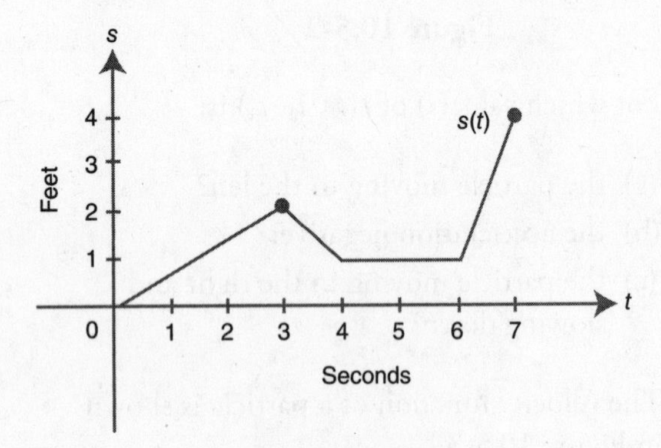

Figure 10.5-4

    (a) What is the particle's position at $t = 5$?
    (b) When is the particle moving to the left?
    (c) When is the particle standing still?
    (d) When does the particle have the greatest speed?

Part B—Calculators are allowed.

13. The position function of a particle moving on a line is $s(t) = t^3 - 3t^2 + 1,\ t \geq 0$ where t is measured in seconds and s in meters. Describe the motion of the particle.

14. Find the linear approximation of $f(x) = \sin x$ at $x = \pi$. Use the equation to find the approximate value of $f\left(\frac{181\pi}{180}\right)$.

15. Find the linear approximation of $f(x) = \ln(1 + x)$ at $x = 2$.

16. Find the coordinates of each point on the graph of $y^2 = 4 - 4x^2$ at which the tangent line is vertical. Write an equation of each vertical tangent.

17. Find the value(s) of x at which the graphs of $y = \ln x$ and $y = x^2 + 3$ have parallel tangents.

18. The position functions of two moving particles are $s_1(t) = \ln t$ and $s_2(t) = \sin t$ and the domain of both functions is $1 \leq t \leq 8$. Find the values of t such that the velocities of the two particles are the same.

19. The position function of a moving particle on a line is $s(t) = \sin(t)$ for $0 \leq t \leq 2\pi$. Describe the motion of the particle.

20. A coin is dropped from the top of a tower and hits the ground 10.2 seconds later. The position function is given as $s(t) = -16t^2 - v_0 t + s_0$, where s is measured in feet, t in seconds, and v_0 is the initial velocity and s_0 is the initial position. Find the approximate height of the building to the nearest foot.

10.6 Cumulative Review Problems

(Calculator) indicates that calculators are permitted.

21. Find $\frac{dy}{dx}$ if $y = x \sin^{-1}(2x)$.

22. Given $f(x) = x^3 - 3x^2 + 3x - 1$ and the point (1, 2) is on the graph of $f^{-1}(x)$. Find the slope of the tangent line to the graph of $f^{-1}(x)$ at (1, 2).

23. Evaluate $\lim_{x \to 100} \frac{x - 100}{\sqrt{x} - 10}$.

24. A function f is continuous on the interval $(-1, 8)$ with $f(0) = 0$, $f(2) = 3$, and $f(8) = 1/2$ and has the following properties:

INTERVALS	(–1, 2)	x = 2	(2, 5)	x = 5	(5, 8)
f'	+	0	–	–	–
f''	–	–	–	0	+

(a) Find the intervals on which f is increasing or decreasing.
(b) Find where f has its absolute extrema.
(c) Find where f has the points of inflection.
(d) Find the intervals on which f is concave upward or downward.
(e) Sketch a possible graph of f.

25. The graph of the velocity function of a moving particle for $0 \le t \le 8$ is shown in Figure 10.6-1. Using the graph:

(a) Estimate the acceleration when $v(t) = 3$ ft/sec.
(b) Find the time when the acceleration is a minimum.

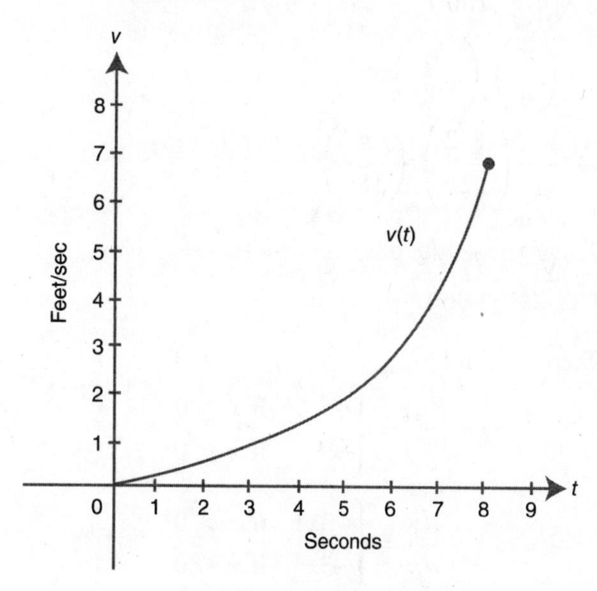

Figure 10.6-1

10.7 Solutions to Practice Problems

Part A—The use of a calculator is not allowed.

1. Equation of tangent line:

$y = f(a) + f'(a)(x - a)$

$f'(x) = \frac{1}{4}(1 + x)^{-3/4}(1) = \frac{1}{4}(1 + x)^{-3/4}$

$f'(0) = \frac{1}{4}$ and $f(0) = 1$;

thus, $y = 1 + \frac{1}{4}(x - 0) = 1 + \frac{1}{4}x$.

$f(0.1) = 1 + \frac{1}{4}(0.1) = 1.025$

2. $f(a + \Delta x) \approx f(a) + f'(a)\Delta x$

Let $f(x) = \sqrt[3]{x}$ and $f(28) = f(27 + 1)$.

Then $f'(x) = \frac{1}{3}(x)^{-2/3}$,

$f'(27) = \frac{1}{27}$, and $f(27) = 3$.

$f(27 + 1) \approx f(27) + f'(27)(1) \approx 3 + \left(\frac{1}{27}\right)(1) \approx 3.\overline{037}$

3. $f(a + \Delta x) \approx f(a) + f'(a)\Delta x$

Convert to radians:

$\frac{46}{180} = \frac{a}{\pi} \Rightarrow a = \frac{23\pi}{90}$ and $1° = \frac{\pi}{180}$;

$45° = \frac{\pi}{4}$.

Let $f(x) = \cos x$ and $f(45°) =$

$f\left(\frac{\pi}{4}\right) = \cos\left(\frac{\pi}{4}\right) = \frac{\sqrt{2}}{2}$.

Then $f'(x) = -\sin x$ and

$$f'(45°) = f'\left(\frac{\pi}{4}\right) = -\frac{\sqrt{2}}{2}$$

$$f(46°) = f\left(\frac{23\pi}{90}\right) = f\left(\frac{\pi}{4} + \frac{\pi}{180}\right)$$

$$f\left(\frac{\pi}{4} + \frac{\pi}{180}\right) \approx f\left(\frac{\pi}{4}\right) +$$

$$f'\left(\frac{\pi}{4}\right)\left(\frac{\pi}{180}\right) \approx$$

$$\frac{\sqrt{2}}{2} - \left(\frac{\sqrt{2}}{2}\right)\left(\frac{\pi}{180}\right)$$

$$\approx \frac{\sqrt{2}}{2} - \frac{\pi\sqrt{2}}{360}$$

4. Step 1: Find m_{tangent}.

$$y = |x^3| = \begin{cases} x^3 & \text{if } x \geq 0 \\ -x^3 & \text{if } x < 0 \end{cases}$$

$$\frac{dy}{dx} = \begin{cases} 3x^2 & \text{if } x > 0 \\ -3x^2 & \text{if } x < 0 \end{cases}$$

Step 2: Set m_{tangent} = slope of line $y - 12x = 3$.
Since $y - 12x = 3 \Rightarrow y = 12x + 3$, then $m = 12$.
Set $3x^2 = 12 \Rightarrow x = \pm 2$ since $x \geq 0$, $x = 2$.
Set $-3x^2 = 12 \Rightarrow x^2 = -4$. Thus, $\varnothing$.

Step 3: Find the point on the curve. (See Figure 10.7-1.)

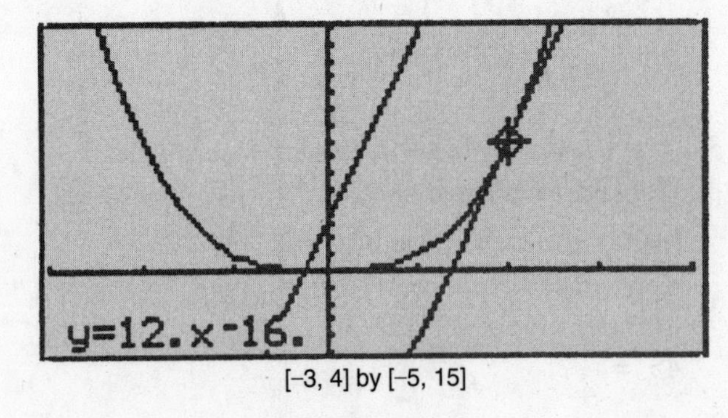

[−3, 4] by [−5, 15]

Figure 10.7-1

At $x = 2$, $y = x^3 = 2^3 = 8$.
Thus, the point is (2, 8).

5. Step 1: Find m_{tangent}.

$$y = e^x; \quad \frac{dy}{dx} = e^x$$

$$\left.\frac{dy}{dx}\right|_{x=\ln 2} = e^{\ln 2} = 2$$

Step 2: Find m_{normal}.
At $x = \ln 2$, $m_{\text{normal}} = \dfrac{-1}{m_{\text{tangent}}} = -\dfrac{1}{2}$.

Step 3: Write equation of the normal line.
At $x = \ln 2$, $y = e^x = e^{\ln 2} = 2$. Thus, the point of tangency is (ln 2, 2).
The equation of the normal line:

$$y - 2 = -\frac{1}{2}(x - \ln 2) \text{ or}$$

$$y = -\frac{1}{2}(x - \ln 2) + 2.$$

6. Step 1: Find m_{tangent}.

$$y = -x^2 + 4; \quad \frac{dy}{dx} = -2x.$$

Step 2: Find the slope of line $y - 2x = b$
$y - 2x = b \Rightarrow y = 2x + b$ or $m = 2$.

Step 3: Find point of tangency.
Set m_{tangent} = slope of line $y - 2x = b \Rightarrow -2x = 2 \Rightarrow x = -1$.
At $x = -1$, $y = -x^2 + 4 = -(-1)^2 + 4 = 3$; $(-1, 3)$.

Step 4: Find b.
Since the line $y - 2x = b$ passes through the point $(-1, 3)$, thus $3 - 2(-1) = b$ or $b = 5$.

7. $v(t) = s'(t) = t^2 - 6t$;
$a(t) = v'(t) = s''(t) = 2t - 6$
Set $a(t) = 0 \Rightarrow 2t - 6 = 0$ or $t = 3$.
$v(3) = (3)^2 - 6(3) = -9$;

$$s(3) = \frac{(3)^3}{3} - 3(3)^2 + 4 = -14.$$

8. On the interval (0, 1), the slope of the line segment is 2. Thus, the velocity $v(t) = 2$ ft/sec. On the interval (1, 3), $v(t) = 0$ and on (3, 5), $v(t) = -1$. (See Figure 10.7-2.)

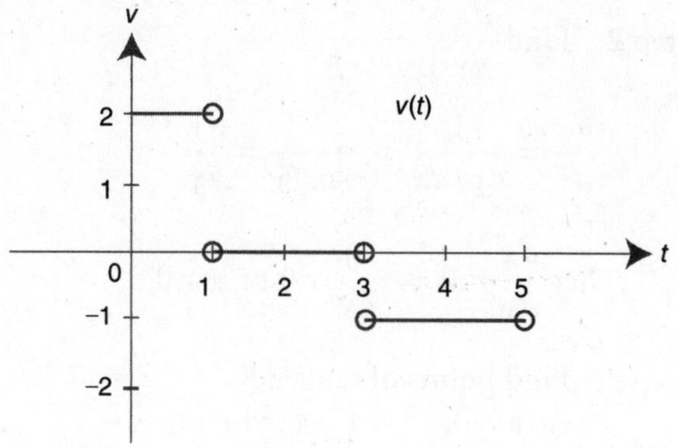

Figure 10.7-2

9. (a) At $t = t_2$, the slope of the tangent is negative. Thus, the particle is moving to the left.
   (b) At $t = t_1$, and at $t = t_2$, the curve is concave downward $\Rightarrow \frac{d^2s}{dt^2} =$ acceleration is negative.
   (c) At $t = t_1$, the slope > 0, and thus, the particle is moving to the right. The curve is concave downward $\Rightarrow$ the particle is slowing down.

10. (a) At $t = 2$, $v(t)$ changes from positive to negative. Thus, the particle reverses its direction.
   (b) At $t = 1$, and at $t = 3$, the slope of the tangent to the curve is 0. Thus, the acceleration is 0.
   (c) At $t = 3$, speed is equal to $|-5| = 5$ and 5 is the greatest speed.

11. (a) $s(4) = -16(4)^2 + 640 = 384$ ft
   (b) $v(t) = s'(t) = -32t$
   $v(4) = -32(4)$ ft/s $= -128$ ft/sec
   (c) Average velocity $= \frac{s(4) - s(0)}{4 - 0}$
   $= \frac{384 - 640}{4} = -64$ ft/sec.
   (d) Set $s(t) = 0 \Rightarrow -16t^2 + 640 = 0 \Rightarrow$ $16t^2 = 640$ or $t = \pm 2\sqrt{10}$.
   Since $t \geq 0$, $t = +2\sqrt{10}$ or $t \approx 6.32$ sec.
   (e) $|v(2\sqrt{10})| = |-32(2\sqrt{10})| = |-64\sqrt{10}|$ ft/s or ≈ 202.39 ft/sec

12. (a) At $t = 5$, $s(t) = 1$.
   (b) For $3 < t < 4$, $s(t)$ decreases. Thus, the particle moves to the left when $3 < t < 4$.
   (c) When $4 < t < 6$, the particle stays at 1.
   (d) When $6 < t < 7$, speed $= 2$ ft/sec, the greatest speed, which occurs where s has the greatest slope.

Part B—Calculators are allowed.

13. Step 1: $v(t) = 3t^2 - 6t$
   $a(t) = 6t - 6$
   Step 2: Set $v(t) = 0 \Rightarrow 3t^2 - 6t = 0 \Rightarrow$ $3t(t - 2) = 0$, or $t = 0$ or $t = 2$
   Set $a(t) = 0 \Rightarrow 6t - 6 = 0$ or $t = 1$.
   Step 3: Determine the directions of motion. (See Figure 10.7-3.)

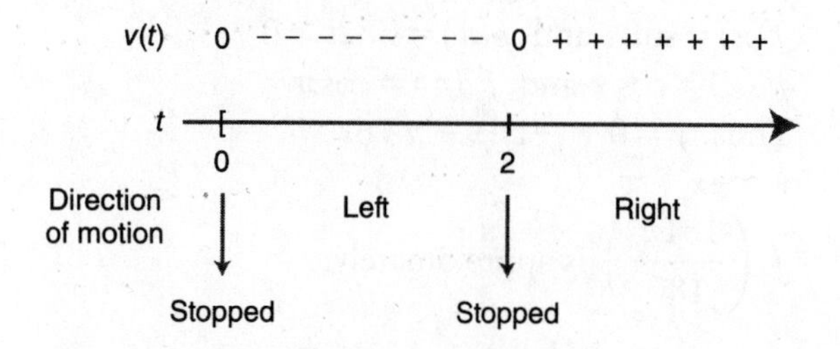

Figure 10.7-3

   Step 4: Determine acceleration. (See Figure 10.7-4.)

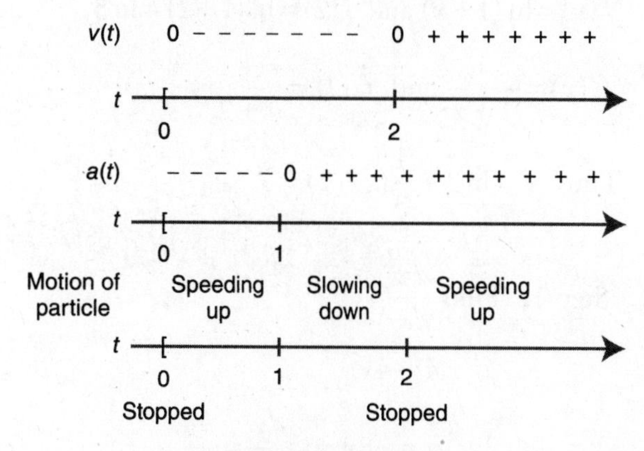

Figure 10.7-4

   Step 5: Draw the motion of the particle. (See Figure 10.7-5.)
   $s(0) = 1$, $s(1) = -1$, and $s(2) = -3$.

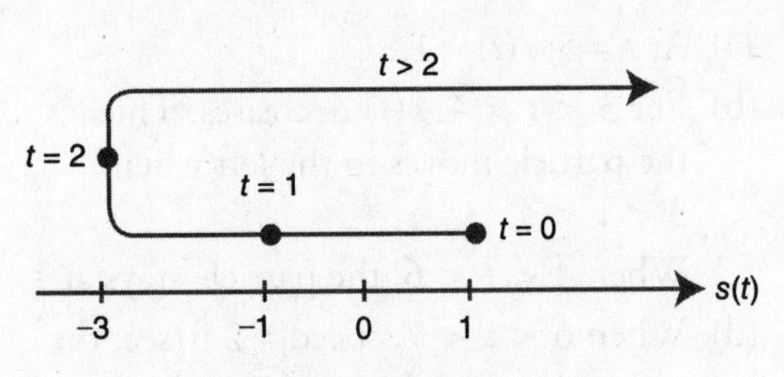

Figure 10.7-5

The particle is initially at 1 ($t = 0$). It moves to the left speeding up until $t = 1$, when it reaches -1. Then it continues moving to the left, but slowing down until $t = 2$ at -3. The particle reverses direction, moving to the right and speeding up indefinitely.

14. Linear approximation:
$y = f(a) + f'(a)(x - a)$ $a = \pi$
$f(x) = \sin x$ and $f(\pi) = \sin \pi = 0$
$f'(x) = \cos x$ and $f'(\pi) = \cos \pi = -1$.
Thus, $y = 0 + (-1)(x - \pi)$ or
$y = -x + \pi$.
$f\left(\dfrac{181\pi}{180}\right)$ is approximately:
$$y = -\left(\frac{181\pi}{180}\right) + \pi = \frac{-\pi}{180} \text{ or } \approx -0.0175.$$

15. $y = f(a) + f'(a)(x - a)$
$f(x) = \ln(1 + x)$ and $f(2) = \ln(1 + 2) = \ln 3$
$$f'(x) = \frac{1}{1+x} \text{ and } f'(2) = \frac{1}{1+2} = \frac{1}{3}.$$
Thus, $y = \ln 3 + \dfrac{1}{3}(x - 2)$.

16. Step 1: Find $\dfrac{dy}{dx}$.
$$y^2 = 4 - 4x^2$$
$$2y\frac{dy}{dx} = -8x \Rightarrow \frac{dy}{dx} = \frac{-4x}{y}$$

Step 2: Find $\dfrac{dx}{dy}$.
$$\frac{dx}{dy} = \frac{1}{dy/dx} = \frac{1}{-4x/y} = \frac{-y}{4x}$$
Set $\dfrac{dx}{dy} = 0 \Rightarrow \dfrac{-y}{4x} = 0$ or $y = 0$.

Step 3: Find points of tangency.
At $y = 0$, $y^2 = 4 - 4x^2$ becomes
$0 = 4 - 4x^2$
$\Rightarrow x = \pm 1$.
Thus, points of tangency are $(1, 0)$ and $(-1, 0)$.

Step 4: Write equations of vertical tangents $x = 1$ and $x = -1$.

17. Step 1: Find $\dfrac{dy}{dx}$ for $y = \ln x$ and $y = x^2 + 3$.
$$y = \ln x;\ \frac{dy}{dx} = \frac{1}{x}$$
$$y = x^2 + 3;\ \frac{dy}{dx} = 2x$$

Step 2: Find the x-coordinate of point(s) of tangency.
Parallel tangents $\Rightarrow$ slopes are equal. Set $\dfrac{1}{x} = 2x$.
Using the [*Solve*] function of your calculator, enter [*Solve*] $\left(\dfrac{1}{x} = 2x,\ x\right)$ and obtain
$x = \dfrac{\sqrt{2}}{2}$ or $x = \dfrac{-\sqrt{2}}{2}$. Since for $y = \ln x$, $x > 0$, $x = \dfrac{\sqrt{2}}{2}$.

18. $s_1(t) = \ln t$ and $s_1'(t) = \dfrac{1}{t}$; $1 \le t \le 8$.
$s_2(t) = \sin(t)$ and
$s_2'(t) = \cos(t)$; $1 \le t \le 8$.
Enter $y1 = \dfrac{1}{x}$ and $y2 = \cos(x)$. Use the [*Intersection*] function of the calculator and obtain $t = 4.917$ and $t = 7.724$.

19. Step 1: $s(t) = \sin t$
$v(t) = \cos t$
$a(t) = -\sin t$

Step 2: Set $v(t) = 0 \Rightarrow \cos t = 0$;
$t = \dfrac{\pi}{2}$ and $\dfrac{3\pi}{2}$.
Set $a(t) = 0 \Rightarrow -\sin t = 0$;
$t = \pi$ and 2π.

Step 3: Determine the directions of motion. (See Figure 10.7-6.)

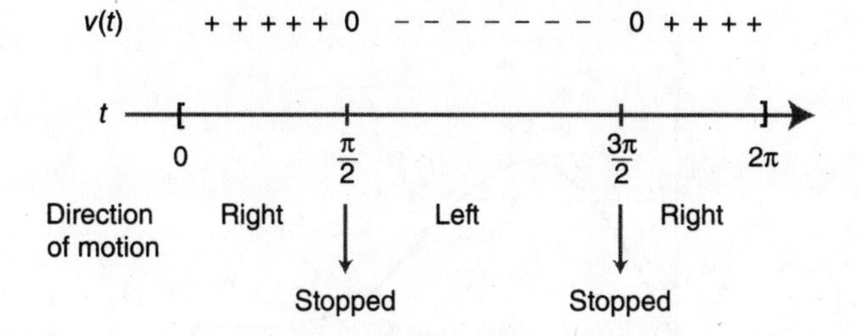

Figure 10.7-6

Step 4: Determine acceleration. (See Figure 10.7-7.)

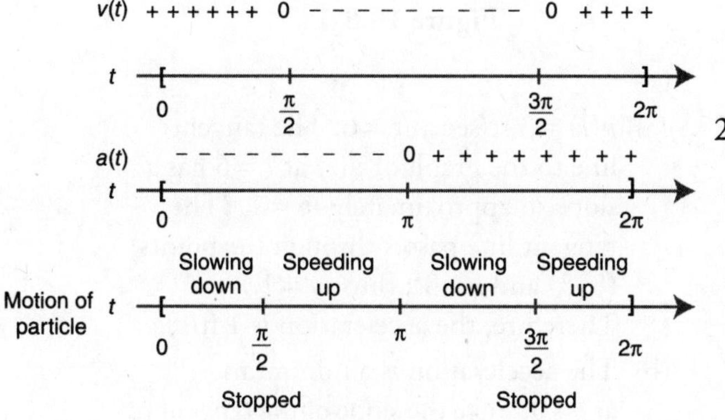

Figure 10.7-7

Step 5: Draw the motion of the particle. (See Figure 10.7-8.)

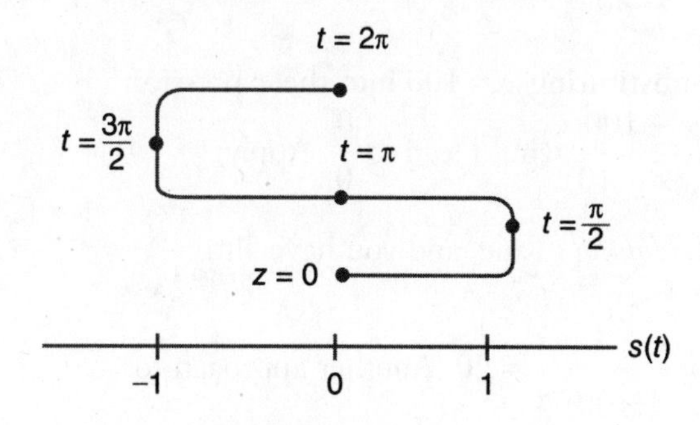

Figure 10.7-8

The particle is initially at 0, $s(0) = 0$. It moves to the right but slows down to a stop at 1 when $t = \dfrac{\pi}{2}$, $s\left(\dfrac{\pi}{2}\right) = 1$. It then turns and moves to the left speeding up until it reaches 0, when $t = \pi$, $s(\pi) = 0$ and continues to the left, but slowing down to a stop at -1 when $t = \dfrac{3\pi}{2}$, $s\left(\dfrac{3\pi}{2}\right) = -1$. It then turns around again, moving to the right, speeding up to 0 when $t = 2\pi$, $s(2\pi) = 0$.

20. $s(t) = -16t^2 + v_0 t + s_0$
s_0 = height of building and $v_0 = 0$.
Thus, $s(t) = -16t^2 + s_0$.
When the coin hits the ground, $s(t) = 0$, $t = 10.2$. Thus, set $s(t) = 0 \Rightarrow$
$-16t^2 + s_0 = 0 \Rightarrow -16(10.2)^2 + s_0 = 0$
$s_0 = 1664.64$ ft. The building is approximately 1665 ft tall.

10.8 Solutions to Cumulative Review Problems

21. Using product rule, let $u = x$; $v = \sin^{-1}(2x)$.

$$\frac{dy}{dx} = (1)\sin^{-1}(2x) + \frac{1}{\sqrt{1-(2x)^2}}(2)(x)$$

$$= \sin^{-1}(2x) + \frac{2x}{\sqrt{1-4x^2}}$$

22. Let $y = f(x) \Rightarrow y = x^3 - 3x^2 + 3x - 1$.
To find $f^{-1}(x)$, switch x and y: $x = y^3 - 3y^2 + 3y - 1$.

$$\frac{dx}{dy} = 3y^2 - 6y + 3$$

$$\frac{dy}{dx} = \frac{1}{dx/dy} = \frac{1}{3y^2 - 6y + 3}$$

$$\left.\frac{dy}{dx}\right|_{y=2} = \frac{1}{3(2)^2 - 6(2) + 3} = \frac{1}{3}$$

23. Substituting $x = 100$ into the expression $\frac{x - 100}{\sqrt{x} - 10}$ would lead to $\frac{0}{0}$. Apply *L'Hôpital's* Rule, and you have $\lim_{x \to 100} \frac{1}{\frac{1}{2}x^{-\frac{1}{2}}}$ or $\frac{1}{\frac{1}{2}(100)^{-\frac{1}{2}}} = 20$. Another approach to solve the problem is as follows. Multiply both numerator and denominator by the conjugate of the denominator $(\sqrt{x} + 10)$:

$$\lim_{x \to 100} \frac{(x - 100)}{(\sqrt{x} - 10)} \cdot \frac{(\sqrt{x} + 10)}{(\sqrt{x} + 10)} =$$

$$\lim_{x \to 100} \frac{(x - 100)(\sqrt{x} + 10)}{(x - 100)}$$

$$\lim_{x \to 100} (\sqrt{x} + 10) = 10 + 10 = 20.$$

An alternative solution is to factor the numerator:

$$\lim_{x \to 10} \frac{(\sqrt{x} - 10)(\sqrt{x} + 10)}{(\sqrt{x} - 10)} = 20.$$

24. (a) $f' > 0$ on $(-1, 2)$, f is increasing on $(-1, 2)$, $f' < 0$ on $(2, 8)$, f is decreasing on $(2, 8)$.
(b) At $x = 2$, $f' = 0$ and $f'' < 0$, thus at $x = 2$, f has a relative maximum. Since it is the only relative extremum on the interval, it is an absolute maximum. Since f is a continuous function on a closed interval and at its endpoints $f(-1) < 0$ and $f(8) = 1/2$, f has an absolute minimum at $x = -1$.
(c) At $x = 5$, f has a change of concavity and f' exists at $x = 5$.
(d) $f'' < 0$ on $(-1, 5)$, f is concave downward on $(-1, 5)$. $f'' > 0$ on $(5, 8)$, f is concave upward on $(5, 8)$.
(e) A possible graph of f is given in Figure 10.8-1.

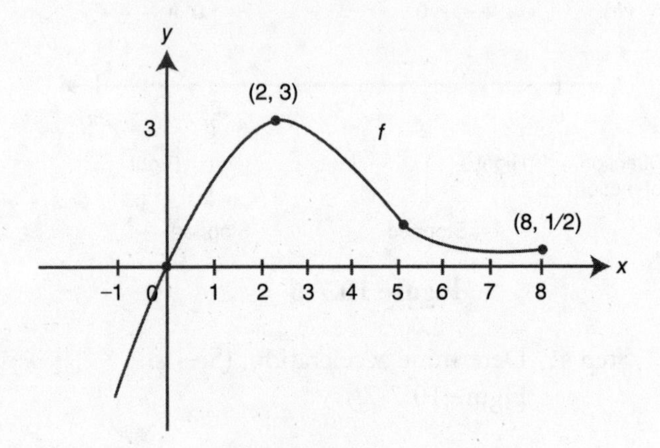

Figure 10.8-1

25. (a) $v(t) = 3$ ft/sec at $t = 6$. The tangent line to the graph of $v(t)$ at $t = 6$ has a slope of approximately $m = 1$. (The tangent line passes through the points (8, 5) and (6, 3); thus $m = 1$.) Therefore, the acceleration is 1 ft/sec^2.
(b) The acceleration is a minimum at $t = 0$, since the slope of the tangent to the curve of $v(t)$ is the smallest at $t = 0$.

Big Idea 3: Integrals and the Fundamental Theorems of Calculus

Integration

IN THIS CHAPTER

Summary: On the AP Calculus AB exam, you will be asked to evaluate integrals of various functions. In this chapter, you will learn several methods of evaluating integrals including U-Substitution. Also, you will be given a list of common integration and differentiation formulas, and a comprehensive set of practice problems. It is important that you work out these problems and check your solutions with the given explanations.

Key Ideas

- Evaluating Integrals of Algebraic Functions
- Integration Formulas
- U-Substitution Method Involving Algebraic Functions
- U-Substitution Method Involving Trigonometric Functions
- U-Substitution Method Involving Inverse Trigonometric Functions
- U-Substitution Method Involving Logarithmic and Exponential Functions

11.1 Evaluating Basic Integrals

Main Concepts: Antiderivatives and Integration Formulas, Evaluating Integrals

- Answer all parts of a question from Section II even if you think your answer to an earlier part of the question might not be correct. Also, if you do not know the answer to part one of a question, and you need it to answer part two, just make it up and continue.

Antiderivatives and Integration Formulas

Definition: A function F is an antiderivative of another function f if $F'(x) = f(x)$ for all x in some open interval. Any two antiderivatives of f differ by an additive constant C. We denote the set of antiderivatives of f by $\int f(x)dx$, called the indefinite integral of f.

Integration Rules:

1. $\int f(x)dx = F(x) + C \Leftrightarrow F'(x) = f(x)$
2. $\int a\, f(x)dx = a \int f(x)dx$
3. $\int -f(x)dx = -\int f(x)dx$
4. $\int [f(x) \pm g(x)]\, dx = \int f(x)dx \pm \int g(x)dx$

Differentiation Formulas:

1. $\frac{d}{dx}(x) = 1$
2. $\frac{d}{dx}(ax) = a$
3. $\frac{d}{dx}(x^n) = nx^{n-1}$
4. $\frac{d}{dx}(\cos x) = -\sin x$
5. $\frac{d}{dx}(\sin x) = \cos x$
6. $\frac{d}{dx}(\tan x) = \sec^2 x$
7. $\frac{d}{dx}(\cot x) = -\csc^2 x$
8. $\frac{d}{dx}(\sec x) = \sec x \tan x$
9. $\frac{d}{dx}(\csc x) = -\csc x(\cot x)$

Integration Formulas:

1. $\int 1dx = x + C$
2. $\int a\,dx = ax + C$
3. $\int x^n dx = \frac{x^{n+1}}{n+1} + C, n \neq -1$
4. $\int \sin x\, dx = -\cos x + C$
5. $\int \cos x\, dx = \sin x + C$
6. $\int \sec^2 x\, dx = \tan x + C$
7. $\int \csc^2 x\, dx = -\cot x + C$
8. $\int \sec x(\tan x)\, dx = \sec x + C$
9. $\int \csc x(\cot x)\, dx = -\csc x + C$

Differentiation Formulas (cont.):

10. $\frac{d}{dx}(\ln x) = \frac{1}{x}$

11. $\frac{d}{dx}(e^x) = e^x$

12. $\frac{d}{dx}(a^x) = (\ln a)a^x$

13. $\frac{d}{dx}(\sin^{-1} x) = \frac{1}{\sqrt{1-x^2}}$

14. $\frac{d}{dx}(\tan^{-1} x) = \frac{1}{1+x^2}$

15. $\frac{d}{dx}(\sec^{-1} x) = \frac{1}{|x|\sqrt{x^2-1}}$

Integration Formulas (cont.):

10. $\int \frac{1}{x} dx = \ln|x| + C$

11. $\int e^x dx = e^x + C$

12. $\int a^x dx = \frac{a^x}{\ln a} + C \quad a > 0, a \neq 1$

13. $\int \frac{1}{\sqrt{1-x^2}} dx = \sin^{-1} x + C$

14. $\int \frac{1}{1+x^2} dx = \tan^{-1} x + C$

15. $\int \frac{1}{|x|\sqrt{x^2-1}} dx = \sec^{-1} x + C$

More Integration Formulas:

16. $\int \tan x \, dx = \ln|\sec x| + C$ or $-\ln|\cos x| + C$

17. $\int \cot x \, dx = \ln|\sin x| + C$ or $-\ln|\csc x| + C$

18. $\int \sec x \, dx = \ln|\sec x + \tan x| + C$

19. $\int \csc x \, dx = \ln|\csc x - \cot x| + C$

20. $\int \ln x \, dx = x \ln|x| - x + C$

21. $\int \frac{1}{\sqrt{a^2-x^2}} dx = \sin^{-1}\left(\frac{x}{a}\right) + C$

22. $\int \frac{1}{a^2+x^2} dx = \frac{1}{a}\tan^{-1}\left(\frac{x}{a}\right) + C$

23. $\int \frac{1}{x\sqrt{x^2-a^2}} dx = \frac{1}{a}\sec^{-1}\left|\frac{x}{a}\right| + C$ or $\frac{1}{a}\cos^{-1}\left|\frac{a}{x}\right| + C$

24. $\int \sin^2 x \, dx = \frac{x}{2} - \frac{\sin(2x)}{4} + C$. Note: $\sin^2 x = \frac{1-\cos 2x}{2}$

Note that after evaluating an integral, always check the result by taking the derivative of the answer (i.e., taking the derivative of the antiderivative).

- Remember that the volume of a right-circular cone is $v = \frac{1}{3}\pi r^2 h$ where r is the radius of the base and h is the height of the cone.

Evaluating Integrals

INTEGRAL	REWRITE	ANTIDERIVATIVE
$\int x^3 dx$		$\frac{x^4}{4} + C$
$\int dx$	$\int 1 dx$	$x + C$
$\int 5 dx$		$5x + C$
$\int \sqrt{x}\, dx$	$\int x^{1/2} dx$	$\frac{x^{3/2}}{3/2} + C$ or $\frac{2x^{3/2}}{3} + C$
$\int x^{5/2} dx$		$\frac{x^{7/2}}{7/2} + C$ or $\frac{2x^{7/2}}{7} + C$
$\int \frac{1}{x^2} dx$	$\int x^{-2} dx$	$\frac{x^{-1}}{-1} + C$ or $\frac{-1}{x} + C$
$\int \frac{1}{\sqrt[3]{x^2}} dx$	$\int \frac{1}{x^{2/3}} dx = \int x^{-2/3} dx$	$\frac{x^{1/3}}{1/3} + C$ or $3\sqrt[3]{x} + C$
$\int \frac{x+1}{x} dx$	$\int \left(1 + \frac{1}{x}\right) dx$	$x + \ln\lvert x\rvert + C$
$\int x(x^5 + 1) dx$	$\int (x^6 + x) dx$	$\frac{x^7}{7} + \frac{x^2}{2} + C$

Example 1

Evaluate $\int (x^5 - 6x^2 + x - 1)\, dx$.

Applying the formula $\int x^n dx = \frac{x^{n+1}}{n+1} + C, n \neq 1$.

$$\int (x^5 - 6x^2 + x - 1) dx = \frac{x^6}{6} - 2x^3 + \frac{x^2}{2} - x + C$$

Example 2

Evaluate $\int \left(\sqrt{x} + \frac{1}{x^3}\right) dx$.

Rewrite $\int \left(\sqrt{x} + \frac{1}{x^3}\right) dx$ as $\int (x^{1/2} + x^{-3})\, dx = \frac{x^{3/2}}{3/2} + \frac{x^{-2}}{-2} + C$

$$= \frac{2}{3}x^{3/2} - \frac{1}{2x^2} + C.$$

Example 3

If $\frac{dy}{dx} = 3x^2 + 2$, and the point (0, −1) lies on the graph of y, find y.

Since $\frac{dy}{dx} = 3x^2 + 2$, then y is an antiderivative of $\frac{dy}{dx}$. Thus,

$y = \int (3x^2 + 2)\, dx = x^3 + 2x + C$. The point (0, −1) is on the graph of y.

Thus, $y = x^3 + 2x + C$ becomes $-1 = 0^3 + 2(0) + C$ or $C = -1$. Therefore, $y = x^3 + 2x - 1$.

Example 4

Evaluate $\int \left(1 - \frac{1}{\sqrt[3]{x^4}}\right) dx$.

Rewrite as $\int \left(1 - \frac{1}{x^{4/3}}\right) dx = \int \left(1 - x^{-4/3}\right) dx$

$$= x - \frac{x^{-1/3}}{-1/3} + C = x + \frac{3}{\sqrt[3]{x}} + C.$$

Example 5

Evaluate $\int \frac{3x^2 + x - 1}{x^2} dx$.

Rewrite as $\int \left(3 + \frac{1}{x} - \frac{1}{x^2}\right) dx = \int \left(3 + \frac{1}{x} - x^{-2}\right) dx$

$$= 3x + \ln|x| - \frac{x^{-1}}{-1} + C = 3x + \ln|x| + \frac{1}{x} + C.$$

Example 6

Evaluate $\int \sqrt{x}\left(x^2 - 3\right) dx$.

Rewrite as See Example 5 $\int x^{1/2}\left(x^2 - 3\right) dx = \int \left(x^{5/2} - 3x^{1/2}\right) dx$

$$= \frac{x^{7/2}}{7/2} - \frac{3x^{3/2}}{3/2} + C = \frac{2}{7}x^{7/2} - 2\sqrt{x^3} + C.$$

Example 7

Evaluate $\int \left(x^3 - 4\sin x\right) dx$.

$$\int \left(x^3 - 4\sin x\right) dx = \frac{x^4}{4} + 4\cos x + C.$$

Example 8

Evaluate $\int (4\cos x - \cot x)\, dx$.

$$\int (4\cos x - \cot x)\, dx = 4\sin x - \ln|\sin x| + C.$$

Example 9

Evaluate $\int \frac{\sin x - 1}{\cos} dx.$

Rewrite as $\int \left(\frac{\sin x}{\cos x} - \frac{1}{\cos x} \right) dx = \int (\tan x - \sec x)\, dx = \int \tan x\, dx - \int \sec x\, dx$

$$= \ln|\sec x| - \ln|\sec x + \tan x| + C = \ln\left|\frac{\sec x}{\sec x + \tan x}\right| + C$$

or $- \ln|\sin x + 1| + C.$

Example 10

Evaluate $\int \frac{e^{2x}}{e^x} dx.$

Rewrite as $\int e^x dx = e^x + C.$

Example 11

Evaluate $\int \frac{3}{1+x^2} dx.$

Rewrite as $3\int \frac{1}{1+x^2} dx = 3\tan^{-1} x + C.$

Example 12

Evaluate $\int \frac{1}{\sqrt{9-x^2}} dx.$

Rewrite as $\int \frac{1}{\sqrt{3^2 - x^2}} dx = \sin^{-1}\left(\frac{x}{3}\right) + C.$

Example 13

Evaluate $\int 7^x dx.$

$\int 7^x\, dx = \frac{7^x}{\ln 7} + C$

Reminder: You can always check the result by taking the derivative of the answer.

- Be familiar with the instructions for the different parts of the exam before the day of the exam. Review the instructions in the practice tests provided at the end of this book.

11.2 Integration by U-Substitution

Main Concepts: The U-Substitution Method, U-Substitution and Algebraic Functions, U-Substitution and Trigonometric Functions, U-Substitution and Inverse Trigonometric Functions, U-Substitution and Logarithmic and Exponential Functions

The U-Substitution Method

The Chain Rule for Differentiation

$$\frac{d}{dx}F(g(x)) = f(g(x))g'(x), \quad \text{where } F' = f$$

The Integral of a Composite Function

If $f(g(x))$ and f' are continuous and $F' = f$, then

$$\int f(g(x))g'(x)dx = F(g(x)) + C.$$

Making a U-Substitution

Let $u = g(x)$, then $du = g'(x)dx$

$$\int f(g(x))g'(x)dx = \int f(u)du = F(u) + C = F(g(x)) + C.$$

Procedure for Making a U-Substitution

Steps:

1. Given $f(g(x))$; let $u = g(x)$.
2. Differentiate: $du = g'(x)dx$.
3. Rewrite the integral in terms of u.
4. Evaluate the integral.
5. Replace u by $g(x)$.
6. Check your result by taking the derivative of the answer.

U-Substitution and Algebraic Functions

Another Form of the Integral of a Composite Function

If f is a differentiable function, then

$$\int (f(x))^n f'(x)dx = \frac{(f(x))^{n+1}}{n+1} + C, n \neq -1.$$

Making a U-Substitution

Let $u = f(x)$; then $du = f'(x)dx$.

$$\int (f(x))^n f'(x)dx = \int u^n du = \frac{u^{n+1}}{n+1} + C = \frac{(f(x))^{n+1}}{n+1} + C, n \neq -1$$

Example 1

Evaluate $\int x(x+1)^{10}dx$.

Step 1: Let $u = x + 1$; then $x = u - 1$.

Step 2: Differentiate: $du = dx$.

Step 3: Rewrite: $\int (u-1)\,u^{10}du = \int \left(u^{11} - u^{10}\right)du$.

Step 4: Integrate: $\frac{u^{12}}{12} - \frac{u^{11}}{11} + C$.

Step 5: Replace u: $\frac{(x+1)^{12}}{12} - \frac{(x+1)^{11}}{11} + C$.

Step 6: Differentiate and Check: $\frac{12\,(x+1)^{11}}{12} - \frac{11\,(x+1)^{10}}{11}$

$= (x+1)^{11} - (x+1)^{10}$

$= (x+1)^{10}(x+1-1)$

$= (x+1)^{10}x$ or $x(x+1)^{10}$.

Example 2

Evaluate $\int x\sqrt{x-2}\,dx$.

Step 1: Let $u = x - 2$; then $x = u + 2$.

Step 2: Differentiate: $du = dx$.

Step 3: Rewrite: $\int (u+2)\sqrt{u}\,du = \int (u+2)u^{1/2}du = \int \left(u^{3/2} + 2u^{1/2}\right)du$.

Step 4: Integrate: $\frac{u^{5/2}}{5/2} + \frac{2u^{3/2}}{3/2} + C$.

Step 5: Replace: $\frac{2\,(x-2)^{5/2}}{5} + \frac{4\,(x-2)^{3/2}}{3} + C$.

Step 6: Differentiate and Check: $\left(\frac{5}{2}\right)\frac{2\,(x-2)^{3/2}}{5} + \left(\frac{3}{2}\right)\frac{4\,(x-2)^{1/2}}{3}$

$= (x-2)^{3/2} + 2(x-2)^{1/2}$

$= (x-2)^{1/2}[(x-2)+2]$

$= (x-2)^{1/2}\,x$ or $x\sqrt{x-2}$.

Example 3

Evaluate $\int (2x-5)^{2/3}dx$.

Step 1: Let $u = 2x - 5$.

Step 2: Differentiate: $du = 2dx \Rightarrow \frac{du}{2} = dx$.

Step 3: Rewrite: $\int u^{2/3}\frac{du}{2} = \frac{1}{2}\int u^{2/3}du.$

Step 4: Integrate: $\frac{1}{2}\left(\frac{u^{5/3}}{5/3}\right) + C = \frac{3u^{5/3}}{10} + C.$

Step 5: Replace u: $\frac{3(2x-5)^{5/3}}{10} + C.$

Step 6: Differentiate and Check: $\left(\frac{3}{10}\right)\left(\frac{5}{3}\right)(2x-5)^{2/3}(2) = (2x-5)^{2/3}.$

Example 4

Evaluate $\int \frac{x^2}{(x^3-8)^5}dx.$

Step 1: Let $u = x^3 - 8.$

Step 2: Differentiate: $du = 3x^2dx \Rightarrow \frac{du}{3} = x^2dx.$

Step 3: Rewrite: $\int \frac{1}{u^5}\frac{du}{3} = \frac{1}{3}\int \frac{1}{u^5}du = \frac{1}{3}\int u^{-5}du.$

Step 4: Integrate: $\frac{1}{3}\left(\frac{u^{-4}}{-4}\right) + C.$

Step 5: Replace u: $\frac{1}{-12}\left(x^3-8\right)^{-4} + C$ or $\frac{-1}{12\,(x^3-8)^4} + C.$

Step 6: Differentiate and Check: $\left(-\frac{1}{12}\right)(-4)\left(x^3-8\right)^{-5}\left(3x^2\right) = \frac{x^2}{(x^3-8)^5}.$

U-Substitution and Trigonometric Functions

Example 1

Evaluate $\int \sin 4x\, dx.$

Step 1: Let $u = 4x.$

Step 2: Differentiate: $du = 4\,dx$ or $\frac{du}{4} = dx.$

Step 3: Rewrite: $\int \sin u\frac{du}{4} = \frac{1}{4}\int \sin u\, du.$

Step 4: Integrate: $\frac{1}{4}(-\cos u) + C = -\frac{1}{4}\cos u + C.$

Step 5: Replace u: $-\frac{1}{4}\cos(4x) + C.$

Step 6: Differentiate and Check: $\left(-\frac{1}{4}\right)(-\sin 4x)(4) = \sin 4x.$

Example 2

Evaluate $\int 3\left(\sec^2 x\right)\sqrt{\tan x}\,dx$.

Step 1: Let $u = \tan x$.

Step 2: Differentiate: $du = \sec^2 x\,dx$.

Step 3: Rewrite: $3\int (\tan x)^{1/2} \sec^2 x\,dx = 3\int u^{1/2} du$.

Step 4: Integrate: $3\dfrac{u^{3/2}}{3/2} + C = 2u^{3/2} + C$.

Step 5: Replace u: $2(\tan x)^{3/2} + C$ or $2\tan^{3/2} x + C$.

Step 6: Differentiate and Check: $(2)\left(\dfrac{3}{2}\right)\left(\tan^{1/2} x\right)\left(\sec^2 x\right) = 3\left(\sec^2 x\right)\sqrt{\tan x}$.

Example 3

Evaluate $\int 2x^2 \cos\left(x^3\right) dx$.

Step 1: Let $u = x^3$.

Step 2: Differentiate: $du = 3x^2 dx \Rightarrow \dfrac{du}{3} = x^2 dx$.

Step 3: Rewrite: $2\int \left[\cos\left(x^3\right)\right] x^2 dx = 2\int \cos u \dfrac{du}{3} = \dfrac{2}{3}\int \cos u\,du$.

Step 4: Integrate: $\dfrac{2}{3}\sin u + C$.

Step 5: Replace u: $\dfrac{2}{3}\sin\left(x^3\right) + C$.

Step 6: Differentiate and Check: $\dfrac{2}{3}\left[\cos\left(x^3\right)\right] 3x^2 = 2x^2 \cos\left(x^3\right)$.

- Remember that the area of a semicircle is $\dfrac{1}{2}\pi r^2$. Do not forget the $\dfrac{1}{2}$. If the cross sections of a solid are semicircles the integral for the volume of the solid will involve $\left(\dfrac{1}{2}\right)^2$, which is $\dfrac{1}{4}$.

U-Substitution and Inverse Trigonometric Functions

Example 1

Evaluate $\int \dfrac{dx}{\sqrt{9 - 4x^2}}$.

Step 1: Let $u = 2x$.

Step 2: Differentiate: $du = 2dx$; $\dfrac{du}{2} = dx$.

Step 3: Rewrite: $\displaystyle\int \frac{1}{\sqrt{9-u^2}}\frac{du}{2} = \frac{1}{2}\int \frac{du}{\sqrt{3^2-u^2}}$.

Step 4: Integrate: $\frac{1}{2}\sin^{-1}\left(\frac{u}{3}\right) + C$.

Step 5: Replace u: $\frac{1}{2}\sin^{-1}\left(\frac{2x}{3}\right) + C$.

Step 6: Differentiate and Check: $\displaystyle\frac{1}{2}\frac{1}{\sqrt{1-\left(\frac{2x}{3}\right)^2}}\cdot\frac{2}{3} = \frac{1}{3}\frac{1}{\sqrt{1-\frac{4x^2}{9}}}$

$$= \frac{1}{\sqrt{9}}\frac{1}{\sqrt{1-\frac{4x^2}{9}}} = \frac{1}{\sqrt{9\left(1-\frac{4x^2}{9}\right)}}$$

$$= \frac{1}{\sqrt{9-4x^2}}.$$

Example 2

Evaluate $\displaystyle\int \frac{1}{x^2+2x+5}dx$.

Step 1: Rewrite: $\displaystyle\int \frac{1}{(x^2+2x+1)+4} = \int \frac{1}{(x+1)^2+2^2}dx$

$$= \int \frac{1}{2^2+(x+1)^2}dx.$$

Let $u = x + 1$.

Step 2: Differentiate: $du = dx$.

Step 3: Rewrite: $\displaystyle\int \frac{1}{2^2+u^2}du$.

Step 4: Integrate: $\frac{1}{2}\tan^{-1}\left(\frac{u}{2}\right) + C$.

Step 5: Replace u: $\frac{1}{2}\tan^{-1}\left(\frac{x+1}{2}\right) + C$.

Step 6: Differentiate and Check: $\displaystyle\left(\frac{1}{2}\right)\frac{1\left(\frac{1}{2}\right)}{1+[(x+1)/2]^2} = \left(\frac{1}{4}\right)\frac{1}{1+(x+1)^2/4}$

$$= \left(\frac{1}{4}\right)\frac{4}{4+(x+1)^2} = \frac{1}{x^2+2x+5}.$$

- If the problem gives you that the diameter of a sphere is 6 and you are using formulas such as $v = \frac{4}{3}\pi r^3$ or $s = 4\pi r^2$, do not forget that $r = 3$.

U-Substitution and Logarithmic and Exponential Functions

Example 1

Evaluate $\int \frac{x^3}{x^4 - 1} dx$.

Step 1: Let $u = x^4 - 1$.

Step 2: Differentiate: $du = 4x^3 dx \Rightarrow \frac{du}{4} = x^3 dx$.

Step 3: Rewrite: $\int \frac{1}{u}\frac{du}{4} = \frac{1}{4}\int \frac{1}{u} du$.

Step 4: Integrate: $\frac{1}{4} \ln|u| + C$.

Step 5: Replace u: $\frac{1}{4} \ln|x^4 - 1| + C$.

Step 6: Differentiate and Check: $\left(\frac{1}{4}\right) \frac{1}{x^4 - 1} (4x^3) = \frac{x^3}{x^4 - 1}$.

Example 2

Evaluate $\int \frac{\sin x}{\cos x + 1} dx$.

Step 1: Let $u = \cos x + 1$.

Step 2: Differentiate: $du = -\sin x dx \Rightarrow -du = \sin x dx$.

Step 3: Rewrite: $\int \frac{-du}{u} = -\int \frac{du}{u}$.

Step 4: Integrate: $-\ln |u| + C$.

Step 5: Replace u: $-\ln |\cos x + 1| + C$.

Step 6: Differentiate and Check: $-\left(\frac{1}{\cos x + 1}\right)(-\sin x) = \frac{\sin x}{\cos x + 1}$.

Example 3

Evaluate $\int \frac{x^2 + 3}{x - 1} dx$.

Step 1: Rewrite $\frac{x^2 + 3}{x - 1} = x + 1 + \frac{4}{x - 1}$; by dividing $(x^2 + 3)$ by $(x - 1)$.

$$\int \frac{x^2 + 3}{x - 1} dx = \int \left(x + 1 + \frac{4}{x - 1}\right) dx = \int (x + 1) dx + \int \frac{4}{x - 1} dx$$

$$= \frac{x^2}{2} + x + 4\int \frac{1}{x - 1} dx$$

Let $u = x - 1$.

Step 2: Differentiate: $du = dx$.

Step 3: Rewrite: $4\int \frac{1}{u}\,du$.

Step 4: Integrate: $4\ln|u| + C$.

Step 5: Replace u: $4\ln|x-1| + C$.

$$\int \frac{x^2+3}{x-1}dx = \frac{x^2}{2} + x + 4\ln|x-1| + C.$$

Step 6: Differentiate and Check:

$$\frac{2x}{2} + 1 + 4\left(\frac{1}{x-1}\right) + C = x + 1 + \frac{4}{x-1} = \frac{x^2+3}{x-1}.$$

Example 4

Evaluate $\int \frac{\ln x}{3x}dx$.

Step 1: Let $u = \ln x$.

Step 2: Differentiate: $du = \frac{1}{x}dx$.

Step 3: Rewrite: $\frac{1}{3}\int u\,dx$.

Step 4: Integrate: $\left(\frac{1}{3}\right)\frac{u^2}{2} + C = \frac{1}{6}u^2 + C$.

Step 5: Replace u: $\frac{1}{6}(\ln x)^2 + C$.

Step 6: Differentiate and Check: $\frac{1}{6}(2)(\ln x)\left(\frac{1}{x}\right) = \frac{\ln x}{3x}$.

Example 5

Evaluate $\int e^{(2x-5)}dx$.

Step 1: Let $u = 2x - 5$.

Step 2: Differentiate: $du = 2dx \Rightarrow \frac{du}{2} = dx$.

Step 3: Rewrite: $\int e^u\left(\frac{du}{2}\right) = \frac{1}{2}\int e^u du$.

Step 4: Integrate: $\frac{1}{2}e^u + C$.

Step 5: Replace u: $\frac{1}{2}e^{(2x-5)} + C$.

Step 6: Differentiate and Check: $\frac{1}{2}e^{2x-5}(2) = e^{2x-5}$.

Example 6

Evaluate $\int \frac{e^x}{e^x+1} dx$.

Step 1: Let $u = e^x + 1$.

Step 2: Differentiate: $du = e^x dx$.

Step 3: Rewrite: $\int \frac{1}{u} du$.

Step 4: Integrate: $\ln|u| + C$.

Step 5: Replace u: $\ln|e^x + 1| + C$.

Step 6: Differentiate and Check: $\frac{1}{e^x + 1} \cdot e^x = \frac{e^x}{e^x + 1}$.

Example 7

Evaluate $\int x e^{3x^2} dx$.

Step 1: Let $u = 3x^2$.

Step 2: Differentiate: $du = 6x\ dx \Rightarrow \frac{du}{6} = x\ dx$.

Step 3: Rewrite: $\int e^u \frac{du}{6} = \frac{1}{6}\int e^u du$.

Step 4: Integrate: $\frac{1}{6}e^u + C$.

Step 5: Replace u: $\frac{1}{6}e^{3x^2} + C$.

Step 6: Differentiate and Check: $\frac{1}{6}\left(e^{3x^2}\right)(6x) = xe^{3x^2}$.

Example 8

Evaluate $\int 5^{(2x)} dx$.

Step 1: Let $u = 2x$.

Step 2: Differentiate: $du = 2dx \Rightarrow \frac{du}{2} = dx$.

Step 3: Rewrite: $\int 5^u \frac{du}{2} = \frac{1}{2}\int 5^u du$.

Step 4: Integrate: $\frac{\frac{1}{2}(5^u)}{\ln 5} + C = \frac{5^u}{(2\ln 5)} + C$.

Step 5: Replace u: $\frac{5^{2x}}{2\ln 5} + C$.

Step 6: Differentiate and Check: $\frac{(5^{2x})(2)\ln 5}{2\ln 5} = 5^{2x}$.

Example 9

Evaluate $\int x^3 5^{(x^4)} dx$.

Step 1: Let $u = x^4$.

Step 2: Differentiate: $du = 4x^3 dx \Rightarrow \frac{du}{4} = x^3 dx$.

Step 3: Rewrite: $\int 5^u \frac{du}{4} = \frac{1}{4}\int 5^u du$.

Step 4: Integrate: $\dfrac{\frac{1}{4}(5^u)}{\ln 5} + C$.

Step 5: Replace u: $\dfrac{5^{x^4}}{4\ln 5} + C$.

Step 6: Differentiate and Check: $\dfrac{5^{(x^4)}\left(4x^3\right)\ln 5}{4\ln 5} = x^3 5^{(x^4)}$.

Example 10

Evaluate $\int (\sin \pi x) e^{\cos \pi x} dx$.

Step 1: Let $u = \cos \pi x$.

Step 2: Differentiate: $du = -\pi \sin \pi x\, dx; -\frac{du}{\pi} = \sin \pi x\, dx$.

Step 3: Rewrite: $\int e^u \left(\frac{-du}{\pi}\right) = -\frac{1}{\pi}\int e^u du$.

Step 4: Integrate: $-\frac{1}{\pi}e^u + C$.

Step 5: Replace u: $-\frac{1}{\pi}e^{\cos \pi x} + C$.

Step 6: Differentiate and Check: $-\frac{1}{\pi}(e^{\cos \pi x})(-\sin \pi x)\pi = (\sin \pi x)e^{\cos \pi x}$.

11.3 Rapid Review

1. Evaluate $\int \frac{1}{x^2} dx$.

   Answer: Rewrite as $\int x^{-2} dx = \frac{x^{-1}}{-1} + C = -\frac{1}{x} + C$.

2. Evaluate $\int \frac{x^3 - 1}{x} dx$.

   Answer: Rewrite as $\int \left(x^2 - \frac{1}{x}\right) dx = \frac{x^3}{3} - \ln|x| + C$.

3. Evaluate $\int x\sqrt{x^2 - 1}\, dx$.

Answer: Rewrite as $\int x(x^2-1)^{1/2}dx$. Let $u = x^2 - 1$.

Thus, $\frac{du}{2} = x\,dx \Rightarrow \frac{1}{2}\int u^{1/2}du = \frac{1}{2}\frac{u^{3/2}}{3/2} + C = \frac{1}{3}(x^2-1)^{3/2} + C.$

4. Evaluate $\int \sin x\,dx$.

Answer: $-\cos x + C$.

5. Evaluate $\int \cos(2x)dx$.

Answer: Let $u = 2x$ and obtain $\frac{1}{2}\sin 2x + C$.

6. Evaluate $\int \frac{\ln x}{x}dx$.

Answer: Let $u = \ln x$; $du = \frac{1}{x}dx$ and obtain $\frac{(\ln x)^2}{2} + C$.

7. Evaluate $\int xe^{x^2}dx$.

Answer: Let $u = x^2$; $\frac{du}{2} = xdx$ and obtain $\frac{e^{x^2}}{2} + C$.

11.4 Practice Problems

Evaluate the following integrals in problems 1 to 20. No calculators are allowed. (However, you may use calculators to check your results.)

1. $\int (x^5 + 3x^2 - x + 1)dx$

2. $\int \left(\sqrt{x} - \frac{1}{x^2}\right)dx$

3. $\int x^3(x^4 - 10)^5 dx$

4. $\int x^3\sqrt{x^2+1}\,dx$

5. $\int \frac{x^2+5}{\sqrt{x-1}}dx$

6. $\int \tan\left(\frac{x}{2}\right)dx$

7. $\int x\csc^2(x^2)dx$

8. $\int \frac{\sin x}{\cos^3 x}dx$

9. $\int \frac{1}{x^2+2x+10}dx$

10. $\int \frac{1}{x^2}\sec^2\left(\frac{1}{x}\right)dx$

11. $\int (e^{2x})(e^{4x})dx$

12. $\int \frac{1}{x\ln x}dx$

13. $\int \ln(e^{5x+1})dx$

14. $\int \frac{e^{4x}-1}{e^x}dx$

15. $\int (9-x^2)\sqrt{x}\,dx$

16. $\int \sqrt{x}\left(1+x^{3/2}\right)^4 dx$

17. If $\frac{dy}{dx} = e^x + 2$ and the point (0, 6) is on the graph of y, find y.

18. $\int -3e^x \sin(e^x)dx$

19. $\int \frac{e^x - e^{-x}}{e^x + e^{-x}} dx$

20. If $f(x)$ is the antiderivative of $\frac{1}{x}$ and $f(1) = 5$, find $f(e)$.

11.5 Cumulative Review Problems

(Calculator) indicates that calculators are permitted.

21. The graph of the velocity function of a moving particle for $0 \leq t \leq 10$ is shown in Figure 11.5-1.

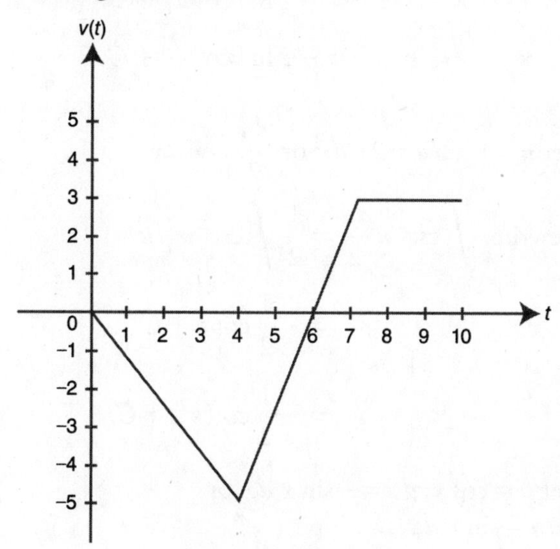

Figure 11.5-1

(a) At what value of t is the speed of the particle the greatest?
(b) At what time is the particle moving to the right?

22. Air is pumped into a spherical balloon, whose maximum radius is 10 meters. For what value of r is the rate of increase of the volume a hundred times that of the radius?

23. Evaluate $\int \frac{\ln^3(x)}{x} dx$.

24. (Calculator) The function f is continuous and differentiable on (0, 2) with $f''(x) > 0$ for all x in the interval (0, 2). Some of the points on the graph are shown below.

x	0	0.50	1	1.50	2
$f(x)$	1	1.25	2	3.25	5

Which of the following is the best approximation for $f'(1)$?

(a) $f'(1) < 2$
(b) $0.5 < f'(1) < 1$
(c) $1.5 < f'(1) < 2.5$
(d) $2.5 < f'(1) < 3.5$
(e) $f'(1) > 2$

25. The graph of the function f'' on the interval [1, 8] is shown in Figure 11.5-2. At what value(s) of t on the open interval (1, 8), if any, does the graph of the function f':

(a) have a point of inflection?
(b) have a relative maximum or minimum?
(c) become concave upward?

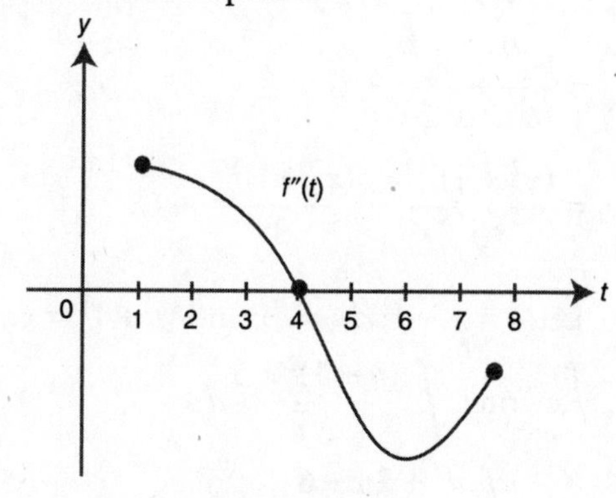

Figure 11.5-2

11.6 Solutions to Practice Problems

1. $\dfrac{x^6}{6}+x^3-\dfrac{x^2}{2}+x+C$

2. Rewrite: $\displaystyle\int (x^{1/2}-x^{-2})dx$

$=\dfrac{x^{3/2}}{3/2}-\dfrac{x^{-1}}{-1}+C$

$=\dfrac{2x^{3/2}}{3}+\dfrac{1}{x}+C.$

3. Let $u=x^4-10$; $du=4x^3dx$ or $\dfrac{du}{4}=x^3dx$.

Rewrite: $\displaystyle\int u^5\frac{du}{4}=\frac{1}{4}\int u^5du$

$=\left(\dfrac{1}{4}\right)\dfrac{u^6}{6}+C$

$=\dfrac{(x^4-10)^6}{24}+C.$

4. Let $u=x^2+1 \Rightarrow (u-1)=x^2$ and $du=2x\,dx$ or $\dfrac{du}{2}=x\,dx$.

Rewrite: $\displaystyle\int x^2\sqrt{x^2+1}(x\,dx)$

$=\displaystyle\int (u-1)\sqrt{u}\frac{du}{2}$

$=\dfrac{1}{2}\displaystyle\int (u-1)u^{1/2}du$

$=\dfrac{1}{2}\displaystyle\int (u^{3/2}-u^{1/2})du$

$=\dfrac{1}{2}\left(\dfrac{u^{5/2}}{5/2}-\dfrac{u^{3/2}}{3/2}\right)+C$

$=\dfrac{u^{5/2}}{5}-\dfrac{u^{3/2}}{3}+C$

$=\dfrac{(x^2+1)^{5/2}}{5}-\dfrac{(x^2+1)^{3/2}}{3}+C.$

5. Let $u=x-1$; $du=dx$ and $(u+1)=x$.

Rewrite: $\displaystyle\int \frac{(u+1)^2+5}{\sqrt{u}}du$

$=\displaystyle\int \frac{u^2+2u+6}{u^{1/2}}du$

$=\displaystyle\int \left(u^{3/2}+2u^{1/2}+6u^{-1/2}\right)du$

$=\dfrac{u^{5/2}}{5/2}+\dfrac{2u^{3/2}}{3/2}+\dfrac{6u^{1/2}}{1/2}+C$

$=\dfrac{2(x-1)^{5/2}}{5}+\dfrac{4(x-1)^{3/2}}{3}$
$+12(x-1)^{1/2}+C.$

6. Let $u=\dfrac{x}{2}$; $du=\dfrac{1}{2}dx$ or $2du=dx$.

Rewrite: $\displaystyle\int \tan u(2\,du)=2\int \tan u\,du$

$=-2\ln|\cos u|+C$

$=-2\ln\left|\cos\dfrac{x}{2}\right|+C.$

7. Let $u=x^2$; $du=2x\,dx$ or $\dfrac{du}{2}=x\,dx$.

Rewrite: $\displaystyle\int \csc^2 u\frac{du}{2}=\frac{1}{2}\int \csc^2 u\,du$

$=-\dfrac{1}{2}\cot u+C$

$=-\dfrac{1}{2}\cot(x^2)+C.$

8. Let $u=\cos x$; $du=-\sin x\,dx$ or $-du=\sin x\,dx$.

Rewrite: $\displaystyle\int \frac{-du}{u^3}=-\int \frac{du}{u^3}$

$=-\dfrac{u^{-2}}{-2}+C=\dfrac{1}{2\cos^2 x}+C.$

9. Rewrite: $\displaystyle\int \frac{1}{(x^2+2x+1)+9}dx$

$=\displaystyle\int \frac{1}{(x+1)^2+3^2}dx.$

Let $u=x+1$; $du=dx$.

Rewrite: $\displaystyle\int \frac{1}{u^2+3^2}du$

$=\dfrac{1}{3}\tan^{-1}\left(\dfrac{u}{3}\right)+C$

$=\dfrac{1}{3}\tan^{-1}\left(\dfrac{x+1}{3}\right)+C.$

10. Let $u = \frac{1}{x}$; $du = \frac{-1}{x^2}dx$ or $-du = \frac{1}{x^2}dx$.

Rewrite: $\int \sec^2 u(-du) = -\int \sec^2 u\, du$

$= -\tan u + C$

$= -\tan\left(\frac{1}{x}\right) + C.$

11. Rewrite: $\int e^{(2x+4x)}dx = \int e^{6x}dx.$

Let $u = 6x$; $du = 6\,dx$ or $\frac{du}{6} = dx.$

Rewrite: $\int e^u \frac{du}{6} = \frac{1}{6}\int e^u du$

$= \frac{1}{6}e^u + C = \frac{1}{6}e^{6x} + C.$

12. Let $u = \ln x$; $du = \frac{1}{x}dx.$

Rewrite: $\int \frac{1}{u}du = \ln|u| + C$

$= \ln|\ln x| + C.$

13. Since e^x and $\ln x$ are inverse functions:

$\int \ln\left(e^{5x+1}\right)dx = \int (5x+1)dx$

$= \frac{5x^2}{2} + x + C.$

14. Rewrite: $\int \left(\frac{e^{4x}}{e^x} - \frac{1}{e^x}\right)dx$

$= \int \left(e^{3x} - e^{-x}\right)dx$

$= \int e^{3x}dx - \int e^{-x}dx.$

Let $u = 3x$; $du = 3dx$;

$\int e^{3x}dx = \int e^u\left(\frac{du}{3}\right) = \frac{1}{3}e^u + C_1$

$= \frac{1}{3}e^{3x} + C.$

Let $v = -x$; $dv = -dx$;

$\int e^{-x}dx = \int e^v(-dv) = e^v + C_2$

$= -e^{-x} + C_2$

Thus, $\int e^{3x}dx - \int e^{-x}dx$

$= \frac{1}{3}e^{3x} + e^{-x} + C.$

Note that C_1 and C_2 are arbitrary constants, and thus, $C_1 + C_2 = C.$

15. Rewrite:

$\int (9 - x^2)x^{1/2}dx = \int \left(9x^{1/2} - x^{5/2}\right)dx$

$= \frac{9x^{3/2}}{3/2} - \frac{x^{7/2}}{7/2} + C$

$= 6x^{3/2} - \frac{2x^{7/2}}{7} + C.$

16. Let $u = 1 + x^{3/2}$; $du = \frac{3}{2}x^{1/2}dx$ or $\frac{2}{3}du = x^{1/2}dx = \sqrt{x}\,dx.$

Rewrite: $\int u^4\left(\frac{2}{3}du\right) = \frac{2}{3}\int u^4 du$

$= \frac{2}{3}\left(\frac{u^5}{5}\right) + C = \frac{2\left(1 + x^{3/2}\right)^5}{15} + C.$

17. Since $\frac{dy}{dx} = e^x + 2$, then $y =$

$\int (e^x + 2)dx = e^x + 2x + C.$

The point (0, 6) is on the graph of y. Thus, $6 = e^0 + 2(0) + C \Rightarrow 6 = 1 + C$ or $C = 5$. Therefore, $y = e^x + 2x + 5.$

18. Let $u = e^x$; $du = e^x dx.$

Rewrite: $-3\int \sin(u)du = -3(-\cos u) + C$

$= 3\cos(e^x) + C.$

19. Let $u = e^x + e^{-x}$; $du = (e^x - e^{-x})\,dx.$

Rewrite: $\int \frac{1}{u}du = \ln|u| + C$

$= \ln|e^x + e^{-x}| + C$

or $= \ln\left|e^x + \frac{1}{e^x}\right| + C$

$= \ln\left|\frac{e^{2x}+1}{e^x}\right| + C$

$= \ln|e^{2x} + 1| - \ln|e^x| + C$

$= \ln|e^{2x} + 1| - x + C.$

20. Since $f(x)$ is the antiderivative of $\frac{1}{x}$,

$f(x) = \int \frac{1}{x} d = \ln|x| + C.$

Given $f(1) = 5$; thus, $\ln(1) + C = 5$
$\Rightarrow 0 + C = 5$ or $C = 5$.

Thus, $f(x) = \ln|x| + 5$ and
$f(e) = \ln(e) + 5 = 1 + 5 = 6.$

11.7 Solutions to Cumulative Review Problems

21. (a) At $t = 4$, speed is 5 which is the greatest on $0 \leq t \leq 10$.
    (b) The particle is moving to the right when $6 < t < 10$.

22. $V = \frac{4}{3}\pi r^3;$

$\frac{dV}{dt} = \left(\frac{4}{3}\right)(3)\pi r^2 \frac{dr}{dt} = 4\pi r^2 \frac{dr}{dt}$

If $\frac{dV}{dt} = 100\frac{dr}{dt}$, then $100\frac{dr}{dt}$
$= 4\pi r^2 \frac{dr}{dt} \Rightarrow 100.$
$= 4\pi r^2$ or $r = \pm\sqrt{\frac{25}{\pi}} = \pm\frac{5}{\sqrt{\pi}}.$

Since $r \geq 0$, $r = \frac{5}{\sqrt{\pi}}$ meters.

23. Let $u = \ln x$; $du = \frac{1}{x}dx$.

Rewrite: $\int u^3 du = \frac{u^4}{4} + C = \frac{(\ln x)^4}{4} + C$
$= \frac{\ln^4(x)}{4} + C.$

24. Label given points as A, B, C, D, and E.
Since $f''(x) > 0 \Rightarrow f$ is concave upward for all x in the interval $[0, 2]$.
Thus, $m_{\overline{BC}} < f'(x) < m_{\overline{CD}}$
$m_{\overline{BC}} = 1.5$ and $m_{\overline{CD}} = 2.5$.
Therefore, $1.5 < f'(1) < 2.5$, choice (c).
(See Figure 11.7-1.)

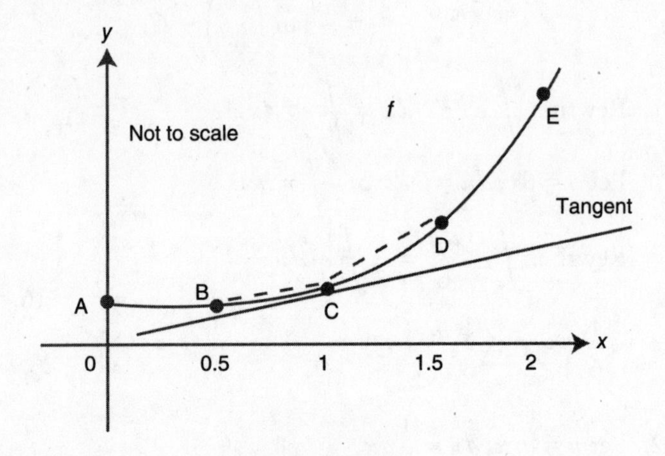

Figure 11.7-1

25. (a) f'' is decreasing on $[1, 6) \Rightarrow$ $f''' < 0 \Rightarrow f'$ is concave downward on $[1, 6)$ and f'' is increasing on $(6, 8]$ $\Rightarrow f'$ is concave upward on $(6, 8]$. Thus, at $x = 6$, f' has a change of concavity. Since f'' exists at $x = 6$, which implies there is a tangent to the curve of f' at $x = 6$, f' has a point of inflection at $x = 6$.
    (b) $f'' > 0$ on $[1, 4] \Rightarrow f'$ is increasing and $f'' < 0$ on $(4, 8] \Rightarrow f'$ is decreasing. Thus at $x = 4$, f' has a relative maximum at $x = 4$. There is no relative minimum.
    (c) f'' is increasing on $[6, 8] \Rightarrow f' > 0$ $\Rightarrow f'$ is concave upward on $[6, 8]$.

Big Idea 3: Integrals and the Fundamental Theorems of Calculus

Definite Integrals

IN THIS CHAPTER

Summary: In this chapter, you will be introduced to the summation notation, the concept of a Riemann Sum, the Fundamental Theorems of Calculus, and the properties of definite integrals. You will also be shown techniques for evaluating definite integrals involving algebraic, trigonometric, logarithmic, and exponential functions. In addition, you will learn how to work with improper integrals. The ability to evaluate integrals is a prerequisite to doing well on the AP Calculus AB exam.

Key Ideas

- Summation Notation
- Riemann Sums
- Properties of Definite Integrals
- The First Fundamental Theorem of Calculus
- The Second Fundamental Theorem of Calculus
- Evaluating Definite Integrals

12.1 Riemann Sums and Definite Integrals

Main Concepts: Sigma Notation, Definition of a Riemann Sum, Definition of a Definite Integral, and Properties of Definite Integrals

Sigma Notation or Summation Notation

$$\sum_{i=1}^{n} a_1 + a_2 + a_3 + \cdots + a_n$$

where i is the index of summation, l is the lower limit, and n is the upper limit of summation. (Note: The lower limit may be any non-negative integer $\leq n$.)

Examples

$$\sum_{i=5}^{7} i^2 = 5^2 + 6^2 + 7^2$$

$$\sum_{k=0}^{3} 2k = 2(0) + 2(1) + 2(2) + 2(3)$$

$$\sum_{i=-1}^{3} (2i + 1) = -1 + 1 + 3 + 5 + 7$$

$$\sum_{k=1}^{4} (-1)^k (k) = -1 + 2 - 3 + 4$$

Summation Formulas

If n is a positive integer, then:

1. $\displaystyle\sum_{i=1}^{n} a = an$

2. $\displaystyle\sum_{i=1}^{n} i = \frac{n(n+1)}{2}$

3. $\displaystyle\sum_{i=1}^{n} i^2 = \frac{n(n+1)(2n+1)}{6}$

4. $\displaystyle\sum_{i=1}^{n} i^3 = \frac{n^2(n+1)^2}{4}$

5. $\displaystyle\sum_{i=1}^{n} i^4 = \frac{n(n+1)(6n^3+9n^2+n-1)}{30}$

Example

Evaluate $\sum_{i=1}^{n} \frac{i(i+1)}{n}$.

Rewrite: $\sum_{i=1}^{n} \frac{i(i+1)}{n}$ as $\frac{1}{n}\sum_{i=1}^{n}(i^2+i) = \frac{1}{n}\left(\sum_{i=1}^{n} i^2 + \sum_{i=1}^{n} i\right)$

$$= \frac{1}{n}\left(\frac{n(n+1)(2n+1)}{6} + \frac{n(n+1)}{2}\right)$$

$$= \frac{1}{n}\left[\frac{n(n+1)(2n+1)+3n(n+1)}{6}\right] = \frac{(n+1)(2n+1)+3(n+1)}{6}$$

$$= \frac{(n+1)\,[(2n+1)+3]}{6} = \frac{(n+1)(2n+4)}{6}$$

$$= \frac{(n+1)(n+2)}{3}.$$

(Note: This question has not appeared in an AP Calculus AB exam in recent years.)

- Remember: In exponential growth/decay problems, the formulas are $\frac{dy}{dx} = ky$ and $y = y_0e^{kt}$.

Definition of a Riemann Sum

Let f be defined on $[a, b]$ and x_i be points on $[a, b]$ such that $x_0 = a$, $x_n = b$, and $a < x_1 < x_2 < x_3 \cdots < x_{n-1} < b$. The points a, x_1, x_2, x_3, $\ldots x_{n-1}$, and b form a partition of f denoted as Δ on $[a, b]$. Let Δx_i be the length of the ith interval $[x_{i-1}, x_i]$ and c_i be any point in the ith interval. Then the Riemann sum of f for the partition is $\sum_{i=1}^{n} f(c_i)\Delta x_i$.

Example 1

Let f be a continuous function defined on [0, 12] as shown below.

x	0	2	4	6	8	10	12
$f(x)$	3	7	19	39	67	103	147

Find the Riemann sum for $f(x)$ over [0, 12] with 3 subdivisions of equal length and the midpoints of the intervals as c_i.

Length of an interval $\Delta x_i = \frac{12-0}{3} = 4$. (See Figure 12.1-1.)

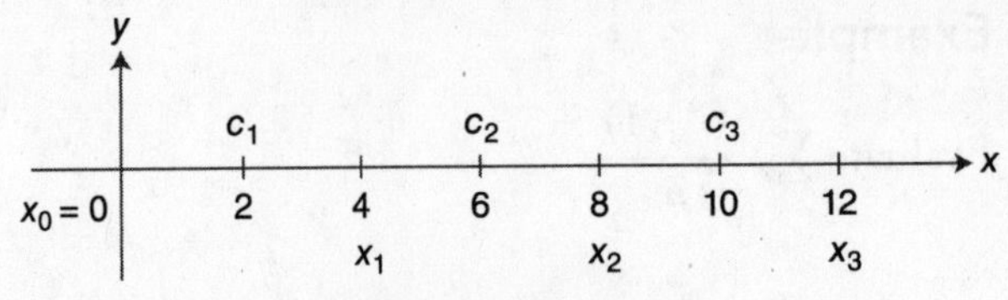

Figure 12.1-1

$$\text{Riemann sum} = \sum_{i=1}^{3} f(c_i)\Delta x_i = f(c_1)\Delta x_1 + f(c_2)\Delta x_2 + f(c_3)\Delta x_3$$

$$= 7(4) + 39(4) + 103(4) = 596$$

The Riemann sum is 596.

Example 2

Find the Riemann sum for $f(x) = x^3 + 1$ over the interval [0, 4] using 4 subdivisions of equal length and the midpoints of the intervals as c_i. (See Figure 12.1-2.)

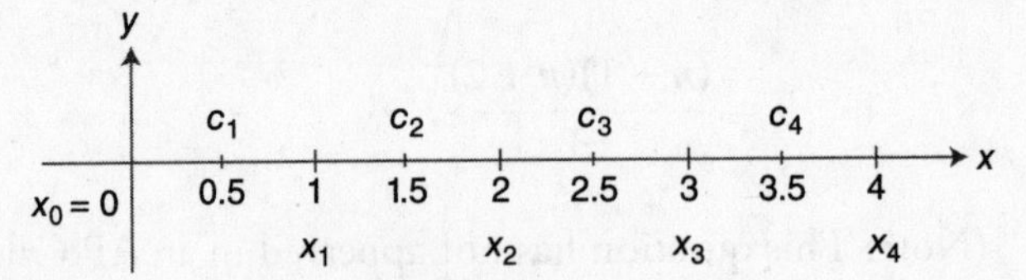

Figure 12.1-2

$$\text{Length of an interval } \Delta x_i = \frac{b-a}{n} = \frac{4-0}{4} = 1; \; c_i = 0.5 + (i-1) = i - 0.5.$$

$$\text{Riemann sum} = \sum_{i=1}^{4} f(c_i)\Delta x_i = \sum_{i=1}^{4} \left[(i-0.5)^3 + 1\right] 1$$

$$= \sum_{i=1}^{4} (i-0.5)^3 + 1.$$

$$\text{Enter} \sum \left((1-0.5)^3 + 1, i, 1, 4\right) = 66.$$

The Riemann sum is 66.

Definition of a Definite Integral

Let f be defined on $[a, b]$ with the Riemann sum for f over $[a, b]$ written as $\sum_{i=1}^{n} f(c_i)\Delta x_i$.

If $\max \Delta x_i$ is the length of the largest subinterval in the partition and the $\lim_{\max \Delta x_i \to 0} \sum_{i=1}^{n} f(c_i)\Delta x_i$ exists, then the limit is denoted by:

$$\lim_{\max \Delta x_i \to 0} \sum_{i=1}^{n} f(c_i)\Delta x_i = \int_a^b f(x)dx.$$

$\int_a^b f(x)dx$ is the definite integral of f from a to b.

Example 1

Use a midpoint Riemann sum with three subdivisions of equal length to find the approximate value of $\int_0^6 x^2 dx$.

$$\Delta x = \frac{6-0}{3} = 2, \; f(x) = x^2$$

midpoints are $x = 1, 3,$ and 5.

$$\int_0^6 x^2 dx \approx f(1)\Delta x + f(3)\Delta x + f(5)\Delta x = 1(2) + 9(2) + 25(2)$$

$$\approx 70$$

Example 2

Using the limit of the Riemann sum, find $\int_1^5 3x dx$.

Using n subintervals of equal lengths, the length of an interval

$$\Delta x_i = \frac{5-1}{n} = \frac{4}{n}; x_i = 1 + \left(\frac{4}{n}\right)i$$

$$\int_1^5 3x dx = \lim_{\max \Delta x_i \to 0} \sum_{i=1}^{n} f(c_i)\Delta x_i.$$

Let $c_i = x_i$; $\max \Delta x_i \to 0 \Rightarrow n \to \infty$.

$$\int_1^5 3x dx = \lim_{n\to\infty} \sum_{i=1}^{n} f\left(1 + \frac{4i}{n}\right)\left(\frac{4}{n}\right) = \lim_{n\to\infty} \sum_{i=1}^{n} 3\left(1 + \frac{4i}{n}\right)\left(\frac{4}{n}\right)$$

$$= \lim_{n\to\infty} \frac{12}{n} \sum_{i=1}^{n} \left(1 + \frac{4i}{n}\right) = \lim_{n\to\infty} \frac{12}{n}\left(n + \frac{4}{n}\left[n\left(\frac{n+1}{2}\right)\right]\right)$$

$$= \lim_{n\to\infty} \frac{12}{n}(n + 2(n+1)) = \lim_{n\to\infty} \frac{12}{n}(3n+2) = \lim_{n\to\infty}\left(36 + \frac{24}{n}\right) = 36$$

Thus, $\int_1^5 3x dx = 36$.

(Note: This question has not appeared in an AP Calculus AB exam in recent years.)

Properties of Definite Integrals

1. If f is defined on $[a, b]$, and the limit $\lim_{\max \Delta x_i \to 0} \sum_{i=1}^{n} f(x_i)\Delta x_i$ exists, then f is integrable on $[a, b]$.
2. If f is continuous on $[a, b]$, then f is integrable on $[a, b]$.

If $f(x)$, $g(x)$, and $h(x)$ are integrable on $[a, b]$, then

3. $\int_a^a f(x)dx = 0$

4. $\int_a^b f(x)dx = -\int_b^a f(x)$

5. $\int_a^b Cf(x)dx = C\int_a^b f(x)dx$ when C is a constant.

6. $\int_a^b [f(x) \pm g(x)]\,dx = \int_a^b f(x)dx \pm \int_a^b g(x)dx$

7. $\int_a^b f(x)dx \geq 0$ provided $f(x) \geq 0$ on $[a, b]$.

8. $\int_a^b f(x)dx \geq \int_a^b g(x)dx$ provided $f(x) \geq g(x)$ on $[a, b]$.

9. $\left|\int_a^b f(x)dx\right| \leq \int_a^b |f(x)|\,dx$

10. $\int_a^b g(x)dx \leq \int_a^b f(x)dx \leq \int_a^b h(x)dx$; provided $g(x) \leq f(x) \leq h(x)$ on $[a, b]$.

11. $m(b-a) \leq \int_a^b f(x)dx \leq M(b-a)$; provided $m \leq f(x) \leq M$ on $[a, b]$.

12. $\int_a^c f(x)dx = \int_a^b f(x)dx + \int_b^c f(x)dx$; provided $f(x)$ is integrable on an interval containing a, b, c.

Examples

1. $\int_\pi^\pi \cos x\,dx = 0$

2. $\int_1^5 x^4 dx = -\int_5^1 x^4 dx$

3. $\int_{-2}^7 5x^2 dx = 5\int_{-2}^7 x^2 dx$

4. $\int_0^4 (x^3 - 2x + 1)dx = \int_0^4 x^3 dx - 2\int_0^4 x dx + \int_0^4 1 dx$

5. $\int_1^5 \sqrt{x}dx = \int_1^3 \sqrt{x}dx + \int_3^5 \sqrt{x}dx$

   Or $\int_1^3 \sqrt{x}dx = \int_1^5 \sqrt{x}dx + \int_5^3 \sqrt{x}dx$

   $\int_a^c = \int_a^b + \int_b^c$ a, b, c do not have to be arranged from smallest to largest.

The remaining properties are best illustrated in terms of the area under the curve of the function, as discussed in the next section.

- Do not forget that $\int_0^{-3} f(x)dx = -\int_{-3}^{0} f(x)dx$.

12.2 Fundamental Theorems of Calculus

Main Concepts: First Fundamental Theorem of Calculus, Second Fundamental Theorem of Calculus

First Fundamental Theorem of Calculus

If f is continuous on $[a, b]$ and F is an antiderivative of f on $[a, b]$, then

$$\int_a^b f(x)dx = F(b) - F(a).$$

Note: $F(b) - F(a)$ is often denoted as $F(x)\big]_a^b$.

Example 1

Evaluate $\int_0^2 (4x^3 + x - 1)dx$.

$$\int_0^2 (4x^3 + x - 1)dx = \frac{4x^4}{4} + \frac{x^2}{2} - x\Big]_0^2 = x^4 + \frac{x^2}{2} - x\Big]_0^2$$

$$= \left(2^4 + \frac{2^2}{2} - 2\right) - (0) = 16$$

Example 2

Evaluate $\int_{-\pi}^{\pi} \sin x dx$.

$$\int_{-\pi}^{\pi} \sin x dx = -\cos x\Big]_{-\pi}^{\pi} = [-\cos \pi] - [-\cos(-\pi)]$$

$$= [-(-1)] - [-(-1)] = (1) - (1) = 0$$

Example 3

If $\int_{-2}^{k} (4x + 1)dx = 30$, $k > 0$, find k.

$$\int_{-2}^{k} (4x + 1)dx = 2x^2 + x\big]_{-2}^{k} = (2k^2 + k) - (2(-2)^2 - 2)$$

$$= 2k^2 + k - 6$$

Set $2k^2 + k - 6 = 30 \Rightarrow 2k^2 + k - 36 = 0$

$$\Rightarrow (2k+9)(k-4) = 0 \text{ or } k = -\frac{9}{2} \text{ or } k = 4.$$

Since $k > 0$, $k = 4$.

Example 4

If $f'(x) = g(x)$, and g is a continuous function for all real values of x, express $\int_2^5 g(3x)dx$ in terms of f.

Let $u = 3x$; $du = 3dx$ or $\frac{du}{3} = dx$.

$$\int g(3x)dx = \int g(u)\frac{du}{3} = \frac{1}{3}\int g(u)du = \frac{1}{3}f(u) + C$$

$$= \frac{1}{3}f(3x) + C$$

$$\int_2^5 g(3x)dx = \frac{1}{3}f(3x)\Big]_2^5 = \frac{1}{3}f(3(5)) - \frac{1}{3}f(3(2))$$

$$= \frac{1}{3}f(15) - \frac{1}{3}f(6)$$

Example 5

Evaluate $\int_0^4 \frac{1}{x-1}dx$.

Note that you cannot evaluate using the First Fundamental Theorem of Calculus since $f(x) = \frac{1}{x-1}$ is discontinuous at $x = 1$.

Example 6

Using a graphing calculator, evaluate $\int_{-2}^2 \sqrt{4-x^2}dx$.

Using a TI-89 graphing calculator, enter $\int\left(\sqrt{(4-x\wedge 2)},\ x,\ -2,\ 2\right)$ and obtain 2π.

Second Fundamental Theorem of Calculus

If f is continuous on $[a, b]$ and $F(x) = \int_a^x f(t)dt$, then $F'(x) = f(x)$ at every point x in $[a, b]$.

Example 1

Evaluate $\int_{\pi/4}^{\pi} \cos(2t)\, dt$.

Let $u = 2t$; $du = 2dt$ or $\frac{du}{2} = dt$.

$$\int \cos(2t)dt = \int \cos u \frac{du}{2} = \frac{1}{2}\int \cos u\, du$$

$$= \frac{1}{2}\sin u + C = \frac{1}{2}\sin(2t) + C$$

$$\int_{\pi/4}^{x} \cos(2t)dt = \frac{1}{2}\sin(2t)\Big]_{\pi/4}^{x}$$

$$= \frac{1}{2}\sin(2x) - \frac{1}{2}\sin\left(2\left(\frac{\pi}{4}\right)\right)$$

$$= \frac{1}{2}\sin(2x) - \frac{1}{2}\sin\left(\frac{\pi}{2}\right)$$

$$= \frac{1}{2}\sin(2x) - \frac{1}{2}$$

Example 2

If $h(x) = \int_{3}^{x} \sqrt{t+1}\, dt$, find $h'(8)$.

$h'(x) = \sqrt{x+1}$; $h'(8) = \sqrt{8+1} = 3$

Example 3

Find $\frac{dy}{dx}$; if $y = \int_{1}^{2x} \frac{1}{t^3} dt$.

Let $u = 2x$; then $\frac{dy}{dx} = 2$.

Rewrite: $y = \int_{1}^{u} \frac{1}{t^3} dt$.

$$\frac{dy}{dx} = \frac{dy}{du} \cdot \frac{du}{dx} = \frac{1}{u^3} \cdot (2) = \frac{1}{(2x)^3} \cdot 2 = \frac{1}{4x^3}$$

Example 4

Find $\frac{dy}{dx}$; if $y = \int_{x^2}^{1} \sin t\, dt$.

Rewrite: $y = -\int_{1}^{x^2} \sin t\, dt$.

Let $u = x^2$; then $\frac{du}{dx} = 2x$.

Rewrite: $y = -\int_1^u \sin t\, dt$.

$$\frac{dy}{dx} = \frac{dy}{du} \cdot \frac{du}{dx} = (-\sin u)2x = (-\sin x^2)2x$$
$$= -2x \sin(x^2)$$

Example 5

Find $\frac{dy}{dx}$; if $y = \int_x^{x^2} \sqrt{e^t + 1}\, dt$.

$$y = \int_x^0 \sqrt{e^t + 1}\, dt + \int_0^{x^2} \sqrt{e^t + 1}\, dt = -\int_0^x \sqrt{e^t + 1}\, dt + \int_0^{x^2} \sqrt{e^t + 1}\, dt$$

$$= \int_0^{x^2} \sqrt{e^t + 1}\, dt - \int_0^x \sqrt{e^t + 1}\, dt$$

Since $y = \int_0^{x^2} \sqrt{e^t + 1}\, dt - \int_0^x \sqrt{e^t + 1}\, dt$

$$\frac{dy}{dx} = \left(\frac{d}{dx}\int_0^{x^2} \sqrt{e^t + 1}\, dt\right) - \left(\frac{d}{dx}\int_0^x \sqrt{e^t + 1}\, dt\right)$$

$$= \left(\sqrt{e^{x^2} + 1}\right)\frac{d}{dx}(x^2) - \left(\sqrt{e^x + 1}\right)$$

$$= 2x\sqrt{e^{x^2} + 1} - \sqrt{e^x + 1}.$$

Example 6

$F(x) = \int_1^x (t^2 - 4)dt$, integrate to find $F(x)$ and then differentiate to find $f'(x)$.

$$F(x) = \frac{t^3}{3} - 4t\Big]_1^x = \left(\frac{x^3}{3} - 4x\right) - \left(\frac{1^3}{3} - 4(1)\right)$$

$$= \frac{x^3}{3} - 4x + \frac{11}{3}$$

$$F'(x) = 3\left(\frac{x^2}{3}\right) - 4 = x^2 - 4$$

12.3 Evaluating Definite Integrals

Main Concepts: Definite Integrals Involving Algebraic Functions; Definite Integrals Involving Absolute Value; Definite Integrals Involving Trigonometric, Logarithmic, and Exponential Functions; Definite Integrals Involving Odd and Even Functions

- If the problem asks you to determine the concavity of f' (not f), you need to know if f'' is increasing or decreasing, or if f''' is positive or negative.

Definite Integrals Involving Algebraic Functions

Example 1

Evaluate $\int_1^4 \frac{x^3-8}{\sqrt{x}}dx$.

Rewrite: $\int_1^4 \frac{x^3-8}{\sqrt{x}}dx = \int_1^4 \left(x^{5/2}-8x^{-1/2}\right)dx$

$$= \frac{x^{7/2}}{7/2} - \frac{8x^{1/2}}{1/2}\Bigg]_1^4 = \frac{2x^{7/2}}{7} - 16x^{1/2}\Bigg]_1^4$$

$$= \left(\frac{2(4)^{7/2}}{7} - 16(4)^{1/2}\right) - \left(\frac{2(1)^{7/2}}{7} - 16(1)^{1/2}\right) = \frac{142}{7}.$$

Verify your result with a calculator.

Example 2

Evaluate $\int_0^2 x(x^2-1)^7 dx$.

Begin by evaluating the indefinite integral $\int x(x^2-1)^7 dx$.

Let $u = x^2 - 1$; $du = 2x\,dx$ or $\frac{du}{2} = x\,dx$.

Rewrite: $\int \frac{u^7 du}{2} = \frac{1}{2}\int u^7 du = \frac{1}{2}\left(\frac{u^8}{8}\right) + C = \frac{u^8}{16} + C = \frac{(x^2-1)^8}{16} + C.$

Thus, the definite integral $\int_0^2 x(x^2-1)^7 dx = \frac{(x^2-1)^8}{16}\Bigg]_0^2$

$$= \frac{(2^2-1)^8}{16} - \frac{(0^2-1)^8}{16} = \frac{3^8}{16} - \frac{(-1)^8}{16} = \frac{3^8-1}{16} = 410.$$

Verify your result with a calculator.

Example 3

Evaluate $\int_{-8}^{-1}\left(\sqrt[3]{y}+\frac{1}{\sqrt[3]{y}}\right)dy.$

Rewrite: $\int_{-8}^{-1}\left(y^{1/3}+\frac{1}{y^{1/3}}\right)dy=\int_{-8}^{-1}\left(y^{1/3}+y^{-1/3}\right)dy$

$$=\left.\frac{y^{4/3}}{4/3}+\frac{y^{2/3}}{2/3}\right]_{-8}^{-1}=\left.\frac{3y^{4/3}}{4}+\frac{3y^{2/3}}{2}\right]_{-8}^{-1}$$

$$=\left(\frac{3(-1)^{4/3}}{4}+\frac{3(-1)^{2/3}}{2}\right)$$

$$-\left(\frac{3(-8)^{4/3}}{4}+\frac{3(-8)^{2/3}}{2}\right)$$

$$=\left(\frac{3}{4}+\frac{3}{2}\right)-(12+6)=\frac{-63}{4}.$$

Verify your result with a calculator.

- You may bring up to 2 (but no more than 2) approved graphing calculators to the exam.

Definite Integrals Involving Absolute Value

Example 1

Evaluate $\int_{1}^{4}|3x-6|\,dx.$

Set $3x-6=0$; $x=2$; thus, $|3x-6|=\begin{cases}3x-6 \text{ if } x\geq 2\\ -(3x-6) \text{ if } x<2\end{cases}.$

Rewrite: $\int_{1}^{4}|3x-6|\,dx=\int_{1}^{2}-(3x-6)dx+\int_{2}^{4}(3x-6)dx$

$$=\left[\frac{-3x^2}{2}+6x\right]_1^2+\left[\frac{3x^2}{2}-6x\right]_2^4$$

$$=\left(\frac{-3(2)^2}{2}-6(2)\right)-\left(\frac{-3(1)^2}{2}-6(1)\right)$$

$$+\left(\frac{3(4)^2}{2}-6(4)\right)-\left(\frac{3(2)^2}{2}-6(2)\right)$$

$$=(-6+12)-\left(-\frac{3}{2}+6\right)+(24-24)-(6-12)$$

$$=6-4\frac{1}{2}+0+6=\frac{15}{2}.$$

Verify your result with a calculator.

Example 2

Evaluate $\int_0^4 |x^2 - 4|dx$.

Set $x^2 - 4 = 0$; $x = \pm 2$.

Thus, $|x^2 - 4| = \begin{cases} x^2 - 4 \text{ if } x \geq 2 \text{ or } x \leq -2 \\ -(x^2 - 4) \text{ if } -2 < x < 2 \end{cases}$.

Thus, $\int_0^4 |x^2 - 4|dx = \int_0^2 -(x^2 - 4)dx + \int_2^4 (x^2 - 4)dx$

$$= \left[\frac{-x^3}{3} + 4x\right]_0^2 + \left[\frac{x^3}{3} - 4x\right]_2^4$$

$$= \left(\frac{-2^3}{3} + 4(2)\right) - (0) + \left(\frac{4^3}{3} - 4(4)\right) - \left(\frac{2^3}{3} - 4(2)\right)$$

$$= \left(\frac{-8}{3} + 8\right) + \left(\frac{64}{3} - 16\right) - \left(\frac{8}{3} - 8\right) = 16.$$

Verify your result with a calculator.

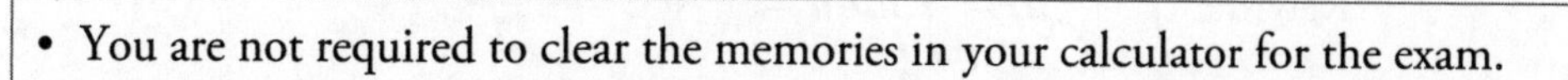

- You are not required to clear the memories in your calculator for the exam.

Definite Integrals Involving Trigonometric, Logarithmic, and Exponential Functions

Example 1

Evaluate $\int_0^\pi (x + \sin x)dx$.

Rewrite: $\int_0^\pi (x + \sin x)dx = \left.\frac{x^2}{2} - \cos x\right]_0^\pi = \left(\frac{\pi^2}{2} - \cos\pi\right) - (0 - \cos 0)$

$$= \frac{\pi^2}{2} + 1 + 1 = \frac{\pi^2}{2} + 2.$$

Verify your result with a calculator.

Example 2

Evaluate $\int_{\pi/4}^{\pi/2} \csc^2(3t)dt$.

Let $u = 3t$; $du = 3dt$ or $\frac{du}{3} = dt$.

Rewrite the indefinite integral: $\int \csc^2 u \frac{du}{3} = -\frac{1}{3} \cot u + c$

$$= -\frac{1}{3} \cot(3t) + c$$

$$\int_{\pi/4}^{\pi/2} \csc^2 (3t) dt = -\frac{1}{3} \cot (3t) \Big]_{\pi/4}^{\pi/2}$$

$$= -\frac{1}{3} \left[\cot\left(\frac{3\pi}{2}\right) - \cot\left(\frac{3\pi}{4}\right) \right]$$

$$= -\frac{1}{3} [0 - (-1)] = -\frac{1}{3}.$$

Verify your result with a calculator.

Example 3

Evaluate $\int_1^e \frac{\ln t}{t} dt$.

Let $u = \ln t,\ du = \frac{1}{t} dt$.

Rewrite: $\int \frac{\ln t}{t} dt = \int u \, du = \frac{u^2}{2} + C = \frac{(\ln t)^2}{2} + C$

$$\int_1^e \frac{\ln t}{t} dt = \frac{(\ln t)^2}{2} \Big]_1^e = \frac{(\ln e)^2}{2} - \frac{(\ln 1)^2}{2}$$

$$= \frac{1}{2} - 0 = \frac{1}{2}.$$

Verify your result with a calculator.

Example 4

Evaluate $\int_{-1}^{2} x e^{(x^2+1)} dx$.

Let $u = x^2 + 1; du = 2x \; dx$ or $\frac{dx}{2} = x \, dx$.

$$\int x e^{(x^2+1)} dx = \int e^u \frac{du}{2} = \frac{1}{2} e^u + C = \frac{1}{2} e^{(x^2+2)} + C$$

Rewrite: $\int_{-1}^{2} x e^{(x^2+1)} dx = \frac{1}{2} e^{(x^2+1)} \Big]_{-1}^{2} = \frac{1}{2} e^5 - \frac{1}{2} e^2 = \frac{1}{2} e^2 (e^3 - 1)$.

Verify your result with a calculator.

Definite Integrals Involving Odd and Even Functions

If f is an even function, that is, $f(-x) = f(x)$, and is continuous on $[-a, a]$, then

$$\int_{-a}^{a} f(x)dx = 2\int_{0}^{a} f(x)dx.$$

If f is an odd function, that is, $F(x) = -f(-x)$, and is continuous on $[-a, a]$, then

$$\int_{-a}^{a} f(x)dx = 0.$$

Example 1

Evaluate $\int_{-\pi/2}^{\pi/2} \cos x \, dx$.

Since $f(x) = \cos x$ is an even function,

$$\int_{-\pi/2}^{\pi/2} \cos x \, dx = 2\int_{0}^{\pi/2} \cos x \, dx = 2\,[\sin x]_0^{\pi/2} = 2\left[\sin\left(\frac{\pi}{2}\right) - \sin(0)\right]$$

$$= 2(1-0) = 2.$$

Verify your result with a calculator.

Example 2

Evaluate $\int_{-3}^{3} \left(x^4 - x^2\right)dx$.

Since $f(x) = x^4 - x^2$ is an even function, i.e., $f(-x) = f(x)$, thus

$$\int_{-3}^{3} \left(x^4 - x^2\right)dx = 2\int_{0}^{3} \left(x^4 - x^2\right)dx = 2\left[\frac{x^5}{5} - \frac{x^3}{3}\right]_0^3$$

$$= 2\left[\left(\frac{3^5}{5} - \frac{3^3}{3}\right) - 0\right] = \frac{396}{5}.$$

Verify your result with a calculator.

Example 3

Evaluate $\int_{-\pi}^{\pi} \sin x \, dx$.

Since $f(x) = \sin x$ is an odd function, i.e., $f(-x) = -f(x)$, thus

$$\int_{-\pi}^{\pi} \sin x \, dx = 0.$$

Verify your result algebraically.

$$\int_{-\pi}^{\pi} \sin x \, dx = -\cos x\Big]_{-\pi}^{\pi} = (-\cos\pi) - [-\cos(-\pi)]$$

$$= [-(-1)] - [-(1)] = (1) - (1) = 0$$

You can also verify the result with a calculator.

Example 4

If $\int_{-k}^{k} f(x)dx = 2\int_{0}^{k} f(x)dx$ for all values of k, then which of the following could be the graph of f? (See Figure 12.3-1.)

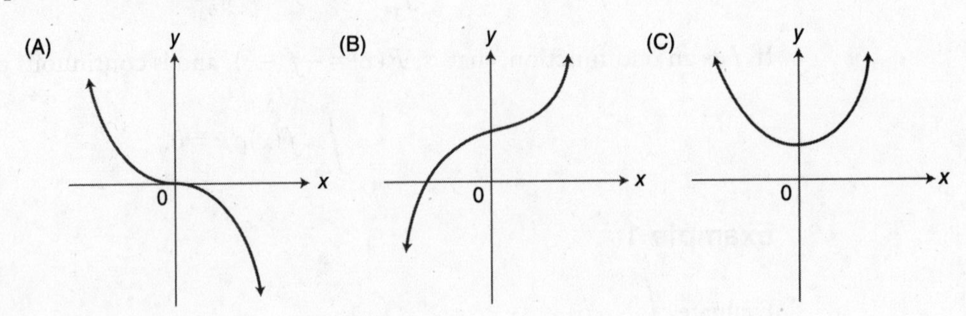

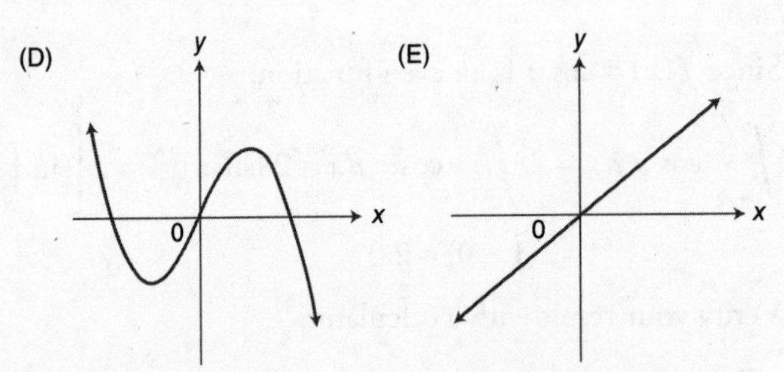

Figure 12.3-1

$$\int_{-k}^{k} f(x)dx = \int_{-k}^{0} f(x)dx + \int_{0}^{k} f(x)dx$$

Since $\int_{-k}^{k} f(x)dx = 2\int_{0}^{k} f(x)dx$, then $\int_{0}^{k} f(x)dx = \int_{-k}^{0} f(x)dx$.

Thus f is an even function. Choice (C).

12.4 Rapid Review

1. Evaluate $\int_{\pi/2}^{x} \cos t\, dt$.

   Answer: $\sin t]_{\pi/2}^{x} = \sin x - \sin\left(\frac{\pi}{2}\right) = \sin x - 1$.

2. Evaluate $\int_{0}^{1} \frac{1}{x+1} dx$.

   Answer: $\ln(x+1)]_{0}^{1} = \ln 2 - \ln 1 = \ln 2$.

3. If $G(x) = \int_{0}^{x} (2t+1)^{3/2} dt$, find $G'(4)$.

   Answer: $G'(x) = (2x+1)^{3/2}$ and $G'(4) = 9^{3/2} = 27$.

4. If $\int_{1}^{k} 2x\, dx = 8$, find k.

   Answer: $x^2]_{1}^{k} = 8 \Rightarrow k^2 - 1 = 8 \Rightarrow k = \pm 3$.

5. If $G(x)$ is an antiderivative of $(e^x + 1)$ and $G(0) = 0$, find $G(1)$.

   Answer: $G(x) = e^x + x + C$
   $G(0) = e^0 + 0 + C = 0 \Rightarrow C = -1.$
   $G(1) = e^1 + 1 - 1 = e.$

6. If $G'(x) = g(x)$, express $\int_0^2 g(4x)dx$ in terms of $G(x)$.

   Answer: Let $u = 4x$; $\frac{du}{4} = dx$.

   $\int g(u)\frac{du}{4} = \frac{1}{4}G(u)$. Thus, $\int_0^2 g(4x)dx = \frac{1}{4}G(4x)\Big]_0^2 = \frac{1}{4}[G(8) - G(0)]$.

12.5 Practice Problems

Part A—The use of a calculator is not allowed.

Evaluate the following definite integrals.

1. $\int_{-1}^{0} (1 + x - x^3)dx$

2. $\int_{6}^{11} (x - 2)^{1/2}\, dx$

3. $\int_{1}^{3} \frac{t}{t+1}dt$

4. $\int_{0}^{6} |x - 3|dx$

5. If $\int_{0}^{k} (6x - 1)dx = 4$, find k.

6. $\int_{0}^{\pi} \frac{\sin x}{\sqrt{1 + \cos x}}dx$

7. If $f'(x) = g(x)$ and g is a continuous function for all real values of x, express $\int_{1}^{2} g(4x)dx$ in terms of f.

8. $\int_{\ln 2}^{\ln 3} 10e^x dx$

9. $\int_{e}^{e^2} \frac{1}{t+3}dt$

10. If $f(x) = \int_{-\pi/4}^{x} \tan^2(t)dt$, find $f'\left(\frac{\pi}{6}\right)$.

11. $\int_{-1}^{1} 4xe^{x^2} dx$

12. $\int_{-\pi}^{\pi} \left(\cos x - x^2\right)dx$

Part B—Calculators are allowed.

13. Find k if $\int_{0}^{2} \left(x^3 + k\right)dx = 10$.

14. Evaluate $\int_{-1.2}^{3.1} 2\theta \cos\theta\, d\theta$ to the nearest 100th.

15. If $y = \int_{1}^{x^3} \sqrt{t^2 + 1}\, dt$, find $\frac{dy}{dx}$.

16. Use a midpoint Riemann sum with four subdivisions of equal length to find the approximate value of $\int_{0}^{8} \left(x^3 + 1\right)dx$.

17. Given $\int_{-2}^{2} g(x)dx = 8$ and $\int_{0}^{2} g(x)dx = 3$, find

(a) $\int_{-2}^{0} g(x)dx$

(b) $\int_{2}^{-2} g(x)dx$

(c) $\int_{0}^{-2} 5g(x)dx$

(d) $\int_{-2}^{2} 2g(x)dx$

18. Evaluate $\int_{0}^{1/2} \frac{dx}{\sqrt{1-x^2}}$.

19. Find $\frac{dy}{dx}$ if $y = \int_{\cos x}^{\sin x} (2t+1)dt$.

20. Let f be a continuous function defined on [0, 30] with selected values as shown below:

x	0	5	10	15	20	25	30
$f(x)$	1.4	2.6	3.4	4.1	4.7	5.2	5.7

Use a midpoint Riemann sum with three subdivisions of equal length to find the approximate value of $\int_{0}^{30} f(x)dx$.

12.6 Cumulative Review Problems

(Calculator) indicates that calculators are permitted.

21. Evaluate $\lim_{x\to-\infty} \frac{\sqrt{x^2-4}}{3x-9}$.

22. Find $\frac{dy}{dx}$ at $x = 3$ if $y = \ln|x^2 - 4|$.

23. The graph of f', the derivative of f, $-6 \le x \le 8$ is shown in Figure 12.6-1.

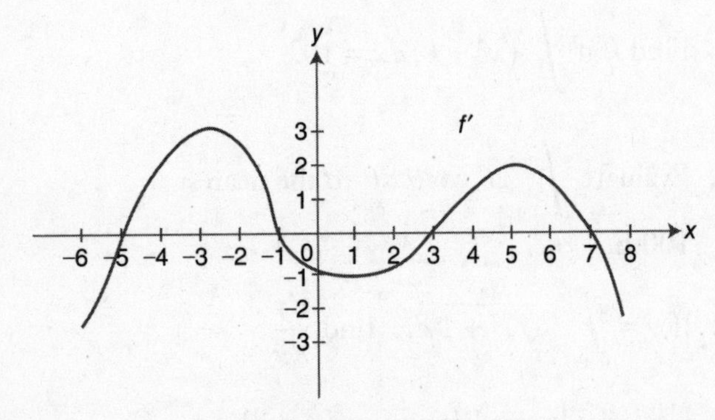

Figure 12.6-1

(a) Find all values of x such that f attains a relative maximum or a relative minimum.

(b) Find all values of x such that f is concave upward.

(c) Find all values of x such that f has a change of concavity.

24. (Calculator) Given the equation $9x^2 + 4y^2 - 18x + 16y = 11$, find the points on the graph where the equation has a vertical or horizontal tangent.

25. (Calculator) Two corridors, one 6 feet wide and another 10 feet wide meet at a corner. (See Figure 12.6-2.) What is the maximum length of a pipe of negligible thickness that can be carried horizontally around the corner?

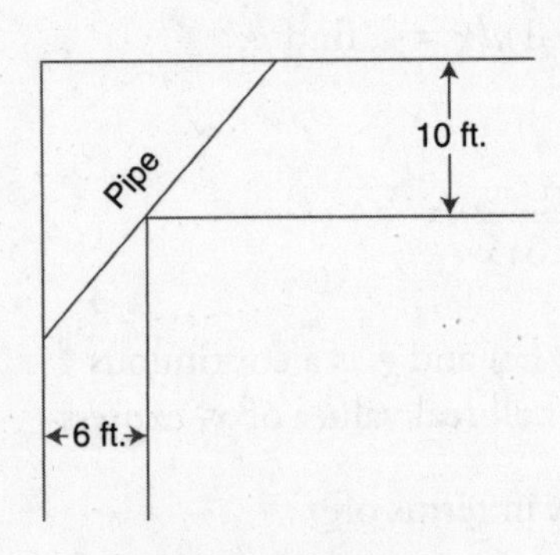

Figure 12.6-2

12.7 Solutions to Practice Problems

Part A—The use of a calculator is not allowed.

1. $\int_{-1}^{0}\left(1+x-x^3\right)dx$

$= x+\frac{x^2}{2}-\frac{x^4}{4}\Big]_{-1}^{0}$

$= 0-\left[(-1)+\frac{(-1)^2}{2}-\frac{(-1)^4}{4}\right]$

$= \frac{3}{4}$

2. Let $u = x-2\ du = dx$.

$$\int (x-2)^{1/2}dx = \int u^{1/2}du$$

$$= \frac{2u^{1/2}}{3}+C$$

$$= \frac{2}{3}(x-2)^{3/2}+C$$

Thus, $\int_{6}^{11}(x-2)^{1/2}dx = \frac{2}{3}(x-2)^{3/2}\Big]_{6}^{11}$

$$= \frac{2}{3}\left[(11-2)^{3/2}\right.$$

$$\left.-(6-2)^{3/2}\right]$$

$$= \frac{2}{3}(27-8) = \frac{38}{3}.$$

3. Let $u = t+1$; $du = dt$ and $t = u-1$.

Rewrite: $\int \frac{t}{t+1}dt = \int \frac{u-1}{u}du$

$$= \int\left(1-\frac{1}{u}\right)du$$

$$= u-\ln|u|+C$$

$$= t+1-\ln|t+1|+C$$

$$\int_{1}^{3}\frac{t}{t+1}dt = [t+1-\ln|t+1|]_{1}^{3}$$

$$= [(3)+1-\ln|3+1|]$$

$$-((1)+1-\ln|1+1|)$$

$$= 4-\ln 4-2+\ln 2$$

$$= 2-\ln 4+\ln 2$$

$$= 2-\ln(2)^2+\ln 2$$

$$= 2-2\ln 2+\ln 2$$

$$= 2-\ln 2.$$

4. Set $x-3=0$; $x=3$.

$$|x-3| = \begin{cases}(x-3) \text{ if } x\geq 3\\ -(x-3) \text{ if } x<3\end{cases}$$

$$\int_{0}^{6}|x-3|\,dx = \int_{0}^{3}-(x-3)dx$$

$$+\int_{3}^{6}(x-3)dx$$

$$= \left[\frac{-x^2}{2}+3x\right]_{0}^{3}+\left[\frac{x^2}{2}-3x\right]_{3}^{6}$$

$$= \left(-\frac{(3)^2}{2}+3(3)\right)-0$$

$$+\left(\frac{6^2}{2}-3(6)\right)-\left(\frac{3^2}{2}-3(3)\right)$$

$$= \frac{9}{2}+\frac{9}{2} = 9$$

5. $\int_{0}^{k}(6x-1)dx = 3x^2-x\Big]_{0}^{k} = 3k^2-k$

Set $3k^2-k=4 \Rightarrow 3k^2-k-4=0$

$\Rightarrow (3k-4)(k+1)=0$

$\Rightarrow k=\frac{4}{3}$ or $k=-1$.

Verify your results by evaluating

$\int_{0}^{4/3}(6x-1)dx$ and $\int_{0}^{-1}(6x-1)dx$.

6. Let $u = 1 + \cos x$; $du = -\sin x\, dx$ or $-du = \sin x\, dx$.

$$\int \frac{\sin x}{\sqrt{1+\cos x}}dx = \int \frac{-1}{\sqrt{u}}(du)$$
$$= -\int \frac{1}{u^{1/2}}du$$
$$= -\int u^{-1/2}du$$
$$= -\frac{u^{1/2}}{1/2} + C$$
$$= -2u^{1/2} + C$$
$$= -2(1+\cos x)^{1/2} + C$$

$$\int_0^{\pi} \frac{\sin x}{\sqrt{1+\cos x}}dx = -2(1+\cos x)^{1/2}\Big]_0^{\pi}$$
$$= -2\left[(1+\cos\pi)^{1/2} - (1+\cos 0)^{1/2}\right]$$
$$= -2\left[0 - 2^{1/2}\right] = 2\sqrt{2}$$

7. Let $u = 4x$; $du = 4\, dx$ or $\frac{du}{4} = dx$.

$$\int g(4x)dx = \int g(u)\frac{du}{4} = \frac{1}{4}\int g(u)du$$
$$= \frac{1}{4}f(u) + C$$
$$= \frac{1}{4}f(4x) + C$$

$$\int_1^2 g(4x)dx = \frac{1}{4}\ f(4x)]_1^2$$
$$= \frac{1}{4}f(4(2)) - \frac{1}{4}f(4(1))$$
$$= \frac{1}{4}f(8) - \frac{1}{4}f(4)$$

8. $$\int_{\ln 2}^{\ln 3} 10e^x dx = 10e^x\Big]_{\ln 2}^{\ln 3}$$
$$= 10\left[\left(e^{\ln 3}\right) - \left(e^{\ln 2}\right)\right]$$
$$= 10(3-2) = 10$$

9. Let $u = t + 3$; $du = dt$.

$$\int \frac{1}{t+3}dt = \int \frac{1}{u}du = \ln|u| + C$$
$$= \ln|t+3| + C$$

$$\int_e^{e^2} \frac{1}{t+3}dt = \ln|t+3|\Big]_e^{e^2}$$
$$= \ln(e^2+3) - \ln(e+3)$$
$$= \ln\left(\frac{e^2+3}{e+3}\right)$$

10. $f'(x) = \tan^2 x$;

$$f'\left(\frac{\pi}{6}\right) = \tan^2\left(\frac{\pi}{6}\right) = \left(\frac{1}{\sqrt{3}}\right)^2 = \frac{1}{3}$$

11. Let $u = x^2$; $du = 2x\, dx$ or $\frac{du}{2} = x\, dx$.

$$\int 4xe^{x^2}dx = 4\int e^u\left(\frac{du}{2}\right)$$
$$= 2\int e^u du = 2e^u + c = 2e^{x^2} + C$$

$$\int_{-1}^{1} 4xe^{x^2}dx = 2e^{x^2}\Big]_{-1}^{1}$$
$$= 2\left[e^{(1)^2} - e^{(-1)^2}\right] = 2(e-e) = 0$$

Note that $f(x) = 4xe^{x^2}$ is an odd function.

Thus, $\int_{-a}^{a} f(x)dx = 0$.

12. $$\int_{-\pi}^{\pi} (\cos x - x^2)dx = \sin x - \frac{x^3}{3}\Big]_{-\pi}^{\pi}$$
$$= \left(\sin\pi - \frac{\pi^3}{3}\right) - \left(\sin(-\pi) - \frac{(-\pi)^3}{3}\right)$$
$$= -\frac{\pi^3}{3} - \left(0 - \frac{-\pi^3}{3}\right)$$
$$= -\frac{2\pi^3}{3}$$

Note that $f(x) = \cos x - x^2$ is an even function. Thus, you could have written $\int_{-\pi}^{\pi} (\cos x - x^2)dx = 2\int_{0}^{\pi} (\cos x - x^2)dx$ and obtained the same result.

Part B—Calculators are allowed.

13. $\int_0^2 (x^3 + k)dx = \frac{x^4}{4} + kx\Big]_0^2$

$= \left(\frac{2^4}{4} + k(2)\right) - 0$

$= 4 + 2k$

Set $4 + 2k = 10$ and $k = 3$.

14. Enter $\int (2x * \cos(x), x, -1.2, 3.1)$ and obtain $-4.70208 \approx -4.702$.

15. $\frac{d}{dx}\left(\int_1^{x^3} \sqrt{t^2 + 1}\, dt\right)$

$= \sqrt{(x^3)^2 + 1}\frac{d}{dx}(x^3)$

$= 3x^2\sqrt{x^6 + 1}.$

16. $\Delta x = \frac{8 - 0}{4} = 2$

Midpoints are $x = 1, 3, 5$, and 7.

$\int_0^8 (x^3 + 1)dx = (1^3 + 1)(2) + (3^3 + 1)(2)$

$+ (5^3 + 1)(2) + (7^3 + 1)(2)$

$= (2)(2) + (28)(2) + (126)(2)$

$+ (344)(2) = 1000$

17. (a) $\int_{-2}^{0} g(x)dx + \int_{0}^{2} g(x)dx$

$= \int_{-2}^{2} g(x)dx \int_{-2}^{0} g(x)dx + 3$

$= 8$. Thus, $\int_{-2}^{0} g(x)dx = 5$.

(b) $\int_{2}^{-2} g(x)dx = -\int_{-2}^{2} g(x)dx = -8$

(c) $\int_{0}^{-2} 5g(x)dx = 5\int_{0}^{-2} g(x)dx$

$= 5\left(-\int_{-2}^{0} g(x)dx\right)$

$= 5(-5) = -25$

(d) $\int_{-2}^{2} 2g(x)dx = 2\int_{-2}^{2} g(x)dx$

$= 2(8) = 16$

18. $\int_0^{1/2} \frac{dx}{\sqrt{1 - x^2}} = \sin^{-1}(x)\Big]_0^{1/2}$

$= \sin^{-1}\left(\frac{1}{2}\right) - \sin^{-1}(0)$

$= \frac{\pi}{6} - 0 = \frac{\pi}{6}$

19. $\int_{\cos x}^{\sin x} (2t + 1)dt = \int_0^{\sin x} (2t + 1)dt$

$- \int_0^{\cos x} (2t + 1)dt$

$\frac{dy}{dx} = \frac{d}{dx}\left(\int_{\cos x}^{\sin x} (2t + 1)dt\right) = (2\sin x + 1)\frac{d}{dx}(\sin x)$

$- (2\cos x + 1)\frac{d}{dx}(\cos x)$

$= (2\sin x + 1)\cos x - (2\cos x + 1)(-\sin x)$

$= 2\sin x\cos x + \cos x + 2\sin x\cos x + \sin x$

$= 4\sin x\cos x + \cos x + \sin x$

20. $\Delta x = \frac{30 - 0}{3} = 10$

Midpoints are $x = 5, 15$, and 25.

$\int_0^{30} f(x)dx = [f(5)]10 + [f(15)]10 + [f(25)]10$

$= (2.6)(10) + (4.1)(10) + (5.2)10$

$= 119$

12.8 Solutions to Cumulative Review Problems

21. As $x \to -\infty$, $x = -\sqrt{x^2}$.

$$\lim_{x \to -\infty} \frac{\sqrt{x^2-4}}{3x-9} = \lim_{x \to -\infty} \frac{\sqrt{x^2-4}/-\sqrt{x^2}}{(3x-9)/x}$$

$$= \lim_{x \to -\infty} \frac{-\sqrt{(x^2-4)/x^2}}{3-(9/x)}$$

$$= \lim_{x \to -\infty} \frac{-\sqrt{1-(4/x)^2}}{3-9/x}$$

$$= \frac{-\sqrt{1-0}}{3-0} = -\frac{1}{3}$$

22. $y = \ln|x^2-4|$, $\dfrac{dy}{dx} = \dfrac{1}{(x^2-4)}(2x)$

$$\left.\frac{dy}{dx}\right|_{x=3} = \frac{2(3)}{(3^2-4)} = \frac{6}{5}$$

23. (a) (See Figure 12.8-1.)

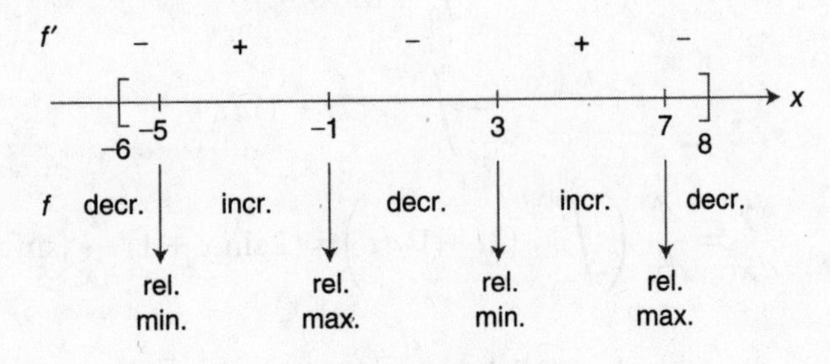

Figure 12.8-1

The function f has a relative minimum at $x = -5$ and $x = 3$, and f has a relative maximum at $x = -1$ and $x = 7$.

(b) (See Figure 12.8-2.)

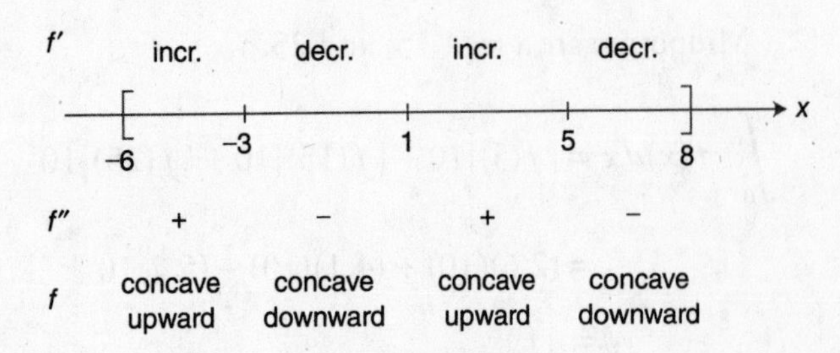

Figure 12.8-2

The function f is concave upward on intervals $(-6, -3)$ and $(1, 5)$.

(c) A change of concavity occurs at $x = -3$, $x = 1$, and $x = 5$.

24. (Calculator) Differentiate both sides of $9x^2 + 4y^2 - 18x + 16y = 11$.

$$18x + 8y\frac{dy}{dx} - 18 + 16\frac{dy}{dx} = 0$$

$$8y\frac{dy}{dx} + 16\frac{dy}{dx} = -18x + 18$$

$$\frac{dy}{dx}(8y + 16) = -18x + 18$$

$$\frac{dy}{dx} = \frac{-18x + 18}{8y + 16}$$

Horizontal tangent $\Rightarrow \dfrac{dy}{dx} = 0$.

Set $\dfrac{dy}{dx} = 0 \Rightarrow -18x + 18 = 0$ or $x = 1$.

At $x = 1$, $9 + 4y^2 - 18 + 16y = 11$

$$\Rightarrow 4y^2 + 16y - 20 = 0.$$

Using a calculator, enter [*Solve*] $(4y\text{^}2 + 16y - 20 = 0, y)$; obtaining $y = -5$ or $y = 1$.

Thus, at each of the points at $(1, 1)$ and $(1, -5)$, the graph has a horizontal tangent.

Vertical tangent $\Rightarrow \dfrac{dy}{dx}$ is undefined.

Set $8y + 16 = 0 \Rightarrow y = -2$.

At $y = -2$, $9x^2 + 16 - 18x - 32 = 11$

$$\Rightarrow 9x^2 - 18x - 27 = 0.$$

Enter [*Solve*] $(9x^2 - 18x - 27 = 0, x)$ and obtain $x = 3$ or $x = -1$.

Thus, at each of the points $(3, -2)$ and $(-1, -2)$, the graph has a vertical tangent. (See Figure 12.8-3.)

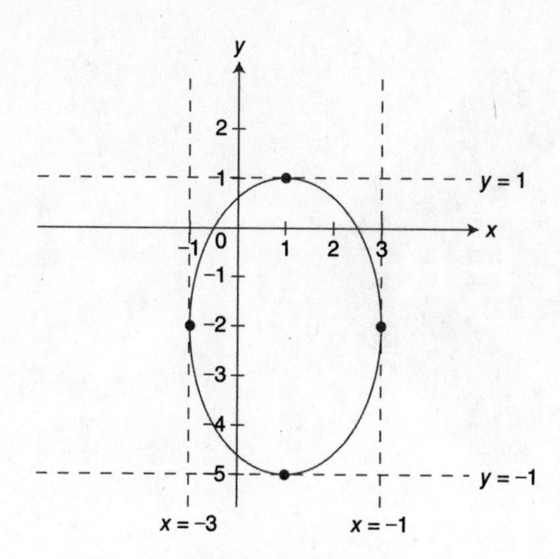

Figure 12.8-3

25. (Calculator)

Step 1: (See Figure 12.8-4.) Let $P = x + y$ where P is the length of the pipe and x and y are as shown. The minimum value of P is the maximum length of the pipe to be able to turn in the corner. By similar triangles, $\frac{y}{10} = \frac{x}{\sqrt{x^2 - 36}}$

and thus, $y = \frac{10x}{\sqrt{x^2 - 36}}$, $x > 6$

$P = x + y = x + \frac{10x}{\sqrt{x^2 - 36}}$.

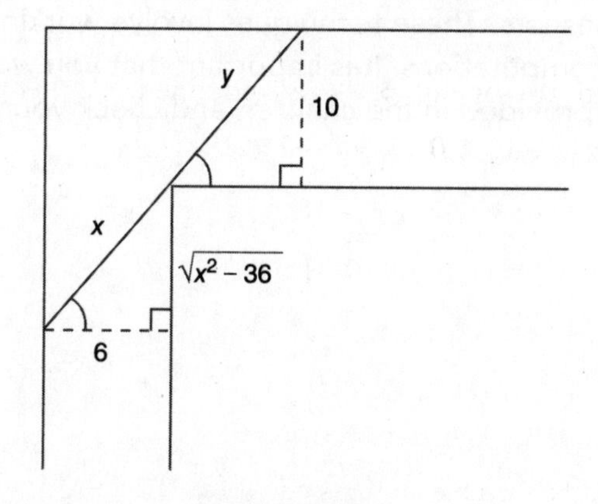

Figure 12.8-4

Step 2: Find the minimum value of P. Enter $y_1 = x + 10 * x / \left(\sqrt{(x\text{^}2 - 36)}\right)$.

Use the [*Minimum*] function of the calculator and obtain the minimum point (9.306, 22.388).

Step 3: Verify with the First Derivative Test. Enter $y2 = (y1(x), x)$ and observe. (See Figure 12.8-5.)

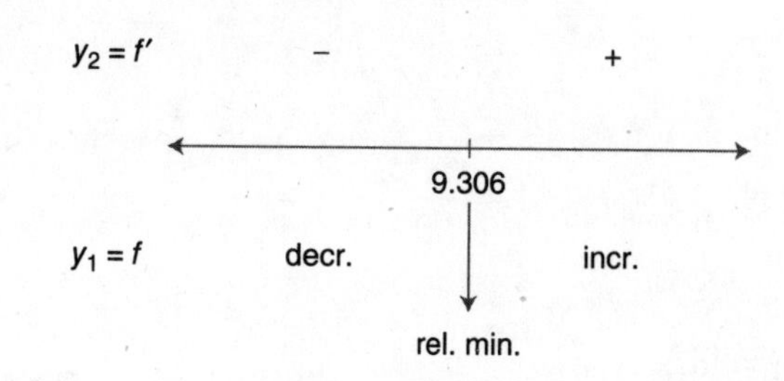

Figure 12.8-5

Step 4: Check the endpoints. The domain of x is $(6, \infty)$. Since $x = 9.306$ is the only relative extremum, it is the absolute minimum. Thus, the maximum length of the pipe is 22.388 feet.

Big Idea 3: Integrals and the Fundamental Theorems of Calculus

Areas and Volumes

IN THIS CHAPTER

Summary: In this chapter, you will be introduced to several important applications of the definite integral. You will learn how to find the area under a curve and the volume of a solid. Some of the techniques that you will be shown include finding area under a curve by using rectangular and trapezoidal approximations and finding the volume of a solid using cross sections, discs, and washers. These techniques involve working with algebraic expressions and lengthy computations. It is important that you work carefully through the practice problems provided in the chapter, and check your solutions with the given explanations.

Key Ideas

- The function $F(x) = \int_a^x f(t)dt$
- Rectangular Approximations
- Trapezoidal Approximations
- Area Under a Curve
- Area Between Two Curves
- Solids with Known Cross Sections
- The Disc Method
- The Washer Method

13.1 The Function $F(x) = \int_a^x f(t)dt$

The Second Fundamental Theorem of Calculus defines

$$F(x) = \int_a^x f(t)dt$$

and states that if f is continuous on $[a, b]$, then $F'(x) = f(x)$ for every point x in $[a, b]$.

If $f \geq 0$, then $F \geq 0$. $F(x)$ can be interpreted geometrically as the area under the curve of f from $t = a$ to $t = x$. (See Figure 13.1-1.)

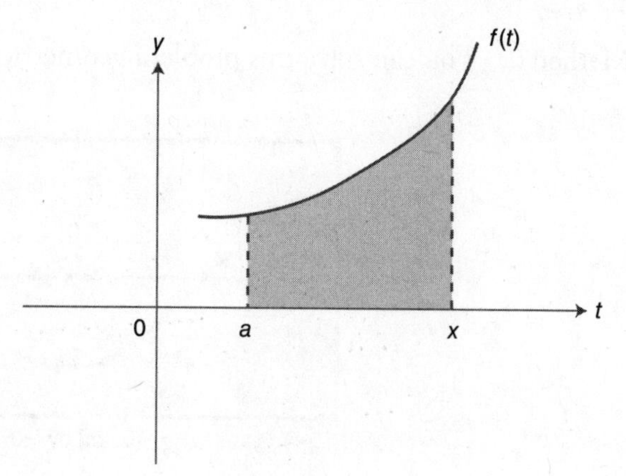

Figure 13.1-1

If $f < 0$, $F < 0$, $F(x)$ can be treated as the negative value of the area between the curve of f and the t-axis from $t = a$ to $t = x$. (See Figure 13.1-2.)

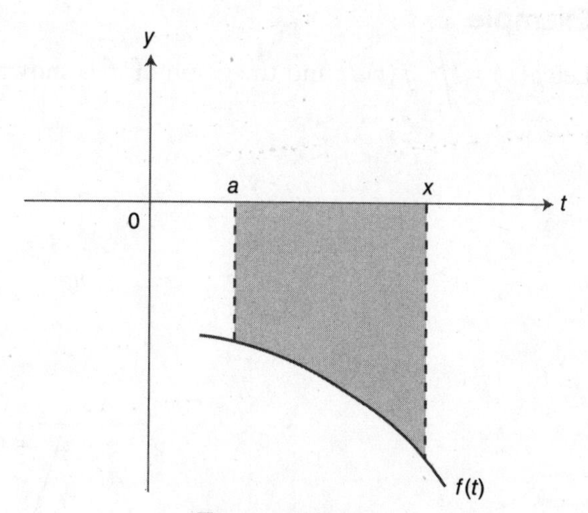

Figure 13.1-2

Example 1

If $F(x)=\int_0^x 2\cos t\, dt$ for $0 \leq x \leq 2\pi$, find the value(s) of x where f has a local minimum.

Method 1: Since $f(x)=\int_0^x 2\cos t\, dt,\ f'(x)=2\cos x.$

Set $f'(x)=0;\ 2\cos x=0,\ x=\dfrac{\pi}{2}$ or $\dfrac{3\pi}{2}$.

$f''(x)=-2\sin x$ and $f''\left(\dfrac{\pi}{2}\right)=-2$ and $f''\left(\dfrac{3\pi}{2}\right)=2.$

Thus, at $x=\dfrac{3\pi}{2}$, f has a local minimum.

Method 2: You can solve this problem geometrically by using area. (See Figure 13.1-3.)

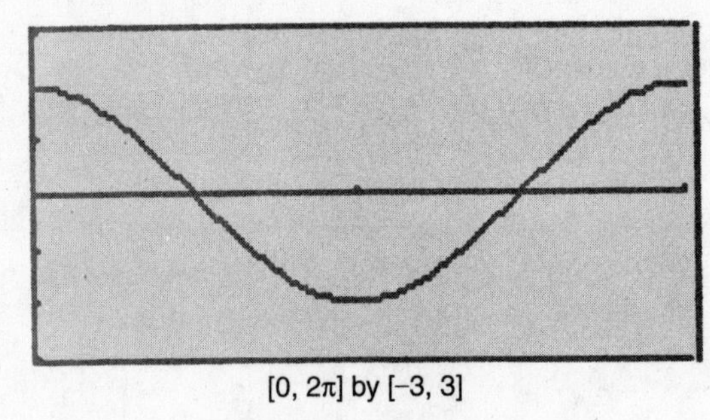

[0, 2π] by [−3, 3]

Figure 13.1-3

The area "under the curve" is above the t-axis on $[0, \pi/2]$ and below the x-axis on $[\pi/2, 3\pi/2]$. Thus, the local minimum occurs at $3\pi/2$.

Example 2

Let $p(x)=\int_0^x f(t)dt$ and the graph of f is shown in Figure 13.1-4.

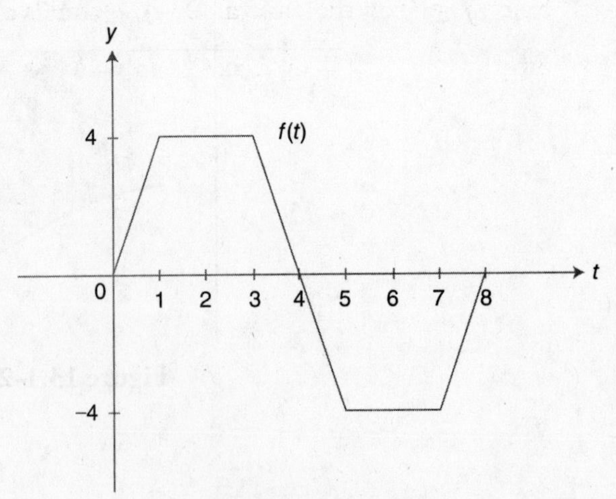

Figure 13.1-4

(a) Evaluate: $p(0)$, $p(1)$, $p(4)$.
(b) Evaluate: $p(5)$, $p(7)$, $p(8)$.
(c) At what value of t does p have a maximum value?
(d) On what interval(s) is p decreasing?
(e) Draw a sketch of the graph of p.

Solution:

(a) $p(0) = \int_0^0 f(t)dt = 0$

$$p(1) = \int_0^1 f(t)dt = \frac{(1)(4)}{2} = 2$$

$$p(4) = \int_0^4 f(t)dt = \frac{1}{2}(2+4)(4) = 12$$

(Note: $f(t)$ forms a trapezoid from $t = 0$ to $t = 4$.)

(b) $p(5) = \int_0^5 f(t)dt = \int_0^4 f(t)dt + \int_4^5 f(t)dt$

$$= 12 - \frac{(1)(4)}{2} = 10$$

$$p(7) = \int_0^7 f(t)dt = \int_0^4 f(t)dt + \int_4^5 f(t)dt + \int_5^7 f(t)dt$$

$$= 12 - 2 - (2)(4) = 2$$

$$p(8) = \int_0^8 f(t)dt = \int_0^4 f(t)dt + \int_4^8 f(t)dt$$

$$= 12 - 12 = 0$$

(c) Since $f \geq 0$ on the interval $[0, 4]$, p attains a maximum at $t = 4$.

(d) Since $f(t)$ is below the x-axis from $t = 4$ to $t = 8$, if $x > 4$,

$$\int_0^x f(t)dt = \int_0^4 f(t)dt + \int_4^x f(t)dt \text{ where } \int_4^x f(t)dt < 0.$$

Thus, p is decreasing on the interval $(4, 8)$.

(e) $p(x) = \int_0^x f(t)dt$. See Figure 13.1-5 for a sketch.

x	0	1	2	3	4	5	6	7	8
$p(x)$	0	2	6	10	12	10	6	2	0

Figure 13.1-5

- Remember that differentiability implies continuity, but the converse is not true, i.e., continuity does not imply differentiability, e.g., as in the case of a cusp or a corner.

Example 3

The position function of a moving particle on a coordinate axis is:

$$s = \int_0^t f(x)dx, \text{ where } t \text{ is in seconds and } s \text{ is in feet.}$$

The function f is a differentiable function, and its graph is shown below in Figure 13.1-6.

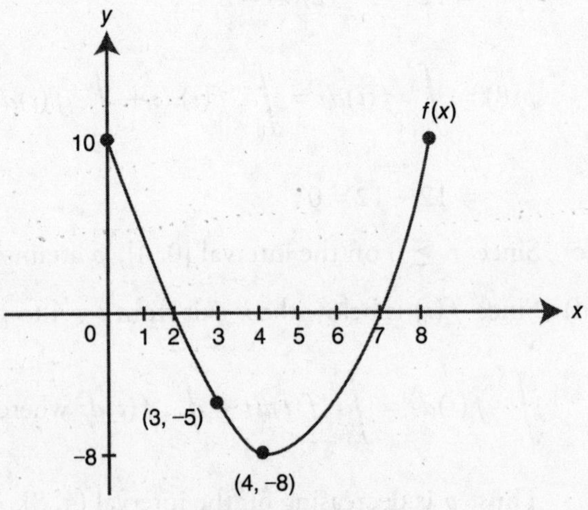

Figure 13.1-6

(a) What is the particle's velocity at $t = 4$?
(b) What is the particle's position at $t = 3$?
(c) When is the acceleration zero?
(d) When is the particle moving to the right?
(e) At $t = 8$, is the particle on the right side or left side of the origin?

Solution:

(a) Since $s = \int_0^t f(x)dx$, then $v(t) = s'(t) = f(t)$.
Thus, $v(4) = -8$ ft/sec.

(b) $s(3) = \int_0^3 f(x)dx = \int_0^2 f(x)dx + \int_2^3 f(x)dx = \frac{1}{2}(10)(2) - \frac{1}{2}(1)(5) = \frac{15}{2}$ ft.

(c) $a(t) = v'(t)$. Since $v'(t) = f'(t)$, $v'(t) = 0$ at $t = 4$. Thus, $a(4) = 0$ ft/sec^2.

(d) The particle is moving to the right when $v(t) > 0$. Thus, the particle is moving to the right on intervals (0, 2) and (7, 8).

(e) The area of f below the x-axis from $x = 2$ to $x = 7$ is larger than the area of f above the x-axis from $x = 0$ to $x = 2$ and $x = 7$ to $x = 8$. Thus, $\int_0^8 f(x)dx < 0$ and the particle is on the left side of the origin.

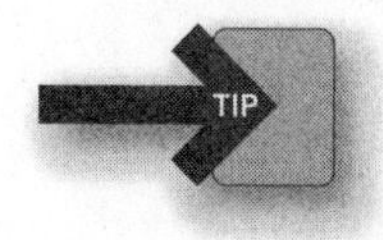

- Do not forget that $(fg)' = f'g + g'f$ and *not* $f'g'$. However, $\lim(fg) = (\lim f)(\lim g)$.

13.2 Approximating the Area Under a Curve

Main Concepts: Rectangular Approximations, Trapezoidal Approximations

Rectangular Approximations

If $f \geq 0$, the area under the curve of f can be approximated using three common types of rectangles: left-endpoint rectangles, right-endpoint rectangles, or midpoint rectangles. (See Figure 13.2-1.)

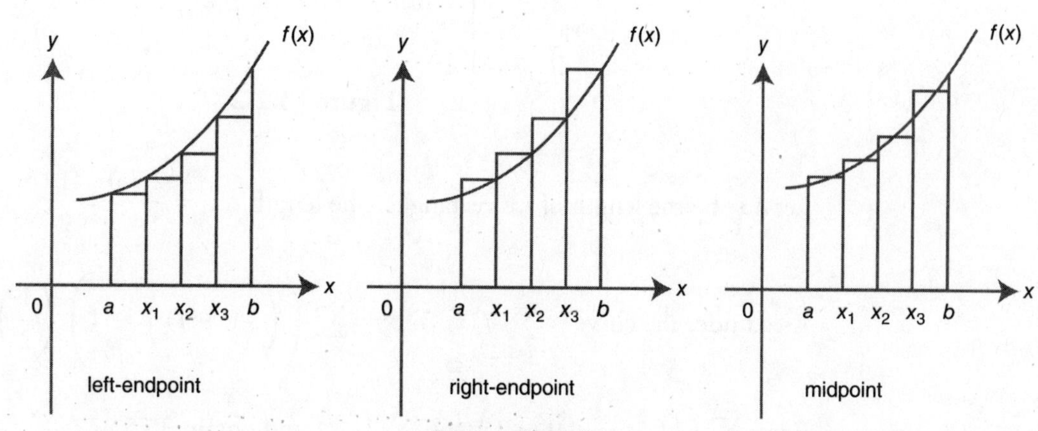

Figure 13.2-1

The area under the curve using n rectangles of equal length is approximately:

$$\sum_{i=1}^{n} (\text{area of rectangle}) = \begin{cases} \sum_{i=1}^{n} f(x_{i-1})\Delta x & \text{left-endpoint rectangles} \\ \sum_{i=1}^{n} f(x_i)\Delta x & \text{right-endpoint rectangles} \\ \sum_{i=1}^{n} f\left(\frac{x_i + x_{i-1}}{2}\right)\Delta x & \text{midpoint rectangles} \end{cases}$$

where $\Delta x = \dfrac{b-a}{n}$ and $a = x_0 < x_1 < x_2 < \cdots < x_n = b$.

If f is increasing on $[a, b]$, then left-endpoint rectangles are inscribed rectangles and the right-endpoint rectangles are circumscribed rectangles. If f is decreasing on $[a, b]$, then left-endpoint rectangles are circumscribed rectangles and the right-endpoint rectangles are inscribed. Furthermore,

$$\sum_{i=1}^{n} \text{inscribed rectangle} \leq \text{area under the curve} \leq \sum_{i=1}^{n} \text{circumscribed rectangle.}$$

Example 1

Find the approximate area under the curve of $f(x) = x^2 + 1$ from $x = 0$ to $x = 2$, using 4 left-endpoint rectangles of equal length. (See Figure 13.2-2.)

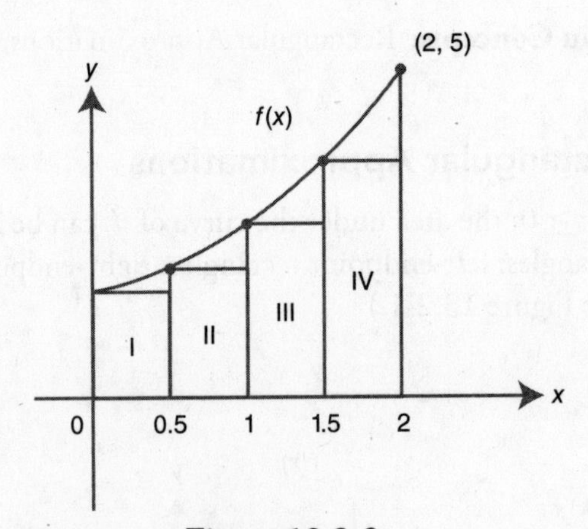

Figure 13.2-2

Let Δx_i be the length of ith rectangle. The length $\Delta x_i = \dfrac{2-0}{4} = \dfrac{1}{2}$; $x_{i-1} = \dfrac{1}{2}(i-1)$.

$$\text{Area under the curve} \approx \sum_{i=1}^{4} f(x_{i-1})\Delta x_i = \sum_{i=1}^{4}\left(\left(\frac{1}{2}(i-1)\right)^2 + 1\right)\left(\frac{1}{2}\right).$$

Enter $\sum\left(\left((.5(x-1))^2 + 1\right) * .5, x, 1, 4\right)$ and obtain 3.75.

Or, find the area of each rectangle:

$$\text{Area of Rect}_{\text{I}} = (f(0))\Delta x_1 = (1)\left(\frac{1}{2}\right) = \frac{1}{2}.$$

$$\text{Area of Rect}_{\text{II}} = f(0.5)\Delta x_2 = ((.5)^2 + 1)\left(\frac{1}{2}\right) = 0.625.$$

$$\text{Area of Rect}_{\text{III}} = f(1)\Delta x_3 = (1^2 + 1)\left(\frac{1}{2}\right) = 1.$$

$$\text{Area of Rect}_{\text{IV}} = f(1.5)\Delta x_4 = (1.5^2 + 1)\left(\frac{1}{2}\right) = 1.625.$$

Area of $(\text{Rect}_{\text{I}} + \text{Rect}_{\text{II}} + \text{Rect}_{\text{III}} + \text{Rect}_{\text{IV}}) = 3.75$.

Thus, the approximate area under the curve of $f(x)$ is 3.75.

Example 2

Find the approximate area under the curve of $f(x) = \sqrt{x}$ from $x = 4$ to $x = 9$ using 5 right-endpoint rectangles. (See Figure 13.2-3.)

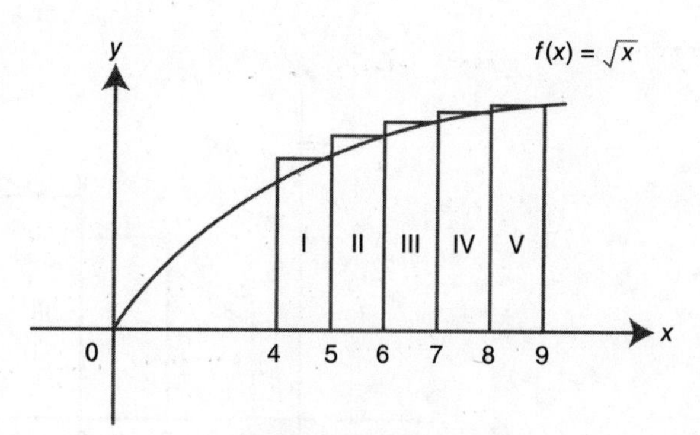

Figure 13.2-3

Let Δx_i be the length of the ith rectangle. The length $\Delta x_i = \dfrac{9-4}{5} = 1$; $x_i = 4 + (1)i = 4 + i$.

$$\text{Area of Rect}_{\text{I}} = f(x_1)\Delta x_1 = f(5)(1) = \sqrt{5}.$$

$$\text{Area of Rect}_{\text{II}} = f(x_2)\Delta x_2 = f(6)(1) = \sqrt{6}.$$

$$\text{Area of Rect}_{\text{III}} = f(x_3)\Delta x_3 = f(7)(1) = \sqrt{7}.$$

$$\text{Area of Rect}_{\text{IV}} = f(x_4)\Delta x_4 = f(8)(1) = \sqrt{8}.$$

$$\text{Area of Rect}_{\text{v}} = f(x_5)\Delta x_5 = f(9)(1) = \sqrt{9} = 3.$$

$$\sum_{i=1}^{5}(\text{Area of Rect}_{\text{I}}) = \sqrt{5} + \sqrt{6} + \sqrt{7} + \sqrt{8} + 3 = 13.160.$$

Or, using $\sum$ notation:

$$\sum_{i=1}^{5} f(x_i)\,\Delta x_i = \sum_{i=1}^{5} f(4+i)(1) = \sum_{i=1}^{5} \sqrt{4+1}.$$

Enter $\sum\left(\sqrt{(4+x)},\, x,\, 1,\, 5\right)$ and obtain 13.160.

Thus, the area under the curve is approximately 13.160.

Example 3

The function f is continuous on [1, 9] and $f > 0$. Selected values of f are given below:

x	1	2	3	4	5	6	7	8	9
$f(x)$	1	1.41	1.73	2	2.37	2.45	2.65	2.83	3

Using 4 midpoint rectangles, approximate the area under the curve of f for $x = 1$ to $x = 9$. (See Figure 13.2-4.)

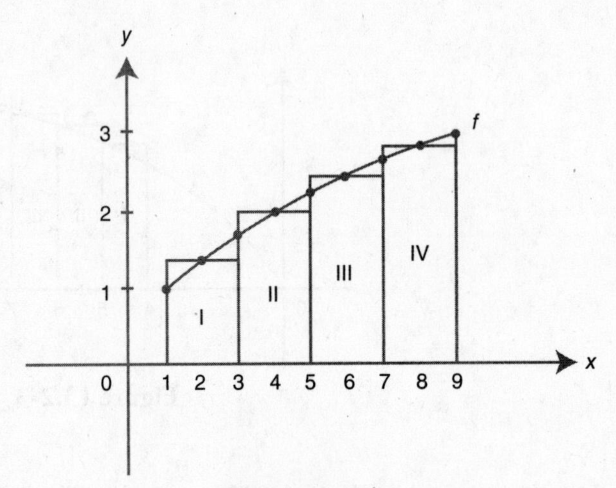

Figure 13.2-4

Let Δx_i be the length of the ith rectangle. The length $\Delta x_i = \dfrac{9-1}{4} = 2.$

$$\text{Area of Rect}_{\text{I}} = f(2)(2) = (1.41)2 = 2.82.$$

$$\text{Area of Rect}_{\text{II}} = f(4)(2) = (2.00)2 = 4.$$

$$\text{Area of Rect}_{\text{III}} = f(6)(2) = (2.45)2 = 4.90.$$

$$\text{Area of Rect}_{\text{IV}} = f(8)(2) = (2.83)2 = 5.66.$$

Area of $(\text{Rect}_{\text{I}} + \text{Rect}_{\text{II}} + \text{Rect}_{\text{III}} + \text{Rect}_{\text{IV}}) = 2.82 + 4.00 + 4.90 + 5.66 = 17.38.$

Thus, the area under the curve is approximately 17.38.

Trapezoidal Approximations

Another method of approximating the area under a curve is to use trapezoids. (See Figure 13.2-5.)

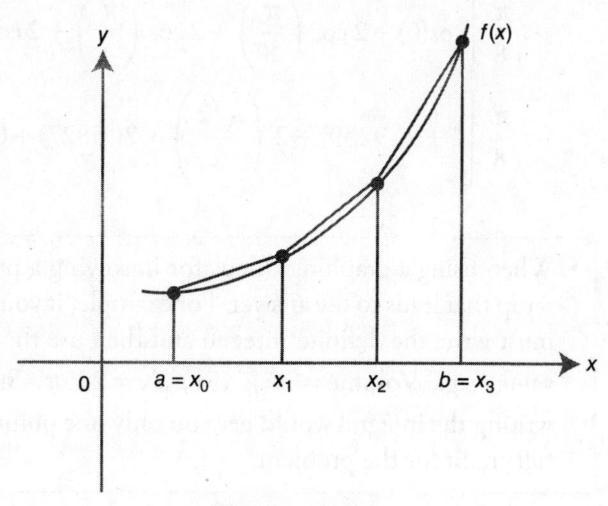

Figure 13.2-5

Formula for Trapezoidal Approximation

If f is continuous, the area under the curve of f from $x = a$ to $x = b$ is:

$$\text{Area} \approx \frac{b-a}{2n}\left[f(x_0) + 2f(x_1) + 2f(x_2) + \cdots + 2f(x_{n-1}) + f(x_n)\right].$$

Example 1

Find the approximate area under the curve of $f(x) = \cos\left(\frac{x}{2}\right)$ from $x = 0$ to $x = \pi$, using 4 trapezoids. (See Figure 13.2-6.)

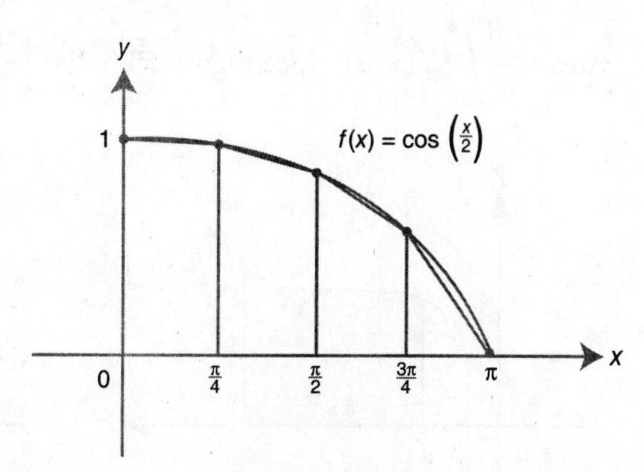

Figure 13.2-6

Since $n = 4$, $\Delta x = \dfrac{\pi - 0}{4} = \dfrac{\pi}{4}$.

Area under the curve:

$$\approx \frac{\pi}{4}\cdot\frac{1}{2}\left[\cos(0)+2\cos\left(\frac{\pi/4}{2}\right)+2\cos\left(\frac{\pi/2}{2}\right)+2\cos\left(\frac{3\pi/4}{2}\right)+\cos\left(\frac{\pi}{2}\right)\right]$$

$$\approx \frac{\pi}{8}\left[\cos(0)+2\cos\left(\frac{\pi}{8}\right)+2\cos\left(\frac{\pi}{4}\right)+2\cos\left(\frac{3\pi}{8}\right)+\cos\left(\frac{\pi}{2}\right)\right]$$

$$\approx \frac{\pi}{8}\left[1+2(.9239)+2\left(\frac{\sqrt{2}}{2}\right)+2(.3827)+0\right] \approx 1.9743.$$

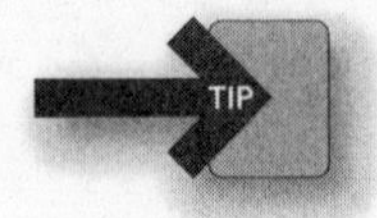

- When using a graphing calculator in solving a problem, you are required to write the setup that leads to the answer. For example, if you are finding the volume of a solid, you must write the definite integral and then use the calculator to compute the numerical value, e.g., Volume $= \pi \int_0^3 (5x)^2 dx = 225\pi$. Simply indicating the answer without writing the integral would get you only one point for the answer. And you will not get full credit for the problem.

13.3 Area and Definite Integrals

Main Concepts: Area Under a Curve, Area Between Two Curves

Area Under a Curve

If $y = f(x)$ is continuous and non-negative on $[a, b]$, then the area under the curve of f from a to b is:

$$\text{Area} = \int_a^b f(x)dx.$$

If f is continuous and $f < 0$ on $[a, b]$, then the area under the curve from a to b is:

$$\text{Area} = -\int_a^b f(x)dx. \text{ (See Figure 13.3-1.)}$$

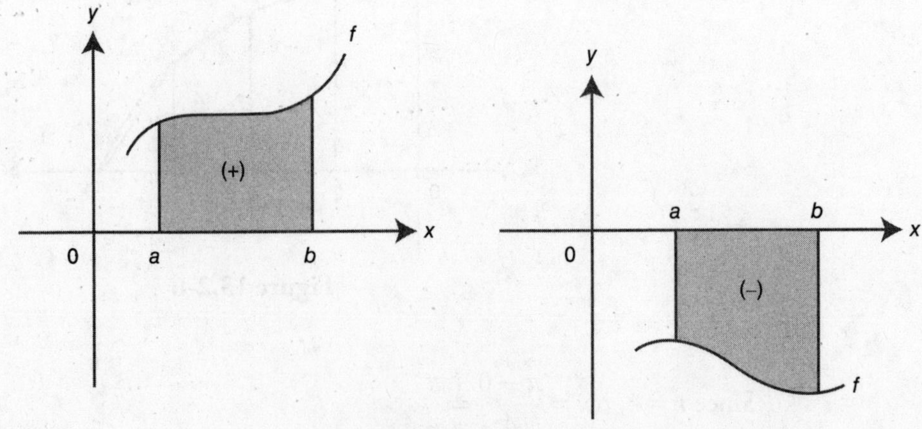

Figure 13.3-1

If $x = g(y)$ is continuous and non-negative on $[c, d]$, then the area under the curve of g from c to d is:

Area $\int_c^d g(y)dy$. (See Figure 13.3-2.)

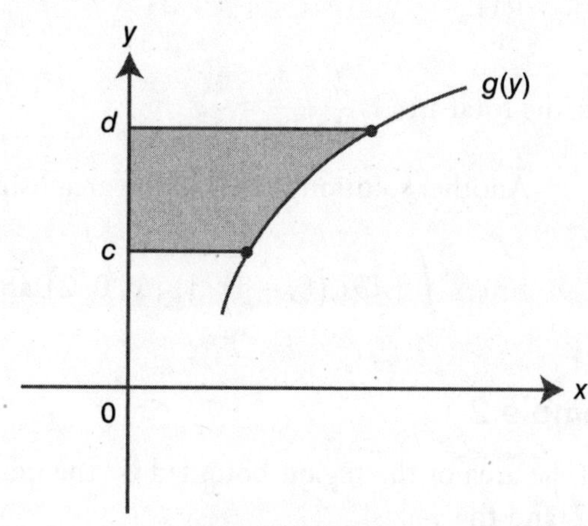

Figure 13.3-2

Example 1

Find the area under the curve of $f(x) = (x - 1)^3$ from $x = 0$ to $x = 2$.

Step 1: Sketch the graph of $f(x)$. (See Figure 13.3-3.)

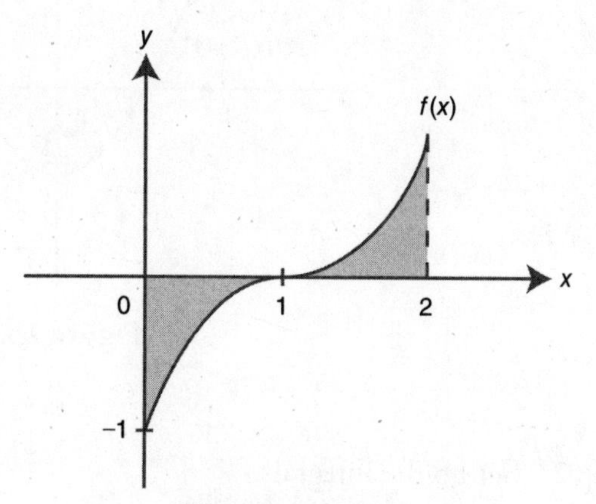

Figure 13.3-3

Step 2: Set up the integrals.

$$\text{Area} = \left|\int_0^1 f(x)dx\right| + \int_1^2 f(x)dx.$$

Step 3: Evaluate the integrals.

$$\left|\int_0^1 (x-1)^3 dx\right| = \left|\left.\frac{(x-1)^4}{4}\right]_0^1\right| = \left|-\frac{1}{4}\right| = \frac{1}{4}$$

$$\int_1^2 (x-1)^3 dx = \left.\frac{(x-1)^4}{4}\right]_1^2 = \frac{1}{4}$$

Thus, the total area is $\frac{1}{4} + \frac{1}{4} = \frac{1}{2}$.

Another solution is to find the area using a calculator:

Enter $\int \left(abs\left((x-1)\verb|^|3\right), x, 0, 2\right)$ and obtain $\frac{1}{2}$.

Example 2

Find the area of the region bounded by the graph of $f(x) = x^2 - 1$, the lines $x = -2$ and $x = 2$, and the x-axis.

Step 1: Sketch the graph of $f(x)$. (See Figure 13.3-4.)

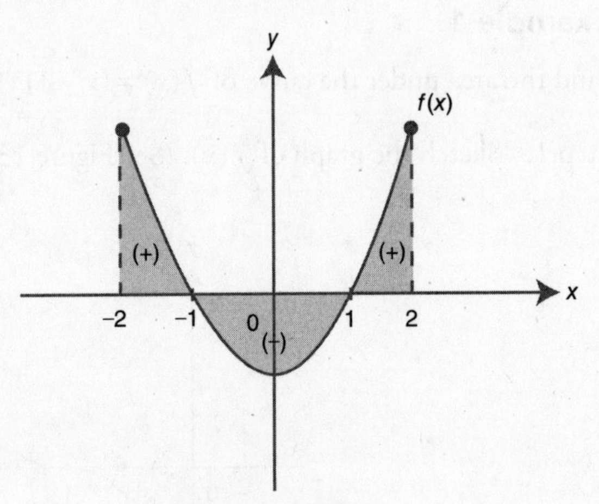

Figure 13.3-4

Step 2: Set up the integrals.

$$\text{Area} = \int_{-2}^{-1} f(x)dx + \left|\int_{-1}^{1} f(x)dx\right| + \int_1^2 f(x)dx.$$

Step 3: Evaluate the integrals.

$$\int_{-2}^{-1} (x^2 - 1)\,dx = \left.\frac{x^3}{3} - x\right]_{-2}^{-1} = \frac{2}{3} - \left(-\frac{2}{3}\right) = \frac{4}{3}$$

$$\left|\int_{-1}^{1} (x^2 - 1)\,dx\right| = \left|\left.\frac{x^3}{3} - x\right]_{-1}^{1}\right| = \left|-\frac{2}{3} - \left(\frac{2}{3}\right)\right| = \left|-\frac{4}{3}\right| = \frac{4}{3}$$

$$\int_{1}^{2} (x^2 - 1)\,dx = \left.\frac{x^3}{3} - x\right]_{1}^{2} = \frac{2}{3} - \left(-\frac{2}{3}\right) = \frac{4}{3}$$

Thus, the total area $= \frac{4}{3} + \frac{4}{3} + \frac{4}{3} = 4.$

Note that since $f(x) = x^2 - 1$ is an even function, you can use the symmetry of the graph and set area $= 2\left(\left|\int_0^1 f(x)dx\right| + \int_1^2 f(x)dx\right)$.

An alternate solution is to find the area using a calculator:

Enter $\int$ (abs (x^2 − 1), x, −2, 2) and obtain 4.

Example 3

Find the area of the region bounded by $x = y^2$, $y = -1$, and $y = 3$. (See Figure 13.3-5.)

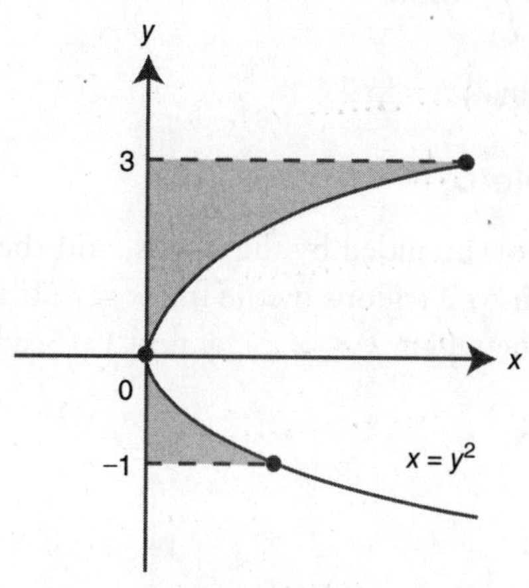

Figure 13.3-5

$$\text{Area} = \int_{-1}^{3} y^2 dy = \left.\frac{y^3}{3}\right]_{-1}^{3} = \frac{3^3}{3} - \frac{(-1)^3}{3} = \frac{28}{3}.$$

Example 4

Using a calculator, find the area bounded by $f(x) = x^3 + x^2 - 6x$ and the x-axis. (See Figure 13.3-6.)

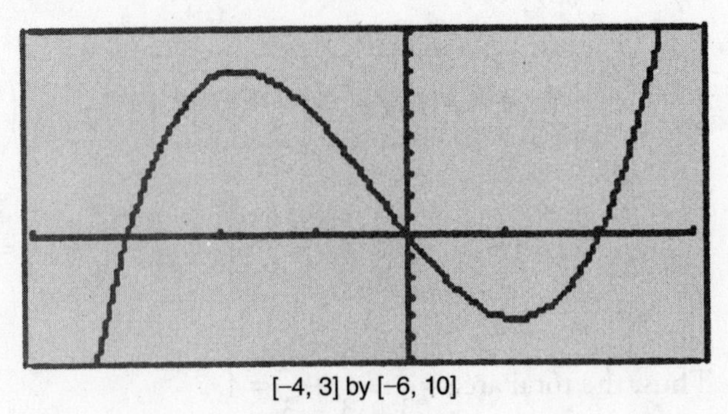

Figure 13.3-6

Step 1: Enter $y_1 = x\^3 + x\^2 - 6x$.

Step 2: Enter $\int (abs\ (x\^3 + x\^2 - 6 * x)\ , x, -3, 2)$ and obtain 21.083.

Example 5

The area under the curve $y = e^x$ from $x = 0$ to $x = k$ is 1. Find the value of k.

Area $= \int_0^k e^x dx = e^x]_0^k = e^k - e^0 = e^k - 1 \Rightarrow e^k = 2$. Take ln of both sides:

$\ln(e^k) = \ln 2$; $k = \ln 2$.

Example 6

The region bounded by the x-axis, and the graph of $y = \sin x$ between $x = 0$ and $x = \pi$ is divided into 2 regions by the line $x = k$. If the area of the region for $0 \leq x \leq k$ is twice the area of the region $k \leq x \leq \pi$, find k. (See Figure 13.3-7.)

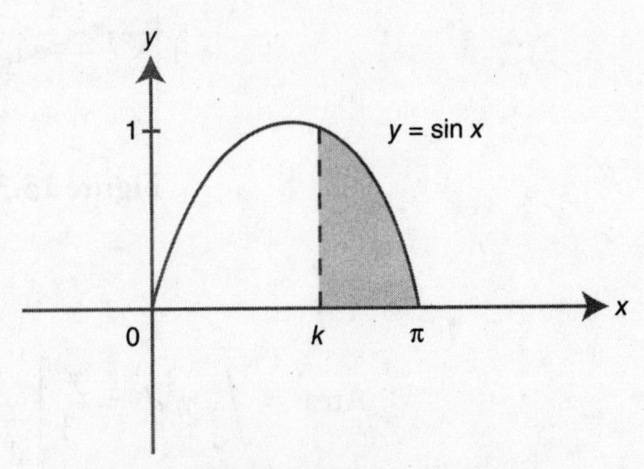

Figure 13.3-7

$$\int_0^k \sin x\, dx = 2\int_k^{\pi} \sin x\, dx$$

$$-\cos x]_0^k = 2\,[-\cos x]_k^{\pi}$$

$$-\cos k - (-\cos(0)) = 2\,(-\cos\pi - (-\cos k))$$

$$-\cos k + 1 = 2(1 + \cos k)$$

$$-\cos k + 1 = 2 + 2\cos k$$

$$-3\cos k = 1$$

$$\cos k = -\frac{1}{3}$$

$$k = \arccos\left(-\frac{1}{3}\right) = 1.91063$$

Area Between Two Curves

Area Bounded by Two Curves. (See Figure 13.3-8.)

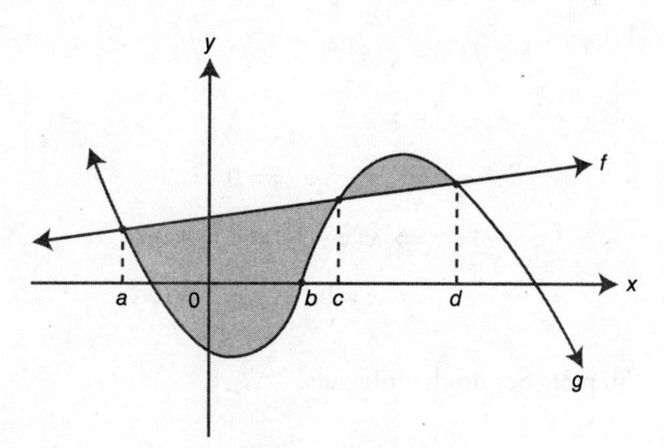

Figure 13.3-8

$$\text{Area} = \int_a^c [f(x) - g(x)]\, dx + \int_c^d [g(x) - f(x)]\, dx.$$

$$\text{Area} = \int_a^d (\text{upper curve} - \text{lower curve})\, dx.$$

Example 1

Find the area of the regions bounded by the graphs of $f(x) = x^3$ and $g(x) = x$. (See Figure 13.3-9.)

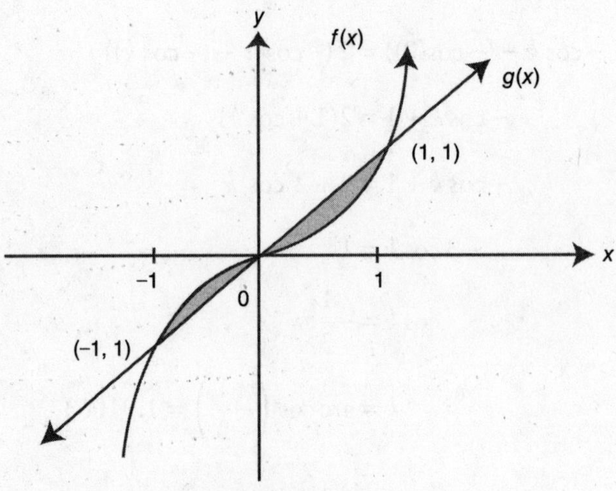

Figure 13.3-9

Step 1: Sketch the graphs of $f(x)$ and $g(x)$.

Step 2: Find the points of intersection.

Set $f(x) = g(x)$

$$x^3 = x$$

$$\Rightarrow x(x^2 - 1) = 0$$

$$\Rightarrow x(x - 1)(x + 1) = 0$$

$$\Rightarrow x = 0,\ 1,\ \text{and}\ -1.$$

Step 3: Set up the integrals.

$$\text{Area} = \int_{-1}^{0} (f(x) - g(x))dx + \int_{0}^{1} (g(x) - f(x))dx$$

$$= \int_{-1}^{0} \left(x^3 - x\right)dx + \int_{0}^{1} \left(x - x^3\right)dx$$

$$= \left[\frac{x^4}{4} - \frac{x^2}{2}\right]_{-1}^{0} + \left[\frac{x^2}{2} - \frac{x^4}{4}\right]_{0}^{1}$$

$$= 0 - \left(\frac{(-1)^4}{4} - \frac{(-1)^2}{2}\right) + \left(\frac{1^2}{2} - \frac{1^4}{4}\right) - 0$$

$$= -\left(-\frac{1}{4}\right) + \frac{1}{4} = \frac{1}{2}.$$

Note that you can use the symmetry of the graphs and let area = $2\int_0^1 \left(x - x^3\right)dx$.

An alternate solution is to find the area using a calculator:

$$\text{Enter } \int (abs\,(x\wedge 3 - x), x, -1, 1) \text{ and obtain } \frac{1}{2}.$$

Example 2

Find the area of the region bounded by the curve $y = e^x$, the y-axis, and the line $y = e^2$.

Step 1: Sketch a graph. (See Figure 13.3-10.)

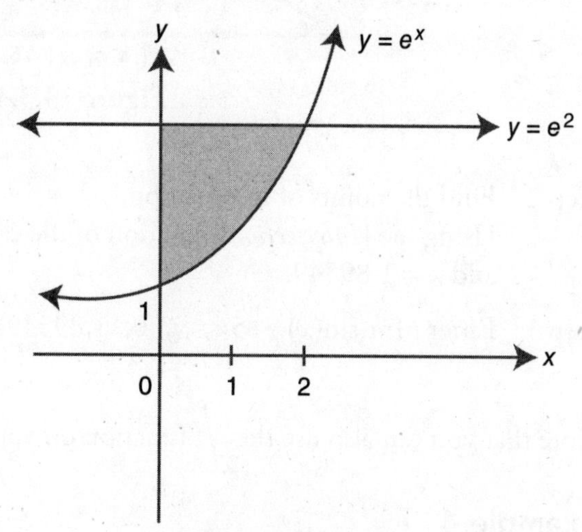

Figure 13.3-10

Step 2: Find the point of intersection. Set $e^2 = e^x \Rightarrow x = 2$.

Step 3: Set up an integral:

$$\text{Area} = \int_0^2 (e^2 - e^x)dx = (e^2)x - e^x]_0^2$$

$$= (2e^2 - e^2) - (0 - e^0)$$

$$= e^2 + 1.$$

Or using a calculator, enter $\int ((e\wedge 2 - e\wedge x), x, 0, 2)$ and obtain $(e^2 + 1)$.

Example 3

Using a calculator, find the area of the region bounded by $y = \sin x$ and $y = \frac{x}{2}$ between $0 \le x \le \pi$.

Step 1: Sketch a graph. (See Figure 13.3-11.)

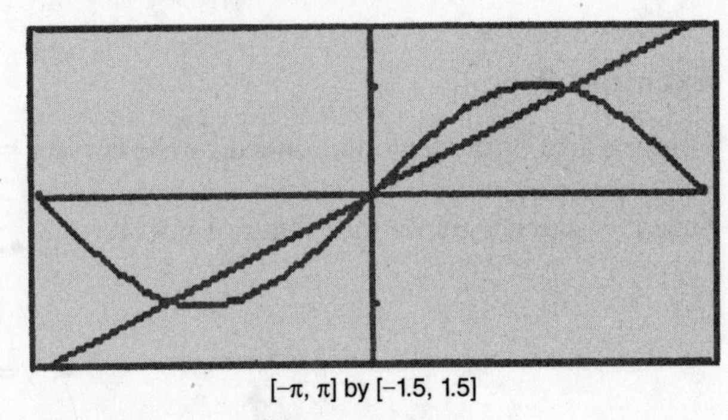

Figure 13.3-11

Step 2: Find the points of intersection.
Using the [*Intersection*] function of the calculator, the intersection points are $x = 0$ and $x = 1.89549$.

Step 3: Enter nInt(sin(x) − .5x, x, 0, 1.89549) and obtain $0.420798 \approx 0.421$.

Note that you can also use the $\int$ function on your calculator and get the same result.

Example 4

Find the area of the region bounded by the curve $xy = 1$ and the lines $y = -5$, $x = e$, and $x = e^3$.

Step 1: Sketch a graph. (See Figure 13.3-12.)

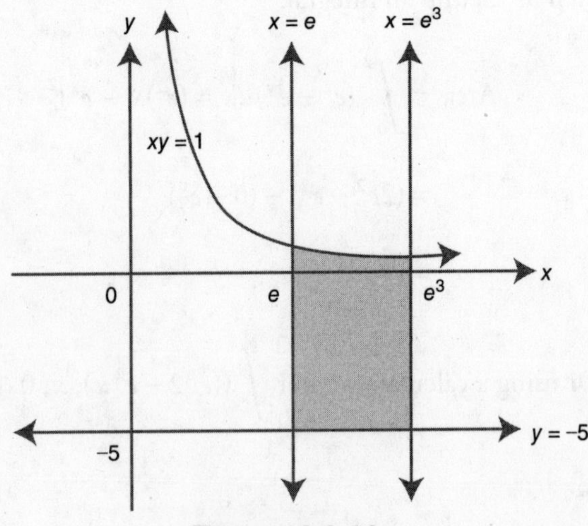

Figure 13.3-12

Step 2: Set up an integral.

$$\text{Area} = \int_{e}^{e^3} \left(\frac{1}{x} - (-5)\right) dx.$$

Step 3: Evaluate the integral.

$$\text{Area} = \int_{e}^{e^3} \left(\frac{1}{x} - (-5)\right) dx = \int_{e}^{e^3} \left(\frac{1}{x} + 5\right) dx$$

$$= \ln|x| + 5x\Big]_{e}^{e^3} = \left[\ln(e^3) + 5(e^3)\right] - \left[\ln(e) + 5(e)\right]$$

$$= 3 + 5e^3 - 1 - 5e = 2 - 5e + 5e^3.$$

- Remember: If $f' > 0$, then f is increasing, and if $f'' > 0$, then the graph of f is concave upward.

13.4 Volumes and Definite Integrals

Main Concepts: Solids with Known Cross Sections, The Disc Method, The Washer Method

Solids with Known Cross Sections

If $A(x)$ is the area of a cross section of a solid and $A(x)$ is continuous on $[a, b]$, then the volume of the solid from $x = a$ to $x = b$ is:

$$V = \int_{a}^{b} A(x)dx.$$

(See Figure 13.4-1.)

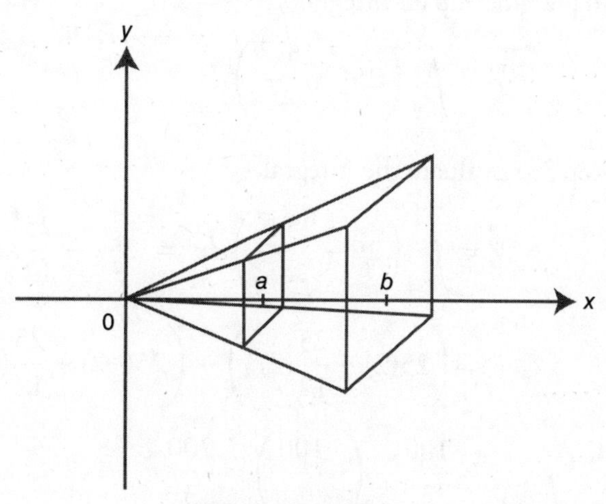

Figure 13.4-1

Note: A cross section of a solid is perpendicular to the height of the solid.

Example 1

The base of a solid is the region enclosed by the ellipse $\frac{x^2}{4} + \frac{y^2}{25} = 1$. The cross sections are perpendicular to the x-axis and are isosceles right triangles whose hypotenuses are on the ellipse. Find the volume of the solid. (See Figure 13.4-2.)

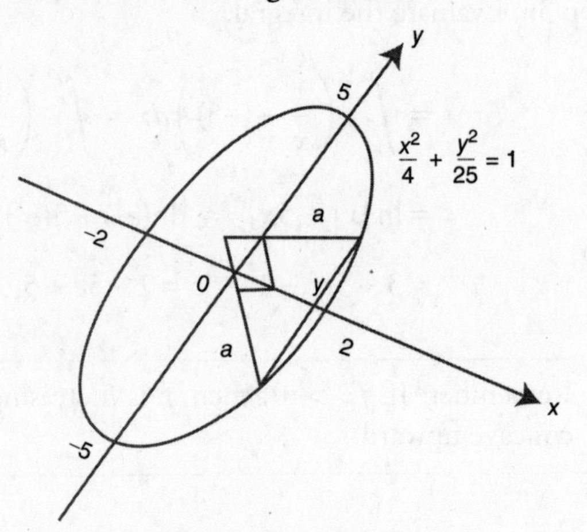

Figure 13.4-2

Step 1: Find the area of a cross section $A(x)$.

Use the Pythagorean Theorem: $a^2 + a^2 = (2y)^2$

$$2a^2 = 4y^2$$

$$a = \sqrt{2}y,\ a > 0.$$

$$A(x) = \frac{1}{2}a^2 = \frac{1}{2}\left(\sqrt{2}y\right)^2 = y^2$$

Since $\frac{x^2}{4} + \frac{y^2}{25} = 1$, $\frac{y^2}{25} = 1 - \frac{x^2}{4}$ or $y^2 = 25 - \frac{25x^2}{4}$,

$$A(x) = 25 - \frac{25x^2}{4}.$$

Step 2: Set up an integral.

$$V = \int_{-2}^{2}\left(25 - \frac{25x^2}{4}\right)dx$$

Step 3: Evaluate the integral.

$$V = \int_{-2}^{2}\left(25 - \frac{25x^2}{4}\right)dx = 25x - \frac{25}{12}x^3\Big]_{-2}^{2}$$

$$= \left(25(2) - \frac{25}{12}(2)^3\right) - \left(25(-2) - \frac{25}{12}(-2)^3\right)$$

$$= \frac{100}{3} - \left(-\frac{100}{3}\right) = \frac{200}{3}$$

The volume of the solid is $\frac{200}{3}$.

Verify your result with a graphing calculator.

Example 2

Find the volume of a pyramid whose base is a square with a side of 6 feet long, and a height of 10 feet. (See Figure 13.4-3.)

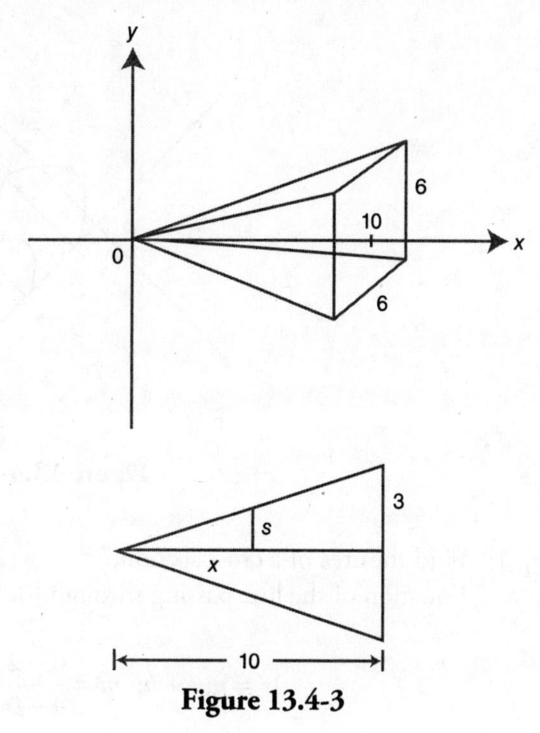

Figure 13.4-3

Step 1: Find the area of a cross section $A(x)$. Note each cross section is a square of side $2s$.

Similar triangles: $\frac{x}{s} = \frac{10}{3} \Rightarrow s = \frac{3x}{10}$.

$$A(x) = (2s)^2 = 4s^2 = 4\left(\frac{3x}{10}\right)^2 = \frac{9x^2}{25}$$

Step 2: Set up an integral.

$$V = \int_0^{10} \frac{9x^2}{25} dx$$

Step 3: Evaluate the integral.

$$V = \int_0^{10} \frac{9x^2}{25} dx = \left.\frac{3x^3}{25}\right]_0^{10} = \frac{3\,(10)^3}{25} - 0 = 120$$

The volume of the pyramid is 120 ft^3.

Example 3

The base of a solid is the region enclosed by a triangle whose vertices are (0, 0), (4, 0), and (0, 2). The cross sections are semicircles perpendicular to the x-axis. Using a calculator, find the volume of the solid. (See Figure 13.4-4.)

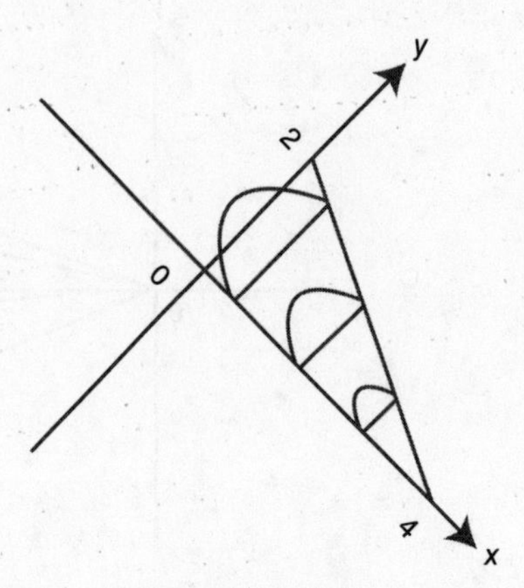

Figure 13.4-4

Step 1: Find the area of a cross section.
Equation of the line passing through (0, 2) and (4, 0):

$$y = mx + b;\ m = \frac{0-2}{4-0} = -\frac{1}{2};\ b = 2$$

$$y = -\frac{1}{2}x + 2.$$

$$\text{Area of semicircle } = \frac{1}{2}\pi r^2;\ r = \frac{1}{2}y = \frac{1}{2}\left(-\frac{1}{2}x + 2\right) = -\frac{1}{4}x + 1.$$

$$A(x) = \frac{1}{2}\pi\left(\frac{y}{2}\right)^2 = \frac{\pi}{2}\left(-\frac{1}{4}x + 1\right)^2$$

Step 2: Set up an integral.

$$V = \int_0^4 A(x)dx = \int_0^4 \frac{\pi}{2}\left(-\frac{1}{4}x + 1\right)^2 dx$$

Step 3: Evaluate the integral.

Enter $\int\left(\left(\frac{\pi}{2}\right) * (-.25x + 1)\text{^}2,\ x,\ 0,\ 4\right)$ and obtain 2.0944.

Thus, the volume of the solid is 2.094.

- Remember: If $f' < 0$, then f is decreasing, and if $f'' < 0$, then the graph of f is concave downward.

The Disc Method

The volume of a solid of revolution using discs:

Revolving about the x-axis:

$$V = \pi \int_a^b (f(x))^2 \, dx, \quad \text{where } f(x) = \text{radius}.$$

Revolving about the y-axis:

$$V = \pi \int_c^d (g(y))^2 \, dy, \quad \text{where } g(y) = \text{radius}.$$

(See Figure 13.4-5.)

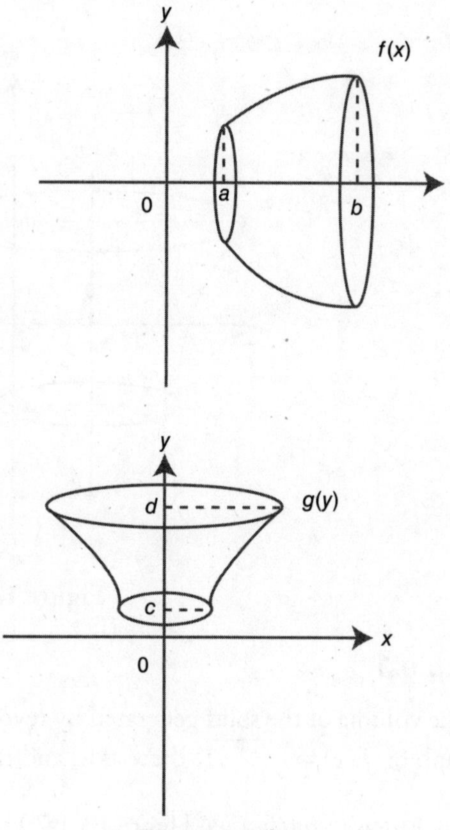

Figure 13.4-5

Revolving about a line $y = k$:

$$V = \pi \int_a^b (f(x) - k)^2 \, dx, \quad \text{where } |f(x) - k| = \text{radius}.$$

Revolving about a line $x = h$:

$$V = \pi \int_c^d (g(y) - h)^2 \, dy, \quad \text{where } |g(y) - h| = \text{radius}.$$

(See Figure 13.4-6.)

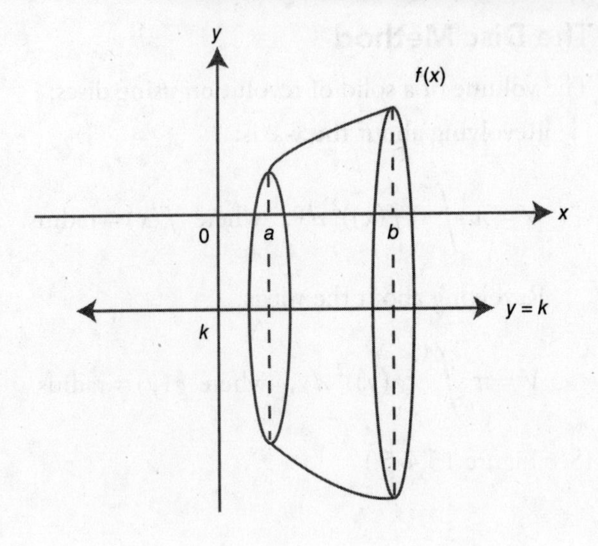

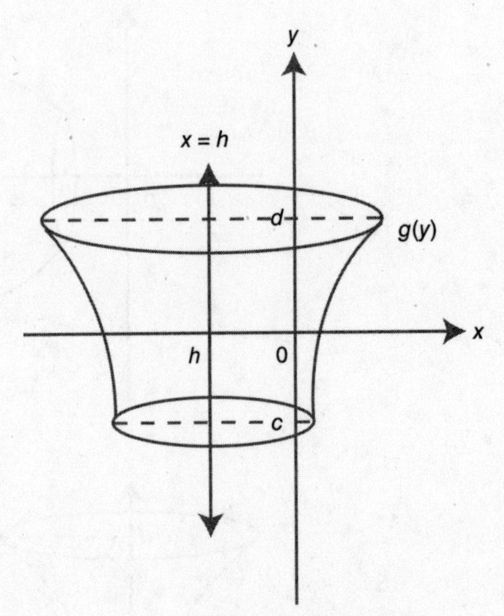

Figure 13.4-6

Example 1

Find the volume of the solid generated by revolving about the x-axis the region bounded by the graph of $f(x) = \sqrt{x-1}$, the x-axis, and the line $x = 5$.

Step 1: Draw a sketch. (See Figure 13.4-7.)

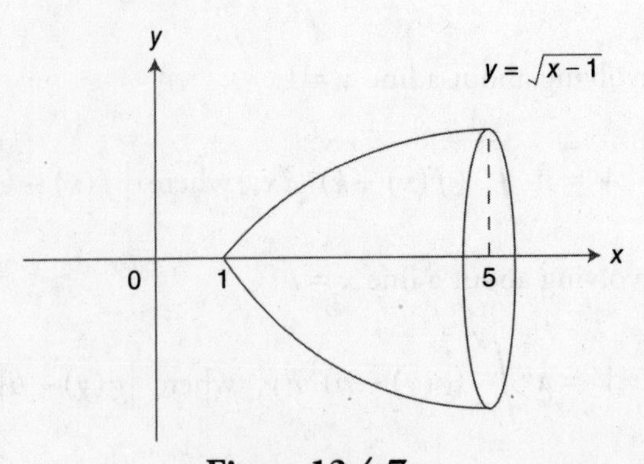

Figure 13.4-7

Step 2: Determine the radius of a disc from a cross section.

$$r = f(x) = \sqrt{x-1}$$

Step 3: Set up an integral.

$$V = \pi \int_1^5 (f(x))^2 dx = \pi \int_1^5 \left(\sqrt{x-1}\right)^2 dx$$

Step 4: Evaluate the integral.

$$V = \pi \int_1^5 \left(\sqrt{x-1}\right)^2 dx = \pi\, [(x-1)]_1^5 = \pi \left[\frac{x^2}{2} - x\right]_1^5$$

$$= \pi \left(\left(\frac{5^2}{2} - 5\right) - \left(\frac{1^2}{2} - 1\right)\right) = 8\pi$$

Verify your result with a calculator.

Example 2

Find the volume of the solid generated by revolving about the x-axis the region bounded by the graph of $y = \sqrt{\cos x}$ where $0 \leq x \leq \frac{\pi}{2}$, the x-axis, and the y-axis.

Step 1: Draw a sketch. (See Figure 13.4-8.)

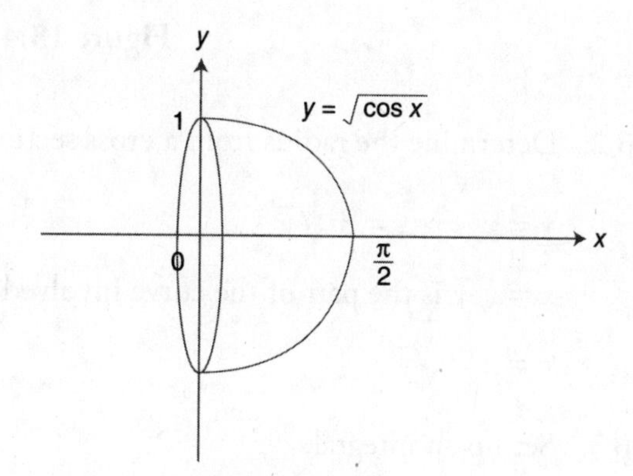

Figure 13.4-8

Step 2: Determine the radius from a cross section.

$$r = f(x) = \sqrt{\cos x}$$

Step 3: Set up an integral.

$$V = \pi \int_0^{\pi/2} \left(\sqrt{\cos x}\right)^2 dx = \pi \int_0^{\pi/2} \cos x \, dx$$

Step 4: Evaluate the integral.

$$V = \pi \int_0^{\pi/2} \cos x dx = \pi \left[\sin x\right]_0^{\pi/2} = \pi \left(\sin\left(\frac{\pi}{2}\right) - \sin 0\right) = \pi$$

Thus, the volume of the solid is π.
Verify your result with a calculator.

Example 3

Find the volume of the solid generated by revolving about the y-axis the region in the first quadrant bounded by the graph of $y = x^2$, the y-axis, and the line $y = 6$.

Step 1: Draw a sketch. (See Figure 13.4-9.)

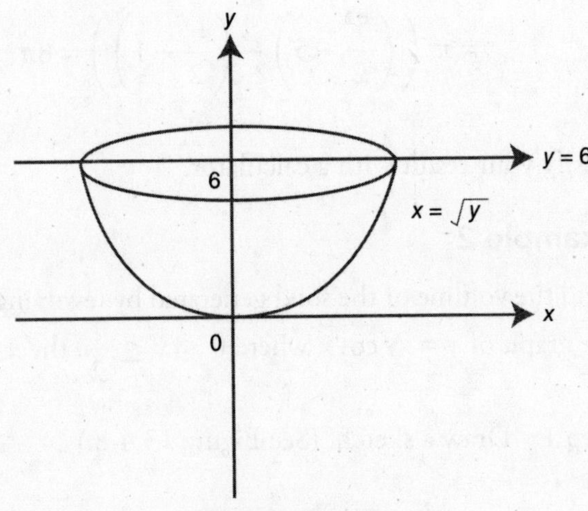

Figure 13.4-9

Step 2: Determine the radius from a cross section.

$y = x^2 \Rightarrow x = \pm\sqrt{y}$

$x = \sqrt{y}$ is the part of the curve involved in the region.

$r = x = \sqrt{y}$

Step 3: Set up an integral.

$$V = \pi \int_0^6 x^2 dy = \pi \int_0^6 (\sqrt{y})^2 dy = \pi \int_0^6 y dy$$

Step 4: Evaluate the integral.

$$V = \pi \int_0^6 y dy = \pi \left[\frac{y^2}{2}\right]_0^6 = 18\pi$$

Thus, the volume of the solid is 18π.
Verify your result with a calculator.

Example 4

Using a calculator, find the volume of the solid generated by revolving about the line $y = 8$, the region bounded by the graph of $y = x^2 + 4$, and the line $y = 8$.

Step 1: Draw a sketch. (See Figure 13.4-10.)

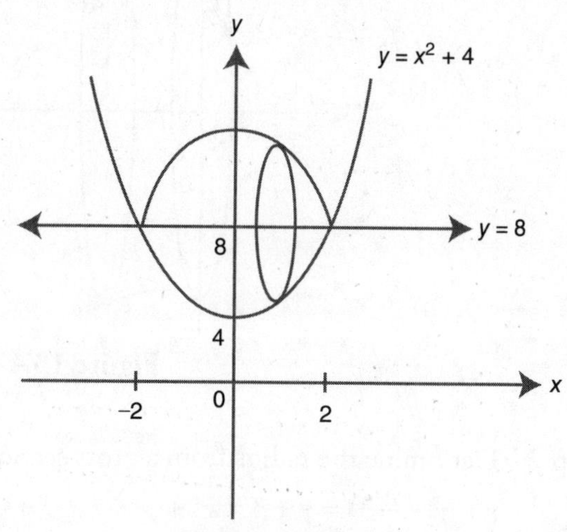

Figure 13.4-10

Step 2: Determine the radius from a cross section.

$$r = 8 - y = 8 - (x^2 + 4) = 4 - x^2$$

Step 3: Set up an integral.
To find the intersection points, set $8 = x^2 + 4 \Rightarrow x = \pm 2$.

$$V = \pi \int_{-2}^{2} \left(4 - x^2\right)^2 dx$$

Step 4: Evaluate the integral.

Enter $\int \left(\pi \left(4 - x^2\right)^2, \ x, \ -2, \ 2\right)$ and obtain $\frac{512}{15}\pi$.

Thus, the volume of the solid is $\frac{512}{15}\pi$.
Verify your result with a calculator.

Example 5

Using a calculator, find the volume of the solid generated by revolving about the line $y = -3$, the region bounded by the graph of $y = e^x$, the y-axis, and the lines $x = \ln 2$ and $y = -3$.

Step 1: Draw a sketch. (See Figure 13.4-11.)

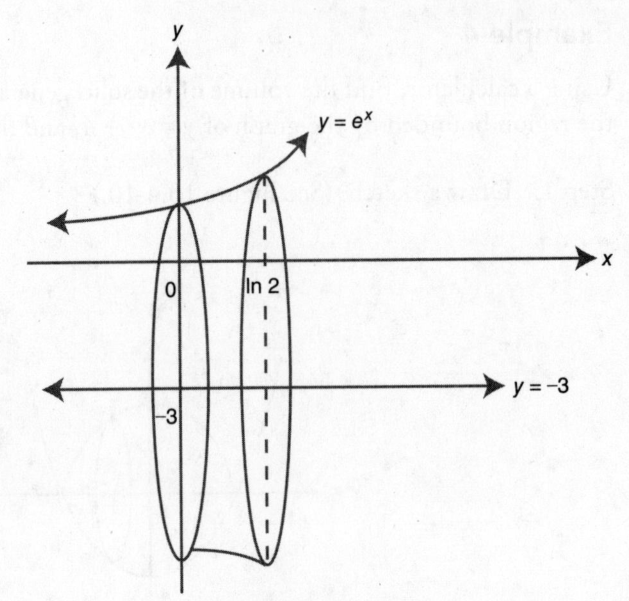

Figure 13.4-11

Step 2: Determine the radius from a cross section.

$r = y - (-3) = y + 3 = e^x + 3$

Step 3: Set up an integral.

$$V = \pi \int_0^{\ln 2} (e^x + 3)^2 dx$$

Step 4: Evaluate the integral.

Enter $\int (\pi\, (e\text{^}(x) + 3)\text{^}2,\ x,\ 0\ \ln(2))$ and obtain $\pi\left(9\ln 2 + \frac{15}{2}\right)$ $= 13.7383\pi$.

Thus, the volume of the solid is approximately 13.7383π.

- Remember: If f' is increasing, then $f'' > 0$ and the graph of f is concave upward.

The Washer Method

The volume of a solid (with a hole in the middle) generated by revolving a region bounded by 2 curves.

About the x-axis:

$$V=\pi \int_a^b \left[(f(x))^2-(g(x))^2\right] dx;$$ where $f(x)$=outer radius and $g(x)$=inner radius.

About the y-axis:

$$V=\pi \int_c^d \left[(p(y))^2-(q(y))^2\right] dy;$$ where $p(y)$=outer radius and $q(y)$=inner radius.

About a line $x = h$:

$$V = \pi \int_a^b \left[(f(x) - h)^2 - (g(x) - h)^2 \right] dx.$$

About a line $y = k$:

$$V = \pi \int_c^d \left[(p(y) - k)^2 - (q(y) - k)^2 \right] dy.$$

Example 1

Using the Washer Method, find the volume of the solid generated by revolving the region bounded by $y = x^3$ and $y = x$ in the first quadrant about the x-axis.

Step 1: Draw a sketch. (See Figure 13.4-12.)

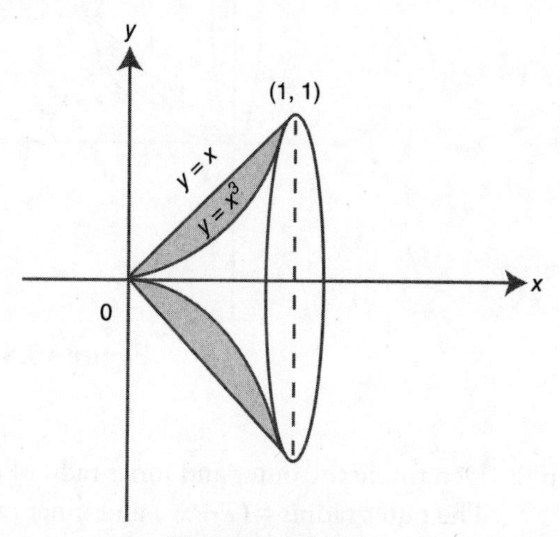

Figure 13.4-12

To find the points of intersection, set $x = x^3 \Rightarrow x^3 - x = 0$ or $x(x^2 - 1) = 0$, or $x = -1, 0, 1$. In the first quadrant $x = 0, 1$.

Step 2: Determine the outer and inner radii of a washer.
The outer radius $= x$ and inner radius $= x^3$.

Step 3: Set up an integral.

$$V = \int_0^1 \left[x^2 - (x^3)^2 \right] dx$$

Step 4: Evaluate the integral.

$$V = \int_0^1 \left(x^2 - x^6 \right) dx = \pi \left[\frac{x^3}{3} - \frac{x^7}{7} \right]_0^1$$

$$= \pi \left(\frac{1}{3} - \frac{1}{7} \right) = \frac{4\pi}{21}$$

Verify your result with a calculator.

Example 2

Using the Washer Method and a calculator, find the volume of the solid generated by revolving the region in Example 1 about the line $y = 2$.

Step 1: Draw a sketch. (See Figure 13.4-13.)

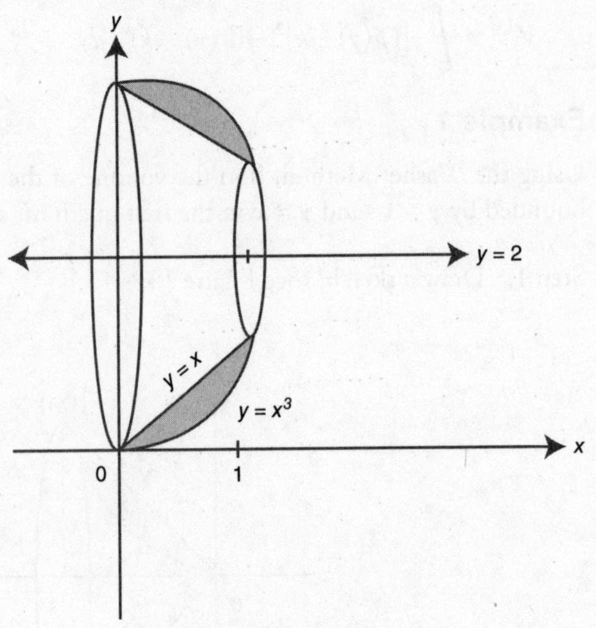

Figure 13.4-13

Step 2: Determine the outer and inner radii of a washer.
The outer radius $= (2 - x^3)$ and inner radius $= (2 - x)$.

Step 3: Set up an integral.

$$V = \pi \int_0^1 \left[\left(2 - x^3\right)^2 - \left(2 - x\right)^2 \right] dx$$

Step 4: Evaluate the integral.

Enter $\int \left(\pi * \left((2 - x\text{^}3)\text{^}2 - (2 - x)\text{^}2\right),\ x,\ 0,\ 1\right)$ and obtain $\frac{17\pi}{21}$.

Thus, the volume of the solid is $\frac{17\pi}{21}$.

Example 3

Using the Washer Method and a calculator, find the volume of the solid generated by revolving the region bounded by $y = x^2$ and $x = y^2$ about the y-axis.

Step 1: Draw a sketch. (See Figure 13.4-14.)

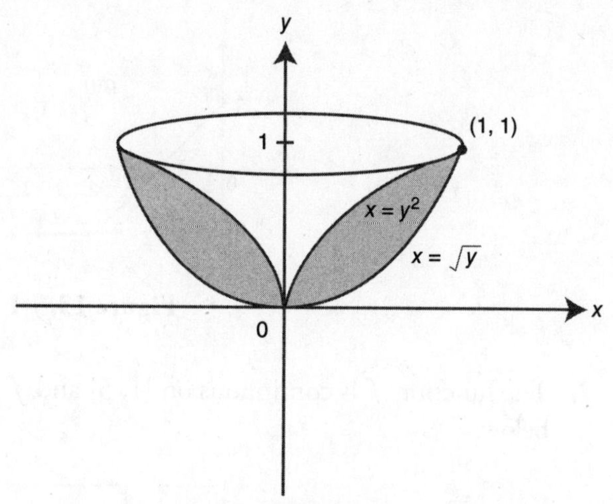

Figure 13.4-14

Intersection points: $y = x^2$; $x = y^2 \Rightarrow y = \pm\sqrt{x}$.

Set $x^2 = \sqrt{x} \Rightarrow x^4 = x \Rightarrow x^4 - x = 0 \Rightarrow x(x^3 - 1) = 0 \Rightarrow x = 0$ or $x = 1$

$x = 0,\ y = 0\ (0, 0)$

$x = 1,\ y = 1\ (1, 1)$.

Step 2: Determine the outer and inner radii of a washer.
The outer radius:

$x = \sqrt{y}$ and inner radius: $x = y^2$.

Step 3: Set up an integral.

$$V = \pi \int_0^1 \left((\sqrt{y})^2 - (y^2)^2 \right) dy$$

Step 4: Evaluate the integral.

Enter $\int \left(\pi * ((\sqrt{y})^\wedge 2 - (y^\wedge 2)^\wedge 2),\ y,\ 0,\ 1\right)$ and obtain $\frac{3\pi}{10}$.

Thus, the volume of the solid is $\frac{3\pi}{10}$.

13.5 Rapid Review

1. If $f(x) = \int_0^x g(t)dt$ and the graph of g is shown in Figure 13.5-1. Find $f(3)$.

Answer: $f(3) = \int_0^3 g(t)dt = \int_0^1 g(t)\,dt + \int_1^3 g(t)\,dt$

$= 0.5 - 1.5 = -1$

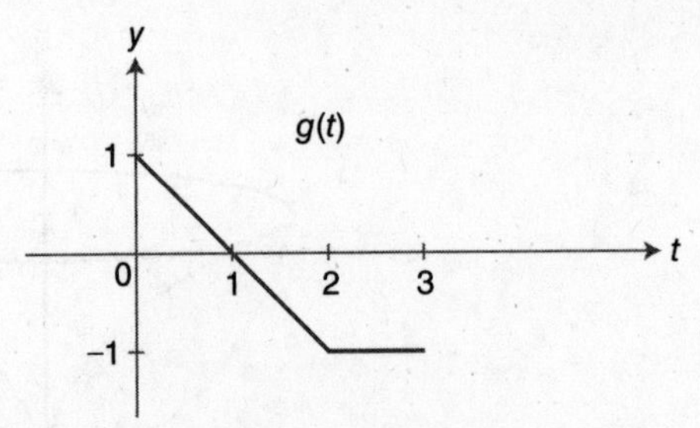

Figure 13.5-1

2. The function f is continuous on [1, 5] and $f > 0$, and selected values of f are given below.

x	1	2	3	4	5
$f(x)$	2	4	6	8	10

Using 2 midpoint rectangles, approximate the area under the curve of f for $x = 1$ to $x = 5$.

Answer: Midpoints are $x = 2$ and $x = 4$, and the width of each rectangle $= \frac{5-1}{2} = 2.$

Area $\approx$ Area of Rect_1 + Area of $\text{Rect}_2 \approx 4(2) + 8(2) \approx 24.$

3. Set up an integral to find the area of the regions bounded by the graphs of $y = x^3$ and $y = x$. Do not evaluate the integral.

Answer: Graphs intersect at $x = -1$ and $x = 1$. (See Figure 13.5-2.)

$$\text{Area} = \int_{-1}^{0} \left(x^3 - x\right) dx + \int_{0}^{1} \left(x - x^3\right) dx.$$

Or, using symmetry, $\text{Area} = 2\int_{0}^{1} \left(x - x^3\right) dx.$

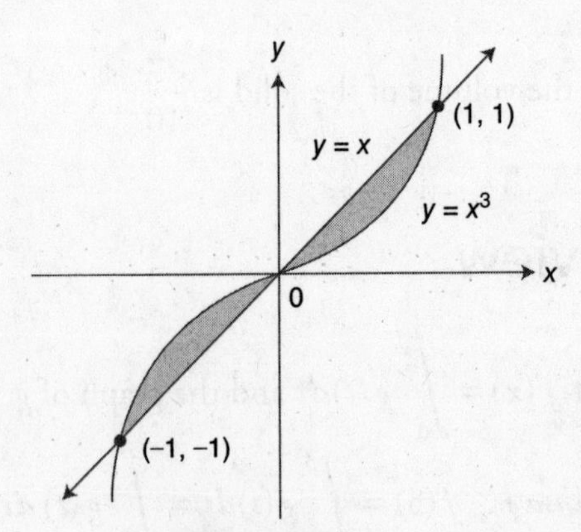

Figure 13.5-2

4. The base of a solid is the region bounded by the lines $y = x$, $x = 1$, and the x-axis. The cross sections are squares perpendicular to the x-axis. Set up an integral to find the volume of the solid. Do not evaluate the integral.

Answer: Area of cross section $= x^2$.

$$\text{Volume of solid} = \int_0^1 x^2 dx.$$

5. Set up an integral to find the volume of a solid generated by revolving the region bounded by the graph of $y = \sin x$, where $0 \leq x \leq \pi$ and the x-axis, about the x-axis. Do not evaluate the integral.

Answer: Volume $= \pi \int_0^{\pi} (\sin x)^2 dx$.

6. The area under the curve of $y = \frac{1}{x}$ from $x = a$ to $x = 5$ is approximately 0.916 where $1 \leq a < 5$. Using your calculator, find a.

Answer: $\int_a^5 \frac{1}{x} dx = \ln x \big|_a^5 = \ln 5 - \ln a = 0.916$

$$\ln a = \ln 5 - 0.916 \approx .693$$

$$a \approx e^{0.693} \approx 2$$

13.6 Practice Problems

Part A—The use of a calculator is not allowed.

1. Let $F(x) = \int_0^x f(t)dt$ where the graph of f is given in Figure 13.6-1.

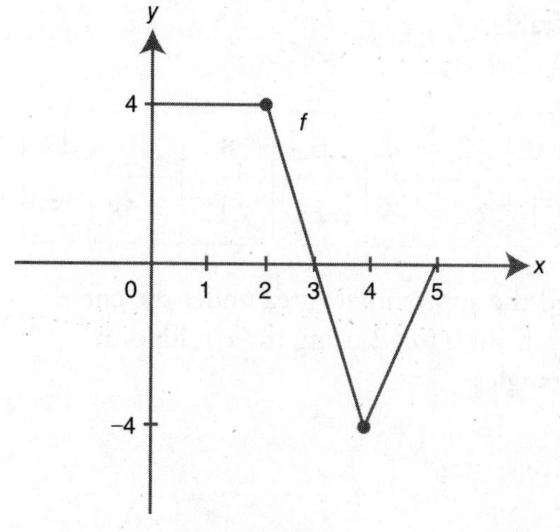

Figure 13.6-1

(a) Evaluate $F(0)$, $F(3)$, and $F(5)$.

(b) On what interval(s) is F decreasing?

(c) At what value of t does F have a maximum value?

(d) On what interval is F concave upward?

2. Find the area of the region(s) enclosed by the curve $f(x) = x^3$, the x-axis, and the lines $x = -1$ and $x = 2$.

3. Find the area of the region(s) enclosed by the curve $y = |2x - 6|$, the x-axis, and the lines $x = 0$ and $x = 4$.

4. Find the approximate area under the curve $f(x) = \frac{1}{x}$ from $x = 1$ to $x = 5$, using four right-endpoint rectangles of equal lengths.

5. Find the approximate area under the curve $y = x^2 + 1$ from $x = 0$ to $x = 3$, using the Trapezoidal Rule with $n = 3$.

6. Find the area of the region bounded by the graphs $y = \sqrt{x}$, $y = -x$, and $x = 4$.

7. Find the area of the region bounded by the curves $x = y^2$ and $x = 4$.

8. Find the area of the region bounded by the graphs of all four equations:
$f(x) = \sin\left(\frac{x}{2}\right)$; x-axis; and the lines, $x = \frac{\pi}{2}$ and $x = \pi$.

9. Find the volume of the solid obtained by revolving about the x-axis, the region bounded by the graphs of $y = x^2 + 4$, the x-axis, the y-axis, and the lines $x = 3$.

10. The area under the curve $y = \frac{1}{x}$ from $x = 1$ to $x = k$ is 1. Find the value of k.

11. Find the volume of the solid obtained by revolving about the y-axis the region bounded by $x = y^2 + 1$, $x = 0$, $y = -1$, and $y = 1$.

12. Let R be the region enclosed by the graph $y = 3x$, the x-axis, and the line $x = 4$. The line $x = a$ divides region R into two regions such that when the regions are revolved about the x-axis, the resulting solids have equal volume. Find a.

Part B—Calculators are allowed.

13. Find the volume of the solid obtained by revolving about the x-axis the region bounded by the graphs of $f(x) = x^3$ and $g(x) = x^2$.

14. The base of a solid is a region bounded by the circle $x^2 + y^2 = 4$. The cross sections of the solid perpendicular to the x-axis are equilateral triangles. Find the volume of the solid.

15. Find the volume of the solid obtained by revolving about the y-axis, the region bounded by the curves $x = y^2$ and $y = x - 2$.

For Problems 16 through 19, find the volume of the solid obtained by revolving the region as described below. (See Figure 13.6-2.)

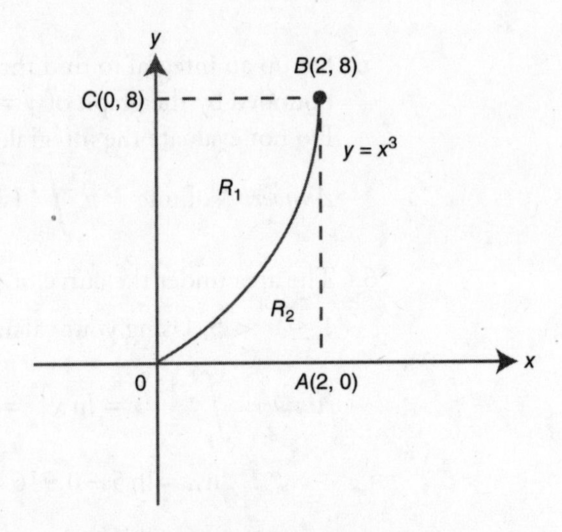

Figure 13.6-2

16. R_1 about the x-axis.

17. R_2 about the y-axis.

18. R_1 about the line $\overleftrightarrow{BC}$.

19. R_2 about the line $\overleftrightarrow{AB}$.

20. The function $f(x)$ is continuous on [0, 12], and the selected values of $f(x)$ are shown in the table.

x	0	2	4	6	8	10	12
$f(x)$	1	2.24	3	3.61	4.12	4.58	5

Find the approximate area under the curve of f from 0 to 12 using three midpoint rectangles.

13.7 Cumulative Review Problems

(Calculator) indicates that calculators are permitted.

21. If $\int_{-a}^{a} e^{x^1} dx = k$, find $\int_{0}^{a} e^{x^2} dx$ in terms of k.

22. A man wishes to pull a log over a 9-foot-high garden wall as shown in Figure 13.7-1. He is pulling at a rate of 2 ft/sec. At what rate is the angle between the rope and the ground changing when there are 15 feet of rope between the top of the wall and the log?

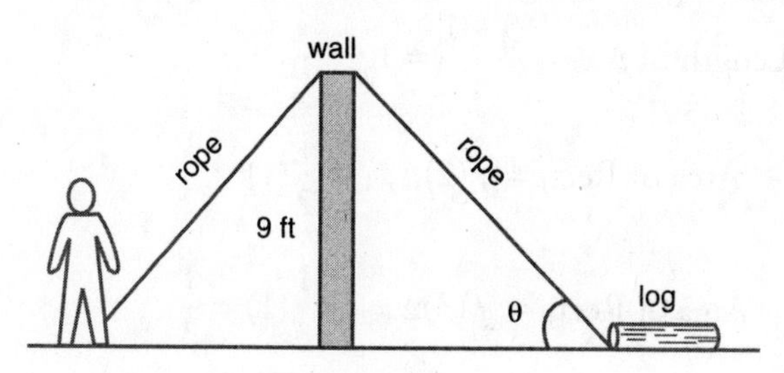

Figure 13.7-1

23. (Calculator) Find a point on the parabola $y = \frac{1}{2}x^2$ that is closest to the point (4, 1).

24. The velocity function of a particle moving along the x-axis is $v(t) = t\cos(t^2 + 1)$ for $t \geq 0$.

   (a) If at $t = 0$, the particle is at the origin, find the position of the particle at $t = 2$.
   (b) Is the particle moving to the right or left at $t = 2$?
   (c) Find the acceleration of the particle at $t = 2$ and determine if the velocity of the particle is increasing or decreasing. Explain why.

25. (Calculator) Given $f(x) = xe^x$ and $g(x) = \cos x$, find:

   (a) The area of the region in the first quadrant bounded by the graphs $f(x)$, $g(x)$, and $x = 0$.
   (b) The volume obtained by revolving the region in part (a) about the x-axis.

13.8 Solutions to Practice Problems

Part A—The use of a calculator is not allowed.

1. (a) $F(0) = \int_0^0 f(t)dt = 0$

$F(3) = \int_0^3 f(t)dt$

$= \frac{1}{2}(3+2)(4) = 10$

$F(5) = \int_0^5 f(t)dt$

$= \int_0^3 f(t)dt + \int_3^5 f(t)dt$

$= 10 + (-4) = 6$

(b) Since $\int_3^5 f(t)dt \leq 0$, F is decreasing on the interval [3, 5].

(c) At $t = 3$, F has a maximum value.

(d) $F'(x) = f(x)$, $F'(x)$ is increasing on (4, 5) which implies $F \leq (x) > 0$. Thus, F is concave upward on (4, 5).

2. (See Figure 13.8-1.)

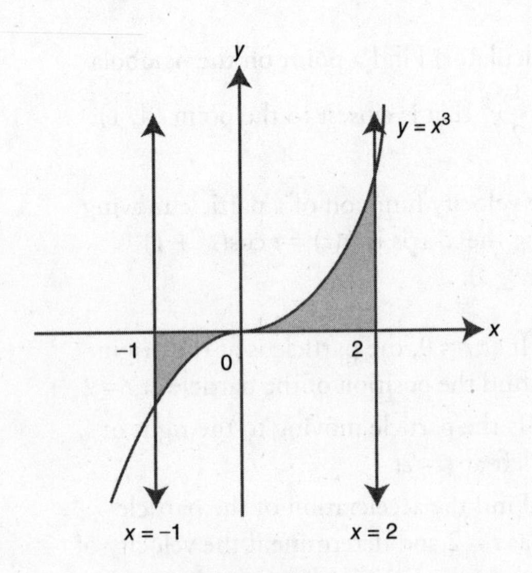

Figure 13.8-1

$$A = \left| \int_{-1}^{0} x^3 dx \right| + \int_{0}^{2} x^3 dx$$

$$= \left| \left[\frac{x^4}{4} \right]_{-1}^{0} \right| + \left[\frac{x^4}{4} \right]_{0}^{2}$$

$$= \left| 0 - \frac{(-1)^4}{4} \right| + \left(\frac{2^4}{4} - 0 \right)$$

$$= \frac{1}{4} + 4 = \frac{17}{4}$$

3. (See Figure 13.8-2.)
Set $2x - 6 = 0$; $x = 3$ and

$$f(x) = \begin{cases} 2x - 6 & \text{if } x \geq 3 \\ -(2x - 6) & \text{if } x < 3 \end{cases}.$$

$$A = \int_{0}^{3} -(2x - 6)dx + \int_{3}^{4} (2x - 6)dx$$

$$= \left[-x^2 + 6x\right]_{0}^{3} + \left[x^2 - 6x\right]_{3}^{4}$$

$$= \left[-(3)^2 + 6(3)\right]$$

$$- 0 + \left[4^2 + 6(4)\right] - \left[3^2 - 6(3)\right]$$

$$= 9 + 1 = 10$$

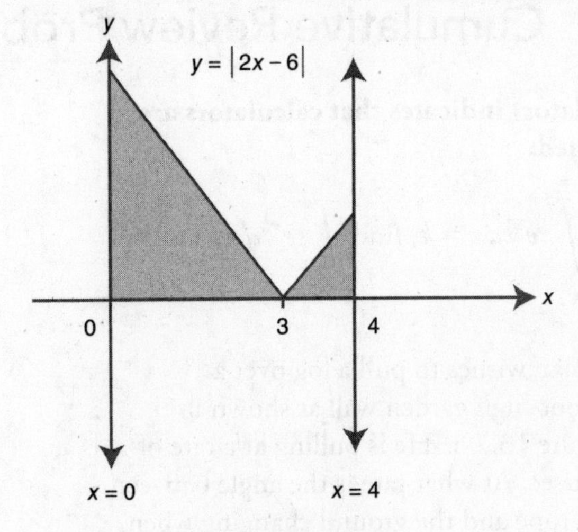

Figure 13.8-2

4. (See Figure 13.8-3.)

$$\text{Length of } \Delta x_1 = \frac{5-1}{4} = 1.$$

$$\text{Area of Rect}_{\text{I}} = f(2)\Delta x_1 = \frac{1}{2}(1) = \frac{1}{2}.$$

$$\text{Area of Rect}_{\text{II}} = f(3)\Delta x_2 = \frac{1}{3}(1) = \frac{1}{3}.$$

$$\text{Area of Rect}_{\text{III}} = f(4)\Delta x_3 = \frac{1}{4}(1) = \frac{1}{4}.$$

$$\text{Area of Rect}_{\text{IV}} = f(5)\Delta x_4 = \frac{1}{5}(1) = \frac{1}{5}.$$

$$\text{Total Area} = \frac{1}{2} + \frac{1}{3} + \frac{1}{4} + \frac{1}{5} = \frac{77}{60}.$$

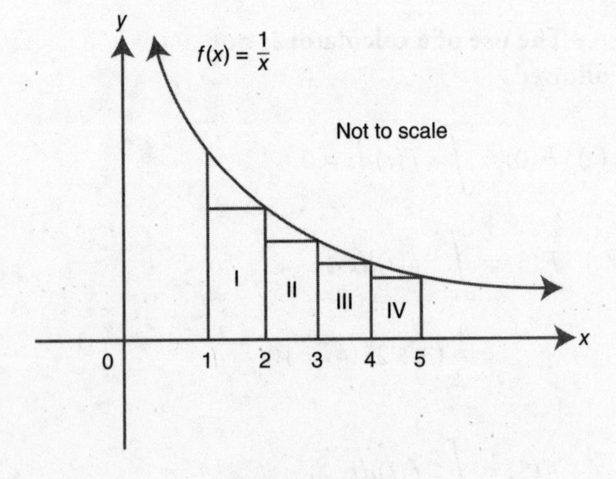

Figure 13.8-3

5. (See Figure 13.8-4.)

$$\text{Trapezoid Rule} = \frac{b-a}{2n}\left(f(a)+2f(x_1)+2f(x_2)+f(b)\right).$$

$$A=\frac{3-0}{2(3)}(f(0)+2f(1)+2f(2)+f(3))$$

$$=\frac{1}{2}(1+4+10+10)=\frac{25}{2}$$

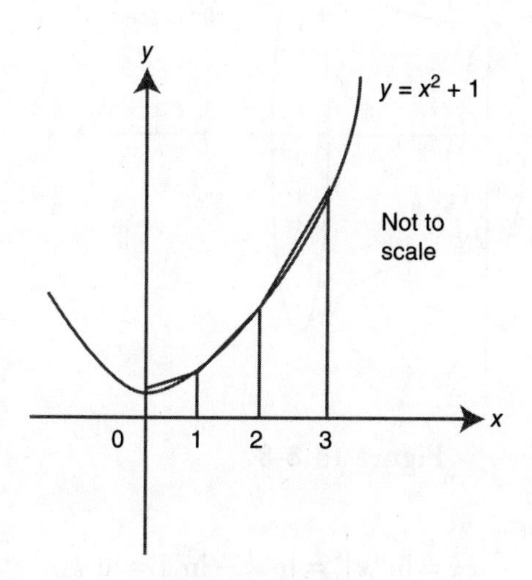

Figure 13.8-4

6. (See Figure 13.8-5.)

$$A=\int_0^4\left(\sqrt{x}-(-x)\right)dx$$

$$=\int_0^4\left(x^{1/2}+x\right)dx=\left[\frac{2x^{3/2}}{3}+\frac{x^2}{2}\right]_0^4$$

$$=\left(\frac{2(4)^{3/2}}{3}+\frac{4^2}{2}\right)-0$$

$$=\frac{16}{3}+8=\frac{40}{3}$$

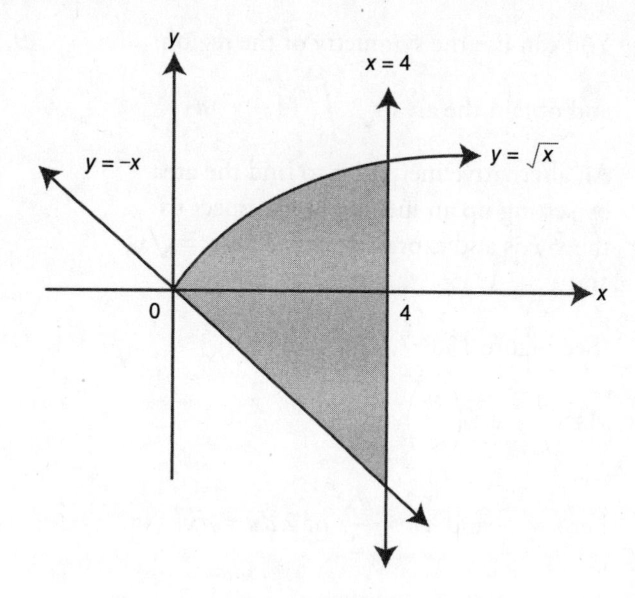

Figure 13.8-5

7. (See Figure 13.8-6.)
Intersection points: $4=y^2 \Rightarrow y=\pm 2$.

$$A=\int_{-2}^{2}\left(4-y^2\right)dy=\left[4y-\frac{y^3}{3}\right]_{-2}^{2}$$

$$=\left(4(2)-\frac{2^3}{3}\right)-\left(4(-2)-\frac{(-2)^3}{3}\right)$$

$$=\left(8-\frac{8}{3}\right)-\left(-8+\frac{8}{3}\right)$$

$$=\frac{16}{3}+\frac{16}{3}=\frac{32}{3}$$

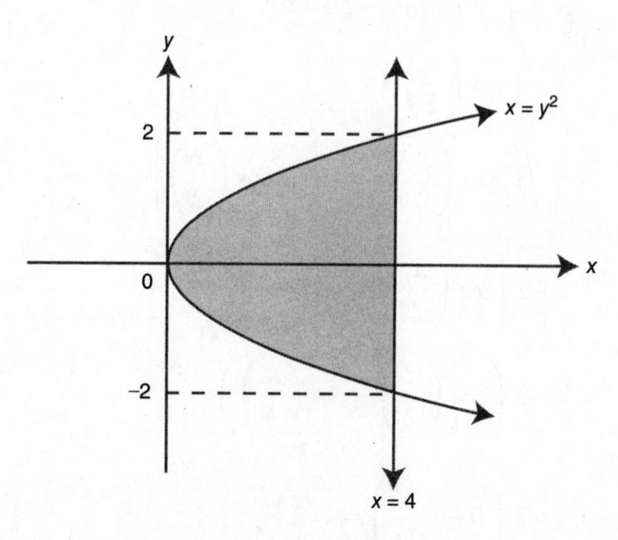

Figure 13.8-6

You can use the symmetry of the region and obtain the area $= 2\int_0^2 (4-y^2)dy$.

An alternative method is to find the area by setting up an integral with respect to the x-axis and expressing $x = y^2$ as $y = \sqrt{x}$ and $y = -\sqrt{x}$.

8. (See Figure 13.8-7.)

$$A = \int_{\pi/2}^{\pi} \sin\left(\frac{x}{2}\right) dx$$

Let $u = \frac{x}{2}$ and $du = \frac{dx}{2}$ or $2\,du = dx$.

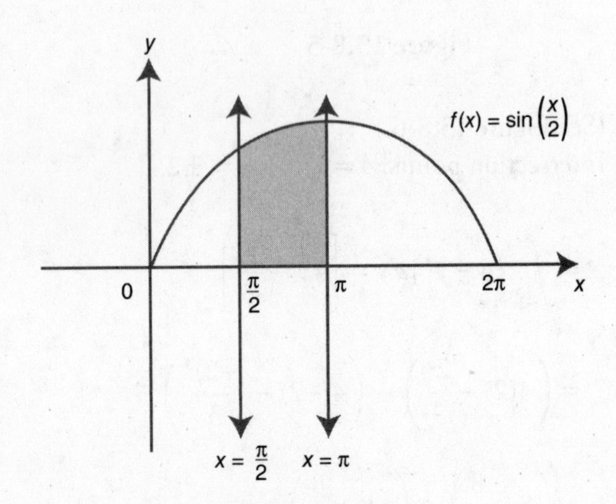

Figure 13.8-7

$$\int \sin\left(\frac{x}{2}\right) dx = \int \sin u(2du)$$

$$= 2\int \sin u\,du = -2\cos u + c$$

$$= -2\cos\left(\frac{x}{2}\right) + c$$

$$A = \int_{\pi/2}^{\pi} \sin\left(\frac{x}{2}\right) dx = \left[-2\cos\left(\frac{x}{2}\right)\right]_{\pi/2}^{\pi}$$

$$= -2\left[\cos\left(\frac{\pi}{2}\right) - \cos\left(\frac{\pi/2}{2}\right)\right]$$

$$= -2\left(\cos\left(\frac{\pi}{2}\right) - \cos\left(\frac{\pi}{4}\right)\right)$$

$$= -2\left(0 - \frac{\sqrt{2}}{2}\right) = \sqrt{2}$$

9. (See Figure 13.8-8.)
Using the Disc Method:

$$V = \pi\int_0^3 \left(x^2+4\right)^2 dx$$

$$= \pi\int_0^3 \left(x^4 + 8x^2 + 16\right) dx$$

$$= \pi\left[\frac{x^5}{5} + \frac{8x^3}{3} + 16x\right]_0^3$$

$$= \pi\left[\frac{3^5}{5} + \frac{8(3)^3}{3} + 16(3)\right] - 0 = \frac{843}{5}\pi$$

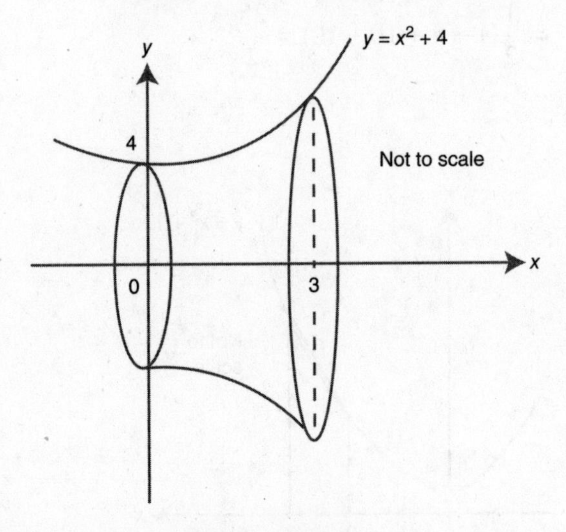

Figure 13.8-8

10. Area

$$= \int_1^k \frac{1}{x}dx = \ln\,x]_1^k = \ln\,k - \ln\,1 = \ln\,k.$$

Set $\ln k = 1$. Thus, $e^{\ln k} = e^1$ or $k = e$.

11. (See Figure 13.8-9.)

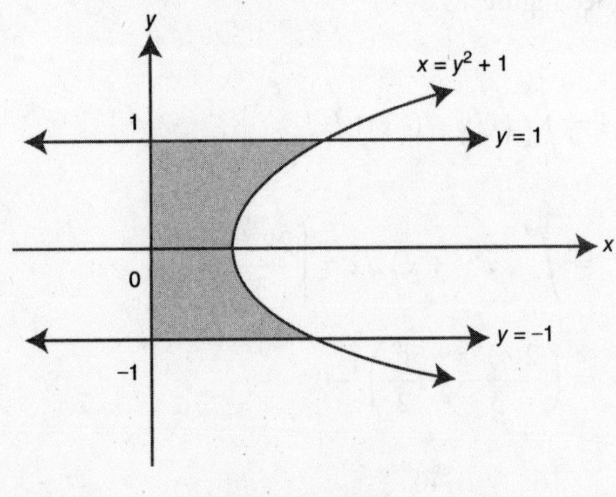

Figure 13.8-9

Using the Disc Method:

$$V = \pi \int_{-1}^{1} (y^2+1)^2 dy$$

$$= \pi \int_{-1}^{1} (y^4+2y^2+1) dy$$

$$= \pi \left[\frac{y^5}{5} + \frac{2y^3}{3} + y\right]_{-1}^{1}$$

$$= \pi \left[\left(\frac{1^5}{5} + \frac{2(1)^3}{3} + 1\right) - \left(\frac{(-1)^5}{5} + \frac{2(-1)^3}{3} + (-1)\right)\right]$$

$$= \pi \left(\frac{28}{15} + \frac{28}{15}\right) = \frac{56\pi}{15}$$

Note that you can use the symmetry of the region and find the volume by $2\pi \int_0^1 (y^2+1)^2 dy$.

12. Volume of solid by revolving R:

$$V_R = \int_0^4 \pi (3x)^2 dx = \pi \int_0^4 9x^2 dx$$

$$= \pi [3x^2]_0^4 = 192\pi$$

Set $\int_0^4 \pi (3x)^2 dx = \frac{192\pi}{2}$

$$\Rightarrow 3a^3\pi = 96\pi$$

$$a^3 = 32$$

$$a = (32)^{1/3} = 2(2)^{2/3}$$

You can verify your result by evaluating $\int_0^{2(2)^{2/3}} \pi (3x)^2 dx$. The result is 96π.

Part B—Calculators are allowed.

13. (See Figure 13.8-10.)

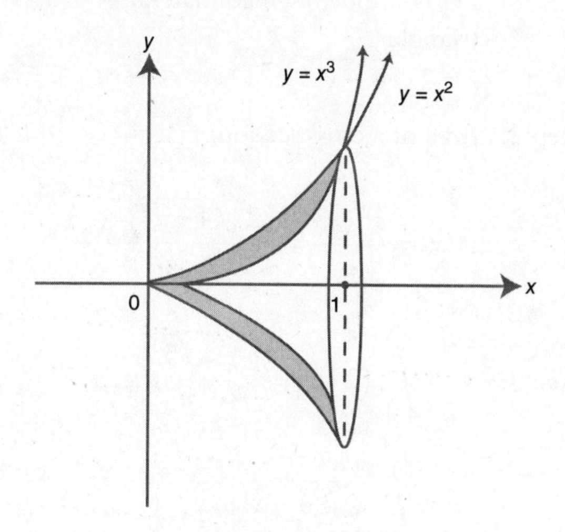

Figure 13.8-10

Step 1: Using the Washer Method:
Points of Intersection: Set $x^3 = x^2 \Rightarrow x^3 - x^2 = 0 \Rightarrow x^2(x-1) = 0$ or $x = 1$.
Outer radius $= x^2$;
Inner radius $= x^3$.

Step 2: $V = \pi \int_0^1 \left((x^2)^2 - (x^3)^2\right) dx$

$$= \pi \int_0^1 (x^4 - x^6) dx$$

Step 3: Enter $\int (\pi (x\hat{\ }4 - x\hat{\ }6), x, 0, 1)$ and obtain $\frac{2\pi}{35}$.

14. (See Figure 13.8-11.)

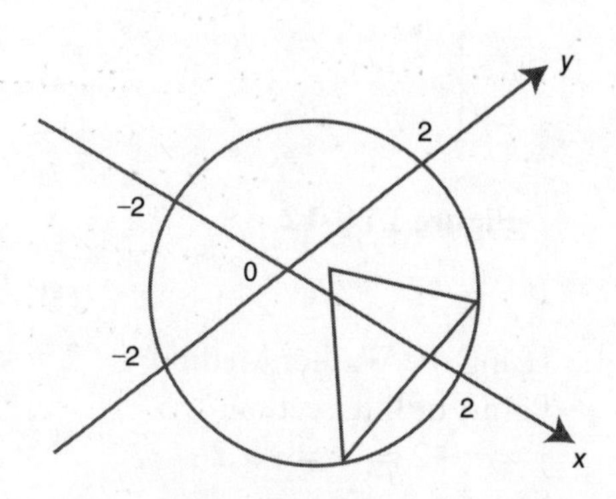

Figure 13.8-11

Step 1: $x^2 + y^2 = 4 \Rightarrow y^2 = 4 - x^2 \Rightarrow$
$y = \pm\sqrt{4 - x^2}$
Let s = a side of an equilateral triangle
$s = 2\sqrt{4 - x^2}$.

Step 2: Area of a cross section:

$$A(x) = \frac{s^2\sqrt{3}}{4} = \frac{\left(2\sqrt{4 - x^2}\right)^2\sqrt{3}}{4}.$$

Step 3: $$V = \int_{-2}^{2} \left(2\sqrt{4 - x^2}\right)^2 \frac{\sqrt{3}}{4} dx$$

$$= \int_{-2}^{2} \sqrt{3}(4 - x^2)dx$$

Step 4: Enter $\int (\sqrt{(3)} * (4 - x^2), x, -2, 2)$
and obtain $\frac{32\sqrt{3}}{3}$.

15. (See Figure 13.8-12.)

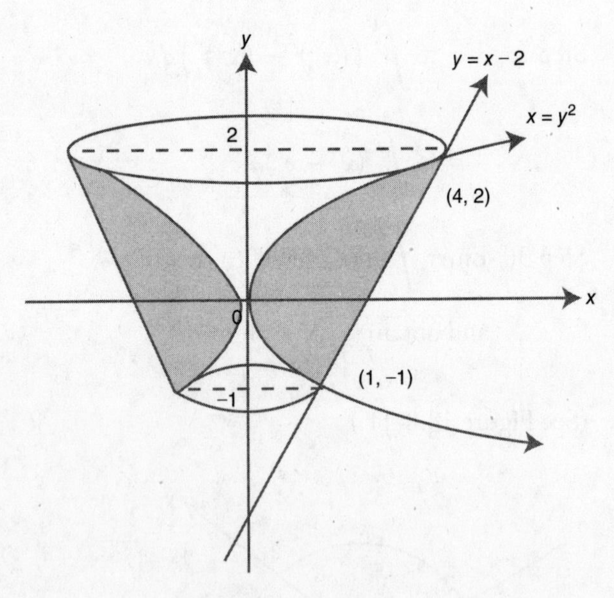

Figure 13.8-12

Step 1: Using the Washer Method:
Points of Intersection:
$y = x - 2 \Rightarrow x = y + 2$
Set $y^2 = y + 2$
$\Rightarrow y^2 - y - 2 = 0$
$\Rightarrow (y - 2)(y + 1) = 0$
or $y = -1$ or $y = 2$.

Outer radius $= y + 2$;
Inner radius $= y^2$.

Step 2: $$V = \pi \int_{-1}^{2} \left((y + 2)^2 - (y^2)^2\right) dy$$

Step 3: Enter
$\pi \int ((y + 2)\text{^}2 - y\text{^}4, -1, 2)$
and obtain $\frac{72}{5}\pi$.

16. (See Figure 13.8-13.)

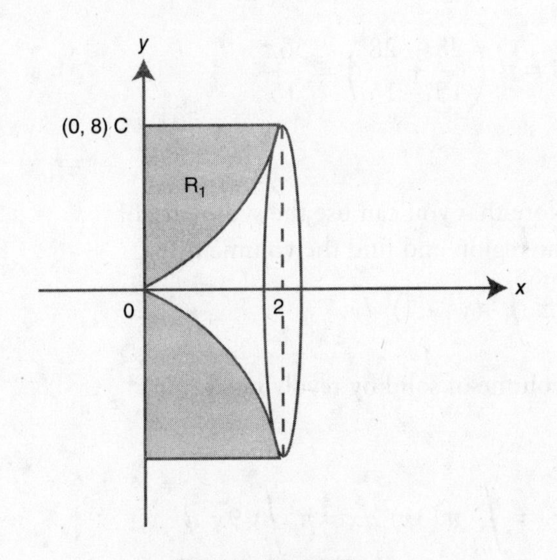

Figure 13.8-13

Step 1: Using the Washer Method:
$y = 8, \ y = x^3$

Outer radius $= 8$;

Inner radius $= x^3$.

$$V = \pi \int_{0}^{2} \left(8^2 - (x^3)^2\right) dx$$

Step 2: Enter $\int \pi \,(8^2 - x^6, x, 0, 2)$
and obtain $\frac{768\pi}{7}$.

17. (See Figure 13.8-14.)

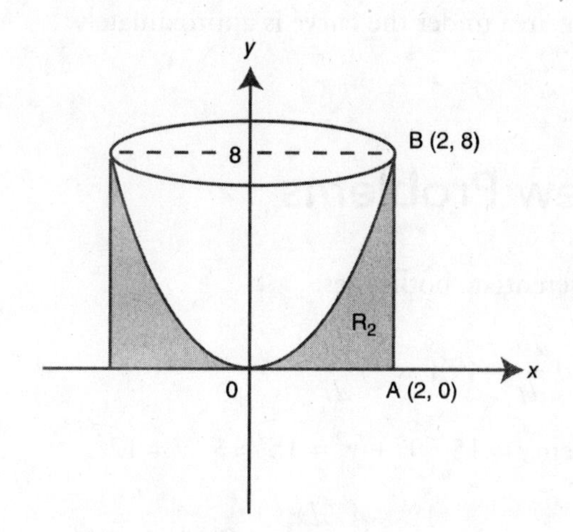

Figure 13.8-14

Using the Washer Method:
Outer radius: $x = 2$;
Inner radius: $x = y^{1/3}$.

$$V = \pi \int_0^8 \left(2^2 - \left(y^{1/3}\right)^2\right) dy$$

Using your calculator, you obtain

$$V = \frac{64\pi}{5}.$$

18. (See Figure 13.8-15.)

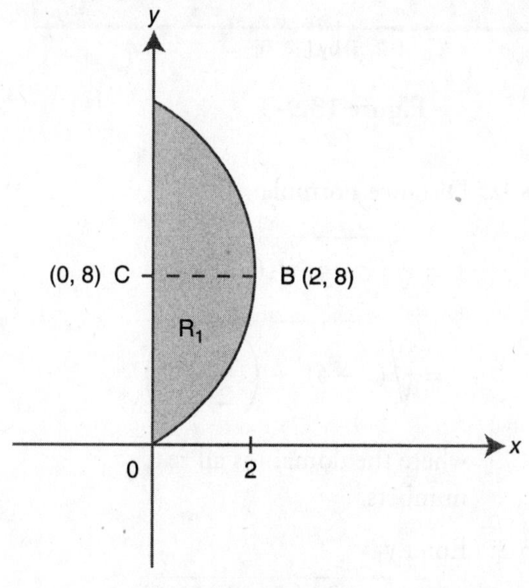

Figure 13.8-15

Step 1: Using the Disc Method:
Radius $= (8 - x^3)$.

$$V = \pi \int_0^2 \left(8 - x^3\right)^2 dx$$

Step 2: Enter $\int \left(\pi * \left(8 - x^3\right)^2, x, 0, 2\right)$ and obtain $\frac{576\pi}{7}$.

19. (See Figure 13.8-16.)
Using the Disc Method:

Radius $= 2 - x = \left(2 - y^{1/3}\right)$.

$$V = \pi \int_0^8 \left(2 - y^{1/3}\right)^2 dy$$

Using your calculator, you obtain

$$V = \frac{16\pi}{5}.$$

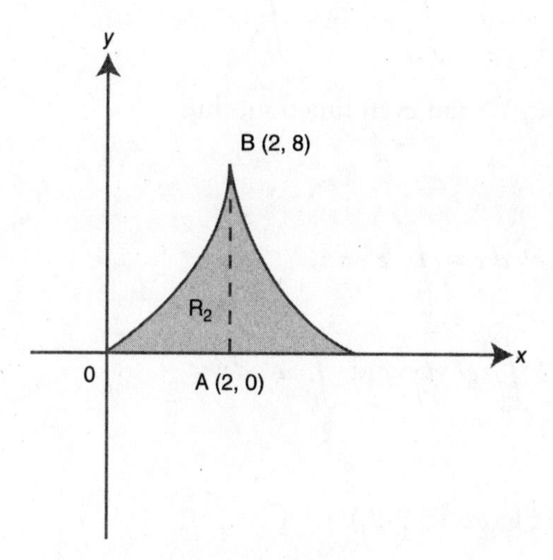

Figure 13.8-16

20. Area $= \sum_{i=1}^{3} f(x_i)\Delta x_i$.

x_i = midpoint of the ith interval.

Length of $\Delta x_i = \frac{12 - 0}{3} = 4$.

Area of $\text{Rect}_\text{I} = f(2)\Delta x_1 = (2.24)(4) = 8.96$.

Area of $\text{Rect}_{\text{II}} = f(6)\Delta x_2 = (3.61)(4) = 14.44.$

Area of $\text{Rect}_{\text{III}} = f(10)\Delta x_3 = (4.58)(4) = 18.32.$

Total Area $= 8.96 + 14.44 + 18.32 = 41.72.$

The area under the curve is approximately 41.72.

13.9 Solutions to Cumulative Review Problems

21. (See Figure 13.9-1.)

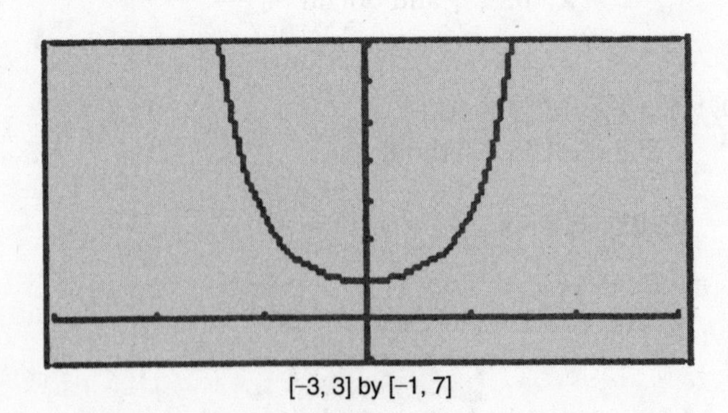

Figure 13.9-1

$$\int_{-a}^{a} e^{x^2}\,dx = \int_{-a}^{0} e^{x^2}\,dx + \int_{0}^{a} e^{x^2}\,dx$$

Since e^{x^2} is an even function, thus

$$\int_{-a}^{0} e^{x^2}\,dx = \int_{0}^{a} e^{x^2}\,dx.$$

$$k = 2\int_{0}^{a} e^{x^2}\,dx \text{ and } \int_{0}^{a} e^{x^2}\,dx = \frac{k}{2}.$$

22. (See Figure 13.9-2.)

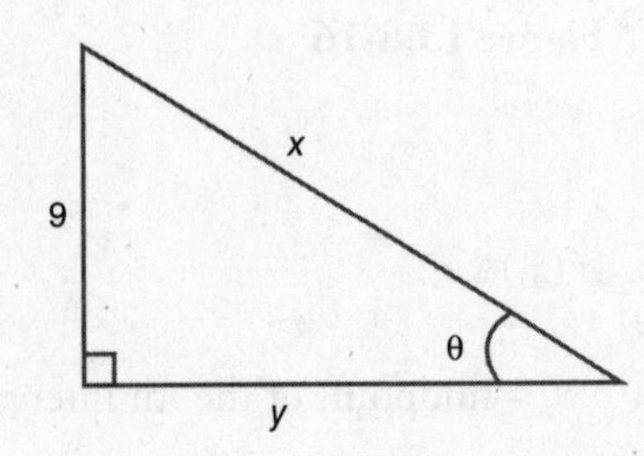

Figure 13.9-2

$$\sin\theta = \frac{9}{x}$$

Differentiate both sides:

$$\cos\theta\frac{d\theta}{dt} = (9)(-x^{-2})\frac{dx}{dt}.$$

When $x = 15,\ 9^2 + y^2 = 15^2 \Rightarrow y = 12.$

Thus, $\cos\theta = \frac{12}{15} = \frac{4}{5};\ \frac{dx}{dt} = -2$ ft/sec.

$$\frac{4}{5}\frac{d\theta}{dt} = 9\left(-\frac{1}{15^2}\right)(-2)$$

$$= \frac{d\theta}{dt} = \frac{18}{15^2}\frac{5}{4} = \frac{1}{10} \text{ radian/sec.}$$

23. (See Figure 13.9-3.)

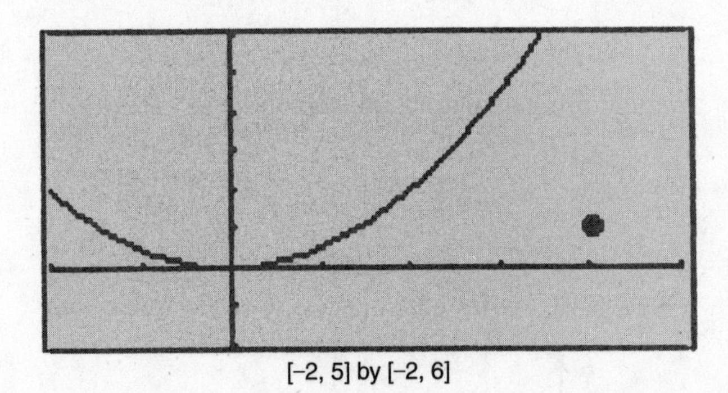

Figure 13.9-3

Step 1: Distance Formula:

$$L = \sqrt{(x-4)^2 + (y-1)^2}$$

$$= \sqrt{(x-4)^2 + \left(\frac{x^2}{2} - 1\right)^2}$$

where the domain is all real numbers.

Step 2: Enter $y_1 = \sqrt{((x-4)\text{^}2+(.5x\text{^}2-1)\text{^}2)}$

Enter $y_2 = d\left(y_1(x)\,,x\right).$

Step 3: Use the [*Zero*] function and obtain $x = 2$ for y_2.

Step 4: Use the First Derivative Test. (See Figures 13.9-4 and 13.9-5.) At $x = 2$, L has a relative minimum. Since at $x = 2$, L has the only relative extremum, it is an absolute minimum.

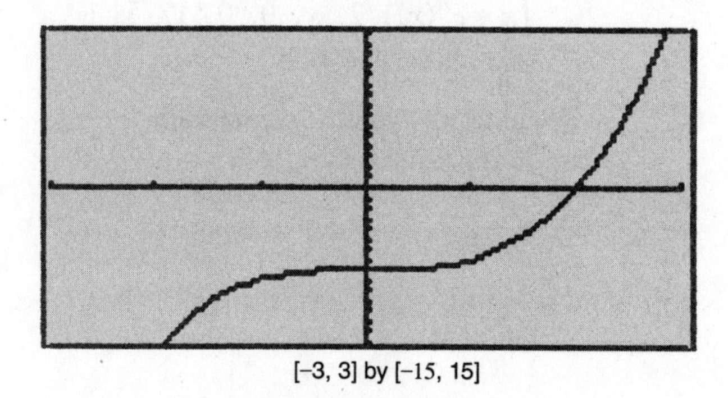

[–3, 3] by [–15, 15]

Figure 13.9-4

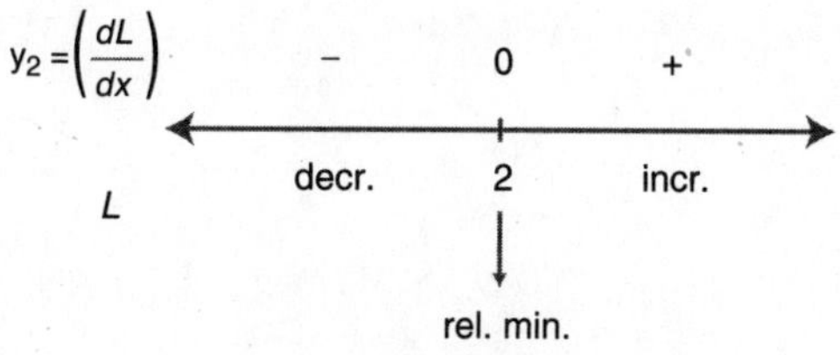

$y_2 = \left(\frac{dL}{dx}\right)$ –   0   +

L decr.   2   incr.

rel. min.

Figure 13.9-5

Step 5: At $x = 2$, $y = \frac{1}{2}(x^2) = \frac{1}{2}(2^2) = 2$. Thus, the point on $y = \frac{1}{2}(x^2)$ closest to the point (4, 1) is the point (2, 2).

24. (a) $s(0) = 0$ and

$$s(t) = \int v(t)dt = \int t\cos(t^2 + 1)dt.$$

Enter $\int (x * \cos(x\char`^2 + 1),\ x)$ and obtain $\frac{\sin(x^2 + 1)}{2}$.

Thus, $s(t) = \frac{\sin(t^2 + 1)}{2} + C.$

Since $s(0) = 0 \Rightarrow \frac{\sin(0^2 + 1)}{2} + C = 0$

$$\Rightarrow \frac{.841471}{2} + C = 0$$

$$\Rightarrow C = -0.420735 = -0.421$$

$$s(t) = \frac{\sin(t^2 + 1)}{2} - 0.420735$$

$$s(2) = \frac{\sin(2^2 + 1)}{2} - 0.420735$$

$$= -0.900197 \approx -0.900.$$

(b) $v(2) = 2\cos(2^2 + 1) = 2\cos(5) = 0.567324$

Since $v(2) > 0$, the particle is moving to the right at $t = 2$.

(c) $a(t) = v'(t)$

Enter $d(x * \cos(x\char`^2 + 1),\ x)|x = 2$ and obtain 7.95506.

Thus, the velocity of the particle is increasing at $t = 2$, since $a(2) > 0$.

25. (See Figure 13.9-6.)

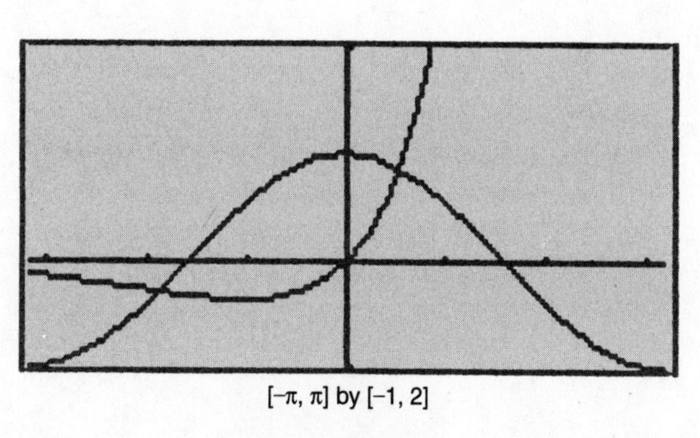

[–π, π] by [–1, 2]

Figure 13.9-6

(a) Point of Intersection: Use the [*Intersection*] function of the calculator and obtain (0.517757, 0.868931).

$$\text{Area} = \int_0^{0.51775} (\cos x - xe^x)dx$$

Enter $\int (\cos(x) - x * e\hat{\ }x,\ x,$ $0,\ 0.51775)$ and obtain 0.304261. The area of the region is approximately 0.304.

(b) Step 1: Using the Washer Method:

Outer radius = $\cos x$;

Inner radius = xe^x.

$$V = \pi \int_0^{0.51775} \left[(\cos x)^2 - (xe^x)^2\right] dx$$

Step 2: Enter $\pi \int \left(\left(\cos(x)\right)\hat{\ }2 - \left(x * e\hat{\ }(x)\right)\hat{\ }2,\ x,\ 0,\ 0.51775\right)$ and obtain 1.16678.

Thus, the volume of the solid is approximately 1.167.

Big Idea 3: Integrals and the Fundamental Theorems of Calculus

More Applications of Definite Integrals

IN THIS CHAPTER

Summary: In this chapter, you will learn to solve problems using a definite integral as accumulated change. These problems include distance traveled problems, temperature problems, and growth problems. You will also learn to work with slope fields and to solve differential equations.

Key Ideas

- Average Value of a Function
- Mean Value Theorem for Integrals
- Distance Traveled Problems
- Definite Integral as Accumulated Change
- Differential Equations
- Slope Fields

14.1 Average Value of a Function

Main Concepts: Mean Value Theorem for Integrals, Average Value of a Function on $[a, b]$

Mean Value Theorem for Integrals

If f is continuous on $[a, b]$, then there exists a number c in $[a, b]$ such that $\int_a^b f(x)\,dx = f(c)(b - a)$. (See Figure 14.1-1.)

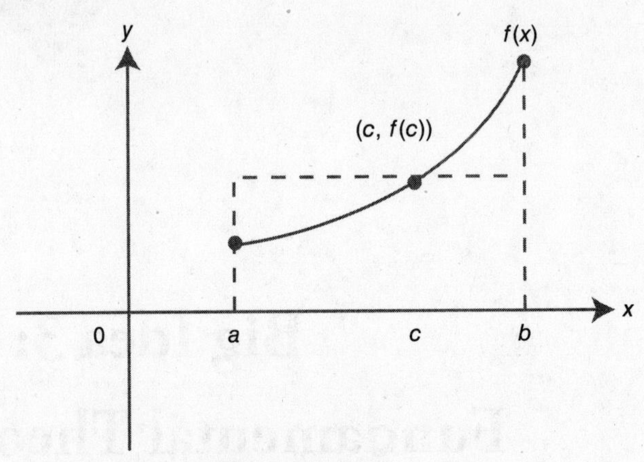

Figure 14.1-1

Example 1

Given $f(x) = \sqrt{x - 1}$, verify the hypotheses of the Mean Value Theorem for Integrals for f on $[1, 10]$ and find the value of c as indicated in the theorem.

The function f is continuous for $x \geq 1$, thus:

$$\int_1^{10} \sqrt{x-1}\,dx = f(c)(10 - 1)$$

$$\left.\frac{2(x-1)^{3/2}}{3}\right]_1^{10} = 9f(c)$$

$$\frac{2}{3}\left[(10-1)^{3/2} - 0\right] = 9f(c)$$

$$18 = 9f(c);$$

$$2 = f(c);$$

$$2 = \sqrt{c-1};$$

$$4 = c - 1$$

$$5 = c.$$

Example 2

Given $f(x)=x^2$, verify the hypotheses of the Mean Value Theorem for Integrals for f on [0, 6] and find the value of c as indicated in the theorem.

Since f is a polynomial, it is continuous and differentiable everywhere, thus,

$$\int_0^6 x^2 dx = f(c)(6-0)$$

$$\left.\frac{x^3}{3}\right]_0^6 = f(c)6$$

$$72 = 6f(c);\ 12 = f(c);\ 12 = c^2$$

$$c = \pm\sqrt{12} = \pm 2\sqrt{3}\left(\pm 2\sqrt{3} \approx \pm 3.4641\right).$$

Since only $2\sqrt{3}$ is in the interval [0, 6], $c = 2\sqrt{3}$.

- Remember: If f' is decreasing, then $f'' < 0$ and the graph of f is concave downward.

Average Value of a Function on [a, b]

Average Value of a Function on an Interval

If f is a continuous function on $[a, b]$, then the Average Value of f on $[a, b]$ $= \frac{1}{b-a}\int_a^b f(x)dx$.

Example 1

Find the average value of $y = \sin x$ between $x = 0$ and $x = \pi$.

$$\text{Average value} = \frac{1}{\pi - 0}\int_0^{\pi} \sin x\, dx$$

$$= \frac{1}{\pi}[-\cos x]_0^{\pi} = \frac{1}{\pi}[-\cos\pi - (-\cos(0))]$$

$$= \frac{1}{\pi}[1+1] = \frac{2}{\pi}.$$

Example 2

The graph of a function f is shown in Figure 14.1-2. Find the average value of f on $[0, 4]$.

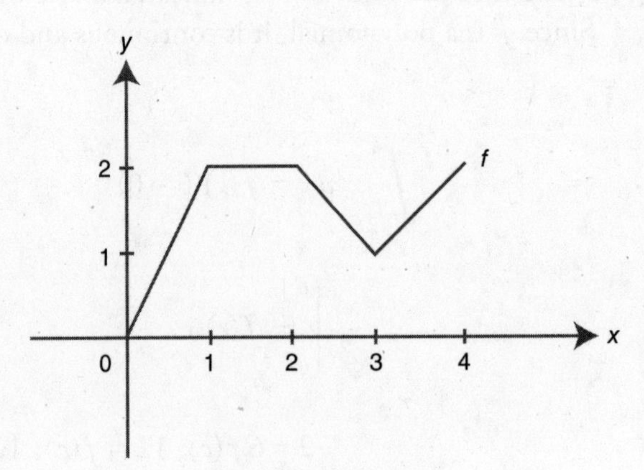

Figure 14.1-2

$$\text{Average value} = \frac{1}{4-0}\int_0^4 f(x)\,dx$$

$$= \frac{1}{4}\left(1 + 2 + \frac{3}{2} + \frac{3}{2}\right) = \frac{3}{2}.$$

Example 3

The velocity of a particle moving on a line is $v(t) = 3t^2 - 18t + 24$. Find the average velocity from $t = 1$ to $t = 3$.

$$\text{Average velocity} = \frac{1}{3-1}\int_1^3 (3t^2 - 18t + 24)\,dt$$

$$= \frac{1}{2}\left[t^3 - 9t^2 + 24t\right]_1^3$$

$$= \frac{1}{2}\left[\left(3^3 - 9(3^2) + 24(3)\right) - \left(1^3 - 9(1^2) + 24(1)\right)\right]$$

$$= \frac{1}{2}(18 - 16) = \frac{1}{2}(2) = 1.$$

Note: The average velocity for $t = 1$ to $t = 3$ is $\dfrac{s(3) - s(1)}{2}$, which is equivalent to the computations above.

14.2 Distance Traveled Problems

Summary of Formulas

Position Function: $s(t)$; $s(t) = \int v(t)\,dt$.

Velocity: $v(t) = \frac{ds}{dt}$; $v(t) = \int a(t)\,dt$.

Acceleration: $a(t) = \frac{dv}{dt}$.

Speed: $|v(t)|$.

Displacement from t_1 to $t_2 = \int_{t_1}^{t_2} v(t)\,dt = s(t_2) - s(t_1)$.

Total Distance Traveled from t_1 to $t_2 = \int_{t_1}^{t_2} |v(t)|\,dt$.

Example 1

(See Figure 14.2-1.)

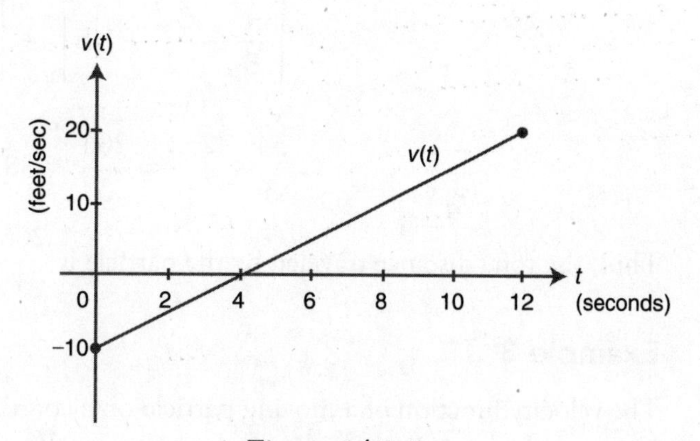

Figure 14.2-1

The graph of the velocity function of a moving particle is shown in Figure 14.2-1. What is the total distance traveled by the particle during $0 \leq t \leq 12$?

$$\text{Total Distance Traveled} = \left| \int_0^4 v(t)dt \right| + \int_4^{12} v(t)dt$$

$$= \frac{1}{2}(4)(10) + \frac{1}{2}(8)(20) = 20 + 80 = 100 \text{ feet.}$$

Example 2

The velocity function of a moving particle on a coordinate line is $v(t) = t^2 + 3t - 10$ for $0 \leq t \leq 6$. Find (a) the displacement by the particle during $0 \leq t \leq 6$, and (b) the total distance traveled by the particle during $0 \leq t \leq 6$.

(a) Displacement $= \int_{t_1}^{t_2} v(t)dt$

$$= \int_0^6 (t^2 + 3t - 10)dt = \left.\frac{t^3}{3} + \frac{3t^2}{2} - 10t\right]_0^6 = 66.$$

(b) Total Distance Traveled $= \int_{t_1}^{t_2} |v(t)|dt$

$$= \int_0^6 |t^2 + 3t - 10|dt.$$

Let $t^2 + 3t - 10 = 0 \Rightarrow (t+5)\ (t-2) = 0 \Rightarrow t = -5$ or $t = 2$

$$|t^2 + 3t - 10| = \begin{cases} -\left(t^2 + 3t - 10\right) & \text{if } 0 \le t \le 2 \\ t^2 + 3t - 10 & \text{if } t > 2 \end{cases}$$

$$\int_0^6 |t^2 + 3t - 10|dt = \int_0^2 -(t^2 + 3t - 10)dt + \int_2^6 (t^2 + 3t - 10)dt$$

$$= \left[\frac{-t^3}{3} - \frac{3t^2}{2} + 10t\right]_0^2 + \left[\frac{t^3}{3} + \frac{3t^2}{2} - 10t\right]_2^6$$

$$= \frac{34}{3} + \frac{232}{3} = \frac{266}{3} \approx 88.667.$$

Thus, the total distance traveled by the particle is $\frac{266}{3}$ or approximately 88.667.

Example 3

The velocity function of a moving particle on a coordinate line is $v(t) = t^3 - 6t^2 + 11t - 6$. Using a calculator, find (a) the displacement by the particle during $1 \le t \le 4$ and (b) the total distance traveled by the particle during $1 \le t \le 4$.

(a) Displacement $= \int_{t_1}^{t_2} v(t)dt$

$$= \int_1^4 (t^3 - 6t^2 + 11t - 6)dt.$$

Enter $\int \left(x^3 - 6x^2 + 11x - 6, x, 1, 4\right)$ and obtain $\frac{9}{4}$.

(b) Total Distance Traveled $= \int_{t_1}^{t_2} |v(t)|\, dt.$

Enter $y1 = x\verb|^|3 - 6x\verb|^|2 + 11x - 6$ and use the [*Zero*] function to obtain x-intercepts at $x = 1, 2, 3$.

$$|v(t)| = \begin{cases} v(t) & \text{if } 1 \le t \le 2 \text{ and } 3 \le t \le 4 \\ -v(t) & \text{if } 2 < t < 3 \end{cases}$$

Total Distance Traveled $\int_1^2 v(t)dt + \int_2^3 -v(t)dt + \int_3^4 v(t)dt.$

Enter $\int (y1(x), x, 1, 2)$ and obtain $\frac{1}{4}$.

Enter $\int (-y1(x), x, 2, 3)$ and obtain $\frac{1}{4}$.

Enter $\int (y1(x), x, 3, 4)$ and obtain $\frac{9}{4}$.

Thus, the total distance traveled is $\left(\frac{1}{4}+\frac{1}{4}+\frac{9}{4}\right)=\frac{11}{4}$.

Example 4

The acceleration function of a moving particle on a coordinate line is $a(t) = -4$ and $v_0 = 12$ for $0 \le t \le 8$. Find the total distance traveled by the particle during $0 \le t \le 8$.

$a(t) = -4$

$v(t) = \int a(t)dt = \int -4dt = -4t + C$

Since $v_0 = 12 \Rightarrow -4(0) + C = 12$ or $C = 12$.

Thus, $v(t) = -4t + 12$.

Total Distance Traveled $= \int_0^8 \left|-4t + 12\right| dt.$

Let $-4t + 12 = 0 \Rightarrow t = 3$.

$$|-4t+12| = \begin{cases} -4t+12 & \text{if } 0 \le t \le 3 \\ -(-4t+12) & \text{if } t > 3 \end{cases}$$

$$\int_0^8 \left|-4t+12\right| dt = \int_0^3 \left|-4t+12\right| dt + \int_3^8 -(-4t+12)dt$$

$$= \left[-12t^2 + 12t\right]_0^3 + \left[2t^2 + 12t\right]_3^8$$

$$= 18 + 50 = 68.$$

Thus, the total distance traveled by the particle is 68.

Example 5

The velocity function of a moving particle on a coordinate line is $v(t) = 3\cos(2t)$ for $0 \le t \le 2\pi$. Using a calculator:

(a) Determine when the particle is moving to the right.

(b) Determine when the particle stops.

(c) Determine the total distance traveled by the particle during $0 \le t \le 2\pi$.

Solution:

(a) The particle is moving to the right when $v(t) > 0$.

Enter $y1 = 3\cos(2x)$. Obtain $y_1 = 0$ when $t = \frac{\pi}{4}, \frac{3\pi}{4}, \frac{5\pi}{4}$, and $\frac{7\pi}{4}$.

The particle is moving to the right when:

$$0 < t < \frac{\pi}{4}, \frac{3\pi}{4} < t < \frac{5\pi}{4}, \frac{7\pi}{4} < t < 2\pi.$$

(b) The particle stops when $v(t) = 0$.

Thus, the particle stops at $t = \frac{\pi}{4}, \frac{3\pi}{4}, \frac{5\pi}{4}$, and $\frac{7\pi}{4}$.

(c) Total distance traveled $\int_0^{2\pi} |3\cos(2t)|\,dt$.

Enter $\int$ (abs(3 cos(2x)), x, 0, 2π) and obtain 12.

Thus, the total distance traveled by the particle is 12.

14.3 Definite Integral as Accumulated Change

Main Concepts: Business Problems, Temperature Problem, Leakage Problem, Growth Problem

Business Problems

Profit = Revenue − Cost	$P(x) = R(x) - C(x)$
Revenue = (price)(items sold)	$R(x) = P(x)$
Marginal Profit	$P'(x)$
Marginal Revenue	$R'(x)$
Marginal Cost	$C'(x)$

$P'(x)$, $R'(x)$, and $C'(x)$ are the instantaneous rates of change of profit, revenue, and cost, respectively.

Example 1

The marginal profit of manufacturing and selling a certain drug is $P'(x) = 100 - 0.005x$. How much profit should the company expect if it sells 10,000 units of this drug?

$$P(t) = \int_0^1 P'(x)dx$$

$$= \int_0^{10,000} (100 - 0.005x)\,dx = 100x - \frac{0.005x^2}{2}\Bigg]_0^{10,000}$$

$$= \left(100(10,000) - \frac{0.005}{2}(10,000)^2\right) = 750,000$$

- If $f''(a) = 0$, f may or may not have a point of inflection at $x = a$, e.g., as in the function $f(x) = x^4$, $f''(0) = 0$ *but* at $x = 0$, f has an absolute minimum.

Example 2

If the marginal cost of producing x units of a commodity is $C'(x) = 5 + 0.4x$,

find (a) the marginal cost when $x = 50$;

(b) the cost of producing the first 100 units.

Solution:

(a) Marginal cost at $x = 50$:

$$C'(50) = 5 + 0.4(50) = 5 + 20 = 25.$$

(b) Cost of producing the first 100 units:

$$C(t) = \int_0^1 C'(x)dx$$

$$= \int_0^{100} (5 + 0.4x)dx$$

$$= 5x + 0.2x^2\Big]_0^{100}$$

$$= \left(5(100) + 0.2(100)^2\right) - 0 = 2500.$$

Temperature Problem

Example

On a certain day, the changes in the temperature in a greenhouse beginning at 12 noon are represented by $f(t) = \sin\left(\frac{t}{2}\right)$ degrees Fahrenheit, where t is the number of hours elapsed after 12 noon. If at 12 noon, the temperature is 95°F, find the temperature in the greenhouse at 5 p.m.

Let $F(t)$ represent the temperature of the greenhouse.

$$F(0) = 95^\circ F$$

$$F(t) = 95 + \int_0^5 f(x)\,dx$$

$$F(5) = 95 + \int_0^5 \sin\left(\frac{x}{2}\right) dx$$

$$= 95 + \left[-2\cos\left(\frac{x}{2}\right)\right]_0^5 = 95 + \left[-2\cos\left(\frac{5}{2}\right) - (-2\cos(0))\right]$$

$$= 95 + 3.602 = 98.602$$

Thus, the temperature in the greenhouse at 5 p.m. is 98.602°F.

Leakage Problem

Example

Water is leaking from a faucet at the rate of $l(t) = 10e^{-0.5t}$ gallons per hour, where t is measured in hours. How many gallons of water will have leaked from the faucet after a 24-hour period?

Let $L(x)$ represent the number of gallons that have leaked after x hours.

$$L(x) = \int_0^x l(t)\,dt = \int_0^{24} 10e^{-0.5t}\,dt$$

Using your calculator, enter $\int(10e\hat{}(-0.5x), x, 0, 24)$ and obtain 19.9999. Thus, the number of gallons of water that have leaked after x hours is approximately 20 gallons.

- You are permitted to use the following four built-in capabilities of your calculator to obtain an answer: plotting the graph of a function, finding the zeros of a function, finding the numerical derivative of a function, and evaluating a definite integral. All other capabilities of your calculator can only be used to *check* your answer. For example, you may *not* use the built-in [*Inflection*] function of your calculator to find points of inflection. You must use calculus using derivatives and showing change of concavity.

Growth Problem

Example

On a farm, the animal population is increasing at a rate which can be approximately represented by $g(t) = 20 + 50\ln(2 + t)$, where t is measured in years. How much will the animal population increase to the nearest tens between the third and fifth years?

Let $G(x)$ be the increase in animal population after x years.

$$G(x) = \int_0^x g(t)\,dt$$

The population increase between the third and fifth year will

$$= G(5) - G(3)$$

$$= \int_0^5 \left(20 + 50\ln(2 + t)\,dt\right) - \int_0^3 \left(20 + 50\ln(2 + t)dt\right)$$

$$= \int_3^5 [20 + 50\ln(2 + t)]\,dt.$$

Enter $\int(20 + 50\ln(2 + x), x, 3, 5)$ and obtain 218.709.

Thus, the animal population will increase by approximately 220 between the third and fifth years.

14.4 Differential Equations

Main Concepts: Exponential Growth/Decay Problems, Separable Differential Equations

Exponential Growth/Decay Problems

1. If $\frac{dy}{dx} = ky$, then the rate of change of y is proportional to y.
2. If y is a differentiable function of t with $y > 0$ $\frac{dy}{dx} = ky$, then $y(t) = y_0 e^{kt}$; where y_0 is initial value of y and k is constant. If $k > 0$, then k is a growth constant and if $k < 0$, then k is the decay constant.

Example 1—Population Growth

If the amount of bacteria in a culture at any time increases at a rate proportional to the amount of bacteria present and there are 500 bacteria after one day and 800 bacteria after the third day:

(a) approximately how many bacteria are there initially, and

(b) approximately how many bacteria are there after 4 days?

Solution:

(a) Since the rate of increase is proportional to the amount of bacteria present, then:

$\frac{dy}{dx} = ky$ where y is the amount of bacteria at any time.

Therefore, this is an exponential growth/decay model: $y(t) = y_0 e^{kt}$.

Step 1: $y(1) = 500$ and $y(3) = 800$

$500 = y_0 e^{k}$ and $800 = y_0 e^{3k}$

Step 2: $500 = y_0 e^{k} \Rightarrow y_0 = \frac{500}{e^{k}} = 500e^{-k}$

Substitute $y_0 = 500e^{-k}$ into $800 = y_0 e^{3k}$.

$$800 = (500)\left(e^{-k}\right)\left(e^{3k}\right)$$

$$800 = 500e^{2k} \Rightarrow \frac{8}{5} = e^{2k}$$

Take the ln of both sides:

$$\ln\left(\frac{8}{5}\right) = \ln\left(e^{2k}\right)$$

$$\ln\left(\frac{8}{5}\right) = 2k$$

$$k = \frac{1}{2}\ln\left(\frac{8}{5}\right) = \ln\sqrt{\frac{8}{5}}.$$

Step 3: Substitute $k = \frac{1}{2}\ln\left(\frac{8}{5}\right)$ into one of the equations.

$$500 = y_0 e^k$$

$$500 = y_0 e^{\ln\left(\sqrt{\frac{8}{5}}\right)}$$

$$500 = y_0\left(\sqrt{\frac{8}{5}}\right)$$

$$y_0 = \frac{500}{\sqrt{8/5}} = 125\sqrt{10} \approx 395.285$$

Thus, there are 395 bacteria present initially.

(b) $y_0 = 125\sqrt{10},\ k = \ln\sqrt{\frac{8}{5}}$

$$y(t) = y_0 e^{kt}$$

$$y(t) = \left(125\sqrt{10}\right) e^{\left(\ln\sqrt{\frac{8}{5}}\right)^t} = \left(125\sqrt{10}\right)\left(\frac{8}{5}\right)^{(1/2)t}$$

$$y(4) = \left(125\sqrt{10}\right)\left(\frac{8}{5}\right)^{(1/2)4} = \left(125\sqrt{10}\right)\left(\frac{8}{5}\right)^2 = 1011.93$$

Thus, there are approximately 1012 bacteria present after 4 days.

- Get a good night's sleep the night before. Have a light breakfast before the exam.

Example 2—Radioactive Decay

Carbon-14 has a half-life of 5750 years. If initially there are 60 grams of carbon-14, how many grams are left after 3000 years?

Step 1: $y(t) = y_0 e^{kt} = 60e^{kt}$

Since half-life is 5750 years, $30 = 60e^{k(5750)} \Rightarrow \frac{1}{2} = e^{5750k}$.

$$\ln\left(\frac{1}{2}\right) = \ln\left(e^{5750k}\right)$$

$$-\ln 2 = 5750k$$

$$\frac{-\ln 2}{5750} = k$$

Step 2: $y(t) = y_0 e^{kt}$

$$y(t) = 60e^{\left[\frac{-\ln 2}{5750}\right]}$$

$$y(t) = 60e^{\left[\frac{-\ln 2}{5750}\right](3000)}$$

$$y(3000) \approx 41.7919$$

Thus, there will be approximately 41.792 grams of carbon-14 after 3000 years.

Separable Differential Equations

General Procedure

Steps:

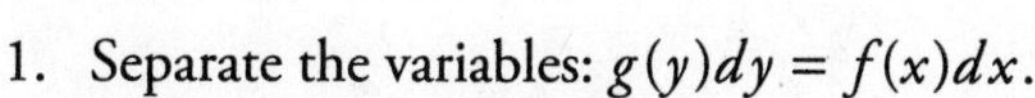

1. Separate the variables: $g(y)dy = f(x)dx$.
2. Integrate both sides: $\int g(y)dy = \int f(x)dx$.
3. Solve for y to get a general solution.
4. Substitute given conditions to get a particular solution.
5. Verify your result by differentiating.

Example 1

Given $\frac{dy}{dx} = 4x^3y^2$ and $y(1) = -\frac{1}{2}$, solve the differential equation.

Step 1: Separate the variables: $\frac{1}{y^2}dy = 4x^3dx$.

Step 2: Integrate both sides: $\int \frac{1}{y^2}dy = \int 4x^3dx; -\frac{1}{y} = x^4 + C$.

Step 3: General solution: $y = \frac{-1}{x^4 + C}$.

Step 4: Particular solution: $-\frac{1}{2} = \frac{-1}{1 + C} \Rightarrow c = 1; y = \frac{-1}{x^4 + 1}$.

Step 5: Verify the result by differentiating.

$$y = \frac{-1}{x^4 + 1} = (-1)(x^4 + 1)^{-1}$$

$$\frac{dy}{dx} = (-1)(-1)(x^4 + 1)^{-2}(4x^3) = \frac{4x^3}{(x^4 + 1)^2}.$$

Note that $y = \frac{-1}{x^4 + 1}$ implies $y^2 = \frac{1}{(x^4 + 1)^2}$.

Thus, $\frac{dy}{dx} = \frac{4x^3}{(x^4 + 1)^2} = 4x^3y^2$.

Example 2

Find a solution of the differentiation equation $\frac{dy}{dx} = x\sin(x^2); y(0) = -1$.

Step 1: Separate the variables: $dy = x\sin(x^2)dx$.

Step 2: Integrate both sides: $\int dy = \int x\sin(x^2)dx; \int dy = y$.

Let $u = x^2; du = 2x\,dx$ or $\frac{du}{2} = x\,dx$.

$$\int x\sin(x^2)dx = \int \sin u\left(\frac{du}{2}\right) = \frac{1}{2}\int \sin u\,du = -\frac{1}{2}\cos u + C$$

$$= -\frac{1}{2}\cos(x^2) + C$$

Thus, $y = -\frac{1}{2}\cos(x^2) + C$.

Step 3: Substitute given condition:

$$y(0) = -1; -1 = -\frac{1}{2}\cos(0) + C; -1 = \frac{-1}{2} + C; -\frac{1}{2} = C.$$

Thus, $y = -\frac{1}{2}\cos(x^2) - \frac{1}{2}$.

Step 4: Verify the result by differentiating:

$$\frac{dy}{dx} = \frac{1}{2}\left[\sin(x^2)\right](2x) = x\sin(x^2).$$

Example 3

If $\frac{d^2y}{dx^2} = 2x + 1$ and at $x = 0$, $y' = -1$, and $y = 3$, find a solution of the differential equation.

Step 1: Rewrite $\frac{d^2y}{dx^2}$ as $\frac{dy'}{dx}$; $\frac{dy'}{dx} = 2x + 1$.

Step 2: Separate the variables: $dy' = (2x + 1)dx$.

Step 3: Integrate both sides: $\int dy' = \int (2x + 1)dx$; $y' = x^2 + x + C_1$.

Step 4: Substitute given condition: At $x = 0$, $y' = -1; -1 = 0 + 0 + C_1 \Rightarrow C_1 = -1$. Thus, $y' = x^2 + x - 1$.

Step 5: Rewrite: $y' = \frac{dy}{dx}$; $\frac{dy}{dx} = x^2 + x - 1$.

Step 6: Separate the variables: $dy = (x^2 + x - 1)dx$.

Step 7: Integrate both sides: $\int dy = \int (x^2 + x - 1)dx$

$$y = \frac{x^3}{3} + \frac{x^2}{2} - x + C_2.$$

Step 8: Substitute given condition: At $x = 0$, $y = 3; 3 = 0 + 0 - 0 + C_2 \Rightarrow C_2 = 3$.

Therefore, $y = \frac{x^3}{3} + \frac{x^2}{2} - x + 3$.

Step 9: Verify the result by differentiating:

$$y = \frac{x^3}{3} + \frac{x^2}{2} - x + 3$$

$$\frac{dy}{dx} = x^2 + x - 1; \frac{d^2y}{dx^2} = 2x + 1.$$

Example 4

Find the general solution of the differential equation $\frac{dy}{dx} = \frac{2xy}{x^2 + 1}$.

Step 1: Separate the variables:

$$\frac{dy}{y} = \frac{2x}{x^2 + 1}dx.$$

Step 2: Integrate both sides: $\int \frac{dy}{y} = \int \frac{2x}{x^2 + 1}dx$ (let $u = x^2 + 1; du = 2x\,dx$)

$$\ln|y| = \ln(x^2 + 1) + C_1.$$

Step 3: General solution: solve for y.

$$e^{\ln|y|} = e^{\ln(x^2+1)+C_1}$$

$$|y| = e^{\ln(x^2+1)} \cdot e^{C_1}; |y| = e^{C_1}(x^2 + 1)$$

$$y = \pm e^{C_1}(x^2 + 1)$$

The general solution is $y = C(x^2 + 1)$.

Step 4: Verify the result by differentiating:

$$y = C(x^2 + 1)$$

$$\frac{dy}{dx} = 2Cx = 2x\frac{C(x^2 + 1)}{x^2 + 1} = \frac{2xy}{x^2 + 1}.$$

Example 5

Write an equation for the curve that passes through the point (3, 4) and has a slope at any point (x, y) as $\frac{dy}{dx} = \frac{x^2+1}{2y}$.

Step 1: Separate the variables: $2y\,dy = (x^2+1)dx$.

Step 2: Integrate both sides: $\int 2y\,dy = \int (x^2+1)\,dx;\ y^2 = \frac{x^3}{3} + x + C.$

Step 3: Substitute given condition: $4^2 = \frac{3^3}{3} + 3 + C \Rightarrow C = 4.$

Thus, $y^2 = \frac{x^3}{3} + x + 4.$

Step 4: Verify the result by differentiating:

$$2y\frac{dy}{dx} = x^2 + 1$$

$$\frac{dy}{dx} = \frac{x^2+1}{2y}.$$

14.5 Slope Fields

Main Concepts: Slope Fields, Solution of Different Equations

A *slope field* (or a *direction field*) for first-order differential equations is a graphic representation of the slopes of a family of curves. It consists of a set of short line segments drawn on a pair of axes. These line segments are the tangents to a family of solution curves for the differential equation at various points. The tangents show the direction which the solution curves will follow. Slope fields are useful in sketching solution curves without having to solve a differential equation algebraically.

Example 1

If $\frac{dy}{dx} = 0.5x$, draw a slope field for the given differential equation.

Step 1: Set up a table of values for $\frac{dy}{dx}$ for selected values of x.

x	–4	–3	–2	–1	0	1	2	3	4
$\frac{dy}{dx}$	–2	–1.5	–1	–0.5	0	0.5	1	1.5	2

Note that since $\frac{dy}{dx} = 0.5x$, the numerical value of $\frac{dy}{dx}$ is independent of the value of y. For example, at the points $(1, -1)$, $(1, 0)$, $(1, 1)$, $(1, 2)$, $(1, 3)$, and at all the points whose x-coordinates are 1, the numerical value of $\frac{dy}{dx}$ is 0.5 regardless of their y-coordinates. Similarly, for all the points, whose x-coordinates are 2 (e.g., (2, − 1), (2, 0), (2, 3), etc.), $\frac{dy}{dx} = 1$. Also, remember that $\frac{dy}{dx}$ represents the slopes of the tangent lines to the curve at various points. You are now ready to draw these tangents.

Step 2: Draw short line segments with the given slopes at the various points. The slope field for the differential equation $\frac{dy}{dx} = 0.5x$ is shown in Figure 14.5-1.

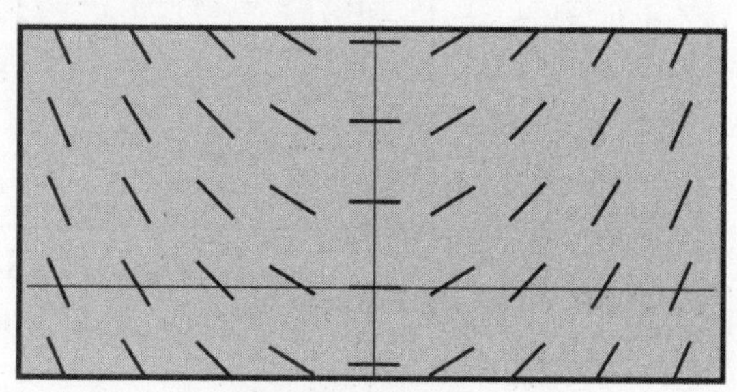

Figure 14.5-1

Example 2

Figure 14.5-2 shows a slope field for one of the differential equations given below. Identify the equation.

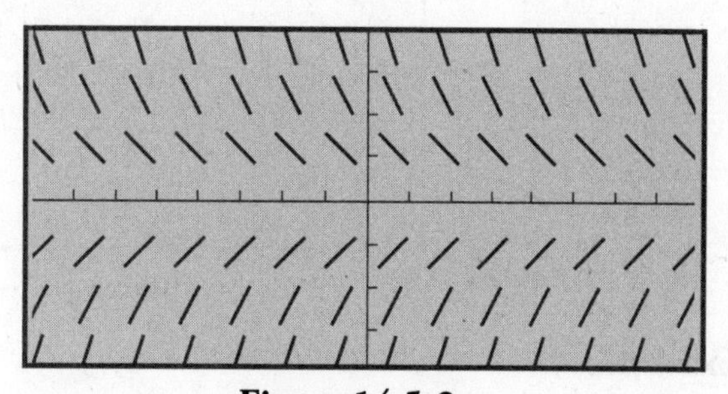

Figure 14.5-2

(a) $\frac{dy}{dx} = 2x$ (b) $\frac{dy}{dx} = -2x$ (c) $\frac{dy}{dx} = y$

(d) $\frac{dy}{dx} = -y$ (e) $\frac{dy}{dx} = x + y$

Solution:

If you look across horizontally at any row of tangents, you'll notice that the tangents have the same slope. (Points on the same row have the same y-coordinate but different x-coordinates.) Therefore, the numerical value of $\frac{dy}{dx}$ (which represents the slope of the tangent) depends solely on the y-coordinate of a point and it is independent of the x-coordinate. Thus, only choice (c) and choice (d) satisfy this condition. Also notice that the tangents have a negative slope when $y > 0$ and have a positive slope when $y < 0$.

Therefore, the correct choice is (d), $\frac{dy}{dx} = -y$.

Example 3

A slope field for a differential equation is shown in Figure 14.5-3. Draw a possible graph for the particular solution $y = f(x)$ to the differential equation function, if (a) the initial condition is $f(0) = -2$ and (b) the initial condition is $f(0) = 0$.

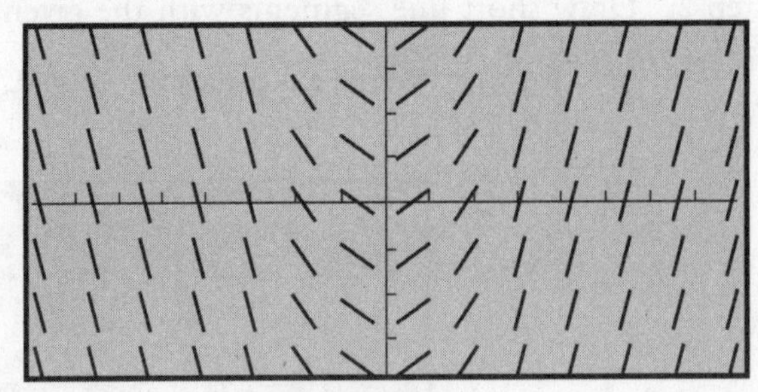

Figure 14.5-3

Solution:

Begin by locating the point (0, −2) as given in the initial condition. Follow the flow of the field and sketch the graph of the function. Repeat the same procedure with the point (0, 0). See the curves as shown in Figure 14.5-4.

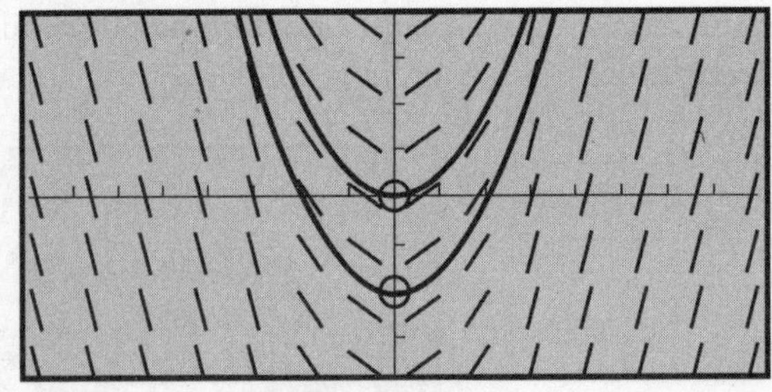

Figure 14.5-4

Example 4

Given the differential equation $\frac{dy}{dx} = -xy$.

(a) Draw a slope field for the differential equation at the 15 points indicated on the provided set of axes in Figure 14.5-5.

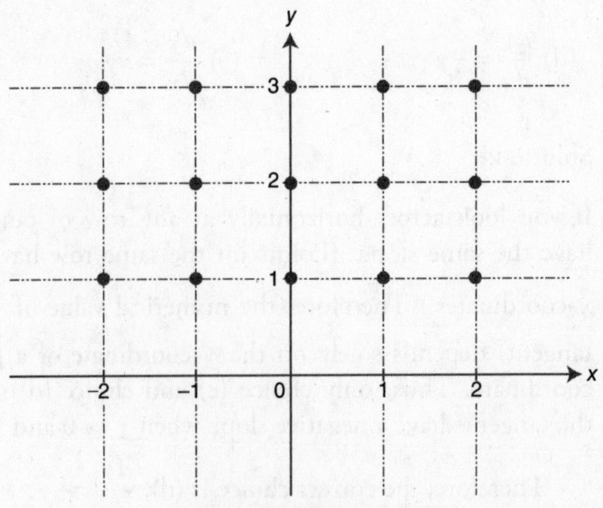

Figure 14.5-5

(b) Sketch a possible graph for the particular solution $y = f(x)$ to the differential equation with the initial condition $f(0) = 3$.

(c) Find, algebraically, the particular solution $y = f(x)$ to the differential equation with the initial condition $f(0) = 3$.

Solution:

(a) Set up a table of values for $\frac{dy}{dx}$ at the 15 given points.

	$x=-2$	$x=-1$	$x=0$	$x=1$	$x=2$
$y=1$	2	1	0	−1	−2
$y=2$	4	2	0	−2	−4
$y=3$	6	3	0	−3	−6

Then sketch the tangents at the various points as shown in Figure 14.5-6.

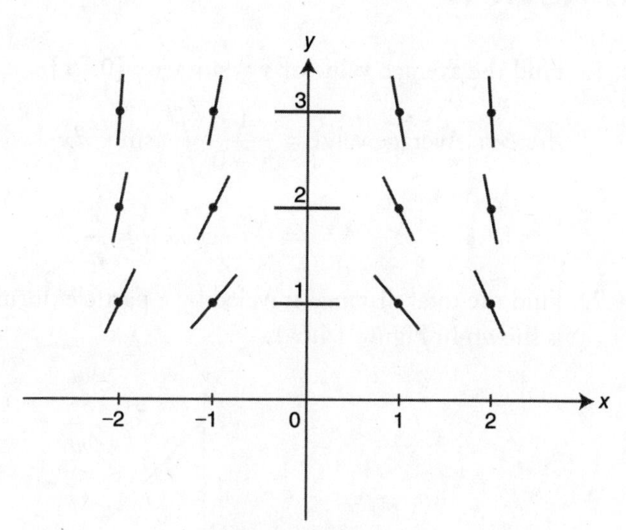

Figure 14.5-6

(b) Locate the point (0, 3) as indicated in the initial condition. Follow the flow of the field and sketch the curve as shown Figure 14.5-7.

(c) Step 1: Rewrite $\frac{dy}{dx} = -xy$ as $\frac{dy}{y} = -x\,dx$.

Step 2: Integrate both sides $\int \frac{dy}{y} = \int -x\,dx$ and obtain $\ln|y| = -\frac{x^2}{2} + C$.

Step 3: Apply the exponential function to both sides and obtain $e^{\ln|y|} = e^{-\frac{x^2}{2}+C}$.

Step 4: Simplify the equation and get $y = \left(e^{\frac{-x^2}{2}}\right)(e^C) = \frac{e^C}{e^{\frac{x^2}{2}}}$.

Let $k = e^C$ and you have $y = \frac{k}{e^{\frac{x^2}{2}}}$.

Step 5: Substitute initial condition (0, 3) and obtain $k = 3$. Thus, you have $y = \frac{3}{e^{\frac{x^2}{2}}}$.

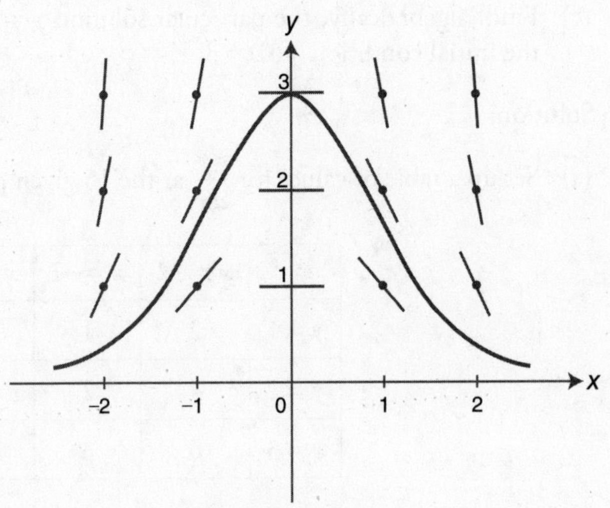

Figure 14.5-7

14.6 Rapid Review

1. Find the average value of $y = \sin x$ on $[0, \pi]$.

   Answer: Average value $= \frac{1}{\pi - 0}\int_0^{\pi} \sin x \, dx$

   $= \frac{1}{\pi}\left[-\cos x\right]_0^{\pi} = \frac{2}{\pi}$.

2. Find the total distance traveled by a particle during $0 \leq t \leq 3$ whose velocity function is shown in Figure 14.6-1.

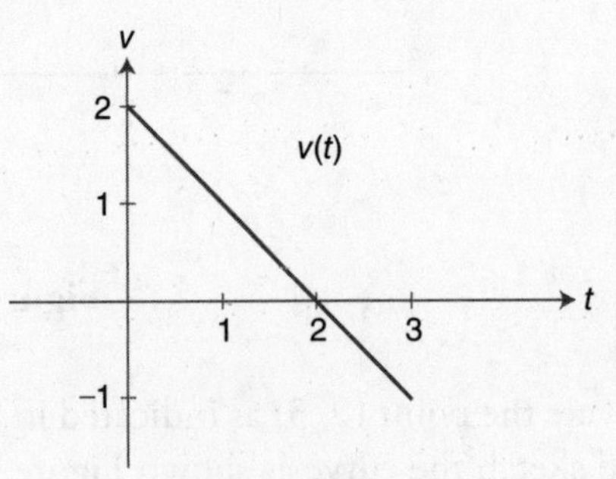

Figure 14.6-1

   Answer: Total distance traveled $= \int_0^2 v(t)dt + \left|\int_2^3 v(t)dt\right|$

   $= 2 + 0.5 = 2.5$.

3. Oil is leaking from a tank at the rate of $f(t) = 5e^{-0.1t}$ gallons/hour, where t is measured in hours. Write an integral to find the total number of gallons of oil that will have leaked from the tank after 10 hours. Do not evaluate the integral.

   Answer: Total number of gallons leaked $= \int_0^{10} 5e^{-0.1t} \, dt$.

4. How much money should Mary invest at 7.5% interest a year compounded continuously, so that she will have $100,000 after 20 years.

   Answer: $y(t) = y_0e^{kt}$, $k = 0.075$, and $t = 20$. $y(20) = 100{,}000 = y_0e^{(0.075)(20)}$. Thus, using a calculator, you obtain $y_0 \approx 22313$, or $22,313.

5. Given $\frac{dy}{dx} = \frac{x}{y}$ and $y(1) = 0$, solve the differential equation.

   Answer: $y\,dy = x\,dx \Rightarrow \int y\,dy = \int x\,dx \Rightarrow \frac{y^2}{2} = \frac{x^2}{2} + c \Rightarrow 0 = \frac{1}{2} + c \Rightarrow c = -\frac{1}{2}$

   Thus, $\frac{y^2}{2} = \frac{x^2}{2} - \frac{1}{2}$ or $y^2 = x^2 - 1$.

6. Identify the differential equation for the slope field shown.

   Answer: The slope field suggests a hyperbola of the form $y^2 - x^2 = k$, so $2y\frac{dy}{dx} - 2x = 0$ and $\frac{dy}{dx} = \frac{x}{y}$.

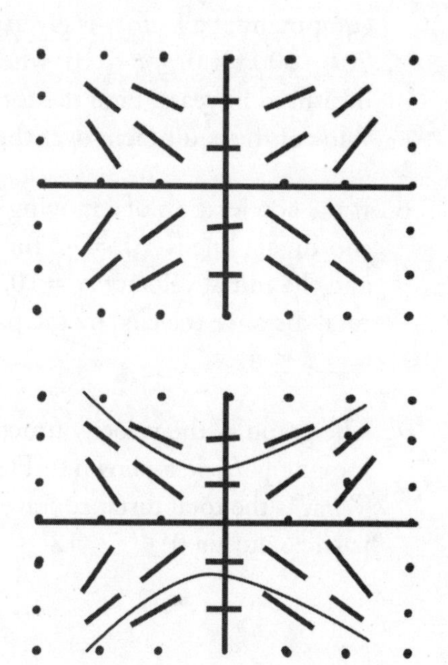

14.7 Practice Problems

Part A—The use of a calculator is not allowed.

1. Find the value of c as stated in the Mean Value Theorem for Integrals for $f(x) = x^3$ on [2, 4].

2. The graph of f is shown in Figure 14.7-1. Find the average value of f on [0, 8].

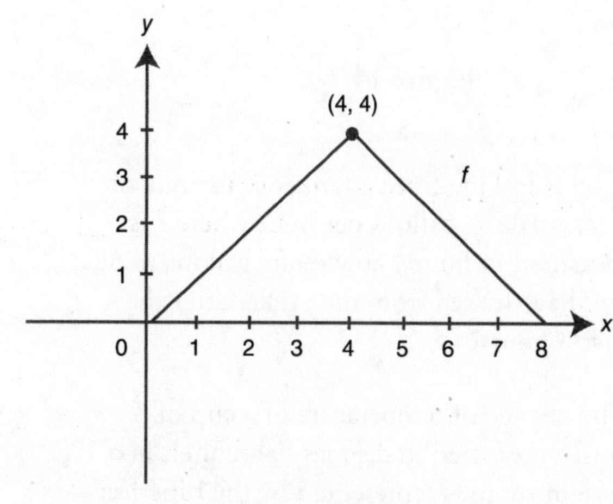

Figure 14.7-1

3. The position function of a particle moving on a coordinate line is given as $s(t) = t^2 - 6t - 7$, $0 \leq t \leq 10$. Find the displacement and total distance traveled by the particle from $1 \leq t \leq 4$.

4. The velocity function of a moving particle on a coordinate line is $v(t) = 2t + 1$ for $0 \leq t \leq 8$. At $t = 1$, its position is -4. Find the position of the particle at $t = 5$.

5. The rate of depreciation for a new piece of equipment at a factory is given as $p(t) = 50t - 600$ for $0 \leq t \leq 10$, where t is measured in years. Find the total loss of value of the equipment over the first 5 years.

6. If the acceleration of a moving particle on a coordinate line is $a(t) = -2$ for $0 \leq t \leq 4$, and the initial velocity $v_0 = 10$, find the total distance traveled by the particle during $0 \leq t \leq 4$.

7. The graph of the velocity function of a moving particle is shown in Figure 14.7-2. What is the total distance traveled by the particle during $0 \leq t \leq 12$?

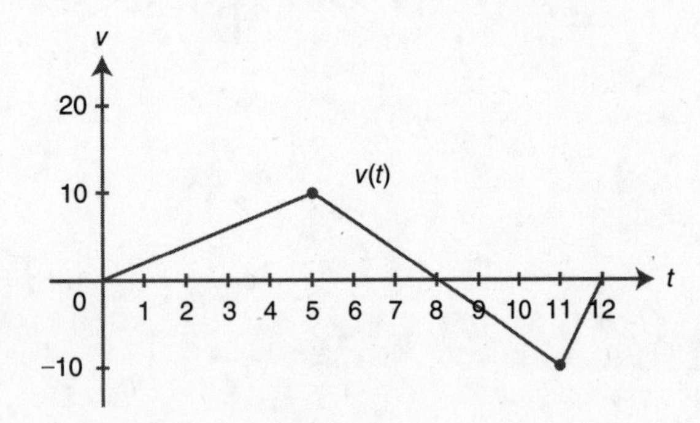

Figure 14.7-2

8. If oil is leaking from a tanker at the rate of $f(t) = 10e^{0.2t}$ gallons per hour where t is measured in hours, how many gallons of oil will have leaked from the tanker after the first 3 hours?

9. The change of temperature of a cup of coffee measured in degrees Fahrenheit in a certain room is represented by the function $f(t) = -\cos\left(\frac{t}{4}\right)$ for $0 \leq t \leq 5$, where t is measured in minutes. If the temperature of the coffee is initially 92°F, find its temperature after the first 5 minutes.

10. If the *half-life* of a radioactive element is 4500 years, and initially, there are 100 grams of this element, approximately how many grams are left after 5000 years?

11. Find a solution of the differential equation:
$$\frac{dy}{dx} = x\cos(x^2);\ y(0) = \pi.$$

12. If $\frac{d^2y}{dx^2} = x - 5$ and at $x = 0$, $y' = -2$ and $y = 1$, find a solution of the differential equation.

Part B—Calculators are allowed.

13. Find the average value of $y = \tan x$ from $x = \frac{\pi}{4}$ to $x = \frac{\pi}{3}$.

14. The acceleration function of a moving particle on a straight line is given by $a(t) = 3e^{2t}$, where t is measured in seconds, and the initial velocity is $\frac{1}{2}$. Find the displacement and total distance traveled by the particle in the first 3 seconds.

15. The sales of an item in a company follow an exponential growth/decay model, where t is measured in months. If the sales drop from 5000 units in the first month to 4000 units in the third month, how many units should the company expect to sell during the seventh month?

16. Find an equation of the curve that has a slope of $\frac{2y}{x+1}$ at the point (x, y) and passes through the point $(0, 4)$.

17. The population in a city was approximately 750,000 in 1980, and grew at a rate of 3% per year. If the population growth

followed an exponential growth model, find the city's population in the year 2002.

18. Find a solution of the differential equation $4e^y = y' - 3xe^y$ and $y(0) = 0$.

19. How much money should a person invest at 6.25% interest compounded continuously so that the person will have \$50,000 after 10 years?

20. The velocity function of a moving particle is given as $v(t) = 2 - 6e^{-t}$, $t \geq 0$ and t is measured in seconds. Find the total distance traveled by the particle during the first 10 seconds.

14.8 Cumulative Review Problems

(Calculator) indicates that calculators are permitted.

21. If $3e^y = x^2 y$, find $\dfrac{dy}{dx}$.

22. Evaluate $\displaystyle\int_0^1 \frac{x^2}{x^3+1}dx$.

23. The graph of a continuous function f, which consists of three line segments on $[-2, 4]$, is shown in Figure 14.8-1. If $F(x) = \displaystyle\int_{-2}^{x} f(t)dt$ for $-2 \leq x \leq 4$,

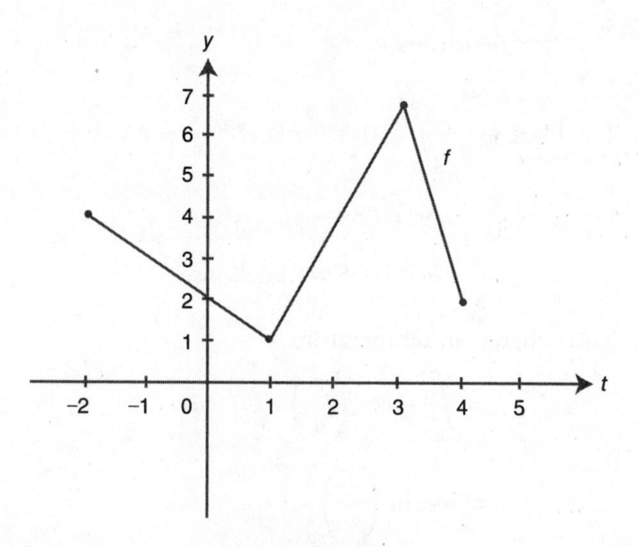

Figure 14.8-1

(a) Find $F(-2)$ and $F(0)$.

(b) Find $F'(0)$ and $F'(2)$.

(c) Find the value of x such that F has a maximum on $[-2, 4]$.

(d) On which interval is the graph of F concave upward?

24. (Calculator) The slope of a function $y = f(x)$ at any point (x, y) is $\dfrac{y}{2x+1}$ and $f(0) = 2$.

(a) Write an equation of the line tangent to the graph of f at $x = 0$.

(b) Use the tangent in part (a) to find the approximate value of $f(0.1)$.

(c) Find a solution $y = f(x)$ for the differential equation.

(d) Using the result in part (c), find $f(0.1)$.

25. (Calculator) Let R be the region in the first quadrant bounded by $f(x) = e^x - 1$ and $g(x) = 3 \sin x$.

(a) Find the area of region R.

(b) Find the volume of the solid obtained by revolving R about the x-axis.

(c) Find the volume of the solid having R as its base and semicircular cross sections perpendicular to the x-axis.

14.9 Solutions to Practice Problems

Part A—The use of a calculator is not allowed.

1. $\int_2^4 x^3 dx = f(c)(4-2)$

$\int_2^4 x^3 dx = \left.\frac{x^4}{4}\right]_2^4 = \left(\frac{4^4}{4}\right) - \left(\frac{2^4}{4}\right) = 60$

$2f(c) = 60 \Rightarrow f(c) = 30$

$c^3 = 30 \Rightarrow C = 30^{(1/3)}$.

2. Average value $= \frac{1}{8-0}\int_0^1 f(x)\,dx$

$= \frac{1}{8}\left(\frac{1}{2}(8)(4)\right) = 2.$

3. Displacement $= s(4) - s(1) = -15 - (-12) = -3.$

Distance traveled $= \int_1^4 |v(t)|dt.$

$v(t) = s'(t) = 2t - 6$

Set $2t - 6 = 0 \Rightarrow t = 3$

$$|2t-6| = \begin{cases} -(2t-6) & \text{if } 0 \le t < 3 \\ 2t-6 & \text{if } 3 \le t \le 10 \end{cases}$$

$\int_1^4 |v(t)|dt = \int_1^3 -(2t-6)dt + \int_3^4 (2t-6)dt$

$= \left[-t^2 + 6t\right]_1^3 + \left[t^2 - 6t\right]_3^4$

$= 4 + 1 = 5.$

4. Position function $s(t) = \int v(t)dt$

$= \int (2t+1)dt$

$= t^2 + t + C$

$s(1) = -4 \Rightarrow (1)^2 + 1 + C = -4$ or $C = -6$

$s(t) = t^2 + t - 6$

$s(5) = 5^2 + 5 - 6 = 24.$

5. Total loss $= \int_0^5 p(t)dt$

$= \int_0^5 (50t - 600)dt$

$= \left.25t^2 - 600t\right]_0^5 = -\$2375.$

6. $v(t) = \int a(t)dt = \int -2\,dt = -2t + C$

$v_0 = 10 \Rightarrow -2(0) + C = 10$ or $C = 10$

$v(t) = -2t + 10$

Distance traveled $= \int_0^4 |v(t)|\,dt.$

Set $v(t) = 0 \Rightarrow -2t + 10 = 0$ or $t = 5.$

$|-2t + 10| = -2t + 10$ if $0 \le t < 5$

$\int_0^4 |v(t)|dt = \int_0^4 (-2t + 10)dt$

$= \left.-t^2 + 10t\right]_0^4 = 24$

7. Total distance traveled

$= \int_0^8 |v(t)|\,dt + \left|\int_8^{12} v(t)\right|$

$= \frac{1}{2}(8)(10) + \frac{1}{2}(4)(10)$

$= 60$ meters.

8. Total leakage $= \int_0^3 10e^{0.2t} = \left.50e^{0.2t}\right]_0^3$

$= 91.1059 - 50$

$= 41.1059 = 41$ gallons.

9. Total change in temperature

$= \int_0^5 -\cos\left(\frac{t}{4}\right)dt$

$= \left.-4\sin\left(\frac{t}{4}\right)\right]_0^5$

$= -3.79594 - 0$

$= -3.79594$°F.

Thus, the temperature of coffee after 5 minutes is $(92 - 3.79594) \approx 88.204$°F.

10. $y(t) = y_0 e^{kt}$

Half-life = 4500 years $\Rightarrow \frac{1}{2} = e^{4500k}$.

Take ln of both sides:

$$\ln\left(\frac{1}{2}\right) = \ln e^{4500k}$$

$$\Rightarrow -\ln 2 = 4500k$$

$$\text{or } k = \frac{-\ln 2}{4500}.$$

$$y(t) = 100e^{\left(\frac{-\ln 2}{4500}\right)(5000)} = 25(2^{2/9})$$
$$\approx 46.293.$$

There are approximately 46.29 grams left.

11. Step 1: Separate the variables:
$dy = x\cos(x^2)\,dx.$

Step 2: Integrate both sides:

$$\int dy = \int x\cos(x^2)\,dx$$

$$\int dy = y$$

$$\int x\cos(x^2)dx : \text{Let } u = x^2;$$

$$du = 2x\,dx,\ \frac{du}{2} = x\,dx$$

$$\int x\cos(x^2)dx = \int \cos u \frac{du}{2}$$

$$= \frac{\sin u}{2} + c = \frac{\sin(x^2)}{2} + C.$$

$$\text{Thus, } y = \frac{\sin(x^2)}{2} + C.$$

Step 3: Substitute given values.

$$y(0) = \frac{\sin(0)}{2} + C = \pi \Rightarrow C = \pi$$

$$y = \frac{\sin(x^2)}{2} + \pi$$

Step 4: Verify result by differentiating:

$$\frac{dy}{dx} = \frac{\cos(x^2)(2x)}{2} = x\cos(x^2).$$

12. Step 1: Rewrite $\frac{d^2y}{dx^2}$ as $\frac{dy'}{dx}$

$$\frac{dy'}{dx} = x - 5.$$

Step 2: Separate variables:
$dy' = (x-5)dx.$

Step 3: Integrate both sides:

$$\int dy' = \int (x-5)\,dx$$

$$y' = \frac{x^2}{2} - 5x + C_1.$$

Step 4: Substitute given values:

$$\text{At } x = 0,\ y' = \frac{0}{2} - 5(0) + C_1$$

$$= -2 \Rightarrow C_1 = -2$$

$$y' = \frac{x^2}{2} - 5x - 2.$$

Step 5: Rewrite:

$$y' = \frac{dy}{dx};\ \frac{dy}{dx} = \frac{x^2}{2} - 5x - 2.$$

Step 6: Separate the variables:

$$dy = \left(\frac{x^2}{2} - 5x - 2\right)dx.$$

Step 7: Integrate both sides:

$$\int dy = \int \left(\frac{x^2}{2} - 5x - 2\right)dx.$$

$$y = \frac{x^3}{6} - \frac{5x^2}{2} - 2x + C_2$$

Step 8: Substitute given values:

$$\text{At } x = 0,\ y = 0 - 0 - 0 + C_2$$

$$= 1 \Rightarrow C_2 = 1$$

$$y = \frac{x^3}{6} - \frac{5x^2}{2} - 2x + 1.$$

Step 9: Verify result by differentiating:

$$\frac{dy}{dx} = \frac{x^2}{2} - 5x - 2$$

$$\frac{d^2y}{dx^2} = x - 5.$$

Part B—Calculators are allowed.

13. Average value $= \dfrac{1}{\pi/3 - \pi/4} \displaystyle\int_{\pi/4}^{\pi/3} \tan x \, dx.$

Enter $= (1/(\pi/3 - \pi/4)) \int (\tan x, x, \pi/4, \pi/3)$

and obtain $\dfrac{6 \ln(2)}{\pi} = 1.32381.$

14. $v(t) = \int a(t)dt$

$$= \int 3e^{2t} = \frac{3}{2}e^{2t} + C$$

$$v(0) = \frac{3}{2}e^0 + C = \frac{1}{2} \Rightarrow \frac{3}{2} + C = \frac{1}{2}$$

or $C = -1$

$$v(t) = \frac{3}{2}e^{2t} - 1$$

Displacement $= \displaystyle\int_0^3 \left(\frac{3}{2}e^{2t} - 1\right) dt.$

Enter $\int (3/2 * e\text{^}(2x) - 1, x, 0, 3)$

and obtain 298.822.

Distance traveled $= \displaystyle\int_0^3 |v(t)| \, dt.$

Since $\frac{3}{2}e^{2t} - 1 > 0$ for $t \geq 0$,

$$\int_0^3 |v(t)| dt = \int_0^3 \left(\frac{3}{2}e^{2t} - 1\right) dt = 298.822.$$

15. Step 1: $y(t) = y_0 e^{kt}$

$y(1) = 5000 \Rightarrow 5000 = y_0 e^k \Rightarrow y_0$

$= 5000e^{-k}$

$y(3) = 4000 \Rightarrow 4000 = y_0 e^{3k}$

Substituting:

$y(0) = 5000e^{-k}$, $4000 = (5000e^{-k})e^{3k}$

$4000 = 5000e^{2k}$

$$\frac{4}{5} = e^{2k}$$

$$\ln\left(\frac{4}{5}\right) = \ln\left(e^{2k}\right) = 2k$$

$$k = \frac{1}{2}\ln\left(\frac{4}{5}\right) \approx -0.111572.$$

Step 2: $5000 = y_0 e^{-0.111572}$

$y(0) = (5000)/e^{-0.111572} \approx 5590.17$

$y(t) = (5590.17)\, e^{-0.111572}$

Step 3: $y(7) = (5590.17)e^{-0.111572(7)} \approx 2560$

Thus, sales for the 7th month are approximately 2560 units.

16. Step 1: Separate the variables:

$$\frac{dy}{dx} = \frac{2y}{x+1}$$

$$\frac{dy}{2y} = \frac{dy}{x+1}.$$

Step 2: Integrate both sides:

$$\int \frac{dy}{2y} = \int \frac{dx}{x+1}$$

$$\frac{1}{2}\ln|y| = \ln|x+1| + C.$$

Step 3: Substitute given value (0, 4):

$$\frac{1}{2}\ln(4) = \ln(1) + C$$

$$\ln 2 = C$$

$$\frac{1}{2}\ln|y| - \ln|x+1| = \ln 2$$

$$\ln\left|\frac{y^{1/2}}{x+1}\right| = \ln 2$$

$$e^{\ln\left|\frac{y^{1/2}}{x+1}\right|} = e^{\ln 2}$$

$$\frac{y^{1/2}}{x+1} = 2$$

$$y^{1/2} = 2(x+1)$$

$$y = (2)^2(x+1)^2$$

$$y = 4(x+1)^2.$$

Step 4: Verify result by differentiating:

$$\frac{dy}{dx} = 4(2)(x+1) = 8(x+1).$$

Compare with $\frac{dy}{dx} = \frac{2y}{x+1}$

$$= \frac{2(4(x+1)^2)}{(x+1)}$$

$$= 8(x+1).$$

17. $y(t) = y_0 e^{kt}$

$y_0 = 750{,}000$

$y(22) = (750{,}000)\,e^{(0.03)(22)}$

$$\approx \begin{cases} 1.45109E6 \approx 1{,}451{,}090 \text{ using a TI-89,} \\ 1{,}451{,}094 \text{ using a TI-85.} \end{cases}$$

18. Step 1: Separate the variables:

$$4e^y = \frac{dy}{dx} - 3xe^y$$

$$4e^y + 3xe^y = \frac{dy}{dx}$$

$$e^y(4+3x) = \frac{dy}{dx}$$

$$(4+3x)dx = \frac{dy}{e^y} = e^{-y}dy.$$

Step 2: Integrate both sides:

$$\int (4+3x)\,dx = \int e^{-y}dy$$

$$4x + \frac{3x^2}{2} = -e^{-y} + C$$

Switch sides: $e^{-y} = -\frac{3x^2}{2} - 4x + C.$

Step 3: Substitute given value: $y(0) = 0$ $\Rightarrow e^0 = 0 - 0 + c \Rightarrow c = 1.$

Step 4: Take ln of both sides:

$$e^{-y} = -\frac{3x^2}{2} - 4x + 1$$

$$\ln(e^{-y}) = \ln\left(-\frac{3x^2}{2} - 4x + 1\right)$$

$$y = -\ln\left(1 - 4x - \frac{3x^2}{2}\right).$$

Step 5: Verify result by differentiating: Enter $d(-\ln(1 - 4x - 3(x\text{-}\wedge 2)/2), x)$ and obtain $\frac{-2(3x+4)}{3x^2+8x-2}$, which is equivalent to $e^y(4+3x)$.

19. $y(t) = y_0 e^{kt}$

$k = 0.0625,\ y(10) = 50{,}000$

$50{,}000 = y_0 e^{10(0.0625)}$

$$y_0 = \frac{50{,}000}{e^{0.625}} \begin{cases} \$26763.1 \text{ using a TI-89,} \\ \$26763.071426 \approx \$26763.07 \\ \text{using a TI-85.} \end{cases}$$

20. Set $v(t) = 2 - 6e^{-t} = 0$. Using the [*Zero*] function on your calculator, compute $t = 1.09861$.

Distance traveled $= \int_0^{10} |v(t)|dt$

$$|2 - 6e^{-t}| = \begin{cases} -(2 - 6e^{-t}) & \text{if } 0 \le t < 1.09861 \\ 2 - 6e^{-t} & \text{if } t \ge 1.09861 \end{cases}$$

$$\int_0^{10} |2 - 6e^{-t}|dt = \int_0^{1.09861} -(2 - 6e^{-t})dt + \int_{1.09861}^{10} (2 - 6e^{-t})dt$$

$$= 1.80278 + 15.803 = 17.606.$$

Alternatively, use the [*nInt*] function on the calculator.

Enter nInt(abs(2 − 6e^(−x)), x, 0, 10) and obtain the same result.

14.10 Solutions to Cumulative Review Problems

21. $3e^y = x^2 y$

$$3e^y \frac{dy}{dx} = 2xy + \frac{dy}{dx}\left(x^2\right)$$

$$3e^y \frac{dy}{dx} - \frac{dy}{dx}x^2 = 2xy$$

$$\frac{dy}{dx}\left(3e^y - x^2\right) = 2xy$$

$$\frac{dy}{dx} = \frac{2xy}{3e^y - x^2}$$

22. Let $u = x^3 + 1$; $du = 3x^2 dx$ or $\frac{du}{3} = x^2 dx$.

$$\int \frac{x^2}{x^3+1}dx = \int \frac{1}{u}\frac{du}{3}$$

$$= \frac{1}{3}\ln|u| + C$$

$$= \frac{1}{3}\ln\left|x^3+1\right| + C$$

$$\int_0^3 \frac{x^2}{x^3+1}dx = \frac{1}{3}\ln\left|x^3+1\right|\Big]_0^3$$

$$= \frac{1}{3}(\ln 2 - \ln 1) = \frac{\ln 2}{3}$$

23. (a) $F(-2) = \int_{-2}^{-2} f(t)\,dt = 0$

$$F(0) = \int_{-2}^{0} f(t)dt = \frac{1}{2}(4+2)2 = 6$$

(b) $F'(x) = f(x)$; $F'(0) = 2$ and $F'(2) = 4$.

(c) Since $f > 0$ on $[-2, 4]$, F has a maximum value at $x = 4$.

(d) The function f is increasing on $(1, 3)$, which implies that $f' > 0$ on $(1, 3)$. Thus, F is concave upward on $(1, 3)$. (Note that f' is equivalent to the 2nd derivative of F.)

24. (a) $\frac{dy}{dx} = \frac{y}{2x+1}$; $f(0) = 2$

$$\left.\frac{dy}{dx}\right|_{x=0} = \frac{2}{2(0)+1} = 2 \Rightarrow m = 2 \text{ at } x = 0.$$

$$y - y_1 = m(x - x_1)$$

$$y - 2 = 2(x - 0) \Rightarrow y = 2x + 2$$

The equation of the tangent to f at $x = 0$ is $y = 2x + 2$.

(b) $f(0.1) = 2(0.1) + 2 = 2.2$

(c) Solve the differential equation: $\frac{dy}{dx} = \frac{y}{2x+1}$.

Step 1: Separate the variables:

$$\frac{dy}{y} = \frac{dx}{2x+1}$$

Step 2: Integrate both sides:

$$\int \frac{dy}{y} = \int \frac{dx}{2x+1}$$

$$\ln|y| = \frac{1}{2}\ln|2x+1| + C.$$

Step 3: Substitute given values (0, 2):

$$\ln 2 = \frac{1}{2}\ln 1 + C \Rightarrow C = \ln 2$$

$$\ln|y| = \frac{1}{2}\left|2x+1\right| + \ln 2$$

$$\ln|y| - \frac{1}{2}\left|2x+1\right| = \ln 2$$

$$\ln\left|\frac{y}{(2x+1)^{1/2}}\right| = \ln 2$$

$$e^{\ln\left|\frac{y}{(2x+1)^{1/2}}\right|} = e^{\ln 2}$$

$$\frac{y}{(2x+1)^{1/2}} = 2$$

$$y = 2(2x+1)^{1/2}.$$

Step 4: Verify result by differentiating

$$y = 2(2x + 1)^{1/2}$$

$$\frac{dy}{dx} = 2\left(\frac{1}{2}\right)(2x + 1)^{-1/2}(2)$$

$$= \frac{2}{\sqrt{2x+1}}.$$

Compare this with:

$$\frac{dy}{dx} = \frac{y}{2x+1} = \frac{2(2x+1)^{1/2}}{2x+1}$$

$$= \frac{2}{\sqrt{2x+1}}.$$

Thus, the function is $y = f(x) = 2(2x + 1)^{1/2}$.

(d) $f(x) = 2(2x + 1)^{1/2}$

$f(0.1) = 2(2(0.1) + 1)^{1/2} = 2(1.2)^{1/2}$

≈ 2.191

25. See Figure 14.10-1.

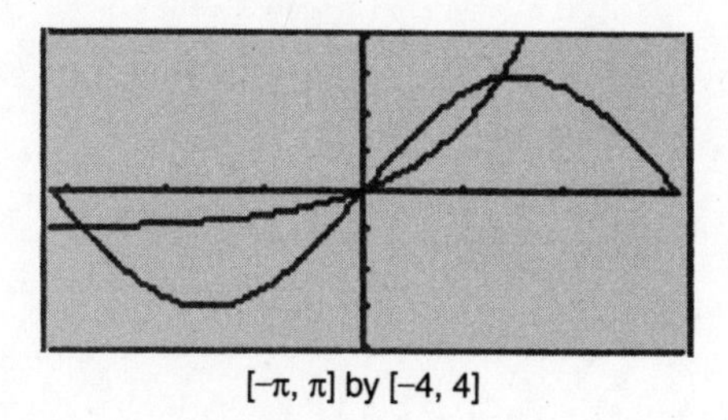

[−π, π] by [−4, 4]

Figure 14.10-1

(a) Intersection points: Using the [*Intersection*] function on the calculator, you have $x = 0$ and $x = 1.37131$.

$$\text{Area of } R = \int_0^{1.37131} [3 \sin x - (e^x - 1)]dx.$$

Enter $\int$ (3 sin(x)) − (e^(x) − 1), x, 0, 1.37131 and obtain 0.836303.

The area of region R is approximately 0.836.

(b) Using the Washer Method, volume of

$$R = \pi \int_0^{1.37131} \left[(3 \sin x)^2 - (e^x - 1)^2\right]dx.$$

Enter $\pi \int$ ((3 sin(x))^2 − (e^(x) − 1)^2, x, 0, 1.37131) and obtain 2.54273π or 7.98824.

Thus, the volume of the solid is 7.988.

(c) Volume of solid $= \pi \int_0^{1.37131}$ (Area of Cross Section)dx.

Area of Cross Section $= \frac{1}{2}\pi r^2$

$$= \frac{1}{2}\pi\left(\frac{1}{2}(3 \sin x - (e^x - 1))\right)^2.$$

Enter $\left(\frac{\pi}{2}\right)\frac{1}{4} * \int$((3 sin(x) − (e^(x) − 1))^2, x, 0, 1.37131) and obtain $0.077184\,\pi$ or 0.24248. Thus, the volume of the solid is approximately $0.077184\,\pi$ or 0.242.

Build Your Test-Taking Confidence

AP Calculus AB Practice Exam 1

ANSWER SHEET FOR MULTIPLE-CHOICE QUESTIONS

Part A	Part B
1 Ⓐ Ⓑ Ⓒ Ⓓ	76 Ⓐ Ⓑ Ⓒ Ⓓ
2 Ⓐ Ⓑ Ⓒ Ⓓ	77 Ⓐ Ⓑ Ⓒ Ⓓ
3 Ⓐ Ⓑ Ⓒ Ⓓ	78 Ⓐ Ⓑ Ⓒ Ⓓ
4 Ⓐ Ⓑ Ⓒ Ⓓ	79 Ⓐ Ⓑ Ⓒ Ⓓ
5 Ⓐ Ⓑ Ⓒ Ⓓ	80 Ⓐ Ⓑ Ⓒ Ⓓ
6 Ⓐ Ⓑ Ⓒ Ⓓ	81 Ⓐ Ⓑ Ⓒ Ⓓ
7 Ⓐ Ⓑ Ⓒ Ⓓ	82 Ⓐ Ⓑ Ⓒ Ⓓ
8 Ⓐ Ⓑ Ⓒ Ⓓ	83 Ⓐ Ⓑ Ⓒ Ⓓ
9 Ⓐ Ⓑ Ⓒ Ⓓ	84 Ⓐ Ⓑ Ⓒ Ⓓ
10 Ⓐ Ⓑ Ⓒ Ⓓ	85 Ⓐ Ⓑ Ⓒ Ⓓ
11 Ⓐ Ⓑ Ⓒ Ⓓ	86 Ⓐ Ⓑ Ⓒ Ⓓ
12 Ⓐ Ⓑ Ⓒ Ⓓ	87 Ⓐ Ⓑ Ⓒ Ⓓ
13 Ⓐ Ⓑ Ⓒ Ⓓ	88 Ⓐ Ⓑ Ⓒ Ⓓ
14 Ⓐ Ⓑ Ⓒ Ⓓ	89 Ⓐ Ⓑ Ⓒ Ⓓ
15 Ⓐ Ⓑ Ⓒ Ⓓ	90 Ⓐ Ⓑ Ⓒ Ⓓ
16 Ⓐ Ⓑ Ⓒ Ⓓ	
17 Ⓐ Ⓑ Ⓒ Ⓓ	
18 Ⓐ Ⓑ Ⓒ Ⓓ	
19 Ⓐ Ⓑ Ⓒ Ⓓ	
20 Ⓐ Ⓑ Ⓒ Ⓓ	
21 Ⓐ Ⓑ Ⓒ Ⓓ	
22 Ⓐ Ⓑ Ⓒ Ⓓ	
23 Ⓐ Ⓑ Ⓒ Ⓓ	
24 Ⓐ Ⓑ Ⓒ Ⓓ	
25 Ⓐ Ⓑ Ⓒ Ⓓ	
26 Ⓐ Ⓑ Ⓒ Ⓓ	
27 Ⓐ Ⓑ Ⓒ Ⓓ	
28 Ⓐ Ⓑ Ⓒ Ⓓ	
29 Ⓐ Ⓑ Ⓒ Ⓓ	
30 Ⓐ Ⓑ Ⓒ Ⓓ	

Section I—Part A

Number of Questions	Time	Use of Calculator
30	60 Minutes	No

Directions:

Tear out the answer sheet provided on the previous page and mark your answers on it. All questions are given equal weight. Points are *not* deducted for incorrect answers, and no points are given to unanswered questions. Unless otherwise indicated, the domain of a function f is the set of all real numbers. The use of a calculator is *not* permitted in this part of the exam.

1. 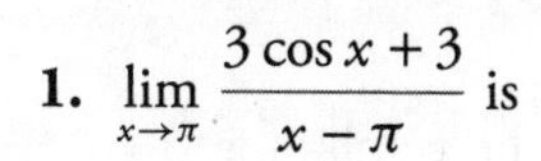
$\lim_{x\to\pi} \dfrac{3\cos x + 3}{x - \pi}$ is

(A) -3 (B) $-\dfrac{1}{3}$

(C) 0 (D) 3

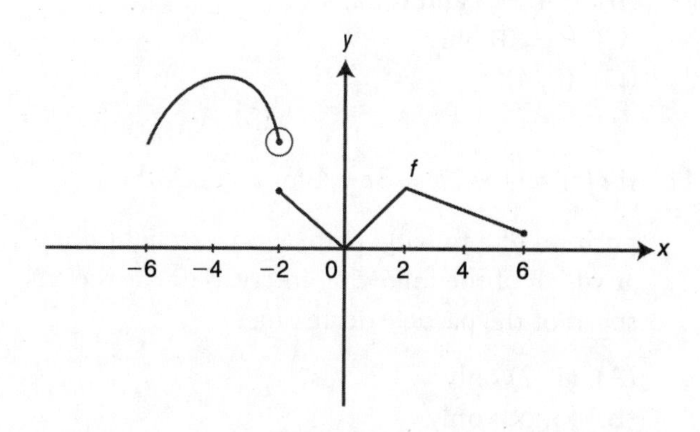

2. The figure above shows the graph of a piecewise function. The graph has a horizontal tangent at $x = -4$. What are all the values of x, such that $-6 < x < 6$, at which f is continuous but not differentiable?

(A) $x = -2$
(B) $x = -4$ and $x = 0$
(C) $x = -2$ and $x = 0$
(D) $x =$ 0, and $x = 2$

3. 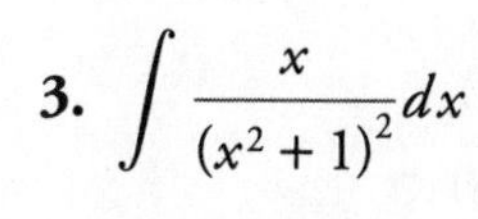
$\displaystyle\int \frac{x}{(x^2+1)^2}dx$

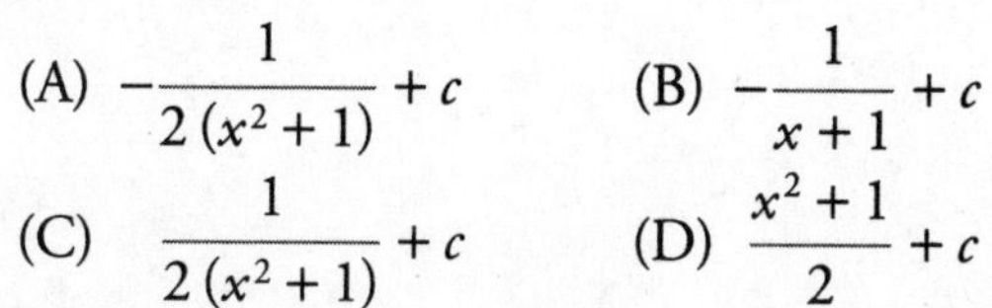
(A) $-\dfrac{1}{2(x^2+1)} + c$ (B) $-\dfrac{1}{x+1} + c$

(C) $\dfrac{1}{2(x^2+1)} + c$ (D) $\dfrac{x^2+1}{2} + c$

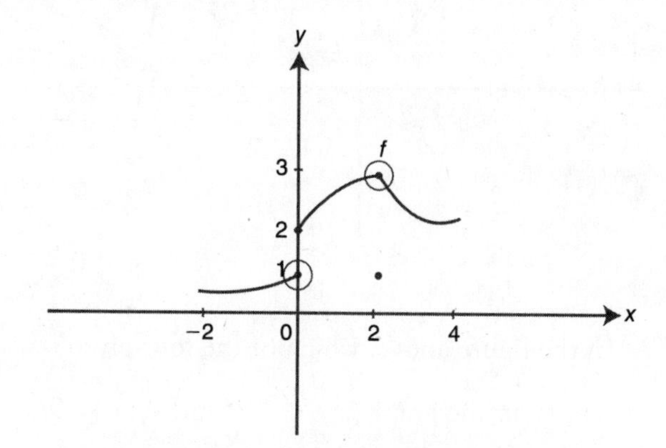

4. The diagram above shows the graph of a function f for $-2 \le x \le 4$. Which of the following statements is/are true?

I. $\lim_{t\to 2} f(x)$ exists.

II. $\lim_{x\to 0^-} f(x)$ exists.

III. $\lim_{x\to 0} f(x)$ exists.

(A) I only
(B) II only
(C) I and II only
(D) I, II, and III

5. If $y = 10\log_2(x^4+1)$, $\dfrac{dy}{dx} =$

(A) $\dfrac{10}{x^4+1}$

(B) $\dfrac{10}{\ln(2)\cdot(x^4+1)}$

(C) $\dfrac{40x^3}{\ln(2)\cdot(x^4+1)}$

(D) $\dfrac{40\ln(2)\cdot x^3}{x^4+1}$

GO ON TO THE NEXT PAGE

6. $\int x(3^{x^2-4})dx =$

(A) $\frac{3^{x^2-4}}{2\ln 3} + c$ (B) $\frac{\ln 3(3^{x^2-4})}{2} + c$

(C) $\frac{3^{x^2-3}}{2} + c$ (D) $3^{x^2-3} + c$

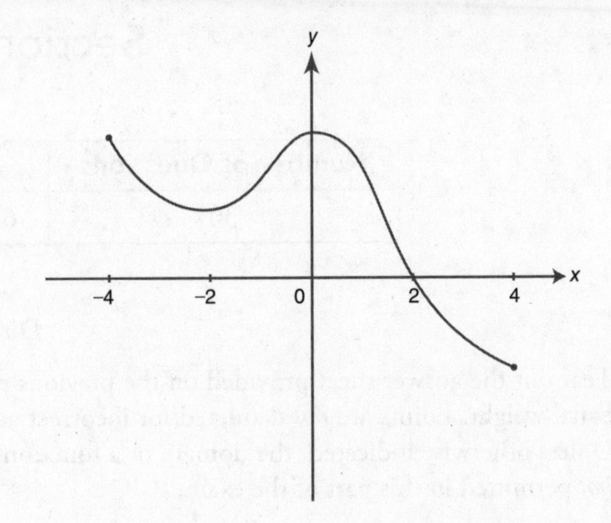

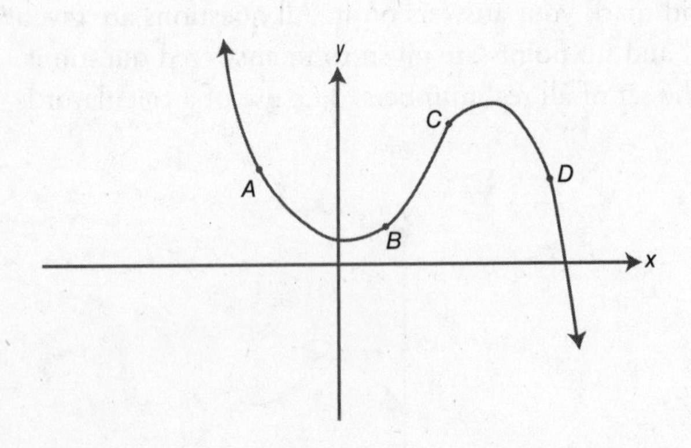

7. In the figure above, which of the four given points on the graph is $\frac{dy}{dx} > 0$ and $\frac{d^2y}{dx^2} < 0$?

(A) A (B) B
(C) C (D) D

x	$h(x)$	$k(x)$	$h'(x)$	$k'(x)$
4	6	4	1	3
8	12	8	2	5

8. The table above shows the values of the functions h and k and their derivatives at $x = 4$ and $x = 8$. What is the value of $\left.\frac{d}{dx}h(k(x))\right|_{x=4}$?

(A) 3 (B) 4
(C) 12 (D) 24

9. If $f(x) = \int_{x^2}^{1} \frac{1}{2+t^2}dt$, then $f'(2)=$

(A) $-\frac{2}{9}$ (B) $\frac{1}{18}$

(C) $\frac{1}{6}$ (D) $\frac{2}{9}$

10. The figure above shows the graph of a continuous function f for $-4 \le x \le 4$. On what intervals is $f'(x) > 0$ for all values of x?

(A) $(-4, 2)$ only
(B) $(-4, -2)$ and $(0, 2)$
(C) $(-2, 0)$ only
(D) $(0, 4)$

11. If $s(t) = \frac{t^3}{3} - 2t^2 + 3t + 4$ for $t \ge 0$ is the position of a moving particle on a straight line, at which of the following intervals of t is the speed of the particle decreasing?

(A) $(1, 2)$ only
(B) $(3, \infty)$ only
(C) $(0, 1)$ and $(2, 3)$
(D) $(1, 2)$ and $(3, \infty)$

12. $\lim_{x \to -\infty} \frac{\sqrt{4x^2+3}}{2x-1}$ is

(A) -2 (B) -1
(C) 1 (D) nonexistent

13. $\int_{-1}^{1} \frac{2}{1+x^2}dx =$

(A) $-\pi$ (B) 0

(C) $\frac{\pi}{2}$ (D) π

GO ON TO THE NEXT PAGE

14. If $f(x) = 8\sin^3(x)$, $f'\left(\frac{\pi}{3}\right)$ is

(A) 6 (B) 9
(C) $6\sqrt{3}$ (D) 18

15. Given the function $f(x) = x^3 - 3x^2 + 3x - 1$, which of the following statements is true?

(A) The graph of f has a relative minimum at $x = 1$.
(B) The graph of f has a relative maximum at $x = 1$.
(C) The graph of f has a point of inflection at $x = 1$.
(D) The graph of f has a relative minimum at $x = 0$.

16. If $\int_{8}^{-8} f(x)dx = -10$ and $\int_{-8}^{-2} f(x)dx = 4$, what is the value of $\int_{-2}^{8} f(x)dx$?

(A) −14 (B) −6
(C) 6 (D) 8

17. What is the slope of the tangent line to the curve $2x^2y - 3xy^2 - 2 = 0$ at the point (2, 1)?

(A) $-\frac{5}{4}$ (B) $-\frac{2}{5}$
(C) $\frac{2}{3}$ (D) $\frac{5}{4}$

18. If f is a piecewise function defined as $f(x) = \begin{cases} x^2 & \text{for } x \le 2 \\ 4x & \text{for } x > 2 \end{cases}$, which of the following statements is/are true?

I. $\lim_{h \to 0^-} \frac{f(2+h) - f(2)}{h} = 4$

II. $\lim_{h \to 0^+} \frac{f(2+h) - f(2)}{h} = 4$

III. $f'(2) = 4$

(A) None
(B) I only
(C) II only
(D) I, II, and III

19. Air is pumped into a spherical balloon at the rate of 10 cubic centimeters per second. How fast is the diameter, in centimeters per second, increasing when the radius is 5 cm? (The volume of a sphere is $V = \frac{4}{3}\pi r^3$.)

(A) $\frac{1}{10\pi}$ (B) $\frac{1}{5\pi}$
(C) $\frac{1}{\pi}$ (D) 5π

20. $\lim_{h \to 0} \frac{\sin\left(\frac{\pi}{3} + h\right) - \sin\left(\frac{\pi}{3}\right)}{h}$ is

(A) 0 (B) $\frac{1}{2}$
(C) 1 (D) $\frac{\sqrt{3}}{2}$

21. The area of the region bounded by the curves of $y = -x^2$ and $y = x$ is

(A) $-\frac{5}{6}$ (B) $\frac{1}{6}$
(C) $\frac{5}{6}$ (D) 6

22. $\int \frac{4}{e^{2x}} dx =$

(A) $-\frac{4}{e^{2x}} + c$ (B) $-\frac{2}{e^{2x}} + c$
(C) $\frac{2}{e^{2x}} + c$ (D) $4e^{2x} + c$

23. If $f(x) = \frac{2x}{(2x-1)^2}$, then $f'(x) =$

(A) $\frac{2}{2x-1}$ (B) $\frac{1}{2(2x+1)}$
(C) $\frac{2x+1}{(2x-1)^3}$ (D) $\frac{-2(2x+1)}{(2x-1)^3}$

GO ON TO THE NEXT PAGE

24. Let f and g be differentiable functions such that $f^{-1}(x) = g(x)$ for all x. The table below shows selected values of $f(x)$ and $f'(x)$. What is the value of $g'(2)$?

x	2	4
$f(x)$	6	2
$f'(x)$	−5	−3

(A) −5 (B) −3

(C) $-\frac{1}{3}$ (D) 2

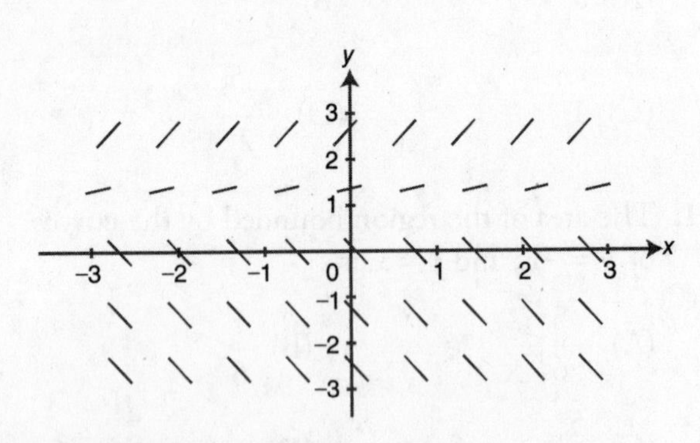

25. The figure above shows the slope field for which of the following differential equations?

(A) $\frac{dy}{dx} = x - 1$ (B) $\frac{dy}{dx} = y - 1$

(C) $\frac{dy}{dx} = y + 1$ (D) $\frac{dy}{dx} = xy$

26. Let f be a continuous and twice differentiable function and $f''(x) > 0$ for all x in the interval [4, 5]. Some of the values of f are shown in the table below.

x	4.5	4.6	4.7	4.8
$f(x)$	10.0	10.2	10.8	11.9

Which of the following is true about $f'(4.6)$?

(A) $f'(4.6) < 2$

(B) $f'(4.6) > 6$

(C) $0.2 < f'(4.6) < 0.6$

(D) $2 < f'(4.6) < 6$

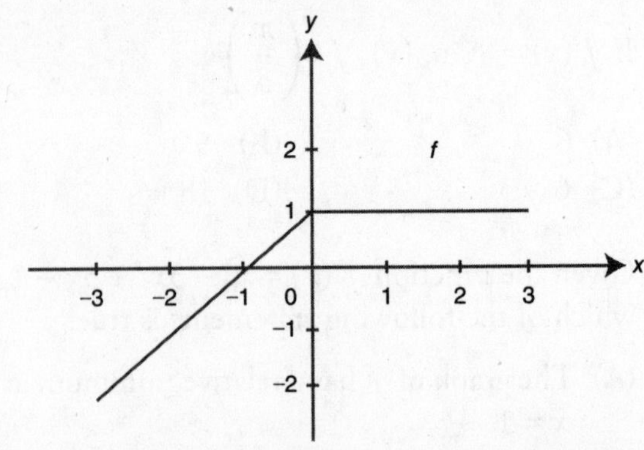

27. The graph of a function f is shown in the figure above. The graph of f consists of two line segments, and f is continuous on the interval [−3, 3]. If $g(x) = \int_0^x f(t)dt$, then $g(-3)$ is

(A) $-\frac{5}{2}$ (B) 1

(C) $\frac{3}{2}$ (D) $\frac{5}{2}$

28. The solution to the differential equation $\frac{dy}{dx} = 3x^2y$ where $y(0) = e$ is

(A) $y = \frac{1}{3x^2}$ (B) $y = e^{3x}$

(C) $y = -e^{(3x+1)}$ (D) $y = e^{(x^3+1)}$

29. The graph of a continuous function $y = f(x)$ passes through the point $(e, 4)$ and $\frac{dy}{dx} = \frac{3}{x}$. Which of the following is the function $f(x)$?

(A) $y = 3\ln x + e$ (B) $y = 3\ln x + 1$

(C) $y = \ln x + 4$ (D) $y = e^x + 4$

30. The definite integral $\int_0^3 (x^2 + 1)dx$ is equivalent to which of the following limits?

(A) $\lim_{x\to\infty} \sum_{k=1}^{n} \left(\frac{1}{n}\right)\left(\frac{k}{n}\right)^2$

(B) $\lim_{n\to\infty} \sum_{k=1}^{n} \frac{3}{n}\left(\left(\frac{k}{n}\right)^2 + 1\right)$

(C) $\lim_{n\to\infty} \sum_{k=1}^{n} \frac{3}{n}\left(\frac{3k}{n}\right)^2$

(D) $\lim_{n\to\infty} \sum_{k=1}^{n} \frac{3}{n}\left(\left(\frac{3k}{n}\right)^2 + 1\right)$

STOP. AP Calculus AB Practice Exam 1, Section I—Part A.

Section I—Part B

Number of Questions	Time	Use of Calculator
15	45 Minutes	Yes

Directions:

Continue using the same answer sheet. *Please note that the questions begin with number 76.* This is not an error; it was done to be consistent with the numbering system of the actual AP Calculus AB exam. All questions are given equal weight. Points are *not* deducted for incorrect answers, and no points are given to unanswered questions. Unless otherwise indicated, the domain of a function f is the set of all real numbers. If the exact numerical value does not appear among the given choices, select the best approximate value. The use of a calculator is *permitted* in this part of the exam.

76. Given that $f(x) = \begin{cases} x^3 + x + a & \text{for } x \le 1 \\ 2bx - 1 & \text{for } x > 1 \end{cases}$, if f is differentiable at $x = 1$, what is the value of $a + 2b$?

(A) 2
(B) 3
(C) 4
(D) 5

77. If $g'(x) = \cos\left(\pi x^2 + 1\right)$ and $g(1) = \frac{1}{2}$, then $g(2) =$

(A) −0.681
(B) −0.319
(C) 0.181
(D) 0.681

78. The point (−2, 5) is on the graph of a differential function f, and $f'(-2) = 3$. What is the estimate for $f(-1.9)$, using local linear approximation for f at $x = -2$?

(A) −16.7
(B) −6.7
(C) −4.7
(D) 5.3

79. If a function f has its derivative $f'(x)$ given as $f'(x) = -3e^{-x}\cos x + 2$ for $-2 < x < 2$, on what interval(s) is f increasing?

(A) (−1.0407, 0.345)
(B) (0, 0.345)
(C) (−2, 0)
(D) (−2, − 1.407) and (0.345, 2)

80. A particle moving in a straight line has a velocity of $v(t) = 2^{-t}\cos(2t)$ feet per minute. What is the total distance in feet traveled by the particle when $0 \le t \le 3$ minutes?

(A) 0.096
(B) 0.121
(C) 0.732
(D) 0.813

81. How many points of inflection does the graph of $y = 5e^{\left(\frac{-x}{2}\right)}\sin\left(\frac{3x}{2}\right)$ have on the intervals $(0, 2\pi)$?

(A) 0
(B) 1
(C) 2
(D) 3

82. The table below shows selected values of a continuous function f on the closed interval [1,16].

x	1	6	11	16
$f(x)$	1.1	2.1	2.6	2.9

Using three left endpoint rectangles of equal length (*LRAM*), what is the approximate value of $\int_1^{16} f(x)dx$?

(A) 16.5
(B) 29
(C) 33.5
(D) 38

GO ON TO THE NEXT PAGE

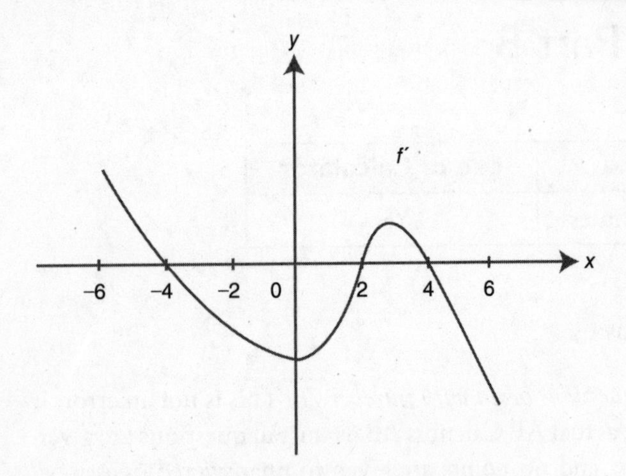

83. The figure above shows the graph of the derivative of a function f. Which of the following could be the graph of f?

(A)

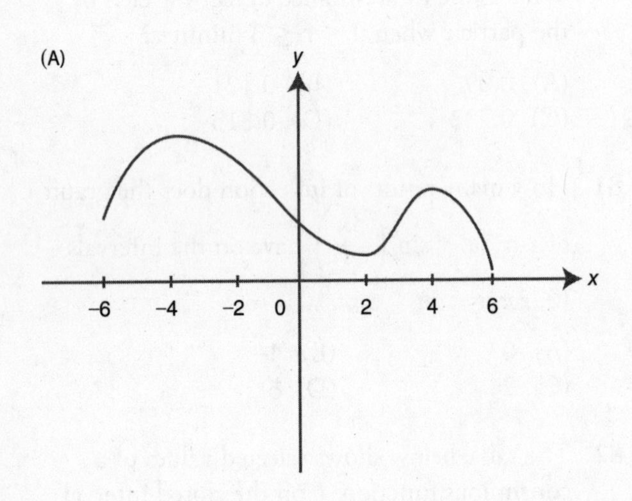

(B)

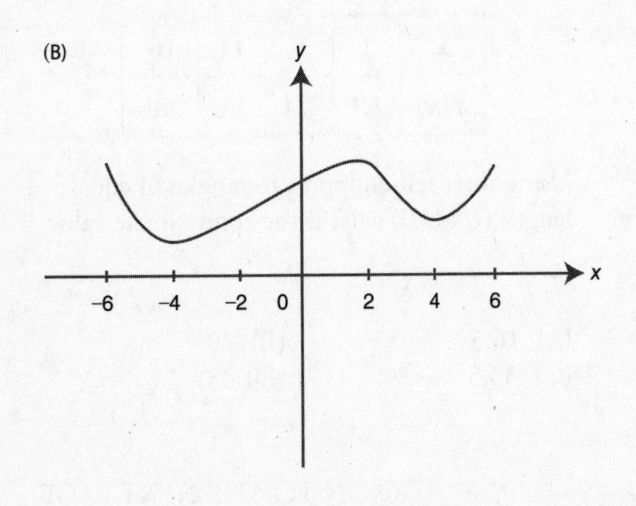

(C)

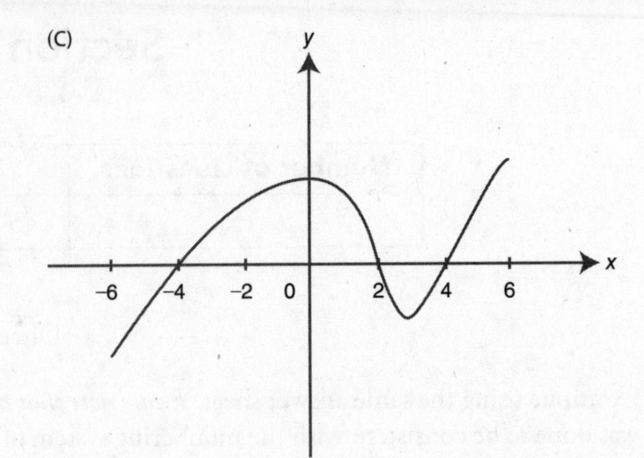

(D)

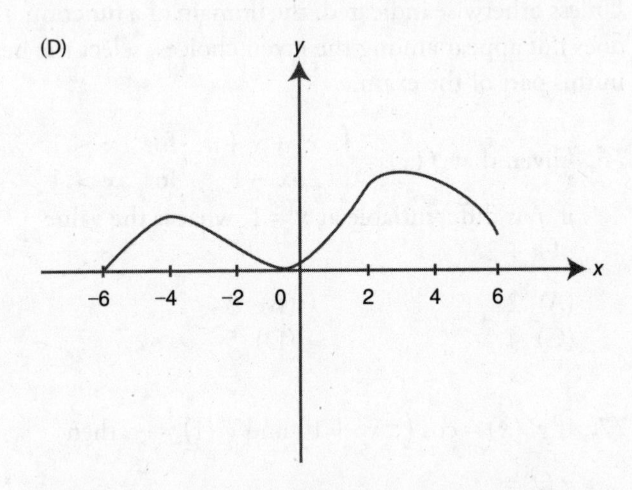

84. A wire 40 centimeters long is being cut into two pieces to make a circle and a square. If the total area of the circle and the square is a minimum, how much wire, in centimeters, should be used for the circle?

(A) 17.596 (B) 22.404
(C) 40 (D) 56.010

85. Given that $f(x) = \arccos(2x)$, what is the slope of the tangent to the graph of f at $x = 0$?

(A) −2 (B) 0
(C) 1 (D) 2

86. A solid is generated by revolving about the x-axis, the region bounded by the curves $y = x^3 - 2x + 1$ and $y = x + 2$ in the second quadrant. The volume of the solid is

(A) 2.241 (B) 7.039
(C) 10.348 (D) 16.279

GO ON TO THE NEXT PAGE

87. A function f is given as $f(x) = x^3 + 1$ for $1 \leq x \leq 3$. If $1 \leq c \leq 3$ and $f'(c)$ is equal to the average rate of change of f on the interval $[1, 3]$, what is the value of c?

(A) $\sqrt{\dfrac{13}{3}}$ (B) $\dfrac{3\sqrt{2}}{2}$
(C) 6.5 (D) 13

88. The velocity of a particle moving on a straight line is given as $v(t) = \cos(t)(e^{\sin(t)})$ for $t \geq 0$. At $t = \dfrac{\pi}{4}$, what is the acceleration of the particle?

(A) −2.718 (B) −1.346
(C) −0.420 (D) 1.434

89. The region, bounded by the graph of $y = \cos(x)$, the y-axis, and the x-axis, for $0 \leq x \leq \dfrac{\pi}{2}$, is divided by the line $x = k$ into two regions of equal area. What is the value of k?

(A) $\dfrac{\pi}{6}$ (B) $\dfrac{\pi}{4}$
(C) $\dfrac{\pi}{3}$ (D) $\dfrac{\pi}{2}$

90. In a given petri dish, the number of bacteria grows at approximately the rate of $r(t) = 2\sqrt{t} \cdot e^{1.02t}$ bacteria per day, where t is the number of days since the culture in the petri dish began. At day two, there are 150 bacteria. Which of the following is the best approximation for the number of bacteria at day five?

(A) 623 (B) 638
(C) 773 (D) 788

STOP. End of AP Calculus AB Practice Exam, Section I—Part B.

Section II—Part A

Number of Questions	Time	Use of Calculator
2	30 Minutes	Yes

Directions:

Show all your work. You may *not* receive any credit for correct answers without supporting work. You may use an approved calculator to help solve a problem. However, you must clearly indicate the setup of your solution using mathematical notations and *not* calculator syntax. Calculators may be used to find the derivative of a function at a point, compute the numerical value of a definite integral, or solve an equation. Unless otherwise indicated, you may assume the following: (a) the numeric or algebraic answers need not be simplified; (b) your answer, if expressed in approximation, should be correct to three places after the decimal point; and (c) the domain of a function f is the set of all real numbers.

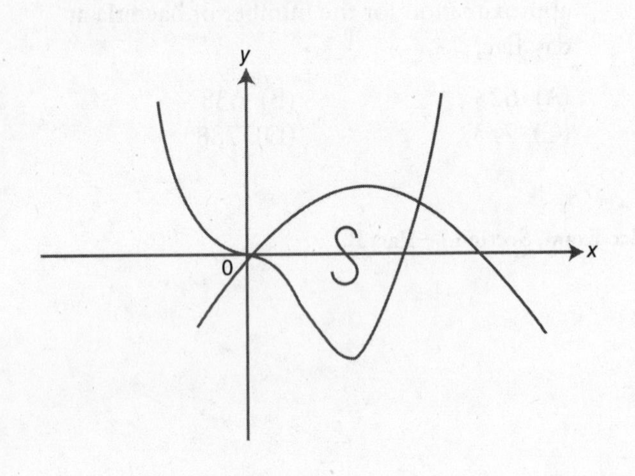

1. As shown in the figure above, S is the region enclosed by the graphs of $y = 2x^4 - 5x^3$ and $y = 4x - x^2$.

(A) Find the area of the region S.
(B) Find the volume of the solid generated when the region S is revolved about the line $y = 4$.
(C) If region S is the base of a solid, the height of the solid for all points in S at a distance x from the y-axis is given as $h(x) = 0.5 + 2x$. Find the volume of the solid.

2. A particle is moving along the x-axis. For $0 \leq t \leq 10$, the acceleration of the particle is given by $a(t) = \dfrac{\sin(\ln(x+1))}{x+1}$. At $t = 0$, the velocity of the particle is $v(0) = -1.1$.

(A) Find the velocity of the particle at $t = 4$.
(B) For $0 \leq t \leq 10$, at what time interval is the speed of the particle increasing?
(C) For $0 \leq t \leq 10$, what is the average velocity of the particle?
(D) For $0 \leq t \leq 10$, find the total distance traveled by the particle.

STOP. End of AP Calculus AB Practice Exam, Section II—Part A.

Section II—Part B

Number of Questions	Time	Use of Calculator
4	60 Minutes	No

Directions:

The use of a calculator is *not* permitted in this part of the exam. You should *show all work*. You may *not* receive any credit for correct answers without supporting work. Unless otherwise indicated, the numeric or algebraic answers need not be simplified, and the domain of a function f is the set of all real numbers. When you have finished this part of the exam, you may return to the problems in Part A of Section II and continue to work on them. However, you may not use a calculator.

3. Given the differential equation $\frac{dy}{dx} - 2x = xy$.

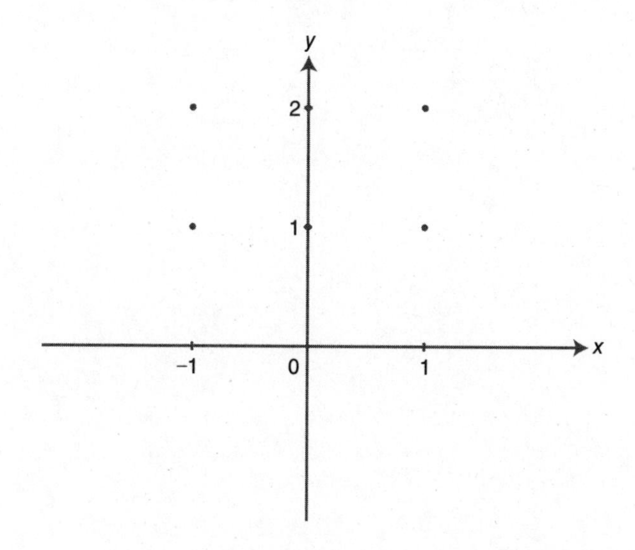

(A) On the axes provided in the figure above, sketch a slope field at the six indicated points.

(B) Let $y = f(x)$ be the particular solution to the differential equation with the initial condition $f(0) = 1$. Write an equation of the tangent line to the solution curve at the point (0, 1).

(C) Using the equation of the tangent line in part (b), find the approximate value of $f(0.02)$.

(D) Find $y = f(x)$ with $f(0) = 1$.

4. Let f be a continuous function on the interval $[-2, 6]$, and f is twice differentiable with $f''(x) < 0$ on the open interval $(-2, 6)$.

Selected values of f are shown below.

x	−2	−1	0	2	6
$f(x)$	1	4	6	7	8

(A) Using trapezoidal approximation with four trapezoids, find the approximate value of $\int_{-2}^{6} f(x)dx$.

(B) Using the result in part (A), find the average value of f on $[-2, 6]$.

(C) Find the average rate of change of f over the interval $[-2, 6]$.

(D) Examine $f'(0)$ and $f'(2)$, and determine if $f'(0)$ is greater than, less than, or equal to $f'(2)$. Explain.

GO ON TO THE NEXT PAGE

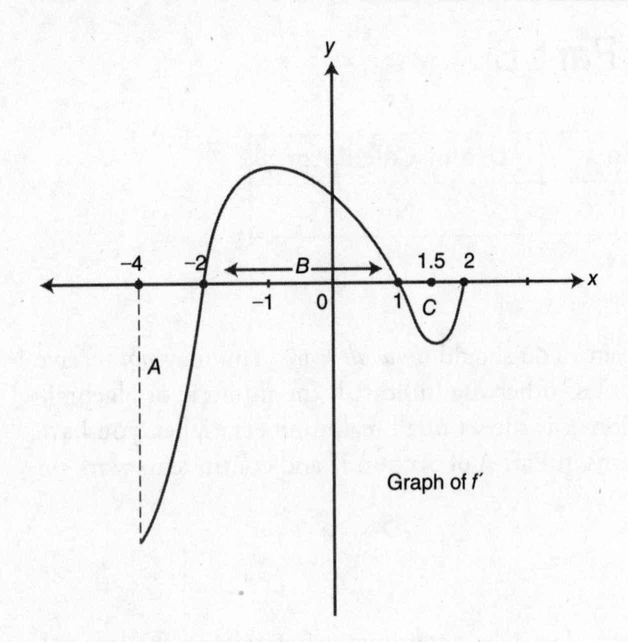

5. A function f is continuous and twice differentiable for all values of x. The figure above shows the graph of f', the derivative of function f on the closed interval $[-4, 2]$. The graph of f' has horizontal tangents at $x = -1$ and $x = 1.5$. The areas of regions A, B, and C are 20, 10, and 6, respectively, and $f(2) = 3$.

(A) Find all x-values on $(-4, 2)$ such that the function f has a local minimum. Justify your answer.

(B) Find all x-values on $(-4, 2)$ such that the graph of f has a point of inflection. Justify your answer.

(C) Evaluate $\lim_{x \to 1} \frac{2f(x)}{x+1}$. Explain your reasoning.

(D) Evaluate $\lim_{x \to -2} \frac{f(x)+1}{x+2}$. Explain your reasoning.

6. The equation of a curve is given as $14y - y^2 = x^2 + 24$.

(A) Find $\frac{dy}{dx}$ in terms of x and y.

(B) Write an equation of the line tangent to the curve at the point $(-3, 11)$.

(C) Write an equation of each vertical tangent to the curve.

(D) Evaluate $\frac{d^2y}{dx^2}$ at the point $(0, 2)$.

STOP. End of AP Calculus AB Practice Exam, Section II—Part B.

Answers to AB Practice Exam 1–Section I

Part A

1. C
2. D
3. A
4. C
5. C
6. A
7. C
8. A
9. A
10. C
11. C
12. B
13. D
14. B
15. C
16. C
17. D
18. B
19. B
20. B
21. B
22. B
23. D
24. C
25. B
26. D
27. C
28. D
29. B
30. D

Part B

76. D
77. D
78. D
79. D
80. D
81. D
82. B
83. A
84. A
85. A
86. B
87. A
88. C
89. A
90. C

Answers to AB Practice Exam 1—Section II

Part A

1. (A) 17.258 (2 pts.)
(B) 172.896 π or 543.17 (4 pts.)
(C) 64.334 (3 pts.)

2. (A) −0.061 (3 pts.)
(B) $4.317 < t \leq 10$ (2 pts.)
(C) −0.018 (2 pts.)
(D) 4.447 (2 pts.)

Part B

3. (A) See solution. (2 pts.)
(B) $y = 1$ (2 pts.)
(C) 1 (2 pts.)
(D) $y = 3e^{\left(\frac{x^2}{2}\right)} - 2$ (3 pts.)

4. (A) 50.5 (3 pts.)
(B) 6.313 (2 pts.)
(C) 0.875 (2 pts.)
(D) $f'(0) > f'(2)$ (2 pts.)

5. (A) $x = -2$ (2 pts.)
(B) $x = -1$ and $x = 1.5$ (2 pts.)
(C) 9 (2 pts.)
(D) 0 (3 pts.)

6. (A) $\frac{dy}{dx} = \frac{x}{7 - y}$ (3 pts.)
(B) $y = \frac{3}{4}x + \frac{53}{4}$ (2 pts.)
(C) $x = -5$ and $x = 5$ (2 pts.)
(D) $\frac{1}{5}$ (2 pts.)

Solutions for AP Calculus AB Practice Exam 1—Section I

Section I—Part A

1. The correct answer is (C).

   Note that $\lim_{x\to\pi}(3\cos x + 3) = 3\cos\pi + 3 = 0$, and $\lim_{x\to\pi}(x - \pi) = 0$. Therefore, the $\lim_{x\to\pi}\frac{3\cos x + 3}{x - \pi}$ is an indeterminate form of the type $\frac{0}{0}$. Applying *L'Hôpital's* Rule, you have

   $\lim_{x\to\pi}\frac{-3\sin x}{1} = \frac{-3\sin\pi}{1} = 0.$

2. The correct answer is (D).

   At $x = -4$, the graph of f has a horizontal tangent, which implies that f is differentiable and $f'(-4) = 0$.

   At $x = -2$, f is discontinuous and thus not differentiable. At each of $x = 0$ and $x = 2$, the one-sided derivatives are different, so f is not differentiable. (Note that the graphs at $x = 0$ and $x = 2$ are sometimes referred to as "corners".)

3. The correct answer is (A).

   Using the u-substitution method, let $u = x^2 + 1$, and obtain $du = 2x\,dx$, or $\frac{du}{2} = x\,dx$. Therefore,

   $$\int \frac{x}{(x^2+1)^2}dx = \frac{1}{2}\int\frac{du}{u^2} = \frac{1}{2}\int u^{-2}\,du$$

   $$= \frac{1}{2}\left(\frac{u^{-1}}{-1}\right) + c = -\frac{1}{2u} + c = -\frac{1}{2(x^2+1)} + c.$$

4. The correct answer is (C).

   Examining the graph, note that $\lim_{x\to 2^+} f(x) = \lim_{x\to 2^-} f(x) = 3$. Since the two one-sided limits are equal, $\lim_{x\to 2} f(x)$ exists. Statement I is true. Also, note that $\lim_{x\to 0^-} f(x) = 1$ and $\lim_{x\to 0^+} f(x) = 2$. Therefore, statement II is true, but statement III is false because the two one-sided limits are not the same.

5. The correct answer is (C).

   Recall the formula $\frac{d}{dx}(\log_a u) = \frac{1}{(\ln a)(u)}\cdot\frac{du}{dx}$. In this case, $y = 10\log_2(x^4+1)$, and you have $\frac{dy}{dx} = (10)\frac{1}{\ln(2)\cdot(x^4+1)}\cdot 4x^3 = \frac{40x^3}{\ln(2)\cdot(x^4+1)}$.

6. The correct answer is (A).

   Let $u = x^2 - 4$. Differentiating, you have $\frac{du}{dx} = 2x$, or $\frac{du}{2} = x\,dx$. Using u-substitution, rewrite $\int x(3^{x^2-4})dx$ as $\frac{1}{2}\int 3^u\,du$. Recall the formula $\int a^x\,dx = \frac{a^x}{\ln a} + c$. Therefore, $\frac{1}{2}\int 3^u\,du = \frac{3^u}{2\ln 3} + c$, or $\frac{3^{x^2-4}}{2\ln 3} + c$.

7. The correct answer is (C).

   If a function is increasing, then $\frac{dy}{dx} > 0$, and if a function is decreasing, $\frac{dy}{dx} < 0$. Also, if a function is concave up, $\frac{d^2y}{dx^2} > 0$, and if a function is concave down, $\frac{d^2y}{dx^2} < 0$. Note the following:

Point	$\frac{dy}{dx}$	$\frac{d^2y}{dx^2}$
A	Decreasing −	Concave up +
B	Increasing +	Concave up +
C	Increasing +	Concave down −
D	Decreasing −	Concave down −

   Thus, at point C, $\frac{dy}{dx} > 0$ and $\frac{d^2y}{dx^2} < 0$.

8. The correct answer is (A).

Note that the notation $\left.\frac{d}{dx}h(k(x))\right|_{x=4}$ is equivalent to $(h \circ k)'(4)$, the derivative of a composite function, at $x = 4$. Begin with the Chain Rule: $\frac{d}{dx}h(k(x)) = h'(k(x))k'(x)$.

Therefore, $\left.\frac{d}{dx}h\,(k(x))\right|_{x=4} = h'(k(4))k'(4) =$ $h'(4)k'(4) = (1)(3) = 3$.

9. The correct answer is (A).

Note that

$$f(x) = \int_{x^2}^{1} \frac{1}{2+t^2}dt = -\int_{1}^{x^2} \frac{1}{2+t^2}dt.$$

Applying the Fundamental Theorem of Calculus, you have

$f'(x) = \frac{-1}{2+(x^2)^2}(2x) = \frac{-2x}{2+x^4}$. Thus,

$f'(2) = \frac{-2(2)}{2+2^4} = \frac{-4}{18} = \frac{-2}{9}$.

10. The correct answer is (C).

Remember that if f is increasing, then $f'(x) > 0$, and that if f is decreasing, then $f'(x) < 0$. Examining the graph of f, you see that $f(x)$ is increasing on the interval $(-2, 0)$. Therefore, $f'(x) > 0$ on the interval $(-2, 0)$.

11. The correct answer is (C).

The speed of a moving particle decreases when its velocity and acceleration have opposite signs, e.g., $v(t) > 0$ and $a(t) < 0$, or $v(t) < 0$ and $a(t) > 0$. The velocity and acceleration functions are $v(t) = t^2 - 4t + 3$ and $a(t) = 2t - 4$. Set $v(t) = 0$, and obtain $t^2 - 4t + 3 = 0$ or $(t - 3)(t - 1) = 0$, which yields $t = 1$ and $t = 3$. Setting $a(t) = 0$, you have $2t - 4 = 0$, or $t = 2$.

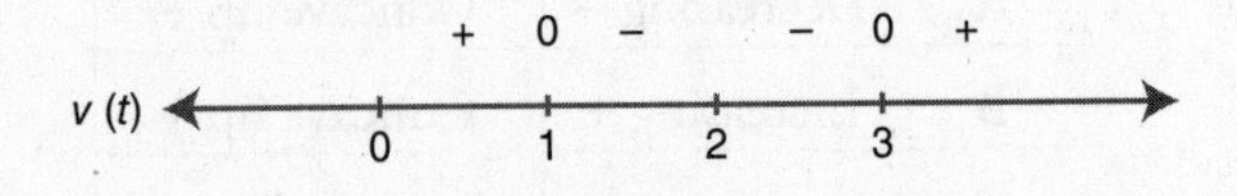

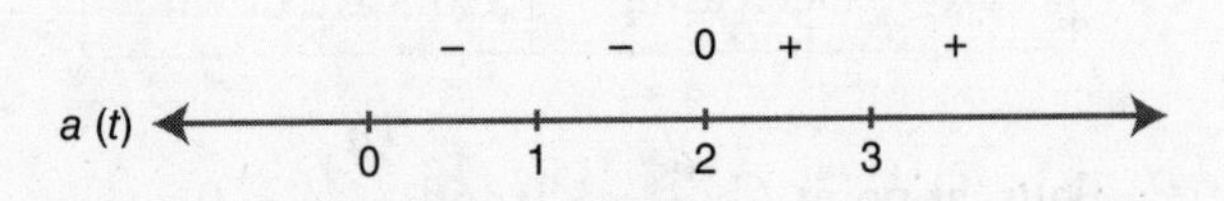

Note that $v(t)$ and $a(t)$ have opposite signs on $(0, 1)$ and $(2, 3)$. Thus, the speed of the particle decreases on $(0, 1)$ and $(2, 3)$.

12. The correct answer is (B).

Note that as x approaches $-\infty$, $\sqrt{x^2} = -x$. Therefore,

$$\lim_{x\to-\infty} \frac{\sqrt{4x^2+3}}{2x-1} = \lim_{x\to-\infty} \frac{\frac{\sqrt{4x^2+3}}{\sqrt{x^2}}}{\frac{2x-1}{\sqrt{x^2}}}$$

$$= \lim_{x\to-\infty} \frac{\frac{\sqrt{4x^2+3}}{\sqrt{x^2}}}{\frac{2x-1}{-x}} = \lim_{x\to-\infty} \frac{\sqrt{4+\frac{3}{x^2}}}{-2+\frac{1}{x}}.$$

Also, $\lim_{x\to-\infty} \frac{3}{x^2} = 0$ and $\lim_{x\to-\infty} \frac{1}{x} = 0$.

Thus, $\lim_{x\to-\infty} \frac{\sqrt{4+\frac{3}{x^2}}}{-2+\frac{1}{x}} = \frac{\sqrt{4}}{-2} = -1$.

13. The correct answer is (D).

Note that the formula

$\int \frac{1}{1+x^2}dx = \tan^{-1}(x) + c$. Therefore,

$$\int_{-1}^{1} \frac{2}{1+x^2}dx = 2\tan^{-1}(x)\Big]_{-1}^{1} =$$

$$2\tan^{-1}(1) - 2\tan^{-1}(-1) = 2\left(\frac{\pi}{4}\right) - 2\left(-\frac{\pi}{4}\right) = \pi.$$

14. The correct answer is (B).

Using the Chain Rule, you have $f' = 24\sin^2(x)\cos x$. Therefore,

$$f'\left(\frac{\pi}{3}\right) = 24\sin^2\left(\frac{\pi}{3}\right)\cos\left(\frac{\pi}{3}\right) =$$

$$24\left(\frac{\sqrt{3}}{2}\right)^2\left(\frac{1}{2}\right) = 9.$$

15. The correct answer is (C).

The first and second derivatives of f are $f'(x) = 3x^2 - 6x + 3$ and $f''(x) = 6x - 6$. Set $f'(x) = 0$, and you have $3x^2 - 6x + 3 = 0$. Solve for x and obtain $3(x^2 - 2x + 1) = 0$, or $3(x - 1)(x - 1) = 0$, which yields $x = 1$. Setting $f''(x) = 0$, you have $6x - 6 = 0$, or $x = 1$.

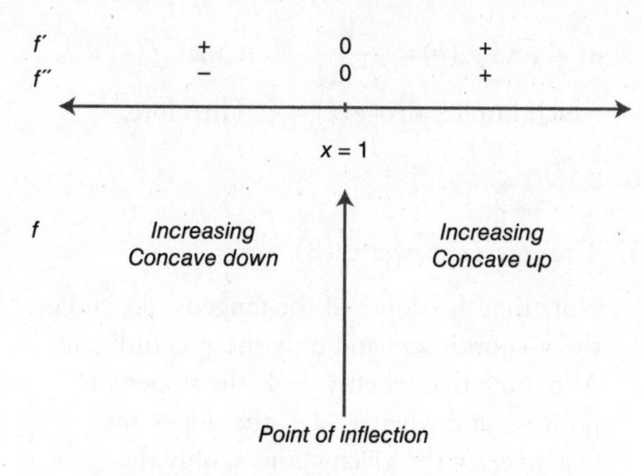

Note that the graph of f changes concavity from down to up. Since f is a differentiable function, it has a point of inflection at $x = 1$.

16. The correct answer is (C).

Since $\int_{8}^{-8} f(x)dx = -10$, $\int_{-8}^{8} f(x)dx = 10$.

Also, $\int_{-8}^{-2} f(x)dx + \int_{-2}^{8} f(x)dx = \int_{-8}^{8} f(x)dx$, so $4 + \int_{-2}^{8} f(x)dx = 10$ or $\int_{-2}^{8} f(x)dx = 6$.

17. The correct answer is (D).

Using implicit differentiation, you have $4xy + 2x^2\frac{dy}{dx} - 3y^2 - 3x\left(2y\frac{dy}{dx}\right) = 0$. Substitute $x = 2$ and $y = 1$, and obtain $8 + 8\frac{dy}{dx} - 3 - 12\frac{dy}{dx} = 0$, or $-4\frac{dy}{dx} = -5$, or $\frac{dy}{dx} = \frac{5}{4}$.

18. The correct answer is (B).

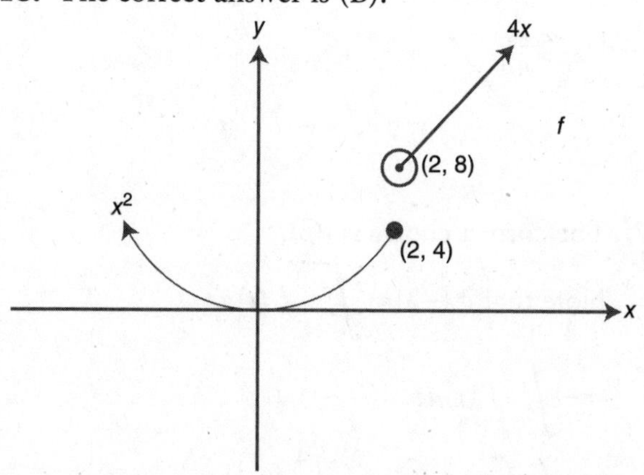

Since $f(x)$ is discontinuous at $x = 2$, $f'(2)$ does not exist. Statement III is false. The limit in statement II represents the right-hand derivative $(f^+)'(x)$, and $(f^+)'(2)$ does not exist because the point (2, 8) is not part of the function. Statement II is false. The limit in statement I represents the left-hand derivative $(f^-)'(x)$, and for $x \leq 2$, $(f^-)'(x) = 2x$. Thus, $(f^-)'(2) = 4$. Statement I is true.

19. The correct answer is (B).

The volume of a sphere is $V = \frac{4}{3}\pi r^3$, so $\frac{dV}{dt} = 4\pi r^2\frac{dr}{dt}$, or $10 = 4\pi r^2\frac{dr}{dt}$, which yields $\frac{dr}{dt} = \frac{5}{2\pi r^2}$. The diameter D is twice the radius; $D = 2r$, and $\frac{dD}{dt} = 2\frac{dr}{dt}$. Therefore, $\frac{dD}{dt} = 2\left(\frac{5}{2\pi r^2}\right) = \frac{5}{\pi r^2}$, and $\left.\frac{dD}{dt}\right|_{r=5} = \left(\frac{5}{\pi\,(5^2)}\right) = \frac{1}{5\pi}$ centimeters per second.

20. The correct answer is (B).

Note that the definition of the derivative $f'(x) = \lim_{h \to 0}\frac{f(x+h) - f(x)}{h}$. Therefore, the $\lim_{h \to 0}\frac{\sin\left(\frac{\pi}{3} + h\right) - \sin\left(\frac{\pi}{3}\right)}{h}$ is equivalent to $\left.\frac{d}{dx}(\sin x)\right|_{x=\frac{\pi}{3}}$ or $\cos\left(\frac{\pi}{3}\right) = \frac{1}{2}$. Note that there are other approaches to solving this problem, e.g., applying *L'Hôpital's* Rule or expanding $\sin\left(\frac{\pi}{3} + h\right)$.

21. The correct answer is (B).

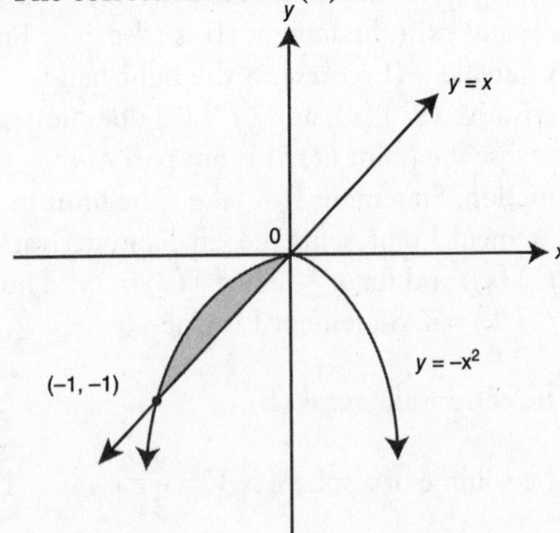

Find the intersection points by setting $x = -x^2$. Rewrite $x = -x^2$ as $x^2 + x = 0$ or $x(x+1) = 0$ and obtain $x = 0$ and $x = -1$. The area of the region bounded by the curves is

$$\int_{-1}^{0} (-x^2 - x)\,dx = \left[\frac{-x^3}{3} - \frac{x^2}{2}\right]_{-1}^{0} =$$

$$0 - \left[\frac{-(-1)^3}{3} - \frac{(-1)^2}{2}\right] = \frac{1}{6}.$$

22. The correct answer is (B).

Rewrite $\int \frac{4}{e^{2x}}\,dx$ as $4\int e^{-2x}\,dx$. Let $u = -2x$ and obtain $du = -2dx$, or $dx = \frac{-du}{2}$.

Therefore, $4\int e^{-2x}\,dx = 4\int e^{u}\left(-\frac{du}{2}\right)$

$= -2\int e^{u}\,du = -2e^{u} + c = -2e^{-2x} + c = \frac{-2}{e^{2x}} + c.$

23. The correct answer is (D).

Applying the quotient rule for derivatives, you have $f'(x) = \frac{(2)(2x-1)^2 - 2(2x-1)(2)(2x)}{(2x-1)^4}$

$= \frac{2(2x-1)[((2x-1) - 4x)]}{(2x-1)^4}$

$= \frac{2(2x-1)(-2x-1)}{(2x-1)^4}$

$= \frac{-2(2x-1)(2x+1)}{(2x-1)^4} = \frac{-2(2x+1)}{(2x-1)^3}.$

24. The correct answer is (C).

Since f and g are inverse functions, if (a, b) is on the graph of f, then (b, a) is on the graph of g, and $g'(b) = \frac{1}{f'(a)}$. Note that $f(4) = 2$, which implies that $g(2) = 4$. Therefore, $g'(2) = \frac{1}{f'(4)} = -\frac{1}{3}$.

25. The correct answer is (B).

Note that the slopes of the tangents depend on the y-coordinates and only the y-coordinates. Also, note that when $y > 1$, the slopes are positive, and when $y < 1$, the slopes are negative. Of the given choices, only the differential equation in choice B satisfies these conditions.

26. The correct choice is (D).

Since $f''(x) > 0$, $f'(x)$ is increasing, and in this case, $f'(4.5) < f'(4.6) < f'(4.7)$. Note that $f'(4.5) \approx \frac{10.2 - 10}{4.6 - 4.5} \approx 2$, and

$f'(4.7) \approx \frac{10.8 - 10.2}{4.7 - 4.6} \approx 6$. Also note that slope of tangent $= f'(4.6)$; therefore, $2 < f'(4.6) < 6$.

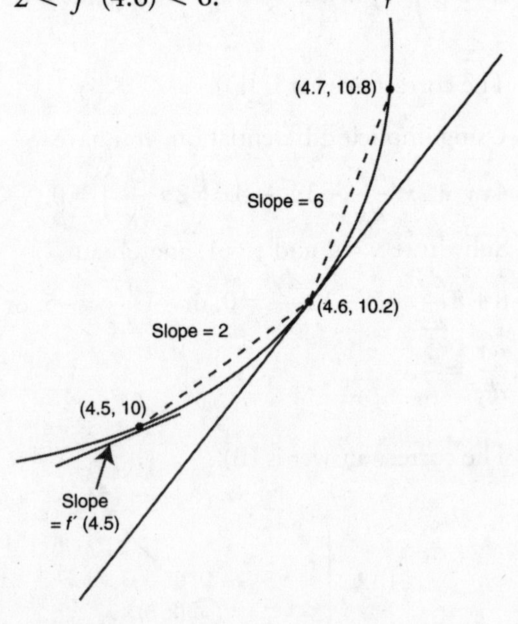

27. The correct choice is (C).

Note that $g(-3) = \int_{0}^{-3} f(t)\,dt$

$= -\int_{-3}^{0} f(t)\,dt$

$= -\left(\int_{-3}^{-1} f(t)\,dt + \int_{-1}^{0} f(t)\,dt\right).$

Using area of a triangle,
$\int_{-3}^{-1} f(t)dt = -\frac{(2)(2)}{2} = -2$, and
$\int_{-1}^{0} f(t)dt = \frac{(1)(1)}{2} = \frac{1}{2}$. Therefore,
$g(-3) = -\left(-2+\frac{1}{2}\right) = \frac{3}{2}$.

28. The correct choice is (D).

Begin with $\frac{dy}{dx} = 3x^2y$ and obtain $\frac{dy}{y} = 3x^2dx$. Integrating both sides of the equation, you have $\ln|y| = x^3 + c$. Substitute $x = 0$ and $y = e$, and obtain $\ln|e| = (0)^3 + c$, or $c = 1$, and the equation becomes $\ln|y| = x^3 + 1$. Solving for y, you have $e^{\ln|y|} = e^{(x^3+1)}$, or $|y| = e^{(x^3+1)}$, which yields $y = \pm e^{(x^3+1)}$. Since $y(0) = e$, which means the graph passes through the point $(0, e)$, $y = e^{(x^3+1)}$.

29. The correct choice is (B).

Begin by solving the differential equation. Rewrite $\frac{dy}{dx} = \frac{3}{x}$ as $dy = \frac{3}{x}dx$. Integrating both sides, you have $\int dy = \int \frac{3}{x}dx$, or $y = 3\ln x + c$. Substitute $x = e$ and $y = 4$ in the equation, and obtain $4 = 3\ln e + c$, or $c = 1$. Since, the point $(e, 4)$ is on the graph of $f(x)$, $y = 3\ln x + 1$.

30. The correct choice is (D).

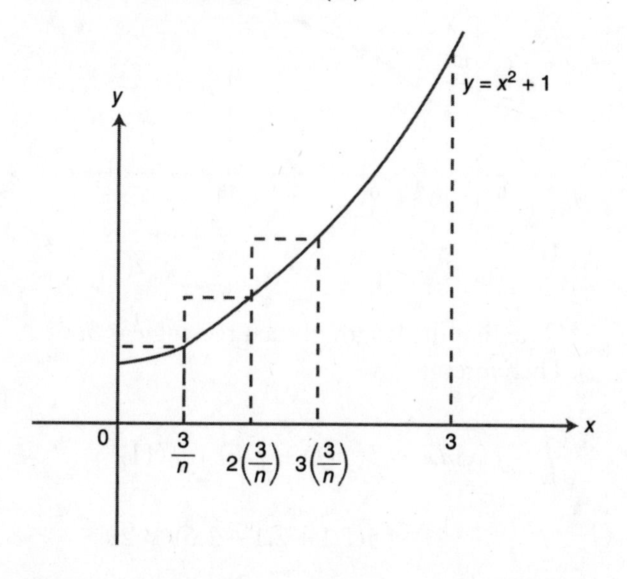

The width of each rectangle is $\frac{3-0}{n} = \frac{3}{n}$, and the heights of the rectangles are as follows:
$f\left(\frac{3}{n}\right), f\left(2\left(\frac{3}{n}\right)\right), f\left(3\left(\frac{3}{n}\right)\right) \ldots$ or
$\left(\frac{3}{n}\right)^2 + 1, \left(2\left(\frac{3}{n}\right)\right)^2 + 1, \left(3\left(\frac{3}{n}\right)\right)^2 + 1, \ldots$

The area under the curve is approximately $\sum_{k=1}^{n} \frac{3}{n}\left(\left(\frac{3k}{n}\right)^2 + 1\right)$, and the
$\int_0^3 (x^2+1)dx = \lim_{n\to\infty} \sum_{k=1}^{n} \frac{3}{n}\left(\left(\frac{3k}{n}\right)^2 + 1\right)$.

Section I—Part B

76. The correct answer is (D).

Since the function f is differentiable at $x = 1$, f is also continuous at $x = 1$. At $x = 1$, set $x^3 + x + a = 2bx - 1$, and obtain $(1)^3 + (1) + a = 2b(1) - 1$, or $a = 2b - 3$. Also, $f'(x) = 3x^2 + 1$ for $x \leq 1$, and $f'(x) = 2b$ for $x > 1$. At $x = 1$, set $3x^2 + 1 = 2b$, and you have $3(1)^2 + 1 = 2b$, or $b = 2$. Substituting $b = 2$ in $a = 2b - 3$, you have $a = 1$. Therefore, $a + 2b = 1 + 2(2) = 5$.

77. The correct answer is (D).

Note that $\int_1^2 g'(x)dx = g(2) - g(1)$.

Therefore, $\int_1^2 \cos(\pi x^{2+1})dx = g(2) - \frac{1}{2}$, or
$g(2) = \int_1^2 \cos(\pi x^{2+1})dx + \frac{1}{2}$. Using your graphing calculator, you have
$\int_1^2 \cos(\pi x^{2+1})dx \approx 0.18126$. Thus,

$g(2) \approx 0.18126 + \frac{1}{2} \approx 0.681$.

78. The correct answer is (D).

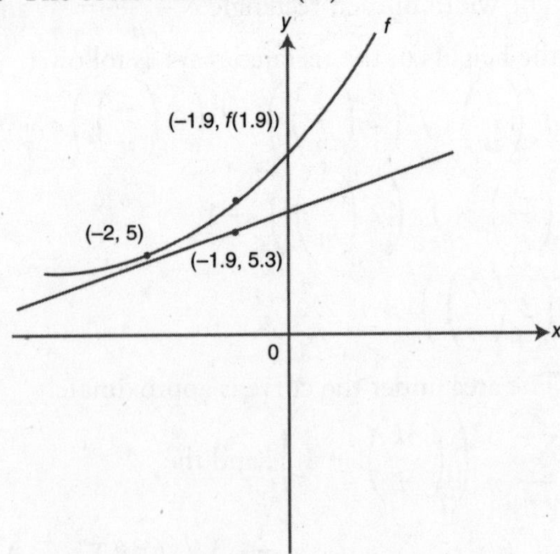

The tangent line for local linearization approximation passes through the point (–2, 5) and has a slope of $m = 3$. The equation of the tangent is $y - 5 = 3\,(x + 2)$. At $x = -1.9$, $y - 5 = 3\,(-1.9 + 2)$, or $y = 5.3$. Therefore, $f\,(-1.9) \approx 5.3$.

79. The correct answer is (D).

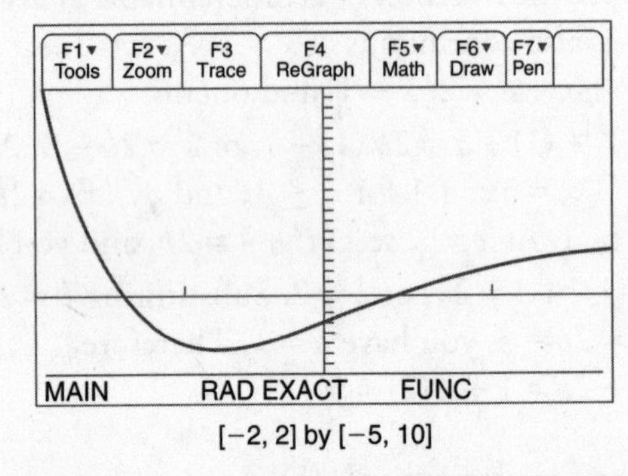

[–2, 2] by [–5, 10]

Using your graphing calculator, examine the graph of $f'\,(x)$. Note that the graph of $f'\,(x)$ crosses the x-axis at $x = -1.407$ and $x = 0.345$. The graph also shows the following:

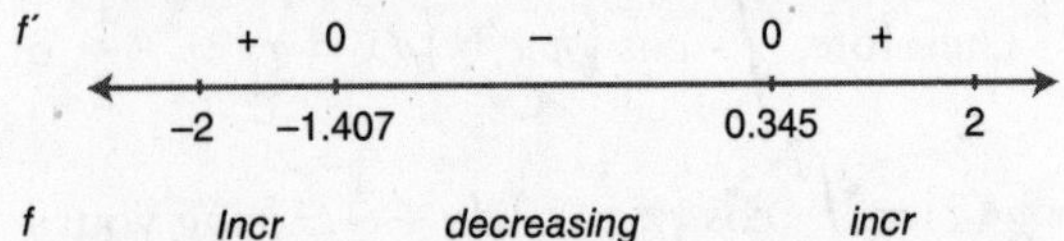

Therefore, the function f increases on the intervals (–2, – 1.407) and (0.345, 2).

80. The correct answer is (D).

The total distance traveled is

$\int_0^3 |v\,(t)|\,dt = \int_0^3 |2^{(-t)} \cos\,(2t)|\,dt$. Using the TI-89 graphing calculator, enter

$\int (abs(2 \wedge (-t) * \cos(2t)), t, 0, 3)$ and obtain $0.812884 \approx 0.813$.

81. The correct answer is (D).

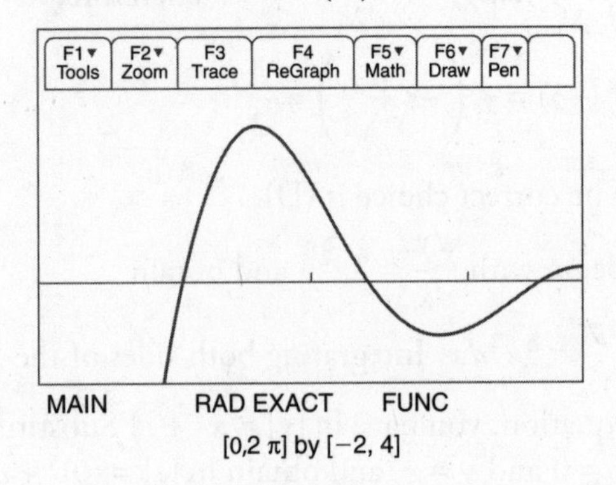

[0,2 π] by [–2, 4]

Using the TI-89 graphing calculator, enter the second derivative $\frac{d^2y}{dx^2}$ into y_1 as follows:

$$y_1 = d\left(5 * e \wedge \left(\frac{-x}{2}\right) * \sin\left(\frac{3x}{2}\right), x, 2\right)$$

Inspecting the graph of y_1, you see that the curve crosses the x-axis three times, which means the second derivative changes sign three times. Since the function y is a twice differentiable function, it has three points of inflection.

82. The correct answer is (B).

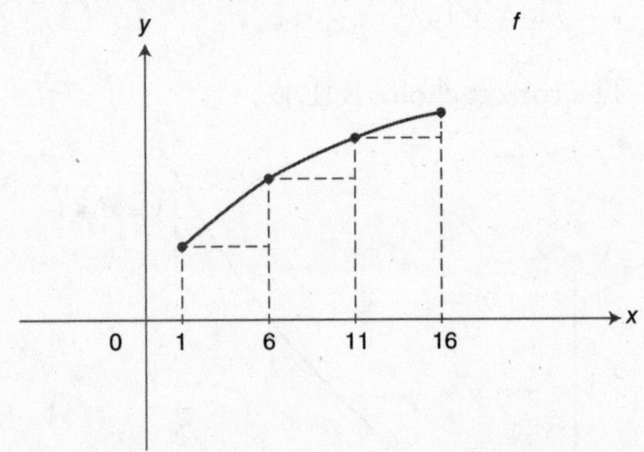

Note that the length of each rectangle is 5. Therefore:

$$\int_1^{16} f(x)dx \approx 5(f(1) + f(6) + f(11))$$

$$\approx 5(1.1 + 2.1 + 2.6) \approx 29$$

83. The correct answer is (A).

Note the following:

f'	+ 0 −		−		− 0 + 0 −			
	−6	−4	−2	0	2	4	6	
f	incr	decr	decr	decr	incr	decr		

Also note that $f'(x) = 0$ at $x = -4, 2, 4$, so the graph of f has a horizontal tangent at each of these points. The only graph that satisfies these conditions is choice (A).

84. The correct answer is (A).

Let x be the length of the wire used to make a circle. Therefore, the circumference of the circle is $2\pi r = x$, or $r = \dfrac{x}{2\pi}$. The area of the circle is $\pi r^2 = \pi\left(\dfrac{x}{2\pi}\right)^2$. The remaining wire has a length of $(40 - x)$ and is made into a square. Therefore, a side of the square is $\left(\dfrac{40-x}{4}\right)$, and the area of the square is $\left(\dfrac{40-x}{4}\right)^2$. Let A be the total area of the two figures. Thus, $A = \pi\left(\dfrac{x}{2\pi}\right)^2 + \left(\dfrac{40-x}{4}\right)^2$.

Enter this equation into the graphing calculator by letting $A = y_1$. Examine the graph of y_1.

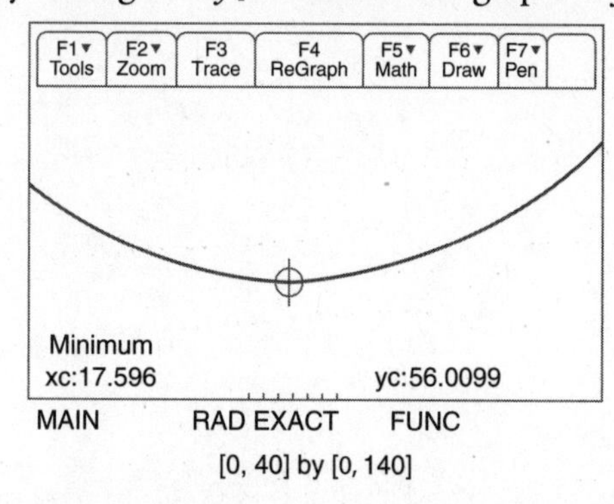

[0, 40] by [0, 140]

Use the [*Minimum*] function on the calculator, and obtain $x = 17.596$ centimeters.

85. The correct answer is (A).

Remember that $\dfrac{d}{dx}\arccos(x) = \dfrac{-1}{\sqrt{1-x^2}}$.

Using the chain rule,

$f'(x) = \dfrac{-1}{\sqrt{1-(2x)^2}} \cdot 2 = \dfrac{-2}{\sqrt{1-4x^2}}$, and $f'(0) = -2$. Therefore, the slope of the tangent to the graph of f at $x = 0$ is -2. Note that you could also find $f(x)$ and $f'(0)$ using your graphing calculator.

86. The correct answer is (B).

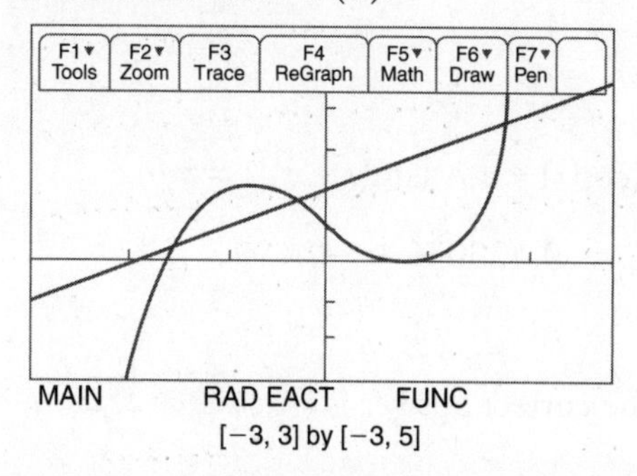

[−3, 3] by [−3, 5]

Using the TI-89 graphing calculator, graph the two given equations. Find their intersection points in the second quadrant: $x = -1.5321$ and $x = -0.3473$. Find the volume of the solid of revolution:

$$V = \pi \int_{-1.5321}^{-0.3473} ((x^3 - 2x + 1)^2 - (x+2)^2)\,dx$$

$$\approx 2.2405\pi \approx 7.039$$

87. The correct answer is (A).

Since $f(x) = x^3 + 1$, $f'(x) = 3x^2$, and $f'(c) = 3(c)^2$. The average rate of change of f on $[1, 3]$ is $\dfrac{f(3) - f(1)}{3 - 1} = \dfrac{28 - 2}{2} = 13$. Set $3c^2 = 13$, and obtain $c = \pm\sqrt{\dfrac{13}{3}} \approx \pm 2.082$.

Note that only $\sqrt{\dfrac{13}{3}}$ is on the interval $[1, 3]$.

Thus, $c = \sqrt{\dfrac{13}{3}}$.

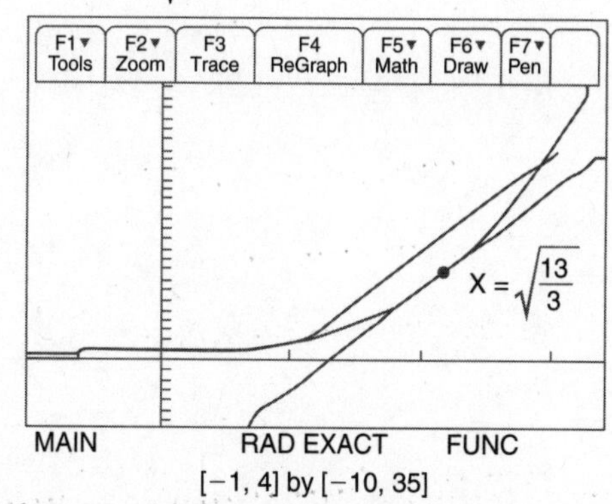

[−1, 4] by [−10, 35]

See the figure above for a geometric representation of the problem.

88. The correct answer is (C).

Remember, $a(t) = v'(t)$. Using the TI-89 graphing calculator, find $a(t)$, and evaluate it at $t = \dfrac{\pi}{4}$:

$d(\cos(t) * e \wedge (\sin(t)), t)|t = \pi/4$

$= -0.420036 \approx -0.420$

89. The correct answer is (A).

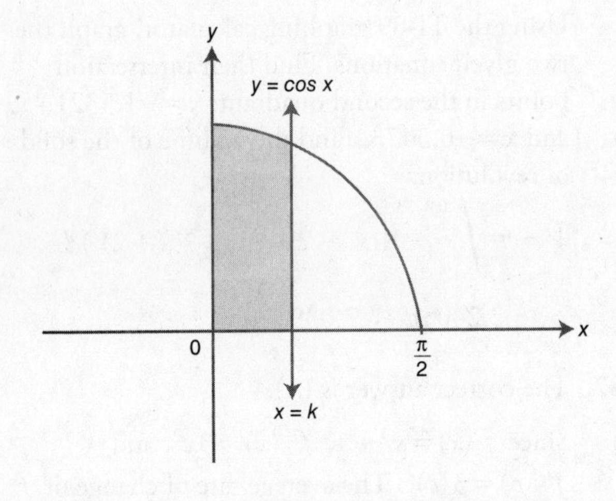

Since the areas of the two regions are equal, set $\int_0^k \cos(x)dx = \int_k^{\frac{\pi}{2}} \cos(x)dx$, and obtain $-\sin(x)]_0^k = -\sin(x)]_k^{\frac{\pi}{2}}$. Evaluating the definite integrals, you have $-\sin(k) - (-\sin(0)) = -\sin\left(\frac{\pi}{2}\right) - (-\sin(k))$, or $2\sin(k) = 1$, which yields $\sin(k) = \frac{1}{2}$, or

$$k = \sin^{-1}\left(\frac{1}{2}\right) = \frac{\pi}{6}.$$

90. The correct answer is (C).

Let $R(t)$ be the function representing the number of bacteria in the petri dish, so $\int r(t)dt = R(t)$. The net change in the number of bacteria between day two and day five is $\int_2^5 r(t)dt = R(5) - R(2)$, or $\int_2^5 2\sqrt{t}(e^{1.02t})dt = R(5) - 150$. Thus,

$623.227 = R(5) - 150$, or

$R(5) = 773.227 \approx 773.$

Solutions for AP Calculus AB Practice Exam 1–Section II

Section II—Part A

1. (A) Begin by finding the intersection points of the two curves, and obtain $x = 0$ and $x = 2.60308$.
Area of a region enclosed by two curves = $\int_a^b$ (Upper Curve – Lower Curve)dx.
In this case, the area of region $S = \int_0^{2.60308} ((4x - x^2) - (2x^4 - 5x^3))\,dx$
$= 17.258$.

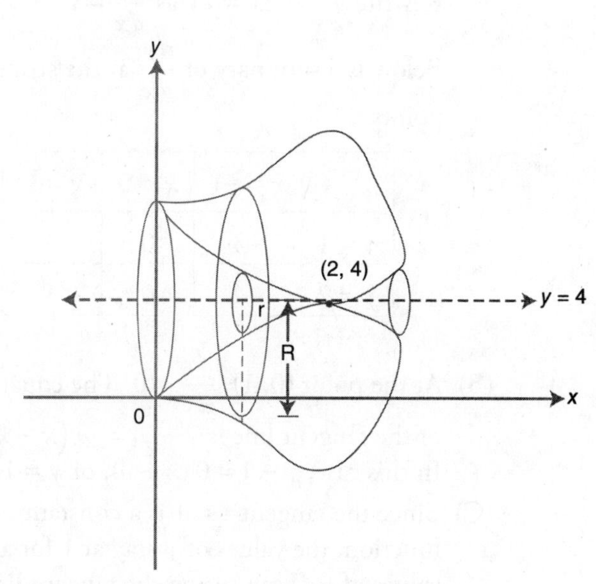

(B) Volume of a solid of revolution with hole $= \pi \int_a^b (R^2 - r^2)dx$, where R is the outer radius and r is the inner radius.
In this case, $R = 4 - (2x^4 - 5x^3)$, and $r = 4 - (4x - x^2)$. (Note that the highest point in region S is $(2, 4)$.) Therefore, the volume of the solid revolving about the line $y = 4$ is:

$$\pi \int_0^{2.60308} [(4 - (2x^4 - 5x^3))^2 - (4 - (4x - x^2))^2]dx = 172.896\pi \text{ or } 543.17$$

(C) Volume of a solid with a given base $= \int_a^b$ (area of cross section)dx. In this case, the area of the cross section $= (0.5 + 2x)((4x - x^2) - (2x^4 - 5x^3))$.
Thus, the volume of the solid with base S is $\int_0^{2.60308} (0.5 + 2x)((4x - x^2) - (2x^4 - 5x^3))dx = 64.334$.

2. (A) Remember that $v(t) = \int a(t)dt$, and in this case, $v(t) = \int \frac{\sin(\ln(t+1))}{t+1}dt$.
Using your TI-89 graphing calculator, you have $v(t) = -\cos(\ln(t+1)) + c$.
Since $v(0) = -1.1$, you have $-1.1 = -\cos(\ln(t+1)) + c$, or $-1.1 = -\cos(0) + c$, which yields $c = -0.1$.
Therefore, $v(t) = -\cos(\ln(t+1)) - 0.1$, and $v(4) = -\cos(\ln(4+1)) - 0.1 = -0.061368$, or -0.061.
Another approach to find $v(4)$ is as follows. Note that $\int_0^4 a(t)dt = v(4) - v(0)$ $= v(4) - (-1.1) = v(4) + 1.1$. Therefore, $v(4) = \int_0^4 a(t)dt - 1.1$. Also note that $\int_0^4 a(t)dt = \int_0^4 \frac{\sin(\ln(t+1))}{t+1}dt =$ 1.03863. Therefore, $v(4) = 1.03863 - 1.1 = -0.061368$, or -0.061.

(B) The speed of the particle is increasing when $a(t)$ and $v(t)$ are both positive or both negative. Examine the graphs of $a(t)$ and $v(t)$.

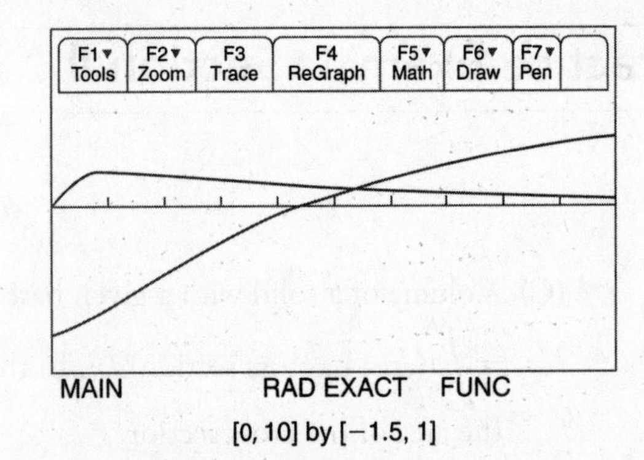

[0, 10] by [−1.5, 1]

Note that for $0 \leq t \leq 10$, $a(t) > 0$, and for $4.31729 < t \leq 10$, $v(t) > 0$. Therefore, for $0 \leq t \leq 10$, $a(t)$ and $v(t)$ have the same sign, both positive on the interval $4.317 < t \leq 10$. Thus, the speed of the particle is increasing on the interval $4.31729 < t \leq 10$.

(C) The average value of a function f on a closed interval $[a, b]$ is $\frac{\int_a^b f(x)dx}{b-a}$. In this case, the average velocity of the particle on the interval [0, 10] is $\frac{\int_0^{10} v(t)dt}{(10-0)} = \frac{1}{10}\int_0^{10}(-\cos(\ln(t+1)) - 0.1)dt = \frac{1}{10}(-0.175738) = -0.0175738$, or -0.018.

(D) The total distance traveled by the particle for $0 \leq t \leq 10$ is $\int_0^{10} |v(t)|dt = \int_0^{10} |-\cos(\ln(t+1)) - 0.1|dt = 4.44663$, or 4.447. Another approach to find the total distance traveled by the particle is to add the distance traveled to the left with the distance traveled to the right. From the graph of $v(t)$, the particle changes direction at $t = 4.317$. Therefore, the total distance $= \left|\int_0^{4.31729} v(t)dt\right| + \int_{4.31729}^{10} v(t)dt$, which yields the same result, 4.447.

Section II—Part B

3. (A)

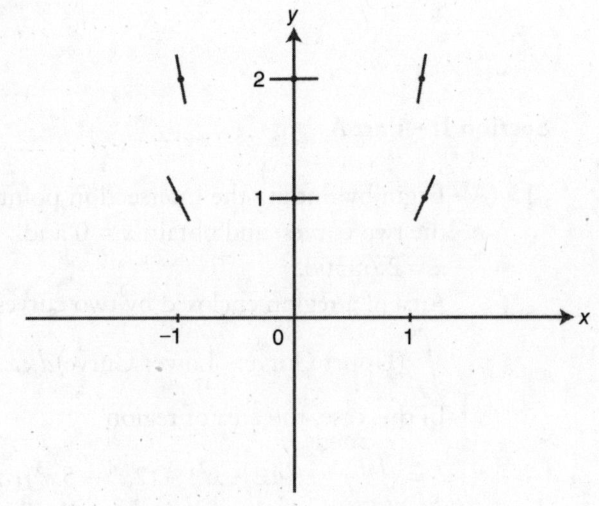

Rewrite $\frac{dy}{dx} - 2x = xy$ as $\frac{dy}{dx} = 2x + xy$.

Below is a summary of $\frac{dy}{dx}$ at the six given points.

	$x = -1$	$x = 0$	$x = 1$
$y = 1$	−3	0	3
$y = 2$	−4	0	4

(B) At the point (0, 1), $\frac{dy}{dx} = 0$. The equation of the tangent line is $y - y_1 = m(x - x_1)$. In this case, $y - 1 = 0(x - 0)$, or $y = 1$.

(C) Since the tangent $y = 1$ is a constant function, the values of y stay at 1 for all values of x. Thus, using the tangent line at (0, 1) to approximate $f(0.02)$, you have $f(0.02) \approx 1$.

(D) Rewrite $\frac{dy}{dx} - 2x = xy$ as $\frac{dy}{dx} = 2x + xy = x(2 + y)$, and obtain $\frac{dy}{2+y} = xdx$. Integrating both sides of the equation, you have $\int \frac{dy}{2+y} = \int xdx$.

Integrate $\int \frac{dy}{2+y}$ by letting $u = 2 + y$ and therefore $du = dx$. You obtain $\int \frac{du}{u} = \ln|u| + c = \ln|2 + y| + c$. Thus,

$\ln|2+y| = \frac{x^2}{2} + c$. Applying the exponential function to both sides, you have $e^{\ln|2+y|} = e^{\left(\frac{x^2}{2}+c\right)}$ or $|2+y| = \left(e^{\frac{x^2}{2}}\right)(e^c)$. Since e^c is an arbitrary constant, let $k = e^c$ and rewrite: $|2+y| = \left(e^{\frac{x^2}{2}}\right)(e^c)$ as $|2+y| = (k)\left(e^{\frac{x^2}{2}}\right)$. Substituting (0, 1) into the equation, you have $|2+1| = (k)(e^0)$, which yields $k = 3$. Since the point (0, 1) is on the curve of $y = f(x)$, $|2+y|$ becomes $2+y$, and you have $2+y = 3e^{\left(\frac{x^2}{2}\right)}$, or $y = 3e^{\left(\frac{x^2}{2}\right)} - 2$.

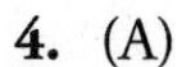

4. (A)

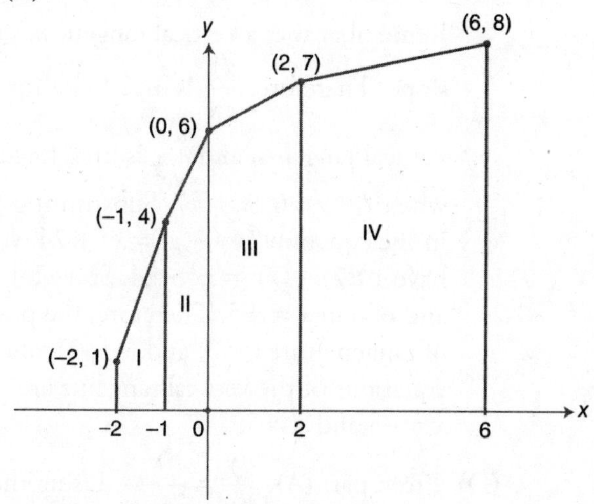

Area of a Trapezoid $= \frac{1}{2}(b_1 + b_2)h$

$$\text{Area}_{\text{TrapI}} = \frac{1}{2}(1+4)[-1-(-2)] = \frac{5}{2} = 2.5$$

$$\text{Area}_{\text{TrapII}} = \frac{1}{2}(4+6)[0-(-1)] = 5$$

$$\text{Area}_{\text{TrapIII}} = \frac{1}{2}(6+7)(2-0) = 13$$

$$\text{Area}_{\text{TrapIV}} = \frac{1}{2}(7+8)(4) = 30$$

Therefore,

$$\int_{-2}^{6} f(x)dx \approx 2.5 + 5 + 13 + 30 \approx 50.5.$$

Note that you cannot use the formula

$$\int_a^b f(x)dx \approx \left(\frac{b-a}{n}\right)[y_0 + 2y_1 + \cdots 2y_{n-1} + y_n],$$

since the four given intervals have different lengths.

(B) The average value of an integrable function on $[a, b]$ is $f_{\text{average}} = \frac{1}{b-a}\int_a^b f(x)dx$. In this case, $f_{\text{average}} = \frac{1}{6-(-2)}\int_{-2}^{6} f(x)dx \approx \frac{1}{8}(50.5) \approx 6.313$.

(C) The average rate of change of a continuous function on $[a, b]$ is $\frac{f(b)-f(a)}{b-a}$. In this case, the average rate of change of f is $\frac{f(6)-f(-2)}{6-(-2)} = \frac{8-1}{8} = \frac{7}{8}$, or 0.875.

(D) Since $f''(x) < 0$ for all $x \in (-2, 6)$, $f'(x)$ is decreasing on the interval. Thus, $f'(0) > f'(2)$.

5. (A) Begin by summarizing $f'(x)$ on a number line.

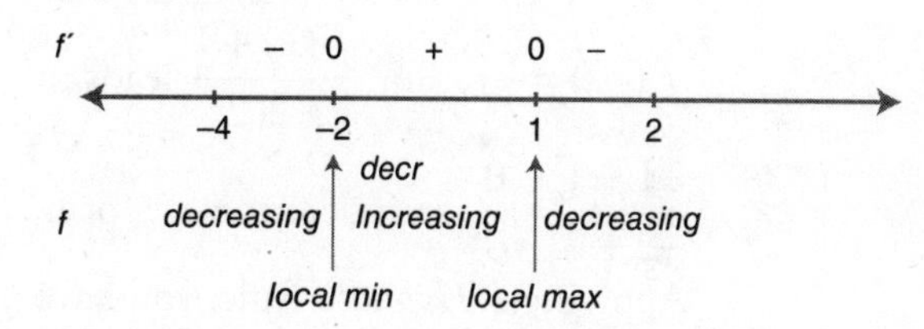

The graph of f has a local minimum at $x = -2$ on the interval $(-4, 2)$.

(B) Summarize $f''(x)$ on the number line.

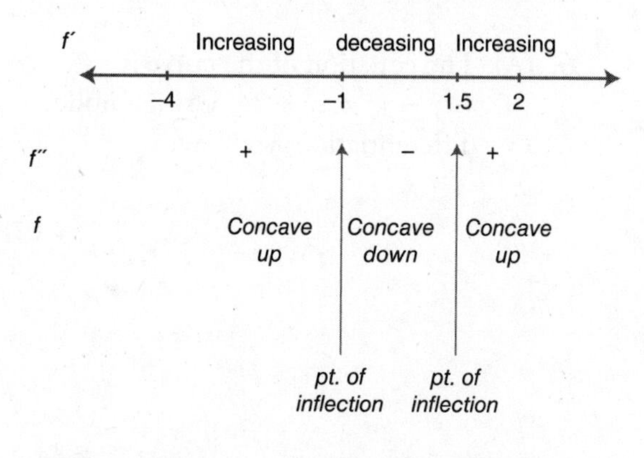

Since the graph of f changes concavity at $x = -1$ and $x = 1.5$, the graph of f has a point of inflection at $x = -1$ and $x = 1.5$.

(C) Note that
$\lim_{x\to 1} \frac{2f(x)}{x+1} = \frac{2f(1)}{1+1} = \frac{2f(1)}{2} = f(1)$.
Also, $\int_1^2 f'(x)dx = f(2) - f(1)$, with $f(2) = 3$. Since the area of region C is 6, and region C is below the x-axis, you have
$\int_1^2 f'(x)dx = -6$. Therefore,
$f(2) - f(1) = -6$, or $3 - f(1) = -6$, which yields $f(1) = 9$. Thus,
$\lim_{x\to 1} \frac{2f(x)}{x+1} = f(1) = 9$.

(D) By substituting $x = -2$, you have
$\lim_{x\to -2} \frac{f(x)+1}{x+2} = \frac{f(-2)+1}{0}$. To find the value of $f(-2)$, note that
$\int_{-2}^2 f'(x)dx = f(2) - f(-2) = 3 - f(-2)$.
Also, using regions B and C, you have
$\int_{-2}^2 f'(x)dx = (10) + (-6) = 4$. Therefore,
$3 - f(-2) = 4$, or $f(-2) = -1$. Since
$f(-2) = -1$, $\lim_{x\to -2} \frac{f(x)+1}{x+2}$ leads to
$\frac{-1+1}{-2+2} = \frac{0}{0}$, an indeterminate form.
Applying *L'Hôpital's* Rule, you have
$\lim_{x\to -2} \frac{f'(x)}{1} = f'(-2)$. Note that
$f'(-2) = 0$, and thus $\lim_{x\to -2} \frac{f(x)+1}{x+2} = 0$.

6. (A) The equation of the curve is
$14y - y^2 = x^2 + 24$. Using implicit differentiation, you have
$14\frac{dy}{dx} - 2y\frac{dy}{dx} = 2x$. Solve for $\frac{dy}{dx}$ and obtain $\frac{dy}{dx} = \frac{2x}{14-2y}$, or $\frac{dy}{dx} = \frac{x}{7-y}$.

(B) The slope of the tangent at the point
$(-3, 11)$ is $\left.\frac{dy}{dx}\right|_{(-3,11)} = \frac{-3}{7-11} = \frac{3}{4}$. Using the formula of a line, $y - y_1 = m(x - x_1)$, you have $y - 11 = \frac{3}{4}(x+3)$, or
$y = \frac{3}{4}x + \frac{53}{4}$.

(C) The slope of all tangents to the curve is derived in part (A) as $\frac{dy}{dx} = \frac{x}{7-y}$.
Remember that a vertical tangent has no slope. Therefore, $\frac{dy}{dx}$ is undefined for all vertical tangents, and $\frac{dy}{dx}$ is undefined when $7 - y = 0$ or $y = 7$. Substituting $y = 7$ in the equation $14y - y^2 = x^2 + 24$, you have $14(7) - (7)^2 = x^2 + 24$. Solve for x and obtain $x = \pm 5$. Therefore, the points of tangency are $(5, 7)$ and $(-5, 7)$, and the equations of the vertical tangents are $x = -5$ and $x = 5$.

(D) From part (A), $\frac{dy}{dx} = \frac{x}{7-y}$. Using the quotient rule, you have:
$$\frac{d^2y}{dx^2} = \frac{(1)(7-y) - (x)\left(-\frac{dy}{dx}\right)}{(7-y)^2}$$
At the point $(0, 2)$ on the curve,
$\frac{dy}{dx} = \frac{0}{7-(2)} = 0$. Therefore,
$\left.\frac{d^2y}{dx^2}\right|_{(0,2)} = \frac{(7-2)-0}{(7-2)^2} = \frac{1}{5}$.

Scoring Sheet for AB Practice Exam 1

Section I—Part A

_____________ × 1.2 = _____________
No. Correct Subtotal A

Section I—Part B

_____________ × 1.2 = _____________
No. Correct Subtotal B

Section II—Part A (Each question is worth 9 points.)

_____________ + _____________ = _____________
Q1 Q2 Subtotal C

Section II—Part B (Each question is worth 9 points)

_____________ + _____________ + _____________ + _____________ = _____________
Q1 Q2 Q3 Q4 Subtotal D

Total Raw Score (Subtotals A + B + C + D) = ☐

Approximate Conversion Scale:

Total Raw Score	Approximate AP Grade
80-108	5
65-79	4
50-64	3
36-49	2
0-35	1

AP Calculus AB Practice Exam 2

ANSWER SHEET FOR MULTIPLE-CHOICE QUESTIONS

Part A

1 (A) (B) (C) (D)
2 (A) (B) (C) (D)
3 (A) (B) (C) (D)
4 (A) (B) (C) (D)
5 (A) (B) (C) (D)
6 (A) (B) (C) (D)
7 (A) (B) (C) (D)
8 (A) (B) (C) (D)
9 (A) (B) (C) (D)
10 (A) (B) (C) (D)
11 (A) (B) (C) (D)
12 (A) (B) (C) (D)
13 (A) (B) (C) (D)
14 (A) (B) (C) (D)
15 (A) (B) (C) (D)
16 (A) (B) (C) (D)
17 (A) (B) (C) (D)
18 (A) (B) (C) (D)
19 (A) (B) (C) (D)
20 (A) (B) (C) (D)
21 (A) (B) (C) (D)
22 (A) (B) (C) (D)
23 (A) (B) (C) (D)
24 (A) (B) (C) (D)
25 (A) (B) (C) (D)
26 (A) (B) (C) (D)
27 (A) (B) (C) (D)
28 (A) (B) (C) (D)
29 (A) (B) (C) (D)
30 (A) (B) (C) (D)

Part B

76 (A) (B) (C) (D)
77 (A) (B) (C) (D)
78 (A) (B) (C) (D)
79 (A) (B) (C) (D)
80 (A) (B) (C) (D)
81 (A) (B) (C) (D)
82 (A) (B) (C) (D)
83 (A) (B) (C) (D)
84 (A) (B) (C) (D)
85 (A) (B) (C) (D)
86 (A) (B) (C) (D)
87 (A) (B) (C) (D)
88 (A) (B) (C) (D)
89 (A) (B) (C) (D)
90 (A) (B) (C) (D)

Section I—Part A

Number of Questions	Time	Use of Calculator
30	60 Minutes	No

Directions:

Use the answer sheet provided on the previous page. All questions are given equal weight. Points will *not* be deducted for incorrect answers, and no points will be given to unanswered questions. Unless otherwise indicated, the domain of a function f is the set of all real numbers. The use of a calculator is *not* permitted in this part of the exam.

1. $\int_0^8 x^{2/3}\,dx$

(A) $\frac{1}{3}$ (B) $\frac{96}{5}$

(C) $\frac{4}{3}$ (D) $-\frac{1}{3}$

2. The $\lim_{x\to 0} \frac{2-2\cos x}{x}$ is:

(A) -1 (B) 0

(C) $\frac{1}{2}$ (D) 2

3. What is the $\lim_{x\to -2} f(x)$, if

$$f(x)=\begin{cases} |x-1| & \text{if } x > -2 \\ 2x+7 & \text{if } x \le -2 \end{cases}?$$

(A) -3 (B) 1

(C) 3 (D) 11

4. The graph of f' is shown below.

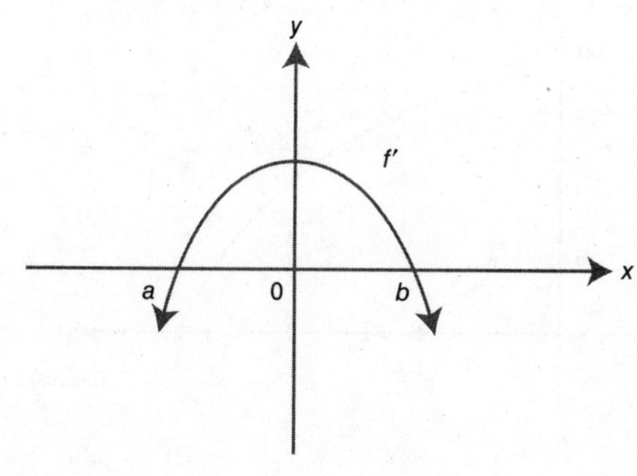

Which of the graphs below is a possible graph of f?

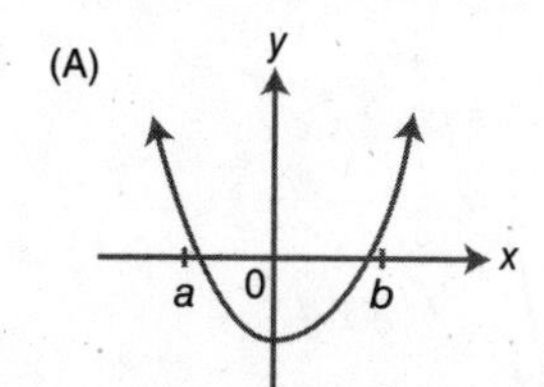

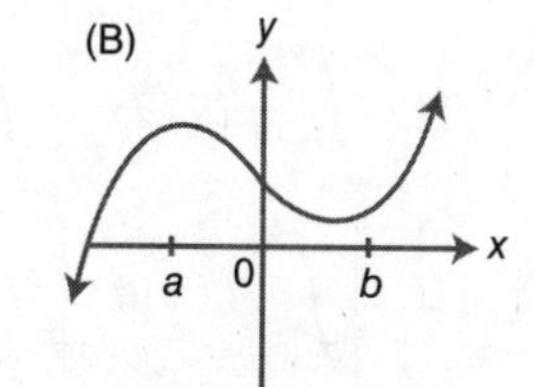

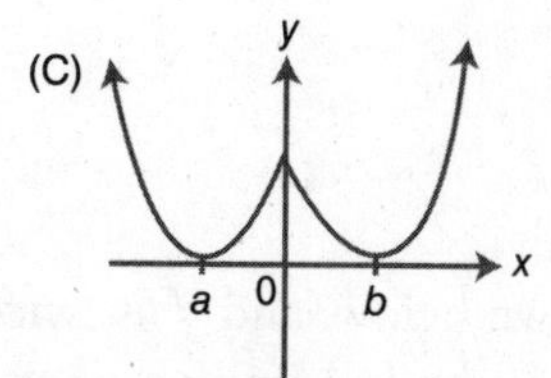

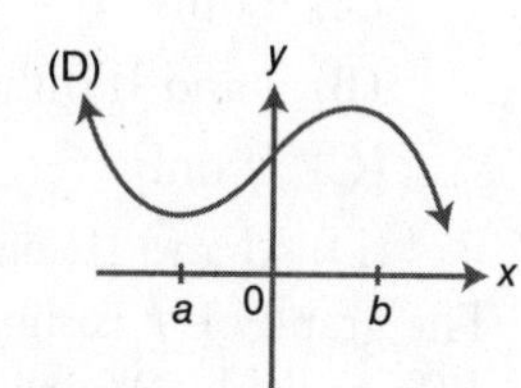

5. $\int_{\pi/2}^{x} 2\cos t\,dt =$

(A) $2\cos x$ (B) $-2\cos x$

(C) $2\sin x$ (D) $2\sin x - 2$

6. Given the equation $y = 3e^{-2x}$, what is an equation of the normal line to the graph at $x = \ln 2$?

(A) $y = \frac{2}{3}(x - \ln 2) + \frac{3}{4}$

(B) $y = \frac{2}{3}(x + \ln 2) - \frac{3}{4}$

(C) $y = -\frac{3}{2}(x - \ln 2) + \frac{3}{4}$

(D) $y = -\frac{3}{2}(x - \ln 2) - \frac{3}{4}$

GO ON TO THE NEXT PAGE

7. What is the $\lim_{h\to 0} \dfrac{\csc(\pi/4 + h) - \csc(\pi/4)}{h}$?

(A) $\sqrt{2}$ (B) $-\sqrt{2}$

(C) 0 (D) $-\dfrac{\sqrt{2}}{2}$

8. If $f(x)$ is an antiderivative of $x^2\sqrt{x^3+1}$ and $f(2) = 0$, then $f(0) =$

(A) -6 (B) 6

(C) $\dfrac{2}{9}$ (D) $\dfrac{-52}{9}$

9. If a function f is continuous for all values of x, which of the following statements is/are always true?

I. $2\int_a^b f(x)dx = \int_{2a}^{2b} f(x)dx$

II. $\int_a^b f(x)dx = \int_b^a -f(x)dx$

III. $\left|\int_a^b f(x)dx\right| = \int_a^b |f(x)|dx$

(A) I only
(B) I and II only
(C) II only
(D) II and III only

10. The graph of f is shown below, and f is twice differentiable. Which of the following has the largest value: $f(0)$, $f'(0)$, $f''(0)$?

(A) $f(0)$
(B) $f'(0)$
(C) $f''(0)$
(D) $f(0)$ and $f'(0)$

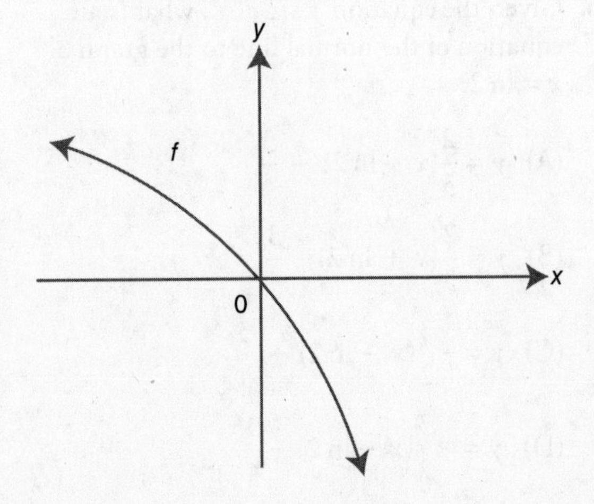

11. $\int \dfrac{x^4-1}{x^2}dx =$

(A) $\dfrac{x^3}{3} + x + C$

(B) $\dfrac{x^3}{3} - x + C$

(C) $\dfrac{x^3}{3} - \dfrac{1}{x} + C$

(D) $\dfrac{x^3}{3} + \dfrac{1}{x} + C$

12. If $p'(x) = q(x)$ and q is a continuous function for all values of x, then $\int_{-1}^{0} q(4x)dx$ is

(A) $p(0) - p(-4)$

(B) $4p(0) - 4p(-4)$

(C) $\dfrac{1}{4}p(0) - \dfrac{1}{4}p(-4)$

(D) $\dfrac{1}{4}p(0) + \dfrac{1}{4}p(-4)$

13. Water is leaking from a tank at a rate represented by $f(t)$ whose graph is shown below. Which of the following is the best approximation of the total amount of water leaked from the tank for $1 \leq t \leq 3$?

(A) $\dfrac{9}{2}$ gallons

(B) 5 gallons

(C) 175 gallons

(D) 350 gallons

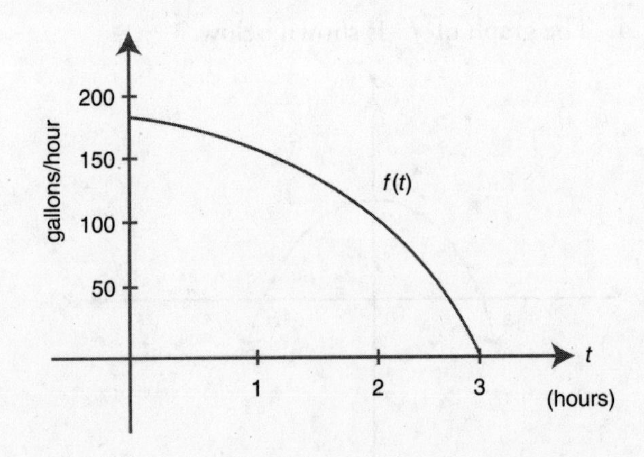

GO ON TO THE NEXT PAGE

14. If $f(x) = 5\cos^2(\pi - x)$, then $f'\left(\frac{\pi}{2}\right)$ is

(A) 0 (B) -5
(C) 5 (D) -10

15. $g(x) = \int_1^x \frac{3t}{t^3 + 1}dt$, then $g'(2)$ is

(A) 0 (B) $-\frac{2}{3}$
(C) $\frac{2}{3}$ (D) $\frac{5}{6}$

16. If $\int_k^2 (2x - 2)dx = -3$, a possible value of k is

(A) -2 (B) 0
(C) 1 (D) 3

17. If $\int_0^a f(x)dx = -\int_{-a}^0 f(x)dx$ for all positive values of a, then which of the following could be the graph of f? (See below.)

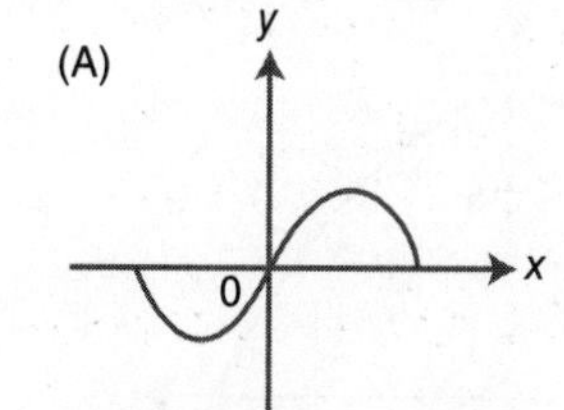

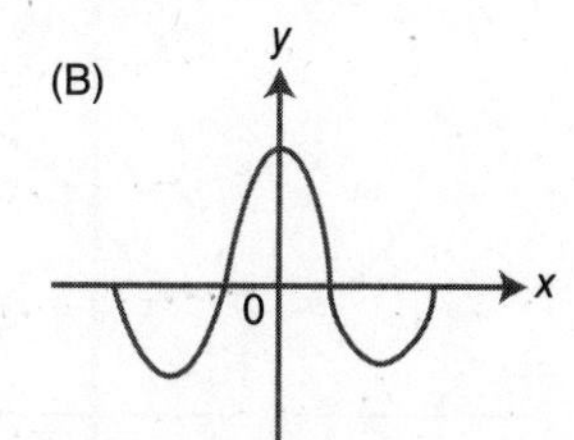

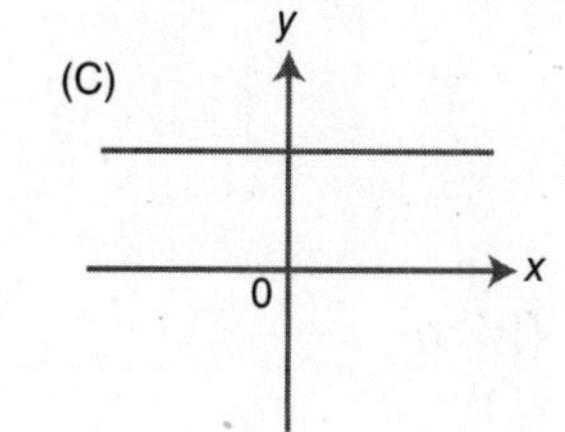

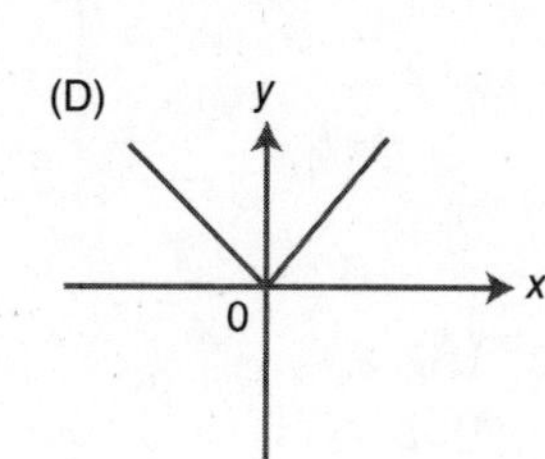

18. A function f is continuous on [1, 5], and some of the values of f are shown below:

x	1	3	5
$f(x)$	-2	b	-1

If f has only one root, r, on the closed interval [1, 5], and $r \neq 3$, then a possible value of b is

(A) -1 (B) 0
(C) 1 (D) 3

19. Given the equation $V = \frac{1}{3}\pi r^2(5 - r)$, what is the instantaneous rate of change of V with respect to r at $r = 5$?

(A) $-\frac{25\pi}{3}$ (B) $\frac{25\pi}{3}$
(C) $\frac{50\pi}{3}$ (D) 25π

20. What is the slope of the tangent to the curve $x^3 - y^2 = 1$ at $x = 1$?

(A) 0 (B) $\frac{3}{2\sqrt{2}}$
(C) $\frac{3}{2}$ (D) Undefined

21. The graph of function f is shown below. Which of the following is true for f on the interval (a, b)?

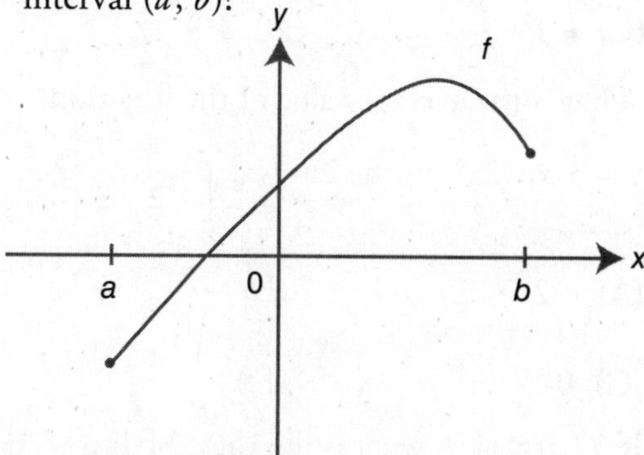

I. The function f is differentiable on (a, b).
II. There exists a number k on (a, b) such that $f'(k) = 0$.
III. $f'' > 0$ on (a, b).

(A) I only
(B) II only
(C) I and II only
(D) II and III only

22. The velocity function of a moving particle on the x-axis is given as $v(t) = t^2 - 3t - 10$. For what positive values of t is the particle's speed increasing?

(A) $0 < t < \frac{3}{2}$ only

(B) $t > \frac{3}{2}$ only

(C) $t > 5$ only

(D) $0 < t < \frac{3}{2}$ and $t > 5$ only

GO ON TO THE NEXT PAGE

23. The graph of f consists of two line segments and a semicircle for $-2 \leq x \leq 2$ as shown below. What is the value of $\int_{-2}^{2} f(x)dx$?

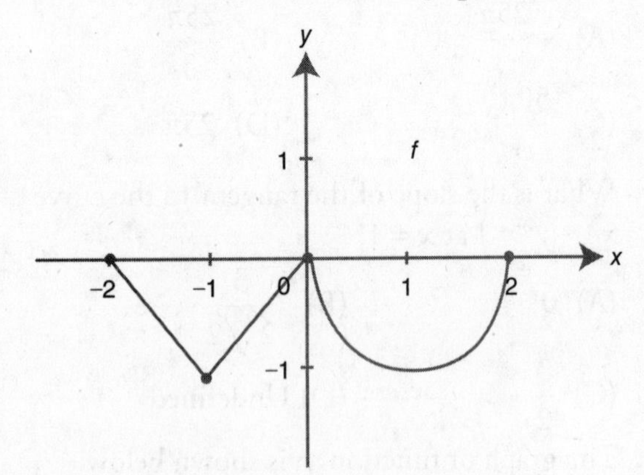

(A) $-2-2\pi$ (B) $-2-\pi$
(C) $-1-\frac{\pi}{2}$ (D) $1+\frac{\pi}{2}$

24. What is the average value of the function $y = 3\cos(2x)$ on the interval $\left[-\frac{\pi}{2}, \frac{\pi}{2}\right]$?

(A) -2 (B) $-\frac{2}{\pi}$
(C) 0 (D) $\frac{1}{\pi}$

25. If $f(x) = |x^3|$, what is the value of $\lim_{x\to -1} f'(x)$?

(A) -3 (B) 0
(C) 1 (D) 3

26. A spherical balloon is being inflated. At the instant when the rate of increase of the volume of the sphere is four times the rate of increase of the radius, the radius of the sphere is

(A) $\frac{1}{4\sqrt{\pi}}$

(B) $\frac{1}{\sqrt{\pi}}$

(C) $\frac{1}{\pi}$

(D) $\frac{1}{16\pi}$

27. If $\frac{dy}{dx} = \frac{x^2}{y}$ and at $x = 0$, $y = 4$, a solution to the differential equation is

(A) $y = \frac{x^3}{3}$

(B) $y = \frac{x^3}{3} + 4$

(C) $\frac{y^2}{2} = \frac{x^3}{3}$

(D) $\frac{y^2}{2} = \frac{x^3}{3} + 8$

28. The area of the region enclosed by the graph of $x = y^2 - 1$ and the y-axis is

(A) $-\frac{4}{3}$ (B) 0
(C) $\frac{2}{3}$ (D) $\frac{4}{3}$

29. The graph of f', the derivative of f, is shown below. At which value of x does the graph f have a horizontal tangent?

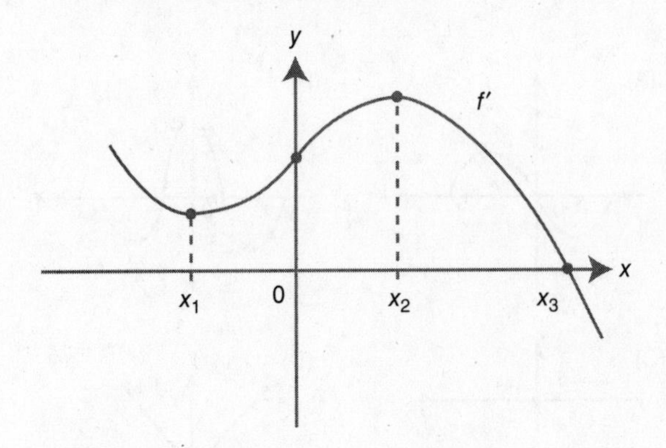

(A) x_1 (B) 0
(C) x_2 (D) x_3

30. If $f(x) = -|x-3|$, which of the following statements about f is true?

I. f is differentiable at $x = 3$.
II. f has an absolute minimum at $x = 3$.
III. f has a point of inflection at $x = 3$.

(A) II only
(B) III only
(C) II and III only
(D) None

STOP. AP Calculus AB Practice Exam 2 Section I—Part A

Section I—Part B

Number of Questions	Time	Use of Calculator
15	45 Minutes	Yes

Directions:

Use the same answer sheet from Part A. *Please note that the questions begin with number* 76. This is not an error. It is done to be consistent with the numbering system of the actual AP Calculus AB exam. All questions are given equal weight. Points will *not* be deducted for incorrect answers, and no points will be given to unanswered questions. Unless otherwise indicated, the domain of a function f is the set of all real numbers. If the exact numerical value does not appear among the given choices, select the best approximate value. The use of a calculator is *permitted* in this part of the exam.

76. The position function of a moving particle is $s(t) = 5 + 4t - t^2$ for $0 \le t \le 10$, where s is in meters and t is measured in seconds. What is the maximum speed in m/sec of the particle on the interval $0 \le t \le 10$?

(A) -16 (B) 2
(C) 4 (D) 16

77. How many points of inflection does the graph of $y = \cos(x^2)$ have on the interval $(0, \pi)$?

(A) 0 (B) 1
(C) 2 (D) 3

78. Let f be a continuous function on [4, 10] and have selected values as shown below:

x	4	6	8	10
$f(x)$	2	2.4	2.8	3.2

Using three right endpoint rectangles of equal length, what is the approximate value of

$$\int_4^{10} f(x)dx?$$

(A) 8.4 (B) 9.6
(C) 14.4 (D) 16.8

79. Given a differentiable function f with $f(-1) = 2$ and $f'(-1) = \frac{1}{2}$. Using a tangent line to the graph of f at $x = -1$, find an approximate value of $f(-1.1)$?

(A) -3.05 (B) -1.95
(C) 0.95 (D) 1.95

80. If area under the curve of $y = \frac{\ln x}{x}$ is 0.66 from $x = 1$ to $x = b$, where $b > 1$, then the value of b is approximately

(A) 1.93 (B) 2.25
(C) 3.15 (D) 3.74

81. The base of a solid is a region enclosed by the circle $x^2 + y^2 = 4$. What is the approximate volume of the solid if the cross sections of the solid perpendicular to the x-axis are semicircles?

(A) 8π (B) $\frac{16\pi}{3}$
(C) $\frac{32\pi}{3}$ (D) $\frac{64\pi}{3}$

82. The temperature of a cup of coffee is dropping at the rate of $f(t) = 4\sin\left(\frac{t}{4}\right)$ degrees for $0 \le t \le 5$, where f is measured in Fahrenheit and t in minutes. If initially, the coffee is 95°F, find its temperature to the nearest degree Fahrenheit 5 minutes later.

(A) 84 (B) 85
(C) 91 (D) 92

83. The graphs of f', g', p', and q' are shown below. Which of the functions f, g, p, or q have a relative minimum on (a, b)?

(A) f only (B) g only
(C) p only (D) q only

GO ON TO THE NEXT PAGE

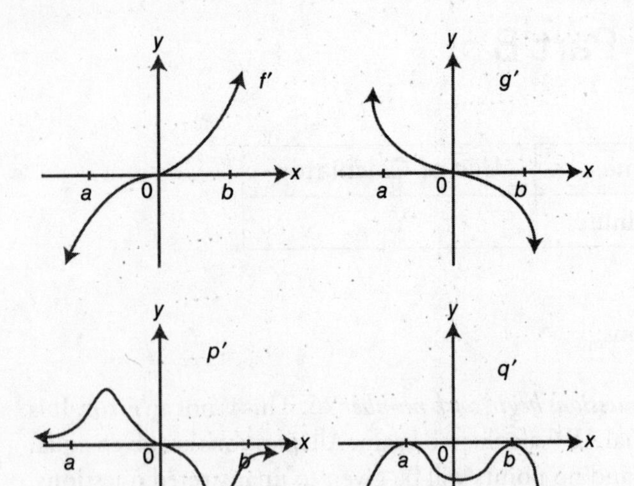

84. What is the volume of the solid obtained by revolving the region enclosed by the graphs of $x=y^2$ and $x=9$ about the y-axis?

(A) 36π (B) $\frac{81\pi}{2}$
(C) $\frac{486\pi}{2}$ (D) $\frac{1944\pi}{5}$

85. At what value(s) of x do the graphs of $y=e^x$ and $y=x^2+5x$ have parallel tangent lines?

(A) -2.5 (B) 0
(C) 0 and 5 (D) -2.45 and 2.25

86. Let y represent the population in a town. If y decreases according to the equation $\frac{dy}{dt}=ky$, with t measured in years, and the population decreases by 25% in 6 years, then $k=$

(A) -8.318 (B) -1.726
(C) -0.231 (D) -0.048

87. If $h(x)=\int_4^x (t-5)^3 dt$ on [4, 8], then h has a local minimum at $x=$

(A) 4 (B) 5
(C) 6 (D) 7

88. The volume of the solid generated by revolving the region bounded by the graph of $y=x^3$, the line $y=1$, and the y-axis about the y-axis is

(A) $\frac{\pi}{4}$ (B) $\frac{2\pi}{5}$
(C) $\frac{3\pi}{5}$ (D) $\frac{2\pi}{3}$

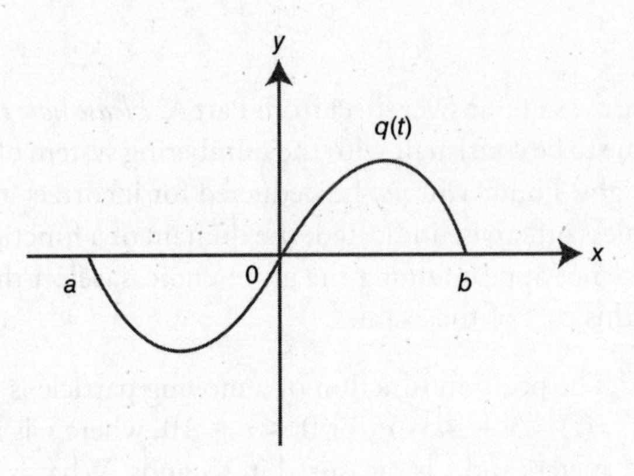

89. If $p(x)=\int_a^x q(t)dt\ a \le x \le b$ and the graph of q is shown above, which of the graphs shown below is a possible graph of p?

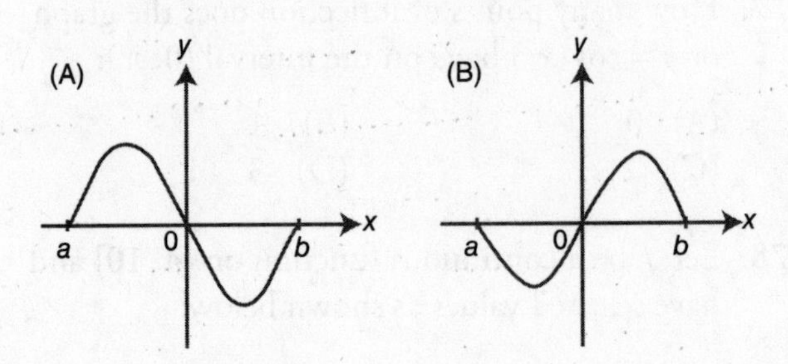

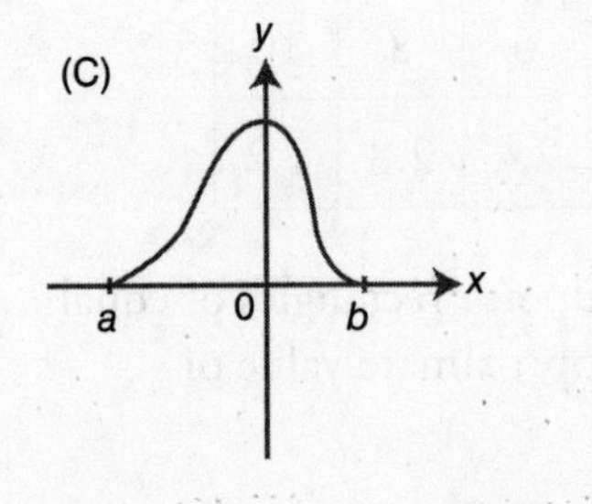

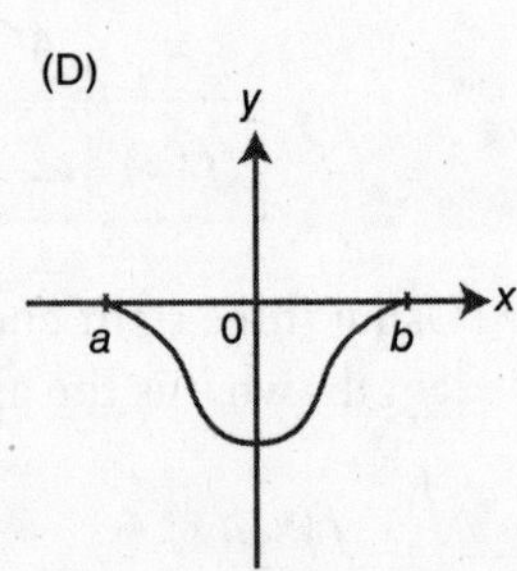

GO ON TO THE NEXT PAGE

90. The equation of the tangent line to the graph of $y = \sin x$ for $0 \leq x \leq \pi$ at the point where $\frac{dy}{dx} = \frac{1}{2}$ is

(A) $y = \frac{1}{2}\left(x - \frac{\pi}{3}\right) - \frac{\sqrt{3}}{2}$

(B) $y = \frac{1}{2}\left(x - \frac{\pi}{3}\right) + \frac{\sqrt{3}}{2}$

(C) $y = \frac{1}{2}\left(x - \frac{1}{2}\right) + \frac{\pi}{3}$

(D) $y = \frac{1}{2}\left(x - \frac{1}{2}\right) - \frac{\pi}{3}$

STOP. AP Calculus AB Practice Exam 2 Section I—Part B

Section II—Part A

Number of Questions	Time	Use of Calculator
2	30 Minutes	Yes

Directions:

Show all work. You may *not* receive any credit for correct answers without supporting work. You may use an approved calculator to help solve a problem. However, you must clearly indicate the setup of your solution using mathematical notations and *not* calculator syntax. Calculators may be used to find the derivative of a function at a point, compute the numerical value of a definite integral, or solve an equation. Unless otherwise indicated, you may assume the following: (a) the numeric or algebraic answers need not be simplified; (b) your answer, if expressed in approximation, should be correct to 3 places after the decimal point; and (c) the domain of a function f is the set of all real numbers.

1. Let R be the region in the first quadrant enclosed by the graph of $y = 2\cos x$, the x-axis, and the y-axis.

(A) Find the area of the region R.

(B) If the line $x = a$ divides the region R into two regions of equal area, find a.

(C) Find the volume of the solid obtained by revolving region R about the x-axis.

(D) If R is the base of a solid whose cross sections perpendicular to the x-axis are semicircles, find the volume of the solid.

2. The temperature of a liquid at a chemical plant during a 20-minute period is given as $g(t) = 90 - 4\tan\left(\frac{t}{20}\right)$, where $g(t)$ is measured in degrees Fahrenheit, $0 \le t \le 20$ and t is measured in minutes.

(A) Sketch the graph of g on the provided grid. What is the temperature of the liquid to the nearest hundredth of a degree Fahrenheit when $t = 10$? (See below.)

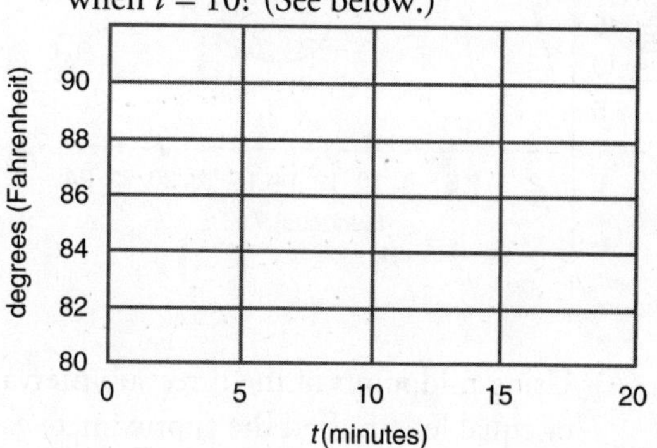

(B) What is the instantaneous rate of change of the temperature of the liquid to the nearest hundredth of a degree Fahrenheit at $t = 10$?

(C) At what values of t is the temperature of the liquid below 86°F?

(D) During the time within the 20-minute period when the temperature is below 86°F, what is the average temperature to the nearest hundredth of a degree Fahrenheit?

STOP. AP Calculus AB Practice Exam 2, Section II—Part A.

Section II—Part B

Number of Questions	Time	Use of Calculator
4	60 Minutes	No

Directions:

The use of a calculator is *not* permitted in this part of the exam. When you have finished this part of the exam, you may return to the problems in Part A of Section II and continue to work on them. However, you may not use a calculator. You should *show all work.* You may *not* receive any credit for correct answers without supporting work. Unless otherwise indicated, the numeric or algebraic answers need not be simplified, and the domain of a function f is the set of all real numbers.

3. A particle is moving on a coordinate line. The graph of its velocity function $v(t)$ for $0 \le t \le 24$ seconds is shown below.

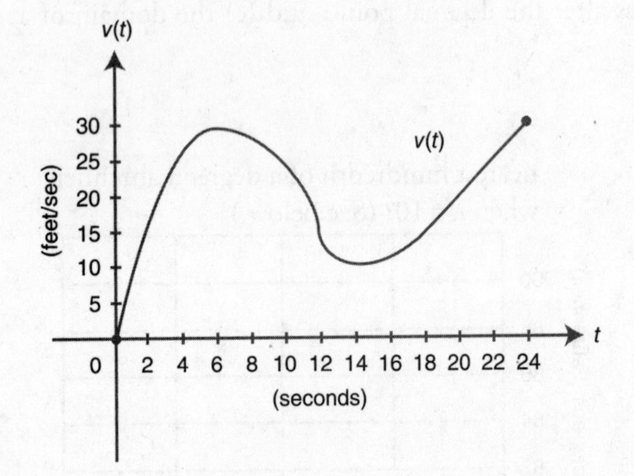

(A) Using midpoints of the three subintervals of equal length, find the approximate value of $\int_0^{24} v(t)dt$.

(B) Using the result in part (A), find the average velocity over the interval $0 \le t \le 24$ seconds.

(C) Find the average acceleration over the interval $0 \le t \le 24$ seconds.

(D) When is the acceleration of the particle equal to zero?

(E) Find the approximate acceleration at $t = 20$ seconds.

4. Given the function $f(x) = 3e^{-2x^2}$,

(A) at what value(s) of x, if any, is $f'(x) = 0$?

(B) at what value(s) of x, if any, is $f''(x) = 0$?

(C) find $\lim_{x\to\infty} f(x)$ and $\lim_{x\to-\infty} f(x)$.

(D) find the absolute maximum value of f and justify your answer.

(E) show that if $f(x) = ae^{-bx^2}$ where $a > 0$ and $b > 0$, the absolute maximum value of f is a.

5. The function f is defined as $f(x) = \int_0^x g(t)dt$ where the graph of g consists of five line segments as shown below.

(A) Find $f(-3)$ and $f(3)$.

(B) Find all values of x on $(-3, 3)$ such that f has a relative maximum or minimum. Justify your answer.

(C) Find all values of x on $(-3, 3)$ such that the graph f has a change of concavity. Justify your answer.

(D) Write an equation of the line tangent to the graph to f at $x = 1$.

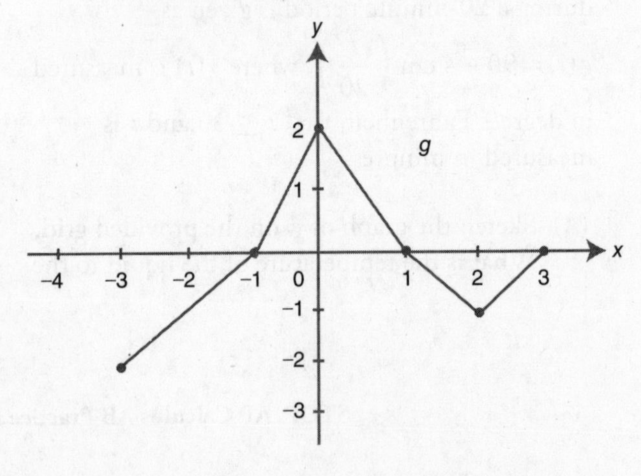

GO ON TO THE NEXT PAGE

6. The slope of a function f at any point (x, y) is $\frac{y}{2x^2}$. The point (2, 1) is on the graph of f.

(A) Write an equation of the tangent line to the graph of f at $x = 2$.

(B) Use the tangent line in part (A) to approximate $f(2.5)$.

(C) Solve the differential equation $\frac{dy}{dx} = \frac{y}{2x^2}$ with the initial condition $f(2) = 1$.

(D) Use the solution in part (C) and find $f(2.5)$.

STOP. AP Calculus AB Practice Exam 2, Section II—Part B.

Answers to AB Practice Exam 2–Section I

Part A

1. B
2. B
3. C
4. D
5. D
6. A
7. B
8. D
9. C
10. A
11. D
12. C
13. C
14. A
15. C
16. D
17. A
18. A
19. A
20. D
21. C
22. D
23. C
24. C
25. A
26. B
27. D
28. D
29. D
30. D

Part B

76. D
77. D
78. D
79. D
80. C
81. B
82. A
83. A
84. D
85. D
86. D
87. B
88. C
89. D
90. B

Answers to AB Practice Exam 2–Section II

Part A

1. (A) 2 (2 pts.)

(B) $a = \frac{\pi}{6}$ (3 pts.)

(C) π^2 (2 pts.)

(D) $\frac{\pi^2}{8}$ (2 pts.)

2. (A) See figure in solution and $g(10) = 87.82°$. (3 pts.)

(B) $-0.26°$ (2 pts.)

(C) $15.708 < t \leq 20$ (2 pts.)

(D) $84.99°$ (2 pts.)

Part B

3. (A) 480 (3 pts.)

(B) 20 ft/s (2 pts.)

(C) 1.25 ft/s^2 (1 pt.)

(D) $t = 6$ and $t = 14$ (2 pts.)

(E) 2.5 ft/s^2 (1 pt.)

4. (A) $x = 0$ (1 pt.)

(B) $x = \pm\frac{1}{2}$ (2 pts.)

(C) $\lim_{x\to\infty} f(x) = 0$ and $\lim_{x\to-\infty} f(x) = 0$ (2 pts.)

(D) 3 (2 pts.)

(E) See solution. (2 pts.)

5. (A) $f(-3) = 1$ and $f(3) = 0$ (2 pts.)

(B) $x = -1, 1$ (3 pts.)

(C) $x = 0$ and $x = 2$ (2 pts.)

(D) $y = 1$ (2 pts.)

6. (A) $y = \frac{1}{8}(x - 2) + 1$ (3 pts.)

(B) 1.063 (1 pt.)

(C) $y = e^{(-1/2x)+(1/4)}$ (4 pts.)

(D) $e^{1/20}$ (or 1.051) (1 pt.)

Solutions to AP Calculus AB Practice Exam 2—Section I

Section I—Part A

1. The correct answer is (B).

$$\int_0^8 x^{2/3}dx = \left.\frac{x^{5/3}}{5/3}\right]_0^8 = \left.\frac{3x^{5/3}}{5}\right]_0^8$$

$$= \frac{3(8)^{5/3}}{5} - 0 = \frac{3(32)}{5} = \frac{96}{5}$$

2. The correct answer is (B).
The $\lim_{x\to 0} 2 - 2\cos x = 2 - 2\cos(0) = 0$, and $\lim_{x\to 0} x = 0$. Therefore, $\lim_{x\to 0} \frac{2 - 2\cos x}{x}$ is an indeterminate form of $\frac{0}{0}$. Applying *L'Hôpital's* Rule, you have $\lim_{x\to 0} \frac{2\sin x}{1} = \frac{2\sin 0}{1} = 0$.

3. The correct answer is (C).

$$\lim_{x\to -2^+} |x-1| = |-2-1| = 3$$

$$\lim_{x\to -2^-} (2x+7) = 2(-2) + 7 = 3$$

Thus, $\lim_{x\to -2} f(x) = 3$.

4. The correct answer is (D).

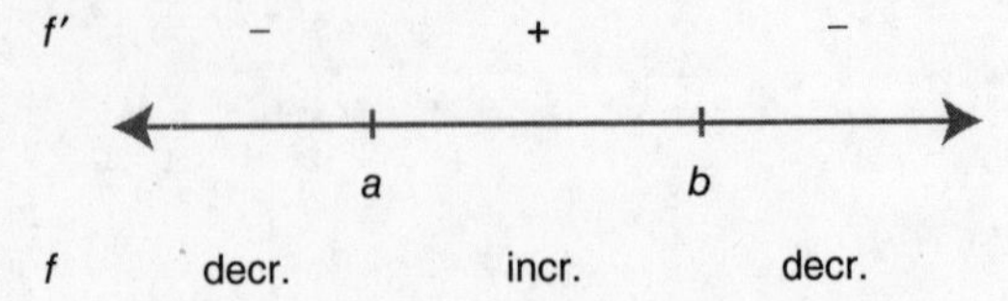

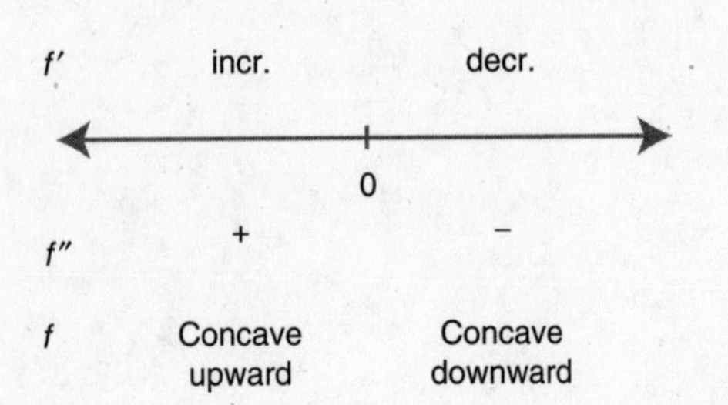

5. The correct answer is (D).

$$\int_{\pi/2}^{x} 2\cos t\,dt = 2\sin t\Big]_{\pi/2}^{x} = 2\sin x - 2(1)$$

$$= 2\sin x - 2$$

6. The correct answer is (A).

$$y = 3e^{-2x}; \quad \frac{dy}{dx} = 3e^{-2x}(-2) = -6e^{-2x}$$

$$\left.\frac{dy}{dx}\right|_{x=\ln 2} = -6e^{-2\ln 2} = -6\left(e^{\ln 2}\right)^{-2}$$

$$= -6(2)^{-2}$$

$$= -6\left(\frac{1}{4}\right) = -\frac{3}{2}$$

Slope of normal line at $x = \ln 2$ is $\frac{2}{3}$.

At $x = \ln 2$, $y = 3e^{-2\ln 2} = \frac{3}{4}$; point $\left(\ln 2, \frac{3}{4}\right)$.

Equation of normal line:

$$y - \frac{3}{4} = \frac{2}{3}(x - \ln 2) \text{ or } y = \frac{2}{3}(x - \ln 2) + \frac{3}{4}.$$

7. The correct answer is (B).

$$f'(x) = \lim_{h\to 0} \frac{f(x+h) - f(x)}{h}$$

Thus, $\lim_{h\to 0} \frac{\csc\left(\frac{\pi}{4} + h\right) - \csc\left(\frac{\pi}{4}\right)}{h}$

$$= \left.\frac{d(\csc x)}{dx}\right|_{x=\frac{\pi}{4}} = -\csc\left(\frac{\pi}{4}\right)\cot\left(\frac{\pi}{4}\right)$$

$$= -\sqrt{2}(1) = -\sqrt{2}.$$

8. The correct answer is (D).
Let $u = x^3 + 1$, $du = 3x^2dx$ or $\frac{du}{3} = x^2dx$.

$$f(x) = \int x^2\sqrt{x^3+1}\,dx = \int \sqrt{u}\frac{du}{3}$$

$$= \frac{1}{3}\int u^{1/2}du$$

$$= \frac{1}{3}\frac{u^{3/2}}{3/2} + C$$

$$= \frac{2}{9}(x^3 + 1)^{3/2} + C$$

$$f(2) = 0 \Rightarrow \frac{2}{9}(2^3 + 1)^{3/2} + C = 0$$

$$\Rightarrow \frac{2}{9}(9)^{3/2} + C = 0$$

$$\Rightarrow 6 + C = 0 \text{ or } C = -6$$

$$f(x) = \frac{2}{9}(x^3 + 1)^{3/2} - 6$$

$$f(0) = \frac{2}{9} - 6 = \frac{-52}{9}$$

9. The correct answer is (C).

Statement I is *not true*, e.g.,

$$2\int_0^4 x\,dx \neq 2\int_0^8 x\,dx.$$

Statement II is always *true* since

$\int_a^b f(x)dx = -\int_b^a f(x)dx$ by properties of definite integrals.

Statement III is *not true*, e.g.,

$$\left|\int_{-2}^{2} x\,dx\right| \neq \int_{-2}^{2} |x|dx.$$

10. The correct answer is (A).

$f(0) = 0$; $f'(0) < 0$ since f is decreasing and $f'' < 0$, f is concave downward. Thus, $f(0)$ has the largest value.

11. The correct answer is (D).

$$\int \frac{x^4 - 1}{x^2}dx = \int \left(x^2 - \frac{1}{x^2}\right)dx$$

$$= \int (x^2 - x^{-2})dx$$

$$= \frac{x^3}{3} - \frac{x^{-1}}{-1} + C = \frac{x^3}{3} + \frac{1}{x} + C$$

12. The correct answer is (C).

Let $u = 4x$; $du = 4dx$ or $\frac{du}{4} = dx$.

$$\int q(4x)dx = \int q(u)\frac{du}{4}$$

$$= \frac{1}{4}p(u) + c = \frac{1}{4}p(4x) + C$$

Thus, $\int_{-1}^{0} q(4x)dx = \frac{1}{4}p(4x)\Big]_{-1}^{0}$

$$= \frac{1}{4}p(0) - \frac{1}{4}p(-4).$$

13. The correct answer is (C).

The total amount of water leaked from the tank for

$$1 \le t \le 3 = \int_1^3 f(t)dt$$

$$\approx 100 + 25 + 50 \approx 175 \text{ gallons.}$$

14. The correct answer is (A).

$$f'(x) = 10[\cos(\pi - x)][-\sin(\pi - x)](-1)$$

$$= 10\cos(\pi - x)\sin(\pi - x)$$

$$f'\left(\frac{\pi}{2}\right) = 10\cos\left(\pi - \frac{\pi}{2}\right)\sin\left(\pi - \frac{\pi}{2}\right)$$

$$= 10\cos\left(\frac{\pi}{2}\right)\sin\left(\frac{\pi}{2}\right)$$

$$= 10(0)(1) = 0$$

15. The correct answer is (C).

$$g'(x) = \frac{3x}{x^3 + 1};\ g'(2) = \frac{3(2)}{2^3 + 1} = \frac{6}{9} = \frac{2}{3}$$

16. The correct answer is (D).

$$\int_k^2 (2x - 2)dx = x^2 - 2x\Big]_k^2$$

$$= \left(2^2 - 2(2)\right) - \left(k^2 - 2k\right)$$

$$= 0 - \left(k^2 - 2k\right) = -k^2 + 2k$$

$$\text{Set } -k^2 + 2k = -3 \Rightarrow 0 = k^2 - 2k - 3$$

$$0 = (k - 3)(k + 1)$$

$$\Rightarrow k = 3 \text{ or } k = -1.$$

17. The correct answer is (A).

$\int_0^a f(x)dx = -\int_{-a}^{0} f(x)dx \Rightarrow f(x)$ is an odd function. The function whose graph is shown in (A) is the only odd function.

18. The correct answer is (A).

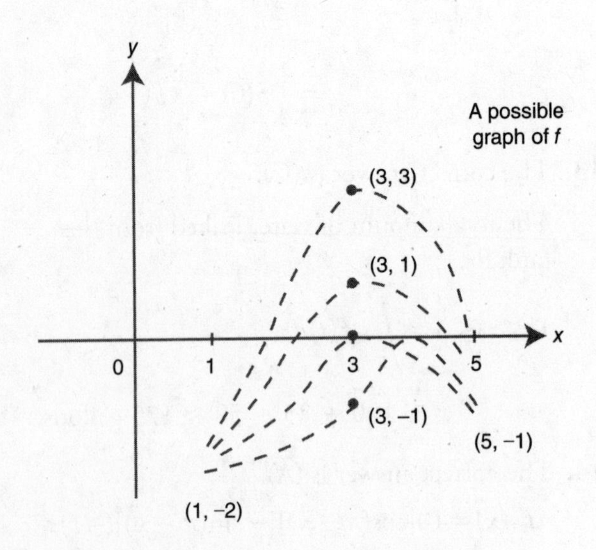

If $b = 0$, then $r = 3$, but r cannot be 3. If $b = 1$, or 3, f would have more than one root. Thus, of all the choices, the only possible value for b is -1.

19. The correct answer is (A).

$$V = \frac{1}{3}\pi r^2(5) - \frac{1}{3}\pi r^2(r)$$

$$= \frac{5}{3}\pi r^2 - \frac{1}{3}\pi r^3$$

$$\frac{dV}{dr} = \frac{10}{3}\pi r - \pi r^2$$

$$\left.\frac{dV}{dr}\right|_{r=5} = \frac{10}{3}\pi(5) - \pi(25) = \frac{-25\pi}{3}$$

20. The correct answer is (D).

$$x^3 - y^2 = 1; 3x^2 - 2y\frac{dy}{dx} = 0 \Rightarrow \frac{dy}{dx} = \frac{3x^2}{2y}$$

At $x = 1$, $1^3 - y^2 = 1 \Rightarrow y = 0 \Rightarrow (1, 0)$

$$\left.\frac{dy}{dx}\right|_{x=1} = \frac{3(1^2)}{2(0)} \text{ undefined.}$$

21. The correct answer is (C).

I. f is differentiable on (a, b) since the graph is a smooth curve.

II. There exists a horizontal tangent to the graph on (a, b); thus, $f'(k) = 0$ for some k on (a, b).

III. The graph is concave downward; thus, $f'' < 0$.

22. The correct answer is (D).

$v(t) = t^2 - 3t - 10$; set $v(t) = 0$

$\Rightarrow (t - 5)(t + 2) = 0$

$\Rightarrow t = 5$ or $t = -2$

$a(t) = 2t - 3$; set $a(t) = 0$

$\Rightarrow 2t - 3 = 0$ or $t = \frac{3}{2}$.

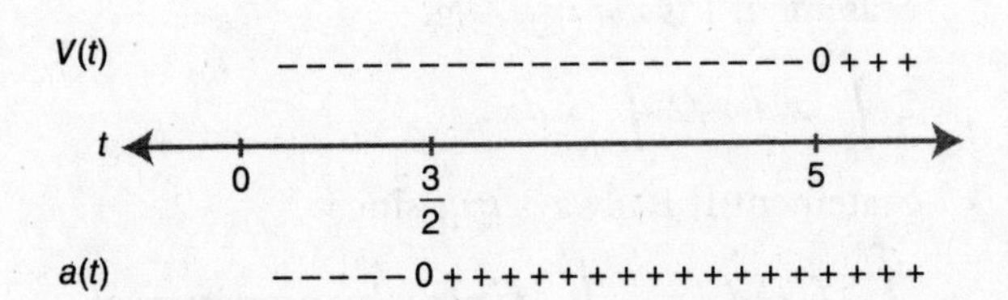

Since $v(t)$ and $a(t)$ are both negative on $(0, 3/2)$, and are both positive on $(5, \infty)$, the particle's speed is increasing on these intervals.

23. The correct answer is (C).

$$\int_{-2}^{2} f(x)\,dx = \int_{-2}^{0} f(x)\,dx + \int_{0}^{2} f(x)\,dx$$

$$= \frac{1}{2}(2)(-1) + \left(-\left(\frac{1}{2}\right)\pi(1)^2\right)$$

$$= -1 - \frac{\pi}{2}$$

24. The correct answer is (C).

$$\text{Average value} = \frac{1}{\frac{\pi}{2} - \left(-\frac{\pi}{2}\right)}\int_{-\pi/2}^{\pi/2} 3\cos(2x)dx$$

$$= \frac{1}{\pi}\left[\frac{3\sin(2x)}{2}\right]_{-\pi/2}^{\pi/2}$$

$$= \frac{3}{2\pi}[\sin\pi - (\sin[-\pi])] = 0.$$

25. The correct answer is (A).

$$f(x) = |x^3| = \begin{cases} x^3 & \text{if } x \geq 0 \\ -x^3 & \text{if } x < 0 \end{cases}$$

$$f'(x) = \begin{cases} 3x^2 & \text{if } x \geq 0 \\ -3x^2 & \text{if } x < 0 \end{cases}$$

$$\lim_{x \to -1} f'(x) = \lim_{x \to -1} (-3x^2) = -3$$

26. The correct answer is (B).

$$V = \frac{4}{3}\pi r^3; \quad \frac{dV}{dt} = 4\pi r^2 \frac{dr}{dt}$$

Since $\frac{dV}{dt} = 4\frac{dr}{dt} \Rightarrow 4 = 4\pi r^2$ or

$r^2 = \frac{1}{\pi}$ or $r = \frac{1}{\sqrt{\pi}}$.

27. The correct answer is (D).

$$\frac{dy}{dx} = \frac{x^2}{y}; \ y\,dy = x^2 dx$$

$$\int y\,dy = \int x^2 dx$$

$\frac{y^2}{2} = \frac{x^3}{3} + C$. Substituting (0, 4)

$$\frac{4^2}{2} = 0 + C \Rightarrow C = 8.$$

Thus, a solution is $\frac{y^2}{2} = \frac{x^3}{3} + 8$.

28. The correct answer is (D).

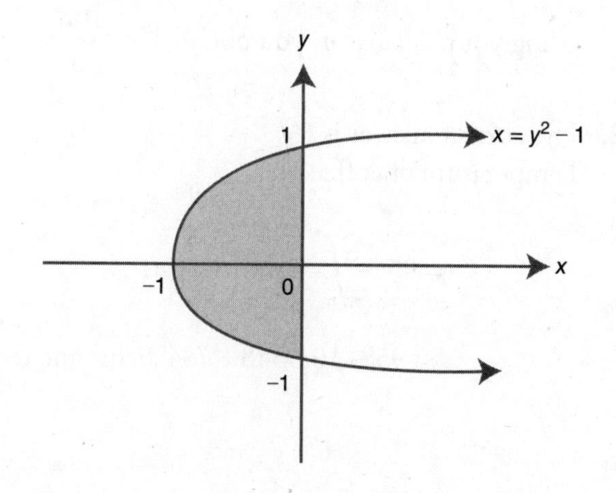

$$A = \left| \int_{-1}^{1} (y^2 - 1)\,dy \right| = \left| \left[\frac{y^3}{3} - y \right]_{-1}^{1} \right|$$

$$= \left| \left(\frac{1}{3} - 1\right) - \left(-\frac{1}{3} - (-1)\right) \right| = \frac{4}{3}$$

29. The correct answer is (D).

At $x = x_3$, $f' = 0$. Thus, the tangent to the graph of f at $x = x_3$ is horizontal.

30. The correct answer is (D).

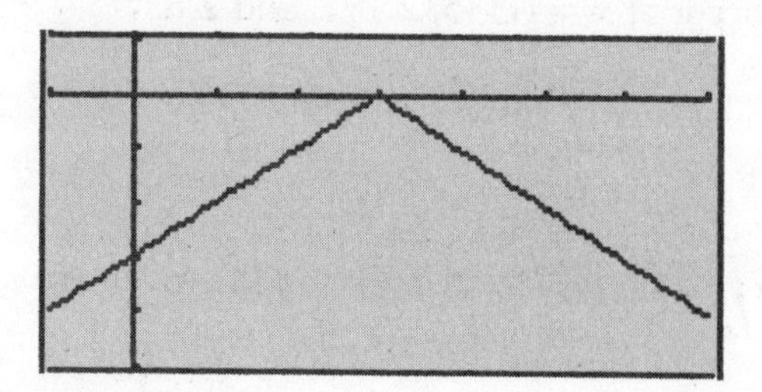

The function f is not differentiable at $x = 3$, has an absolute maximum at $x = 3$, and has no point of inflection. Thus, all three statements are not true.

Section I—Part B

76. The correct answer is (D).

$s(t) = 5 + 4t - t^2$; $v(t) = s'(t) = 4 - 2t$

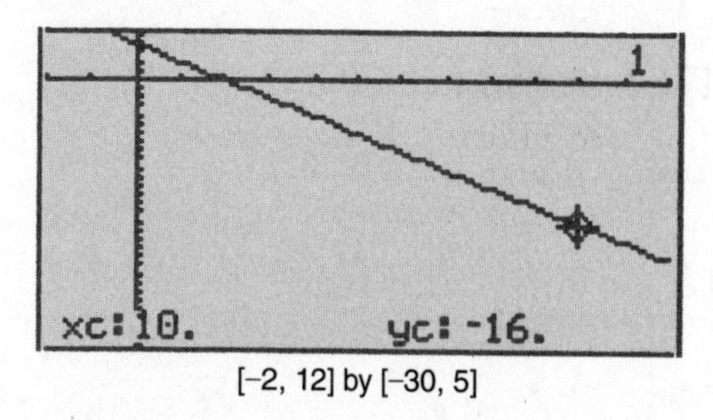

[–2, 12] by [–30, 5]

Since $v(t)$ is a straight line with a negative slope, the maximum speed for $0 \leq t \leq 10$ occurs at $t = 10$ where $v(t) = 4 - 2(10) = -16$. Thus, the maximum speed $=16$.

77. The correct answer is (D).

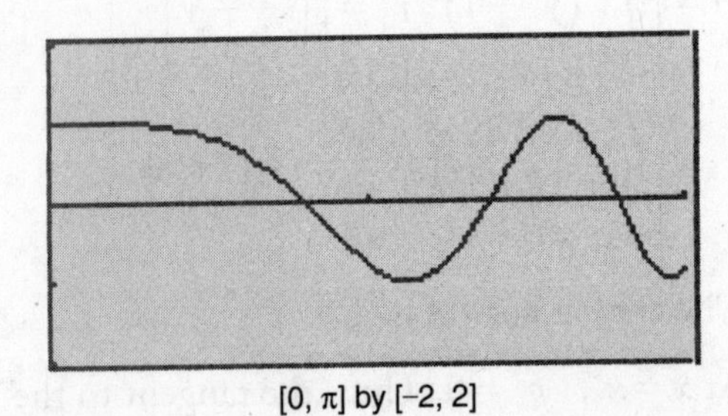

Using the [*Inflection*] function of your calculator, you will find three points of inflection. They occur at $x = 1.335$, 2.195, and 2.8.

78. The correct answer is (D).

$$\int_4^{10} f(x)\,dx \approx 2(f(6) + f(8) + f(10))$$

$$\approx 2(2.4 + 2.8 + 3.2) \approx 16.8$$

79. The correct answer is (D).

$f(-1) = 2 \Rightarrow$ a point $(-1, 2)$.

$f'(-1) = \frac{1}{2} \Rightarrow$ the slope at $x = -1$ is $\frac{1}{2}$.

Equation of tangent at $x = -1$ is

$y - 2 = \frac{1}{2}(x + 1)$ or $y = \frac{1}{2}(x + 1) + 2$.

Thus, $f(-1.1) \approx \frac{1}{2}(-1.1 + 1) + 2 \approx 1.95$.

80. The correct answer is (C).

$\text{Area} = \int_1^b \frac{\ln x}{x}dx = 0.66.$

Let $u = \ln x$; $du = \frac{1}{x}dx$.

$$\int \frac{\ln x}{x}dx = \int u\,du = \frac{u^2}{2} + C = \frac{(\ln x)^2}{2} + C$$

$$\int_1^b \frac{\ln x}{x}dx = \left.\frac{(\ln x)^2}{2}\right]_1^b = \frac{(\ln b)^2}{2} - \frac{(\ln 1)^2}{2}$$

$$= \frac{(\ln b)^2}{2}$$

Let $\frac{(\ln b)^2}{2} = 0.66$, $(\ln b)^2 = 1.32$.

$\ln b = \sqrt{1.32}$

$e^{\ln b} = e^{\sqrt{1.32}}$

$b \approx 3.15$

81. The correct answer is (B).

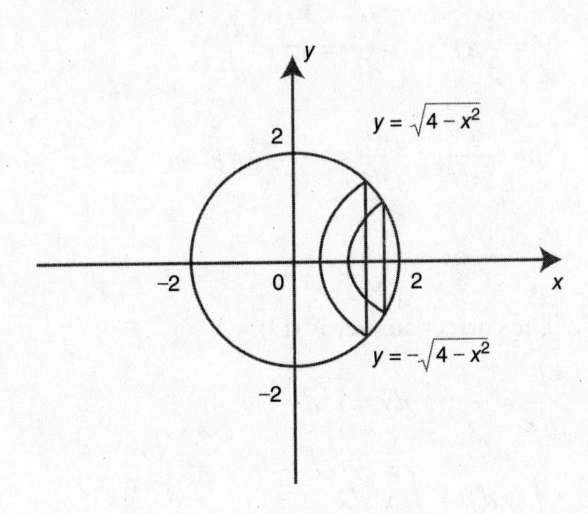

$\text{Area of a cross section} = \frac{1}{2}\pi\left(\sqrt{4 - x^2}\right)^2$

$= \frac{1}{2}\pi(4 - x^2).$

$\text{Volume of the solid} = \int_{-2}^{2} \frac{1}{2}\pi(4 - x^2)\,dx.$

Using your calculator, you obtain $V = \frac{16\pi}{3}$.

82. The correct answer is (A).
Temperature of coffee

$$= 95 - \int_0^5 4\sin\left(\frac{t}{4}\right)dt$$

$$\approx 95 - 10.9548 \approx 84°\text{Fahrenheit}.$$

83. The correct answer is (A).

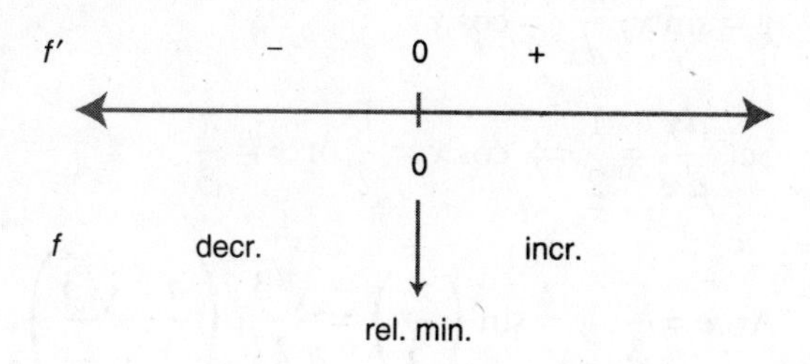

Only f has a relative minimum on (a, b).

84. The correct answer is (D).

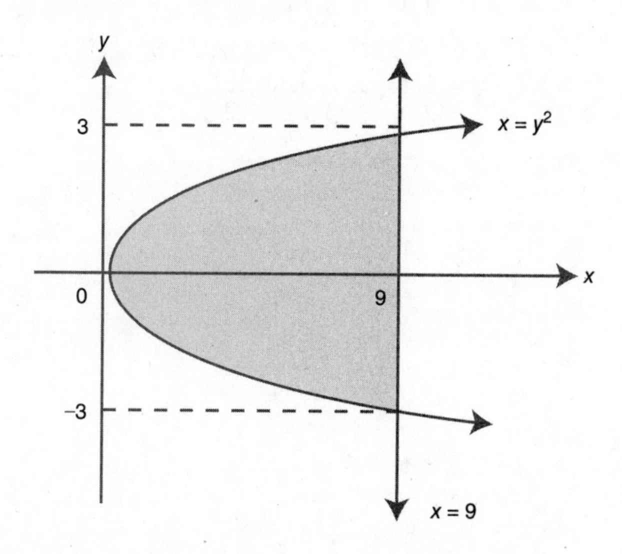

$$\text{Volume} = \pi\int_{-3}^{3}\left(9^2 - \left(y^2\right)^2\right)dy$$

$$= \frac{1944\pi}{5}.$$

85. The correct answer is (D).

$$y = e^x;\ \frac{dy}{dx} = e^x$$

$$y = x^2 + 5x;\ \frac{dy}{dx} = 2x + 5$$

If the graphs have parallel tangents at a point, then the slopes of the tangents are equal. Enter $y1 = e^x$ and $y2 = 2x + 5$. Using the [*Intersection*] function on your calculator, you obtain $x = -2.45$ and $x = 2.25$.

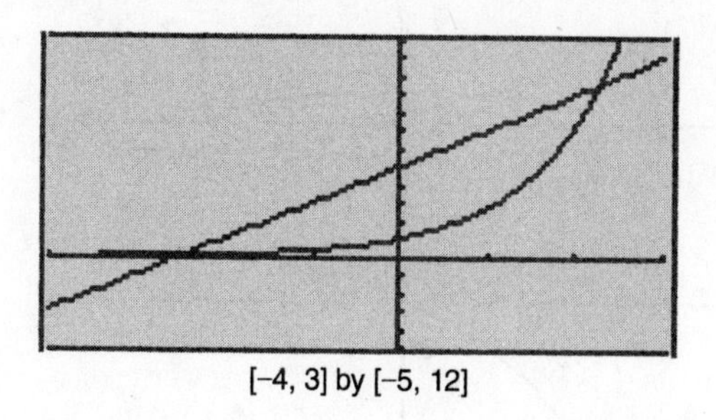
[-4, 3] by [-5, 12]

86. The correct answer is (D).

$$\text{Since } \frac{dy}{dx} = ky \Rightarrow y = y_0e^{kt}$$

$$\frac{3}{4}y_0 = y_0e^{k(6)} \Rightarrow \frac{3}{4} = e^{6k}$$

$$\Rightarrow \ln\left(\frac{3}{4}\right) = \ln\left(e^{6k}\right)$$

$$\Rightarrow \ln\frac{3}{4} = 6k \text{ or } k = \frac{\ln\left(\frac{3}{4}\right)}{6} = -0.048.$$

87. The correct answer is (B).

$$h'(x) = (x - 5)^3$$

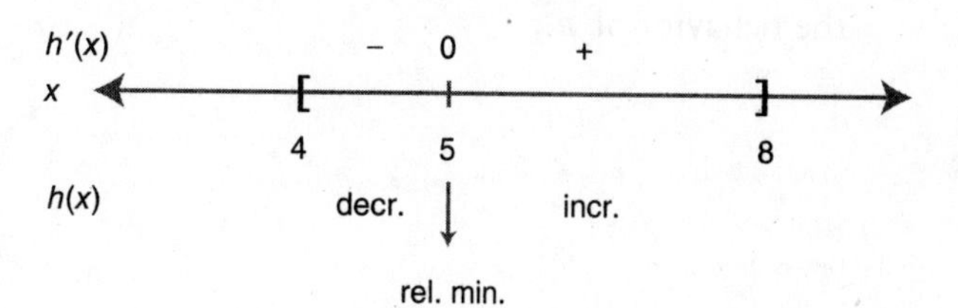

Thus, h has a relative minimum at $x = 5$.

88. The correct answer is (C).

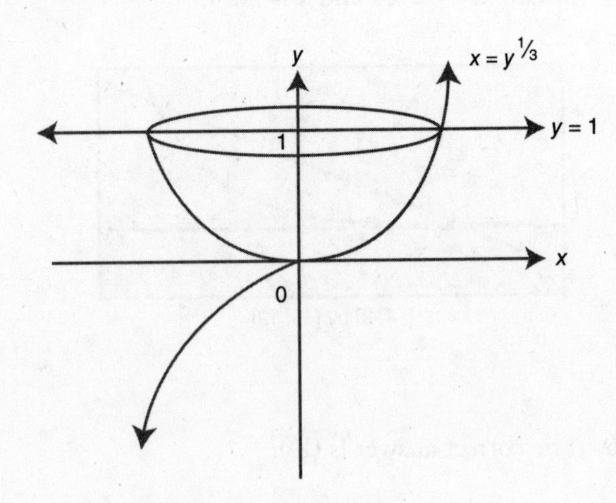

Volume $= \pi \int_0^1 \left(y^{1/3}\right)^2 dy$.

Using your calculator, you obtain $V = 3\pi/5$.

89. The correct answer is (D).

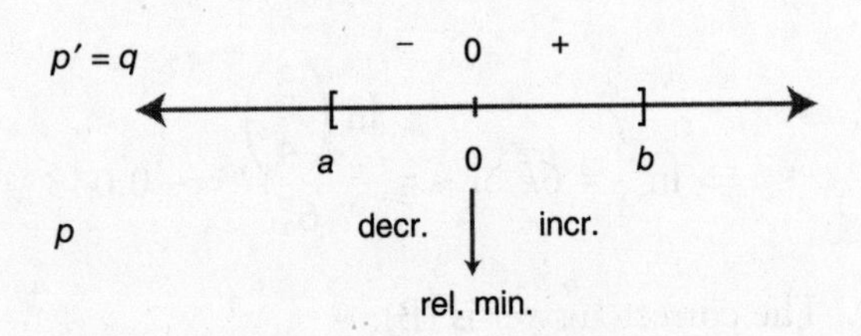

The graph in choice (D) is the one that satisfies the behavior of p.

90. The correct answer is (B).

$y = \sin x; \frac{dy}{dx} = \cos x$

Set $\frac{dy}{dx} = \frac{1}{2} \Rightarrow \cos x = \frac{1}{2}$ or $x = \frac{\pi}{3}$.

At $x = \frac{\pi}{3}$, $y = \sin\left(\frac{\pi}{3}\right) = \frac{\sqrt{3}}{2}$; $\left(\frac{\pi}{3}, \frac{\sqrt{3}}{2}\right)$.

Equation of tangent line at $x = \frac{\pi}{3}$:

$$y - \frac{\sqrt{3}}{2} = \frac{1}{2}\left(x - \frac{\pi}{3}\right) \text{ or}$$

$$y = \frac{1}{2}\left(x - \frac{\pi}{3}\right) + \frac{\sqrt{3}}{2}.$$

Solutions to AP Calculus AB Practice Exam 2—Section II

Section II—Part A

1.

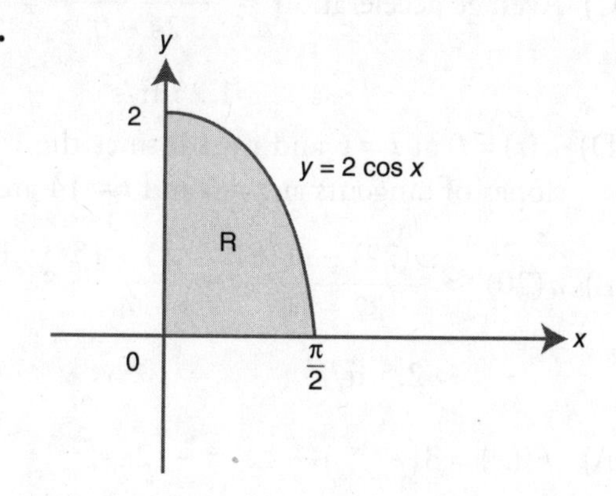

(A) Area of $R = \int_0^{\pi/2} 2\cos x\, dx$

$= 2\sin x]_0^{\pi/2}$

$= 2\sin\left(\frac{\pi}{2}\right) - 2\sin(0) = 2.$

(B) $\int_0^a 2\cos x\, dx = 1$

$2\sin x]_0^a = 2\sin a - 2\sin(0) = 2\sin a$

$2\sin a = 1 \Rightarrow \sin a = \frac{1}{2}$

$\Rightarrow a = \sin^{-1}\left(\frac{1}{2}\right) = \frac{\pi}{6}$

(C) Volume $= \pi \int_0^{\pi/2} (2\cos x)^2 dx$

$= \pi \int_0^{\pi/2} 4\cos^2 x\, dx$

$= 4\pi \int_0^{\pi/2} \cos^2 dx$

$= 4\pi \int_0^{\pi/2} \frac{1 + \cos(2x)}{2} dx$

$= 2\pi \int_0^{\pi/2} [1 + \cos(2x)]\, dx$

$= 2\pi \left[x + \frac{\sin(2x)}{2}\right]_0^{\pi/2}$

$= 2\pi \left[\left(\frac{\pi}{2} + \frac{\sin \pi}{2}\right) - 0\right] = \pi^2.$

(D) Area of cross section $= \frac{1}{2}\pi \left(\frac{2\cos x}{2}\right)^2$

$= \frac{1}{2}\pi \cos^2 x.$

$V = \int_0^{\pi/2} \frac{1}{2}\pi \cos^2 x\, dx$

$= \frac{1}{2}\pi \int_0^{\pi/2} \cos^2 x\, dx$

$= \frac{1}{2}\pi \int_0^{\pi/2} \frac{1 + \cos(2x)}{2} dx$

$= \frac{\pi}{4} \int_0^{\pi/2} (1 + \cos(2x) dx)$

$= \frac{\pi}{4} \left[x + \frac{\sin(2x)}{2}\right]_0^{\pi/2}$

$= \frac{\pi}{4} \left[\left(\frac{\pi}{2} + \frac{\sin \pi}{2}\right) - 0\right] = \frac{\pi^2}{8}$

2. (A)

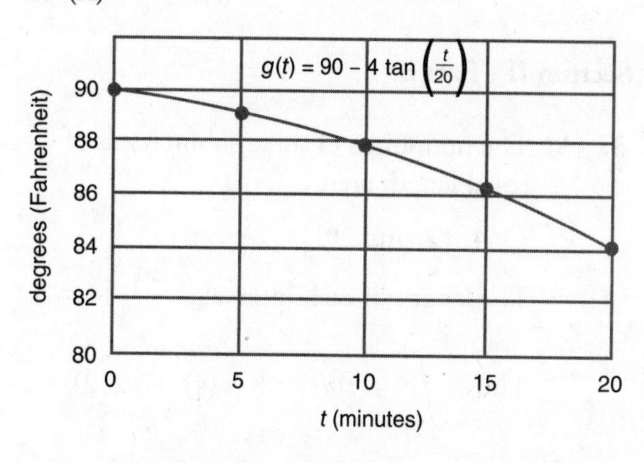

$g(10) = 90 - 4\tan\left(\frac{10}{20}\right) = 90 - 4\tan\left(\frac{1}{2}\right)$

$\approx 87.81°\text{F or } 87.82°\text{ F}$

(B) $g'(t) = -4\sec^2\left(\frac{t}{20}\right)\left(\frac{1}{20}\right)$

$g'(10) = -\frac{1}{5}\sec^2\left(\frac{10}{20}\right) \approx -0.26$

(C) Set the temperature of the liquid equal to 86°F. Using your calculator, let

$y1 = 90 - 4\tan\left(\frac{x}{20}\right)$; and $y2 = 86$.

To find the intersection point of y_1 and y_2, let $y3 = y1 - y2$ and find the zeros of y_3. Using the [*Zero*] function of your calculator, you obtain $x = 15.708$. Since $y_1 < y_2$ on the interval $15.708 < x \le 20$, the temperature of the liquid is below 86°F when $15.708 < t \le 20$.

(D) Average temperature below 86°F

$$= \frac{1}{20 - 15.708}\int_{15.708}^{20}\left(90 - 4\tan\left(\frac{x}{20}\right)\right)dx.$$

Using your calculator, you obtain:

Average temperature $= \frac{1}{4.292}(364.756)$

≈ 84.9851

$\approx 84.99°$ F.

Section II—Part B

3. (A) The midpoints of three subintervals of equal length are:

$t = 4$, 12, and 20.

The length of each interval is $\frac{24-0}{3} = 8$.

Thus, $\int_0^{24} v(t)dt \approx 8\,[v(4) + v(12) + v(20)]$

$\approx 8[25 + 15 + 20]$

$= 8(60) = 480.$

(B) Average velocity $= \frac{1}{24}\int_0^{24} v(t)dt$

$\approx \frac{1}{24}(480) = 20$ ft/s.

(C) Average acceleration $= \frac{v(24) - v(0)}{24 - 0} = \frac{30}{24}$

$= 1.25$ ft/s².

(D) $a(t) = 0$ at $t = 6$ and $t = 14$, since the slopes of tangents at $t = 6$ and $t = 14$ are 0.

(E) $a(20) \approx \frac{v(22) - v(18)}{22 - 18} \approx \frac{25 - 15}{4} \approx \frac{10}{4}$

≈ 2.5 ft/s²

4. (A) $f'(x) = 3\left(e^{-2x^2}\right)(-4x) = -12xe^{-2x^2}$

Setting $f'(x) = 0$, $-12xe^{-2x^2} = 0$

$\Rightarrow x = 0$.

(B) $f''(x) = (-12)\left(e^{-2x^2}\right)$

$+ (-12x)\left(e^{-2x^2}\right)(-4x)$

$= -12e^{-2x^2} + 48x^2e^{-2x^2}$

Setting $f''(x) = 0$, $12e^{-2x^2} + 48x^2e^{-2x^2} = 0$

$\Rightarrow 48x^2e^{-2x^2} = 12e^{-2x^2}$

$\Rightarrow 48x^2 = 12 \Rightarrow x^2 = \frac{1}{4}$ or

$x = \pm\frac{1}{2}$.

(C) $\lim_{x\to\infty} 3e^{-2x^2} = \lim_{x\to\infty} \frac{3}{e^{2x^2}} = 0$

$\lim_{x\to-\infty} 3e^{-2x^2} = \lim_{x\to-\infty} \frac{3}{e^{2x^2}} = 0$

(D)

f'	+	0	–
x	←	0	→
f'	incr.	rel. max.	decr.

$f'(x) = -12xe^{-2x^2} = \dfrac{-12x}{e^{2x^2}}$

$f(0) = 3$, since f has only one critical point (at $x = 0$), thus at $x = 0$, f has an absolute maximum. The absolute maximum value is 3.

(E) $f(x) = ae^{-bx^2}$, $a > 0$, $b > 0$

$f'(x) = ae^{-bx^2}(-2bx) = -2abxe^{-bx^2}$

Setting

$f'(x) = 0$, $-2abxe^{-bx^2} = 0 \Rightarrow x = 0$

$f'(x) = \dfrac{-2abx}{e^{bx^2}}$.

$f'(x) > 0$ if $x < 0$ and $f'(x) < 0$ if $x > 0$. Thus, f has a relative maximum at $x = 0$, and since it is the only critical point, f has an absolute maximum at $x = 0$. Since $f(0) = a$, the absolute maximum for f is a.

5. (A) $f(-3) = \displaystyle\int_0^{-3} g(t)dt = -\int_{-3}^{0} g(t)dt$

$= -\displaystyle\int_{-3}^{-1} g(t)dt - \int_{-1}^{0} g(t)dt$

$= -\left(-\dfrac{1}{2}(2)(2)\right) - \left(\dfrac{1}{2}(1)(2)\right)$

$= 2 - 1 = 1$

$f(3) = \displaystyle\int_0^{3} g(t)dt$

$= \displaystyle\int_0^{1} g(t)dt + \int_1^{3} g(t)dt$

$= \dfrac{1}{2}(1)(2) + \left(-\dfrac{1}{2}(1)(2)\right)$

$= 1 - 1 = 0$

(B) Note that $f'(x) = g(x)$, and $g(x) < 0$ on $(-3, -1)$ and $(1, 3)$ and that $g(x) > 0$ on $(-1, 1)$. The function f increases on $(-1, 1)$ and decreases on $(1, 3)$. Thus f has a relative maximum at $x = 1$. Also, f decreases on $(-3, -1)$ and increases on $(-1, 1)$. Thus, f has a relative minimum at $x = -1$.

(C) $f'(x) = g(x)$ and $f''(x) = g'(x)$

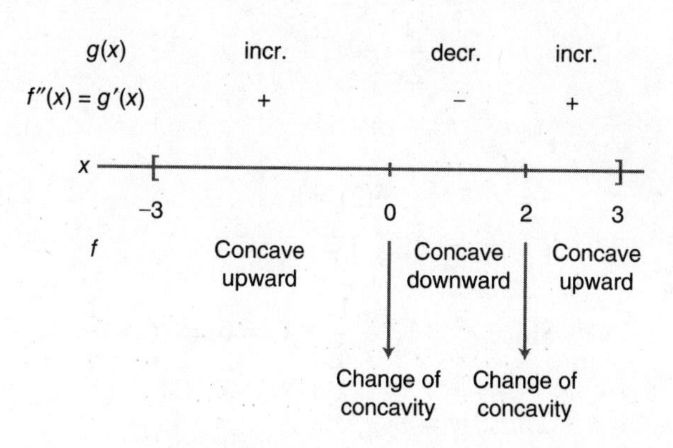

The function f has a change of concavity at $x = 0$ and $x = 2$.

(D) $f(1) = \displaystyle\int_0^{1} g(t)dt = \dfrac{1}{2}(1)(2) = 1$

$f'(1) = g(1) = 0$

Thus, $m = 0$, point $(1, 1)$; the equation of the tangent line to $f(x)$ at $x = 1$ is $y = 1$.

6. (A) $\dfrac{dy}{dx} = \dfrac{y}{2x^2}$; $(2,1)$

$\left.\dfrac{dy}{dx}\right|_{x=2, y=1} = \dfrac{1}{2(2)^2} = \dfrac{1}{8}$

Equation of tangent:

$y - 1 = \dfrac{1}{8}(x - 2)$ or

$y = \dfrac{1}{8}(x - 2) + 1.$

(B) $f(2.5) \approx \dfrac{1}{8}(2.5 - 2) + 1 = 1.0625$

≈ 1.063

(C) $\dfrac{dy}{dx} = \dfrac{y}{2x^2}$

$\dfrac{dy}{y} = \dfrac{dx}{2x^2}$ and $\displaystyle\int \dfrac{dy}{y} = \int \dfrac{dx}{2x^2}$

$$\ln|y| = \int \frac{1}{2}x^{-2}dx = \frac{1}{2}\frac{(x^{-1})}{-1} + C$$

$$= -\frac{1}{2x} + C$$

$$e^{\ln|y|} = e^{\left(-\frac{1}{2x} + C\right)}$$

$y = e^{-\frac{1}{2x} + C}$; $f(2) = 1$

$1 = e^{-\frac{1}{2(2)} + C} \Rightarrow 1 = e^{-\frac{1}{4} + C}$

Since $e^0 = 1$, $-\frac{1}{4} + C = 0 \Rightarrow C = \frac{1}{4}$.

Thus, $y = e^{-\frac{1}{2x} + \frac{1}{4}}$.

(D) $f(2.5) = e^{\left(-\frac{1}{2(2.5)} + \frac{1}{4}\right)}$

$= e^{\left(-\frac{1}{5} + \frac{1}{4}\right)} = e^{\frac{1}{20}}$

Scoring Sheet for AB Practice Exam 2

Section I—Part A

__________ × 1.2 = __________
No. Correct Subtotal A

Section I—Part B

__________ × 1.2 = __________
No. Correct Subtotal B

Section II—Part A (Each question is worth 9 points.)

__________ + __________ = __________
Q1 Q2 Subtotal C

Section II—Part B (Each question is worth 9 points.)

__________ + __________ + __________ + __________ = __________
Q1 Q2 Q3 Q4 Subtotal D

Total Raw Score (Subtotals A + B + C + D) = ☐

Approximate Conversion Scale:

Total Raw Score	Approximate AP Grade
80–108	5
65–79	4
50–64	3
36–49	2
0–35	1

180 Activities and Questions in

5 Minutes a Day

INTRODUCTION

Welcome to *5 Minutes to a 5: 180 Questions and Activities*! This bonus section is another tool for you to use as you work toward your goal of achieving a 5 on the AP exam in May. It includes 180 AP questions and activities that cover the most essential course materials and are meant to be completed in conjunction with the *5 Steps* book.

One of the secrets to excelling in your AP class is spending a bit of time *each day* studying the subject(s). The questions and activities offered here are designed to be done one per day, and each should take 5 minutes or so to complete. (Although there might be exceptions depending on the exam—some exercises may take a little longer, some a little less.) You will encounter stimulating questions to make you think about a topic in a big way and some very subject-specific activities that cover the main book's chapters; some science subjects will offer at-home labs, and some humanities subjects will offer ample chunks of text to be read on one day, with questions and activities for follow-up on the following day(s). There will also be suggestions for relevant videos for you to watch, websites to visit, or both. Most questions and activities are linked to the specific chapters of your book so that you are constantly fortifying your knowledge.

Remember—approaching this section for 5 minutes a day is much more effective than binging a week's worth in one sitting! So if you practice all the extra exercises in this section and reinforce the main content of this book, we are certain you will build the skills and confidence needed to succeed on your exam. Good luck!

—Editors of McGraw Hill

Check off each activity as it is completed.

1. ❑
2. ❑
3. ❑
4. ❑
5. ❑
6. ❑
7. ❑
8. ❑
9. ❑
10. ❑
11. ❑
12. ❑
13. ❑
14. ❑
15. ❑
16. ❑
17. ❑
18. ❑
19. ❑
20. ❑
21. ❑
22. ❑
23. ❑
24. ❑
25. ❑
26. ❑
27. ❑
28. ❑
29. ❑
30. ❑
31. ❑
32. ❑
33. ❑
34. ❑
35. ❑
36. ❑
37. ❑
38. ❑
39. ❑
40. ❑
41. ❑
42. ❑
43. ❑
44. ❑
45. ❑
46. ❑
47. ❑
48. ❑
49. ❑
50. ❑
51. ❑
52. ❑
53. ❑
54. ❑
55. ❑
56. ❑
57. ❑
58. ❑
59. ❑
60. ❑
61. ❑
62. ❑
63. ❑
64. ❑
65. ❑
66. ❑
67. ❑
68. ❑
69. ❑
70. ❑
71. ❑
72. ❑
73. ❑
74. ❑
75. ❑
76. ❑
77. ❑
78. ❑
79. ❑
80. ❑
81. ❑
82. ❑
83. ❑
84. ❑
85. ❑
86. ❑
87. ❑
88. ❑
89. ❑
90. ❑
91. ❑
92. ❑
93. ❑
94. ❑
95. ❑
96. ❑
97. ❑
98. ❑
99. ❑
100. ❑
101. ❑
102. ❑
103. ❑
104. ❑
105. ❑
106. ❑
107. ❑
108. ❑
109. ❑
110. ❑
111. ❑
112. ❑
113. ❑
114. ❑
115. ❑
116. ❑
117. ❑
118. ❑
119. ❑
120. ❑
121. ❑
122. ❑
123. ❑
124. ❑
125. ❑
126. ❑
127. ❑
128. ❑
129. ❑
130. ❑
131. ❑
132. ❑
133. ❑
134. ❑
135. ❑
136. ❑
137. ❑
138. ❑
139. ❑
140. ❑
141. ❑
142. ❑
143. ❑
144. ❑
145. ❑
146. ❑
147. ❑
148. ❑
149. ❑
150. ❑
151. ❑
152. ❑
153. ❑
154. ❑
155. ❑
156. ❑
157. ❑
158. ❑
159. ❑
160. ❑
161. ❑
162. ❑
163. ❑
164. ❑
165. ❑
166. ❑
167. ❑
168. ❑
169. ❑
170. ❑
171. ❑
172. ❑
173. ❑
174. ❑
175. ❑
176. ❑
177. ❑
178. ❑
179. ❑
180. ❑

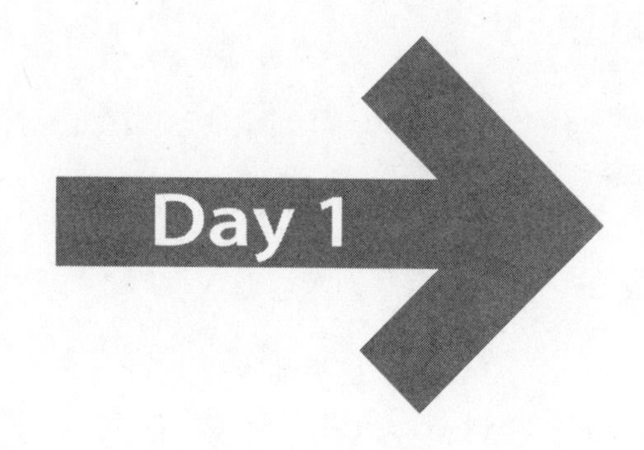

The start of your preparation, before you dig deep into calculus, is a good time to review earlier ideas that you'll call upon repeatedly in calculus, such as coordinate geometry. The formula for the distance between (x_1, y_1) and (x_2, y_2), $d = \sqrt{(x_2 - x_1)^2 + (y_2 - y_1)^2}$, is a modification of the Pythagorean Theorem. The equation of a circle with center (h, k) and radius r is the distance formula applied to the center (h, k) and any point (x, y) on the circle, or $(x - h)^2 + (y - k)^2 = r^2$.

Find the equation of a circle that passes through the point (3, −1), if the lines $3x - 2y = 12$ and $y = 5 - \frac{x}{3}$ contain diameters.

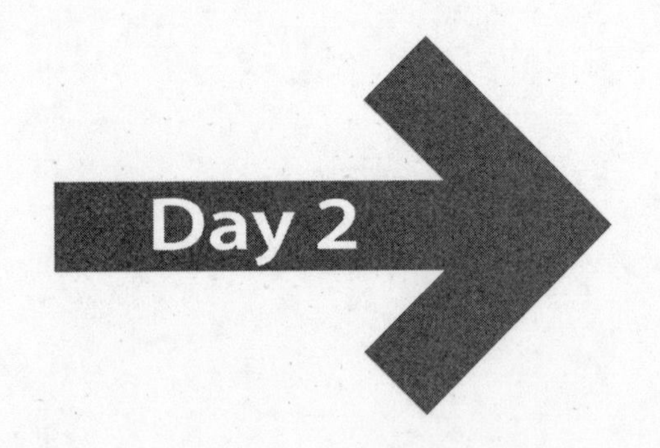

The slope of a line will appear throughout calculus. You'll find the slope, and the equation, of a tangent line to a curve and of a normal line, which is perpendicular to the tangent. Back in geometry you learned that a tangent to a circle is perpendicular to a radius drawn to the point of tangency, so the line containing a radius (or a diameter) is a normal line of the circle.

Complete the table with the missing information.

EQUATION OF CIRCLE	POINT OF TANGENCY	EQUATION OF TANGENT	EQUATION OF NORMAL LINE
$(x-6)^2+(y-3)^2=5^2$	$(3,-1)$		
$x^2+y^2=25$			$y-3=-\frac{3}{4}(x+4)$
$(x+2)^2+(y-1)^2=9$		$x=1$	

Throughout calculus, you'll hear references to intervals. An interval is simply a subset of the real numbers; visually, it's a piece of the real number line. The addition of the label "open" or "closed" serves to tell you whether or not the endpoints are included.

Write the inequalities that describe the graphs shown, and give the interval notation.

NAME	INEQUALITY	INTERVAL	GRAPH
Open interval			1. x: −4 −3 −2 −1 0 1 2 3 4 5 6 7 8
Infinite Interval			2. x: 2 3 4 5 6 7 8 9 10
Closed interval			3. x: −6 −5 −4 −3 −2 −1 0 1 2 3
Infinite Interval			4. x: −10 −9 −8 −7 −6 −5 −4 −3 −2 −1 0 1
Half-open or half-closed			5. x: 5 6 7 8 9 10
			6. x: −10 −9 −8 −7

The absolute value of a real number a is casually thought of as the distance of a from zero, without regard to direction. Formally, the absolute value of a number a is defined by means of a piecewise function:

$$|a| = \begin{cases} a & \text{if } a \geq 0 \\ -a & \text{if } a < 0 \end{cases}$$

Inequalities involving absolute value are best dealt with by translating into compound inequalities. Given that c is a positive real number:

MNEMONIC	ABSOLUTE VALUE INEQUALITY	TRANSLATES TO
GreatOR	$\lvert x \rvert > c$	$-c > x$ OR $x > c$
Less ThAND	$\lvert x \rvert < c$	$-c < x$ AND $x < c$

Solve each inequality.

1. $|7 - 4x| > 9$
2. $|3x - 5| + 2 < 3$
3. $|6 - 2(x - 4)| \leq 10$

In differential calculus, the first and second derivatives will provide important information about the behavior of a function, but you'll need to identify intervals on which each derivative is greater than zero or less than zero. Use the test point method to solve each inequality.

1. $x^2 - x - 6 > 0$
2. $(3 - x)(2 + x)(1 - x) \leq 0$
3. $\dfrac{x^2 - 3x - 10}{x - 2} \leq 0$

A function is defined as a mapping or pairing of elements from one set, called the domain, with elements of another, called the range. When not specified, the domain is assumed to be all real numbers except obvious discontinuities. New functions defined by arithmetic are generally simple to construct, but exist only where the two domains overlap.

If $f(x)=\frac{1}{x}$, $x\neq 0$ and $g(x)=\sqrt{x}$, $x\geq 0$, find the equation, in simplest form of the new function, and give its domain.

1. $(f+g)(x)$
2. $(fg)(x)$
3. $(f-g)(x)$
4. $\left(\frac{f}{g}\right)(x)$

5 Minutes to a 5

When confronted with a new function, you may find it helpful to do a quick analysis of the behavior of its graph. Choose the correct interval, point, or equation from Column B for each of these characteristics of the function $f(x)=\frac{x^3-1}{x}$. You may use items from Column B more than once, and you may use more than one item from Column B for a single item from Column A.

COLUMN A	**COLUMN B**
1. Domain	**(A)** $(2, -1)$
2. Range	**(B)** $(-\infty, 0)$
3. Increasing	**(C)** $(0, 1)$
4. Decreasing	**(D)** $(-\infty, -0.8)$
5. Constant	**(E)** $(-0.8, 1.9)$
6. Relative Maxima	**(F)** $(0, \infty)$
7. Relative Minima	**(G)** $x=0$
8. Intercepts	**(H)** $(-0.8, 0)$
9. Asymptotes	**(I)** $y=0$
	(J) $(1, 0)$
	(K) $(-\infty, 0)\cup(0, \infty)$
	(L) $(-\infty, \infty)$
	(M) None

Recognizing when the graph of a function exhibits symmetry, and what type of symmetry, can provide you quicker ways to approach problems like the area-under-a-curve problems from integral calculus. A function f is an even function, or has even symmetry, if for every x in the domain of f, $f(-x) = f(x)$. A function g is an odd function or has odd symmetry, if for every x in the domain of g, $g(-x) = -g(x)$. Tell whether each graph is Even, Odd, or Neither.

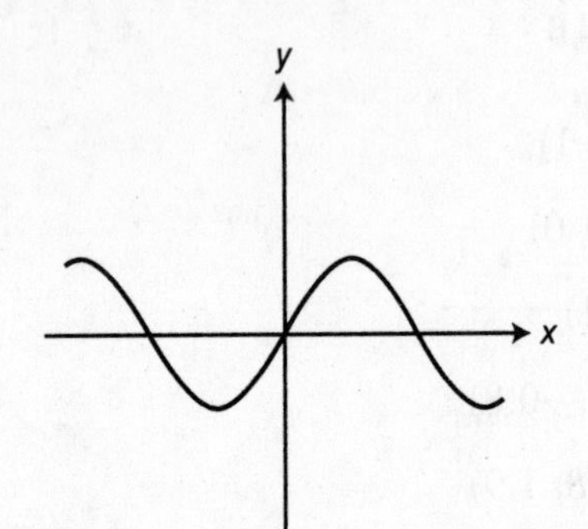

1.

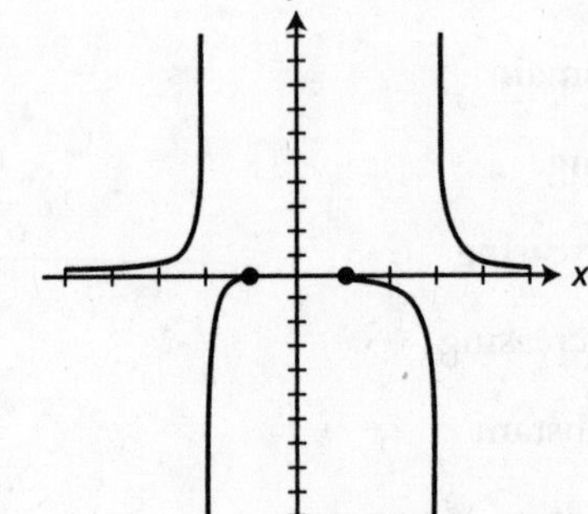

3.

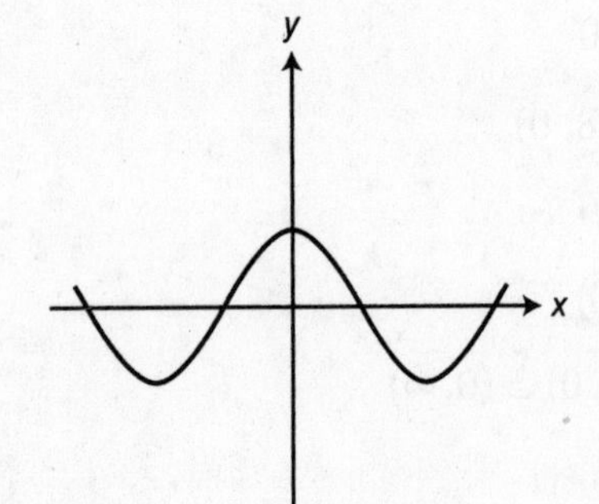

2.

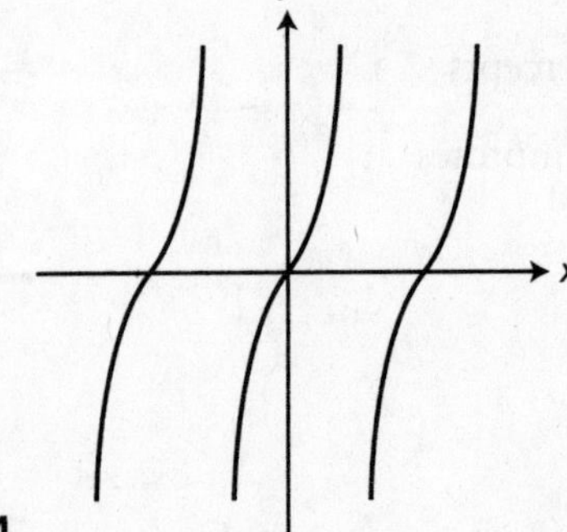

4.

Day 9

Before calculus provides a formal definition of what it means for a function to be continuous, we often talk about "a function you can draw without lifting your pencil." Identify any discontinuities in the functions below, and classify each discontinuity as Essential, Removable, or Jump.

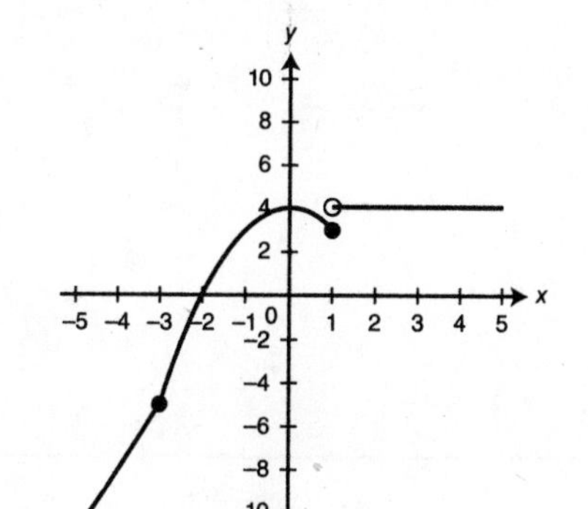

1.

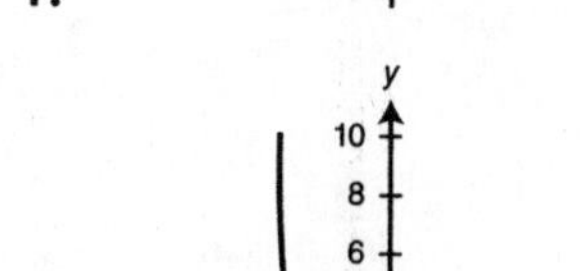

2.

3. $f(x) = 2x^3 - 5x^2 + 8x + 1$

4. $f(x) = \dfrac{x^2 - 9}{x - 3}$

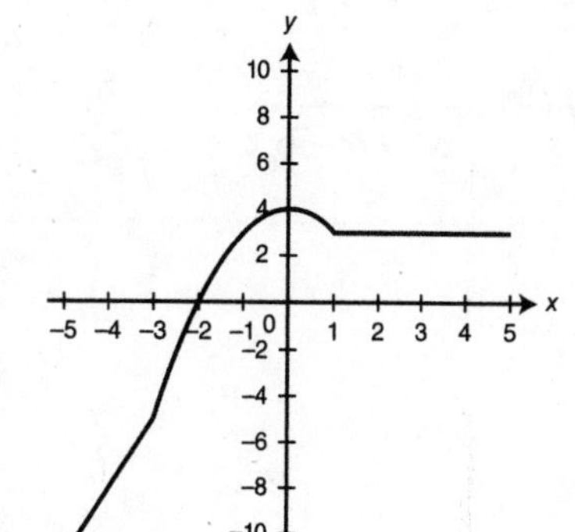

5.

Removable discontinuities or holes appear when one or more of the discontinuities of a rational function are caused by factors of the denominator that also appear in the numerator. Choose the functions that have removable discontinuities (or holes); locate and remove the hole.

1. $f(x) = \dfrac{x^2 - 2x - 35}{x + 5}$

2. $f(x) = \dfrac{x^2 - 3x - 1}{x^2 - 1}$

3. $f(x) = \dfrac{x^2 - 3x}{x^2 - x - 6}$

Vertical asymptotes give information about the behavior of the graph of a rational function near essential discontinuities. Horizontal and oblique asymptotes, on the other hand, provide information about the end behavior of the graph. Find the equation of a horizontal or oblique asymptote by dividing the numerator by the denominator and ignoring the remainder.

Match each function in Column A with its asymptote(s) in Column B. You may use an asymptote once, more than once, or not at all.

COLUMN A	COLUMN B
1. $f(x)=\dfrac{2x+1}{x-3}$	**(A)** $y=1$
2. $f(x)=\dfrac{x-3}{2x^2-x-1}$	**(B)** $y=2x-3$
3. $f(x)=\dfrac{2x^2-5x-3}{x-1}$	**(C)** $y=1-\dfrac{x}{2}$
4. $f(x)=\dfrac{3-x}{2x+1}$	**(D)** $y=2$
5. $f(x)=\dfrac{1-x}{3-x}$	**(E)** $y=-1$
6. $f(x)=\dfrac{3-x}{x-1}$	**(F)** $y=0$
7. $f(x)=\dfrac{x^2-4x+3}{2x+1}$	**(G)** $y=\dfrac{2x-9}{4}$
8. $f(x)=\dfrac{2x^2-5x-3}{x^2-4x+3}$	**(H)** $y=-\dfrac{1}{2}$
9. $f(x)=\dfrac{x^2-2x+1}{2x+1}$	**(I)** $y=2x$
10. $f(x)=\dfrac{2x+1}{2x-6}$	**(J)** $y=\dfrac{x}{2}-\dfrac{5}{4}$

Recognizing when a function has been formed by combining two simpler functions can give you an easier path through some calculus tasks. One rule in particular, the Chain Rule, is based on viewing a function as the composition of two simpler functions.

Given the functions $f(x) = 4x^2$ and $g(x) = \sqrt{x} + 3$, $x \geq 0$, find each composition and give its domain.

1. $(f \circ g)(x)$

2. $(g \circ f)(x)$

3. $(f \circ f)(x)$

4. $(g \circ g)(x)$

When the composition of two functions, regardless of order, produces just x, the two functions are inverses of one another. In order for a function to have an inverse function, both functions must be 1–1. Find the inverse of each function, restricting the domain of the function if necessary to ensure that it is a 1–1 function.

1. $f(x) = 4 - x^2$
2. $f(x) = \sqrt{2x+3}$
3.

x	−3	−2	−1	0	1	2	3
$f(x)$	1	−3	2	−2	3	−1	0

4.

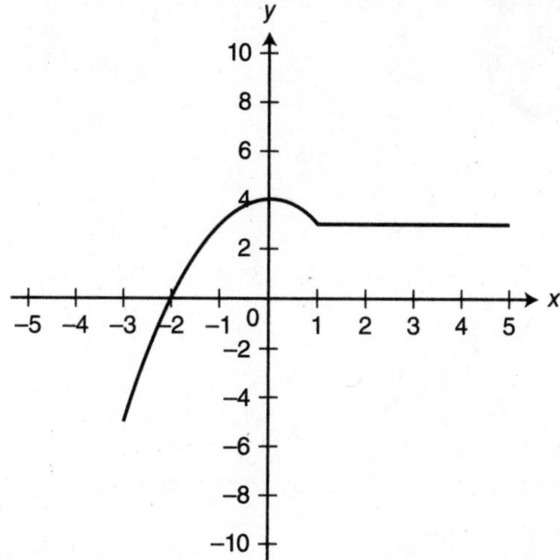

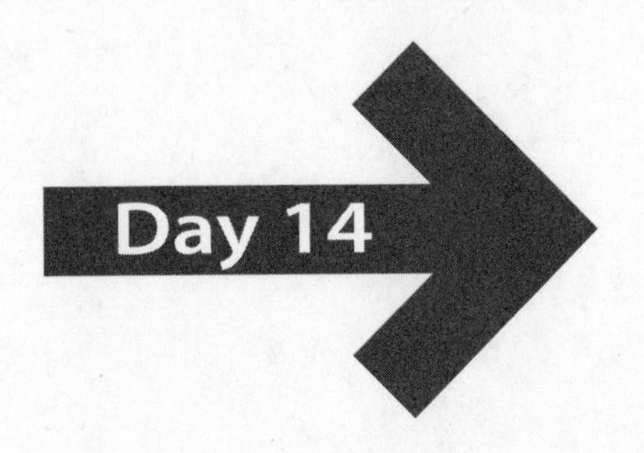

There are a few key strategies to deal with situations when you can't just evaluate a limit directly, and the most common of these is factoring and cancelling. Use this strategy to find the following limits.

1. $\lim_{x\to 1}\left(\frac{4x^2-3x-1}{x^2-3x+2}\right)$

2. $\lim_{x\to 2}\frac{x^2-4}{x^3-8}$

3. $\lim_{x\to 2}\frac{x^3-8}{x^2+3x-10}$

The trigonometric functions, because of their periodic nature, are not 1–1 functions, and to look at inverse trig functions, it's essential to restrict the domain of the function, and so the range of the inverse. $f^{-1}(x) = \sin^{-1} x$ only returns values in the interval $\left[-\frac{\pi}{2}, \frac{\pi}{2}\right]$, $f^{-1}(x) = \cos^{-1} x$ returns values in $[0, \pi]$, and $f^{-1}(x) = \tan^{-1} x$ returns values in $\left(-\frac{\pi}{2}, \frac{\pi}{2}\right)$.

Match each composition in Column A with its value from Column B.

COLUMN A	COLUMN B
1. $\sin^{-1}\left(\cos\left(\frac{2\pi}{3}\right)\right)$	**(A)** $\frac{3\pi}{4}$
2. $\cos^{-1}\left(\tan\left(\frac{7\pi}{4}\right)\right)$	**(B)** π
3. $\tan^{-1}\left(\tan\left(\frac{5\pi}{6}\right)\right)$	**(C)** $-\frac{\pi}{2}$
4. $\tan^{-1}(\sin(-\pi))$	**(D)** 0
5. $\cos^{-1}\left(\sin\left(\frac{5\pi}{4}\right)\right)$	**(E)** $\frac{\pi}{4}$
6. $\sin^{-1}\left(\sin\left(-\frac{\pi}{3}\right)\right)$	**(F)** $-\frac{\pi}{4}$
7. $\cos^{-1}\left(\cos\left(-\frac{\pi}{4}\right)\right)$	**(G)** $\frac{\pi}{2}$
8. $\tan^{-1}\left(\cos\left(-\frac{\pi}{2}\right)\right)$	**(H)** $-\frac{\pi}{3}$
9. $\sin^{-1}\left(\tan\left(\frac{7\pi}{4}\right)\right)$	**(I)** $-\frac{\pi}{6}$
10. $\tan^{-1}\left(\sin\left(-\frac{\pi}{2}\right)\right)$	**(J)** $\frac{\pi}{6}$

The formal definition of a limit is shown below.

Let f be a function defined on an open interval containing a, except possibly at a itself. $\lim_{x \to a} f(x) = L$ if for any real number $\varepsilon > 0$, there exists a real number $\delta > 0$ such that $|f(x) - L| < \varepsilon$ whenever $|x - a| < \delta$.

Apply the definition by answering the following questions for $\lim_{x \to 7}\left(x^2 - 7x + 12\right)$.

- What is the function f?
- What is the value of a?
- What is an appropriate open interval?
- Is f defined at a?
- What is the value of L?

Many of the properties of limits seem obvious. The limit of a constant is the constant. The limit of x, as x approaches a, is a. Add an exponent or a root and $\lim_{x \to a} x^n = a^n$ and $\lim_{x \to a} \sqrt[n]{x} = \sqrt[n]{a}$. The limit of a sum is the sum of the limits, and likewise for the difference, the product, and as long as $\lim_{x \to a} g(x) \neq 0$, the quotient.

Which of the properties are used in evaluating each of these limits?

$$\lim_{x \to 4} \left(\frac{11 - x}{3\sqrt{x}} \right)$$

Day 18

Evaluate

1. $\lim_{x \to -3} \frac{x^2 - 9}{x^2 - x - 12}$

2. $\lim_{x \to 3} \left(\frac{2x^2 - 12x + 18}{4x^2 - 24x + 36} \right)$

Rationalizing denominators is common throughout algebra, but when working on limits, you'll sometimes want to rationalize a numerator instead. Multiply both the denominator and the numerator by the conjugate of the expression you want to rationalize.

Use rationalization to find the limits.

1. $\lim_{x \to 0} \frac{x}{\sqrt{x+9}-3}$

2. $\lim_{x \to 4} \frac{\sqrt{x}-2}{x-4}$

3. $\lim_{x \to -3} \frac{\sqrt{x+3}}{\sqrt{x+3}-\sqrt{3}}$

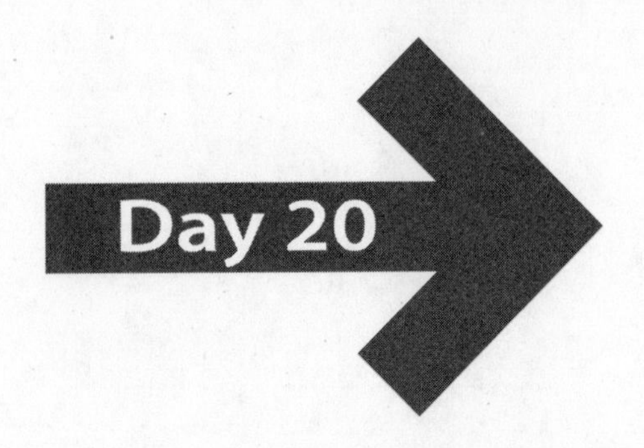

The Squeeze Theorem says that if you are able to find two functions $g(x)$ and $h(x)$ such that $g(x) \le f(x) \le h(x)$ and $\lim_{x \to a} g(x) = \lim_{x \to a} h(x) = L$, then $\lim_{x \to a} f(x)$ will also = L. It's common to use the Squeeze Theorem on expressions involving sin x or cos x because both can be squeezed between −1 and 1.

Use the Squeeze Theorem to find:

1. $\lim_{x \to \infty} \frac{\cos(\pi x)}{x - 1}$

2. $\lim_{x \to -\infty} \frac{2 - \sin\left(x - \frac{\pi}{2}\right)}{e^{-x}}$

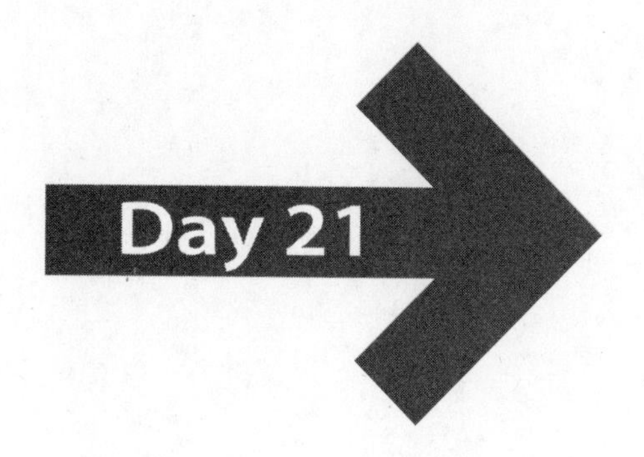

One of the most powerful results obtained by using the Squeeze Theorem is that $\lim\limits_{x\to 0}\dfrac{\sin x}{x} = 1$.

It can be applied to similar problems as long as the denominator matches the argument of the sine function. Use $\lim\limits_{x\to 0}\dfrac{\sin x}{x} = 1$ with appropriate adjustments to find the following limits.

1. $\lim\limits_{x\to 0}\dfrac{\sin(2x)}{2x}$

2. $\lim\limits_{x\to 0}\dfrac{\sin x}{2x}$

3. $\lim\limits_{x\to 0}\dfrac{\sin(3x)}{x}$

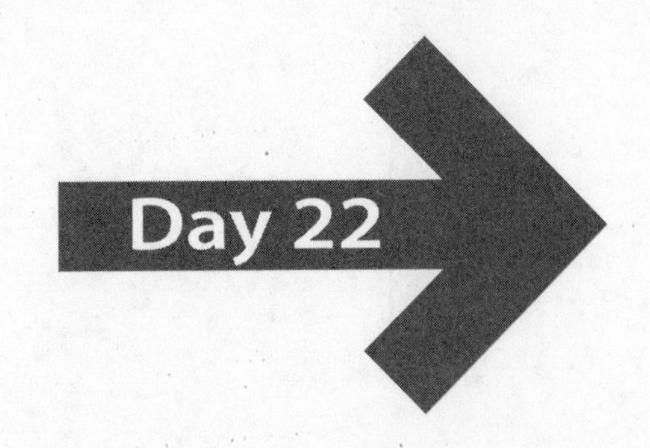

The notation $\lim\limits_{x \to a^-} f(x)$ or $\lim\limits_{x \to a^+} f(x)$ denotes a one-sided limit, the limit as x approaches a "from the left" or "from the right," respectively. If $\lim\limits_{x \to a^-} f(x) = \lim\limits_{x \to a^+} f(x)$, then $\lim\limits_{x \to a} f(x)$ exists and is equal to the one-sided limits.

Find each of the following limits:

1. $\lim\limits_{x \to 0^-} \frac{|x|}{x}$

2. $\lim\limits_{x \to 0^+} \frac{|x|}{x}$

3. $\lim\limits_{x \to 0} \frac{|x|}{x}$

Given $f(x) = \begin{cases} -2x - 7 & \text{if } x \le -2 \\ 4 - x^2 & \text{if } -2 < x \le 3 \\ -5 & \text{if } x > 3 \end{cases}$

4. $\lim\limits_{x \to -2^+} f(x)$

5. $\lim\limits_{x \to 3^-} f(x)$

6. $\lim\limits_{x \to 3} f(x)$

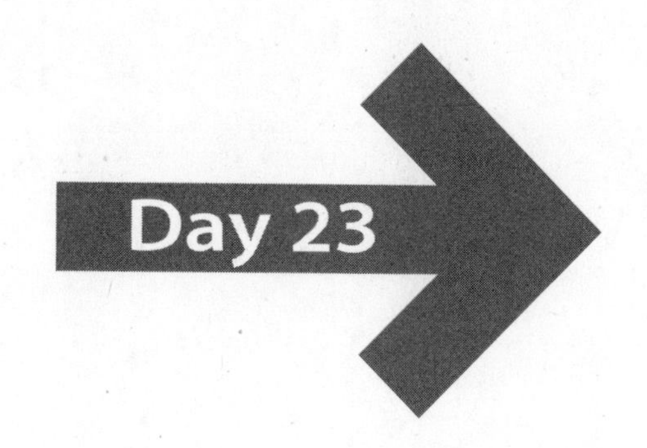

If a function f increases without bound as x approaches a, $\lim_{x \to a} f(x) = \infty$. If, as x approaches a, the function decreases without bound, $\lim_{x \to a} f(x) = -\infty$. Those limits only exist if the limit from the left and the limit from the right both exist and are equal.

Use the graph of $f(x)$ shown to find each of these limits.

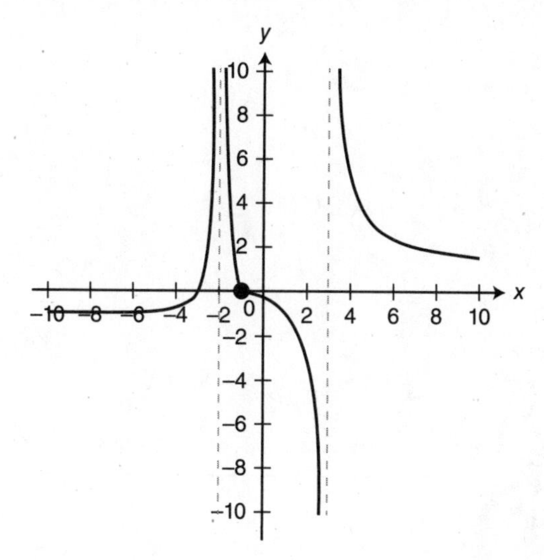

1. $\lim_{x \to 3^-} f(x)$
2. $\lim_{x \to 3^+} f(x)$
3. $\lim_{x \to 3} f(x)$
4. $\lim_{x \to -2^-} f(x)$
5. $\lim_{x \to -2^+} f(x)$
6. $\lim_{x \to -2} f(x)$
7. $\lim_{x \to -1} f(x)$

Finding that $\lim_{x \to a} f(x) = \infty$ or $-\infty$ makes it clear that the function has a vertical asymptote $x = a$. It's also possible to consider what happens to the function as x increases without bound or decreases without bound. There may be a horizontal or an oblique asymptote or the function may increase or decrease without bound.

Find each limit.

1. $\lim_{x \to \infty} \frac{2x-3}{x+4}$
2. $\lim_{x \to -\infty} \left(4 - 3x + 2x^2\right)$
3. $\lim_{x \to \infty} \frac{2x-3}{x^2 - 3x + 1}$
4. $\lim_{x \to -\infty} \frac{4}{x^2}$
5. $\lim_{x \to -\infty} \frac{3x^5}{2x^2}$

Because $\lim_{x\to0}\left(\frac{x}{\sin x}\right)\cdot\lim_{x\to0}\left(\frac{\sin x}{x}\right)=\lim_{x\to0}\left(\frac{x}{\sin x}\cdot\frac{\sin x}{x}\right)=\lim_{x\to0}1=1$, it's possible to show that $\lim_{x\to0}\frac{x}{\sin x}=1$. Use $\lim_{x\to0}\left(\frac{x}{\sin x}\right)$ and $\lim_{x\to0}\frac{\sin x}{x}$ to find the following limits.

1. $\lim_{x\to0}\frac{x\cos x}{\sin 2x}$
2. $\lim_{x\to0}\left(\frac{-\pi\sin(-x)}{\sin(\pi x)}\right)$
3. $\lim_{x\to0}\frac{\sin(2x)}{\sin(3x)}$

The formal definition of continuity at a point, *a*, requires three things: the function exists at *a*; the limit of the function, as *x* approaches *a*, exists; and that limit is equal to the function value.

Determine if each of the functions shown below is continuous at each of the given points by checking each of three requirements of the definition.

$$f(x)=\begin{cases} 7-2|x+4| & \text{if } x\le -1 \\ \dfrac{1}{x^2} & \text{if } -1<x<4 \\ \dfrac{x-4}{5-x} & \text{if } x\ge 4 \end{cases}$$

1. $x=-4$

2. $x=-1$

3. $x=0$

4. $x=4$

5. $x=5$

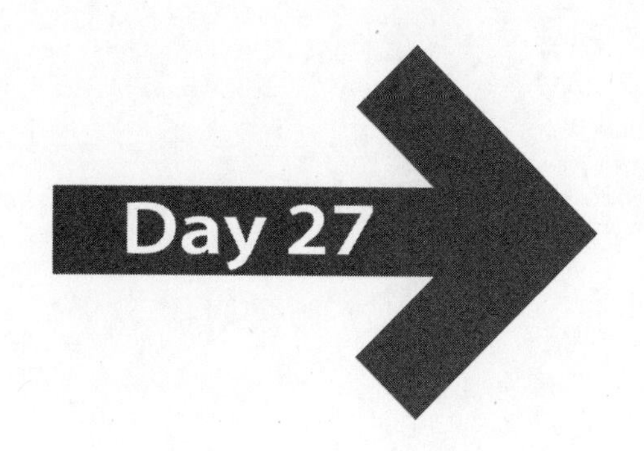

Each of the functions below has at least one discontinuity. Locate each discontinuity and classify it as Essential, Removable or Jump.

1. $f(x)=\dfrac{|x|}{x}$

2. $f(x)=\dfrac{x^2-4}{x^2-5x+6}$

3. $f(x)=\begin{cases} x^2+10x+21 & \text{if } x \le -2 \\ 6-x & \text{if } -2<x<3 \\ -x^2+8x-12 & \text{if } x \ge 3 \end{cases}$

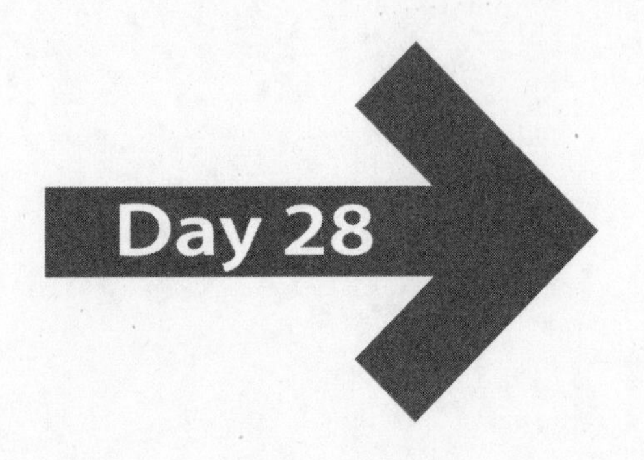

Find a value for the constant k, if possible, that will make the function continuous.

1. $f(x)=\begin{cases} 1-4x & x\le 0 \\ x^2+k & x>0 \end{cases}$

2. $f(x)=\begin{cases} kx^2 & x\le -1 \\ 3x-k & x>-1 \end{cases}$

3. $f(x)=\begin{cases} \dfrac{\sin 2x}{x} & x\ne 0 \\ k-2x & x=0 \end{cases}$

If f is continuous on a closed interval $[a, b]$, and c is any number between $f(a)$ and $f(b)$ inclusive, then there is at least one number x in the closed interval such that $f(x) = c$.

The Intermediate Value Theorem above is often used to assert that a continuous function has a zero in a closed interval, if the conditions of the theorem are met, and $f(a)$ and $f(b)$ have opposite signs. For each of the following functions, determine if the function has a zero in the given interval.

1. $f(x)=\dfrac{7-2x}{x}$ on $[3, 4]$
2. $f(x)=\dfrac{7-2x}{x}$ on $[-1, 1]$
3. $f(x)=\dfrac{2x^2-3}{x^2+1}$ on $[-2, -1]$
4. $f(x)=\dfrac{2x^2-3}{x^2+1}$ on $[-1, 1]$

For a function f, which passes through the points $(x, f(x))$ and $(x+\Delta x, f(x+\Delta x))$, where Δx is a small number, the slope of the line connecting the points, $\dfrac{f(x+\Delta x)-f(x)}{(x+\Delta x)-x}$, is called the difference quotient.

Find the difference quotient for each function, in simplest form. Then find $\lim\limits_{\Delta x \to 0} \dfrac{f(x+\Delta x)-f(x)}{(x+\Delta x)-x}$.

1. $f(x)=2x+5$
2. $f(x)=3-2x^2$
3. $f(x)=\dfrac{1}{x}$

The derivative f' of a function f is defined as $f'(x)=\lim_{h\to 0}\frac{f(x+h)-f(x)}{h}$. Use the definition to find the derivative of each function.

1. $f(x)=x^3+8$
2. $f(x)=\frac{2x}{x-3}$
3. $f(x)=\sqrt{x-2}$

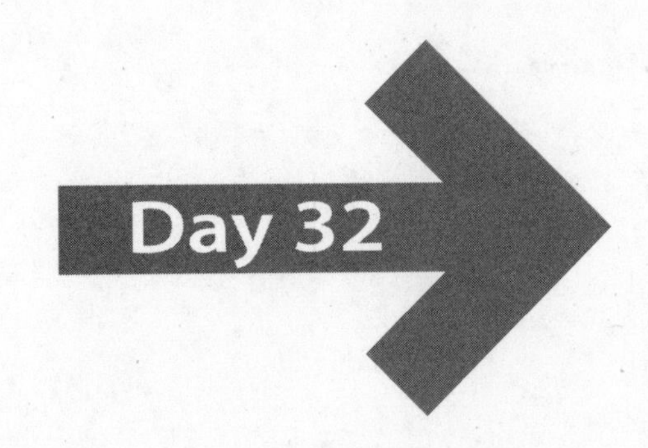

The derivative of a constant is zero. The derivative of a sum (or difference) is the sum (or difference) of the derivative of the individual terms. The Power Rule asserts that the derivative of x^n is nx^{n-1}. Use these fundamental rules to find the derivative of each of the polynomial functions.

1. $f(x) = 3x^7 - 4x^5 + x^2 - 1$
2. $f(x) = 12 - 8x^9 + 5x^6$
3. $f(x) = \frac{2}{3}x^{12} + \frac{1}{5}x^{10} - \frac{3}{4}x^8 - \frac{1}{3}x^6$
4. $f(x) = x^7 - x^5 + x^3 - x$
5. $f(x) = 2 - x^4 + 4x^6 - 3x^8$

The Product Rule says that the derivative of a product of two functions, u and v is $uv' + vu'$, and the Quotient Rule ensures that the derivative of $\frac{u}{v}$ is $\frac{vu' - uv'}{v^2}$.

Use the Product and Quotient Rules to find the derivative of each function.

1. $f(x) = (3x - 7)(x^2 - 4)$
2. $f(x) = 2x^4\sqrt{x}$
3. $f(x) = (x^2 + 1)(2x^3 + 5)$
4. $f(x) = \dfrac{4 - 3x^2}{2x + 5}$

Use the table below to estimate each of the following derivatives.

x	−2	−1	0	1	2	3
f	10.99	13.86	16.09	17.92	19.46	20.79

1. $f'(-1)$
2. $f'(0)$
3. $f'(2)$

The Chain Rule provides a method of differentiating a function that is formed by composing two (or more) simpler functions. Use the Chain Rule to find the derivative of each of the following functions.

1. $f(x) = \left(x^2 - 3x + 2\right)^2$
2. $f(x) = \sqrt{4 + 5x^2}$
3. $f(x) = \dfrac{1}{x^3 - 2x + 5}$
4. $f(x) = (3x - 1)^2(2x + 3)^3$
5. $f(x) = \dfrac{x^2 - 4}{(2x + 1)^2}$

Given functions f and g continuous and differentiable on $[-1, 2]$. Let $u(x) = f(g(x))$, $v(x) = g\left(f(x)\right)$ and $w(x) = f\left(f(x)\right)$. Use the information in the table to evaluate each derivative.

x	-1	0	1	2
f	0	-2	2	-1
g	3	1	0	2
f'	-1	4	-3	1
g'	-2	-1	2	3

1. $\left.\dfrac{d}{dx}(f+g)\right|_{x=2}$
2. $\left.\dfrac{d}{dx}\left(\dfrac{g}{f}\right)\right|_{x=1}$
3. $\left.\dfrac{d}{dx}u\right|_{x=0}$
4. $\left.\dfrac{d}{dx}v\right|_{x=-1}$
5. $\left.\dfrac{d}{dx}w\right|_{x=2}$

Use the derivatives of the six trig functions and the Product, Quotient, and Chain Rules to find the derivatives.

1. $\frac{d}{dx}(4-\cos(2x))$
2. $\frac{d}{dx}\left(\tan^2\left(\frac{\pi}{2}x\right)\right)$
3. $\frac{d}{dx}\left(3x^2\left(\sin^2 x\right)\right)$
4. $\frac{d}{dx}\left(\frac{\csc x}{1-1\cot x}\right)$
5. $\frac{d}{dx}(\sin(3x)\cos(2x))$

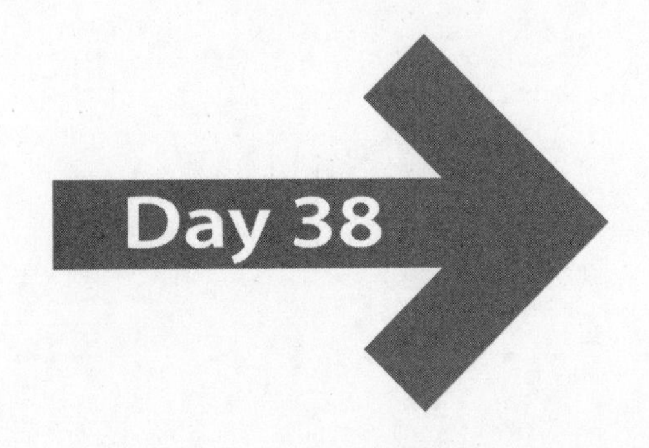

Derivatives of inverse trig functions demand some memory work. They are not intuitive, but practice will help fix them in memory. Find the derivatives of the following functions that contain inverse trig functions.

1. $f(x) = x^3 \arccos x$
2. $f(x) = \dfrac{2 + \arccos x}{3 - \arcsin x}$
3. $f(x) = (\arctan x)(\arcsin x)$
4. $f(x) = x^2 (\arcsin x)^2$
5. $f(x) = \left(\arctan\left(x^2\right)\right)^3$

Don't let the simplicity of the derivative of e^x lead you into errors. Apply the Chain Rule as appropriate and remember that for bases other than e, $\frac{d\left(b^x\right)}{dx} = b^x \ln b, b > 0$.

Find each derivative.

1. $\frac{d}{dx}\left(e^{x^2}\right)$
2. $\frac{d}{dx}\left(e^{x^2}\sqrt{x}\right)$
3. $\frac{d}{dx}\left(e^{\cos(2x)}\right)$
4. $\frac{d}{dx}\left(2^x \cdot \sin^2 x\right)$
5. $\frac{d}{dx}\left(\frac{e^{2x+1}}{2^{1-3x}}\right)$

As is the case with derivatives of exponential functions, the derivative of $\ln x$ is simple: $\frac{1}{x}$. For other bases, the derivative involves the log of the base: $\frac{d}{dx}\log_b x = \frac{1}{x \ln b}, x > 0, b > 0,$ and $b \neq 1$. The Chain Rule may be necessary as well, depending on the argument of the log.

For each logarithmic function in column A, choose the correct derivative from column B. Derivatives may be used more than once.

COLUMN A

1. $f(x) = \ln(x+3)$
2. $f(x) = \ln\left(x^2+3\right)$
3. $f(x) = \ln(2x+3)$
4. $f(x) = \ln\left(x^2+3x\right)$
5. $f(x) = 2\ln(x)+3$
6. $f(x) = 2\log_3(x)$
7. $f(x) = \log_3\left(x^2\right)$
8. $f(x) = \log_2\left(x^3\right)$
9. $f(x) = \log_3\left(x^2+3x\right)$
10. $f(x) = \log_3(2x+3)$

COLUMN B

(A) $f'(x) = \frac{2}{2x+3}$

(B) $f'(x) = \frac{1}{x+3}$

(C) $f'(x) = \frac{2}{x}$

(D) $f'(x) = \frac{2}{x \ln 3}$

(E) $f'(x) = \frac{3}{x \ln 2}$

(F) $f'(x) = \frac{2x}{x^2+3}$

(G) $f'(x) = \frac{2x+3}{x^2+3x}$

(H) $f'(x) = \frac{2}{(\ln 3)(2x+3)}$

(I) $f'(x) = \frac{2x+3}{(\ln 3)(x)(x+3)}$

In many cases, it's a simple matter to solve an equation to express y explicitly as a function of x. The equation $4x^2 - 2y = 12$ can be transformed to $y = 2x^2 - 6$. When that is not possible, or when the explicit form is difficult to differentiate, implicit differentiation may be the better choice. In each of the following, assume y is a function of x and find the derivative of y with respect to x.

1. $y^3 - x^2 = 4$
2. $2x - 3y + 1 = (x + y)^2$
3. $x^2 y + xy^2 = x$
4. $e^{x+y} = e^x + e^y$

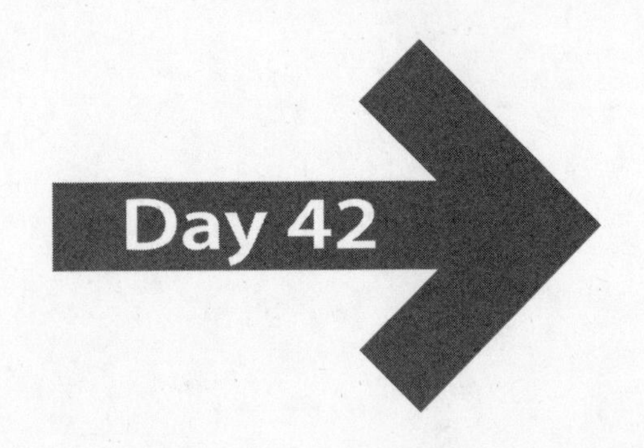

Implicit differentiation can be useful when functions involve trigonometric expressions, exponentials, or logs. Find the derivative of y with respect to x for each of the following.

1. $x \sin y - y \cos x = x^2$
2. $\sin(x + y) + \cos(x - y) = 1$
3. $e^{x+y} = x^2 - 3y$
4. $\ln\left(\frac{x}{y}\right) = x + y$

The rule for the derivative of an inverse is a valuable tool in those situations for which it is difficult or impossible to find the derivative of the inverse function directly. If f is a function that has an inverse $f^{-1}(x)$, and if f is differentiable at $f^{-1}(x)$, and if $f'\left(f^{-1}(x)\right) \neq 0$, then $\frac{d}{dx} f^{-1}(x) = \frac{1}{f'\left(f^{-1}(x)\right)}$.

Let $f(x)$ and $g(x)$ be inverse functions. Values of $f(x)$ and $f'(x)$ are given in the table below. Find the values of $g'(x)$ requested below.

x	−3	−2	−1	0	1	2	3
f(x)	9	4	1	−1	−2	−3	−10
f′(x)	−6	−4	−2	−1	−1	−3	−14

1. $g'(1)$
2. $g'(-1)$
3. $g'(-3)$

The derivative of a function, depending on context, may give information about instantaneous rate of change, velocity, or the increasing or decreasing behavior of the function graph. The second derivative, taken in context, may talk about acceleration or concavity. Finding higher-order derivatives uses the techniques of differentiation repetitively.

Find the requested derivatives.

1. $f(x) = x^5 - 2x^2 + 3,\ f'''(x)$
2. $f(x) = \dfrac{3x-1}{x+2},\ f''(x)$
3. $f(x) = e^{x^2-4},\ f'''(x)$
4. $f(x) = 3\sin^2 x,\ f'''(x)$
5. $f(x) = \cos x,\ f^{(4)}(x)$

Rolle's Theorem states: If $f(x)$ is continuous on the closed interval $[a, b]$ and differentiable on the open interval (a, b), and if $f(a) = f(b)$, then there is a number c such that $a < c < b$ and $f'(c) = 0$.

Check if Rolle's Theorem applies in each of the following situations, and if so, find the value of c.

1. $f(x) = x^3 - 3x^2 + 1$ on $[-1, 2]$

2. $f(x) = 2 - 5\cos\left(\frac{x}{4}\right)$ on $[-\pi, \pi]$

The Mean Value Theorem states that if $f(x)$ is continuous on the closed interval $[a, b]$, and differentiable on the open interval (a, b), then there is a number c such that $a < c < b$ and $f'(c) = \frac{f(b) - f(a)}{b - a}$.

For each of the following functions, verify that the conditions of the Mean Value Theorem are met, and find a value, c, at which a tangent line is parallel to the line containing $(a, f(a))$ and $(b, f(b))$.

1. $f(x) = x^3 - 2x^2 - x + 2$ on $[-2, 2]$
2. $f(x) = \sin^2 x + 2\cos x$ on $[0, 2\pi]$
3. $f(x) = 1 + e^{-x}$ on $[0, 1]$

The Extreme Value Theorem guarantees that if a function $f(x)$ is continuous on a closed interval $[a, b]$, it has both a maximum and minimum value on $[a, b]$. Note that the extrema will occur on the closed interval, so it is important to remember to examine the endpoints.

Locate the maximum and minimum values of the function on the given interval.

1. $f(x) = \dfrac{3 - x^2}{x^2 + 3}$ on $[-1, 2]$
2. $f(x) = x^3 - 5x^2 + 3x - 4$ on $[-2, 5]$
3. $f(x) = x^4 - 2x^2$ on $\left[-\dfrac{3}{2}, \dfrac{1}{2}\right]$

The first derivative is positive when the function is increasing, negative when the function is decreasing, and zero at critical points. Use the first derivative test to determine where each function is increasing and where it is decreasing.

1. $f(x) = 7 + 6x - x^3$
2. $f(x) = x^3 - 6x^2 + 12x + 5$
3. $f(x) = -x^3 + 6x - 1$
4. $f(x) = x^3 - 4x^2 + 4x + 6$

A critical point is a relative maximum if at that point the function changes from increasing to decreasing, and a relative minimum if the function changes from decreasing to increasing. Use the first derivative test to determine whether the given critical point is a relative maximum or a relative minimum.

1. $f(x) = x^4 - 8x^2 + 3$, critical point: $x = -2$
2. $f(x) = e^{-x^2/2}$, critical point: $x = 0$
3. $f(x) = x^2 \ln x$, critical point: $x = \dfrac{1}{\sqrt{e}} \approx 0.607$
4. $f(x) = \cos^2 x - 2\sin x$, critical point: $x = \dfrac{\pi}{2}$
5. $f(x) = \cos^2 x - 2\sin x$, critical point: $x = \dfrac{3\pi}{2}$

The sign of the first derivative indicates whether the function is increasing or decreasing. In a similar fashion, the sign of the second derivative communicates the concavity of the graph, whether it holds water or spills water. For the second derivative, a positive value indicates the graph is concave upward, holding water, while a negative result occurs when the graph is concave downward, spilling water. At a point of inflection, a change in concavity must occur, and the graph of the function has a tangent line.

Find any inflection points for the function, and then determine where the function is concave upward and where it is concave downward.

1. $f(x) = (x-1)^3$
2. $f(x) = \ln\left(4+x^2\right)$

Testing the sign of the first derivative to the left and right of a critical point can identify the point as a relative maximum or relative minimum. The same information can be obtained by letting the second derivative tell whether the graph is concave upward, indicating a minimum, or concave downward, at a relative maximum.

Find the critical points and use the second derivative test to identify each as a relative maximum or a relative minimum.

1. $f(x) = \sin^2 x\ ,\ 0 < x < 2\pi$
2. $f(x) = \frac{2}{3}x^3 - \frac{1}{2}x^2 - 3x + 4$
3. $f(x) = \frac{2x - 1}{x^2}\ ,\ x > 0$

The first and second derivatives combine to provide quite a bit of information about the graph of a function: relative extrema, points of inflection, increasing, decreasing, constant, concavity. Add domain and range, nature of discontinuities, behavior near asymptotes, end behavior, and intercepts, and you have a good image of the graph (without the help of technology).

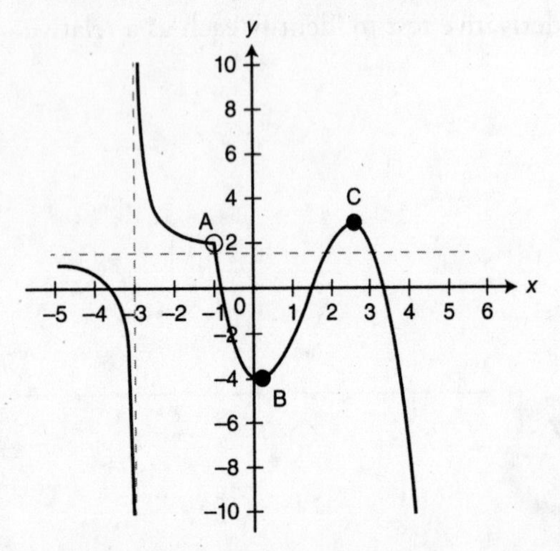

Use the graph of $f(x)$ shown above to complete each of the statements.

1. On the interval from A to B, $f'(x)$ is __________________.

2. On $[-1,1]$, $f''(x)$ is __________________.

3. At point C, $f'(x)$ is ____________ and $f''(x)$ is __________________.

4. $\lim_{x\to -3^-} f(x) =$ ____________ and $\lim_{x\to -3^+} f(x) =$ ____________, therefore,

$\lim_{x\to -3} f(x) =$ ____________.

5. $\lim_{x\to -\infty} f(x) =$ ____________ and $\lim_{x\to \infty} f(x) =$ ____________.

Day 53

The graph of the function $f(x)$ is shown. Sketch a graph of $f'(x)$.

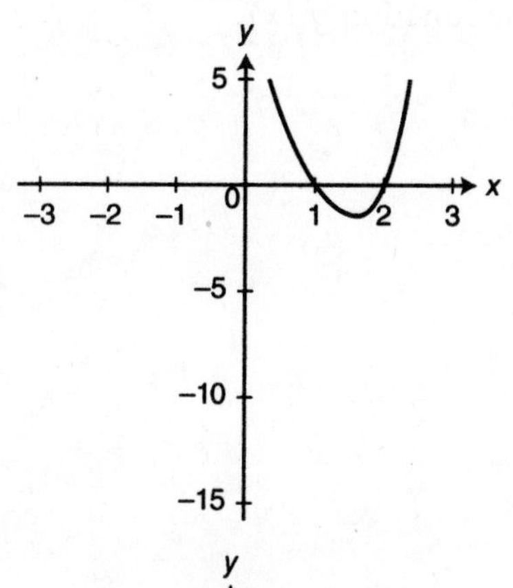

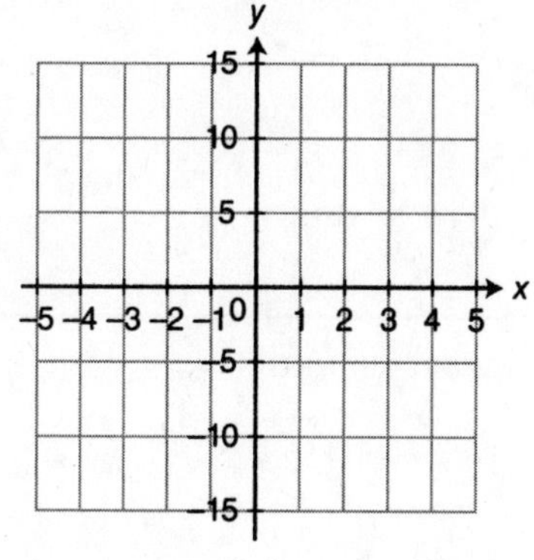

The graph shown is the first derivative, $f'(x)$, of a function $f(x)$. Use information derived from the first derivative to sketch a graph of the function $f(x)$.

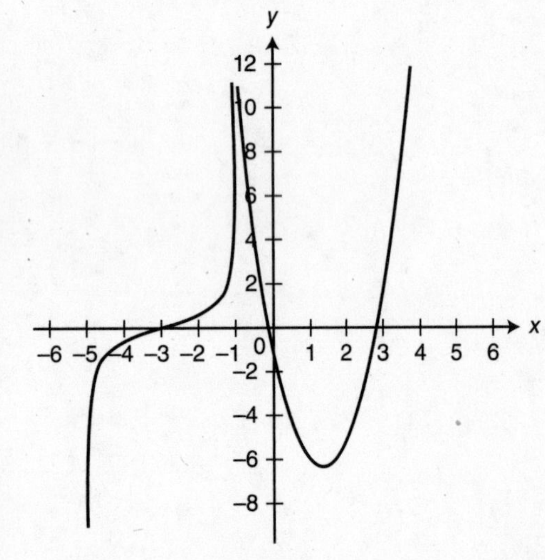

Day 55

The graphs below show three functions, their first derivatives, and their second derivatives. Arrange the graphs into three groups, by function, and order them from function to first derivative to second derivative.

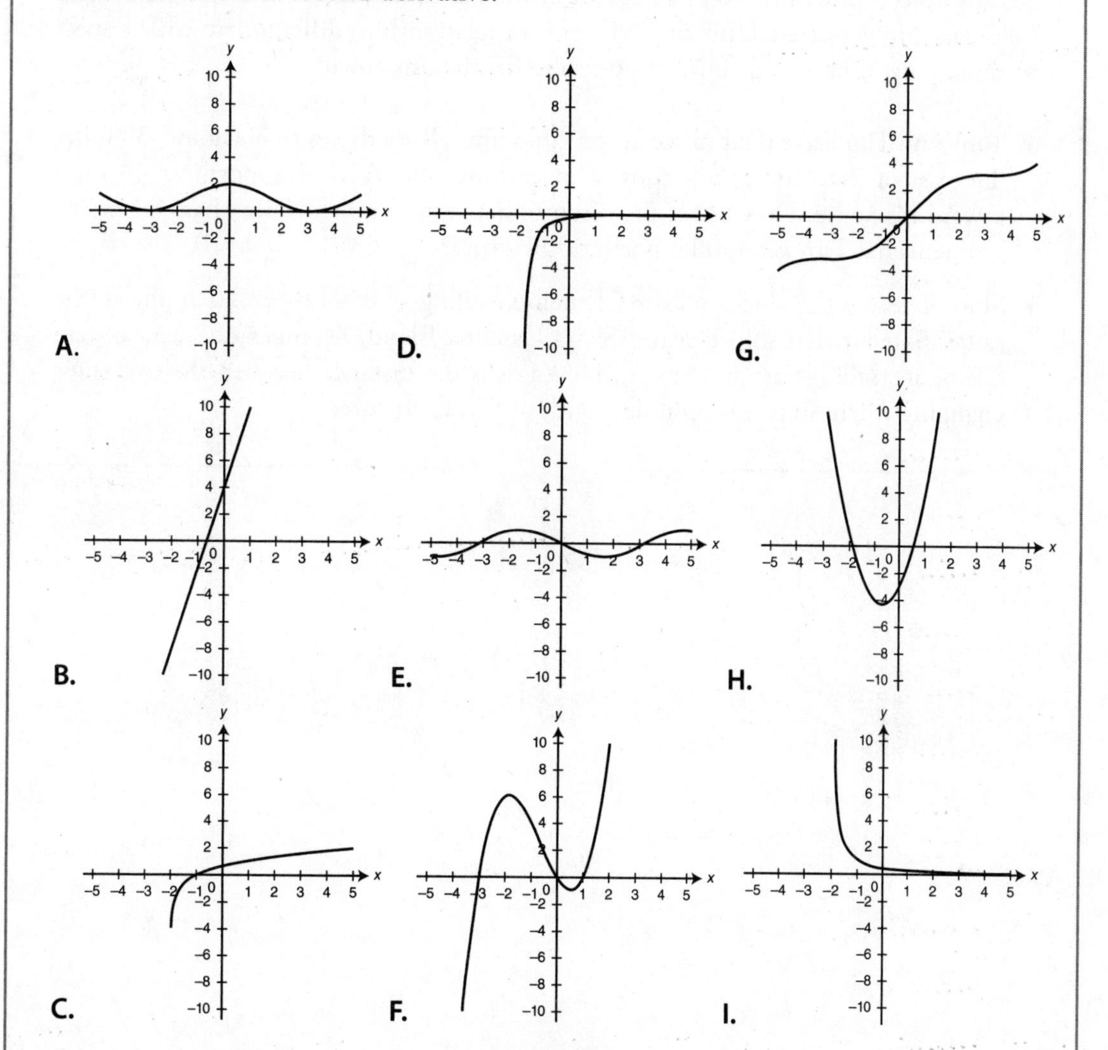

Every Related Rates problem has a defining equation or formula. One of the most common is the Pythagorean Theorem, when lengths of the sides of the triangle created by objects moving at right angles are all seen as functions of time. For each of the following problems, establish the Pythagorean relationship, differentiate with respect to time, plug in known quantities, and solve for the unknown.

1. Tony and Tim leave their office at the same time. Tony drives to his home, 30 miles due east of the office, and stops. Tim continues to drive due north at 40 mph toward his home. How fast is the distance between Tony and Tim changing at the moment that Tim is 80 miles north of the office?
2. Ship A leaves the dock in San Clemente, sailing due south at 20 mph, at the same moment that ship B leaves Santa Catalina Island, 46 miles due west of San Clemente, sailing east at 15 mph. How fast is the distance between the two ships changing when Ship A is 50 miles south of San Clemente?

Related Rates problems about filling or draining containers are structured by volume formulas. For each of the following problems, set up the appropriate volume formula, differentiate assuming each dimension is a function of time, plug in known quantities, and solve for the unknown.

A water container in the shape of an inverted cone is 12 feet in diameter and 16 feet deep at its deepest point. Water leaks out of the container at a constant rate of 2 cubic feet per minute. How fast is the water level dropping when the depth of water remaining in the tank is 4 feet?

For any Related Rates problem, even one that doesn't fit the common patterns, there's a formula or rule to tie the situation together. Look for that equation and solve the Related Rates problems below.

A photographer covering the launch of a new satellite sets up a camera 2,000 feet from the base of the launch pad across level ground. When the rocket carrying the satellite is 3,000 feet off the ground, it rises vertically at 1,200 feet per second. At what rate must the angle of elevation of the camera change to keep the rocket in view?

To solve problems that ask for the maximum or the minimum value of a function, use the first derivative to identify the critical values of the function, and then either the first derivative test or the second derivative test to determine at which x values will the maxima (or minima) occur.

1. $f(x) = x^{1/2}(x-3)^2$

2. $f(x) = \sin 2x - 2\cos x,\ 0 < x < 2\pi$

3. $f(x) = -2x + (x+1)^{3/2}$

1. Find the dimensions of the rectangular garden of largest area that can be enclosed with 100 feet of fencing.
2. Find the dimensions of the rectangular garden of largest area that can be enclosed with 100 feet of fencing if one side of the garden is the wall of a barn and so only three sides need to be fenced.
3. Find the shortest distance from the point $(3, -1)$ to the line $y = 2x - 1$.

Sketch a graph of a function $f(x)$ that has all of these characteristics.

Domain: $[-5, 5]$, Range: $[-5, 5]$

5 x-intercepts

y-intercept: $(0, -2)$

x	-4	-3	-2	1	3.5
$f'(x)$	0	DNE	0	0	0
$f''(x)$	$-$		$-$	$+$	$-$

L'Hôpital's Rule provides another tool for finding limits. When $\lim_{x \to a} \frac{f(x)}{g(x)}$ is an indeterminate form $\frac{0}{0}$ or $\frac{\pm\infty}{\pm\infty}$, *L'Hôpital's* Rule says $\lim_{x \to a} \frac{f(x)}{g(x)} = \lim_{x \to a} \frac{f'(x)}{g'(x)}$. It doesn't solve every problem but it can help. Use *L'Hôpital's* Rule to find these limits, if possible.

1. $\lim_{x \to 0} \frac{\sin x}{2x}$
2. $\lim_{x \to 3} \frac{\ln(x-2)}{x^2 - 9}$
3. $\lim_{x \to \infty} \frac{e^{2x}}{2x^2}$

Day 63

The derivative of a function gives a formula for the slope of a tangent line to the graph at any point. When evaluated for a particular value of *x*, it gives the slope of a tangent drawn to that point. With the slope and a point, it's possible to find the equation of the tangent line. Match each function with the tangent to the curve at the point (–1, –2).

COLUMN A	COLUMN B
1. $f(x) = x^3 - 2x^2 + 1$	**(A)** $y = -x - 3$
2. $f(x) = \dfrac{1-x}{x}$	**(B)** $y = -2$
3. $f(x) = -1 - e^{-x^2-2x-1}$	**(C)** $y = -2x - 1$
4. $x^2 + 2y^2 = 9$	**(D)** $y = 7x + 5$
5. $f(x) = 2\sin\left(\dfrac{\pi x}{2}\right)$	**(E)** $y = -\dfrac{1}{2}x - \dfrac{5}{2}$

A normal line is one that is perpendicular to a tangent line at the point of tangency. For a circle, any normal line is a line that contains a radius, because a radius drawn to the point of tangency is perpendicular to the tangent. For other curves, however, the normal lines are not so predictable. For each curve, find an equation of the normal line at the given point.

1. $f(x) = 2x^4 - 5x^3 + 2$ at $(1, -1)$
2. $f(x) = \sin(2x) + \cos x$ at $(-\pi, -1)$

A tangent line to the graph of a function runs so close to the curve that it can be used to approximate values of the function close to the point of tangency. It is also possible to tell if the estimate is larger or smaller than the actual value by examining the concavity of the function graph. For each function below, use the equation of a tangent line to estimate the requested function value and tell whether it is an underestimate or an overestimate.

1. $f(x) = 5 + 2x - x^3$; $f(1.01)$
2. $f(x) = 3 - \ln(2x + 1)$; $f(0.9)$

Use tangent line to estimate indicated function value and tell whether it is an underestimate or an overestimate.

$f(x) = 2 + 3e^{-x}$; $f(-0.01)$

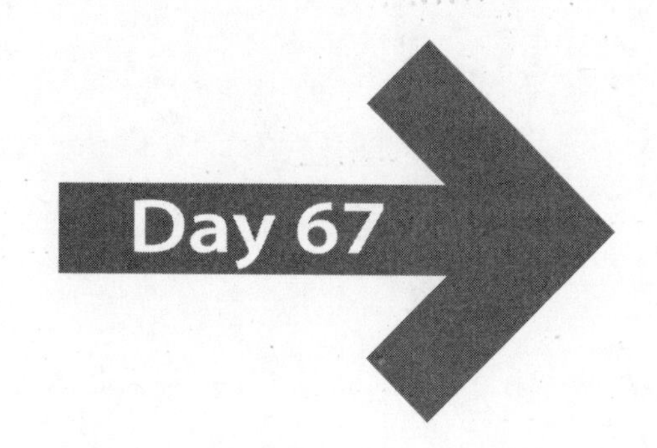

Finding approximate values for roots of integers is one case in which linear approximation can be helpful. Of course, how close to the actual value your estimate is will depend on how far you are from the point of tangency. Use linear approximation to estimate each of the following roots, and calculate the difference from the actual value.

Desired root	$\sqrt{17}$	$\sqrt{20}$	$\sqrt[3]{-26}$	$\sqrt[3]{-25}$	$\sqrt[5]{33}$
Linear Approximation					
Actual Value (to nearest thousandth)	4.123	4.472	−2.962	−2.924	2.012
Actual minus Approximate					

Estimating the values of trig functions for angles near those commonly memorized is another application of linear approximation. Measures should be in radians, but conversion is simple if you remember that one degree is $\frac{\pi}{180}$ radians. Complete the chart for these approximations.

Desired	cos 58°	sin 50°	tan 170°
Function	$y = \cos x$		
Point of tangency		$\left(\frac{\pi}{4}, \frac{\sqrt{2}}{2}\right)$	
Derivative			
Derivative Evaluated			
Distance from point of tangency			$-10° = -\frac{\pi}{18}$
(Derivative)(distance) + y-value of point of tangency			
Approximate Value	cos 58° ≈	sin 50° ≈	tan 170° ≈
Actual (to nearest thousandth)	0.530	0.766	−0.176

Velocity is the change in position over time. Acceleration is the change in velocity over time. If $s(t)$ is a function that gives the position of an object over time, then the velocity is $v(t) = s'(t)$ and acceleration is $a(t) = v'(t) = s''(t)$. The 12 expressions below represent the position, velocity, and acceleration functions for four different situations. Organize the equations into the appropriate four groups.

-10	$2\cos t - \sin t$	$-2t^3 + 15t^2 + 5t - 3$	$-10t + 30$
$\sin(2t)$	$-4\sin(2t)$	$-2\sin t - \cos t$	$-5t^2 + 30t + 5$
$-12t + 30$	$2\sin t + \cos t$	$2\cos(2t)$	$-6t^2 + 30t + 5$

	SITUATION 1	SITUATION 2	SITUATION 3	SITUATION 4
Position				
Velocity				
Acceleration				

The phrase "vertical motion" refers to the movement of an object that has been dropped from a height or thrown upward (and ultimately comes down). The position of the object is usually modeled by the equation $s(t) = -\frac{1}{2}gt^2 + v_0t + s_0$, where g is the acceleration due to gravity (32 ft/sec^2 in customary units or 9.8 m/sec^2 in metric), v_0 is the initial velocity, and s_0 is the initial position. The graph shows the position of an object launched upward from a height of 3 feet, with an initial velocity of 100 ft/sec.

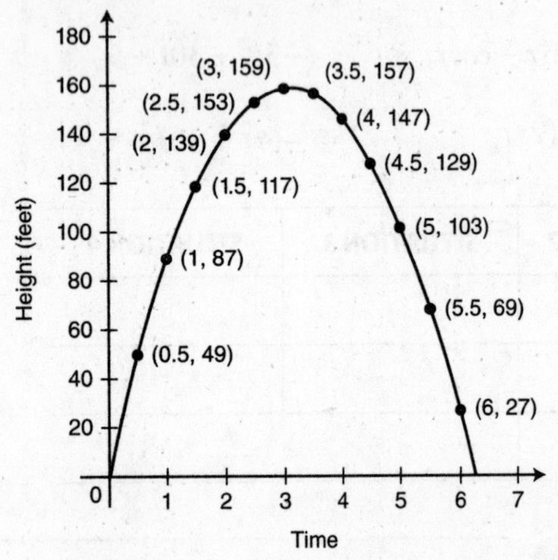

1. Approximate the velocity of the object 1.5 seconds after launch.
2. Approximate the velocity of the object 5 seconds after launch.
3. Use the derivative of the position function to calculate those velocities.
4. At what time does the object reach maximum height?
5. What is the average velocity of the object from $t = 2$ to $t = 5.5$?
6. At what time is the instantaneous velocity equal to this average velocity?
7. True or False: The acceleration is greater at $t = 1$ than at $t = 2$.
8. When is the object slowing down?

Horizontal motion examines movement to the left and to the right along a line. Imagine a particle moving along the x-axis, with its position at any time $t \geq 0$ given by the function $s(t) = 1 - \cos\left(\frac{\pi t}{2}\right)$.

1. Chart the position of the particle each second from $t = 0$ to $t = 4$.

time (seconds)	0	1	2	3	4
position	(,0)	(,0)	(,0)	(,0)	(,0)

2. When does the particle change direction?

3. When is the particle speeding up? When is it slowing down?

4. Find the total distance traveled by the particle.

5 Minutes to a 5

A water container in the shape of an inverted cone is 6 feet in diameter and 8 feet deep at its deepest point. Water leaks out of the container at a constant rate of 2 cubic feet per minute.

The water leaking from the cone drips into a tank that is 4 feet high and has a square base 3 feet on a side. How fast is the depth of water in this tank changing?

Find the maximum and minimum values of the functions.

1. $f(x) = x^3 + \dfrac{2}{x^2}$
2. $f(x) = x^4 - 3x^3 + 3x^2 + 1$

The long-running television game show *Jeopardy* is based on giving contestants the answer and asking them to come up with the correct question. Antiderivatives are a lot like *Jeopardy*. You're given the derivative, and asked for the function it came from. The derivative of a constant is zero, so remember there may have been a constant in the original function. Find a function that could have each of these derivatives.

1. $f'(x) = 5$
2. $f'(x) = 3x^2 + 8x + 3$
3. $f'(x) = \dfrac{5}{x}$
4. $f'(x) = \sin x$
5. $f'(x) = \sec^2 x$

Evaluate each of the following summations, using the properties of summations to simplify the calculation.

1. $\sum_{i=1}^{5} i$
2. $\sum_{i=1}^{10} \left(\frac{i}{5}\right)$
3. $\sum_{i=2}^{4} (i^2)(i+3)$
4. $\sum_{i=-3}^{1} (i^3 + 2i^2 - 5i + 4)$

Use geometric reasoning to estimate the area between the x-axis and the graph on the interval $[a, b]$. Then find the antiderivative $F(x)$ and evaluate $F(b) - F(a)$.

1. $f(x) = 4$ on $[0, 4]$

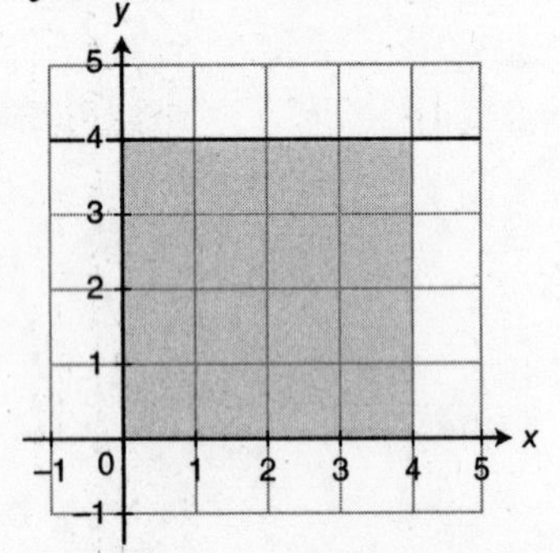

2. $f(x) = 4 - 2x$ on $[0, 2]$

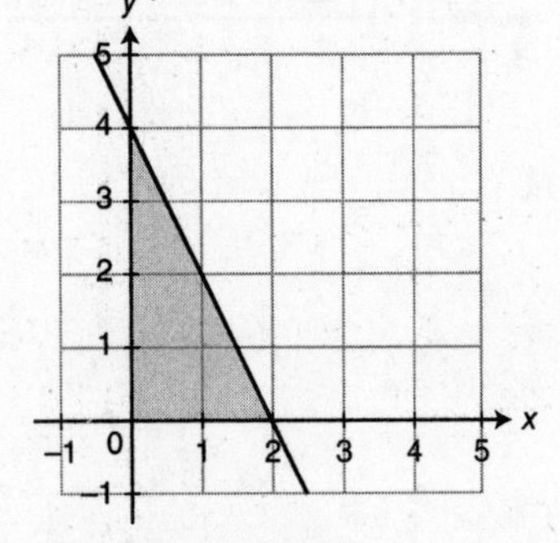

3. $f(x) = 3x$ on $[0, 1]$

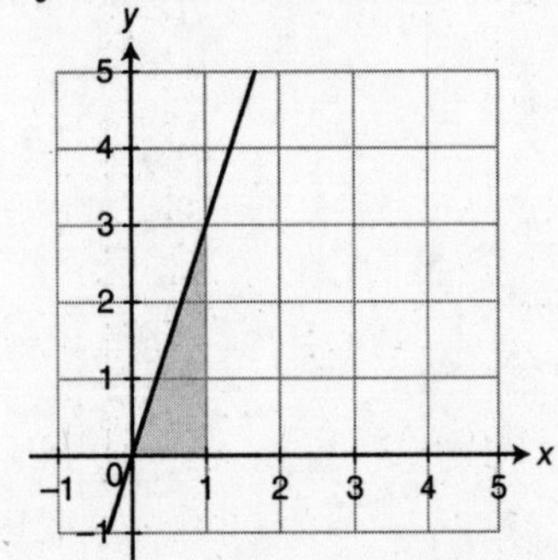

4. $f(x) = 3x + 1$ on $[0, 1]$

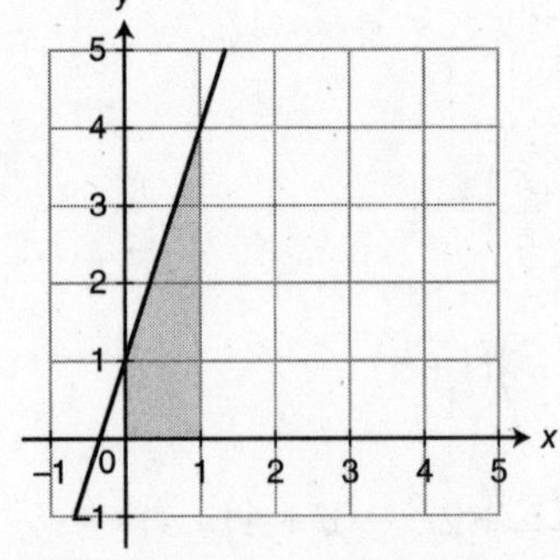

5. $f(x) = |x - 1|$ on $[0, 4]$

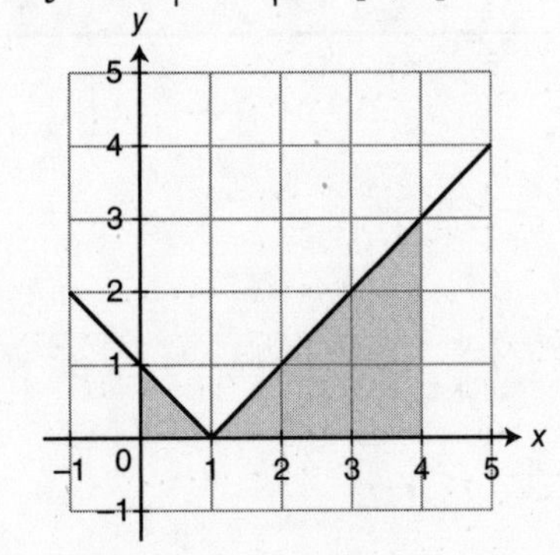

Day 77

Given $\int_5^9 f(x)\ dx = 12$, $\int_2^5 f(x)\ dx = -3$, $\int_5^9 g(x)\ dx = 12$, and $\int_2^5 g(x)\ dx = -1$, find:

1. $\int_8^8 g(x)\ dx$

2. $\int_9^5 f(x)\ dx$

3. $\int_2^5 7g(x)\ dx$

4. $\int_2^9 f(x)\ dx$

5. $\int_2^5 [f(x)+g(x)]\ dx$

Day 78

Evaluate each of the definite integrals.

1. $\int_{-3}^{1}\left(6x^2 - 5x + 2\right)dx$

2. $\int_{1}^{2}\frac{x^5 + x^2 + x}{x}\,dx$

3. $\int_{1}^{3}\frac{dx}{x^3}$

4. $\int_{1}^{4}\sqrt{x}\;dx$

5. $\int_{1}^{32}\sqrt[5]{x^2}\;dx$

Sum, Difference, and Power Rules are not always enough to find the integrals of exponential, logarithmic, or trigonometric functions. Thoughtful application of memorized forms will help with finding these definite integrals.

1. $\int_{0}^{\pi/3} (2\sin\theta - 5\cos\theta)\,d\theta$

2. $\int_{\pi/4}^{\pi/3} 5\sec^2 x \; dx$

3. $\int_{1}^{e} \left(\frac{3}{z}\right) dz$

4. $\int_{-\pi}^{\pi/2} (2\sin\theta\cos\theta)\,d\theta$

5. $\int_{0}^{\ln(1+\pi)} \left(-e^x \cos\left(1-e^x\right)\right) dx$

Recognizing ways to rewrite functions is a key skill for integration. Match each integral with an equivalent form that is easier to integrate.

COLUMN A

1. $\int x\sqrt{x}\,dx$
2. $\int \frac{x^3+x}{x}\,dx$
3. $\int \left(\sqrt{x}(x-2)\right)dx$
4. $\int \left(\frac{2x^5-x+3}{x^2}\right)dx$
5. $\int \sqrt[4]{x^3}\,dx$

COLUMN B

(A) $\int \sqrt{x^2-2x}\,dx$

(B) $\int x^{3/2}\,dx$

(C) $\int \frac{1}{x}\left(x^3+x\right)dx$

(D) $\int x^{3/4}\,dx$

(E) $\int \sqrt{x^3}\,dx$

(F) $\int \left(2x^3-x^{-1}+3x^{-2}\right)dx$

(G) $\int x^{4/3}\,dx$

(H) $\int \left(x^2+1\right)dx$

(I) $\int \left(x^{3/2}-2x^{1/2}\right)dx$

(J) $\int \left(2x^3-x+3x^2\right)dw$

You can find $\int x^3\,dx$ by just applying the Power Rule to get $\int x^3\,dx = \frac{x^4}{4} + C$ but $\int (x^2+5x-3)^3\,dx \neq \frac{(x^2+5x-3)^4}{4} + C$. To find the derivative of $\frac{(x^2+5x-3)^4}{4} + C$, you'd use the Chain Rule, and get $\frac{d}{dx}\left[\frac{(x^2+5x-3)^4}{4} + C\right] = (x^2+5x-3)^3(2x+5)$.

The integral $\int (x^2+5x-3)^3\,dx$ is missing the $(2x+5)$ that is the derivative of the quantity being raised to the third power. Fill in the missing piece of each of the integrals, and then integrate.

1. $\int \left[(2x+x^2)^5(\quad)\right]dx$
2. $\int \left[\sqrt{x^3-1}(\quad)\right]dx$
3. $\int \left[\frac{(\quad)}{x^2+4}\right]dx$

5 Minutes to a 5

Each of these integrals has the form $\int u \cdot du$ (give or take a constant multiplier). For each, identify u and du, then evaluate the integral.

1. $\displaystyle\int_{-\pi}^{\pi/2} \left(\cos\theta \cos(\sin\theta)\right) d\theta$

2. $\displaystyle\int_{0}^{\ln(1+\pi)} \left(e^{x} \cos\left(1-e^{x}\right)\right) dx$

3. $\displaystyle\int_{0}^{1/4} \left(\frac{2t}{1-4t^{2}}\right) dt$

Day 83

Let f be continuous and differentiable on $(-\infty, \infty)$.

1. For which values of k does $\int_{-2}^{k} x^2\, dx = 24$?

a) 1 **b)** 2 **c)** 3 **d)** 4

2. Which of the following is an expression for $f'(x)$ when $f(x) = \int_{0}^{x} (3t^2 - t)\, dt$?

a) $x^3 - \frac{x^2}{2}$ **b)** $t^3 - \frac{t^2}{2}$ **c)** $3x^2 - x$ **d)** $3t^2 - t$

3. $\int_{\pi}^{\pi} \frac{\sin^2(3x)}{\cos(2x)} dx =$

a) 0 **b)** $\frac{1}{2}$ **c)** 1 **d)** $\frac{3}{2}$

4. If $\int_{2}^{5} f(x)\, dx = 8$, then $\int_{5}^{2} f(x)\, dx =$

a) -8 **b)** 0 **c)** $\frac{1}{8}$ **d)** 8

5. $\int_{1}^{5} 2\, dx =$

a) 4 **b)** 5 **c)** 8 **d)** 10

Whenever you find an antiderivative, you must contend with the invisible constant term, but if you have the right bit of information, you can actually find that constant. Once the constant is identified, you can apply antiderivatives again.

Find $y = f(x)$ if $f''(x) = 6x - 4$, $f'(0) = 1$, and $f(1) = -3$

Day 85

Arrange the expressions below to form five equations showing integrals evaluated by u-substitution.

2	$2x$	$6x$	$2x+3$	$2x+3$
x^2+3	$3x^2-3$	$3x^2-3$	x^2+3x-2	$(3x^2-3)^2$
$(x^3-3x-2)^2$	$\frac{1}{3}(x^3-3x-2)^3$	$\frac{1}{2}(x^2+3)^2$	$\frac{1}{2}(2x+3)^2$	$\frac{1}{2}(x^2+3x-2)^2$

$\int(\qquad\qquad)(\qquad\qquad)dx = \qquad\qquad + C$

$\int(\qquad\qquad)(\qquad\qquad)dx = \qquad\qquad + C$

$\int(\qquad\qquad)(\qquad\qquad)dx = \qquad\qquad + C$

$\int(\qquad\qquad)(\qquad\qquad)dx = \qquad\qquad + C$

$\int(\qquad\qquad)(\qquad\qquad)dx = \qquad\qquad + C$

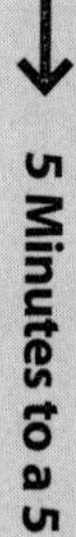

For each integral, identify u and du, make any necessary adjustments, and integrate.

1. $\int \frac{x}{2x^2-3}\,dx$
2. $\int \frac{x^2}{x^3-5}\,dx$
3. $\int \frac{x}{4-x^2}\,dx$
4. $\int e^{3x}\,dx$
5. $\int x^2 e^{x^3}\,dx$

Day 87

Complete each integral with the appropriate du, and integrate.

1. $\int \sin^2 x(\qquad\qquad) \, dx$
2. $\int \cos\sqrt{2x}\,(\qquad\qquad)\, dx$
3. $\int \sec^2(2x)(\qquad\qquad)\, dx$
4. $\int \frac{(\qquad\qquad)}{1+\sin 3x}\, dx$
5. $\int \tan^4 3x(\qquad\qquad)\, dx$

Recognizing an integrand as the derivative of an inverse trig function can be challenging, so be sure you've done the necessary memory work, and practice recognizing the forms as you work problems.

1. $\int \frac{-2dx}{1+4x^2}$

2. $\int \frac{-3dx}{\sqrt{1-9x^2}}$

3. $\int \frac{1}{\sqrt{4-x^2}} = \int \frac{\frac{1}{2}}{\sqrt{1-\frac{x^2}{4}}}$

4. $\int \frac{\cos x}{1+\sin^2 x}\, dx$

Find the dimensions of the rectangle of greatest area that can be inscribed in a semicircle with the radius of 10 centimeters.

The graph of a piecewise linear function *f*, on [–4, 3], is shown.

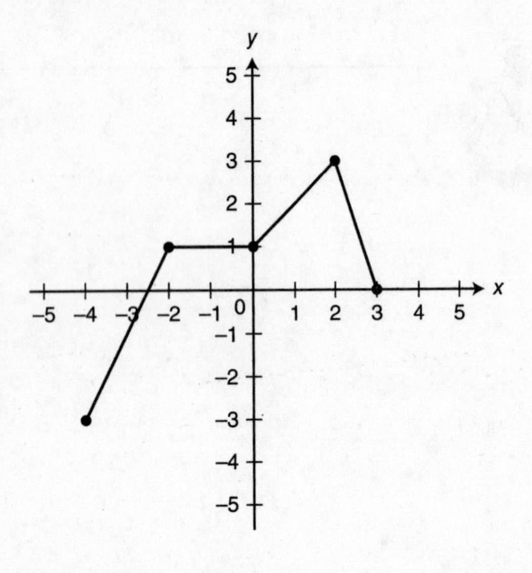

What is the value of

1. $\int_{0}^{2} f(x)\,dx$

2. $\int_{0}^{-2} f(x)\,dx$

3. $\int_{-4}^{-2} f(x)\,dx$

4. $\int_{-4}^{3} f(x)\,dx$

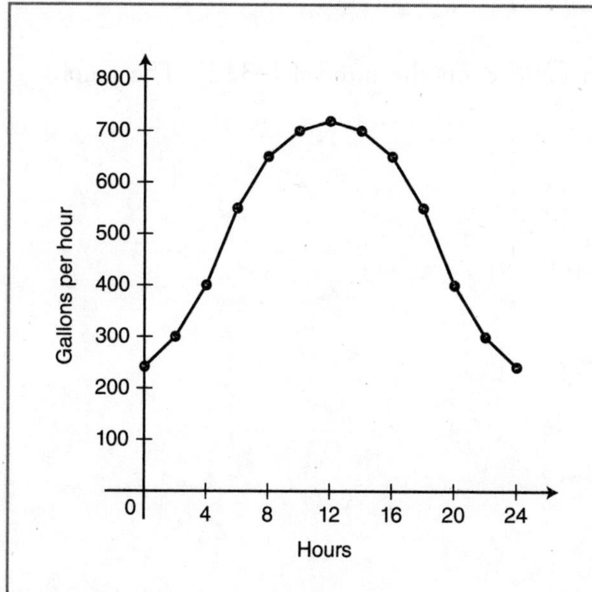

The graph shows the rate of flow of water, in gallons per hour, through a pipe over the course of 24 hours.

Estimate the number of gallons of water that flowed through the pipe

1. in the first 4 hours

2. between hour 8 and hour 16

3. over the 24-hour period

$f(x)$ is a twice differentiable function defined on the interval $[-3, 2]$. The graph below is the graph of $f'(x)$.

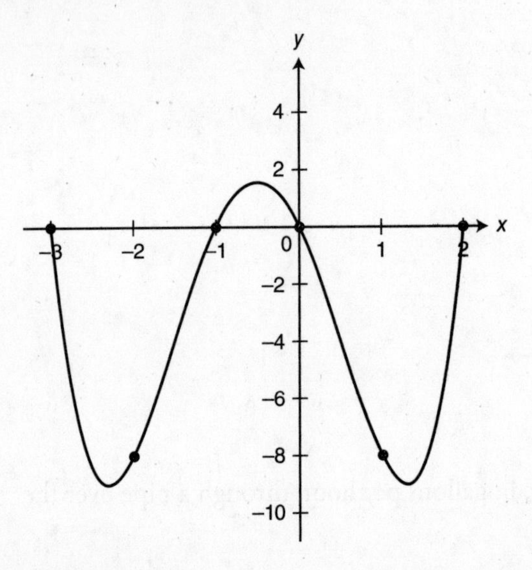

Label each of the following statements True or False.

1. $f'(-2) < f''(-2)$
2. $f(0) < f'(0) < f''(0)$
3. $f'(-1) < f''(-1)$
4. $f(-2) < f(-1)$
5. $f(-1) < f(0)$

A function f is continuous and twice differentiable on the interval (a, b). The graph of the derivative $f'(x)$ is shown below. Sketch the graphs of $f(x)$ and $f''(x)$ on [a, b].

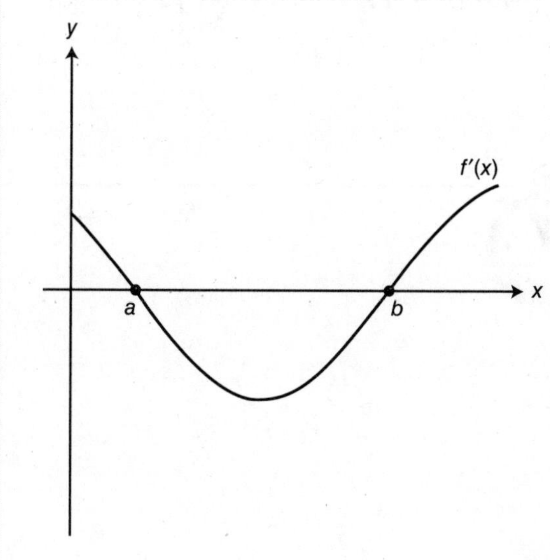

The function f is continuous on the closed interval [2, 8] and has values as shown in the table. Using four subintervals, find the trapezoidal approximation of $\int_2^8 f(x)\,dx$.

x	2	3	5	6	8
f(x)	1	4	7	3	2

Set up and evaluate Riemann sums for the area under the graph of $f(x)$ shown, on the interval $[-3, 2]$, using the partitions given.

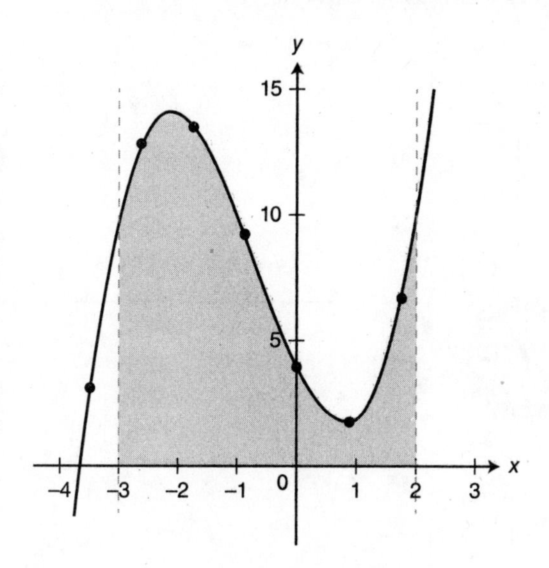

1. Use $\Delta x = 1$ and LRAM (Left Rectangular Approximation Method).
2. Use $\Delta x = 1$ and RRAM (Right Rectangular Approximation Method).

Evaluate each definite integral.

1. $\int_{-2}^{2} \left(4 - x^2\right) dx$

2. $\int_{-1}^{7} \sqrt{x+2}\ dx$

3. $\int_{-3}^{2} \left(x^3 + 2x^2 - 5x + 4\right) dx$

The first part of the Fundamental Theorem of Calculus says that if a function $f(x)$ is continuous on $[a, b]$ and $F(x)$ is the indefinite integral of $f(x)$ on $[a, b]$, the definite integral $\int_a^b f(x)\,dx = F(b) - F(a)$. Match each definite integral with the appropriate evaluation.

COLUMN A

1. $\int_3^5 \left(x^2 - 3x + 2\right) dx$
2. $\int_3^5 \left(3x^2 + 2x - 1\right) dx$
3. $\int_5^3 \left(x^3 + x^2 - x\right) dx$
4. $\int_3^5 \left(x^3 - 3x^2 + 2x\right) dx$
5. $\int_3^5 \left(x^3 + x^2 - x\right) dx$

COLUMN B

(A) $\left(\frac{625}{4} - 125 + 25\right) - \left(\frac{81}{4} - 27 + 9\right)$

(B) $(125 + 25 - 5) - (27 + 9 - 3)$

(C) $\left(\frac{125}{3} - \frac{3(25)}{2} + 2(5)\right) - \left(\frac{27}{3} - \frac{3(9)}{2} + 2(3)\right)$

(D) $\left(\frac{81}{4} - 27 + 9\right) - \left(\frac{625}{4} - 125 + 25\right)$

(E) $\left(\frac{625}{4} + \frac{125}{3} - \frac{25}{2}\right) - \left(\frac{81}{4} + \frac{27}{3} - \frac{9}{2}\right)$

(F) $\left(\frac{81}{4} + \frac{27}{3} - \frac{9}{2}\right) - \left(\frac{625}{4} + \frac{125}{3} - \frac{25}{2}\right)$

(G) $(27 + 9 - 3) - (125 + 25 - 5)$

Use second part of the Fundamental Theorem of Calculus to complete the chart.

1.	$F(x)=\int_{5}^{x}\left(2t\sqrt{t}\right)dt$	$F'(x)=$	$F'(9)=$
2.	$F(x)=\int_{0}^{4x}t\left(1+t^{3}\right)dt$	$F'(x)=$	$F'(1)=$
3.	$F(x)=\int_{4}^{2x}\left(\frac{t-3}{t}\right)dt$	$F'(x)=$	$F'(10)=$
4.	$F(x)=\int_{-1}^{x}t\left(t^{2}-4\right)dt$	$F'(x)=$	$F'(2)=$
5.	$F(x)=\int_{0}^{x}e^{t^{2}}\,dt$	$F'(x)=$	$F'(1)=$

The second part of the Fundamental Theorem of Calculus says that if $f(x)$ is continuous on an open interval and a is any value in that interval, and $F(x)=\int_a^x f(t)\,dt$, then at every point in that interval, $F'(x)=f(x)$. State $F'(x)$ if:

1. $F(x)=\int_0^x \left(t^3-7t+1\right)dt$

2. $F(x)=\int_{-3}^x \sqrt{t+5}\ dt$

3. $F(x)=\int_1^{2x} \left(t^2-2t+5\right)dt$

Each graph shows the same function $f(x)$. Use the rectangles provided to estimate the area under the curve from $x = -4$ to $x = 4$.

1.

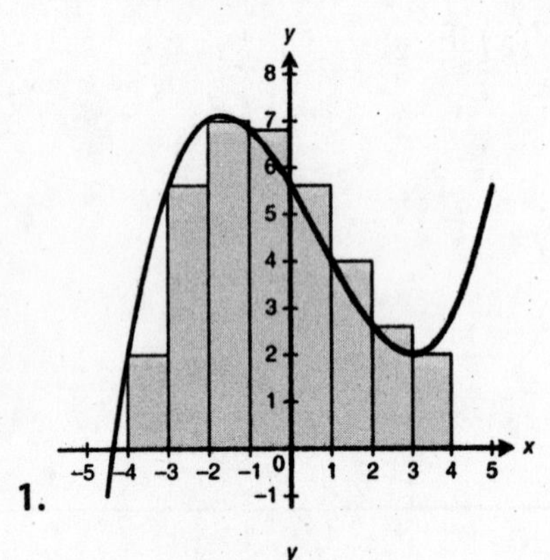

2.

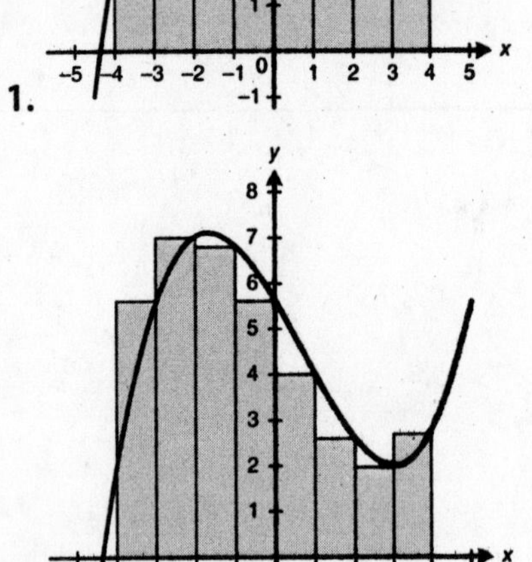

3.

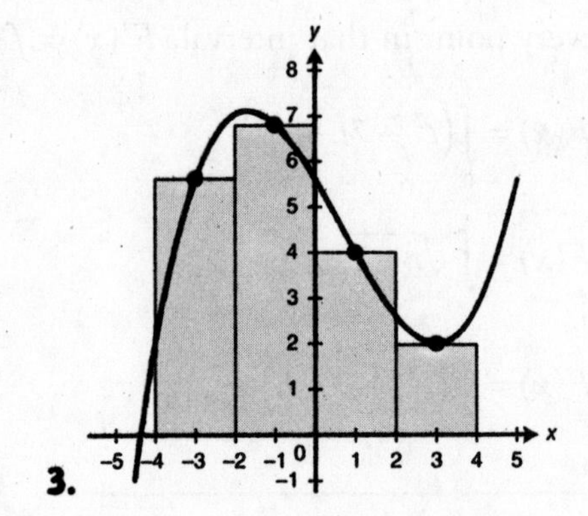

If you're working on an area-under-the-curve problem and want to check your work, or you're facing an integral that can't be evaluated by hand, your graphing calculator may help. In the math menu, you'll find a function fnInt(. Give it a function, the variable you want to integrate with respect to, and the lower and upper limits of integration. It will return the value of the definite integral.

But if you want to see the area under the curve actually shaded, as well as evaluated, graph the function, and then go to the CALC menu (2nd TRACE). Option 7, $\int f(x)\,dx$, will ask for lower and upper limits, then shade the region and calculate the area.

Use your graphing calculator to evaluate each integral.

1. $\int_0^4 \sqrt{x^2+1}\ dx$

2. $\int_{-3}^3 \sqrt{9-x^2}\ dx$

3. $\int_{-2}^5 \ln\left(x^2+2x+3\right)dx$

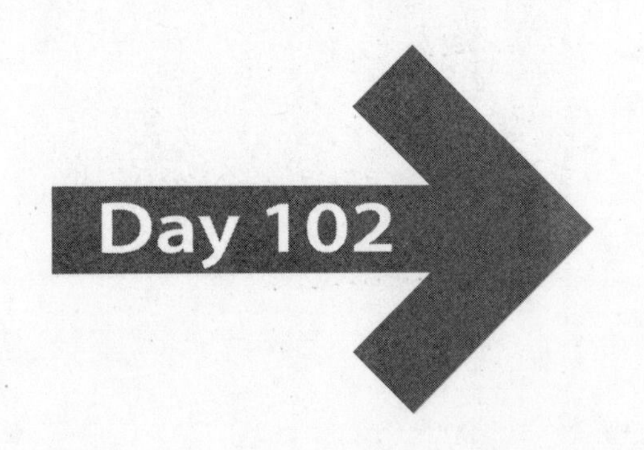

x	−4	−3.5	−3	−2.5	−2	−1.5	−1	−0.5	0	0.5	1	1.5	2	2.5	3
$f(x)$	0	4.5	6	5.5	4	2	0	−1.5	−2.5	−2.5	−2	−1	0	0.5	0

Let $f(x)$ be a continuous function on [−4, 3] that takes the values shown in the table. Write and evaluate an approximation of the area under the curve using the conditions described.

1. From $x = -4$ to $x = -1$ using 6 subintervals of equal width and left-hand approximation
2. From $x = -1$ to $x = 2$ using 6 subintervals of equal width and right-hand approximation
3. From $x = -4$ to $x = -1$ using 3 subintervals of equal width and midpoint approximation

x	–4	–3.5	–3	–2.5	–2	–1.5	–1	–0.5	0	0.5	1	1.5	2	2.5	3
$f(x)$	0	4.5	6	5.5	4	2	0	–1.5	–2.5	–2.5	–2	–1	0	0.5	0

Let $f(x)$ be a continuous function on [–4, 3] that takes the values shown in the table.

Use a trapezoidal approximation to estimate the area under the curve from $x = -1$ to $x = 2$, using 3 subintervals of equal width.

5 Minutes to a 5

Find the area of the region(s) bounded by:

1. $f(x) = x^3 + 3x^2$ and the x-axis.

2. $f(x) = 2x^3 + 3x^2 + 4$, the x-axis, and the vertical lines $x = -3$ and $x = -1$.

Day 105

Evaluate each of the definite integrals.

1. $\int_{0}^{\pi} (1 + \cos x)\, dx$

2. $\int_{\pi/6}^{\pi/3} (-\sin x)\, dx$

3. $\int_{-\pi/2}^{\pi/2} (x + \sin x)\, dx$

The definite integrals below involve exponential functions and log functions. Find each definite integral.

1. $\int_{-2}^{2} \left(xe^{-x^2} \right) dx$

2. $\int_{1}^{4} \frac{x+2}{x^2+4x} dx$

When evaluating a definite integral that involves an absolute value, rewrite the integrand as a piecewise function. Then rewrite the integral as a sum of two integrals, or as a single integral that does not include an absolute value if the limits of integration are both in the same portion of the domain.

1. $\int_{1}^{5} |x-3|\,dx$

2. $\int_{-1}^{3} |2x+1|\,dx$

3. $\int_{-2}^{2} |x^3|\,dx$

Let $f(x) = 3 + 2x - x^2$, $g(x) = x^2 - 2x - 3$, $h(x) = 9 + 9x - x^2 - x^3$, and $s(x) = \sin\left(\dfrac{\pi x}{3}\right)$. Find the area between the curves on the specified interval.

1. $f(x)$ and $g(x)$ between $x = -1$ and $x = 3$.

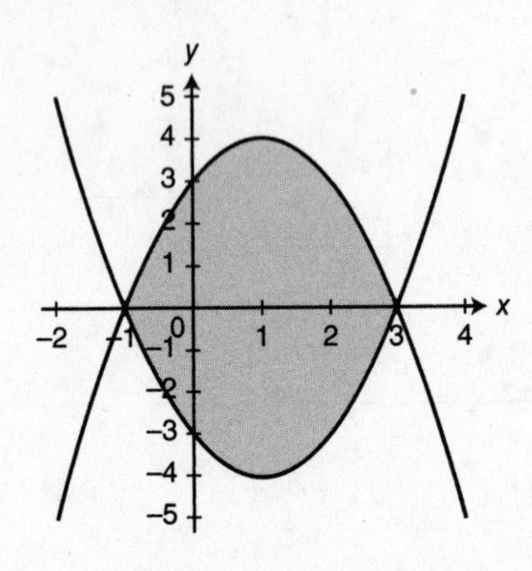

2. $f(x)$ and $h(x)$ between $x = -1$ and $x = 3$.

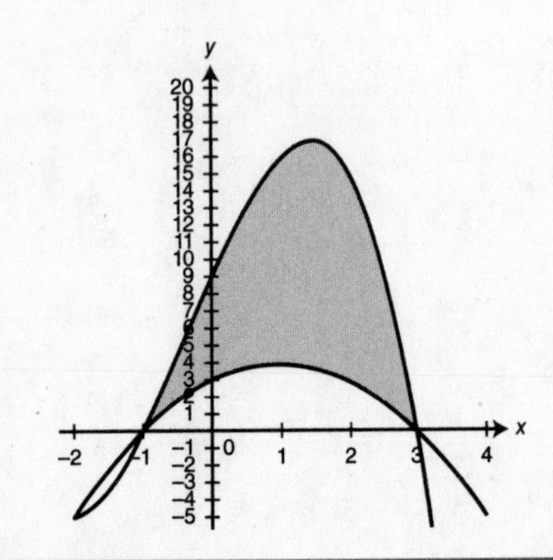

3. $f(x)$ and $s(x)$ between $x = 0$ and $x = 3$.

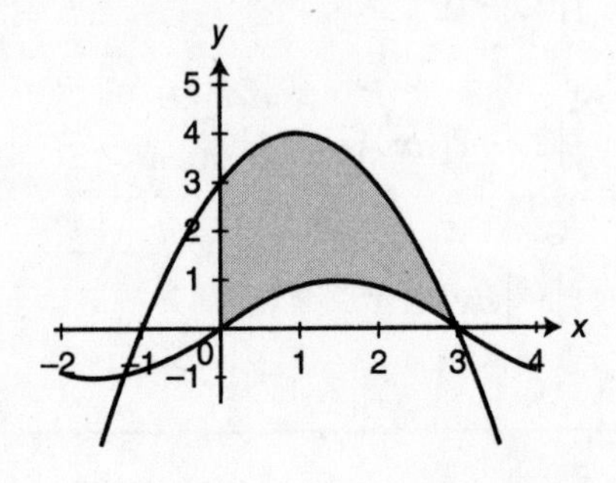

$y_1 = 4x - x^2$	$y_2 = 2x - 3$	$y_3 = 2\sin x$	$y_4 = 2\cos x$
$y_5 = e^{-x}$	$y_6 = e^x$	$y_7 = 4$	

Using the equations in the table above, write and evaluate an integral that yields the area shown.

1.

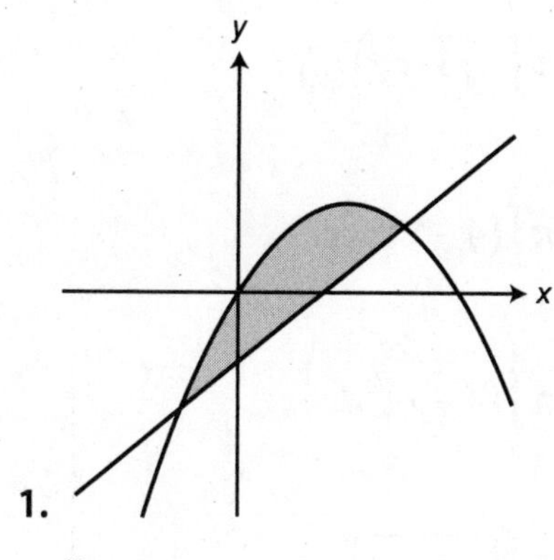

3.

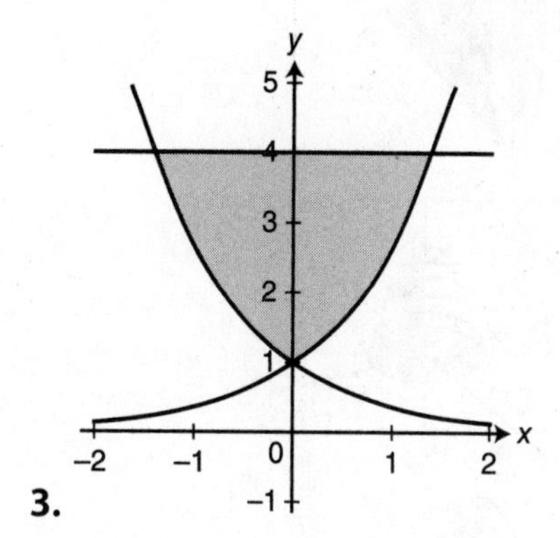

2.

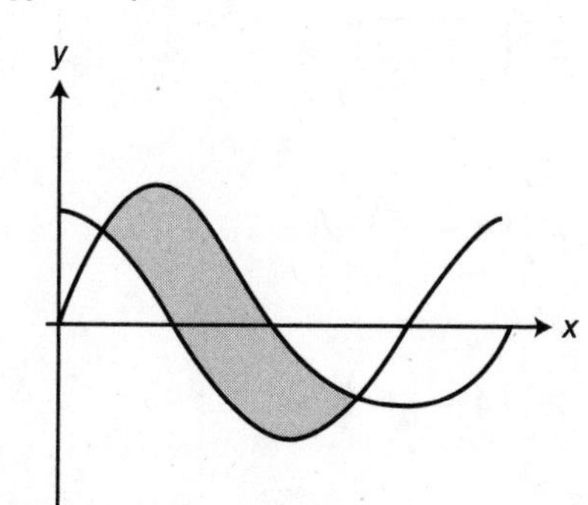

5 Minutes to a 5

Match each solid of revolution with the integral that will produce its volume, and then find the volume.

COLUMN A

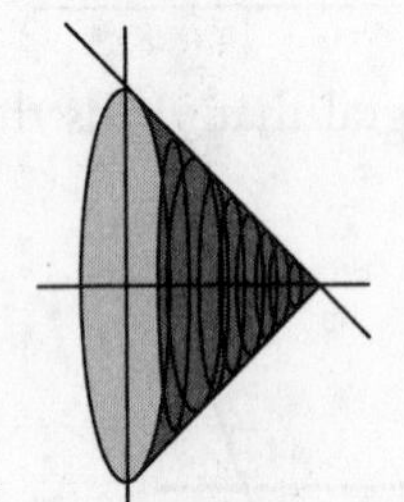

1.

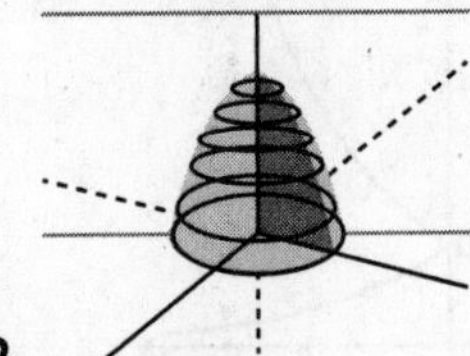

2.

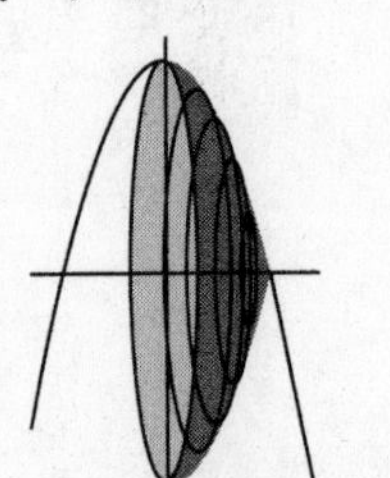

3.

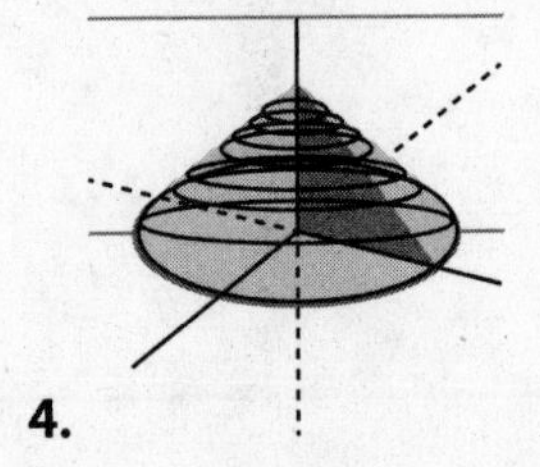

4.

COLUMN B

(A) $\pi \int_0^2 (4-x^2)^2 \, dx$

(B) $\pi \int_0^2 \left(\sqrt{4-y}\right)^2 \, dy$

(C) $\pi \int_0^2 (4-x)^2 \, dx$

(D) $\pi \int_0^4 (4-x)^2 \, dx$

(E) $\pi \int_0^4 \left(\sqrt{4-y}\right)^2 \, dy$

(F) $\pi \int_0^2 (4-y^2)^2 \, dy$

(G) $\pi \int_0^4 (4-y)^2 \, dy$

Let R be the region in the first quadrant enclosed by the graphs of $y = x^2$ and $y = 2x$, as shown. Use the Method of Washers to find the volume of each solid of revolution formed when the region R is revolved about the given line.

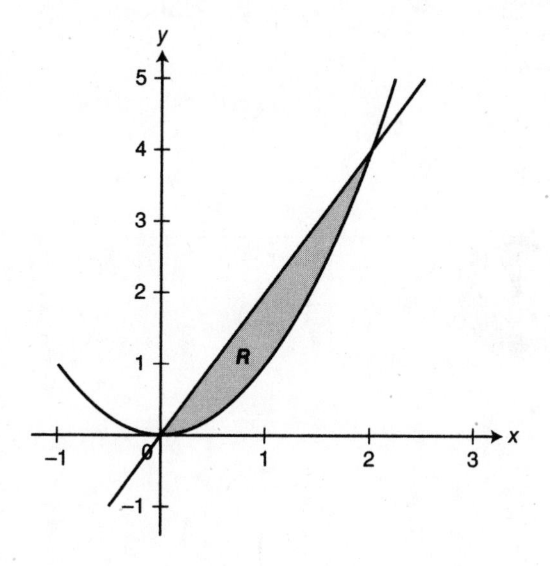

1. About the y-axis

2. About the x-axis

3. About the line $y = 4$

4. About the line $x = 2$

Let R be the region in the first quadrant enclosed by the graphs of $y = x^2$, $y = 4$, and $x = 0$. Let A be the region enclosed by the graphs of $y = -x$, $y = 3$, and $x = 1$. Find the volume of each solid of revolution described.

1. Region R revolved around the line $y = 4$

2. Region A revolved around the line $y = 3$

3. Region A revolved around the line $x = 1$

Let R be the region enclosed by the graphs of $y = x^2$, $y = 1$, and $x = 3$. Find the volume of each solid generated when region R is revolved about the given line.

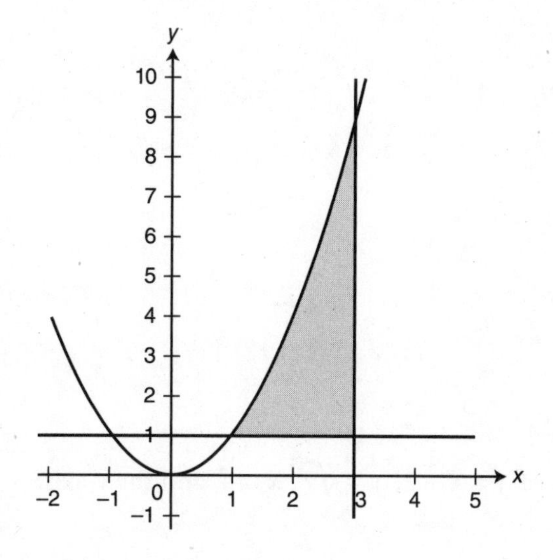

1. R revolved about $x = 4$

2. R revolved about $y = -1$

3. R revolved about $y = 10$

4. R revolved about $x = -1$

Day 114

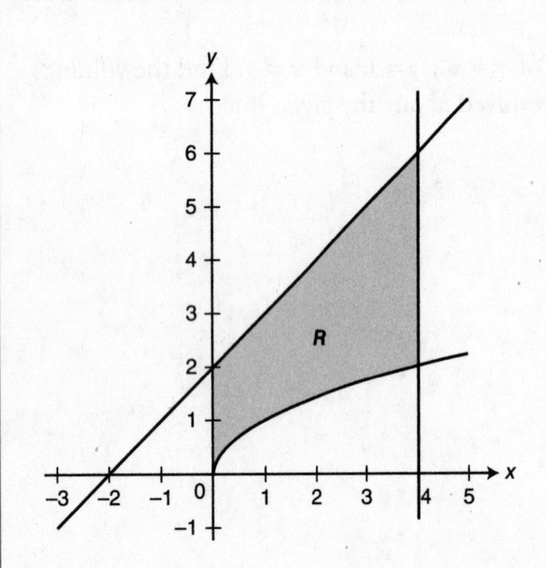

Let R be the region enclosed by the graphs of $y = x + 2$, $y = \sqrt{x}$, $x = 4$, and the y-axis.

1. Find the area of region R.

2. Find the volume of a solid with R as its base and square cross sections perpendicular to the x-axis.

3. Find the volume of a solid with R as its base and cross sections that are isosceles right triangles with hypotenuse perpendicular to the x-axis.

Let A be the region enclosed by the graphs of $y = \sin(x)$, $y = \cos(x)$, and the y-axis, as shown. Let B be the region under the graphs of $y = \sin(x)$ and $y = \cos(x)$ and above the x-axis, as marked.

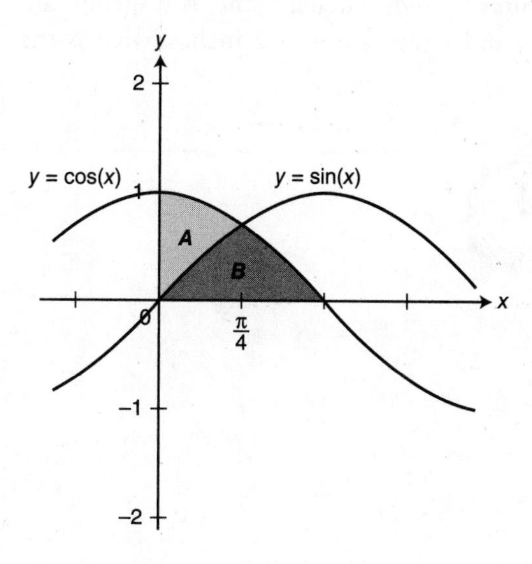

1. Write, but do not integrate, an expression for the volume of the solid created when region B is revolved about the y-axis.
2. Write, but do not integrate, an expression for the volume of the solid created when region A is revolved about the y-axis.
3. Find the volume of the solid created when region A is revolved about the x-axis.

For many Americans (probably too many), breakfast is often a cup of coffee and a pastry.

If your coffee cup, in the shape of a frustum of a right circular cone, is 8 inches tall, has a circular base with a radius of 1 inch, and a top radius of 2 inches, what is the volume of the cup?

Find the dimensions of a right - circular cylindrical container with minimum surface area if it must have a volume of 1,000 cubic centimeters.

If your pastry has a circular base with a radius of 3 inches and has parallel cross sections that are semicircles, what is the volume of the pastry?

Slope Fields allow you to visualize what the antiderivative might look like. Create a Slope Field for each of the differential equations on the axes provided. Then sketch the graph of the antiderivative that meets the given initial condition.

1. $\frac{dy}{dx} = 3x^2 + 2x,\ y(0) = 1$

2. $\frac{dy}{dx} = \frac{1}{x+2},\ y(-1) = 0$

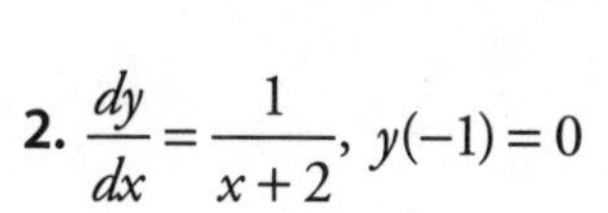

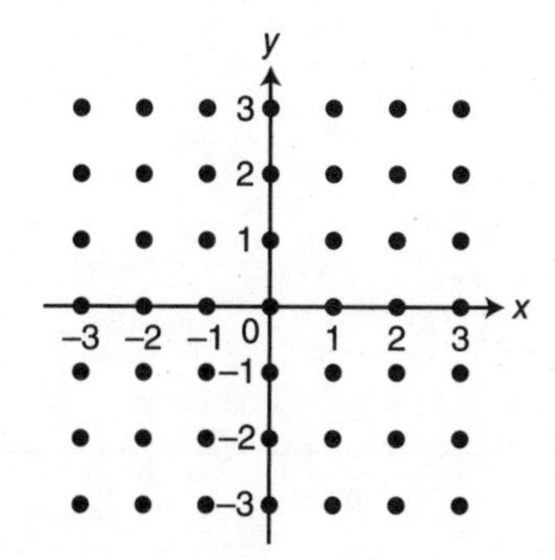

3. $\frac{dy}{dx} = x\cos x + \sin x,\ y(0) = -2$

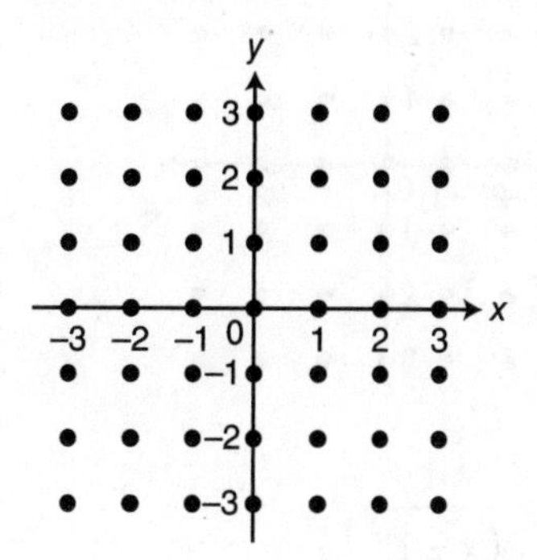

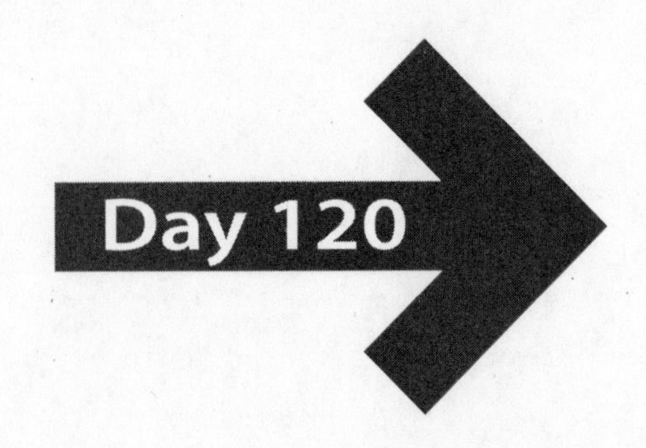

Create a Slope Field for each of the given differential equations.

1. $\dfrac{dy}{dx} = 3x^2 - 1$

3. $\dfrac{dy}{dx} = 2x + y$

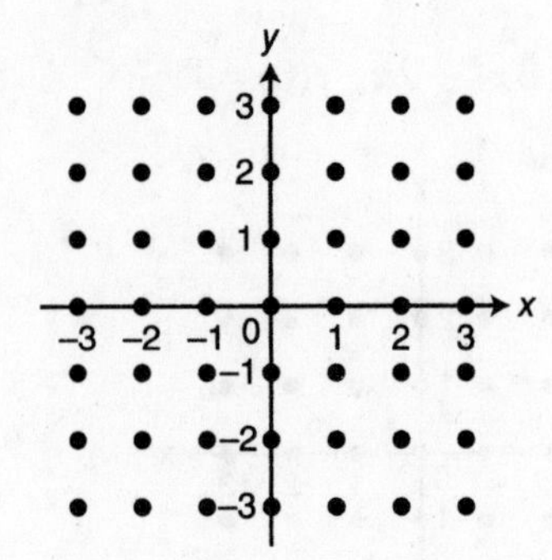

2. $\dfrac{dy}{dx} = \dfrac{1}{(x-1)^2}$

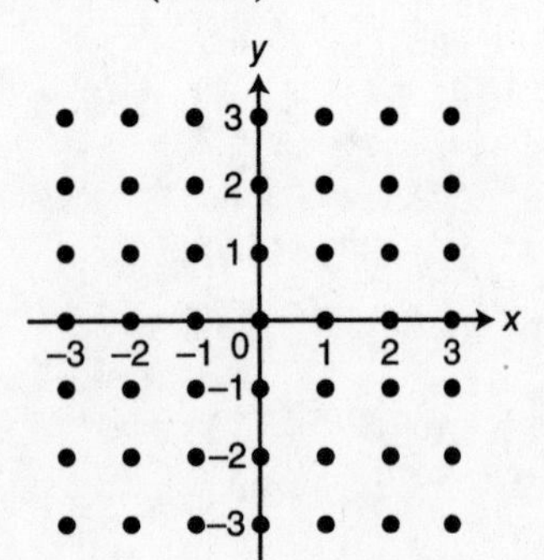

A differential equation, one containing derivatives, may be simple or quite complex. If the differential equation can be rearranged so that one variable is on the left side and the other is on the right, finding the antiderivative of each side can lead to a solution. Solve each of the following separable differential equations.

1. $\frac{dy}{dx} = \frac{2\cos x}{y^2}$
2. $\frac{dy}{dx} = x(1+2y)$

For a discrete function, finding the average value of the function on an interval is a simple task: add the values the function takes on that interval and divide by the number of values. For a continuous function, however, an integral is necessary to find the sum of the function values. The divisor is the length of the interval. Find the average value of each function on the specified interval.

1. $f(x) = \sqrt{x-2}$ on [2, 6]
2. $g(x) = e^{1-x}$ on [0, 1]
3. $f(x) = x^3 - 2x + 1$ on [−1, 2]

Day 123

Find the average value of the functions shown, over the interval [0, 2].

1. $f(x) = x^3 - 4x$
2. $f(x) = x + \sin x$
3. $f(x) = 2xe^{-x^2}$
4. $f(x) = 1 - xe^{3x^2}$

Mean Value Theorem for integrals says that if $f(x)$ is a continuous function on $[a, b]$, then there is a number c in $[a, b]$ such that $\int_a^b f(x)\,dx = f(c)(b-a)$. Find the value of c that satisfies the Mean Value Theorem for each of the following functions on the interval $[0, 2]$.

1. $f(x) = 4 - x^2$
2. $f(x) = x^2 - x - 2$
3. $f(x) = 5\sin\left(\frac{x-1}{2}\right)$

The graph shows the velocity of a car traveling on a straight road.

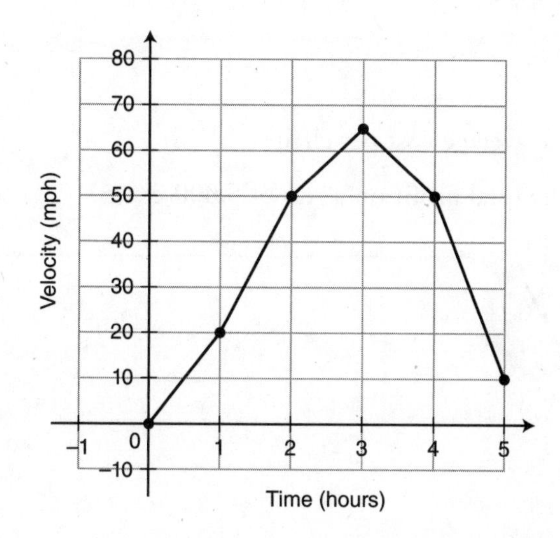

1. Over what period of time is the acceleration of the car negative?

2. What is the average velocity of the car over the 5-hour period?

3. What is the total distance traveled by the car over the 5-hour period.

5 Minutes to a 5

A furniture company produces a line of chairs that has cost function given by the function $C(x) = 17x + 0.025x^2$.

1. Find the marginal cost function.
2. If 5,000 chairs are produced, what is the average cost per chair?
3. If each chair is sold for $250, what is the total profit on a run of 5,000 chairs?

A new business estimates that their costs can be approximated by the function $C(x) = x^2 + 6x + 1,200$, $0 \leq x \leq 125$, and the their revenue by the function $R(x) = 120x$.

1. Find the profit function.

2. Find the marginal cost and marginal profit when production is at $x = 80$.

3. At what level of production does the business make maximum profit?

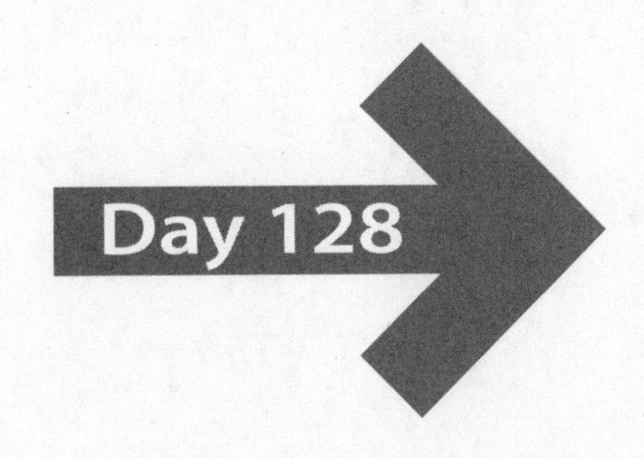

Find the maximum and minimum values of the function on the given interval.

1. $f(x) = \dfrac{x-1}{2x+1}$ on $[0,3]$

2. $f(x) = x + 4x^{-2}$ on $[1,4]$

Oil is pumped into a tank at a constant rate of 1 gallon per minute but leaks out at $3\sqrt{t}$ gallons per minute, $t \geq 0$. At time $t = 0$, the tank contains 50 gallons of oil.

1. How many gallons of oil leak out of the tank from time $t = 0$ until $t = 9$ minutes?

2. How many gallons of oil are in the tank at time $t = 9$ minutes?

A colony of a certain bacterium initially has a population of 5 million bacteria. Suppose that the colony grows at a rate of $f(t) = e^{(t+1)/2}$ million bacteria per hour.

1. Find the total change in the bacteria population during the time from $t = 0$ to $t = 3$.

2. Find the bacteria population at time $t = 3$.

At time $t = 0$, a cup of coffee at 200°F is put into a 72°F room. The coffee is left to cool, with its temperature changing at a rate of $r(t) = -12e^{-t/6}$ degrees per minute.

1. Estimate the temperature of the coffee when $t = 5$.
2. At what time will the temperature of the coffee drop to 140°F, which some say is the perfect drinking temperature?

The same types of plants are planted in two comparable garden plots to investigate the effects of different methods of watering. The rates of growth for the two plots are shown in the following figure.

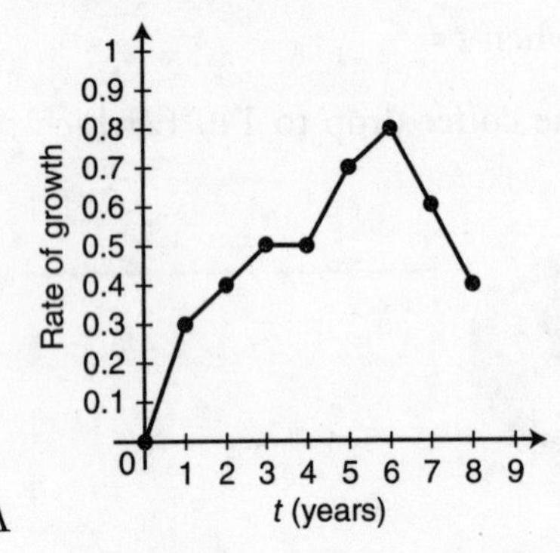

Plot A

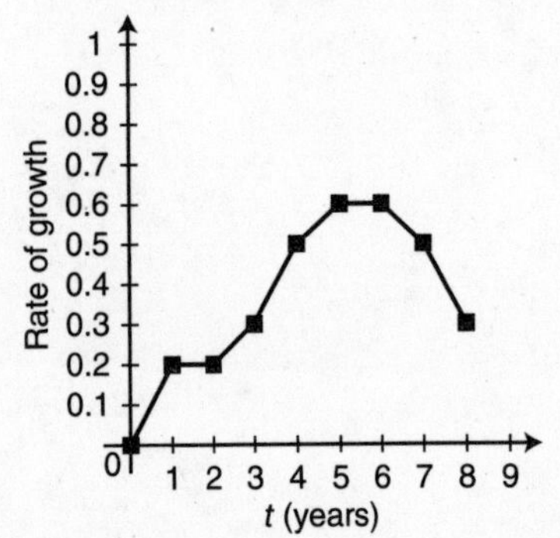

Plot B

Assume that the populations of the two plots are equal at time $t = 0$. Using RRAM (Right Rectangular Approximation Method):

1. Estimate the change in the population of plot A during the first 3 years.

2. Estimate the change in the population of plot B during years 4 to 8.

3. Which population is larger after 8 years?

At time $t = 0$, a cup of coffee at 200°F is put into a 72°F room. The temperature of the coffee is measured every 3 minutes and recorded in the table below.

time (min)	0	3	6	9	12	15	18	21	24
temp (°F)	200	153	124	107	96	90	86	84	82

1. Estimate the average temperature of the coffee over the first 15 minutes.

2. Estimate the rate of change of the temperature at $t = 15$.

5 Minutes to a 5

A cylindrical container with a diameter of 12 inches contains 900 in^3 of water when a leak forms. Let $t = 0$ represent the moment the leak forms. Water leaks out of the container at a rate modeled by $r(t) = -45e^{-t/20}$.

1. How many in^3 of water leak out of the container in the first 10 minutes?

2. How many in^3 of water remain in the container after 30 minutes?

3. Write an equation that relates the change in volume over time to the change in the height of the water in the container over time.

A woman 168 cm tall is walking at a rate of 90 cm per second toward a streetlight 610 cm tall. How fast is the length of her shadow changing?

Find an equation of the normal line at the given point.

$x^3 - 2xy - y^2 + 4 = 0$ at $(0,-2)$

Let f be the function given by $f(x)=\dfrac{x}{\sqrt{x^2+2x+1}}$.

1. Find the domain and range of f.

2. Sketch a graph of f.

3. Write an equation for each horizontal asymptote of f.

A particle moves along the y-axis so that its velocity at any time $t \geq 0$ is given by $v(t) = t^2 - \sin t$. At time $t = 0$, the position of the particle is $y = 1$.

1. Write an expression for the acceleration of the particle in terms of t.
2. Write an expression for the position $y(t)$ of the particle.
3. Find the position of the particle the first time, $t > 0$, when the velocity of the particle is zero.

Consider the curve defined by $2y^2 - xy + x^3 = 16$.

1. Find $\frac{dy}{dx}$.
2. Write an equation for the line tangent to the curve at the point (–2, 3).
3. There is a number k such that the point (–2.2, k) is on the curve. Use the tangent line to approximate the value of k.

Day 140

The graphs of $f(x) = x^3 + 1$ and $g(x) = 4x - x^3$ enclose two regions, labeled $R1$ and $R2$, as shown. The graphs intersect at $x = a$, $x = b$, and $x = c$ with $a < b < c$.

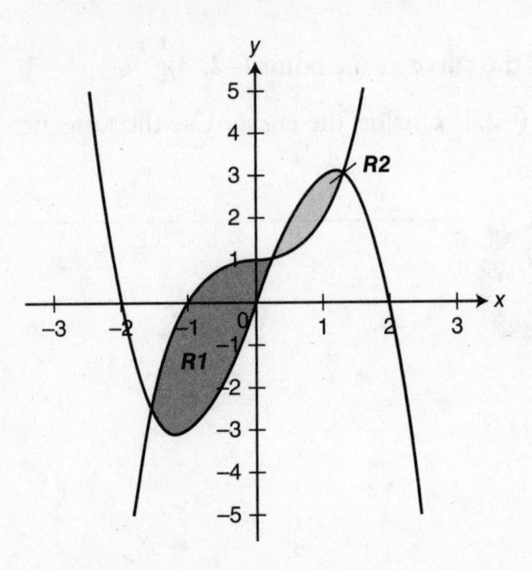

1. Without using absolute value, set up an expression involving one or more integrals that gives the area enclosed by the graphs of f and g.

2. Assume $b = 0.3$ and $c = 1.3$. Set up an expression involving one or more integrals that gives the volume of the solid generated by revolving the region $R2$ about the x-axis. Find the volume of this solid.

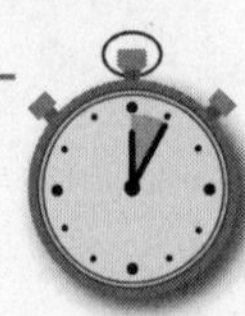

Water is draining from a conical tank, completely filled with water, with a height of 8 feet and a diameter of 6 feet into a cylindrical tank that has a base with an area of 100π square feet. The depth, h, in feet, of the water in the conical tank is changing at the rate of $(h - 1)$ feet per minute.

1. Write an expression for the volume of water in the conical tank as a function of h.
2. At what rate is the volume of water in the conical tank changing with $h = 6$?
3. Let y be the depth, in feet, of the water in the cylindrical tank. At what rate is y changing when $h = 6$?

The graph of a differentiable function $f(t)=1+\frac{2t}{t^2+1}$ f on the closed interval $[-1, 5]$ is shown. Let $h(x)=\int_0^x f(t)\,dt$ for $-1 \le x \le 5$.

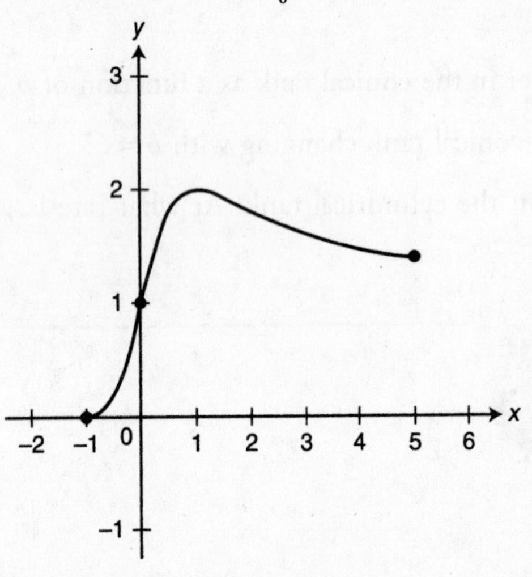

1. Find $h(0)$ and $h'(1)$.
2. On what interval(s) is the graph of h concave upward?
3. Find any inflection points of h on the interval $-1 < x < 5$.

Consider the graph of the function $h(x) = 8 - \frac{x^4}{2}$ for $x \geq 0$. Let R be the region in the first quadrant below the graph of h, and let $A(w)$ be the area of a rectangle in the first quadrant with sides along the x- and y-axis and diagonally opposite vertices at the origin and the point $(w, h(w))$.

1. Find the maximum value of $A(w)$.
2. Find the area included in R but not included in the rectangle with maximum value as described in Question 1.

The cost of a single dose of a particular drug in the United States is approximately given by $S(t) = Ce^{kt}$, where S is measured in dollars and t is measured in years from the beginning of 2007. The single-dose price at the beginning of 2007 was \$57, and in 2015, \$375.

1. Find C and k.

2. Find the average price of a single dose of the drug over the 8-year time period beginning January 1, 2011.

A family of functions has the form $f(x) = \cos(x) - bx$, where b is a positive constant and $-2\pi \leq x \leq 2\pi$.

Find the x-coordinates of all points, $-2\pi \leq x \leq 2\pi$, where the line $y = 1 - bx$ is tangent to the graph of $f(x) = \cos(x) - bx$.

A storage tank has the shape shown, obtained by revolving the curve $y = \frac{x^6}{64}$ from $x = 0$ to $x = 3$ about the y-axis. Both x and y are measured in feet. A lubricant weighing 35 pounds per cubic foot flowed into an initially empty tank at a constant rate of 5 cubic feet per minute. When the depth of the oil reached 8 feet, the flow stopped.

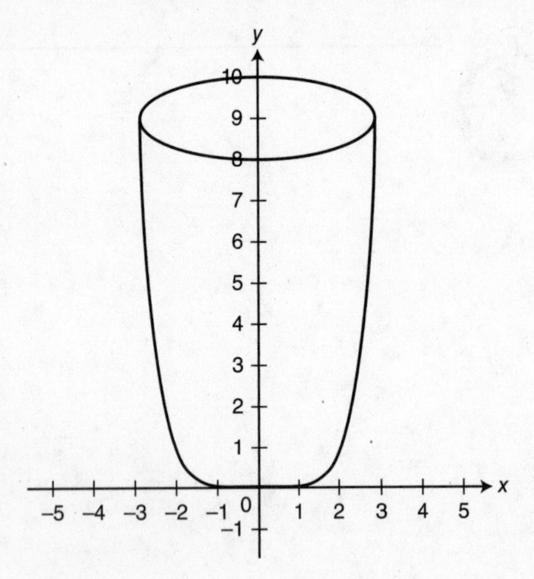

1. Find the volume of the lubricant in the tank, and the total weight of that volume of lubricant.
2. To the nearest minute, how long would it take to fill the tank to a depth of 8 feet?
3. How fast is the depth of lubricant in the tank, h, measured in feet, increasing when $h = 1$ ft?

Distance (cm)	x	0	2	6	10	15
Temperature (°C)	$f(x)$	80	73	65	62	60

A metal wire 15 cm long is heated at one end. The table gives selected values of the temperature $f(x)$ in degrees Celsius of the wire x centimeters from the end where heat was applied. The temperature $f(x)$ is decreasing and twice differentiable.

1. Approximate $f'(4)$.

2. Estimate the average temperature of the wire, using a trapezoidal approximation.

3. Find $\int_0^{15} f'(x)\,dx$ and explain its significance.

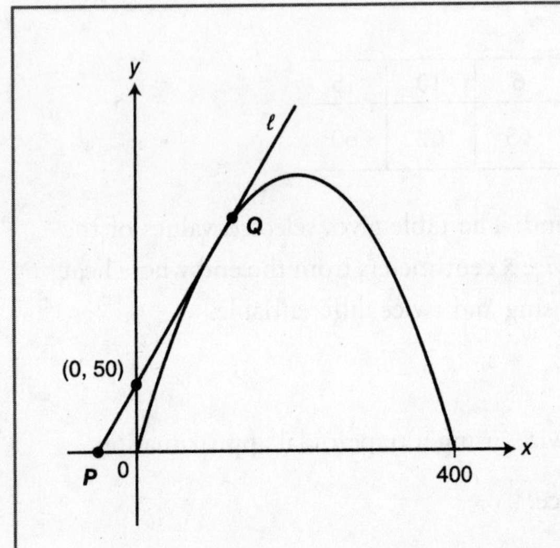

Line ℓ is tangent to the graph of $y = \dfrac{400x - x^2}{200}$ at point Q as shown.

Find the x-coordinate of point Q.

Consider the curve given by $x^3y^2 + 3x^2y^2 + xy^2 = 2$.

1. Show that $\dfrac{dy}{dx} = -\dfrac{y\left(3x^2 + 6x + 1\right)}{2x\left(x^2 + 3x + 1\right)}$.

2. Find all points on the curve whose x-coordinate is -1, and write an equation for the tangent line at each of these points.

Find any inflection points for the function:

$f(x) = 2x + \cos x, \quad -2\pi \leq x \leq 2\pi$

A particle moves along the y-axis with velocity $v(t) = 3 + 5t - t^2$ for $0 \le t \le 10$.

1. In which direction is the particle moving at time $t = 5$? Why?
2. Find the acceleration of the particle at time $t = 5$. Is the particle speeding up or slowing down at $t = 5$? Explain.
3. Find the position function $s(t)$ if $s(0) = 1$, and determine the position of the particle at $t = 6$.

The shaded region, R, is bounded by the graph of $y = \dfrac{x^3}{9}$ and the y-axis and the line $y = 3$, as shown in the figure.

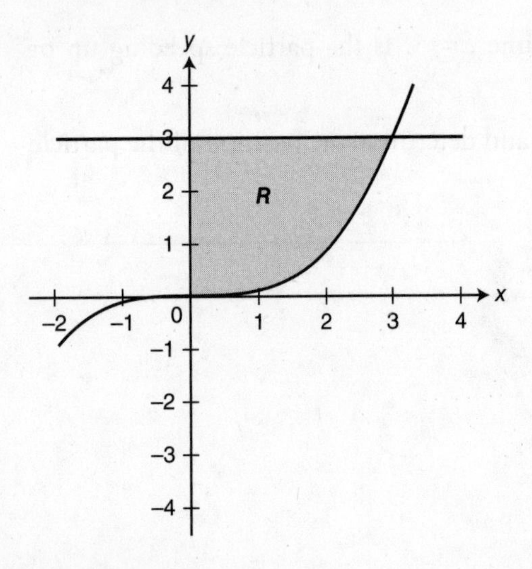

1. Find the area of R.
2. Find the volume of the solid generated by revolving R about the x-axis.
3. There exists a number k, $k > 3$, such that when R is revolved about the line $y = k$, the resulting solid has the same volume as that in question 2. Set up an equation containing an integral such that the solution to the equation will yield the value of k. Do not solve the equation.

t (hours)	0	2	4	6	8	10	12
$R(t)$ (gallons/ hour)	16.67	17.67	18	17.67	16.67	15	12.67

The rate at which water flows out of a pipe, in gallons per hour, is given by a differentiable function R of time t. The table above shows the rate as measured every 2 hours for a 12-hour period.

1. Use a midpoint Riemann sum with 3 subdivisions of equal length to approximate $\int_0^{12} R(t)\,dt$. Explain the meaning of your answer, using correct units.
2. The rate of water flow $R(t)$ can be approximated by $Q(t) = \dfrac{200 + 8t - t^2}{12}$. Use $Q(t)$ to approximate the average rate of water flow during the 12-hour period and indicate units of measure.

Suppose that the function f is at least three times differentiable for all x, and that $f(0)=1$, $f'(0)=2$, and $f''(0)=-1$. Let g be a function whose derivative is given by $g'(x)=6xf(x)+\left(3x^2+2\right)f'(x)+2xf''(x)$ for all x.

1. Write an equation of the line tangent to the graph of f at the point where $x=0$.
2. If $g(0)=-3$, write an equation of the line tangent to the graph of g at the point where $x=0$.
3. Show that $g''(x)=6f(x)+12xf'(x)+\left(3x^2+4\right)f''(x)+2xf'''(x)$.

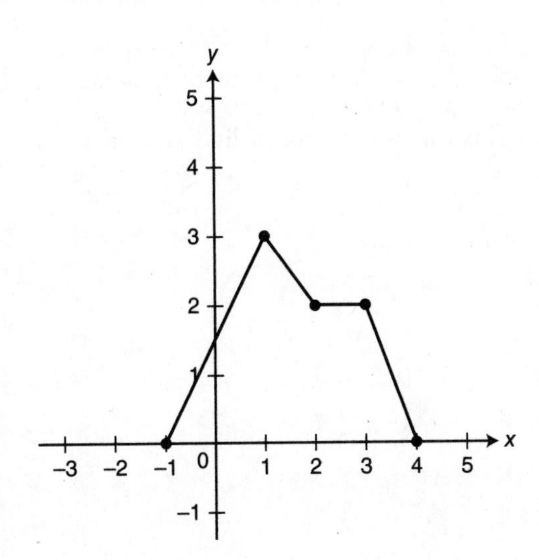

The graph of the function f, consisting of four line segments, is shown. Let $g(x) = \int_0^x f(t)\,dt$.

1. Compute $g(1)$ and $g(-1)$.
2. Find the instantaneous rate of change of g, with respect to x, at $x = 3$.
3. Find the absolute minimum value of g on the closed interval $[-1, 4]$.

Line ℓ is tangent to the graph of $y = \frac{5}{x}$ at point $\left(w, \frac{5}{w}\right)$. This point of tangency is point P. Point Q is the point $(w, 0)$ and point R is the x-intercept of line ℓ, which has coordinates $(k, 0)$.

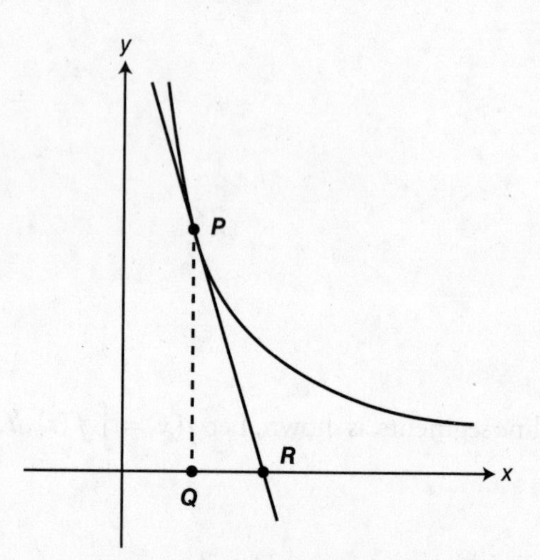

1. Find the value of k when $w = 1$.
2. Express k in terms of w, for all $w > 0$.
3. Suppose that w is increasing at a constant rate of 2 units per second. When $w = 1$, what is the rate of change of k with respect to time?

Let R be the region bounded by the x-axis, the graph of $y = x^2$, and the line $x = 3$.

1. Find the area of the region R.
2. Find the value of k for which the vertical line $x = k$ divides the region R into two regions of equal area.
3. Find the volume of the solid generated when the region R is revolved about the x-axis.
4. The vertical line $x = c$ divides the region R into two regions, such that the volumes of the solids formed when these two regions are revolved about the x-axis are equal. Find the value of c.

Let f be the function given by $f(x)=\dfrac{\ln(x)}{x^2}$

1. Find $\lim\limits_{x\to\infty} f(x)$ and $\lim\limits_{x\to 0^+} f(x)$.
2. Find the absolute maximum value of f. Justify your answer.

t	0	5	10	15	20	25	30
$v(t)$	0	10	25	40	60	50	45

The velocity $v(t)$ in ft/sec of a car traveling on a straight road for $0 \leq t \leq 30$ is shown.

1. During what intervals of time is the acceleration of the car negative?
2. Use a trapezoidal sum of 6 subintervals of equal width to approximate the total distance traveled by the car over the interval $0 \leq t \leq 30$.
3. Find the average velocity of the car in ft/sec, over the interval $0 \leq t \leq 30$.

For a certain equation, the slope of the graph at every point (x, y) is given by $\frac{dy}{dx} = \frac{x - 3x^2}{y}$, and the point (2, 2) is on the graph.

1. Write an equation of the line tangent to the graph at $x = 2$, and use it to approximate y when $x = 2.1$.

2. Solve the differential equation $\frac{dy}{dx} = \frac{x - 3x^2}{y}$ with initial condition $y = 2$ when $x = 2$.

3. Use the result of question 2 to find y when $x = 2.1$.

The temperature outside a house during a 24-hour period is given by $F(t) = 65 + 8\sin\left(\frac{\pi t}{12}\right)$, $0 \le t \le 24$, where $F(t)$ is measured in degrees Fahrenheit and t is measured in hours.

1. At what time, t, does the highest temperature of the day occur? At what time does the lowest temperature occur?
2. Find the average temperature, to the nearest degree Fahrenheit, between the two values of t found in question 1.

Consider the curve defined by $x^3 - 2x^2y - 3x^2 + 2y = 10$

1. Show that $\dfrac{dy}{dx} = \dfrac{3x^2 - 4xy - 6x}{2x^2 - 2}$.

2. Write an equation of the tangent line to the curve at $x = 0$.

t	0	2	4	6	8	10	12	14	16	18	20
$v(t)$	0	22	48	73	98	112	120	117	106	86	62

A sled moves on a straight track with positive velocity $v(t)$, in feet per minute at time t minutes, where v is a differentiable function of t. Selected values of $v(t)$ for $0 \le t \le 20$ are shown in the table.

1. Use a midpoint Riemann sum with five subintervals of equal length and values from the table to approximate $\int_0^{20} v(t)\,dt$ and explain its meaning.

2. The function $v(t) = 20 + 80\sin\left(\frac{t}{8}\right) - 20\cos\left(\frac{t}{4}\right)$ is used to model the velocity of the sled, in feet per minute, for $0 \le t \le 20$. According to this model, what is the acceleration of the sled at $t = 16$?

3. According to the model, what is the average velocity of the sled, in feet per minute, over the interval $0 \le t \le 20$?

5 Minutes to a 5

The figure shows the graph of $f'(x)$, the derivative of the function $f(x)$, on the closed interval $[-3, 7]$. The graph of $f'(x)$ has horizontal tangent lines at $x = -1$ and $x = 5$. The function $f(x)$ is twice differentiable with $f(2) = \frac{13}{3}$.

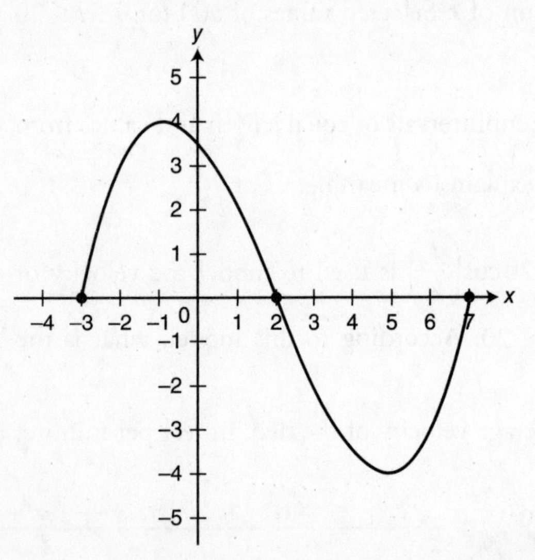

1. Find the coordinates of any relative maxima or minima of the graph of f.
2. Write an equation of the line tangent to the graph of f at $x = 2$.
3. Let g be the function defined by $g(x) = x^3 f(x)$. Find an equation for the line tangent to the graph of g at $x = 2$.

Day 165

Let g be the function given by $g(x)=1+\sin\left(\frac{\pi x}{2}\right)$.

1. Find the area of the region R, in the first quadrant, enclosed by the graph of $y=g(x)$ and the x-axis and y-axis.
2. Find the average value of g on the closed interval $[0, 3]$.
3. Let S by the solid generated when the region R is revolved about the x-axis. Set up an integral that could be used to find the volume of S. Do not evaluate the integral.

Let R be the region in the first quadrant bounded by the y-axis, the graph of $f(x) = 5 - x^2$, and the line $y = \frac{x}{2}$.

1. Find the area of the region R.

2. Find the volume of the solid generated when R is revolved about the y-axis.

Let f be the function given by $f(x) = x\cos(x)$ on the interval $0 \leq x \leq 2\pi$.

1. Find the rate at which f is changing when $x = \dfrac{\pi}{2}$.

2. Write an equation of the tangent line to the graph of f at $x = \dfrac{\pi}{2}$.

3. Find the area enclosed between the x-axis and the graph of f from $x = \dfrac{\pi}{2}$ and $x = \dfrac{3\pi}{2}$.

Day 168

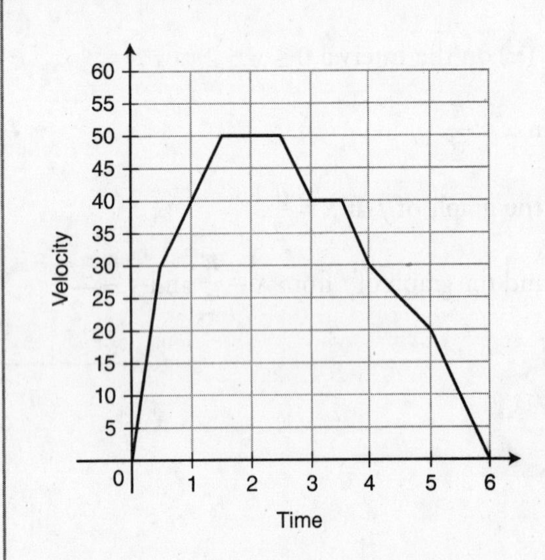

The graph of the velocity $v(t)$, in miles per hour, of a car traveling on a straight road, for $0 \leq t \leq 6$, is shown.

1. During what intervals of time is the acceleration of the car positive?

2. Estimate the average velocity of the car, in miles per hour over the interval from at $t = 0$ to $t = 3$.

Day 169

Let f be a function defined for $x \geq 0$ whose derivative is given by $\frac{dy}{dx} = \frac{3x^2 - x - 2}{4y}$, and let $f(2) = 1$.

1. Find the slope of a tangent line to the graph of f at the point where $x = 2$.

2. Write the equation of the tangent line at $x = 2$ and use it to approximate $f(2.1)$.

3. Find $f(x)$ by solving $\frac{dy}{dx} = \frac{3x^2 - x - 2}{4y}$ with the initial condition $f(2) = 1$.

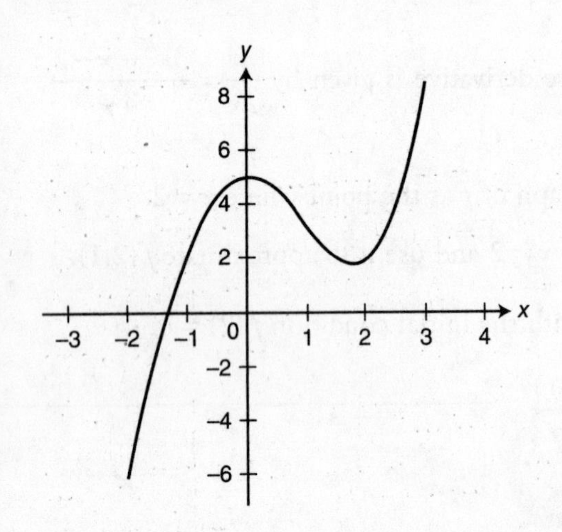

The figure shows the graph of $y = f'(x)$, the derivative of twice-differentiable function $f(x)$ defined on $[-2, 3]$.

Label each of the statements TRUE or FALSE.

1. The graph of $f(x)$ is decreasing on the interval $(0, 2)$.

2. The graph of $f(x)$ has a relative minimum on the interval $(-2, -1)$.

3. The graph of $f(x)$ is concave upward on the interval $(0, 1)$.

4. The graph of $f(x)$ has two points of inflection on the interval $(-2, 3)$.

5. The graph of $f(x)$ has only one horizontal tangent on the interval $(-2, 3)$.

Consider the curve defined by $y^3 + x^2 y + x^2 + 4y = 4$

1. Find the first derivative $\frac{dy}{dx}$.

2. Write the equation of any horizontal tangent lines to the curve.

3. The line with a slope of $-\frac{1}{2}$ and a y-intercept of 1 is tangent to the curve at a zero of the function. Find that point.

A particle moves along the x-axis. The position of the particle is given by the function $s(t) = 3\cos(1-t^2)$ for $0 \le t \le 5$.

1. Find the position of the particle at $t = 1$.
2. Find an expression for the velocity of the particle.
3. What is the first time the particle changes direction?
4. Find the total distance traveled by the particle from $t = 0$ to $t = 5$.

Let R be the region bounded by the graph of $y = 4 - |x - 4|$ and the x-axis.

1. Find the area of R by geometric methods.

2. Find the volume of the solid created when the region R is revolved about the x-axis.

Let f be a twice differentiable function. Selected values of f, its first derivative f', and its second derivative f'' are shown in the table.

x	−3	−1	0	2	4
f	34	9	0	−12	−16
f'	−17	−13	−5	−1	4
f''	2	2	2	2	2

Define $g = xf(x) + f'(x)$ and $h(x) = \dfrac{f(x)}{f'(x)}$.

1. Find $g'(-3)$.

2. Find $h'(2)$.

A function $f(x)$ is twice differentiable, and it is known that $f(0) = 2$, $f'(0) = -3$, and $f''(0) = 1$.

1. Write an equation of the tangent line to the graph of $f(x)$ at $x = 0$.
2. Use your equation to estimate $f(0.1)$.
3. Is your estimate an overestimate or an underestimate?

On the axes provided, sketch a graph of a function f that displays the following properties.

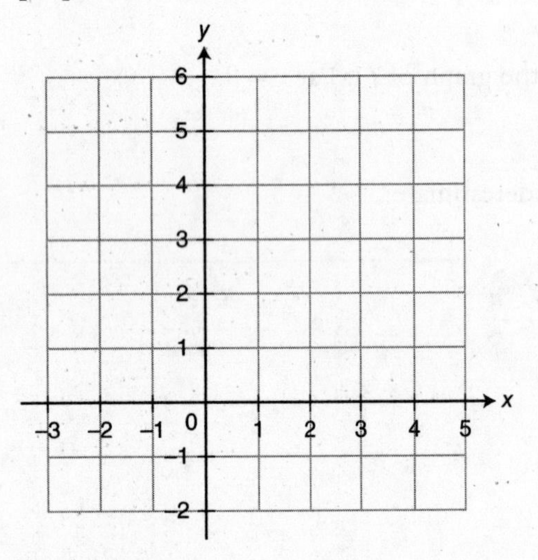

Domain $[-3, 5]$

Range $[-1, 4]$

Intercepts: $(-2, 0)$ $(0, 0)$

$f'(x) < 0$ on $(-3, -1)$ and $(2, 4)$

$f'(x) = 0$ at $x = -1$, $x = 2$, and $x = 4$

$f''(x) < 0$ on $(1, 3)$

$f''(x) = 0$ at $x = 1$ and $x = 3$

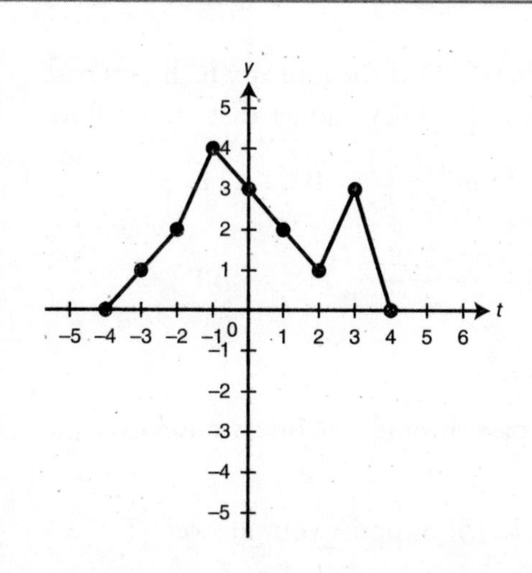

The graph of a function f is shown. Let g be the function defined by $g(x) = \int_0^x f(t)\,dt$. Sketch a graph of g on the axes provided.

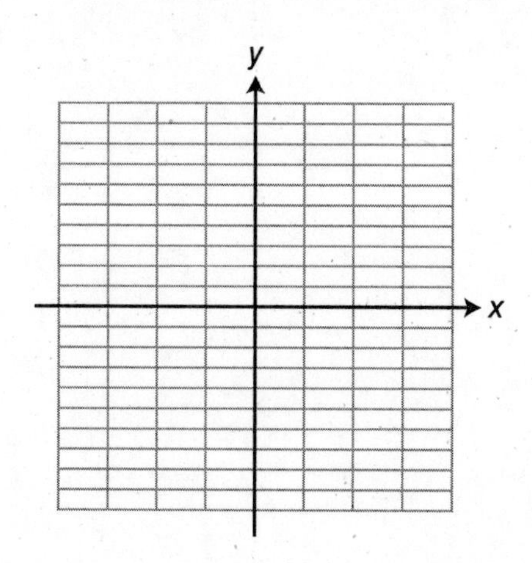

To measure the flow of traffic in an area, researchers record the rate at which cars pass through a particular intersection, in cars per minute. They model that traffic flow, in cars per minute, by the equation $F(t)=\begin{cases} 4+5\sin\left(\dfrac{t}{3}\right) & 0\le t<11 \\ 25+\dfrac{(t-19)^2}{10} & 11\le t<30 \end{cases}$, where t is number of minutes, and $0\le t\le 30$.

1. To the nearest whole number, how many cars pass through the intersection over the first 10 minutes?

2. Is the traffic flow increasing or decreasing at $t = 15$? Support your answer.

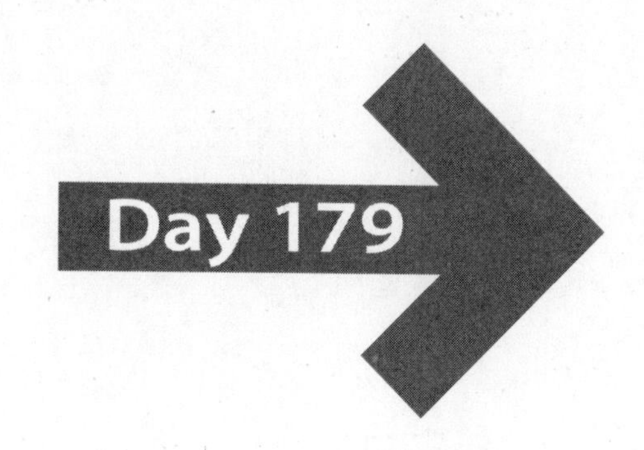

Let f and g be the functions given by $f(x) = 3x - x^2$ and $g(x) = \dfrac{x\sqrt{2x}}{2}$ for $0 \le x \le 2$. The graphs are shown in the figure.

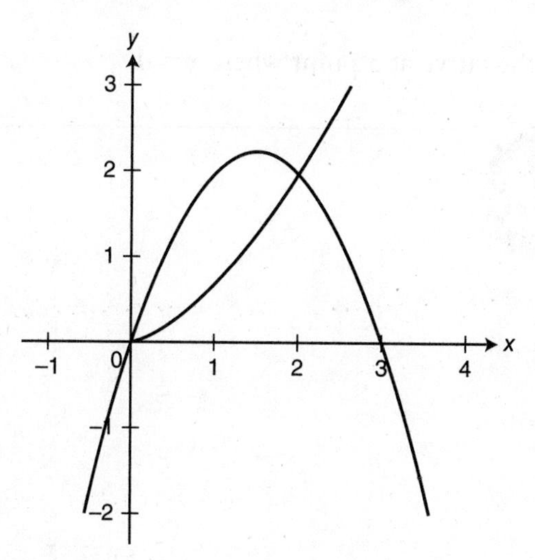

1. Find the area of the shaded region enclosed by the graphs of f and g.
2. The region enclosed by the graphs of f and g is the base of a solid with square cross sections perpendicular to the x-axis. Find the volume of this solid.

Consider the curve given by $5x^2 + 6xy + 5y^2 = 8$.

1. Show that $\frac{dy}{dx} = -\frac{5x+3y}{3x+5y}$.

2. Find an equation for the line tangent to the curve at a point where $x = 0$.

Answers

Day 1

Graph the lines $3x-2y=12$ and $y=5-\frac{x}{3}$ to show the center of the circle. The lines intersect at (6, 3). A circle with center at (6, 3) that passes through the point (3, −1) has a radius equal to $r=\sqrt{(6-3)^2+(3+1)^2}=\sqrt{9+16}=\sqrt{25}=5$. The equation of the circle is $(x-6)^2+(y-3)^2=5^2$.

Day 2

EQUATION OF CIRCLE	POINT OF TANGENCY	EQUATION OF TANGENT	EQUATION OF NORMAL LINE
$(x-6)^2+(y-3)^2=5^2$	(3, −1)	$y+1=-\frac{3}{4}(x-3)$	$y+1=\frac{4}{3}(x-3)$
$x^2+y^2=25$	(−4, 3)	$y-3=\frac{4}{3}(x+4)$	$y-3=-\frac{3}{4}(x+4)$
$(x+2)^2+(y-1)^2=9$	(1, 1)	$x=1$	$y=1$

Day 3

1. $-3<x<7$, $(-3, 7)$
2. $x>4$, $(4, \infty)$
3. $-5\le x\le 2$, $[-5, 2]$
4. $x\le -1$, $(-\infty, -1]$
5. $6<x\le 10$, $(6, 10]$
6. $-10\le x<-8$, $[-10, -8)$

Day 4

1. $|7-4x|>9$ becomes $-9>7-4x$ *OR* $7-4x>9$. $x>4$ *OR* $x<-\frac{1}{2}$

2. $|3x-5|+2<3$ or $|3x-5|<1$ becomes $-1<3x-5$ *AND* $3x-5<1$. $\frac{4}{3}<x$ *AND* $x<2$. The solution can also be written as $\frac{4}{3}<x<2$.

3. $|6-2(x-4)|\le 10$ simplifies to $|14-2x|\le 10$, which becomes $-10\le 14-2x\le 10$. $12\ge x\ge 2$ or $2\le x\le 12$

Day 5

1. $x^2-x-6>0$ on $(-\infty,-2)$ or $(3,\infty)$

2. $(3-x)(2+x)(1-x)\le 0$ on $(-\infty,-2]$ and $[1,3]$

3. $\frac{x^2-3x-10}{x-2}\le 0$ on $(-\infty,-2]$ and $(2,5]$

Day 6

1. $(f+g)(x)=f(x)+g(x)=\frac{1}{x}+\sqrt{x}=\frac{1+x\sqrt{x}}{x}$, Domain: $x>0$

2. $(fg)(x)=f(x)\cdot g(x)=\frac{1}{x}\cdot\sqrt{x}=\frac{\sqrt{x}}{x}$, Domain: $x>0$

3. $(f-g)(x)=f(x)-g(x)=\frac{1}{x}-\sqrt{x}=\frac{1-x\sqrt{x}}{x}$, Domain: $x>0$

4. $\left(\frac{f}{g}\right)(x)=\frac{f(x)}{g(x)}=\frac{1/x}{\sqrt{x}}=\frac{1}{x\sqrt{x}}=\frac{\sqrt{x}}{x^2}$, Domain: $x>0$

Day 7

1. K **2.** L **3.** H, F **4.** D **5.** M **6.** M **7.** E **8.** J **9.** G

Day 8

1. Odd

2. Even

3. Neither

4. Odd

Day 9

1. Jump $x=1$ **2.** Essential $x=-2$ **3.** No discontinuities

4. Removable $x=3$ **5.** No discontinuities

Day 10

1. $f(x)=\frac{x^2-2x-35}{x+5}=\frac{(x-7)(x+5)}{x+5}$ has a hole at $x=-5$. The point $(-5, -12)$ will fill the hole.

2. $f(x)=\frac{x^2-3x-1}{x^2-1}=\frac{x^2-3x-1}{(x+1)(x-1)}$ has two essential discontinuities but no hole.

3. $f(x)=\frac{x^2-3x}{x^2-x-6}=\frac{x(x-3)}{(x+2)(x-3)}$ has a hole at $x=3$. The graph must look like the graph of $g(x)=\frac{x}{x+2}$, so the point $\left(3,\frac{3}{5}\right)$ fills the hole.

Day 11

1. D **2.** F **3.** B **4.** H **5.** A **6.** E **7.** G **8.** D **9.** J **10.** A

Day 12

1. $(f\circ g)(x)=4\left(\sqrt{x}+3\right)^2=4x+24\sqrt{x}+36, [0, \infty)$
2. $(g\circ f)(x)=\sqrt{4x^2}+3=2|x|+3, (-\infty, \infty)$
3. $(f\circ f)(x)=4\left(4x^2\right)^2=64x^4, (-\infty, \infty)$
4. $(g\circ g)(x)=\sqrt{\sqrt{x}+3}+3, [0, \infty)$

Day 13

1. $f(x)=4-x^2$ is not 1-1. Restrict domain to $[0, \infty)$, with range $(-\infty, 4]$. $f^{-1}(x)=\sqrt{4-x}$. Domain of $f^{-1}(x)$ is $(-\infty, 4]$, range is $[0, \infty)$.

2. $f(x)=\sqrt{2x+3}$ is 1-1 on domain $[-1.5, \infty)$, with range $[0, \infty)$. $f^{-1}(x)=\frac{x^2-3}{2}$ with domain $[0, \infty)$ and range $[-1.5, \infty)$.

3.

x	−3	−2	−1	0	1	2	3
$f^{-1}(x)$	−2	0	2	3	−3	−1	1

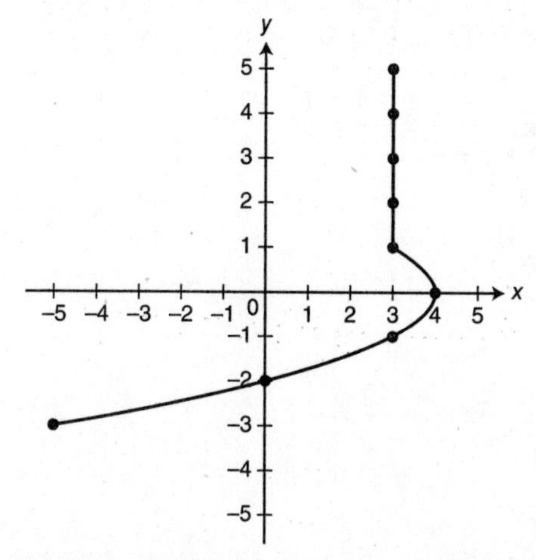

4. Note that $f(x)$ is not 1 – 1 and that $f^{-1}(x)$ is not a function as shown in the accompanying diagram. If we restrict the domain of $f(x)$ to $[-3, 0]$, then $f^{-1}(x)$ would be a function with the domain of $[-5, 4]$.

Answers

Day 14

1. $\lim_{x\to 1}\left(\frac{4x^2-3x-1}{x^2-3x+2}\right)=\lim_{x\to 1}\left[\frac{(4x+1)(x-1)}{(x-2)(x-1)}\right]=\lim_{x\to 1}\left[\frac{4x+1}{x-2}\right]=\frac{4(1)+1}{1-2}=-5$

2. $\lim_{x\to 2}\frac{x^2-4}{x^3-8}=\lim_{x\to 2}\frac{(x-2)(x+2)}{(x-2)(x^2+2x+4)}=\lim_{x\to 2}\frac{x+2}{x^2+2x+4}=\frac{2+2}{4+4+4}=\frac{4}{12}=\frac{1}{3}$

3. $\lim_{x\to 2}\frac{x^3-8}{x^2+3x-10}=\lim_{x\to 2}\frac{(x-2)(x^2+2x+4)}{(x-2)(x+5)}=\lim_{x\to 2}\frac{x^2+2x+4}{x+5}=\frac{4+4+4}{2+5}=\frac{12}{7}$

Day 15

1. I **2.** B

3. I **4.** D

5. A **6.** H

7. E **8.** D

9. C **10.** F

Day 16

$f(x)=x^2-7x+12$, $a=7$, (6, 8), yes, $L=12$,

Day 17

$$\lim_{x\to 4}\left(\frac{11-x}{3\sqrt{x}}\right)=\frac{\lim_{x\to 4}(11-x)}{\lim_{x\to 4}(3\sqrt{x})}=\frac{\lim_{x\to 4}(11)-\lim_{x\to 4}(x)}{\lim_{x\to 4}(3)\cdot\lim_{x\to 4}\sqrt{x}}=\frac{\lim_{x\to 4}(11)-\lim_{x\to 4}(x)}{\lim_{x\to 4}(3)\cdot\sqrt{\lim_{x\to 4}(x)}}=\frac{11-4}{3\cdot 2}=\frac{7}{6}$$

Day 18

1. $\lim_{x\to -3}\frac{x^2-9}{x^2-x-12}=\lim_{x\to -3}\frac{(x+3)(x-3)}{(x+3)(x-4)}=\lim_{x\to -3}\frac{x-3}{x-4}=\frac{-6}{-7}=\frac{6}{7}$

2. $\lim_{x\to 3}\left(\frac{2x^2-12x+18}{4x^2-24x+36}\right)=\lim_{x\to 3}\left(\frac{2(x-3)^2}{4(x-3)^2}\right)=\frac{1}{2}$

Day 19

1. $\lim\limits_{x\to 0}\dfrac{x}{\sqrt{x+9}-3}=\lim\limits_{x\to 0}\dfrac{x}{\sqrt{x+9}-3}\cdot\dfrac{\sqrt{x+9}+3}{\sqrt{x+9}+3}=\lim\limits_{x\to 0}\dfrac{x\left(\sqrt{x+9}+3\right)}{x}=\lim\limits_{x\to 0}\sqrt{x+9}+3=6$

2. $\lim\limits_{x\to 4}\dfrac{\sqrt{x}-2}{x-4}=\lim\limits_{x\to 4}\dfrac{\sqrt{x}-2}{x-4}\cdot\dfrac{\sqrt{x}+2}{\sqrt{x}+2}=\lim\limits_{x\to 4}\dfrac{x-4}{(x-4)\left(\sqrt{x}+2\right)}=\lim\limits_{x\to 4}\dfrac{1}{\sqrt{x}+2}=\dfrac{1}{4}$

3. $\lim\limits_{x\to -3}\dfrac{\sqrt{x+3}}{\sqrt{x+3}-\sqrt{3}}=\lim\limits_{x\to -3}\dfrac{\sqrt{x+3}}{\sqrt{x+3}-\sqrt{3}}\cdot\dfrac{\sqrt{x+3}+\sqrt{3}}{\sqrt{x+3}+\sqrt{3}}=\lim\limits_{x\to -3}\dfrac{\sqrt{x+3}\cdot\left(\sqrt{x+3}+\sqrt{3}\right)}{x+3-3}$

$=\lim\limits_{x\to -3}\dfrac{\sqrt{x+3}\cdot\left(\sqrt{x+3}+\sqrt{3}\right)}{x}=\lim\limits_{x\to -3}\dfrac{x+3+\sqrt{3x+9}}{x}=\dfrac{-3+3+0}{-3}=0$

Day 20

1. $-1\le\cos(\pi x)\le 1$. Because x approaches $+\infty$, so it's safe to consider $x-1$ positive. Divide through by $x-1$. $\dfrac{-1}{x-1}\le\dfrac{\cos(\pi x)}{x-1}\le\dfrac{1}{x-1}$. Then $\lim\limits_{x\to\infty}\dfrac{-1}{x-1}=0$ and $\lim\limits_{x\to\infty}\dfrac{1}{x-1}=0$, so $\lim\limits_{x\to\infty}\dfrac{\cos(\pi x)}{x-1}=0$.

2. $-1\le\sin\left(x-\dfrac{\pi}{2}\right)\le 1$. Multiply by -1. $1\ge-\sin\left(x-\dfrac{\pi}{2}\right)\ge-1$ or $-1\le-\sin\left(x-\dfrac{\pi}{2}\right)\le 1$. Add 2. $1\le 2-\sin\left(x-\dfrac{\pi}{2}\right)\le 3$. As $x\to-\infty$, and e^{-x} can be assumed to be positive. $\dfrac{1}{e^{-x}}\le\dfrac{2-\sin\left(x-\dfrac{\pi}{2}\right)}{e^{-x}}\le\dfrac{3}{e^{-x}}$. Because $\lim\limits_{x\to-\infty}\dfrac{1}{e^{-x}}=\lim\limits_{x\to-\infty}\dfrac{3}{e^{-x}}=0$. $\lim\limits_{x\to-\infty}\dfrac{2-\sin\left(x-\dfrac{\pi}{2}\right)}{e^{-x}}=0$.

Day 21

1. $\lim\limits_{x\to 0}\dfrac{\sin(2x)}{2x}=1$

2. $\lim\limits_{x\to 0}\dfrac{\sin x}{2x}=\lim\limits_{x\to 0}\left(\dfrac{1}{2}\cdot\dfrac{\sin x}{x}\right)=\lim\limits_{x\to 0}\left(\dfrac{1}{2}\right)\cdot\lim\limits_{x\to 0}\left(\dfrac{\sin x}{x}\right)=\dfrac{1}{2}\cdot 1=\dfrac{1}{2}$

3. $\lim\limits_{x\to 0}\dfrac{\sin(3x)}{x}=\lim\limits_{x\to 0}\left[\dfrac{\sin(3x)}{x}\cdot\dfrac{3}{3}\right]=\lim\limits_{x\to 0}\left[3\left(\dfrac{\sin(3x)}{3x}\right)\right]=3$

Day 22

1. -1 **2.** 1 **3.** DNE

4. 0 **5.** -5 **6.** -5

Day 23

1. $-\infty$

2. ∞

3. DNE

4. ∞

5. ∞

6. ∞

7. 0

Day 24

1. $\lim_{x\to\infty}\frac{2x-3}{x+4}=\lim_{x\to\infty}\frac{2-3/x}{1+4/x}=2$

2. $\lim_{x\to-\infty}\left(4-3x+2x^2\right)=\infty$

3. $\lim_{x\to\infty}\frac{2x-3}{x^2-3x+1}=0$

4. $\lim_{x\to-\infty}\frac{4}{x^2}=0$

5. $\lim_{x\to-\infty}\frac{3x^5}{2x^2}=\lim_{x\to-\infty}\frac{3}{2}x^3=-\infty$

Day 25

1. $\lim_{x\to0}\frac{x\cos x}{\sin 2x}=\lim_{x\to0}\left(\frac{x}{\sin 2x}\cdot\frac{\cos x}{1}\right)=\lim_{x\to0}\left(\frac{2x}{\sin 2x}\cdot\frac{\cos x}{2}\right)$

$=\lim_{x\to0}\frac{2x}{\sin 2x}\cdot\lim_{x\to0}\frac{\cos x}{2}=1\cdot\frac{1}{2}=\frac{1}{2}$

2. $\lim_{x\to0}\left(\frac{-\pi\sin(-x)}{\sin(\pi x)}\right)=\lim_{x\to0}\left(\frac{\sin(-x)}{-1}\cdot\frac{\pi}{\sin(\pi x)}\right)=\lim_{x\to0}\left(\frac{\sin(-x)}{-1x}\cdot\frac{\pi x}{\sin(\pi x)}\right)=1$

3. $\lim_{x\to0}\frac{\sin(2x)}{\sin(3x)}=\lim_{x\to0}\left[\frac{\sin(2x)}{1}\cdot\frac{1}{\sin(3x)}\right]=\lim_{x\to0}\left[\frac{\sin(2x)}{1}\cdot\frac{1}{\sin(3x)}\cdot\frac{2x}{2x}\cdot\frac{3x}{3x}\right]$

$=\lim_{x\to0}\left[\frac{\sin(2x)}{2x}\cdot\frac{3x}{\sin(3x)}\cdot\frac{2x}{1}\cdot\frac{1}{3x}\right]=\frac{2}{3}$

Day 26

1. $f(-4)=7,\ \lim_{x\to -4^-} f(x)=7,\ \lim_{x\to -4^+} f(x)=7,\ \lim_{x\to -4} f(x)=7$, continuous

2. $f(-1)=1,\ \lim_{x\to -1^-} 7-2|x+4|=1,\ \lim_{x\to -1^+} \frac{1}{x^2}=1,\ \lim_{x\to -1} f(x)=1$, continuous

3. $f(0)$ is undefined, $\lim_{x\to 0^-} f(x)=\infty,\ \lim_{x\to 0^+} f(x)=\infty,\ \lim_{x\to 0} f(x)=\infty$, discontinuous because $f(0)$ does not exist.

4. $f(4)=0,\ \lim_{x\to 4^-} \frac{1}{x^2}=\frac{1}{16},\ \lim_{x\to 4^+} \frac{x-4}{5-x}=0,\ \lim_{x\to 4} f(x)$ DNE because limit from left is not equal to limit from right, discontinuous because $\lim_{x\to 4} f(x)$ DNE.

5. $f(5)$ is undefined, $\lim_{x\to 5^-} \frac{x-4}{5-x}=\infty,\ \lim_{x\to 5^+} \frac{x-4}{5-x}=-\infty,\ \lim_{x\to 5} f(x)$ DNE because limit from left is not equal to limit from right, discontinuous because $\lim_{x\to 5} f(x)$ DNE or $f(5)$ is undefined.

Day 27

1. Jump discontinuity at $x = 0$.
2. Essential discontinuity at $x = 3$, removable discontinuity at $x = 2$.
3. Jump discontinuity at $x = -2$.

Day 28

1. $\lim_{x\to 0^-} 1-4x=1,\ \lim_{x\to 0^+} x^2+k=k,\ k=1$

2. $\lim_{x\to -1^-} kx^2=k,\ \lim_{x\to -1^+} 3x-k=-3-k,\ k=-3-k,\ 2k=-3,\ k=-1.5$

3. $\lim_{x\to 0^-} \frac{\sin 2x}{x}=\lim_{x\to 0^-}\left[2\left(\frac{\sin 2x}{2x}\right)\right]=2,\ \lim_{x\to 0^+} \frac{\sin 2x}{x}=\lim_{x\to 0^+}\left[2\left(\frac{\sin 2x}{2x}\right)\right]=2,$
$f(0)=k-2(0)=k,\ k=2$

Day 29

1. There is a zero on $[3,4]$. $f(x)=\frac{7-2x}{x}$ is continuous on $[3,4]$, $f(3)=\frac{7-(2\cdot 3)}{3}=\frac{1}{3}$, $f(4)=\frac{7-(2\cdot 4)}{4}=-\frac{1}{4}$

2. $f(x)=\frac{7-2x}{x}$ is not continuous on $[-1,1]$. Cannot apply the Intermediate Value Theorem.

3. There is a zero on $[-2,-1]$. $f(x)=\frac{2x^2-3}{x^2+1}$ is continuous on $[-2,-1]$.

$$f(-2)=\frac{2(-2)^2-3}{(-2)^2+1}=\frac{8-3}{4+1}=1,\ f(-1)=\frac{2(-1)^2-3}{(-1)^2+1}=\frac{-1}{2}$$

4. $f(x)=\frac{2x^2-3}{x^2+1}$ is continuous on $[-1,1]$. However, $f(-1)=\frac{2(-1)^2-3}{(-1)^2+1}=-\frac{1}{2}$ and $f(1)=\frac{2(1)^2-3}{(1)^2+1}=-\frac{1}{2}$. Because $f(-1)$ and $f(1)$ are not opposite in sign, it is not possible to apply the Intermediate Value Theorem to conclude that there is a zero on $[-1,1]$.

Day 30

1. $$\frac{2(x+\Delta x)+5-(2x+5)}{(x+\Delta x)-x}=\frac{2x+2\Delta x-2x+5-5}{\Delta x}=\frac{2\Delta x}{\Delta x}=2.\ \lim_{\Delta x\to 0}2=2$$

2. $$\frac{3-2(x+\Delta x)^2-(3-2x^2)}{(x+\Delta x)-x}=\frac{3-2\left(x^2+2x\Delta x+(\Delta x)^2\right)-3+2x^2}{\Delta x}$$

$$=\frac{-4x\Delta x-2(\Delta x)^2}{\Delta x}=-4x-2\Delta x$$

$$\lim_{\Delta x\to 0}(-4x-2\Delta x)=-4x$$

3. $$\frac{\frac{1}{x+\Delta x}-\frac{1}{x}}{\Delta x}=\frac{1}{\Delta x}\left[\frac{x}{x(x+\Delta x)}-\frac{x+\Delta x}{x(x+\Delta x)}\right]=\frac{1}{\Delta x}\left[\frac{-\Delta x}{x(x+\Delta x)}\right]=\frac{-1}{x(x+\Delta x)}.$$

$$\lim_{\Delta x\to 0}\frac{-1}{x(x+\Delta x)}=-\frac{1}{x^2}$$

Day 31

1. $$f'(x)=\lim_{h\to 0}\frac{\left[(x+h)^3+8\right]-\left[x^3+8\right]}{h}=\lim_{h\to 0}\frac{\left[x^3+3hx^2+3h^2x+h^3+8\right]-\left[x^3+8\right]}{h}$$
$$=\lim_{h\to 0}\frac{3hx^2+3h^2x+h^3}{h}=\lim_{h\to 0}\left(3x^2+3hx+h^2\right)=3x^2$$

2. $$f'(x)=\lim_{h\to 0}\frac{\frac{2(x+h)}{(x+h-3)}-\frac{2x}{x-3}}{h}=\lim_{h\to 0}\frac{1}{h}\cdot\frac{2(x+h)(x-3)-2x(x+h-3)}{(x+h-3)(x-3)}$$
$$=\lim_{h\to 0}\frac{1}{h}\cdot\frac{2x^2+2hx-6x-6h-2x^2-2hx+6x}{(x+h-3)(x-3)}=\lim_{h\to 0}\frac{1}{h}\cdot\frac{-6h}{(x+h-3)(x-3)}$$
$$=\lim_{h\to 0}\frac{-6}{(x+h-3)(x-3)}=\frac{-6}{(x-3)^2}$$

3. $$f'(x)=\lim_{h\to 0}\frac{\sqrt{x+h-2}-\sqrt{x-2}}{h}=\lim_{h\to 0}\frac{\sqrt{x+h-2}-\sqrt{x-2}}{h}\cdot\frac{\sqrt{x+h-2}+\sqrt{x-2}}{\sqrt{x+h-2}+\sqrt{x-2}}$$
$$=\lim_{h\to 0}\frac{x+h-2-(x-2)}{h\left(\sqrt{x+h-2}+\sqrt{x-2}\right)}=\lim_{h\to 0}\frac{h}{h\left(\sqrt{x+h-2}+\sqrt{x-2}\right)}$$
$$=\lim_{h\to 0}\frac{1}{\sqrt{x+h-2}+\sqrt{x-2}}=\frac{1}{2\sqrt{x-2}}$$

Day 32

1. $f'(x)=21x^6-20x^4+2x$

2. $f'(x)=-72x^8+30x^5$

3. $f'(x)=8x^{11}+2x^9-6x^7-2x^5$

4. $f'(x)=7x^6-5x^4+3x^2-1$

5. $f'(x)=-4x^3+24x^5-24x^7$

Day 33

1. $f'(x)=(3x-7)(2x)+\left(x^2-4\right)(3)=9x^2-14x-12$

2. $f'(x)=2x^4\left(\frac{1}{2}x^{-\frac{1}{2}}\right)+x^{\frac{1}{2}}\left(8x^3\right)=x^{\frac{7}{2}}+8x^{\frac{7}{2}}=9x^{\frac{7}{2}}.$

3. $f'(x)=\left(x^2+1\right)\left(6x^2\right)+\left(2x^3+5\right)(2x)=10x^4+6x^2+10x$

4. $$f'(x)=\frac{(2x+5)(-6x)-\left(4-3x^2\right)(2)}{(2x+5)^2}=\frac{-6x^2-30x-8}{(2x+5)^2}$$

Day 34

1. $f'(-1) \approx \frac{f(0)-f(-2)}{0-(-2)} = \frac{16.09-10.99}{2} = \frac{5.10}{2} = 2.55$

2. $f'(0) \approx \frac{f(1)-f(-1)}{1-(-1)} = \frac{17.92-13.86}{2} = \frac{4.06}{2} = 2.03$

3. $f'(2) \approx \frac{f(3)-f(1)}{3-1} = \frac{20.79-17.92}{2} = \frac{2.87}{2} = 1.435$

Day 35

1. $f'(x) = 2(x^2-3x+2)(2x-3)$

2. $f'(x) = \frac{1}{2}(4+5x^2)^{-1/2}(10x)$

3. $f'(x) = -1(x^3-2x+5)^{-2}(3x^2-2)$

4. $f'(x) = \left[(3x-1)^2 \cdot 3(2x+3)^2 \cdot 2\right] + \left[(2x+3)^3 \cdot 2(3x-1) \cdot 3\right]$

5. $f'(x) = \frac{\left[(2x+1)^2 \cdot 2x\right] - \left[(x^2-4) \cdot 2(2x+1) \cdot 2\right]}{(2x+1)^4}$

Day 36

1. $f'(2) + g'(2) = 1 + 3 = 4$

2. $\frac{f(1) \cdot g'(1) - g(1) \cdot f'(1)}{(f(1))^2} = \frac{2 \cdot 2 - 0 \cdot (-3)}{2^2} = 1$

3. $f'(g(0)) \cdot g'(0) = f'(1) \cdot g'(0) = (-3)(-1) = 3$

4. $g'(f(-1)) \cdot f'(-1) = g'(0) \cdot f'(-1) = (-1)(-1) = 1$

5. $f'(f(2)) \cdot f'(2) = f'(-1) \cdot f'(2) = (-1)(1) = -1$

Day 37

1. $\frac{d}{dx}(4-\cos(2x))=0-(-\sin(2x))(2)=2\sin(2x)$

2. $\frac{d}{dx}\left(\tan^2\left(\frac{\pi}{2}x\right)\right)=2\tan\left(\frac{\pi}{2}x\right)\sec^2\left(\frac{\pi}{2}x\right)\left(\frac{\pi}{2}\right)=\pi\tan\left(\frac{\pi}{2}x\right)\sec^2\left(\frac{\pi}{2}x\right)$

3. $\frac{d}{dx}\left(3x^2\left(\sin^2 x\right)\right)=3x^2(2\sin x\cos x)+6x\sin^2 x=6x\sin x(x\cos x+\sin x)$

4. $$\frac{d}{dx}\left(\frac{\csc x}{1-\cot x}\right)=\frac{(1-\cot x)(-\csc x\cot x)-\csc x\left(\csc^2 x\right)}{(1-\cot x)^2}$$
$$=\frac{-\csc x\left(\cot x-\cot^2 x+\csc^2 x\right)}{(1-\cot x)^2}=\frac{-\csc x(\cot x+1)}{(1-\cot x)^2}.$$

   Note that $\cot^2 x+1=\csc^2 x$ and thus $1=\csc^2 x-\cot^2 x$.

5. $$\frac{d}{dx}(\sin(3x)\cos(2x))=\sin(3x)(-2\sin(2x))+\cos(2x)(3\cos(3x))$$
$$=-2\sin(3x)\sin(2x)+3\cos(2x)\cos(3x)$$

Day 38

1. $f'(x)=\frac{-x^3}{\sqrt{1-x^2}}+3x^2\arccos x$

2. $$f'(x)=\frac{(3-\arcsin x)\left(\frac{-1}{\sqrt{1-x^2}}\right)-(2+\arccos x)\left(\frac{-1}{\sqrt{1-x^2}}\right)}{(3-\arcsin x)^2}$$
$$=\left(\frac{-1}{\sqrt{1-x^2}}\right)\frac{1-\arcsin x-\arccos x}{(3-\arcsin x)^2}=\frac{\arcsin x+\arccos x-1}{(3-\arcsin x)^2\sqrt{1-x^2}}$$

3. $f'(x)=\frac{\arctan x}{\sqrt{1-x^2}}+\frac{\arcsin x}{1+x^2}$

4. $$f'(x)=x^2\cdot 2(\arcsin x)\frac{1}{\sqrt{1-x^2}}+(\arcsin x)^2(2x)$$
$$=2x(\arcsin x)\left[\frac{x}{\sqrt{1-x^2}}+\arcsin x\right]$$

5. $$f'(x)=3\left(\arctan\left(x^2\right)\right)^2\left(\frac{1}{1+\left(x^2\right)^2}\right)(2x)=\frac{6x}{1+x^4}\left(\arctan\left(x^2\right)\right)^2$$

Answers

Day 39

1. $\frac{d}{dx}\left(e^{x^2}\right)=e^{x^2}(2x)=2xe^{x^2}$

2. $\frac{d}{dx}\left(e^{x^2}\sqrt{x}\right)=e^{x^2}\left(\frac{1}{2\sqrt{x}}\right)+\sqrt{x}\left(2xe^{x^2}\right)=\frac{e^{x^2}\sqrt{x}}{2x}+\frac{4x^2e^{x^2}\sqrt{x}}{2x}=\frac{e^{x^2}\sqrt{x}}{2x}(1+4x^2)$

3. $\frac{d}{dx}\left(e^{\cos(2x)}\right)=\left(e^{\cos(2x)}\right)(-2\sin(2x))=-2\sin(2x)e^{\cos(2x)}$

4. $\frac{d}{dx}(2^x\cdot\sin^2 x)=2^x(2\sin x\cos x)+(\sin^2 x)(2^x\ln 2)=2^x\sin x[2\cos x+(\ln 2)\sin x]$

5. $$\frac{d}{dx}\left(\frac{e^{2x+1}}{2^{1-3x}}\right)=\frac{2^{1-3x}(2e^{2x+1})-e^{2x+1}(2^{1-3x})(\ln 2)(-3)}{(2^{1-3x})^2}$$
$$=\frac{(2^{1-3x})e^{2x+1}[2+3\ln 2]}{(2^{1-3x})^2}=\frac{e^{2x+1}[2+3\ln 2]}{2^{1-3x}}$$

Day 40

1. B **2.** F **3.** A **4.** G **5.** C

6. D **7.** D **8.** E **9.** I **10.** H

Day 41

1. $3y^2\frac{dy}{dx}-2x=0,\ \frac{dy}{dx}=\frac{2x}{3y^2}$

2. $2-3\frac{dy}{dx}=2(x+y)\left(1+\frac{dy}{dx}\right),\ \frac{2-2x-2y}{3+2x+3y}=\frac{dy}{dx}$

3. $x^2\frac{dy}{dx}+2xy+x\cdot 2y\frac{dy}{dx}+y^2=1,\ \frac{dy}{dx}=\frac{1-2xy-y^2}{x^2+2xy}$

4. $e^{x+y}\left(1+\frac{dy}{dx}\right)=e^x+e^y\frac{dy}{dx},\ \frac{dy}{dx}=\frac{e^x-e^{x+y}}{e^{x+y}-e^y}$

Day 42

1. $x\cos y\frac{dy}{dx}+\sin y-\left(-y\sin x+\cos x\frac{dy}{dx}\right)=2x,\ \frac{dy}{dx}=\frac{2x-\sin y-y\sin x}{x\cos y-\cos x}$

2. $\cos(x+y)\left(1+\frac{dy}{dx}\right)-\sin(x-y)\left(1-\frac{dy}{dx}\right)=0,\ \frac{dy}{dx}=\frac{\sin(x-y)-\cos(x+y)}{\cos(x+y)+\sin(x-y)}$

3. $e^{x+y}\left(1+\frac{dy}{dx}\right)=2x-3\frac{dy}{dx},\ \frac{dy}{dx}=\frac{2x-e^{x+y}}{e^{x+y}+3}$

4. $\frac{y}{x}\cdot\frac{y-x\frac{dy}{dx}}{y^2}=1+\frac{dy}{dx},\ \frac{y-xy}{x+xy}=\frac{dy}{dx}$

Day 43

1. $g'(1) = \dfrac{1}{f'\left(f^{-1}(1)\right)} = \dfrac{1}{f'(-1)} = \dfrac{1}{-2} = -\dfrac{1}{2}$

2. $g'(-1) = \dfrac{1}{f'\left(f^{-1}(-1)\right)} = \dfrac{1}{f'(0)} = \dfrac{1}{-1} = -1$

3. $g'(-3) = \dfrac{1}{f'\left(f^{-1}(-3)\right)} = \dfrac{1}{f'(2)} = \dfrac{1}{-3} = -\dfrac{1}{3}$

Day 44

1. $f'(x) = 5x^4 - 4x,\ f''(x) = 20x^3 - 4,\ f'''(x) = 60x^2$

2. $f'(x) = \dfrac{7}{(x+2)^2},\ f''(x) = \dfrac{-14(x+2)}{(x+2)^4} = \dfrac{-14}{(x+2)^3}$

3. $f'(x) = 2xe^{x^2-4},\ f''(x) = \left(4x^2+2\right)e^{x^2-4},\ f'''(x) = \left(8x^3+12x\right)e^{x^2-4}$

4. $f'(x) = 6\sin x\cos x,\ f''(x) = 6\cos^2 x - 6\sin^2 x,\ f'''(x) = -24\sin x\cos x$

5. $f'(x) = -\sin x,\ f''(x) = -\cos x,\ f'''(x) = \sin x,\ f^{(4)}(x) = \cos x$

Day 45

1. $f(x) = x^3 - 3x^2 + 1$ is a polynomial function, and so continuous everywhere and differentiable on $(-1, 2)$. $f(-1) = -1-3+1 = -3$ and $f(2) = 8-12+1 = -3$. Rolle's Theorem applies. $f'(x) = 3x^2 - 6x = 3x(x-2)$ is equal to zero at $x = 0$ and $x = 2$. $-1 < 0 < 2$, so $c = 0$. The critical point is (0, 1).

2. $f(x) = 2 - 5\cos\left(\dfrac{x}{4}\right)$ is continuous everywhere and differentiable on $(-\pi, \pi)$. $f(-\pi) = 2 - 5\cos\left(\dfrac{-\pi}{4}\right) = \dfrac{4-5\sqrt{2}}{2}$, $f(\pi) = 2 - 5\cos\left(\dfrac{\pi}{4}\right) = \dfrac{4-5\sqrt{2}}{2}$. Rolle's Theorem applies. The derivative $f'(x) = \dfrac{5}{4}\sin\left(\dfrac{x}{4}\right) = 0$, when $\sin\left(\dfrac{x}{4}\right) = 0$, or when $x = 0$, and $-\pi < 0 < \pi$. The critical point is (0, −3).

Day 46

1. $\dfrac{f(2) - f(-2)}{2+2} = \dfrac{0+12}{4} = 3,\ f'(c) = 3c^2 - 4c - 1 = 3\ \ c = -\dfrac{2}{3}$ or 2. But only $c = -\dfrac{2}{3}$ satisfies $-2 < -\dfrac{2}{3} < 2$.

2. $\dfrac{f(b) - f(a)}{b-a} = \dfrac{2-2}{2\pi - 0} = 0,\ f'(c) = 2\sin c\cos c - 2\sin c,\ c = \pi$

3. $\dfrac{f(b) - f(a)}{b-a} = \dfrac{\left(1+e^{-1}\right) - 2}{1-0} = e^{-1} - 1,\ f'(x) = -e^{-x} = e^{-1} - 1,\ c = 1 - \ln(e-1)$

Answers

Day 47

1. $f(x)=\dfrac{3-x^2}{x^2+3}$ on $[-1,2]$

$f'(x)=\dfrac{-12x}{(x^2+3)^2}$; set $f'(x)=0$ and obtain $x=0$.

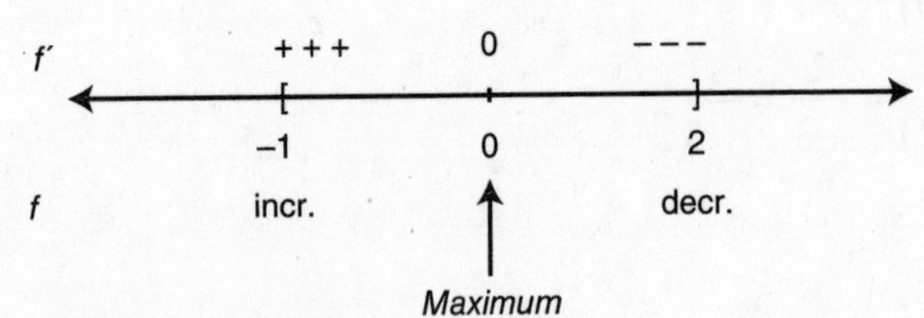

x	$f(x)$
−1	1/2
0	1
2	−1/7

Maximum point: (0, 1), and Minimum point: $\left(2,-\dfrac{1}{7}\right)$

2. Maximum point: (5, 11), Minimum point: (−2, −38)

3. Maximum point: $\left(-\dfrac{3}{2},\dfrac{9}{16}\right)$, Minimum point: (−1, −1)

Day 48

1. $f'(x)=6-3x^2$, Critical values: $x=\pm\sqrt{2}$, Increasing: $(-\sqrt{2},\sqrt{2})$, Decreasing: $(-\infty,-\sqrt{2})$ and $(\sqrt{2},\infty)$

2. $f'(x)=3x^2-12x+12$, Critical values: $x=2$, Increasing: $(-\infty,2)\cup(2,\infty)$

3. $f'(x)=-3x^2+6$, Critical values: $x=\pm\sqrt{2}$, Increasing: $(-\sqrt{2},\sqrt{2})$, Decreasing: $(-\infty,-\sqrt{2})$ and $(\sqrt{2},\infty)$

4. $f'(x)=3x^2-8x+4$, Critical values: $x=\dfrac{2}{3}, x=2$, Increasing: $\left(-\infty,\dfrac{2}{3}\right)$ and $(2,\infty)$, Decreasing: $\left(\dfrac{2}{3},2\right)$

Day 49

1. $f'(x)=4x^3-16x$, $f'(-3)=-60$, $f'(-1)=12$, minimum

2. $f'(x)=-xe^{-x^2/2}$, $f'(-1)=e^{-1/2}$, $f'(1)=-e^{-1/2}$, maximum

3. $f'(x)=x+2x\ln x$, $f'(0.1)=0.1+0.2\ln 0.1\approx-0.36$, $f'(1)=1$, minimum

4. $f'(x)=-2\sin x\cos x-2\cos x$, $f'\left(\dfrac{\pi}{4}\right)=-1-\sqrt{2}$, $f'\left(\dfrac{3\pi}{4}\right)=1+\sqrt{2}$, minimum

5. $f'(x)=-2\sin x\cos x-2\cos x$, $f'\left(\dfrac{5\pi}{4}\right)=-1+\sqrt{2}$, $f'\left(\dfrac{7\pi}{4}\right)=1-\sqrt{2}$, maximum

Day 50

1. $f''(x)=6(x-1)$, $f''(1)=0$, and $f''(x)<0$ for $x<1$, concave downward on $(-\infty, 1)$, and $f''(x)>0$, for $x>1$, concave upward on $(1, \infty)$, point of inflection at $x=1$.

2. $f''(x)=\dfrac{8-2x^2}{(4+x^2)^2}$, $f''(-2)=0$ and $f''(2)=0$, $f''(x)>0$ for $-2<x<2$, concave upward on $(-2, 2)$, and $f''(x)<0$ for $x<-2$ or $x>2$, concave downward on $(-\infty, -2)$ and $(2, \infty)$, point of inflection at $x=-2$ and $x=2$.

Day 51

1. $f'(x)=2\sin x\cos x$, $0<x<2\pi$ critical values: $x=\dfrac{\pi}{2}$, $x=\pi$, $x=\dfrac{3\pi}{2}$,

   $f''(x)=2\cos^2 x-2\sin^2 x$, $f''\left(\dfrac{\pi}{2}\right)=-2$, maximum, $f''(\pi)=2$, minimum,

   $f''\left(\dfrac{3\pi}{2}\right)=-2$, maximum

2. $f'(x)=2x^2-x-3$, critical values: $x=-1$, $x=\dfrac{3}{2}$, $f''(x)=4x-1$, $f''(-1)=-5$, maximum, $f''\left(\dfrac{3}{2}\right)=5$, minimum

3. $f'(x)=\dfrac{2-2x}{x^3}$, critical value: $x=1$, $f''(x)=\dfrac{4x-6}{x^4}$, $f''(1)=-2$, maximum

Day 52

1. negative
2. positive
3. zero, negative
4. $-\infty$, ∞, does not exist
5. 1.5, $-\infty$

Day 53

The function is decreasing for all values $x < 1.5$, so $f'(x) < 0$ for $x < 1.5$. The graph has a horizontal tangent at $x = 1.5$, so $f'(x) = 0$ at $x = 1.5$. The function is increasing for $x > 1.5$, so $f'(x) > 0$. Since the graph of $f(x)$ is concave up, $f''(x) > 0$ and $f'(x)$ is increasing. Therefore, the graph of $f'(x)$ could look like the graph in the accompanying diagram.

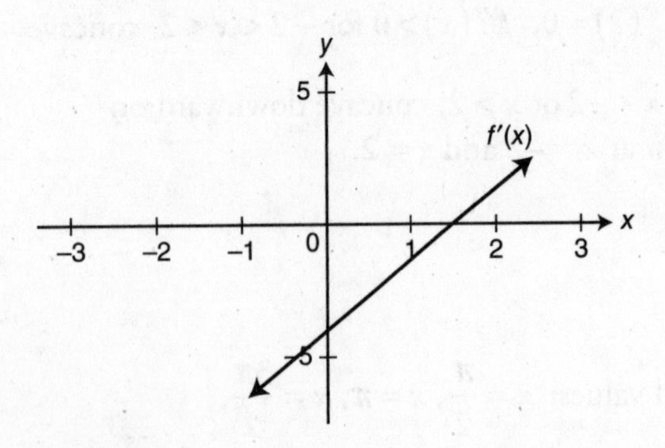

Day 54

The derivative is negative $-5 < x < -3$, so the function is decreasing, and turns at $x = -3$, where the derivative is zero. From -3 to -1, the derivative is positive and the function increases. Undefined at -1, the function continues to increase until just below 0, reaches a max and begins to decrease. There is a point of inflection just past 1, a minimum just below 3, after which the function increases.

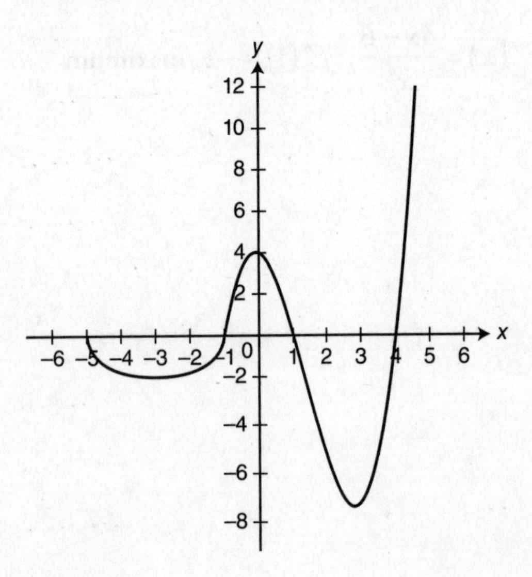

Day 55

1. C, I, D

2. F, H, B

3. G, A, E

Day 56

Assuming a, b, and c are all functions of time, and differentiating $a^2+b^2=c^2$ with respect to time produces $2a\frac{da}{dt}+2b\frac{db}{dt}=2c\frac{dc}{dt}$, or $a\frac{da}{dt}+b\frac{db}{dt}=c\frac{dc}{dt}$

1. $a=30,\ \frac{da}{dt}=0,\ b=80,\ \frac{db}{dt}=40,\ c=\sqrt{30^2+80^2}\approx 85.4,\ 30(0)+80(40)=85.4\frac{dc}{dt},$

$\frac{dc}{dt}\approx 37.5$ mph

2. $a=50,\ \frac{da}{dt}=20,\ t=2.5,\ \frac{db}{dt}=-15,\ b=46-15(2.5)=8.5,\ c=\sqrt{(50)^2+(8.5)^2}\approx 50.7,$

$50(20)+8.5(-15)=50.7\frac{dc}{dt},\ \frac{dc}{dt}\approx 17.2$ mph

Day 57

$\frac{r}{h}=\frac{6}{16},\ r=\frac{3h}{8},\ V=\frac{1}{3}\pi r^2 h=\frac{1}{3}\pi\left(\frac{3h}{8}\right)^2 h=\frac{3\pi h^3}{64},\ \frac{dV}{dt}=\frac{9\pi h^2}{64}\frac{dh}{dt},\ -2=\frac{9\pi(4)^2}{64}\frac{dh}{dt},$

$\frac{dh}{dt}=\frac{-8}{9\pi}\approx -0.283$ ft/min

Day 58

$\tan\theta=\frac{h}{2{,}000},\ \sec^2\theta\frac{d\theta}{dt}=\frac{1}{2{,}000}\frac{dh}{dt},\ \left(\frac{\sqrt{13}}{2}\right)^2\frac{d\theta}{dt}=\frac{1{,}200}{2{,}000},\ \frac{d\theta}{dt}=\frac{1{,}200(4)}{2{,}000(13)}\approx 0.185$ radians per second or approximately 10.578 degrees per second.

Day 59

1. $f'(x)=x^{-1/2}(x-3)\left(\frac{5x-3}{2}\right)$ and the critical points are $x=\frac{3}{5}$ or $x=3$.

$f'\left(\frac{1}{2}\right)=\frac{5\sqrt{2}}{8}>0,\ f'(1)=-2<0$, at $x=\frac{3}{5}$ is a maximum, $f'(4)=\frac{17}{4}$, at $x=3$ is a minimum.

2. $f'(x)=2\cos 2x+2\sin x,\ x=\frac{\pi}{2},\frac{7\pi}{6},\frac{11\pi}{6},\ f''(x)=-4\sin 2x+2\cos x,$

$f''\left(\frac{7\pi}{6}\right)=-3\sqrt{3}<0,\ f''\left(\frac{11\pi}{6}\right)=3\sqrt{3}>0,\ f''\left(\frac{\pi}{2}\right)=0.$ Maximum at $x=\frac{7\pi}{6}$, minimum at $x=\frac{11\pi}{6}$. (at $x=\frac{\pi}{2}$ is an Inflection Point.)

3. $f'(x)=-2+\frac{3}{2}(x+1)^{1/2}=0$ at $x=\frac{7}{9}$. $f''(x)=\frac{3}{4}(x+1)^{-1/2}$ and $f''\left(\frac{7}{9}\right)=\frac{9}{16}>0$.

at $x=\frac{7}{9}$ is a minimum.

Day 60

1. $A = x(50 - x)$, $A' = 50 - 2x = 0$, $x = 25$, $50 - x = 25$. $A'' = -2$.
   Maximum area with a square 25 ft on a side.

2. $A = x(100 - 2x)$, $A' = 100 - 4x = 0$, $x = 25$, $100 - 2x = 50$. $A'' = -4$.
   Maximum area with a 25-by-50-foot rectangle.

3. $d = \sqrt{(x-3)^2 + (2x-1+1)^2}$, $d' = \frac{1}{2}\left[5x^2 - 6x + 9\right]^{-\frac{1}{2}}(10x - 6) = 0$, $x = \frac{3}{5}$,
   $d \approx 2.683$. $d''\left(\frac{3}{5}\right) > 0$. The shortest distance, which is the perpendicular distance, is approximately 2.683 units.

Day 61

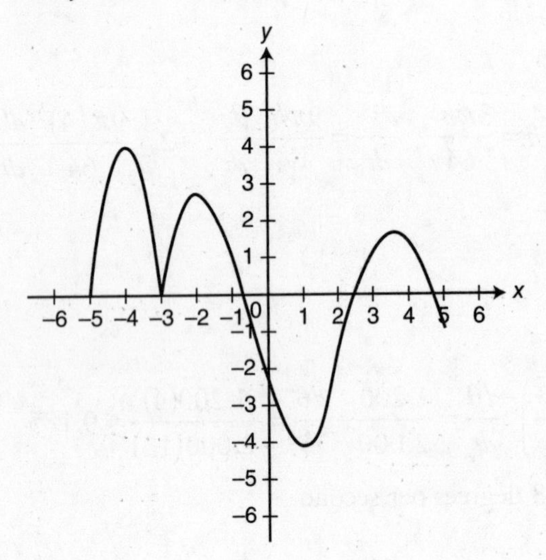

Day 62

1. $\lim\limits_{x\to 0} \frac{\sin x}{2x} = \lim\limits_{x\to 0} \frac{\cos x}{2} = \frac{1}{2}$

2. $\lim\limits_{x\to 3} \frac{\ln(x-2)}{x^2 - 9} = \lim\limits_{x\to 3} \frac{\frac{1}{(x-2)}}{2x} = \lim\limits_{x\to 3} \frac{1}{2x(x-2)} = \frac{1}{6}$

3. $\lim\limits_{x\to \infty} \frac{e^{2x}}{2x^2} = \lim\limits_{x\to \infty} \frac{2e^{2x}}{4x} = \lim\limits_{x\to \infty} \frac{4e^{2x}}{4} = \lim\limits_{x\to \infty} e^{2x} = \infty$

Day 63

1. D $f'(x)=3x^2-4x,\ f'(-1)=7,\ y+2=7(x+1),\ y=7x+5$
2. A $f'(x)=-\frac{1}{x^2},\ f'(-1)=-1,\ y+2=-1(x+1),\ y=-x-3$
3. B $f'(x)=(2x+2)e^{-x^2-2x-1},\ f'(-1)=0,\ y=-2$
4. E $2x+2y\frac{dy}{dx}=0,\ \frac{dy}{dx}=-\frac{x}{y}\Big|_{(-1,-2)}=-\frac{1}{2},\ y+2=-\frac{1}{2}(x+1),\ y=-\frac{1}{2}x-\frac{5}{2}$
5. B $f'(x)=\pi\cos\left(\frac{\pi x}{2}\right),\ f'(-1)=\pi\cos\left(-\frac{\pi}{2}\right)=0,\ y=-2$

Day 64

1. $f'(x)=8x^3-15x^2,\ f'(1)=-7,\ y+1=\frac{1}{7}(x-1),\ y=\frac{1}{7}x-\frac{8}{7}$
2. $f'(x)=2\cos(2x)-\sin x,\ f'(-\pi)=2,\ y+1=-\frac{1}{2}(x+\pi),\ y=-\frac{x}{2}-\frac{\pi+2}{2}$

Day 65

1. $f'(x)=2-3x^2$, $f'(1)=-1$, $y=7-x$; $f(1.01)\approx 5.99$. $f''(x)=-6x$, $f''(1)<0$, overestimate.
2. $f'(x)=\frac{-2}{2x+1},\ f'(1)=-\frac{2}{3},\ y=-\frac{2}{3}x+\frac{11}{3}-\ln 3,\ f(0.9)\approx 1.968.$

   $f''(x)=\frac{4}{(2x+1)^2}$ and $f''(1)=\frac{4}{9}>0$, underestimate.

Day 66

$f'(x)=-3e^{-x}$, $f'(0)=-3$, $y=5-3x$, $f(-0.01)\approx 5.03$. $f''(x)=3e^{-x}$ and $f''(0)=3>0$, underestimate.

Day 67

$f(x)=\sqrt{x}$, $f'(x)=\frac{1}{2\sqrt{x}}$, $f'(16)=\frac{1}{8}$, $y-4=\frac{1}{8}(x-16)$. At $x=17$, $y=4.125$

$\sqrt{17}\approx 4.125$, difference: 0.002

At $x=20$, $y=4.5$; $\sqrt{20}\approx 4.5$; difference: 0.028

$f(x)=\sqrt[3]{x}$, $f'(x)=\frac{x^{-2/3}}{3}$, $f'(-27)=\frac{1}{27}$, $y+3=\frac{1}{27}(x+27)$

$\sqrt[3]{-26}\approx -2.963$, difference: 0.001

$\sqrt[3]{-25}\approx -2.926$, difference: 0.002

$f(x)=\sqrt[5]{x}$, $f'(x)=\frac{1}{5x^{4/5}}$ and $f'(32)=\frac{1}{80}$, $y-2=\frac{1}{80}(x-32)$

$\sqrt[5]{33}\approx 2.0125$, difference: -0.0005

Day 68

Desired	cos 58°	sin 50°	tan 170°
Function	$y=\cos x$	$y=\sin x$	$y=\tan x$
Point of tangency	$\left(\frac{\pi}{3},\frac{1}{2}\right)$	$\left(\frac{\pi}{4},\frac{\sqrt{2}}{2}\right)$	$(\pi,0)$
Derivative	$y'=-\sin x$	$y'=\cos x$	$y'=\sec^2 x$
Derivative Evaluated	$-\frac{\sqrt{3}}{2}$	$\frac{\sqrt{2}}{2}$	1
Distance from point of tangency	$-2^\circ=-\frac{\pi}{90}$	$5^\circ=\frac{\pi}{36}$	$-10^\circ=-\frac{\pi}{18}$
(Derivative)(distance) + y-value of point of tangency	$\left(-\frac{\sqrt{3}}{2}\cdot-\frac{\pi}{90}\right)+\frac{1}{2}$	$\left(\frac{\sqrt{2}}{2}\cdot\frac{\pi}{36}\right)+\frac{\sqrt{2}}{2}$	$1\left(-\frac{\pi}{18}\right)+0$
Approximate Value	cos 58° ≈ 0.530	sin 50° ≈ 0.769	tan 170° ≈ −0.175
Actual (to nearest thousandth)	0.530	0.766	−0.176

Day 69

Position	$-5t^2+30t+5$	$\sin(2t)$	$2\sin t+\cos t$	$-2t^3+15t^2+5t-3$
Velocity	$-10t+30$	$2\cos(2t)$	$2\cos t-\sin t$	$-6t^2+30t+5$
Acceleration	-10	$-4\sin(2t)$	$-2\sin t-\cos t$	$-12t+30$

Day 70

1. $v \approx \dfrac{139-87}{2-1} = 52 \,{}^{\text{ft}}\!/_{\text{sec}}$

2. $v \approx \dfrac{69-129}{5.5-4.5} = -60 \,{}^{\text{ft}}\!/_{\text{sec}}$

3. $s(t) = -16t^2 + 100t + 3$, $v(t) = -32t + 100$, $v(1.5) = -48 + 100 = 52$, $v(5) = -160 + 100 = -60$

4. $v(t) = -32t + 100 = 0$, $t = 3.125$. It reaches a maximum height of 159.25 feet after 3.125 seconds.

5. $v_{average} = \dfrac{69-139}{5.5-2} = \dfrac{-70}{3.5} = -20$

6. $v(t) = -32t + 100 = -20$, $t = 3.75$ seconds

7. False. The acceleration is constant at −32 feet per second per second.

8. The object slows when velocity and acceleration have opposite signs. Acceleration is constant (−32 feet per second per second), so the object is slowing when velocity is positive: $0 < t < 3.125$. It is slowing down as it rises and speeding up as it falls.

Day 71

1.

time (seconds)	0	1	2	3	4
position	(0, 0)	(1, 0)	(2, 0)	(1, 0)	(0, 0)

2. $v(t) = s'(t) = \dfrac{\pi}{2}\sin\left(\dfrac{\pi t}{2}\right) = 0$, $t = 2$

time	0	0.5	1	1.5	2	2.5	3	3.5	4
velocity	0	+	+	+	0	–	–	–	0
acceleration	+	+	0	–	–	–	0	+	+

3. $a(t) = \left(\dfrac{\pi}{2}\right)^2 \cos\left(\dfrac{\pi t}{2}\right)$ Speeding up: $0 < t < 1$ and $2 < t < 3$. Slowing down: $1 < t < 2$ and $3 < t < 4$.

4. Particle travels from (0, 0) to (2, 0), then reverses direction and returns to (0, 0). It travels a total of 4 units.

Day 72

$V = 9h, \quad \frac{dV}{dt} = 9\frac{dh}{dt}, \quad 2 = 9\frac{dh}{dt}, \quad \frac{dh}{dt} = \frac{2}{9} \approx 0.222$ ft/min

Day 73

1. $f'(x) = 3x^2 - 4x^{-3} = 0$ at $x = \left(\frac{4}{3}\right)^{1/5} \approx 1.059$. $f''(x) = 6x + 12x^{-4}$ and

   $f''\left(\frac{4}{3}\right)^{1/5} = 6\left(\frac{4}{3}\right)^{1/5} + 12\left(\frac{4}{3}\right)^{-4/5} > 0$. Relative minimum at $x = \left(\frac{4}{3}\right)^{1/5} \approx 1.059$

   and $f(1.059) \approx 2.971$, relative minimum value of f.

2. $f'(x) = 4x^3 - 9x^2 + 6x = 0$ at $x = 0$. $f''(x) = 12x^2 - 18x + 6$ and $f''(0) = 6 > 0$. Minimum at $x = 0$ and $f(0) = 1$, minimum value of f.

Day 74

1. $f(x) = 5x + C$
2. $f(x) = x^3 + 4x^2 + 3x + C$
3. $f(x) = 5\ln|x| + C$
4. $f(x) = -\cos x + C$
5. $f(x) = \tan x + C$

Day 75

1. $\sum_{i=1}^{5} i = 1 + 2 + 3 + 4 + 5 = 15$
2. $\sum_{i=1}^{10} \left(\frac{i}{5}\right) = \frac{1}{5}\sum_{i=1}^{10} i = \frac{1}{5}(1+2+3+4+5+6+7+8+9+10) = \frac{55}{5} = 11$
3. $\sum_{i=2}^{4} (i^2)(i+3) = (4)(5) + (9)(6) + (16)(7) = 20 + 54 + 112 = 186$
4. $\sum_{i=-3}^{1} (i^3 + 2i^2 - 5i + 4) = \sum_{i=-3}^{1} (i^3) + 2\sum_{i=-3}^{1} (i^2) - 5\sum_{i=-3}^{1} (i) + \sum_{i=-3}^{1} (4)$

   $= -35 + 2(15) - 5(-5) + 20 = 40$

Day 76

1. A = 16, $F(x) = 4x + C,\ F(4) - F(0) = 16 - 0 = 16$

2. A = 4, $F(x) = 4x - x^2 + C,\ F(2) - F(0) = 4 - 0 = 4$

3. A = 1.5, $F(x) = \frac{3}{2}x^2 + C,\ F(1) - F(0) = \frac{3}{2} - 0 = 1.5$

4. A = 2.5, $F(x) = \frac{3}{2}x^2 + x + C,\ F(1) - F(0) = \frac{3}{2} + 1 - 0 = 2.5$

5. A = 5, $F(x) = \begin{cases} -\frac{1}{2}x^2 + x + C_1, & x < 1 \\ \frac{1}{2}x^2 - x + C_2, & x \geq 1 \end{cases}$

$$F(4) - F(0) = (8 - 4) - \left(\frac{1}{2} - 1\right) + \left(-\frac{1}{2} + 1\right) - 0 = 5$$

Alternatively, using your TI-89 calculator, you obtain

$F(x) = \int |x - 1|\, dx = \frac{(x-1)\,|x-1|}{2} + c$. And

$F(4) - F(0) = \frac{9}{2} - \left(-\frac{1}{2}\right) = 5.$

Day 77

1. $\int_8^8 g(x)\ dx = 0$

2. $\int_9^5 f(x)\, dx = -\int_5^9 f(x)\, dx = -12$

3. $\int_2^5 7g(x)\, dx = 7\int_2^5 g(x)\, dx = 7(-1) = -7$

4. $\int_2^9 f(x)\ dx = \int_2^5 f(x)\ dx + \int_5^9 f(x)\ dx = -3 + 12 = 9$

5. $\int_2^5 [f(x) + g(x)]\ dx = \int_2^5 f(x)\ dx + \int_2^5 g(x)\ dx = -3 - 1 = -4$

Day 78

1. $\int_{-3}^{1}\left(6x^2-5x+2\right)dx = 2x^3-\frac{5}{2}x^2+2x\Big|_{-3}^{1} = \left(2-\frac{5}{2}+2\right)-\left(-54-\frac{45}{2}-6\right)=84$

2. $\int_{1}^{2}\frac{x^5+x^2+x}{x}\,dx = \int_{1}^{2}\left(x^4+x+1\right)dx = \frac{x^5}{5}+\frac{x^2}{2}+x\Big|_{1}^{2}$
$= \left(\frac{32}{5}+\frac{4}{2}+2\right)-\left(\frac{1}{5}+\frac{1}{2}+1\right)=8.7$

3. $\int_{1}^{3}\frac{dx}{x^3}=\int_{1}^{3}x^{-3}dx=\frac{x^{-2}}{-2}\Big|_{1}^{3}=\frac{1}{-2(9)}-\frac{1}{-2(1)}=\frac{-8}{-18}=\frac{4}{9}$

4. $\int_{1}^{4}\sqrt{x}\,dx=\int_{1}^{4}x^{1/2}\,dx=\frac{2}{3}x^{3/2}\Big|_{1}^{4}=\frac{2}{3}(8)-\frac{2}{3}(1)=\frac{14}{3}$

5. $\int_{1}^{32}\sqrt[5]{x^2}\,dx=\int_{1}^{32}x^{2/5}\,dx=\frac{5}{7}x^{7/5}\Big|_{1}^{32}=\frac{5}{7}(128)-\frac{5}{7}(1)=\frac{635}{7}$

Day 79

1. $\int_{0}^{\pi/3}(2\sin\theta-5\cos\theta)\,d\theta=(-2\cos\theta-5\sin\theta)\big|_{0}^{\pi/3}=1-5\left(\frac{\sqrt{3}}{2}\right)=\frac{2-5\sqrt{3}}{2}$

2. $\int_{\pi/4}^{\pi/3}5\sec^2 x\,dx=(5\tan x)\big|_{\pi/4}^{\pi/3}=5\sqrt{3}-5=5\left(\sqrt{3}-1\right)$

3. $\int_{1}^{e}\left(\frac{3}{z}\right)dz=(3\ln z)\big|_{1}^{e}=3-0=3$

4. $\int_{-\pi}^{\pi/2}(2\sin\theta\cos\theta)\,d\theta=\sin^2\theta\big|_{-\pi}^{\pi/2}=\sin^2\left(\frac{\pi}{2}\right)-\sin^2(-\pi)=1-0=1$

5. $\int_{0}^{\ln(1+\pi)}\left(-e^x\cos\left(1-e^x\right)\right)dx=\sin\left(1-e^x\right)\big|_{0}^{\ln(1+\pi)}=\sin\left(1-e^{\ln(1+\pi)}\right)-\sin\left(1-e^0\right)$
$=\sin(1-1-\pi)-\sin(0)=\sin(-\pi)-\sin(0)=0$

Day 80

1. B
2. H
3. I
4. F
5. D

Day 81

1. $\int\left[\left(2x+x^2\right)^5(2+2x)\right]dx=\dfrac{\left(2x+x^2\right)^6}{6}+C$

2. $\int\left[\sqrt{x^3-1}\left(3x^2\right)\right]dx=\dfrac{2\left(x^3-1\right)^{3/2}}{3}+C$

3. $\int\left[\dfrac{(2x)}{x^2+4}\right]dx=\ln\left(x^2+4\right)+C$

Day 82

1. Let $u=\sin\,\theta,\ \dfrac{du}{d\theta}=\cos\theta$

$$\int_{-\pi}^{\pi/2}\left(\cos\theta\cos\left(\sin\theta\right)\right)d\theta=\int_{\sin(-\pi)}^{\sin(\pi/2)}\cos(u)\frac{du}{d\theta}d\theta=\sin(u)\Big|_0^1=\sin(1)-\sin(0)\approx 0.841$$

2. Let $u=1-e^x,\ \dfrac{du}{dx}=-e^x$

$$\int_0^{\ln(1+\pi)}\left(e^x\cos\left(1-e^x\right)\right)dx=-\int_{1-e^0}^{1-e^{\ln(1+\pi)}}\cos(u)\frac{du}{dx}dx=-\sin(u)\Big|_0^{-\pi}=0$$

3. Let $u=1-4t^2,\ \dfrac{du}{dt}=-8t$

$$\int_0^{1/4}\left(\frac{2t}{1-4t^2}\right)dt=\int_{1-4\cdot 0^2}^{1-4(1/4)^2}\frac{1}{u}\left(-\frac{1}{4}\cdot\frac{du}{dt}\right)dt=-\frac{1}{4}\ln u\Big|_1^{3/4}=-\frac{1}{4}\ln\left(\frac{3}{4}\right)+\frac{1}{4}\ln 1=\frac{\ln 4-\ln 3}{4}\approx 0.072$$

Answers

Day 83

1. d $\int_{-2}^{k} x^2\,dx = \left.\frac{x^3}{3}\right|_{-2}^{k} = \frac{k^3}{3} + \frac{8}{3} = 24,\ k = 4$

2. c Fundamental Theorem of Calculus

3. a Limits of integration are equal.

4. a Reversing the limits of integration changes the sign.

5. c $\int_{1}^{5} 2\,dx = 2x\Big|_1^5 = 10 - 2 = 8$

Day 84

If $f''(x) = 6x - 4$, then $f'(x) = 3x^2 - 4x + C$. The value of the first derivative at $x = 0$ is known. $f'(0) = 3(0)^2 - 4(0) + C = 1$ indicates $C = 1$, so $f'(x) = 3x^2 - 4x + 1$. Find the antiderivative of $f'(x) = 3x^2 - 4x + 1$ to get $f(x) = x^3 - 2x^2 + x + C$. Use $f(1) = -3$ to find the constant. $f(1) = (1)^3 - 2(1)^2 + 1 + C = -3$, so $C = -3$. Conclude $f(x) = x^3 - 2x^2 + x - 3$.

Day 85

$$\int (x^2+3)(2x)\,dx = \frac{1}{2}(x^2+3)^2 + C$$

$$\int (x^2+3x-2)(2x+3)\,dx = \frac{1}{2}(x^2+3x-2)^2 + C$$

$$\int (2x+3)(2)\,dx = \frac{1}{2}(2x+3)^2 + C$$

$$\int (x^3-3x-2)^2(3x^2-3)\,dx = \frac{1}{3}(x^3-3x-2)^3 + C$$

$$\int (3x^2-3)(6x)\,dx = (3x^2-3)^2 + C$$

Day 86

1. $u = 2x^2 - 3,\ du = 4x,\ \int \frac{x}{2x^2-3}\,dx = \frac{1}{4}\int \frac{4x}{2x^2-3}\,dx = \frac{1}{4}\int \frac{du}{u} = \frac{1}{4}\ln\left|2x^2-3\right| + C$

2. $u = x^3 - 5,\ du = 3x^2,\ \int \frac{x^2}{x^3-5}\,dx = \frac{1}{3}\int \frac{3x^2}{x^3-5}\,dx = \frac{1}{3}\int \frac{du}{u} = \frac{1}{3}\ln\left|x^3-5\right| + C$

3. $u = 4 - x^2,\ du = -2x,\ \int \frac{x}{4-x^2}\,dx = -\frac{1}{2}\int \frac{-2x}{4-x^2}\,dx = -\frac{1}{2}\int \frac{du}{u} = -\frac{1}{2}\ln\left|4-x^2\right| + C$

4. $u = 3x,\ du = 3,\ \int e^{3x}\,dx = \frac{1}{3}\int 3e^{3x}\,dx = \frac{1}{3}\int e^{u}\,du = \frac{1}{3}e^{3x} + C$

5. $u = x^3,\ du = 3x^2,\ \int x^2 e^{x^3}\,dx = \frac{1}{3}\int 3x^2 e^{x^3}\,dx = \frac{1}{3}\int e^{u}\,du = \frac{1}{3}e^{x^3} + C$

Day 87

1. $\int \sin^2 x(\cos x)\,dx = \int u^2\,du = \frac{\sin^3 x}{3} + C$

2. $\int \cos\sqrt{2x}\left(\frac{1}{\sqrt{2x}}\right)dx = \int \cos u\,du = \sin\sqrt{2x} + C$

3. $\int \sec^2(2x)(2)\,dx = \int \sec^2(u)\,du = \tan 2x + C$

4. $\int \frac{(3\cos 3x)}{\sin 3x}\,dx = \int \frac{du}{u} = \ln\left|\sin(3x)\right| + C$

5. $\int \tan^4 3x\left(3\sec^2 3x\right)dx = \int u^4\,du = \frac{\tan^5 3x}{5} + C$

Day 88

1. $\int \frac{-2dx}{1+4x^2} = \cot^{-1}(2x) + C$

2. $\int \frac{-3dx}{\sqrt{1-9x^2}} = \cos^{-1}(3x) + C$

3. $\int \frac{1}{\sqrt{4-x^2}} = \int \frac{\frac{1}{2}}{\sqrt{1-\frac{x^2}{4}}} = \sin^{-1}\left(\frac{x}{2}\right) + C$

4. $\int \frac{\cos x}{1+\sin^2 x}\,dx = \tan^{-1}(\sin x) + C$

Answers

Day 89

$A = 2x\left(\sqrt{100 - x^2}\right)$, $A' = \dfrac{200 - 4x^2}{\sqrt{100 - x^2}} = 0$, $x = 5\sqrt{2} \approx 7.071$; $A''(5\sqrt{2}) < 0$. The rectangle of greatest area is 14.142 by 7.071 units.

Day 90

1. $\int_0^2 f(x)\,dx = \frac{1}{2}(1+3)(2) = 4$

2. $\int_0^{-2} f(x)\,dx = -2$

3. $\int_{-4}^{-2} f(x)\,dx = \frac{-3(1.5)}{2} + \frac{1(0.5)}{2} = \frac{-4}{2} = -2$

4. $\int_{-4}^{3} f(x)\,dx = \int_{-4}^{-2} f(x)\,dx + \int_{-2}^{0} f(x)\,dx + \int_{0}^{2} f(x)\,dx + \int_{2}^{3} f(x)\,dx$
$= -2 + 2 + 4 + 1.5 = 5.5$

Day 91

1. Using (0, 240), and (4, 400) and a trapezoidal approximation, the total number of gallons $\approx \frac{1}{2}(4)(240 + 400)$ = 1,280 gallons.

2. Using (8, 650), (10, 700), (12, 720), (14, 700), and (16, 650), the total number of gallons $\approx 2\left[\frac{1}{2}(2)(650 + 700) + \frac{1}{2}(2)(700 + 720)\right]$ = 2[2,770] = 5,540.

3. Using previous calculations and the symmetry of the graph, the total number of gallons over 24 hours $\approx$ 2[1,280 + $\frac{1}{2}(2)(400 + 550) + \frac{1}{2}(2)(550 + 650)$ + 2,770] = 2[1,280 + 950 + 1,200 + 2,770] = 2[6,240] = 12,400 gallons.

Day 92

1. True
2. False
3. False
4. False
5. True

Day 93

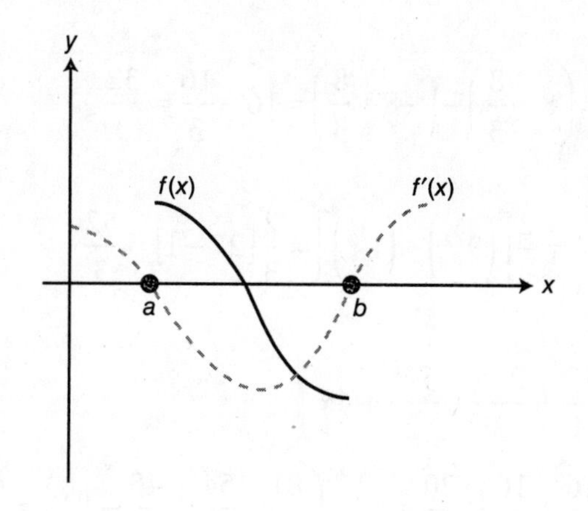

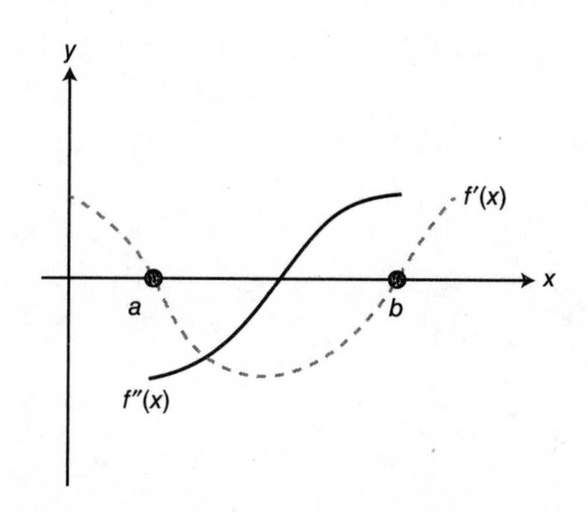

Day 94

$$\int_2^8 f(x)dx \approx \frac{1}{2}(3-2)(1+4)+\frac{1}{2}(5-3)(4+7)+\frac{1}{2}(6-5)(7+3)+\frac{1}{2}(8-6)(3+2)$$

$$\approx \frac{1}{2}(1)(5)+\frac{1}{2}(2)(11)+\frac{1}{2}(1)(10)+\frac{1}{2}(2)(5)$$

$$\approx \frac{1}{2}(5+22+10+10)$$

$$\approx 23.5$$

Day 95

1. $1[f(-3)+f(-2)+f(-1)+f(0)+f(1)]=10+14+10+4+2=40$
2. $1[f(2)+f(1)+f(0)+f(-1)+f(-2)]=(1)[10+2+4+10+14]=40$

Day 96

1. $\int_{-2}^{2}(4-x^2)dx = 4x - \frac{x^3}{3}\Big|_{x=-2}^{x=2} = \left(8-\frac{8}{3}\right)-\left(-8+\frac{8}{3}\right) = 16-\frac{16}{3} = \frac{32}{3}$

2. $\int_{-1}^{7}\sqrt{x+2}\ dx = \frac{2}{3}(x+2)^{3/2}\Big|_{x=-1}^{x=7} = \frac{2}{3}\left[\left(9^{3/2}\right)-\left(1^{3/2}\right)\right] = \frac{2}{3}[27-1] = \frac{52}{3}$

3. $\int_{-3}^{2}(x^3+2x^2-5x+4)dx = \left(\frac{x^4}{4}+\frac{2x^3}{3}-\frac{5x^2}{2}+4x\right)\Bigg|_{x=-3}^{x=2}$

$$= \left(\frac{16}{4}+\frac{16}{3}-\frac{20}{2}+8\right)-\left(\frac{81}{4}-\frac{54}{3}-\frac{45}{2}-12\right) = \frac{475}{12}$$

Day 97

1. C

2. B

3. F

4. A

5. E

Day 98

1.	$F(x)=\int_{5}^{x}\left(2t\sqrt{t}\right)dt$	$F'(x)=2x\sqrt{x}$	$F'(9)=18\cdot 3=54$
2.	$F(x)=\int_{0}^{4x}t(1+t^3)dt$	$F'(x)=16x(1+64x^3)$	$F'(1)=16(65)=1{,}040$
3.	$F(x)=\int_{4}^{2x}\left(\frac{t-3}{t}\right)dt$	$F'(x)=\frac{2x-3}{x}$	$F'(10)=\frac{20-3}{10}=1.7$
4.	$F(x)=\int_{-1}^{x}t(t^2-4)dt$	$F'(x)=x^3-4x$	$F'(2)=8-8=0$
5.	$F(x)=\int_{0}^{x}e^{t^2}\ dt$	$F'(x)=e^{x^2}$	$F'(1)=e^{1^2}=e$

Day 99

1. $\dfrac{d}{dx}\int_{0}^{x}\left(t^3-7t+1\right)dt = x^3-7x+1$

2. $\dfrac{d}{dx}\int_{-3}^{x}\sqrt{t+5}\ dt = \sqrt{x+5}$

3. $\dfrac{d}{dx}\int_{1}^{2x}\left(t^2-2t+5\right)dt = \dfrac{d}{dx}\int_{1}^{u}\left(t^2-2t+5\right)dt = \dfrac{d}{du}\left[\int_{1}^{u}\left(t^2-2t+5\right)dt\right]\dfrac{du}{dx}$
$= \left(u^2-2u+5\right)(2) = 2\left(4x^2-4x+5\right) = 8x^2-8x+10$

Day 100

1. $\sum_{k=-4}^{3} 1\cdot(f(k)) = 2+5.5+7+6.8+5.5+4+2.5+2 = 35.3$

2. $\sum_{k=-3}^{4} 1\cdot(f(k)) = 5.5+7+6.8+5.5+4+2.5+2+2.5 = 35.8$

3. $\sum_{k=-2}^{1} 2\cdot(f(2k+1)) = 2(5.5+6.8+4+2) = 2(18.3) = 36.6$

Day 101

1. $\int_{0}^{4} \sqrt{x^2+1}\ dx \approx 9.294$

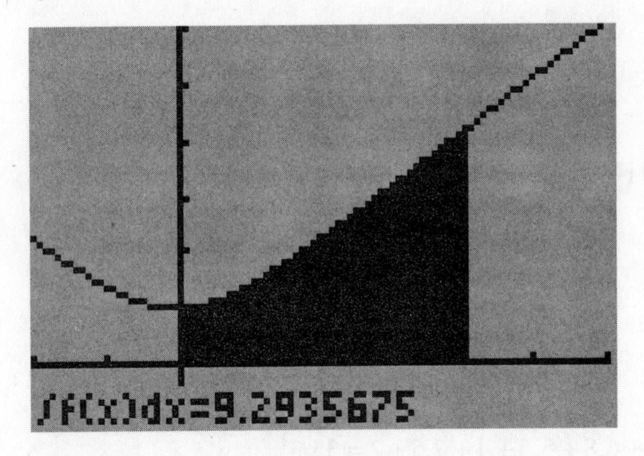

2. $\int_{-3}^{3} \sqrt{9-x^2}\ dx \approx 14.137$

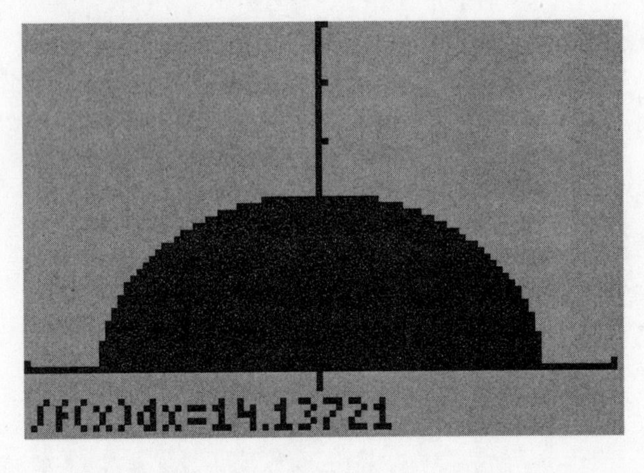

3. $\int_{-2}^{5} \ln\left(x^2+2x+3\right) dx \approx 14.453$

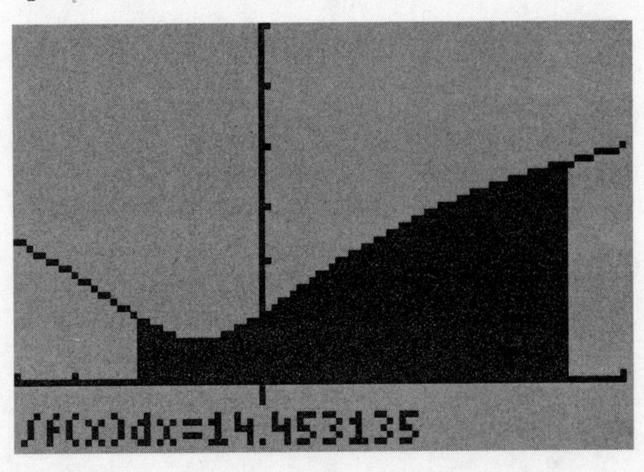

Day 102

1. $\sum_{k=1}^{6} 0.5 f\left(\frac{k-9}{2}\right) = 0.5\left(f(-4)+f(-3.5)+f(-3)+f(-2.5)+f(-2)+f(-1.5)\right)$

$= 0.5(0+4.5+6+5.5+4+2) = 0.5(22) = 11$

2. $\sum_{k=1}^{6} 0.5 f\left(\frac{k}{2}-1\right) = 0.5\left(f(-0.5)+f(0)+f(0.5)+f(1)+f(1.5)+f(2)\right)$

$= 0.5(-1.5-2.5-2.5-2-1+0) = 0.5(-9.5) = -4.75$; Area $= 4.75$

3. $\sum_{k=1}^{3} 1 f(k-4.5) = 1\left(f(-3.5)+f(-2.5)+f(-1.5)\right)$

$= 1(4.5+5.5+2) = 12$

Day 103

$$A = \sum_{k=-1}^{1} \frac{1}{2}\left(f(k)+f(k+1)\right)\Delta x$$

$$= \frac{1}{2}\left[f(-1)+2f(0)+2f(1)+f(2)\right]$$

$$= \frac{1}{2}\left[0+2(-2.5)+2(-2)+0\right] = \frac{1}{2}(-9) = -4.5\text{; Area} = 4.5$$

Day 104

1. $\int_{-3}^{0}\left(x^3+3x^2\right)dx = \left.\frac{x^4}{4}+x^3\right|_{-3}^{0} = 0-\left[\frac{81}{4}-27\right] = \frac{27}{4}$

2. $\int_{-3}^{-2}\left(2x^3+3x^2+4\right)dx = \left.\frac{x^4}{2}+x^3+4x\right|_{-3}^{-2} = -9.5$

$\int_{-2}^{-1}\left(2x^3+3x^2+4\right)dx = \left.\frac{x^4}{2}+x^3+4x\right|_{-2}^{-1} = \left(\frac{1}{2}-1-4\right)-\left(\frac{16}{2}-8-8\right) = -4.5+8 = 3.5$

Total Area $= |-9.5| + 3.5 = 13$.

Day 105

1. $\int_{0}^{\pi}(1+\cos x)\,dx = x+\sin x\Big|_0^{\pi} = (\pi+0)-(0) = \pi$

2. $\int_{\pi/6}^{\pi/3}(-\sin x)\,dx = \cos x\Big|_{\pi/6}^{\pi/3} = \frac{1}{2}-\frac{\sqrt{3}}{2} \approx -0.366$

3. $\int_{-\pi/2}^{\pi/2}(x+\sin x)\,dx = \left(\frac{x^2}{2}-\cos x\right)\Bigg|_{-\pi/2}^{\pi/2} = \left(\frac{\pi^2}{8}-0\right)-\left(\frac{\pi^2}{8}-0\right) = 0$

Day 106

1. $u = e^{-x^2},\ du = -2x\,dx,\ \int_{-2}^{2}\left(xe^{-x^2}\right)dx = -\frac{1}{2}\int_{-2}^{2}\left(e^{-x^2}\right)(-2x)\,dx = -\frac{1}{2}\left[e^{-x^2}\right]_{-2}^{2} = 0$

2. $u = \left(x^2+4x\right),\ du = (2x+4)\,dx,$

$$\frac{1}{2}\int\frac{2x+4}{x^2+4x}\,dx = \frac{1}{2}\ln\left|x^2+4x\right|\Big]_1^4 = \frac{1}{2}(\ln 32-\ln 5) = \frac{\ln\left(\frac{32}{5}\right)}{2} \approx .928$$

Day 107

1. $\int_{1}^{5}|x-3|\,dx = \int_{1}^{3}(3-x)\,dx+\int_{3}^{5}(x-3)\,dx = \left(3x-\frac{x^2}{2}\right)\Bigg|_1^3+\left(\frac{x^2}{2}-3x\right)\Bigg|_3^5$

$$= \left[\left(9-\frac{9}{2}\right)-\left(3-\frac{1}{2}\right)\right]+\left[\left(\frac{25}{2}-15\right)-\left(\frac{9}{2}-9\right)\right]$$

$$= [6-4]+[8-6] = 4$$

2. $\int_{-1}^{3}|2x+1|\,dx = \int_{-1}^{-1/2}(-2x-1)\,dx+\int_{-1/2}^{3}(2x+1)\,dx = \left(-x^2-x\right)\Big|_{-1}^{-1/2}+\left(x^2+x\right)\Big|_{-1/2}^{3}$

$$= \left[\left(-\frac{1}{4}+\frac{1}{2}\right)-(-1+1)\right]+\left[(9+3)-\left(\frac{1}{4}-\frac{1}{2}\right)\right] = \left[\frac{1}{4}\right]+\left[12+\frac{1}{4}\right] = 12\frac{1}{2}$$

3. $\int_{-2}^{2}\left|x^3\right|dx = \int_{-2}^{0}\left(-x^3\right)dx+\int_{0}^{2}x^3\,dx = \left(-\frac{x^4}{4}\right)\Bigg|_{-2}^{0}+\left(\frac{x^4}{4}\right)\Bigg|_0^2 = [0+4]+[4-0] = 8$

Day 108

1. $$\int_{-1}^{3}\left[\left(3+2x-x^2\right)-\left(x^2-2x-3\right)\right]dx=-2\int_{-1}^{3}\left(x^2-2x-3\right)dx=2\left[\frac{x^3}{3}-x^2-3x\right]_{-1}^{3}$$
$$=-2\left[(9-9-9)-\left(-\frac{1}{3}-1+3\right)\right]$$
$$=-2\left[-9-\frac{5}{3}\right]=-2\left[-\frac{32}{3}\right]=\frac{64}{3}$$

2. $$\int_{-1}^{3}\left[\left(9+9x-x^2-x^3\right)-\left(3+2x-x^2\right)\right]dx=\int_{-1}^{3}\left(6+7x-x^3\right)dx$$
$$=6x+\frac{7}{2}x^2-\frac{x^4}{4}\Bigg|_{-1}^{3}$$
$$=\left[\left(18+\frac{63}{2}-\frac{81}{4}\right)-\left(-6+\frac{7}{2}-\frac{1}{4}\right)\right]$$
$$=24+28-20=32$$

3. $$\int_{0}^{3}\left[\left(3+2x-x^2\right)-\sin\left(\frac{\pi x}{3}\right)\right]dx=3x+x^2-\frac{x^3}{3}+\frac{3}{\pi}\cos\left(\frac{\pi x}{3}\right)\Bigg|_{0}^{3}$$
$$=\left(9+9-9+\frac{3}{\pi}\cos(\pi)\right)-\frac{3}{\pi}\cos(0)$$
$$=9+\frac{3}{\pi}(-1)-\frac{3}{\pi}(1)=\frac{9\pi-6}{\pi}$$

Day 109

1. $$\int_{-1}^{3}\left(4x-x^2\right)-(2x-3)\,dx=\int_{-1}^{3}\left(3+2x-x^2\right)dx=3x+x^2-\frac{x^3}{3}\Bigg|_{-1}^{3}=9+\frac{5}{3}=\frac{32}{3}$$

2. $$\int_{\pi/4}^{5\pi/4}(2\sin x-2\cos x)\,dx=2[-\cos x-\sin x]_{\pi/4}^{5\pi/4}=2\left[\sqrt{2}+\sqrt{2}\right]=4\sqrt{2}$$

3. $$\int_{-\ln 4}^{0}\left(4-e^{-x}\right)dx+\int_{0}^{\ln 4}\left(4-e^{x}\right)dx=2\int_{0}^{\ln 4}\left(4-e^{x}\right)dx=2\left(4x-e^{x}\right)\Big|_{0}^{\ln 4}$$
$$=8\ln(4)-6\approx 5.0903549$$

Day 110

1. D $\pi\int_0^4 (4-x)^2\,dx = \pi\int_0^4 (16-8x+x^2)\,dx = \pi\left[16x - 4x^2 + \frac{x^3}{3}\right]_0^4 = \frac{64\pi}{3} \approx 67.021$

2. E $\pi\int_0^4 \left(\sqrt{4-y}\right)^2 dy = \pi\int_0^4 (4-y)\,dy = \pi\left[4y - \frac{y^2}{2}\right]_0^4 = \pi[(16-8)-(0)] = 8\pi \approx 25.133$

3. A $\pi\int_0^2 \left(4-x^2\right)^2 dx = \pi\int_0^2 \left(16-8x^2+x^4\right)dx = \pi\left[16x - \frac{8x^3}{3} + \frac{x^5}{5}\right]_0^2$

$$= \pi\left[\left(32 - \frac{64}{3} + \frac{32}{5}\right) - (0)\right] = \frac{256\pi}{15} \approx 53.617$$

4. G $\pi\int_0^4 (4-y)^2\,dy = \pi\int_0^4 \left(16-8y+y^2\right)dy = \pi\left[16y - 4y^2 + \frac{y^3}{3}\right]_0^4$

$$= \pi\left[\left(64 - 64 + \frac{64}{3}\right) - (0)\right] = \frac{64\pi}{3} \approx 67.021$$

Answers

Day 111

1. Outer: $R=\sqrt{y}$, Inner: $r=\frac{y}{2}$.

$$V=\pi\int_0^4\left[\sqrt{y}^2-\left(\frac{y}{2}\right)^2\right]dy=\pi\int_0^4\left[y-\frac{y^2}{4}\right]dy=\pi\left[\frac{y^2}{2}-\frac{y^3}{12}\right]_0^4=\pi\left(8-\frac{16}{3}\right)=\frac{8\pi}{3}$$

2. Outer: $R=2x$, Inner: $r=x^2$.

$$V=\pi\int_0^2\left[(2x)^2-\left(x^2\right)^2\right]dx=\pi\int_0^2\left[4x^2-x^4\right]dx=\pi\left[\frac{4x^3}{3}-\frac{x^5}{5}\right]_0^2$$

$$=\pi\left(\frac{32}{3}-\frac{32}{5}\right)=\frac{64\pi}{15}$$

3. Outer: $R=4-x^2$, Inner: $r=4-2x$.

$$V=\pi\int_0^2\left[\left(4-x^2\right)^2-(4-2x)^2\right]dx=\pi\int_0^2\left[\left(16-8x^2+x^4\right)-\left(16-16x+4x^2\right)\right]dx$$

$$=\pi\int_0^2\left(16x-12x^2+x^4\right)dx=\pi\left[8x^2-4x^3+\frac{x^5}{5}\right]_0^2=\pi\left(32-32+\frac{32}{5}\right)=\frac{32\pi}{5}$$

4. Outer: $R=2-\frac{y}{2}$, Inner: $r=2-\sqrt{y}$.

$$V=\pi\int_0^4\left[\left(2-\frac{y}{2}\right)^2-\left(2-\sqrt{y}\right)^2\right]dy=\pi\int_0^4\left[\left(4-2y+\frac{y^2}{4}\right)-\left(4-4y^{1/2}+y\right)\right]dy$$

$$=\pi\int_0^4\left(-3y+\frac{y^2}{4}+4y^{1/2}\right)dy=\pi\left[-\frac{3y^2}{2}+\frac{y^3}{12}+\frac{8y^{3/2}}{3}\right]_0^4=\pi\left(-24+\frac{16}{3}+\frac{64}{3}\right)=\frac{8\pi}{3}$$

Day 112

1. $$\pi\int_0^2\left(4-x^2\right)^2dx=\pi\left[16x-\frac{8x^3}{3}+\frac{x^5}{5}\right]_0^2=\pi\left[32-\frac{64}{3}+\frac{32}{5}\right]=\frac{256\pi}{15}\approx 53.617$$

2. $$\pi\int_{-3}^1(3+x)^2dx=\pi\left[9x+3x^2+\frac{x^3}{3}\right]_{-3}^1=\pi\left[\left(9+3+\frac{1}{3}\right)-(-27+27-9)\right]$$

$$=\frac{64\pi}{3}\approx 67.021$$

3. $$\pi\int_{-1}^3(1+y)^2dy=\pi\left[y+y^2+\frac{y^3}{3}\right]_{-1}^3=\pi\left[(3+9+9)-\left(-1+1-\frac{1}{3}\right)\right]$$

$$=\frac{64\pi}{3}\approx 67.021$$

Day 113

1. $$\pi\int_1^9\left[\left(4-\sqrt{y}\right)^2-(4-3)^2\right]dy=\pi\int_1^9\left[15-8y^{1/2}+y\right]dy=\pi\left[15y-\frac{16y^{3/2}}{3}+\frac{y^2}{2}\right]_1^9$$
$$=\pi\left[(135-144+40.5)-\left(15-\frac{16}{3}+\frac{1}{2}\right)\right]$$
$$=\frac{64\pi}{3}\approx 67.021$$

2. $$\pi\int_1^3\left[\left(-1-x^2\right)^2-(-1-1)^2\right]dx=\pi\int_1^3\left[-3+2x^2+x^4\right]dx=\pi\left[-3x+\frac{2x^3}{3}+\frac{x^5}{5}\right]_1^3$$
$$=\pi\left[(-9+18+48.6)-\left(-3+\frac{2}{3}+\frac{1}{5}\right)\right]$$
$$=\frac{896\pi}{15}\approx 187.658$$

3. $$\pi\int_1^3\left[(10-1)^2-\left(10-x^2\right)^2\right]dx=\pi\int_1^3\left(-19+20x^2-x^4\right)dx=\left[-19x+\frac{20x^3}{3}-\frac{x^5}{5}\right]_1^3$$
$$=\left[(-57+180-48.6)-\left(-19+\frac{20}{3}-\frac{1}{5}\right)\right]$$
$$=\pi\left[74.4-\left(\frac{-188}{15}\right)\right]$$
$$=\frac{1304\pi}{15}\approx 273.109$$

4. $$\pi\int_1^9\left[(-1-3)^2-\left(-1-\sqrt{y}\right)^2\right]dy=\pi\int_1^9\left(15-2y^{1/2}-y\right)dy=\pi\left[15y-\frac{4y^{3/2}}{3}-\frac{y^2}{2}\right]_1^9$$
$$=\pi\left[(135-36-40.5)-\left(15-\frac{4}{3}-\frac{1}{2}\right)\right]$$
$$=\frac{272\pi}{6}=\frac{136\pi}{3}\approx 142.419$$

Day 114

1. $A=\int_0^4\left(x+2-\sqrt{x}\right)dx=\frac{x^2}{2}+2x-\frac{2x^{3/2}}{3}\Bigg|_0^4=8+8+\frac{16}{3}=\frac{32}{3}$

2. $V=\int_0^4\left(x+2-\sqrt{x}\right)^2dx=\int_0^4\left(x^2+5x+4-2x^{3/2}-4x^{1/2}\right)dx$

$$=\frac{x^3}{3}+\frac{5x^2}{2}+4x-\frac{4x^{5/2}}{5}-\frac{8x^{3/2}}{3}\Bigg|_0^4=\frac{64}{3}+\frac{80}{2}+16-\frac{128}{5}-\frac{64}{3}=56-25.6=30.4$$

3. $c=x+2-\sqrt{x}$, $a=b=\frac{x+2-\sqrt{x}}{\sqrt{2}}$, $A=\frac{1}{2}ab=\frac{1}{2}\left(\frac{x+2-\sqrt{x}}{\sqrt{2}}\right)^2$

$$=\frac{x^2+5x+4-2x^{3/2}-4x^{1/2}}{4},\ V=\int_0^4\frac{x^2+5x+4-2x^{3/2}-4x^{1/2}}{4}dx$$

$$=\frac{1}{4}\left[\frac{x^3}{3}+\frac{5x^2}{2}+4x-\frac{4x^{5/2}}{5}-\frac{8x^{3/2}}{3}\right]_0^4=\frac{1}{4}(30.4)=7.6$$

Day 115

1. $V=\pi\int_0^{\sqrt{2}/2}\left(\arccos y\right)^2-\left(\arcsin y\right)^2dy$

2. $V=\pi\int_0^{\sqrt{2}/2}\left(\arcsin y\right)^2dy+\int_{\sqrt{2}/2}^{1}\left(\arccos y\right)^2dy$

3. $V=\pi\int_0^{\pi/4}\left(\cos^2x-\sin^2x\right)dx=\frac{\pi}{2}\int_0^{\pi/4}\cos(2x)\cdot 2\,dx=\frac{\pi}{2}\sin(2x)\Big|_0^{\pi/4}=\frac{\pi}{2}(1-0)=\frac{\pi}{2}$

Day 116

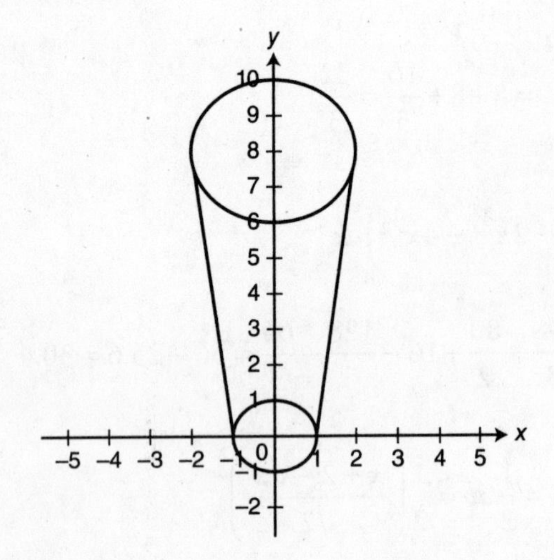

Each horizontal slice is a circle, with a radius from the y-axis to the edge of the cup, which is the line $y = 8(x-1) \Rightarrow x = \frac{y}{8} + 1$

$$V = \pi \int_0^8 \left(\frac{y}{8}+1\right)^2 dy = \pi \int_0^8 \left(\frac{y^2}{64}+\frac{y}{4}+1\right) dy = \pi \left[\frac{y^3}{192}+\frac{y^2}{8}+y\right]_0^8$$

$$= \frac{56\pi}{3} \approx 58.643$$

Day 117

$$V = \pi r^2 h, \quad h = \frac{1{,}000}{\pi r^2}, \quad S = 2\pi r^2 + 2\pi r h = 2\pi r^2 + 2\pi r\left(\frac{1{,}000}{\pi r^2}\right) = 2\pi r^2 + \frac{2{,}000}{r},$$

$$S' = 4\pi r - 2{,}000 r^{-2} = 0, \; r = \sqrt[3]{\frac{500}{\pi}} \approx 5.419, \; h \approx 10.839; \; S''(5.419) > 0, \text{ minimum.}$$

Day 118

Each slice is a semicircle with a radius $r = \sqrt{9 - x^2}$.

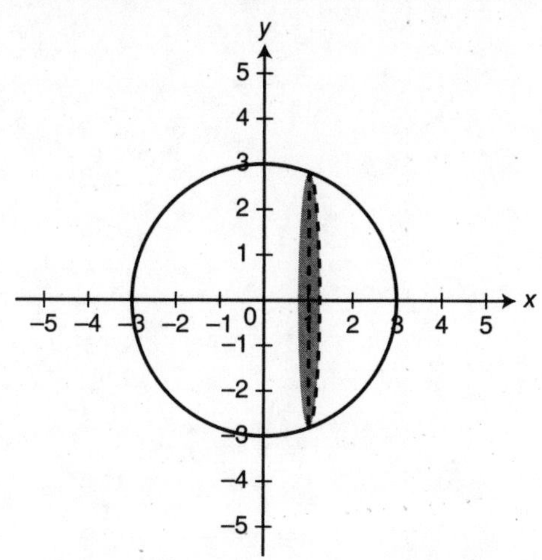

$$V = \int_{-3}^{3} \frac{1}{2}\pi\left(\sqrt{9-x^2}\right)^2 dx = 2\int_{0}^{3} \frac{1}{2}\pi\left(\sqrt{9-x^2}\right)^2 dx$$

$$= \pi\int_{0}^{3}\left(9-x^2\right)dx = \pi\left[9x - \frac{x^3}{3}\right]_0^3$$

$$= \pi\left[(27-9)-0\right] = 18\pi$$

Day 119

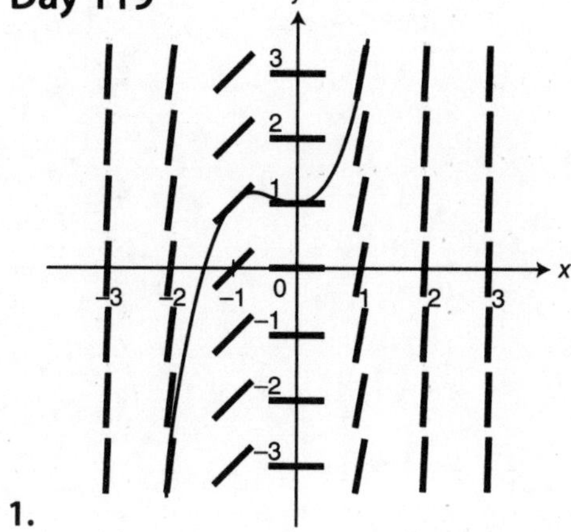

1.

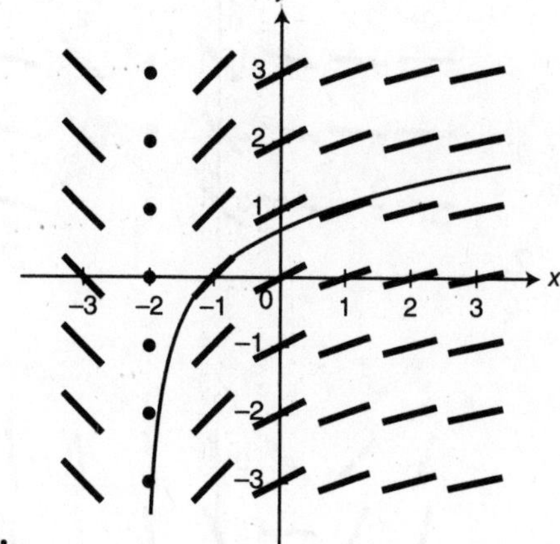

2.

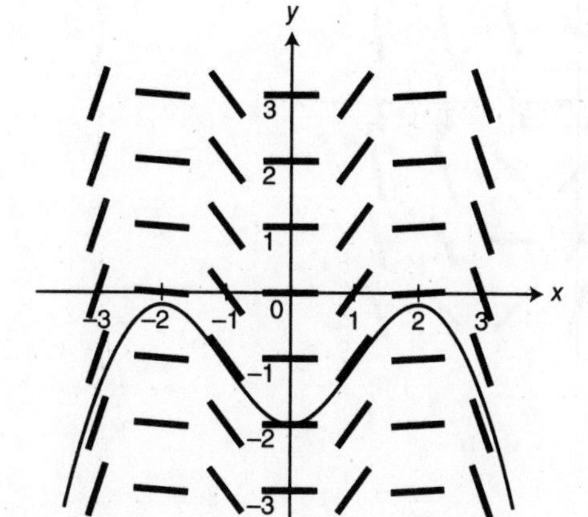

3.

Day 120

1.

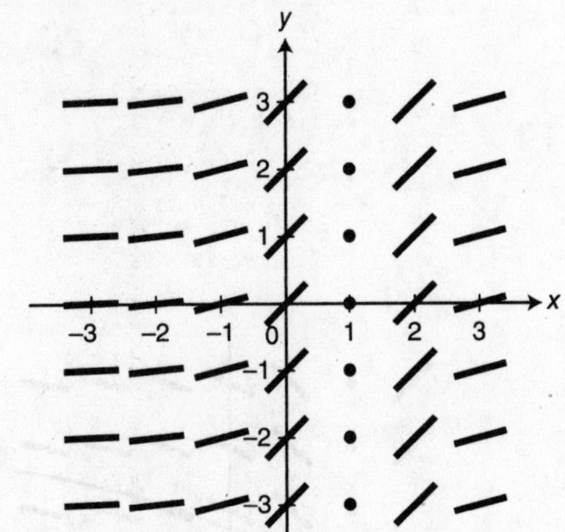

2.

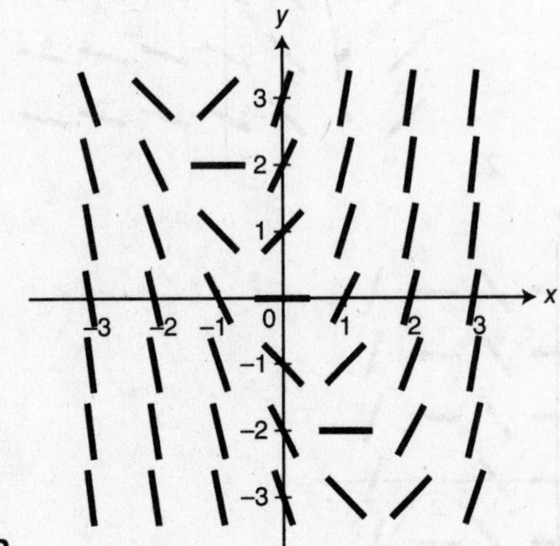

3.

Day 121

1. $\frac{dy}{dx}=\frac{2\cos x}{y^2} \Rightarrow y^2dy=2\cos xdx$

$\int y^2\,dy=2\int \cos xdx$

$\frac{y^3}{3}=2\sin x+C_1$

$y^3=6\sin x+3C_1$

$y=\sqrt[3]{6\sin x+3C_1}=\sqrt[3]{6\sin x+C}.$

Note that C_1 and C are arbitrary constants.

2. $\frac{dy}{dx}=x(1+2y) \Rightarrow \frac{dy}{1+2y}=x\,dx$

$\frac{1}{2}\int\frac{2dy}{1+2y}=\int x\,dx$

$\frac{1}{2}\ln|1+2y|=\frac{x^2}{2}+C_1$

$\ln|1+2y|=x^2+2C_1$

$|1+2y|=e^{(x^2+2C_1)}=C_2e^{x^2}$

$1+2y=\pm C_2e^{x^2}$

$y=\frac{\pm C_2e^{x^2}-1}{2}=\frac{Ce^{x^2}-1}{2}.$

Note that C_1, C_2, and C are arbitrary constants.

Day 122

1. $\frac{1}{4}\int_2^6(x-2)^{1/2}\,dx=\frac{1}{4}\cdot\frac{2}{3}(x-2)^{3/2}\Big|_2^6=\frac{1}{6}(8-0)=\frac{4}{3}$

2. $\frac{1}{1}\int_0^1 e^{1-x}\,dx=-\int_0^1 e^{1-x}(-1)dx=-e^{1-x}\Big|_0^1=-(1-e)=e-1\approx 1.718$

3. $\frac{1}{3}\int_{-1}^{2}\left(x^3-2x+1\right)dx=\frac{1}{3}\left(\frac{x^4}{4}-x^2+x\right)\Bigg|_{-1}^{2}=\frac{1}{3}\left[(4-4+2)-\left(\frac{1}{4}-1-1\right)\right]=\frac{1}{3}\left(\frac{15}{4}\right)=\frac{5}{4}$

Day 123

Order: 4, 1, 3, 2

1. $\frac{1}{2}\int_0^2 (x^3 - 4x)\,dx = \frac{1}{2}\left[\frac{x^4}{4} - 2x^2\right]_0^2 = \frac{1}{2}[(4-8)] = -2$

2. $\frac{1}{2}\int_0^2 (x + \sin x)\,dx = \frac{1}{2}\left[\frac{x^2}{2} - \cos x\right]_0^2$

$= \frac{1}{2}[(2 - \cos 2) - (0 - \cos(0))] = \frac{3 - \cos 2}{2} \approx 1.708$

3. $\frac{1}{2}\int_0^2 \left(2xe^{-x^2}\right)dx = -\frac{1}{2}\int_0^2 \left(e^{-x^2}\right)(-2x)\,dx = -\frac{1}{2}\left(e^{-x^2}\right)\Big|_0^2 = \frac{1 - e^{-4}}{2} \approx 0.491$

4. $\frac{1}{2}\int_0^2 \left(1 - xe^{3x^2}\right)dx = \frac{1}{2}\int_0^2 (1)\,dx - \frac{1}{12}\int_0^2 \left(e^{3x^2}\right)(6x)\,dx$

$= \frac{x}{2}\Big|_0^2 - \frac{e^{3x^2}}{12}\Big|_0^2 = 1 - \left(\frac{e^{12}}{12} - \frac{1}{12}\right) = \frac{13 - e^{12}}{12} \approx -13{,}561.816$

Day 124

1. $\int_0^2 (4 - x^2)\,dx = \left[4x - \frac{x^3}{3}\right]_0^2 = 8 - \frac{8}{3} = \frac{16}{3} = f(c)(2-0)$

$4 - c^2 = \frac{8}{3} \Rightarrow c^2 = \frac{4}{3} \Rightarrow c = \pm\frac{2\sqrt{3}}{3}; c = \frac{2\sqrt{3}}{3}$ on the interval $[0,2]$

2. $\int_0^2 (x^2 - x - 2)\,dx = \left[\frac{x^3}{3} - \frac{x^2}{2} - 2x\right]_0^2 = \frac{8}{3} - 2 - 4 = -\frac{10}{3} = f(c)(2-0)$

$f(c) = \frac{1}{2}\left(-\frac{10}{3}\right) = -\frac{5}{3}$

$x^2 - x - 2 = -\frac{5}{3} \Rightarrow 3x^2 - 3x - 1 = 0 \Rightarrow x = \frac{3 \pm \sqrt{9 - 4(3)(-1)}}{2(3)} = \frac{3 \pm \sqrt{21}}{6};$

$c = \frac{3 + \sqrt{21}}{6} \approx 1.264$ on $[0,2]$.

3. $\int_0^2 \left(5\sin\left(\frac{x-1}{2}\right)\right)dx = 10\int_0^2 \sin\left(\frac{x-1}{2}\right)\cdot\frac{1}{2}\,dx = -10\cos\left(\frac{x-1}{2}\right)\Big|_0^2$

$= -10\left[\cos\left(\frac{1}{2}\right) - \cos\left(-\frac{1}{2}\right)\right] = 0$

$\frac{5}{2}\sin\left(\frac{c-1}{2}\right) = 0 \Rightarrow \sin\left(\frac{c-1}{2}\right) = 0 \Rightarrow \frac{c-1}{2} = 0 \Rightarrow c = 1$

Day 125

1. Acceleration is the derivative of velocity and will be negative from $t = 3$ to $t = 5$.

2. Average velocity

$$\frac{1}{5}\left[\frac{(1)(20)}{2}+\frac{(20+50)(1)}{2}+\frac{(50+65)(1)}{2}+\frac{(50+65)(1)}{2}+\frac{(50+10)(1)}{2}\right]=38 \text{ mph.}$$

3. Total distance is the area under the curve which is 190 miles.

Day 126

1. $C'(x)=17+0.05x.$

2. $$\text{Average cost} = \frac{\text{total cost}}{\text{quantity}}=\frac{17(5{,}000)+0.025(5{,}000)^2}{5{,}000}$$
$$=17+0.025(5{,}000)=17+125=\$142$$

3. $$\text{Total profit} = \text{total revenue} - \text{total cost}$$
$$=250(5{,}000)-\left(17(5{,}000)+0.025(5{,}000)^2\right)$$
$$=1{,}250{,}000-710{,}000=\$540{,}000$$

Day 127

1. $P(x)=R(x)-C(x)=120x-\left(x^2+6x+1{,}200\right)=-x^2+114x-1{,}200$

2. $C'(x)=2x+6, C'(80)=166$ per additional item. $P'(x)=-2x+114, P'(80)=-46$ per additional item.

3. Maximum profit occurs when marginal profit is zero. $P'(x)=-2x+114=0$ when $x=57$. Maximum profit will occur at the production of 57 items.

Day 128

1. $f(x)=\dfrac{x-1}{2x+1}$ on $[0,3]$

$f'(x)=\dfrac{3}{(2x+1)^2}$; note that $f'(x)>0$ on $[0,3]$.

Thus, $f(x)$ is increasing on $[0,3]$.

x	$f(x)$
0	–1
3	2/7

Maximum value is $\frac{2}{7}$ occurring at point $\left(3,\frac{2}{7}\right)$.

Minimum value is –1 occurring at point (0, –1).

2. Maximum value is 5 occurring at point (1, 5).

Minimum value is 3 occurring at point (2, 3).

Day 129

1. $\int_0^9 3\sqrt{t}\,dt = 2t^{3/2}\Big|_0^9 = 2(27) = 54$ gallons

2. $50 + \int_0^9 \left(1 - 3\sqrt{t}\right)dt = 50 + \left[t - 2t^{3/2}\right]_0^9 = 50 + 9 - 54 = 5$ gallons

Day 130

1. $2\int_0^3 \left(e^{(t+1)/2}\right)\left(\frac{1}{2}dt\right) = 2e^{(t+1)/2}\Big|_0^3 = 2\left(e^2 - e^{1/2}\right) \approx 11.481$ million bacteria

2. $5 + \int_0^3 f(t)\,dt \approx 5 + 11.481 \approx 16.481$ million bacteria

Day 131

1. $200 - 12\int_0^5 \left(e^{-t/6}\right)dt = 200 + 72\int_0^5 \left(e^{-t/6}\right)\left(-\frac{1}{6}dt\right) = 200 + 72\left[e^{-t/6}\right]_0^5$

$$= 200 + 72\left[e^{-5/6} - 1\right] \approx 200 + 72\left[e^{-5/6} - 1\right] \approx 159.291^\circ$$

2. $-12\int_0^T \left(e^{-t/6}\right)dt = -60 \Rightarrow -6\int_0^T \left(e^{-t/6}\right)\left(-\frac{1}{6}dt\right) = 5 \Rightarrow -6\left(e^{-T/6} - 1\right) = 5$

$$\Rightarrow e^{-T/6} - 1 = -\frac{5}{6} \Rightarrow e^{-T/6} = \frac{1}{6} \Rightarrow -\frac{T}{6} = \ln\left(\frac{1}{6}\right) \Rightarrow T = -6\ln\left(\frac{1}{6}\right) \approx 10.751 \text{ min}$$

Day 132

1. 1(0.3 + 0.4 + 0.5) = 1.2

2. 1(0.6 + 0.6 + 0.5 + 0.3) = 2.0

3. Plot A: 1(0.3 + 0.4 + 0.5 + 0.5 + 0.7 + 0.8 + 0.6 + 0.4) = 4.2

   Plot B: 1(0.2 + 0.2 + 0.3 + 0.5 + 0.6 + 0.6 + 0.5 + 0.3) = 3.0

   Assuming populations were equal at $t = 0$, Plot A is larger after 8 years.

Day 133

1. Using RRAM $\frac{1}{15}[3(153)+3(124)+3(107)+3(96)+3(90)]=\frac{1}{5}(570)=114°\text{F}$ or

Using LRAM $\frac{1}{15}[3(200)+3(153)+3(124)+3(107)+3(96)]=\frac{1}{5}(680)=136°\text{F}$

You could also use MRAM or Trapezoidal approximation.

2. $\frac{90-96}{15-12}=-2$ or $\frac{86-90}{18-15}=-\frac{4}{3}$ or $\frac{86-96}{18-12}=-\frac{5}{3}$ degree F/min

Day 134

1. $\int_0^{10} r(t)\,dt=-45\int_0^{10} e^{-t/20}\,dt=900\int_0^{10} e^{-t/20}\left(-\frac{1}{20}\right)dt$

$$=900e^{-t/20}\Big|_0^{10}=900e^{-1/2}-900\approx-354.122\text{ in}^3.$$

2. $900-45\int_0^{30} e^{-t/20}\,dt=900+900\int_0^{30} e^{-t/20}\left(\frac{-1}{20}\right)dt=900+900\left[e^{-t/20}\right]_0^{30}$

$$=900+900\left[e^{-3/2}-1\right]\approx 200.817$$

3. Volume of a cylinder is $V=\pi r^2h$, and the radius is constant at $r=6$, so $V=36\pi h$. Differentiate with respect to time. $\frac{dV}{dt}=36\pi\frac{dh}{dt}$.

Day 135

Let x be her distance from the lamp, and s the length of her shadow. $\frac{s}{s+x}=\frac{168}{610}$.

$84x=221s$, $84\frac{dx}{dt}=221\frac{ds}{dt}$, $84(-90)=221\frac{ds}{dt}$, $\frac{ds}{dt}\approx-34.21$ cm per second.

Day 136

$\frac{dy}{dx}=\frac{3x^2-2y}{2x+2y}\Big|_{(0,-2)}=-1$, $y+2=1(x-0)$, $y=x-2$

Day 137

1. $\sqrt{x^2+2x+1}=0 \Rightarrow x^2+2x+1=0 \Rightarrow x=-1$ The domain is all reals except $x=-1$. $\lim_{x\to\infty}\frac{x}{\sqrt{x^2+2x+1}}=1$, $\lim_{x\to-\infty}\frac{x}{\sqrt{x^2+2x+1}}=-1$, $\frac{d}{dx}\frac{x}{\sqrt{x^2+2x+1}}=$ $\frac{1}{(x+1)\sqrt{x^2+2x+1}}\neq 0$ so the graph has no turning point. Range is $(-\infty,1)$.

2. 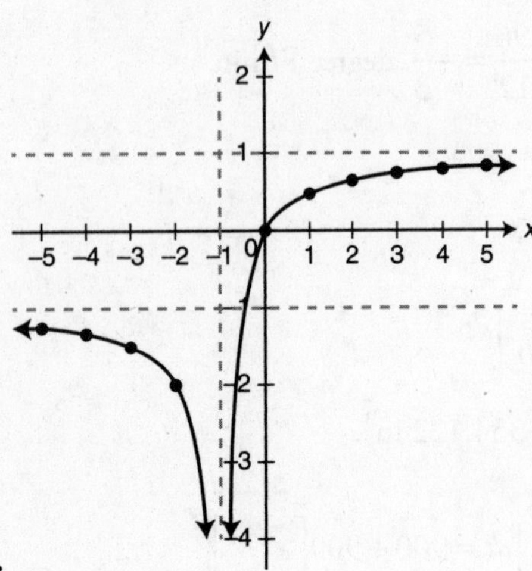

3. $y=1$ and $y=-1$

Day 138

1. $a(t)=v'(t)=2t-\cos t$

2. $y(t)=\int v(t)\,dt=\int\left(t^2-\sin t\right)dt=\frac{t^3}{3}+\cos t+C$. At $t=0$, $0+\cos 0+C=1+C$, so $C=0$, and $y(t)=\frac{t^3}{3}+\cos t$.

3. $v(t)=t^2-\sin t=0$ when $t^2=\sin t$, which is when $t=0$ or $t\approx .8767$. When $t\approx .8767$, $y\approx\frac{(.8767)^3}{3}+\cos(.8767)\approx .864$.

Day 139

1. $4y\frac{dy}{dx} - x\frac{dy}{dx} - y + 3x^2 = 0$, $\frac{dy}{dx}(4y - x) = y - 3x^2$, $\frac{dy}{dx} = \frac{y - 3x^2}{4y - x}$

2. At $(-2,3)$, $\frac{dy}{dx} = \frac{y - 3x^2}{4y - x} = \frac{3 - 12}{12 + 2} = \frac{-9}{14}$, so the equation of the tangent line is $y - 3 = -\frac{9}{14}(x + 2)$ or $9x + 14y = 24$.

3. $$\begin{aligned} 9(-2.2) + 14k &= 24 \\ -19.8 + 14k &= 24 \\ 14k &= 43.8 \\ k &\approx 3.129 \end{aligned}$$

Day 140

1. $\int_a^b \left[\left(x^3 + 1\right) - \left(4x - x^3\right)\right]dx + \int_b^c \left[\left(4x - x^3\right) - \left(x^3 + 1\right)\right]dx$

2. $$\pi \int_b^c \left[\left(4x - x^3\right)^2 - \left(x^3 + 1\right)^2\right]dx$$
$$\approx \pi \int_{0.3}^{1.3} \left[16x^2 - 8x^4 + x^6 - \left(x^6 + 2x^3 + 1\right)\right]dx$$
$$\approx \pi \int_{0.3}^{1.3} \left[-8x^4 - 2x^3 + 16x^2 - 1\right]dx$$
$$\approx \pi \left[-\frac{8}{5}x^5 - \frac{1}{2}x^4 + \frac{16}{3}x^3 - x\right]_{0.3}^{1.3}$$
$$\approx \pi\left[3.048595 - (-0.163938)\right] \approx 3.213\pi$$

Day 141

1. $V = \frac{1}{3}\pi \cdot r^2 h, \frac{h}{r} = \frac{8}{3}$ thus $r = \frac{3h}{8}$ and
$V = \frac{\pi}{3}\left(\frac{3h}{8}\right)^2 h = \frac{3\pi}{64}h^3$.

2. $V = \frac{3\pi}{64}h^3$, $\frac{dv}{dt} = \frac{9\pi}{64}(h^2)\frac{dh}{dt}$.
At $h = 6$, $\frac{dv}{dt} = \frac{9\pi}{64}(36)(5) \approx 79.522\ \text{ft}^3/\text{min}$

3. $V = 100\pi y$, so $\frac{dV}{dt} = 100\pi\frac{dy}{dt}$. $79.523 = 100\pi\frac{dy}{dt}$, so $\frac{dy}{dt} = 0.253$ ft/min

Day 142

1. $h(0)=\int_0^0 f(t)\,dt=0,\ h'(1)=f(1)=2$

2. The function h will be concave upward when $f'(t)>0$ or when $f(t)$ is increasing. This occurs when $-1<x<1$.

3. The point of inflection will occur when the second derivative of h, $f'(t)$, is equal to zero, and h changes from concave upward to concave downward. The point of inflection occurs at $x=1$. $h(1)=\int_0^1 f(t)\,dt=\int_0^1\left(1+\frac{2t}{t^2+1}\right)dt=t+\ln\left(t^2+1\right)\Big|_0^1=$ $(1+\ln 2)-\ln 1=1+\ln 2$. The point of inflection is $(1,1+\ln 2)\approx(1,1.693)$.

Day 143

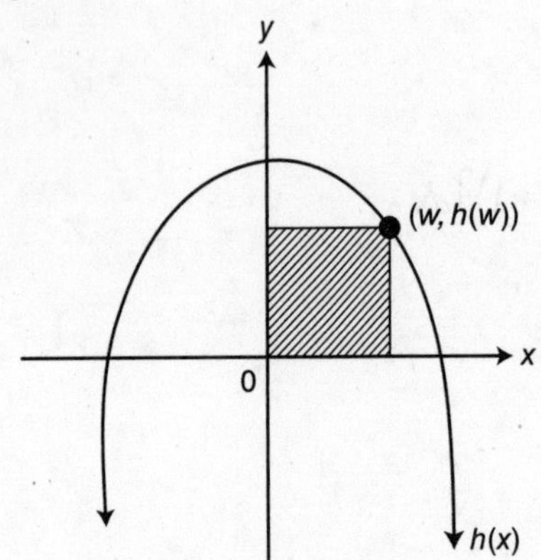

1. $A(w)=w\left(8-\frac{w^4}{2}\right)=8w-\frac{w^5}{2}$

$A'(w)=8-\frac{5}{2}w^4=0\Rightarrow\frac{5}{2}w^4=8\Rightarrow w^4=\frac{16}{5}\Rightarrow w=\frac{2}{\sqrt[4]{5}}\approx 1.337.$

$A''(w)=-10w^3$ and $A''(1.337)<0\Rightarrow$ maximum.

$h(1.337)=8-\frac{(1.337)^4}{2}=8-\frac{16/5}{2}=\frac{40-8}{5}=\frac{32}{5}=6.4$

$A(1.337)\approx 1.337(6.4)=8.560$

2. $\int_0^2\left(8-\frac{x^4}{2}\right)dx=\left[8x-\frac{x^5}{10}\right]_0^2=[(16-3.2)-(0)]=12.8$, $12.8-8.560=4.240$

Day 144

1. $S(t)=Ce^{kt}\Rightarrow 57=Ce^{k0}\Rightarrow C=57$

$375=57e^{8k}\Rightarrow 6.57895\approx e^{8k}\Rightarrow\ln(6.57895)=8k$

$k=\frac{\ln(6.57895)}{8}\approx 0.235$

2. Average Price $=\frac{1}{8}\int_4^{12}57e^{0.235t}\,dt$

$=\frac{3448.34701}{8}=\$431.04$

Day 145

1. If $y=1-bx$ is tangent to the graph of $f(x)=\cos(x)-bx$, $\cos(x)-bx=1-bx$, and $f'(x)=-\sin(x)-b=-b$. This occurs when $\cos(x)=1$ and $\sin(x)=0$, or when $\cos(x)=1$, $x=-2\pi,0,2\pi$ and when $\sin(x)=0$, $x=-2\pi,-\pi,0,\pi,2\pi$.

   Thus, when $\cos(x)=1$ and $\sin(x)=0$, $x=-2\pi,0,2\pi$.

Day 146

1. $V=\pi\int_0^8\left(\sqrt[6]{64y}\right)^2 dy=4\pi\int_0^8 y^{1/3}\,dy=4\pi\left[\frac{3}{4}y^{4/3}\right]_0^8=3\pi\cdot\left(8^{4/3}\right)=48\pi\approx 150.796\text{ ft}^3.$

   $W=35(150.796)\approx 5{,}277.876$ lb.

2. 30 min

3. $V(h)=4\pi\int_0^h y^{1/3}\,dy=3\pi h^{4/3}$

$$\frac{dV}{dt}=4\pi h^{1/3}\frac{dh}{dt}$$

$$5=4\pi\left(1^{1/3}\right)\frac{dh}{dt}$$

$$\frac{dh}{dt}=\frac{5}{4\pi}\approx 0.398$$

Day 147

1. $f'(4)\approx\dfrac{65-73}{6-2}\approx\dfrac{-8}{4}\approx -2$

2. $$\frac{1}{15}\int_0^{15} f(x)\,dx\approx\frac{1}{15}\left[\frac{1}{2}((80+73)\cdot 2)+\frac{1}{2}((73+65)\cdot 4)+\frac{1}{2}((65+62)\cdot 4)+\frac{1}{2}((62+60)\cdot 5)\right]$$

$$\approx\frac{1}{30}\left[((153)\cdot 2)+((138)\cdot 4)+((127)\cdot 4)+((122)\cdot 5)\right]$$

$$\approx\frac{1}{30}[306+552+508+610]$$

$$\approx\frac{1}{30}[1976]\approx 65.867$$

3. $\int_0^{15} f'(x)\,dx=f(15)-f(0)=60-80=-20$. There is a total change of $-20\,^\circ\text{C}$ from one end of the wire to the other.

Day 148

The derivative of $y=\frac{400x-x^2}{200}=2x-\frac{x^2}{200}$ is $y'=2-\frac{x}{100}$. The equation of the line takes the form $y=\left(2-\frac{x}{100}\right)x+50$. The intersection of the line and the parabola can be found by solving $2x-\frac{x^2}{200}=\left(2-\frac{x}{100}\right)x+50$. $400x-x^2=400x-2x^2+10{,}000$, gives $x^2=10{,}000$, so $x=100$ is the x-coordinate of point Q.

Day 149

1. $2x^3y\frac{dy}{dx}+3x^2y^2+6x^2y\frac{dy}{dx}+6xy^2+2xy\frac{dy}{dx}+y^2=0$

$$2x^3y\frac{dy}{dx}+6x^2y\frac{dy}{dx}+2xy\frac{dy}{dx}=-\left(3x^2y^2+6xy^2+y^2\right)$$

$$\frac{dy}{dx}=-\frac{3x^2y^2+6xy^2+y^2}{2x^3y+6x^2y+2xy}=-\frac{y^2\left(3x^2+6x+1\right)}{2xy\left(x^2+3x+1\right)}=-\frac{y\left(3x^2+6x+1\right)}{2x\left(x^2+3x+1\right)}$$

2. $-y^2+3y^2-y^2=2$, $y^2=2$, $y=\pm\sqrt{2}$, At $\left(-1,\sqrt{2}\right)$, $\frac{dy}{dx}=-\frac{\sqrt{2}(3-6+1)}{-2(1-3+1)}=\sqrt{2}$ and the tangent line is $y-\sqrt{2}=\sqrt{2}(x+1)$. At $\left(-1,-\sqrt{2}\right)$, $\frac{dy}{dx}=-\frac{-\sqrt{2}(3-6+1)}{-2(1-3+1)}=-\sqrt{2}$ and the tangent line is $y+\sqrt{2}=-\sqrt{2}(x+1)$.

Day 150

$f''(x)=-\cos x$, critical points at $x=\pm\frac{\pi}{2},\pm\frac{3\pi}{2}$, $f''\left(-\frac{7\pi}{4}\right)=-\frac{\sqrt{2}}{2}$, concave downward on $\left(-2\pi,-\frac{3\pi}{2}\right)$, $f''(-\pi)=1$, concave upward on $\left(-\frac{3\pi}{2},-\frac{\pi}{2}\right)$, $f''(0)=-1$, concave downward on $\left(-\frac{\pi}{2},\frac{\pi}{2}\right)$, $f''(\pi)=1$, concave upward on $\left(\frac{\pi}{2},\frac{3\pi}{2}\right)$, $f''\left(\frac{7\pi}{4}\right)=-\frac{\sqrt{2}}{2}$, concave downward on $\left(\frac{3\pi}{2},2\pi\right)$.

Thus, points of inflection at $x=\pm\frac{\pi}{2}$ and at $x=\pm\frac{3\pi}{2}$.

Day 151

1. $v(5)=3+5(5)-(5)^2=3>0$. At $t=5$, the particle is moving upward.

2. $v(t)=3+5t-t^2 \Rightarrow a(t)=5-2t \Rightarrow a(5)=5-2(5)=-5<0$. The velocity and acceleration have opposite signs at $t=5$, so the particle is slowing down.

3. $v(t)=3+5t-t^2 \Rightarrow s(t)=3t+\frac{5t^2}{2}-\frac{t^3}{3}+C$. If $s(0)=1$, $s(t)=3t+\frac{5t^2}{2}-\frac{t^3}{3}+1$.

   At t = 6, $s(6)=3(6)+\frac{5(6)^2}{2}-\frac{(6)^3}{3}+1=18+90-72+1=37$.

Day 152

1. $\int_0^3\left(3-\frac{x^3}{9}\right)dx=\left[3x-\frac{x^4}{36}\right]_0^3=9-\frac{9}{4}=\frac{27}{4}=6.75$

2. $\pi\int_0^3\left(3^2-\left(\frac{x^3}{9}\right)^2\right)dx=\pi\int_0^3\left(9-\frac{x^6}{81}\right)dx=\pi\left[9x-\frac{x^7}{567}\right]_0^3$

   $=\pi\left[27-\frac{27}{7}\right]=\frac{162\pi}{7}\approx 72.705$

3. $\pi\int_0^3\left[\left(k-\frac{x^3}{9}\right)^2-(k-3)^2\right]dx=\frac{162\pi}{7}$

Day 153

1. $\int_0^{12} R(t)\,dt\approx 4(17.67+17.67+15)\approx 4(50.34)\approx 201.36$ Over the 12-hour period, approximately 201.36 gallons of water flowed through the pipe.

2. $\frac{1}{12}\int_0^{12}\left(\frac{200+8t-t^2}{12}\right)dt=\frac{1}{144}\left[200t+4t^2-\frac{t^3}{3}\right]_0^{12}$

   $=\frac{1}{144}(2{,}400+576-576)=\frac{2{,}400}{144}=16\frac{2}{3}$ gallons/hour.

Day 154

1. $y-1=2(x-0)$

2. $g'(0)=2f'(0)=2(2)=4$, $y+3=4(x-0)$

3. $g'(x)=6xf(x)+(3x^2+2)f'(x)+2xf''(x)$

$$g''(x)=[6xf'(x)+6f(x)]+[(3x^2+2)f''(x)+6xf'(x)]+[2xf'''(x)+2f''(x)]$$

$$g''(x)=6f(x)+6xf'(x)+6xf'(x)+(3x^2+2)f''(x)+2f''(x)+2xf'''(x)$$

$$g''(x)=6f(x)+12xf'(x)+(3x^2+4)f''(x)+2xf'''(x)$$

Day 155

1. $g(1)=\int_0^1 f(t)dt=\frac{1}{2}(1.5+3)(1)=2.25$, $g(-1)=\int_0^{-1} f(t)dt=-\frac{1}{2}(1.5)(1)=-0.75$

2. Instantaneous rate of change of g, with respect to x, at $x = 3$ is $g'(3)=f(3)=2$

3. The derivative of g is equal to 0 only at the endpoints of the closed interval $[-1, 4]$. The function g has no turning points on the interval. $g(-1)=-0.75$,

$$g(4)=\int_0^4 f(t)dt=\int_0^1 f(t)dt+\int_1^4 f(t)dt$$

$$=2.25+\frac{1}{2}(3+2)(1)+\frac{1}{2}(2+2)(1)+\frac{1}{2}(2)(1)$$

$$=2.25+\frac{1}{2}(5+4+2)=2.25+5.5=7.75$$

The minimum value of g is $g(-1)=-0.75$.

Alternatively, $g'(x)=f(x)>0$ on the open interval $(-1, 4)$. Thus, $g(x)$ is strictly increasing on the interval $[-1, 4]$ and the minimum value of g is $g(-1)=-0.75$.

Day 156

1. If $w = 1$, $P = (1,5)$, $y'=-\frac{5}{x^2}\Big|_{x=1}=-5$. The tangent line is $y-5=-5(x-1)$ or $y=10-5x$. The x-intercept is $x = 2$.

2. For all w, $P=\left(w,\frac{5}{w}\right)$, $y'=-\frac{5}{x^2}\Big|_{x=w}=-\frac{5}{w^2}$. The tangent line is $y-\frac{5}{w}=-\frac{5}{w^2}(x-w)$ or $y=\frac{10}{w}-\frac{5}{w^2}x$. When $y = 0$, $0=\frac{10}{w}-\frac{5}{w^2}k$, so $k=\frac{10}{w}\cdot\frac{w^2}{5}=2w$.

3. If $k=2w$, $\frac{dk}{dt}=2\frac{dw}{dt}=2(2)=4$. Note that k is increasing at a constant rate of 4 units per second for all values of w in the domain.

Day 157

1. $\int_0^3 x^2\,dx = \left[\frac{x^3}{3}\right]_0^3 = 9$

2. If $\int_0^k x^2\,dx = \int_k^3 x^2\,dx$, then $\left[\frac{x^3}{3}\right]_0^k = \left[\frac{x^3}{3}\right]_k^3$ or $\frac{k^3}{3} = 9 - \frac{k^3}{3}$. Solving $\frac{2k^3}{3} = 9$ gives $k^3 = \frac{27}{2}$ or $k = \frac{3}{\sqrt[3]{2}} \approx 2.381$.

3. $\pi\int_0^3 \left(x^2\right)^2 dx = \pi\int_0^3 x^4\,dx = \pi\left[\frac{x^5}{5}\right]_0^3 = \frac{243\pi}{5} = 48.6\pi \approx 152.681$

4. If $\pi\int_0^c x^4\,dx = \pi\int_c^3 x^4\,dx$, then $\left.\frac{\pi x^5}{5}\right|_0^c = \left.\frac{\pi x^5}{5}\right|_c^3$ or $\frac{\pi c^5}{5} = \frac{243\pi}{5} - \frac{\pi c^5}{5}$. Solving $\frac{2\pi c^5}{5} = \frac{243\pi}{5}$ gives $2c^5 = 243$ or $c = \sqrt[5]{121.5} \approx 2.612$.

Day 158

1. $\lim_{x\to\infty} f(x) = \lim_{x\to\infty}\frac{\ln(x)}{x^2} = \lim_{x\to\infty}\frac{(1/x)}{2x} = \lim_{x\to\infty}\frac{1}{2x^2} = 0$, $\lim_{x\to 0^+} f(x) = \lim_{x\to 0^+}\frac{\ln(x)}{x^2} = -\infty$

2. $f'(x) = \frac{x^2\left(\frac{1}{x}\right) - 2x\ln(x)}{x^4} = \frac{x - 2x\ln(x)}{x^4} = \frac{1-2\ln(x)}{x^3}$

$$\frac{1-2\ln(x)}{x^3} = 0$$

$$1 - 2\ln(x) = 0$$

$$\ln(x) = \frac{1}{2}$$

$x = \sqrt{e} \approx 1.649$, $f''(x) = \frac{6\ln(x) - 5}{x^4}$ and $f''(\sqrt{e}) = -\frac{2}{e^2} < 0$.

$f\left(\sqrt{e}\right) = \frac{\ln\left(\sqrt{e}\right)}{\sqrt{e}^2} = \frac{\frac{1}{2}}{e} = \frac{1}{2e} \approx 0.184$, maximum value.

Day 159

1. $a(t) = v'(t) < 0$ when $20 < t < 30$.

2. $\int_0^{30} v(t)\,dt \approx \frac{5}{2}[0 + (2\cdot 10) + (2\cdot 25) + (2\cdot 40) + (2\cdot 60) + (2\cdot 50) + 45]$

$$\approx \frac{5}{2}(20 + 50 + 80 + 120 + 100 + 45)$$

$$\approx \frac{5}{2}(415) \approx 1{,}037.5 \text{ feet.}$$

3. $\frac{1}{30-0}\int_0^{30} v(t)\,dt = \frac{1{,}037.5}{30} \approx 34.583$ feet/second.

Day 160

1. At $x = 2$, $\frac{dy}{dx} = \frac{x - 3x^2}{y} = \frac{2 - 12}{2} = -5$. The tangent line is $y - 2 = -5(x - 2)$ or $y = -5x + 12$. $f(2.1) \approx -5(2.1) + 12 \approx -10.5 + 12 \approx 1.5$

2. $\frac{dy}{dx} = \frac{x - 3x^2}{y}$ becomes $\int y\, dy = \int (x - 3x^2)\, dx$ or $\frac{y^2}{2} = \frac{x^2}{2} - x^3 + C$. If $x = 2$ and $y = 2$, $\frac{4}{2} = \frac{4}{2} - 8 + C$ gives $C = 8$. The equation is $\frac{y^2}{2} = \frac{x^2}{2} - x^3 + 8$ or $y^2 = x^2 - 2x^3 + 16$. Or $y = \sqrt{x^2 - 2x^3 + 16}$. Note that $y \neq -\sqrt{x^2 - 2x^3 + 16}$ since the point $(2,2)$ is on the graph.

3. Evaluating $y^2 = x^2 - 2x^3 + 16$ at $x = 2.1$ gives $y^2 = 4.41 - 2(9.261) + 16 = 1.888$ and $y \approx 1.374$.

Day 161

1. Setting the first derivative $F'(t) = \frac{2\pi}{3}\cos\left(\frac{\pi t}{12}\right) = 0$ and solving gives $t = 6$ and $t = 18$.

Using a second derivative test, $F''(x) = \frac{-\pi^2 \sin\left(\frac{\pi x}{12}\right)}{18}$, $F''(6) = -\frac{\pi^2}{18}\sin\left(\frac{\pi}{2}\right) < 0$, indicates that at $t = 6$, a maximum, and $F''(18) = -\frac{\pi^2}{18}\sin\left(\frac{3\pi}{2}\right) > 0$, indicates a minimum at $t = 18$.

2.

$$\begin{aligned}
\frac{1}{18-6}\int_6^{18} 65 + 8\sin\left(\frac{\pi t}{12}\right) dt &= \frac{1}{12}\int_6^{18} 65\, dt + \frac{8}{12}\cdot\frac{12}{\pi}\int_6^{18} \sin\left(\frac{\pi t}{12}\right)\left(\frac{\pi}{12}dt\right) \\
&= \left[\frac{65}{12}t - \frac{8}{\pi}\cos\left(\frac{\pi t}{12}\right)\right]_6^{18} \\
&= \left(\frac{195}{2} - \frac{8}{\pi}\cos\left(\frac{3\pi}{2}\right)\right) - \left(\frac{65}{2} - \frac{8}{\pi}\cos\left(\frac{\pi}{2}\right)\right) \\
&= \frac{195}{2} - \frac{65}{2} = 65
\end{aligned}$$

Day 162

1. Differentiate $x^3 - 2x^2 y - 3x^2 + 2y = 10$ implicitly to get $3x^2 - 2x^2\frac{dy}{dx} - 4xy - 6x + 2\frac{dy}{dx} = 0$. Solve for $\frac{dy}{dx} = \frac{3x^2 - 4xy - 6x}{2x^2 - 2}$.

2. At $x = 0$, $y = 5$, and $\frac{dy}{dx} = \frac{3x^2 - 4xy - 6x}{2x^2 - 2} = 0$. The equation of the tangent line is $y = 5$.

Day 163

1. $\int_0^{20} v(t)\,dt \approx 4(22+73+112+117+86) \approx 4(410) \approx 1{,}640$. The sled travels a total of 1,640 feet in 20 minutes.

2. $a(t)=v'(t)=10\cos\left(\frac{t}{8}\right)+5\sin\left(\frac{t}{4}\right)$ and $a(16)=10\cos(2)+5\sin(4)\approx -7.945$ ft/min^2.

3. $\frac{1}{20}\int_0^{20}\left[20+80\sin\left(\frac{t}{8}\right)-20\cos\left(\frac{t}{4}\right)\right]dt$

$=\frac{1}{20}\int_0^{20} 20\,dt+\frac{640}{20}\int_0^{20}\left[\sin\left(\frac{t}{8}\right)\right]\left(\frac{1}{8}dt\right)-\frac{80}{20}\int_0^{20}\left[\cos\left(\frac{t}{4}\right)\right]\left(\frac{1}{4}dt\right)$

$=t\Big|_0^{20}-32\cos\left(\frac{t}{8}\right)\Big|_0^{20}-4\sin\left(\frac{t}{4}\right)\Big|_0^{20}$

$\approx 20-32(-1.80114)-4(-.95892)\approx 81.477$ ft/min

Day 164

1. Relative extrema, or turning points, occur when $f'(x)=0$. From the graph, $f'(x)=0$ at the endpoints of the closed interval [−3, 7] and at $x = 2$. At $x = 2$, the derivative changes from positive to negative, so $f(x)$ changes from increasing to decreasing. The graph of f has a relative maximum at $\left(2, \frac{13}{3}\right)$.
2. Because the first derivative is equal to zero at $x = 2$, the tangent line is a horizontal line $y=\frac{13}{3}$.
3. If $g(x)=x^3 f(x)$, then $g(2)=8f(2)=8\cdot\frac{13}{3}=\frac{104}{3}$. $g'(x)=x^3 f'(x)+3x^2 f(x)$, so $g'(2)=8f'(2)+3(4)f(2)=8(0)+12\left(\frac{13}{3}\right)=52$. The equation of the tangent line is $y-\frac{104}{3}=52(x-2)$.

Answers

Day 165

1. $A(R)=\int_0^3\left[1+\sin\left(\frac{\pi x}{2}\right)\right]dx=\int_0^3[1]\,dx+\frac{2}{\pi}\int_0^3\left[\sin\left(\frac{\pi x}{2}\right)\right]\left(\frac{\pi}{2}\,dx\right)$

$$=x\Big|_0^3-\frac{2}{\pi}\cos\left(\frac{\pi x}{2}\right)\Big|_0^3=3-\frac{2}{\pi}\left[\cos\left(\frac{3\pi}{2}\right)-\cos(0)\right]=3+\frac{2}{\pi}\approx 3.637$$

2. Average value $=\frac{1}{3}\int_0^3\left[1+\sin\left(\frac{\pi x}{2}\right)\right]dx\approx\frac{1}{3}(3.637)\approx 1.212\left(\text{Or }1+\frac{2}{3\pi}\right)$

3. Volume $=\pi\int_0^3\left[1+\sin\left(\frac{\pi x}{2}\right)\right]^2dx$ or $\pi\int_0^3\left[1+2\sin\left(\frac{\pi x}{2}\right)+\sin^2\left(\frac{\pi x}{2}\right)\right]dx$

Day 166

1. $A=\int_0^2\left(5-x^2-\frac{x}{2}\right)dx=\left[5x-\frac{x^3}{3}-\frac{x^2}{4}\right]_0^2=10-\frac{8}{3}-1=\frac{19}{3}$

2. $y=5-x^2\Rightarrow x=\sqrt{5-y},\ y=\frac{x}{2}\Rightarrow x=2y.$

$$V=\pi\int_0^1(2y)^2dy+\pi\int_1^5\left(\sqrt{5-y}\right)^2dy$$

$$=4\pi\int_0^1 y^2\,dy+\pi\int_1^5(5-y)\,dy=4\pi\cdot\frac{y^3}{3}\Big|_0^1+\pi\left(5y-\frac{y^2}{2}\right)\Big|_1^5$$

$$=\frac{4\pi}{3}+\pi\left[\left(25-\frac{25}{2}\right)-\left(5-\frac{1}{2}\right)\right]=\frac{4\pi}{3}+8\pi=\frac{28\pi}{3}\approx 29.322$$

Day 167

1. $f'(x)=\cos x-x\sin x,\ f'\left(\frac{\pi}{2}\right)=\cos\frac{\pi}{2}-\frac{\pi}{2}\sin\frac{\pi}{2}=0-\frac{\pi}{2}(1)=-\frac{\pi}{2}$

2. $f\left(\frac{\pi}{2}\right)=\frac{\pi}{2}\cos\left(\frac{\pi}{2}\right)=0$, tangent line $y-0=-\frac{\pi}{2}\left(x-\frac{\pi}{2}\right)$ or $y=-\frac{\pi}{2}x+\frac{\pi^2}{4}$

3. $A=\int_{\pi/2}^{3\pi/2}(x\cos x)\,dx=x\sin x\Big|_{\pi/2}^{3\pi/2}-\int_{\pi/2}^{3\pi/2}\sin x\,dx=[x\sin x+\cos x]\Big|_{\pi/2}^{3\pi/2}$

$$=\left(\frac{3\pi}{2}\sin\frac{3\pi}{2}+\cos\frac{3\pi}{2}\right)-\left(\frac{\pi}{2}\sin\frac{\pi}{2}+\cos\frac{\pi}{2}\right)$$

$$=\left(-\frac{3\pi}{2}\right)-\left(\frac{\pi}{2}\right)=-2\pi$$

Day 168

1. Acceleration is the change in velocity over time, or the slope of the velocity curve. The acceleration is positive for $0 \le t < 0.5$ and $0.5 < t < 1.5$.

2. $$\frac{1}{3-0}\int_0^3 v(t)\,dt \approx \frac{1}{3}\left[\frac{1}{2}\cdot\frac{1}{2}(30)+\frac{1}{2}(1)(30+50)+(1)(50)+\frac{1}{2}\cdot\frac{1}{2}(50+40)\right]$$
$$\approx \frac{1}{3}[7.5+40+50+22.5] \approx \frac{1}{3}[120] \approx 40$$

Day 169

1. $$m = \left.\frac{3x^2-x-2}{4y}\right|_{(2,1)} = \frac{3(4)-2-2}{4(1)} = 2$$

2. The tangent line at (2, 1) is $y-1=2(x-2)$ or $y=2x-3$. $f(2.1)\approx 2(2.1)-3 \approx 1.2$.

3. $\frac{dy}{dx}=\frac{3x^2-x-2}{4y}$ becomes $\int 4y\,dy=\int (3x^2-x-2)\,dx$, which integrates to $2y^2=x^3-\frac{x^2}{2}-2x+C$. With the given initial condition, $2(1^2)=2^3-\frac{2^2}{2}-2(2)+C$ means that $C=0$. The function is $2y^2=x^3-\frac{x^2}{2}-2x$ or $f(x)=\sqrt{\frac{x^3}{2}-\frac{x^2}{4}-x}$ or $f(x)=\frac{1}{2}\sqrt{2x^3-x^2-4x}$.

Day 170

1. FALSE. $f'(x)$ is positive on the interval (0, 2), which means $f(x)$ is increasing.
2. TRUE. In the interval (–2, –1), $f'(x)=0$ and $f(x)$ changes from decreasing to increasing.
3. FALSE. On the interval (0, 1), $f'(x)$ is decreasing, so $f''(x)$ is negative, indicating that $f(x)$ is concave downward.
4. TRUE. $f''(x)$ is equal to zero twice, at the two turning points of $f'(x)$.
5. TRUE. $f'(x)$ is equal to zero only once.

Day 171

1. $y^3+x^2y+x^2+4y=4$

$$3y^2\frac{dy}{dx}+x^2\frac{dy}{dx}+2xy+2x+4\frac{dy}{dx}=0$$

$$3y^2\frac{dy}{dx}+x^2\frac{dy}{dx}+4\frac{dy}{dx}=-2xy-2x$$

$$\frac{dy}{dx}=\frac{-2xy-2x}{3y^2+x^2+4}$$

2. $\frac{dy}{dx}=\frac{-2xy-2x}{3y^2+x^2+4}=0$ when $-2xy-2x=0$. $-2x(y+1)=0$ when $x=0$ or $y=-1$.

If $x=0$, $y^3+4y-4=0$ and $y\approx 0.848$. If $y=-1$, $-1-x^2+x^2-4=4$, which has no solution. Equation of the horizontal tangent is $y=0.848$.

3. The tangent line is $y=-\frac{1}{2}x+1$ and the point of tangency is (k, 0).

$0=-\frac{1}{2}k+1\Rightarrow k=2$. The point of tangency is (2, 0).

Day 172

1. $s(1)=3\cos(0)=3$. The position of the particle at $t=1$ is (3, 0).
2. $v(t)=s'(t)=-3\sin(1-t^2)(-2t)=6t\sin(1-t^2)$
3. $v(t)=6t\sin(1-t^2)=0$ when $t=0$ or when $\sin(1-t^2)=0$, which first occurs when $t=1$. And $v(0.9)>0$ and $v(1.1)<0$. The first change of direction occurs when $t=1$.
4. Initial position $s(0)=3\cos(1)\approx 1.621$. Final position $s(5)=3\cos(-24)\approx 1.273$. The particle moves between extremes of 3 and −3, changing direction whenever $\cos(1-t^2)=\pm 1$ or $1-t^2=n\pi$, where $0\le n\le\frac{24}{\pi}$. Total distance = (3−1.621) + 7(6) + (1.273 + 3) = 1.379 + 42 + 4.273 = 47.652. Alternatively, using the TI-89 graphing calculator, enter $\int_0^5|v(t)|\,dt$ which is $\int_0^5|6t\sin(1-t^2)|\,dt$ and obtain 47.652.

Day 173

1. R is a triangle with a base of 8 and a height of 4. $A=\frac{1}{2}bh=\frac{1}{2}(8)(4)=16$.
2. Note that when $0\le x\le 4$, $(x-4)\le 0$ and $|x-4|=-(x-4)=-x+4$. Thus $4-|x-4|=4-(-x+4)=x$.

Volume $=\pi\int_0^8(4-|x-4|)^2\,dx=2\pi\int_0^4 x^2\,dx=2\pi\left[\frac{x^3}{3}\right]_0^4=\frac{128\pi}{3}$.

Day 174

1. $g'(x) = f(x) + xf'(x) + f''(x)$, so $g'(-3) = 34 + (-3)(-17) + 2 = 87$

2. $h'(x) = \dfrac{(f'(x))^2 - f(x)f''(x)}{(f'(x))^2}$ so $h'(2) = \dfrac{(-1)^2 - (-12)(2)}{(-1)^2} = 25$

Day 175

1. Point of tangency is (0, 2). The slope of the tangent line at (0, 2) is –3, so the equation of the tangent line is $y - 2 = -3\,(x - 0)$ or $y = 2 - 3x$.

2. $f(0.1) \approx 2 - 3\,(0.1) \approx 1.7$

3. If $f''(0) = 1$, the graph of $f(x)$ is concave upward at $x = 0$. The tangent line will, therefore, be an underestimate.

Day 176

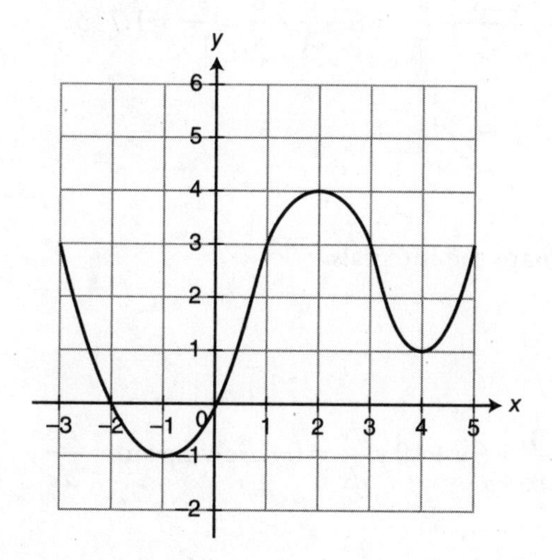

Day 177

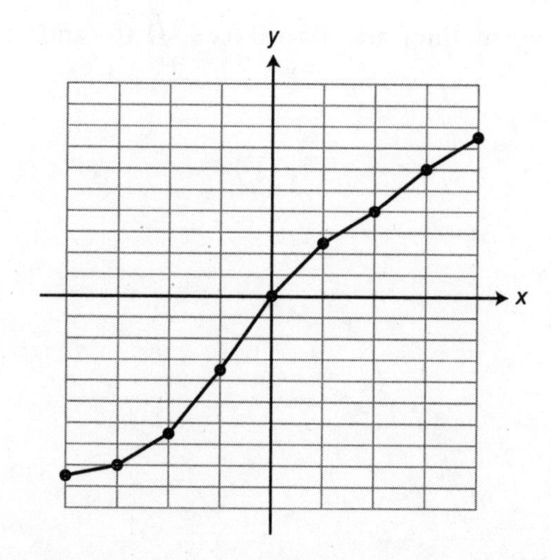

Answers

Day 178

1. Approximately 70 cars pass through in the first 10 minutes.

$$\int_0^{10} F(t)dt = \int_0^{10}\left[4+5\sin\left(\frac{t}{3}\right)\right]dt = \left[4t-15\cos\left(\frac{t}{3}\right)\right]_0^{10}$$

$$=\left[\left(40-15\cos\left(\frac{10}{3}\right)\right)-(-15\cos 0)\right]$$

$$=40+14.725+15=69.725\approx 70$$

2. When $t=15$, $F(t)=25+\frac{(t-19)^2}{10}$. $F'(t)=\frac{(t-19)}{5}$, so $F'(15)=-\frac{4}{5}$. Traffic flow is decreasing.

Day 179

1. $A=\int_0^2\left(3x-x^2-\frac{x\sqrt{2x}}{2}\right)dx=\left[\frac{3x^2}{2}-\frac{x^3}{3}-\frac{\sqrt{2}x^{5/2}}{5}\right]_0^2=6-\frac{8}{3}-\frac{8}{5}=\frac{26}{15}\approx 1.733$

2. $V=\int_0^2\left(3x-x^2-\frac{x\sqrt{2x}}{2}\right)^2 dx\approx 1.797.$

Note: use your graphing calculator to evaluate the integrals.

Day 180

1. If $5x^2+6xy+5y^2=8$, then $10x+6x\frac{dy}{dx}+6y+10y\frac{dy}{dx}=0$. Solving for $\frac{dy}{dx}$,

$\frac{dy}{dx}=-\frac{10x+6y}{6x+10y}=-\frac{5x+3y}{3x+5y}$.

2. The slope of the tangent when $x = 0$ is $\frac{dy}{dx}=-\frac{3y}{5y}=-0.6$. When $x=0, 5y^2=8$ and $y=\pm\sqrt{1.6}$. The equations of the tangent lines are $y=-0.6x+\sqrt{1.6}$ and $y=-0.6x-\sqrt{1.6}$.

APPENDIX

1. Quadratic Formula:

   $ax^2 + bx + c = 0 \ (a \neq 0)$

   $$x = \frac{-b \pm \sqrt{b^2 - 4ac}}{2a}$$

2. Distance Formula:

   $$d = \sqrt{(x_2 - x_1)^2 + (y_2 - y_1)^2}$$

3. Equation of a Circle:

   $x^2 + y^2 = r^2$ center at (0, 0) and radius $= r$.

4. Equation of an Ellipse:

   $\frac{x^2}{a^2} + \frac{y^2}{b^2} = 1$ center at (0, 0).

   $\frac{(x-h)^2}{a^2} + \frac{(y-k)^2}{b^2} = 1$ center at (h, k).

5. Area and Volume Formulas:

FIGURE	AREA FORMULA
Trapezoid	$\frac{1}{2}[\text{base}_1 + \text{base}_2]$ (height)
Parallelogram	(base)(height)
Equilateral triangle	$\frac{s^2\sqrt{3}}{4}$
Circle	πr^2 (circumference $= 2\pi r$)

SOLID	VOLUME	SURFACE AREA (S.A.)
Sphere	$\frac{4}{3}\pi r^3$	$4\pi r^2$
Right circular cylinder	$\pi r^2 h$	Lateral S.A.: $2\pi rh$ Total S.A.: $2\pi rh + 2\pi r^2$
Right circular cone	$\frac{1}{3}\pi r^2 h$	Lateral S.A.: $\pi r\sqrt{r^2 + h^2}$ Total S.A.: $\pi r^2 + \pi r\sqrt{r^2 + h^2}$

6. Special Angles:

ANGLE FUNCTION	0°	$\pi/6$ 30°	$\pi/4$ 45°	$\pi/3$ 60°	$\pi/2$ 90°	π 180°	$3\pi/2$ 270°	2π 360°
Sin	0	$1/2$	$\sqrt{2}/2$	$\sqrt{3}/2$	1	0	-1	0
Cos	1	$\sqrt{3}/2$	$\sqrt{2}/2$	$1/2$	0	-1	0	1
Tan	0	$\sqrt{3}/3$	1	$\sqrt{3}$	Undefined	0	Undefined	0

7. Double Angles:
   - $\sin 2\theta = 2\sin\theta\cos\theta$
   - $\cos 2\theta = \cos^2\theta - \sin^2\theta$ or $1 - 2\sin^2\theta$ or $2\cos^2\theta - 1$.
   - $\cos^2\theta = \dfrac{1+\cos 2\theta}{2}$
   - $\sin^2\theta = \dfrac{1-\cos 2\theta}{2}$
8. Pythagorean Identities:
   - $\sin^2\theta + \cos^2\theta = 1$
   - $1 + \tan^2\theta = \sec^2\theta$
   - $1 + \cot^2\theta = \csc^2\theta$
9. Limits:

$$\lim_{x\to\infty}\frac{1}{x} = 0 \qquad \lim_{x\to 0}\frac{\cos x - 1}{x} = 0$$

$$\lim_{x\to 0}\frac{\sin x}{x} = 1 \qquad \lim_{h\to\infty}\left(1+\frac{1}{h}\right)^h = e$$

$$\lim_{h\to 0}\frac{e^h - 1}{h} = 1 \qquad \lim_{x\to 0}(1+x)^{\frac{1}{x}} = e$$

10. *L'Hôpital's* Rule for Indeterminate Forms: Let *lim* represent one of the limits $\lim_{x\to c}$, $\lim_{x\to c^+}$, $\lim_{x\to c^-}$, $\lim_{x\to\infty}$, or $\lim_{x\to -\infty}$. Suppose $f(x)$ and $g(x)$ are differentiable and $g'(x) \neq 0$ near c, except possibly at c, and suppose $\lim f(x) = 0$ and $\lim g(x) = 0$. Then the $\lim \dfrac{f(x)}{g(x)}$ is an indeterminate form of the type $\dfrac{0}{0}$. Also, if $\lim f(x) = \pm\infty$ and $\lim g(x) = \pm\infty$, then the $\lim \dfrac{f(x)}{g(x)}$ is an indeterminate form of the type $\dfrac{\infty}{\infty}$. In both cases, $\dfrac{0}{0}$ and $\dfrac{\infty}{\infty}$, *L'Hôpital's* Rule states that $\lim \dfrac{f(x)}{g(x)} = \lim \dfrac{f'(x)}{g'(x)}$.
11. Rules of Differentiation:

   a. Definition of the Derivative of a Function:

$$f'(x) = \lim_{h\to 0}\frac{f(x+h) - f(x)}{h}$$

   b. Power Rule: $\dfrac{d}{dx}(x^n) = nx^{n-1}$

   c. Sum and Difference Rules:

$$\frac{d}{dx}(u \pm v) = \frac{du}{dx} \pm \frac{dv}{dx}$$

   d. Product Rule:

$$\frac{d}{dx}(uv) = v\frac{du}{dx} + u\frac{dv}{dx}$$

   e. Quotient Rule:

$$\frac{d}{dx}\left(\frac{u}{v}\right) = \frac{v\dfrac{du}{dx} - u\dfrac{dv}{dx}}{v^2},\ v \neq 0$$

   Summary of Sum, Difference, Product, and Quotient Rules:

$$(u \pm v)' = u' \pm v' \qquad (uv)' = u'v + v'u$$

$$\left(\frac{u}{v}\right)' = \frac{u'v - v'u}{v^2}$$

   f. Chain Rule:

$$\frac{d}{dx}[f(g(x))] = f'(g(x)) \cdot g'(x)$$

   or $\dfrac{dy}{dx} = \dfrac{dy}{du} \cdot \dfrac{du}{dx}$

12. Inverse Function and Derivatives:

$$(f^{-1})'(x) = \frac{1}{f'(f^{-1}(x))} \quad \text{or} \quad \frac{dy}{dx} = \frac{1}{dx/dy}$$

13. Differentiation and Integration Formulas, Integration Rules:

   a. $\int f(x)dx = F(x) + C \Rightarrow F'(x) = f(x)$

   b. $\int af(x)dx = a\int f(x)dx$

   c. $\int -f(x)dx = -\int f(x)dx$

d. $\int [f(x) \pm g(x)]\, dx$

$= \int f(x)dx \pm \int g(x)dx$

Differentiation Formulas:

a. $\frac{d}{dx}(x) = 1$

b. $\frac{d}{dx}(ax) = a$

c. $\frac{d}{dx}(x^n) = nx^{n-1}$

d. $\frac{d}{dx}(\cos x) = -\sin x$

e. $\frac{d}{dx}(\sin x) = \cos x$

f. $\frac{d}{dx}(\tan x) = \sec^2 x$

g. $\frac{d}{dx}(\cot x) = -\csc^2 x$

h. $\frac{d}{dx}(\sec x) = \sec x \tan x$

i. $\frac{d}{dx}(\csc x) = -\csc x \cot x$

j. $\frac{d}{dx}(\ln x) = \frac{1}{x}$

k. $\frac{d}{dx}(e^x) = e^x$

l. $\frac{d}{dx}(a^x) = (\ln a)\, a^x$

m. $\frac{d}{dx}(\sin^{-1} x) = \frac{1}{\sqrt{1-x^2}}$

n. $\frac{d}{dx}\left(\tan^{-1} x\right) = \frac{1}{1+x^2}$

o. $\frac{d}{dx}\left(\sec^{-1} x\right) = \frac{1}{|x|\sqrt{x^2-1}}$

Integration Formulas:

a. $\int 1dx = x + C$

b. $\int a dx = ax + C$

c. $\int x^n dx = \frac{x^{n+1}}{n+1} + C,\ n \neq -1$

d. $\int \sin x dx = -\cos x + C$

e. $\int \cos x dx = \sin x + C$

f. $\int \sec^2 x dx = \tan x + C$

g. $\int \csc^2 x dx = -\cot x + C$

h. $\int \sec x\ (\tan x)\, dx = \sec x + C$

i. $\int \csc x\ (\cot x)\ dx = -\csc x + C$

j. $\int \frac{1}{x} dx = \ln|x| + C$

k. $\int e^x dx = e^x + C$

l. $\int a^x dx = \frac{a^x}{\ln a} + C\ \ a > 0,\ a \neq 1$

m. $\int \frac{1}{\sqrt{1-x^2}} dx = \sin^{-1} x + C$

n. $\int \frac{1}{1+x^2} dx = \tan^{-1} x + C$

o. $\int \frac{1}{|x|\sqrt{x^2-1}} dx = \sec^{-1} x + C$

More Integration Formulas:

a. $\int \tan x dx = \ln|\sec x| + C$ or $-\ln|\cos x| + C$

b. $\int \cot x dx = \ln|\sin x| + C$ or $-\ln|\csc x| + C$

c. $\int \sec x dx = \ln|\sec x + \tan x| + C$

d. $\int \csc x dx = \ln|\csc x - \cot x| + C$

e. $\int \ln x dx = x \ln|x| - x + C$

f. $\int \frac{1}{\sqrt{a^2 - x^2}} dx = \sin^{-1}\left(\frac{x}{a}\right) + C$

g. $\int \frac{1}{a^2 + x^2} dx = \frac{1}{a} \tan^{-1}\left(\frac{x}{a}\right) + C$

h. $\int \frac{1}{x\sqrt{x^2 - a^2}} dx = \frac{1}{a} \sec^{-1}\left|\frac{x}{a}\right| + C$ or

$\frac{1}{a} \cos^{-1}\left|\frac{a}{x}\right| + C$

i. $\int \sin^2 x dx = \frac{x}{2} - \frac{\sin(2x)}{4} + C$

Note that $\sin^2 x = \frac{1 - \cos 2x}{2}$ and

$\cos^2 x = \frac{1 + \cos(2x)}{2}$

Note that after evaluating an integral, always check the result by taking the derivative of the answer (i.e., taking the derivative of the antiderivative).

14. Intergration by parts $\int u dv = uv - \int v du$ (and follow LIPET Rule).

15. The Fundamental Theorem of Calculus

$$\int_a^b f(x)dx = F(b) - F(a),$$

where $F'(x) = f(x)$.

If $F(x) = \int_a^x f(t)dt$, then $F'(x) = f(x)$.

16. Trapezoidal Approximation:

$$\int_a^b f(x)dx = \frac{b-a}{2n}\left[\begin{array}{l} f(x_0) + 2f(x_1) + 2f(x_2) \ldots \\ +2f(x_{n-1}) + f(x_n) \end{array}\right]$$

17. Average Value of a Function:

$$f(c) = \frac{1}{b-a} \int_a^b f(x)dx$$

18. Mean Value Theorem:

$f'(c) = \frac{f(b) - f(a)}{b - a}$ for some c in (a, b).

Mean Value Theorem for Integrals:

$\int_a^b f(x)\, dx = f(c)\,(b - a)$ for some c in (a, b).

19. Area Bounded by 2 Curves:

$$\text{Area} = \int_{x_1}^{x_2} (f(x) - g(x))dx,$$

where $f(x) \geq g(x)$.

20. Volume of a Solid with Known Cross Section:

$$V = \int_a^b A(x)dx,$$

where $A(x)$ is the cross section.

21. Disc Method:

$V = \pi \int_a^b (f(x))^2\, dx$, where $f(x) = \text{radius}$.

22. Using the Washer Method:

$$V = \pi \int_a^b \left((f(x))^2 - (g(x))^2\right) dx,$$

where $f(x) = \text{outer radius}$ and $g(x) = \text{inner radius}$.

23. Distance Traveled Formulas:

- Position Function: $s(t)$; $s(t) = \int v(t)dt$
- Velocity: $v(t) = \frac{ds}{dt}$; $v(t) = \int a(t)dt$
- Acceleration: $a(t) = \frac{dv}{dt}$
- Speed: $|v(t)|$
- Displacement from t_1 to $t_2 = \int_{t_1}^{t_2} v(t)$ $= s(t_2) - s(t_1)$.

- Total Distance Traveled from t_1 to

$$t_2 = \int_{t_1}^{t_2} |v(t)|dt.$$

24. Business Formulas:

Profit = Revenue − Cost	$P(x) = R(x) - C(x)$
Revenue = (price) (items sold)	$R(x) = px$
Marginal Profit	$P'(x)$
Marginal Revenue	$R'(x)$
Marginal Cost	$C'(x)$

$P'(x)$, $R'(x)$, $C'(x)$ are the instantaneous rates of change of profit, revenue, and cost, respectively.

25. Exponential Growth/Decay Formulas:

$\frac{dy}{dt} = ky,\ y > 0$ and $y(t) = y_0 e^{kt}$.

BIBLIOGRAPHY

Advanced Placement Program Course Description. New York: The College Board, 2010.

Anton, H., Bivens, I. & Davis, S. *Calculus*, 7th edition. New York: John Wiley & Sons, 2001.

Apostol, Tom M. *Calculus*. Waltham, MA: Blaisdell Publishing Company, 1967.

Berlinski, David. *A Tour of the Calculus*. Colorado Springs: Vintage, 1997.

Boyer, Carl B. *The History of the Calculus and Its Conceptual Development*. New York: Dover, 1959.

Finney, R., Demana, F. D., Waits, B. K. & Kennedy, D. *Calculus Graphical, Numerical, Algebraic*, 3rd edition. Boston: Pearson Prentice Hall, 2002.

Larson, R. E., Hostetler, R. P. & Edwards, B. H. *Calculus*, 8th edition. New York: Brooks Cole, 2005.

Leithold, Louis. *The Calculus with Analytic Geometry*, 5th edition. New York: Longman Higher Education, 1986.

Sawyer, W. W. *What Is Calculus About*? Washington, DC: Mathematical Association of America, 1961.

Spivak, Michael. *Calculus*, 4th edition. New York: Publish or Perish, 2008.

Stewart, James. *Calculus*, 4th edition. New York: Brooks/Cole Publishing Company, 1999.

NOTES

NOTES

NOTES

NOTES

NOTES

NOTES

NOTES

NOTES

NOTES

NOTES

NOTES

NOTES

5 Steps to Teaching AP Calculus AB

Emily Pillar

AP Calculus teacher at
Schreiber High School, in Port Washington, New York

Thanks to Greg Jacobs, an AP Physics teacher at Woodberry Forest School in Virginia, for developing the 5-step approach used in this teaching manual. Thanks also to Courtney Mayer, an AP Environmental Science teacher, for creating a sample teacher's manual that AP teachers could use to create their own manual.

Introduction to the Teacher's Manual

Nowadays, teachers have no shortage of resources for the AP Calculus AB class. Classes are no longer limited to just the teacher and the textbook; today's teachers can utilize online simulations, apps, computer-based homework, video lectures, and so on. Even the College Board itself provides so much material related to the AP Calculus AB exam that the typical teacher—and student—can easily become overwhelmed by an excess of teaching materials and resources.

One vital resource for you and your class is this book. It explains in straightforward language exactly what a student needs to know for the AP Calculus AB exam. It also provides a complete review for the test, including explanatory materials, questions to check student understanding, and test-like practice exams.

This teacher's manual will take you through the five steps of teaching AP Calculus AB. These five steps are:

1. **Prepare a strategic plan for the course**
2. **Hold an interesting class every day**
3. **Evaluate your students' progress**
4. **Get students ready to take the AP exam**
5. **Become a better teacher every year**

I'll discuss each of these steps, providing suggestions and ideas of things that I use in my class. I present them here because over the years I found that *they work*. There are many different course strategies, teaching activities, and evaluation techniques that help students succeed; each teacher must find what works in their classroom. I hope you find in this teacher's manual something that will be useful to you.

STEP 1

Prepare a Strategic Plan for the Course

The Course and Exam Description (CED) from the College Board, which can be found at **https://apcentral.collegeboard.org/courses/ap-calculus-ab/course**, lays out a suggested scope and sequence for the AP Calculus AB class. The College Board has set it up in a way that topics and skills build as the year goes on. Especially if you are new to the course, I recommend that you follow that scope and sequence.

The chart below shows the units and the time suggested for each unit. The number of class periods is based on a typical 45-minute class. If your school is on a form of block schedule or other atypical schedule, you will need to adjust the pacing to fit your class needs.

TOPICS	PACING	5 STEPS TO A 5
Unit 1: Limits and Continuity	22–23 class periods	Chapter 6, pp. 84–108
Unit 2: Differentiation: Definition and Fundamental Properties	13–14 class periods	Chapter 7, pp. 109–115, 116–118, 119–121
Unit 3: Differentiation: Composite, Implicit, and Inverse Functions	10–11 class periods	Chapter 7, pp. 115–116, 118–119, 121–136
Unit 4: Contextual Applications of Differentiation	10–11 class periods	Chapter 9, pp. 177–182, Chapter 10, pp. 202–206
Unit 5: Analytical Applications of Differentiation	15–16 class periods	Chapter 8, pp. 137–174, Chapter 9, pp. 183–188
Unit 6: Integration and Accumulation of Change	18–20 class periods	Chapter 11–13, pp. 227–280
Unit 7: Differential Equations	8–9 class periods	Chapter 14, pp. 325–334
Unit 8: Applications of Integration	19–20 class periods	Chapter 13–14, pp. 280–324

After you have taught the course a few times and feel comfortable with the material, you may want to move topics and units around to better meet your classroom needs. While the overall units should be primarily taught in the sequence outlined by the College Board, there is some flexibility with Units 6, 7, and 8, as well as some subtopics in Units 4 and 5.

For example, there is flexibility in the order in which you teach derivative rules; below is the order that I have found works for me. It allows me to have students derive derivative rules based on prior knowledge from this sequence. The order is as follows:

Derivatives:

- **Power Rule**
- **Constant, sum, difference, constant multiple**
- **Product Rule**
- **Quotient Rule**
- **High-order derivatives**
- **Trigonometry derivatives**
- **Chain Rule**
- **Exponential and logarithmic derivatives**
- **Implicit differentiation**
- **Differentiating inverse functions**
- **Differentiating inverse trigonometry functions**

After derivative rules, I teach applications of derivatives, including particle motion and related rates. I suggest waiting to teach linearization until after Unit 5 so students can also analyze if the approximation is an over- or underestimation.

At the end of the year there is also flexibility in integrals. Unit 7 is a shorter, stand-alone unit, so you can choose to teach it after Unit 8. I have found that within Unit 8 (especially 8.2 and 8.3), there are some great areas for AP review and practice. There are plenty of past free-response questions (FRQs) that can serve as a segue into AP review.

As you plan your year, make sure to leave plenty of time for review. It is important to give students practice with actual AP exam questions and call upon them to synthesize material from the entire year. I like to leave two to three weeks of dedicated review time just before the test.

STEP 2

Hold an Interesting Class Every Day

AP Calculus students should love coming to your class. Why? Because you should offer many opportunities and strategies to help students understand the material and internalize what they learn. I follow the same schedule daily but with different activities to keep it interesting.

- **Challenge question.** Each day in class students walk in to a "Challenge Question" on the board. They are each given a very small slip of paper on which they must write their solution. I collect this paper and take a quick look at each to get a fast assessment of where students stand on a particular topic that will feed into that day's lesson. While this activity takes only five minutes, it allows me the opportunity to focus on a precalculus, algebra, or calculus skill required for that lesson. If you have the Elite Edition of the *5 Steps to a 5: Calculus AB*, the "5 Minutes to a 5" section at the end provides great challenge questions. The breakdown of the questions by unit is provided later in this manual.
- **Homework presentation.** I assign homework each night to students, which includes a short problem set focusing on material from that lesson; it can include a worksheet, problems from the *5 Steps to a 5*, or FRQs from the AP Classroom. I pick three of the most important or challenging problems and assign them to three different students. The next day in class, these students are asked to present their solutions to the class. They are responsible for answering questions from their peers regarding their work. Being asked to present and explain to their peers affords students the opportunity to get another level of understanding of the material.
- **Flipped classroom.** A flipped classroom can be a great way to change things up. There are many excellent videos available on the web, including those from the Khan Academy and AP Classroom. You can assign a video that teaches the topic and then have students work on practice problems in class. This allows you to spend class time correcting any misconceptions and assisting students with the more challenging AP problems.

- **Creative problem sets.** There are various ways to shake up problem sets for class time. You can try these different ways of presenting and completing problems, including:

  - **Calculus maze:** Students answer a question and must search for the answer to find which problem to complete next.

Ans: $-3\sin 3x$ # <u>1</u> Find $f''(x)$. $f(x)=3x^3+2x^2+x$	Ans: $-\dfrac{\sin x + y}{x}$ # ______ Find $f'(x)$. $f(x)=\dfrac{4}{x^3}$
Ans: $-\dfrac{12}{x^4}$ # ______ Find $\dfrac{dy}{dx}$. $y=\cos 3x$	Ans: $18x+4$ # ______ Find $\dfrac{dy}{dx}$. $xy=\cos x$

  - **I Have, Who Has? game:** A class activity in which each student has a function and a derivative, for example. The first person reads their function out loud, and the person with that function's derivative says the derivative out loud and then asks for the derivative of their function. This continues until you return to the first person.

STUDENT A EXAMPLE:	STUDENT B EXAMPLE:
I have: $\dfrac{2x}{x^2+7}$ Who has? $\dfrac{d}{dx}(e^{x^2})$	I have: $2xe^{x^2}$ Who has? $\dfrac{d}{dx}(\cos(3x^2-4))$

  - **Pair and share activity:** Students are grouped in pairs. Within the pair one student completes column A and one student column B of a worksheet. Each column, while providing different questions, has the same answers for corresponding questions. After completing the problems, pairs work together to discuss and check answers.

COLUMN A	COLUMN B
Evaluate the definite integral. $\displaystyle\int_{-1}^{3}(-x^3+3x^2+1)dx$	Evaluate the definite integral. $\displaystyle\int_{-1}^{0}12x^2(3x^3+3)^2\,dx$

The rigor of the AP exam can be of a different level than prior math classes. Having students work together, whether in groups or partners, will help them capitalize on their strengths. I have found that at this level students really can find their strengths, be it algebra skills, applying the rules of calculus, or understanding modeling/applied problems. Thoughtfully grouping students and having them work together allows each student to bring something to the table, as this course is a synthesis of material from their entire high school career.

If you have a classroom set of the *5 Steps to a 5: AP Calculus AB*, you can assign your students homework of reading a few pages of the book that correlates to the topics you are teaching next. You can also assign them the review questions at the end of each chapter as homework. The book covers each unit and topic in the CED, and all the review questions are aligned to the math practices. This book is also available online; follow the instructions on the back cover of this book to access the Cross-Platform Edition.

Additionally, *5 Steps to a 5: AP Calculus AB* provides cumulative review problems at the end of each section. You can assign these problems for homework occasionally to blend prior material. The book provides solutions to these problems, so you can efficiently have students check homework prior to class and come in only with questions.

5 Steps to a 5: AP Calculus AB Elite Edition: The Elite Edition of this book provides additional questions that can be used in your class. It contains 180 activities and questions that require five minutes a day. While they are primarily intended to be used by students studying for the test, you can use these as daily warm-ups in your course. To do this, you will need the table below that organizes these questions and activities by unit, since they do not follow the course in chronological order.

UNIT	QUESTIONS/ACTIVITIES IN THE ELITE EDITION
Precalculus Review	1–8, 11–13, 15, 137
Unit 1: Limits and Continuity	9–10, 14, 16–29
Unit 2: Differentiation: Definition and Basic Derivative Rules	30–34, 145, 148
Unit 3: Differentiation: Composite, Implicit, and Inverse Functions	35–44, 63, 64, 136, 139, 149, 154, 162, 171, 174, 180
Unit 4: Contextual Applications of Differentiation	56–58, 62, 65–72, 135, 156, 175
Unit 5: Analytical Applications of Differentiation	45–55, 59–61, 73, 89, 126–128, 143, 150, 176
Unit 6: Integration and Accumulation of Change	74–88, 90, 92–103, 105–107, 132, 141, 142, 155, 164, 170, 177
Unit 7: Differential Equations	119–121, 160, 169
Unit 8: Applications of Integration	91, 104, 108–118, 122–125, 129–131, 133, 134, 138, 140, 144, 146, 147, 151–153, 157–159, 161, 163, 165–168, 172, 173, 178, 179

STEP 3

Evaluate Your Students' Progress

I assess progress on the majority of the AP topics with larger unit exams. I recommend incorporating released questions from the College Board (found on AP Classroom) on these exams. While you may incorporate questions from other sources such as your textbook or from *5 Steps to a 5: AP Calculus AB*, it is important to expose students to the rigor and style of AP questions throughout the year. You may select questions on AP Classroom by filtering by unit, subtopic, question type (multiple choice or free response), and calculator/noncalculator.

I use a variety of multiple-choice AP questions on my exam and adapt some free-response questions (FRQs) to include as well. FRQs may be difficult to include in the fall semester, as many questions include topics from various units, but once I get to derivative applications, I have found these questions to be appropriate. Whether they are AP level or not, I make sure that each exam has both multiple-choice and FRQs.

If I include an FRQ on my exam, I allow students to leave their answers in nonsimplified form to practice for the AP exam. When students do calculator-active FRQs, they must have the answer correct to three decimal places, like on the AP exam. I encourage students to truncate answers at four decimal places so they don't make a rounding error. Let students focus on the calculus and eliminate any chance for an algebra mistake!

On the first two units, students are not allowed to use a calculator. Remember, the AP exam is two-thirds noncalculator, so it is important for students to limit dependency on their calculators. Once I introduce calculator-active topics and questions, in Unit 3: Derivative Applications, I give both calculator and noncalculator sections for exams. Students are given both sections at the beginning of the exam and are only able to take out their calculators once they hand in the noncalculator section. Printing one of the sections on colored paper allows for an easier monitoring process.

AP Classroom is a valuable resource, allowing to students to practice with released AP questions from previous exams. Just before the unit exam, I give an AP Classroom assignment of multiple-choice questions that covers all the topics for that exam. These assignments are due the day of the test and graded on a curve, as the AP exam will be. The assignments give students a chance to practice with AP-style questions and rigor, as well as to review for the upcoming exam.

#FlashBackFriday. On selected Fridays I give students a #FBF quiz. This is a short quiz focused on algebra and precalculus material that is essential for success on the AP exam. I review the topic for five to ten minutes in a previous class or give students notes and/or example problems on the topic. Each quiz is no more than ten minutes and forces students to refresh these imperative skills.

Some topics for the quiz include:

- **Factoring**
- **Simplifying rational functions**
- **Domain and range**
- **Graphs of natural log and exponential functions**
- **The unit circle and exact trigonometric values**
- **Reciprocal trigonometric functions**
- **Inverse trigonometric functions**
- **Evaluating and manipulating functions with fractional and negative exponents**
- **Long division with polynomials**

STEP 4

Get Students Ready to Take the AP Exam

I leave two or three weeks to dedicate to reviewing for the exam in class. By this point, we have covered all units in the curriculum and are taking time just to synthesize the material and practice with exam questions. The best way for students to succeed on the AP exam is getting exposure to real AP questions and to *practice, practice, practice.* I consider this last unit to be the most important unit of the year. I love to get students to buy in at the end of their senior year with a team challenge I call "Review-O-Mania."

For my review unit, I place students in heterogeneous groups of three (or four) students. The teams cooperatively compete against one another. Each team earns points each class, as delineated at the end of the manual. Each member of the first-place team wins all the pride and glory that comes with being the Review-O-Mania champion, points on their final, and gets lunch with me! The challenge is outlined in an introduction sheet I give my students; this handout is attached at the end of this teacher's manual.

I give students a schedule in advance with all assignments and due dates listed. Each assignment is typically 15 to 25 multiple-choice questions or 3 to 4 FRQs.

I have found that offering students the opportunity to grade their own work or their peer's work on FRQs is very valuable. I have done this in various ways; one way is to give a group quiz of one AP FRQ in class. I then have groups exchange papers and have them grade their peers based on the AP rubric provided by the College Board. Having students work with the rubric allows them to become familiar with how they can earn points on the exam.

I provide my students with a list of tips for succeeding on the AP exam during the review unit. A sample of this list is also provided at the end of this teacher's manual.

Schedule

ASSIGNMENT #	ASSIGNMENT	DUE DATE	COMPLETED?	QUESTIONS?

STEP 5

Become a Better Teacher Every Year

A good AP teacher tries to do better, regardless of how they measure success. If there is anything that didn't work as well as you had hoped this year, there's always next year to try something different. The message is the same whether you are a novice AP teacher or a veteran: your goal is to become a better teacher every year.

My advice for a teacher new to AP Calculus would be to focus on the content and let go of the nuances of the test. It is challenging to teach a rigorous course for the first time, sprinkle in interesting activities, *and* prepare for the "tricks" of the AP exam. There are a ton of resources out there, but focus on only two or three. After you have conquered the first year, start adding activities, maybe one unit at a time. Finally, the most important task for a teacher new to AP Calculus is to practice AP problems yourself! My first year, in preparation for my review unit, I did about a decade's worth of FRQs. You will start to notice patterns.

Be reflective. After a unit is complete, assess what your students understood and what you need to change. Each year I typically change the sequence a bit or focus on different aspects of each unit. For example, ever since the 2020–2021 school year, I have found students had weak precalculus skills, so I am integrating more limits and precalculus topics throughout my lessons and units.

It is difficult to judge your success as a teacher solely based on AP scores. Each year, your group of students comes to class with different backgrounds. Their math courses prior to AP Calculus offered different areas of focus, and with the ever-changing curriculum, their strengths and weaknesses will vary from year to year.

Prior to reviewing for the AP Exam, I ask students to be honest about their goals for the exam. You can use these goals to help motivate students when they need it, as well as to assess your success in helping students achieve their goals once scores are in. My goal each year is a combination of affording every student the opportunity to "pass" the exam, aiding students in obtaining their goals, and pushing students to aim high and work hard to achieve what they set out for themselves.

A "passing" score on the AP exam is considered a 3 or better. This means students have understood the material at a college level. Most colleges and universities will accept this score and give college credit for Calculus 1. Students may then choose to place into Calculus 2 or simply not have to take Calculus 1 again. Students who score a 2 on the exam, while not earning credit for the course, will be very prepared to take Calculus 1 in college and can use this background knowledge to succeed. As teachers all we can do is offer our students the best chance to earn what they are capable of achieving and then celebrate that achievement with them, be it a 2 or a 5!

For all AP teachers, both new and experienced, the best thing you can do to improve is to use the Instructional Planning Report that you receive after student scores are calculated. You can access this document in the AP Classroom. You will get a breakdown of scores by unit, by question type, and so on. With this information, you can adjust your course for the next school year. If you notice that students as a whole struggled with a particular unit, this is where you make changes. Perhaps you'll spend a little extra time on this unit or maybe you'll blend questions from that unit into later units throughout the year. Maybe you'll review this unit in class before next year's AP exam. If students did better on either multiple choice or free response, you can choose to focus more on the question type they found challenging. With reflection and analysis of the Instructional Planning Report, you can adjust your class effectively.

Another great way to learn new teaching tips and tactics is to attend an AP Summer Institute. Both new and experienced teachers can gain ideas and insights into the course and the AP exam. If you take more than one Institute over the course of your career, make sure to attend courses from different instructors. Check with peers on some of the respected instructors in your area.

Additionally, you can see if a local school hosts a symposium after the exam. In my area, a local high school holds a short meeting following the exam each year to discuss the FRQs. Different teachers present each one of the six FRQs. It is a great way to converse with other teachers about their experience teaching the course that year, chat about their assessment of the questions from the exam, and discuss how the exam informs any changes to their focus in class for the next year.

Finally, another great way to gain insight into the exam is to become an AP reader for the exam. After three years of teaching the course, you may apply. Grading the exam is one of the best ways to truly understand student errors, misconceptions, and how the exam is graded. You can apply online through the College Board website.

Additional Resources for Teachers

COLLEGE BOARD

Make sure to always use the College Board's CED for the course that is found at https://apcentral.collegeboard.org/courses/ap-calculus-ab/course. If the topic is in the CED, it will be on the AP test. If the topic is not in the CED, it is out of the scope for the course and will not be tested.

In addition, the College Board has created AP Curriculum Modules. These include Volumes of Solid of Revolution, Extrema, Motion, Reasoning from Tabular Data, Fundamental Theorem of Calculus, and Functions Defined by Integrals. The College Board also has available some "Special Focus Material" on the following topics: Approximation, The Fundamental Theorem of Calculus, and Differential Equations. Visit https://apcentral.collegeboard.org/courses/ap-calculus-ab/classroom-resources?course=ap-calculus-ab.

WEBSITES

Some of these might be good to use with students, and others are good for you as the teacher as a reference.

Khan Academy: AP Calculus AB
Provides videos and practice problems
https://www.khanacademy.org/math/ap-calculus-ab

Patrick JMT
Provides videos for calculus topics (and beyond)

- First-semester videos: Limits, Continuity and Derivatives:
  https://www.youtube.com/playlist?list=PL58C7BA6C14FD8F48
- Second-semester videos: Integration
  https://www.youtube.com/playlist?list=PLD371506BCA23A437

Mathispower4u Videos
Provides videos for various topics
http://www.mathispower4u.com/calculus.php

Calculus Maximus
Provides notes, worksheets, and solutions broken down by topic
http://www.korpisworld.com/Mathematics/Calculus%20Maximus/Calculus%20Maximus%20Splash.htm

Paul's Online Notes
Includes notes (very helpful for first-time teacher), practice problems with solutions, and problems for assignments
https://tutorial.math.lamar.edu/Problems/CalcI/CalcI.aspx

Flipped Math Calculus
Provides video lessons, including practice and answers
https://calculus.flippedmath.com/

Review-O-Mania

How Do I Earn Points?

Homework:

- **Multiple choice (complete and on time): 4 points**
- **Free response (complete and on time): 4 points**

Classwork

- **Presentation of a problem/explanation to the class: 2 points**

Attendance

- **Class attendance (all members of group present): 1 point/group**
- **Attendance of a review session: 4 points/person**

Quizzes

- **Points vary depending on quiz (group and individual quizzes)**

Further Explanation

Homework: Each multiple homework assignment is worth 4 points for your team. *You are expected to come in with work corrected and highlight questions you still need clarification on.* All assignments must be completed on time to receive credit.

Classwork: If you are on a team where a teammate got a question wrong and you got it right, it is your duty to explain that question thoroughly. After a teammate explains it to them, each member should be able to explain a question they once got wrong, unless all team members got it wrong.

Participation: Any problems that an entire team needs clarification on will be put up on the board by a representative from another team. This representative must rotate each time that team takes a turn, and we will make sure all teams are given an equal chance of putting up a problem. The entire team will earn 2 team points for their board work and possible explanation.

Attendance: Each group will receive 1 point per class if every member of the group is in attendance. Missing class for another AP exam is excused.

Quizzes: There will be a few short quizzes that students will complete in groups and individually.

How is the winning team decided upon?

The winning team is the team with the greatest number of points. You will get a group grade that counts toward your quarter average.

How will this affect my average?

You will also get an individual score based on homework completion/participation.

Can I really pass the AP exam?!

Yes! You have been practicing all year with AP-style questions on your exams, *without* a curve! The AP exam is curved, and with effort, your goals are attainable. Practice, practice, practice. Want to estimate your score on a practice exam? Go to **http://appass.com/calculators/calculusab** to calculate your score out of 5.

This is the final stretch. Work hard and cooperatively on your team! Let's get ready to derive . . . and integrate! [Crowd goes wild.]

Tips for Success on the AP Calculus Exam

Multiple Choice

1. Answer *every* question—there is no penalty for a wrong answer. If you are running out of time and have some questions blank, just guess! Points are not deducted for incorrect answers.
2. Some multiple-choice questions can be answered by working backward. Plug the answer choices into the problem and see which answer works out.
3. If you do not find your solution as one of the choices, plug in a number to your answer and to all choices to see if it appears in a different form.
4. Don't rush through the test, but be mindful of timing. Work slowly enough so that you don't make careless mistakes on the easier questions. Easy questions count the same as difficult ones. Yet don't take too much time on very difficult questions. Skip them and come back.
5. Know your trig values of special angles, especially sin and cos of 0, 30, 45, 60, 90, 180, 270 (and in radians) and $\tan^{-1}(1) = 45°$.

Free Response

1. FRQs must be correct to three decimal places. Show four to ensure you don't make a rounding error! And don't round until the end of the problem. Don't forget units.
2. Show *all work* on FRQs; even if you are using your calculator, you must show the integral or derivative you are taking.
3. Don't waste time simplifying answers. It's OK to leave $\frac{93}{6}$, $\cos 30$, $\ln 1$, or $e^{\ln 2}$.
4. When questions say "justify your answer," you must explain with a sentence or two. A graph or picture or arrows on a number line are not counted as a justification.
5. Don't use words like "it"; specifically refer to the function: $f(x)$, $f'(x)$, $f''(x)$, $g(x)$, and so on.
6. If you change your mind about an answer, cross it out lightly. *Do not waste time erasing.*
7. Never let the beginning of a problem keep you from the points at the end. You can often answer (b) or (c) parts of a problem without even doing part (a).